OXFORD **IB DIPLOMA PROGRAMME**

2014 EDITION

BIOLOGY

COURSE COMPANION

Andrew Allott
David Mindorff

OXFORD
UNIVERSITY PRESS

OXFORD
UNIVERSITY PRESS

Great Clarendon Street, Oxford, OX2 6DP, United Kingdom

Oxford University Press is a department of the University of Oxford. It furthers the University's objective of excellence in research, scholarship, and education by publishing worldwide. Oxford is a registered trade mark of Oxford University Press in the UK and in certain other countries

British Library Cataloguing in Publication Data
Data available

978-0-19-839211-8

20 19 18

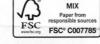

Paper used in the production of this book is a natural, recyclable product made from wood grown in sustainable forests. The manufacturing process conforms to the environmental regulations of the country of origin.

Printed in Great Britain by Bell and Bain Ltd, Glasgow

Acknowledgements

The publishers would like to thank the following for permissions to use their photographs:

Cover image: © Paul Souders/Corbis; p1: Sulston & Horvitz; p2: DR YORGOS NIKAS/SCIENCE PHOTO LIBRARY; p3a: DR.JEREMY BURGESS/SCIENCE PHOTO LIBRARY; p3b: Shutterstock; p6: Ferran Garcia-Pichel, Max Planck Institute of Marine Biology, Bermen Germany; p7a: Prof. P.Motta & T. Naguro/ SPL; p7b: Andrew Allot; p7c: Andrew Allot; p7d: MICHAEL ABBEY/ SCIENCE PHOTO LIBRARY; p8a: Carolina Biological Supply Co/Visuals Unlimited, Inc.; p8b: ASTRID & HANNS-FRIEDER MICHLER/SCIENCE PHOTO LIBRARY; p9: MICHAEL ABBEY/SCIENCE PHOTO LIBRARY; p10a: DR. PETER SIVER, VISUALS UNLIMITED /SCIENCE PHOTO LIBRARY; p10b: Sulston & Horvitz; p12: JAMES CAVALLINI/SCIENCE PHOTO LIBRARY; p14a: CHRIS BARRY/VISUALS UNLIMITED, INC. /SCIENCE PHOTO LIBRARY; p14b: SIMON FRASER/DEPARTMENT OF HAEMATOLOGY, RVI, NEWCASTLE/SCIENCE PHOTO LIBRARY; p16a: TEK IMAGE/SCIENCE PHOTO LIBRARY; p17: LAWRENCE BERKELEY NATIONAL LABORATORY/ SCIENCE PHOTO LIBRARY; p19: A B Dowsett/SPL; p20a: Eye of Science/SPL; p20b: CNRI/SCIENCE PHOTO LIBRARY; p21a: BIOPHOTO ASSOCIATES/SCIENCE PHOTO LIBRARY; p21b: MICROSCAPE/SCIENCE PHOTO LIBRARY; p22a: BIOPHOTO ASSOCIATES/ SCIENCE PHOTO LIBRARY; p22b: DR GOPAL MURTI/SCIENCE PHOTO LIBRARY; p22c: DR GOPAL MURTI/SCIENCE PHOTO LIBRARY; p22d: MICROSCAPE/SCIENCE PHOTO LIBRARY; p22e: DR KARI LOUNATMAA/ SCIENCE PHOTO LIBRARY; p22f: MICROSCAPE/SCIENCE PHOTO LIBRARY; p23a: DON W. FAWCETT/SCIENCE PHOTO LIBRARY; p23b: DR. GOPAL MURTI/SCIENCE PHOTO LIBRARY; p23c: Andrew Allot; p24a: STEVE GSCHMEISSNER/SCIENCE PHOTO LIBRARY; p24b: DR.JEREMY BURGESS/ SCIENCE PHOTO LIBRARY; p25a: STEVE GSCHMEISSNER/SCIENCE PHOTO LIBRARY; p25b: DAVID M. PHILLIPS/SCIENCE PHOTO LIBRARY; p25c: STEVE GSCHMEISSNER/SCIENCE PHOTO LIBRARY; p27: Author Image; p28: NIBSC/ SCIENCE PHOTO LIBRARY; p29: Author Image; p32: Janaka Dharmasena/ Shutterstock; p43a: OUP; p43b: Andrew Allot; p44: Herve Conge/SPL; p45: David Mayer, Consultant and CSL Liver Surgery, Queen Elizabeth Hospital, Birmingham; p46a: THOMAS DEERINCK, NCMIR/SCIENCE PHOTO LIBRARY; p46b: The VRoma Project (www.vroma.org); p48: GEORGETTE DOUWMA/ SCIENCE PHOTO LIBRARY; p49: DAVID MCCARTHY/SCIENCE PHOTO LIBRARY; p51: M.I. Walker/SPL; p53a,b,c,d: STEVE GSCHMEISSNER/SCIENCE PHOTO LIBRARY; p54a,b: STEVE GSCHMEISSNER/SCIENCE PHOTO LIBRARY; p55a: Dharam M Ramnani; p55b: MANFRED KAGE/SCIENCE PHOTO LIBRARY; p55c: MANFRED KAGE/SCIENCE PHOTO LIBRARY; p57: MOREDUN ANIMAL HEALTH LTD/SCIENCE PHOTO LIBRARY; p58: OUP; p54: Andrew Allot; p60: J Herve Conge, ISM/ SPL; p61: OUP; p62: Vasiliy Koval/Shutterstock; p66: LAGUNA DESIGN/SCIENCE PHOTO LIBRARY; p69a-p69b: OUP; p70: CLAIRE PAXTON & JACQUI FARROW/SCIENCE PHOTO LIBRARY; p71: DR KEITH WHEELER/SCIENCE PHOTO LIBRARY; p72: OUP; p73a: Dr. Elena Kiseleva/SPL; p73b: Dr. Gopal Murti/SPL; p73c: Dr. Elena Kiseleva/SPL; p75a: LAGUNA DESIGN/SCIENCE PHOTO LIBRARY; p75b: LAGUNA DESIGN/ SCIENCE PHOTO LIBRARY; p75c: LAGUNA DESIGN/SCIENCE PHOTO LIBRARY; p79: OUP; p80a: Andrew Allot; p80b-81: OUP; p83a: OUP; p83b: Giles Bell; p90a: OUP; p90b: www.rcsb.org; p91: www.rcsb.org; p92a: Yikrazuul/Wikipedia; p92b: OUP; p95: JAMES KING-HOLMES/SCIENCE

PHOTO LIBRARY; p101-102: OUP; p110: SPL; p116: Author Image; p122: © Tony Rusecki / Alamy; p123a: OUP; p123b: Glenn Tattersall; p124a: MATTHEW OLDFIELD/SCIENCE PHOTO LIBRARY; p124b: Author Image; p152: OUP; p126a: OUP; p126b: Petrov Andrey/Shutterstock; p130a: OUP; p130b: OUP; p130c: Andrew Allott; p131c: Andrew Allott; p132a: OUP; p133: William Allott; p134: OUP; p141: OUP; p143a: Jax.org; p143b: Jax.org; p143c: Jax.org; p144: www.ncbi.nlm.nih.gov/pubmed; p146a: Eye of Science/SPL; p146b: Eye of Science/SPL; p148: MAURO FERMARIELLO/SCIENCE PHOTO LIBRARY; p150a: M .Wurtz/Biozentrum/University o fBasel/SPL; p150b: Kwangshin Kim/SPL; p151: www.ncbi.nlm.nih.gov; p152: Dr. Oscar Lee Miller, Jr of the University of Virginia; p155a: OUP; p155b: Andrew Allot; p156: OUP; p158a: DEPT. OF CLINICAL CYTOGENETICS, ADDENBROOKES HOSPITAL/SCIENCE PHOTO LIBRARY; p158b: Tomasz Markowski/ Dreamstime; p159: L. WILLATT, EAST ANGLIAN REGIONAL GENETICS SERVICE/SCIENCE PHOTO LIBRARY; p160-161b: OUP; p162a: Andrew Allot; p164a,b,c,d: Andrew Allot; p165a,b,c,d: Andrew Allot; p166a: OUP; p166b: OUP; p166c: OUP; p169: OUP; p171a: OUP; p171b: OUP; p172: William Allott; p176: Enrico Coen; p177-184a: OUP; p184b: OUP; p186: RIA NOVOSTI/ SCIENCE PHOTO LIBRARY; p188: VOLKER STEGER/SCIENCE PHOTO LIBRARY; p189: OUP; p190a: Andrew Allot; p190b: DAVID PARKER/SCIENCE PHOTO LIBRARY; p190c-196c: OUP; p197: WALLY EBERHART, VISUALS UNLIMITED /SCIENCE PHOTO LIBRARY; p198a: GERARD PEAUCELLIER, ISM /SCIENCE PHOTO LIBRARY; p198b: GERARD PEAUCELLIER, ISM /SCIENCE PHOTO LIBRARY; p198c: Author Image; p199: PHILIPPE PLAILLY/SCIENCE PHOTO LIBRARY; p201: OUP; p202: Parinya Hirunthitima/Shutterstock; p203a: OUP; p203b: OUP; p203c: ERIC GRAVE/SCIENCE PHOTO LIBRARY; p203d: OUP; p204a,b,c,d: Andrew Allot; p205a: Author Image; p205b: CreativeNature.nl/Shutterstock; p205c: Author Image; p206: OUP; p207: OUP; p207b: Author Image; p209: Author Image; p210: OUP; p211: OUP; p212a: OUP; p212b: Andrew Allott; p214: Andrew Allott; p215a: OUP; p215b: Andrew Allott; p215c: Andrew Allott; p215d: Rich Lindie/Shutterstock; p215e: OUP; p217a: OUP; p217b: Andrew Allott; p217d: OUP; p221: Giorgiogp2/Wikipedia; p223a: Andrew Allott; p223b: Andrew Allott; p224: OUP; p225a: OUP; p225b: Andrew Allott; p225c: Andrew Allott; p228-242b: OUP; p243: Erik Lam/Shutterstock; p244: Sinclair Stammers/SPL; p246a: Wikipedia; p246b: Daiju AZUMA; p246c: Wikipedia; p246d: Shutterstock; p248a: Andrew Allott; p248b Andrew Allott; p250a: OUP; p250b: OUP; p251a: OUP; p251b: OUP; p251c: OUP; p251d: OUP; p251e: PETER CHADWICK/SCIENCE PHOTO LIBRARY; p253: OUP; p259: Author Image; p261: OUP; p262a: OUP; p262b: OUP; p264: Andrew Allot; p265: Kipling Brock/Shutterstock; p270a: Author Image; p270b: Author Image; p272: OUP; p276a: OUP; p276b: BOB GIBBONS/SCIENCE PHOTO LIBRARY; p279: BSIP VEM/SCIENCE PHOTO LIBRARY; p281: Dennis Kunkel/Photolibrary; p282: Author Image; p283a: Andrew Allot; p283b: OUP; p286: Author Image; p290: Public Domain/Wikipedia; p292a: OUP; p292b: OUP; p294a: OUP; p294b: BIOPHOTO ASSOCIATES/SCIENCE PHOTO LIBRARY; p298: Andrew Allot; p299: OUP; p302: OUP; p303a: OUP; p303b: Andrew Allot; p304a: OUP; p304b: OUP; p305: JAMES CAVALLINI/SCIENCE PHOTO LIBRARY; p306: ST MARY'S HOSPITAL MEDICAL SCHOOL/SCIENCE PHOTO LIBRARY; p307: OUP; p308: Wikipedia; p309: OUP; p315: OUP; p317: DU CANE MEDICAL IMAGING LTD/SCIENCE PHOTO LIBRARY; p318: OUP; p320a: OUP; p320b: THOMAS DEERINCK, NCMIR/SCIENCE PHOTO LIBRARY; p323: OUP; p325: BSIP VEM/SCIENCE PHOTO LIBRARY; p327: OUP; p328a: SCIENCE VU, VISUALS UNLIMITED /SCIENCE PHOTO LIBRARY; p328b: OUP; p330: J. ZBAEREN/EURELIOS/SCIENCE PHOTO LIBRARY; p331: OUP; p332: OAK RIDGE NATIONAL LABORATORY/US DEPARTMENT OF ENERGY/SCIENCE PHOTO LIBRARY; p333: OUP; p334: POWER AND SYRED/SCIENCE PHOTO LIBRARY; p339: CHASSENET/BSIP/SCIENCE PHOTO LIBRARY; p340: Author Image; p343: SIMON FRASER/SCIENCE PHOTO LIBRARY; p344: LEE D. SIMON/SCIENCE PHOTO LIBRARY; p346: SPL; p348: Image of PDB ID 1aoi (K. Luger, A.W. Mader, R.K. Richmond, D.F. Sargent, T.J. Richmond (1997) structure of the core particle at 2.8 A resolution Nature 389: 251-260) created with Chimera (UCSF Chimera–a visualization system for exploratory research and analysis. Pettersen EF, Goddard TD, Huang CC, Couch GS, Greenblatt DM, Meng EC, Ferrin TE. J Comput Chem. 2004 Oct;25(13):1605-12.); p349: Public Domain/Wikipedia; p351: SCIENCE PHOTO LIBRARY; p352: Andrew Allot; p353: Charvosi/Wikipedia; p357: Axel Bueckert/ Shutterstock; p358: PNAS.Org; p359: DR ELENA KISELEVA/SCIENCE PHOTO LIBRARY; p363a: Jmol; p363b: RCSB.org; p367: © 1970 American Association for the Advancement of Science. Miller, O. L. et al. Visualization of bacterial genes in action. Science 169,392–395 (1970). All rights reserved; p368a: Nobelprize.org; p368b: POWER AND SYRED/SCIENCE PHOTO LIBRARY; p368c: SINCLAIR STAMMERS/SCIENCE PHOTO LIBRARY; p370a: Andrew Allot; p373: Shutterstock; p375: RAMON ANDRADE 3DCIENCIA/SCIENCE PHOTO LIBRARY; p387a: CNRI/SCIENCE PHOTO LIBRARY; p387b: Petrov Andrey/Shutterstock; p387c: Prof. Kenneth R Miller/ SPL; p387d: Andrew Allot; p387e: Andrew Allot; p388: Dr. Carmen Manella, Wadsworth Center,New York State Department of Health; p390: Prof. Kenneth R Miller/ SPL; p392: Andrew Allot; p398: Andrew Allot; p399: Barrie Juniper; p403: POWER AND SYRED/SCIENCE PHOTO LIBRARY; p404: SINCLAIR STAMMERS/ SCIENCE PHOTO LIBRARY; p405a: Smugmug.Com; p405b: SCIENCE PHOTO LIBRARY; p406a: POWER AND SYRED/SCIENCE PHOTO LIBRARY; p406b: DR KEITH WHEELER/SCIENCE PHOTO LIBRARY; p410: SIDNEY MOULDS/ SCIENCE PHOTO LIBRARY; p411: DR KEITH WHEELER/SCIENCE PHOTO

Continued on back page.

Contents

Course book definition

The IB Diploma Programme course books are resource materials designed to support students throughout their two-year Diploma Programme course of study in a particular subject. They will help students gain an understanding of what is expected from the study of an IB Diploma Programme subject while presenting content in a way that illustrates the purpose and aims of the IB. They reflect the philosophy and approach of the IB and encourage a deep understanding of each subject by making connections to wider issues and providing opportunities for critical thinking.

The books mirror the IB philosophy of viewing the curriculum in terms of a whole-course approach; the use of a wide range of resources, international mindedness, the IB learner profile and the IB Diploma Programme core requirements, theory of knowledge, the extended essay, and creativity, action, service (CAS).

Each book can be used in conjunction with other materials and indeed, students of the IB are required and encouraged to draw conclusions from a variety of resources. Suggestions for additional and further reading are given in each book and suggestions for how to extend research are provided.

In addition, the course companions provide advice and guidance on the specific course assessment requirements and on academic honesty protocol. They are distinctive and authoritative without being prescriptive.

IB mission statement

The International Baccalaureate aims to develop inquiring, knowledgeable and caring young people who help to create a better and more peaceful world through intercultural understanding and respect.

To this end the organization works with schools, governments and international organizations to develop challenging programmes of international education and rigorous assessment.

These programmes encourage students across the world to become active, compassionate and lifelong learners who understand that other people, with their differences, can also be right.

The IB Learner Profile

The aim of all IB programmes to develop internationally minded people who work to create a better and more peaceful world. The aim of the programme is to develop this person through ten learner attributes, as described below.

Inquirers: They develop their natural curiosity. They acquire the skills necessary to conduct inquiry and research and snow independence in learning. They actively enjoy learning and this love of learning will be sustained throughout their lives.

Knowledgeable: They explore concepts, ideas, and issues that have local and global significance. In so doing, they acquire in-depth knowledge and develop understanding across a broad and balanced range of disciplines.

Thinkers: They exercise initiative in applying thinking skills critically and creatively to recognize and approach complex problems, and make reasoned, ethical decisions.

Communicators: They understand and express ideas and information confidently and creatively in more than one language and in a variety of modes of communication. They work effectively and willingly in collaboration with others.

Principled: They act with integrity and honesty, with a strong sense of fairness, justice and respect for the dignity of the individual, groups and communities. They take responsibility for their own action and the consequences that accompany them.

Open-minded: They understand and appreciate their own cultures and personal histories, and are open to the perspectives, values and traditions of other individuals and communities. They are accustomed to seeking and evaluating a range of points of view, and are willing to grow from the experience.

Caring: They show empathy, compassion and respect towards the needs and feelings of others. They have a personal commitment to service, and to act to make a positive difference to the lives of others and to the environment.

Risk-takers: They approach unfamiliar situations and uncertainty with courage and forethought, and have the independence of spirit to explore new roles, ideas, and strategies. They are brave and articulate in defending their beliefs.

Balanced: They understand the importance of intellectual, physical and emotional ballance to achieve personal well-being for themselves and others.

Reflective: They give thoughtful consideration to their own learning and experience. They are able to assess and understand their strengths and limitations in order to support their learning and personal development.

A note on academic honesty

It is of vital importance to acknowledge and appropriately credit the owners of information when that information is used in your work. After all, owners of ideas (intellectual property) have property rights. To have an authentic piece of work, it must be based on your individual and original ideas with the work of others fully acknowledged. Therefore, all assignments, written or oral, completed for assessment must use your own language and expression. Where sources are used or referred to, whether in the form of direct quotation or paraphrase, such sources must be appropriately acknowledged.

How do I acknowledge the work of others?

The way that you acknowledge that you have used the ideas of other people is through the use of footnotes and bibliographies.

Footnotes (placed at the bottom of a page) or endnotes (placed at the end of a document) are to be provided when you quote or paraphrase from another document, or closely summarize the information provided in another document. You do not need to provide a footnote for information that is part of a 'body of knowledge'. That is, definitions do not need to be footnoted as they are part of the assumed knowledge.

Bibliographies should include a formal list of the resources that you used in your work. 'Formal' means that you should use one of the several accepted forms of presentation. This usually involves separating the resources that you use into different categories (e.g. books, magazines, newspaper articles, internet-based resources, Cds and works of art) and providing full information as to how a reader or viewer of your work can find the same information. A bibliography is compulsory in the Extended Essay.

What constitutes malpractice?

Malpractice is behaviour that results in, or may result in, you or any student gaining an unfair advantage in one or more assessment component. Malpractice includes plagiarism and collusion.

Plagiarism is defined as the representation of the ideas or work of another person as your own. The following are some of the ways to avoid plagiarism:

- words and ideas of another person to support one's arguments must be acknowledged

- passages that are quoted verbatim must be enclosed within quotation marks and acknowledged

- CD-Roms, email messages, web sites on the Internet and any other electronic media must be treated in the same way as books and journals

- the sources of all photographs, maps, illustrations, computer programs, data, graphs, audio-visual and similar material must be acknowledged if they are not your own work

- works of art, whether music, film dance, theatre arts or visual arts and where the creative use of a part of a work takes place, the original artist must be acknowledged.

Collusion is defined as supporting malpractice by another student. This includes:

- allowing your work to be copied or submitted for assessment by another student

- duplicating work for different assessment components and/or diploma requirements.

Other forms of malpractice include any action that gives you an unfair advantage or affects the results of another student. Examples include, taking unauthorized material into an examination room, misconduct during an examination and falsifying a CAS record.

Using your IB Biology  Online Resources

What is Kerboodle?

Kerboodle is an online learning platform. If your school has a subscription to IB Biology Kerboodle Online Resources you will be able to access a bank of resources and assessments to guide you through this course.

What is in your Kerboodle Online Resources?

There are three main areas on the IB Biology Kerboodle: planning, resources, and assessment.

Resources

There a hundreds of extra resources available on the IB Biology Kerboodle Online. You can use these at home or in the classroom to develop your skills and knowledge as you progress through the course.

- ○ Hundreds of worksheets – read articles, perform experiments and simulations, practice your skills, or use your knowledge to answer questions.

- ○ Find out more by looking at links to recommended sites on the Internet, answer questions, or do more research.

- ○ Plus more to come in regular updates to Kerboodle!

Planning

This area is for your teacher so you won't have access to material in here.

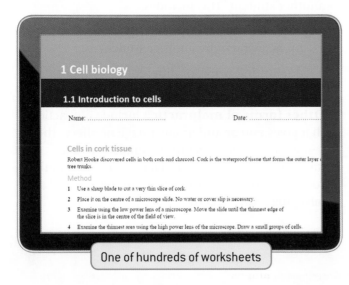

One of hundreds of worksheets

Practical skills presentation

Assessment

Click on the assessment tab to check your knowledge or revise for your examinations. Here you will find lots of interactive quizzes and exam-style practice questions.

- Formative tests: use these to check your comprehension. Evaluate how confident you feel about a sub-topic, then complete the test. You will have two attempts at each question and get feedback after every question. The marks are automatically reported in the markbook, so you can see how you progress throughout the year.

- Summative tests: use these to practice for your exams or as revision. Work through the test as if it were an examination – go back and change any questions you aren't sure about until you are happy, then submit the test for a final mark. The marks are automatically reported in the markbook, so you can see where you may need more practice.

- Assessment practice: use these to practice answering the longer written questions you will come across when you are examined. These worksheets can be printed out and performed as a timed test.

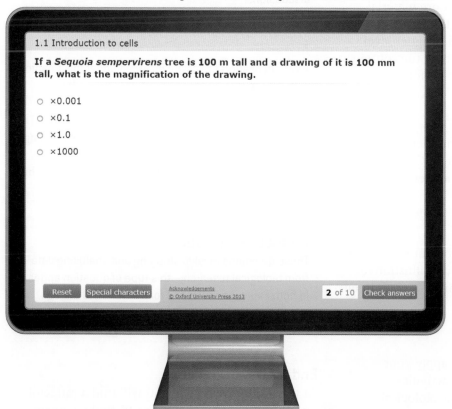

1.1 Introduction to cells

If a *Sequoia sempervirens* tree is 100 m tall and a drawing of it is 100 mm tall, what is the magnification of the drawing.

- ×0.001
- ×0.1
- ×1.0
- ×1000

Reset | Special characters | Acknowledgements © Oxford University Press 2013 | **2** of 10 | Check answers

Don't forget!
You can also find all of the textbook answers on our free website
www.oxfordsecondary.co.uk/ib-biology

Introduction

This book is a companion for students of Biology in the International Baccalaureate Diploma Programme.

Biology is the most popular choice of science subject as part of the IB diploma. The study of biology should lead students to appreciate the interconnectedness of life within the biosphere. With a focus on understanding the nature of science, IB Biology will allow you to develop a level of scientific literacy that will better prepare you to act on issues of local and global concern, with a full understanding of the scientific point of view.

The structure of this book is closely based on the biology programme in the Subject Guide. Sub-headings restate the specific assessment statements.

Topics 1 – 6 explain in detail the Core material that is common to both SL and HL courses. Topics 7 – 11 explain the AHL (additional higher level material). Topics A, B, C and D cover the content of the options. All topics include the following elements:

Understanding

The specifics of the content requirements for each sub-topic are covered in detail. Concepts are presented in ways that will promote enduring understanding.

Applications

These sections help you to develop your understanding by studying a specific illustrative example or learning about a significant experiment in the history of biology.

Skills topics

These sections encourage you to apply your understanding through practical activities and analysis of results from classic biological research. In some cases this involves instructions for handling data from experiments and also use of ICT. Some of the skills sections involve experiments with known outcomes, aimed at promoting understanding through "doing and seeing." Others involve ideas for experimental work with unknown outcomes, where you can define the problem and the methods. These are a valuable opportunities to build the skills that are assessed in IA (see page 708).

Nature of science

Here you can explore the methods of science and some of the knowledge issues that are associated with scientific endeavour. This is done using carefully selected examples, including biological research that led to paradigm shifts in our understanding of the natural world.

Theory of Knowledge

These short sections have headings that are equivocal `knowledge questions´. The text that follows often details one possible answer to the knowledge question. We encourage you draw on these examples of knowledge issues in your TOK essays. Of course, much of the material elsewhere in the book, particularly in the nature of science sections, can be used to prompt TOK discussions.

Activity

A variety of short topics are included under this heading with the focus in all cases on active learning. We encourage you research these topics yourself, using information available in textbooks or on the Internet. The aim is to promote an independent approach to learning. We believe that the optimal approach to learning is to be active – the more that you do for yourself, guided by your teacher, the better you will learn.

Data-based questions

These questions involve studying and analysing data from biological research – this type of question appears in both Paper 2 and Paper 3 for SL and HL IB Biology. Answers to these questions can be found at www.oxfordsecondary.co.uk/ib-biology

End-of-Topic Questions

At the end of each topic you will find a range of questions, including both past IB Biology exam questions and new questions. Answers can be found at www.oxfordsecondary.co.uk/ib-biology

1 CELL BIOLOGY

Introduction

There is an unbroken chain of life from the first cells on Earth to all cells found in organisms alive today. Eukaryotes have a much more complex cell structure than prokaryotes. The evolution of multicellular organisms allowed cell specialization and cell replacement. Cell division is essential but is carried out differently in prokaryotes and eukaryotes. While evolution has resulted in a biological world of enormous diversity, the study of cells shows us that there are also universal features. For example, the fluid and dynamic structure of biological membranes allows them to control the composition of cells.

1.1 Introduction to cells

Understanding

→ According to the cell theory, living organisms are composed of cells.

→ Organisms consisting of only one cell carry out all functions of life in that cell.

→ Surface area to volume ratio is important in the limitation of cell size.

→ Multicellular organisms have properties that emerge from the interaction of their cellular components.

→ Specialized tissues can develop by cell differentiation in multicellular organisms.

→ Differentiation involves the expression of some genes and not others in a cell's genome.

→ The capacity of stem cells to divide and differentiate along different pathways is necessary in embryonic development. It also makes stem cells suitable for therapeutic uses.

Applications

→ Questioning the cell theory using atypical examples, including striated muscle, giant algae and aseptate fungal hyphae.

→ Investigation of functions of life in *Paramecium* and one named photosynthetic unicellular organism.

→ Use of stem cells to treat Stargardt's disease and one other named condition.

→ Ethics of the therapeutic use of stem cells from specially created embryos, from the umbilical cord blood of a new-born baby and from an adult's own tissues.

Nature of science

→ Looking for trends and discrepancies: although most organisms conform to cell theory, there are exceptions.

→ Ethical implications of research: research involving stem cells is growing in importance and raises ethical issues.

Skills

→ Use of a light microscope to investigate the structure of cells and tissues.

→ Drawing cell structures as seen with the light microscope.

→ Calculation of the magnification of drawings and the actual size of structures shown in drawings or micrographs. (Practical 1)

The cell theory

Living organisms are composed of cells.

The internal structure of living organisms is very intricate and is built up from very small individual parts. Organs such as the kidney and the eye are easily visible. If they are dissected we can see that large organs are made of a number of different tissues, but until microscopes were invented little or nothing was discovered about the structure of tissues. From the 17th century onwards biologists examined tissues from both plants and animals using microscopes. Although there was much variation, certain features were seen again and again. A theory was developed to explain the basic features of structure – the cell theory. This states that cells are the fundamental building blocks of all living organisms. The smallest organisms are unicellular – they consist of just one cell. Larger organisms are multicellular – they are composed of many cells.

Cells vary considerably in size and shape but they share certain common features:

- Every living cell is surrounded by a membrane, which separates the cell contents from everything else outside.

- Cells contain genetic material which stores all of the instructions needed for the cell's activities.

- Many of these activities are chemical reactions, catalysed by enzymes produced inside the cell.

- Cells have their own energy release system that powers all of the cell's activities.

So, cells can be thought of as the smallest living structures – nothing smaller can survive.

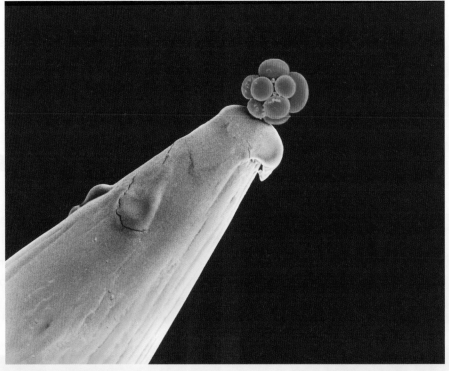

▲ Figure 1 Coloured scanning electron micrograph (SEM) of a human embryo on the tip of a pin

Exceptions to the cell theory

Looking for trends and discrepancies: although most organisms conform to cell theory, there are exceptions.

An early stage in scientific investigation is to look for trends – things that appear to be found generally rather than just in specific cases. These trends can lead to the development of a theory. A scientific theory is a way of interpreting the natural world. Theories allow us to make predictions. Sometimes exceptions to a general trend are found. These are called discrepancies. Scientists have to judge whether the discrepancies are common or serious enough to make predictions too unreliable to be useful. The theory is then discarded.

The cell theory is an example of where scientists have looked for trends and discrepancies. Robert Hooke was the first to use the word cell for structures in living organisms. He did this in 1665 after examining cork and other parts of plants. After describing cells in cork he wrote this:

> *Nor is this kind of texture peculiar to cork only, for upon examination with my microscope I have found that the pith of the Elder or almost any other tree, the inner pith of the Cany hollow stems of several other vegetables: as of Fennel, Carrets, Daucus, Bur-docks, Teasels, Fearn, some kind of Reeds etc. have much such a kind of Schematisme, as I have lately shown that of cork.*

So Hooke wasn't content with looking at just one type of plant tissue – he looked at many and discovered a general trend. Since Hooke's day biologists have looked at tissues from a huge variety of living organisms. Many of these tissues have been found to consist of cells, so the cell theory has not been discarded. However, some discrepancies have been discovered – organisms or parts of organisms that do not consist of typical cells. More discrepancies may be discovered, but it is extremely unlikely that the cell theory will ever be discarded, because so many tissues do consist of cells.

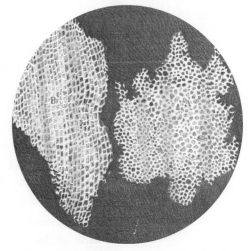

▲ Figure 2 Robert Hooke's drawing of cork cells

Activity

▲ Figure 3 What is the unit of life: the boy or his cells?

These two answers represent the holistic and the reductionist approach in biology.

 ## Using light microscopes

Use of a light microscope to investigate the structure of cells and tissues.

Try to improve your skill at using microscopes as much as you can.

- Learn the names of parts of the microscope.
- Understand how to focus the microscope to get the best possible image.
- Look after your microscope so it stays in perfect working order.
- Know how to troubleshoot problems.

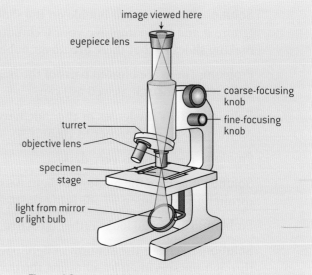

▲ Figure 4 Compound light microscope

Focusing

- Put the slide on the stage, with the most promising region exactly in the middle of the hole in the stage that the light comes through.

- Always focus at low power first even if eventually you need high power magnification.

- Focus with the larger coarse-focusing knobs first, then when you have nearly got the image in focus make it really sharp using the smaller fine-focusing knobs.

- If you want to increase the magnification, move the slide so the most promising region is exactly in the middle of the field of view and then change to a higher magnification lens.

Looking after your microscope

- Always focus by moving the lens and the specimen further apart, never closer to each other.

- Make sure that the slide is clean and dry before putting it on the stage.

- Never touch the surfaces of the lenses with your fingers or anything else.

- Carry the microscope carefully with a hand under it to support its weight securely.

Troubleshooting

Problem: Nothing is visible when I try to focus.

Solution: Make sure the specimen is actually under the lens, by carefully positioning the slide. It is easier to find the specimen if you focus at low power first.

Problem: A circle with a thick black rim is visible.

Solution: There is an air bubble on the slide. Ignore it and try to improve your technique for making slides so that there are no air bubbles.

Problem: There are blurred parts of the image even when I focus it as well as I can.

Solution: Either the lenses or the slide have dirt on them. Ask your teacher to clean it.

Problem: The image is very dark.

Solution: Increase the amount of light passing through the specimen by adjusting the diaphragm.

Problem: The image looks rather bleached.

Solution: Decrease the amount of light passing through the specimen by adjusting the diaphragm.

Types of slide

The slides that we examine with a microscope can be permanent or temporary.

Making permanent slides is very skilled and takes a long time, so these slides are normally made by experts. Permanent slides of tissues are made using very thin slices of tissue.

Making temporary slides is quicker and easier so we can do this for ourselves.

Examining and drawing plant and animal cells

Almost all cells are too small to be seen with the naked eye, so a microscope is needed to study them.

It is usually easy to see whether a cell is from a plant or an animal, even though there are many different cell types in both the plant and animal kingdoms.

- Place the cells on the slide in a layer not more than one cell thick.

- Add a drop of water or stain.

- Carefully lower a cover slip onto the drop. Try to avoid trapping any air bubbles.

- Remove excess fluid or stain by putting the slide inside a folded piece of paper towel and pressing lightly on the cover slip.

It is best to examine the slide first using low power. Move the slide to get the most promising areas in the middle of the field of view and then move up to high power. Draw a few cells, so you remember their structure.

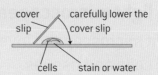

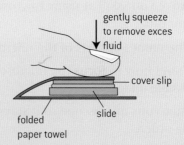

▲ Figure 5 Making a temporary mount

1 Moss leaf

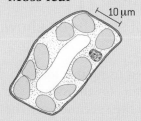

Use a moss plant with very thin leaves. Mount a single leaf in a drop of water or methylene blue stain.

2 Banana fruit cell

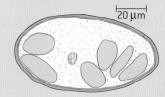

Scrape a small amount of the soft tissue from a banana and place on a slide. Mount in a drop of iodine solution.

3 Mammalian liver cell

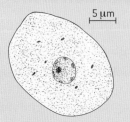

Scrape cells from a freshly cut surface of liver (not previously frozen). Smear onto a slide and add methylene blue to stain.

4 Leaf lower epidermis

Peel the lower epidermis off a leaf. The cell drawn here was from *Valeriana*. Mount in water or in methylene blue.

5 Human cheek cell

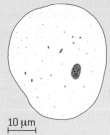

Scrape cells from the inside of your cheek with a cotton bud. Smear them on a slide and add methylene blue to stain.

6 White blood cell

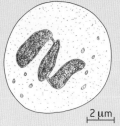

A thin layer of mammalian blood can be smeared over a slide and stained with Leishman's stain.

▲ Figure 6 Plant and animal cell drawings

🜂 Drawing cells

Drawing cell structures as seen with the light microscope.

Careful drawings are a useful way of recording the structure of cells or other biological structures. Usually the lines on the drawing represent the edges of structures. Do not show unnecessary detail and only use faint shading. Drawings of structures seen using a microscope will be larger than the structures actually are – the drawing shows them magnified. On page 6 the method for calculating the magnification of a drawing is explained. Everything on a drawing should be shown to the same magnification.

a) Use a sharp pencil with a hard lead to draw single sharp lines.

b) Join up lines carefully to form continuous structures such as cells

c) Draw lines freehand, but use a ruler for labelling lines.

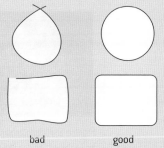

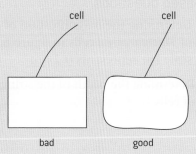

▲ Figure 7 Examples of drawing styles

Calculation of magnification and actual size (Practical 1)

Calculation of the magnification of drawings and the actual size of structures shown in drawings or micrographs.

When we look down a microscope the structures that we see appear larger than they actually are. The microscope is magnifying them. Most microscopes allow us to magnify specimens by two or three different factors. This is done by rotating the turret to switch from one objective lens to another. A typical school microscope has three levels of magnification:

- × 40 (low power)
- × 100 (medium power)
- × 400 (high power)

If we take a photo down a microscope, we can magnify the image even more. A photo taken down a microscope is called a micrograph. There are many micrographs in this book, including electron micrographs taken using an electron microscope.

When we draw a specimen, we can make the drawing larger or smaller, so the magnification of the drawing isn't necessarily the same as the magnification of the microscope.

To find the magnification of a micrograph or a drawing we need to know two things: the size of the image (in the drawing or the micrograph) and the actual size of the specimen. This formula is used for the calculation:

$$\text{magnification} = \frac{\text{size of image}}{\text{actual size of specimen}}$$

If we know the size of the image and the magnification, we can calculate the actual size of a specimen.

It is very important when using this formula to make sure that the units for the size of the image and actual size of the specimen are the same. They could both be millimetres (mm) or micrometres (μm) but they must not be different or the calculation will be wrong. Millimetres can be converted to micrometres by multiplying by one thousand. Micrometres can be converted to millimetres by dividing by one thousand.

Scale bars are sometimes put on micrographs or drawings, or just alongside them. These are straight lines, with the actual size that the scale bar represents. For example, if there was a 10 mm long scale bar on a micrograph with a magnification of ×10,000 the scale bar would have a label of 1 μm.

EXAMPLE:

The length of an image is 30 mm. It represents a structure that has an actual size of 3 μm. Determine the magnification of the image.

Either:

$30\,\text{mm} = 30 \times 10^{-3}\,\text{m}$

$3\,\mu\text{m} = 3 \times 10^{-6}\,\text{m}$

$$\text{Magnification} = \frac{30 \times 10^{-3}}{3 \times 10^{-6}}$$

$= 10{,}000 \times$

Or:

$30\,\text{mm} = 30{,}000\,\mu\text{m}$

$$\text{Magnification} = \frac{30{,}000}{3}$$

$= 10{,}000 \times$

Data-based questions

1   **a)**   Determine the magnification of the string of *Thiomargarita* cells in figure 8, if the scale bar represents 0.2 μm   [3]

   b)   Determine the width of the string of cells.   [2]

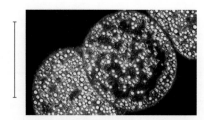

▲ Figure 8 *Thiomargarita*

2 In figure 9 the actual length of the mitochondrion is 8 μm.

   a) Determine the magnification of this electron micrograph. [2]

   b) Calculate how long a 5 μm scale bar would be on this electron micrograph. [2]

   c) Determine the width of the mitochondrion. [1]

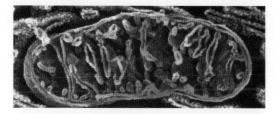

▲ Figure 9 Mitochondrion

3 The magnification of the human cheek cell from a compound microscope (figure 10) is 2,000 ×.

   a) Calculate how long a 20 μm scale bar would be on the image. [2]

   b) Determine the length of the cheek cell. [2]

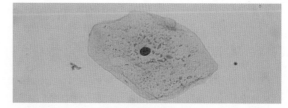

▲ Figure 10 Human cheek cell

4 a) Using the width of the hen's egg as a guide, estimate the actual length of the ostrich egg (figure 11). [2]

   b) Estimate the magnification of the image. [2]

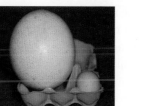

▲ Figure 11 Ostrich egg

 ## Testing the cell theory

Questioning the cell theory using atypical examples, including striated muscle, giant algae and aseptate fungal hyphae.

To test the cell theory you should look at the structure of as many living organisms as you can, using a microscope. Instructions for microscope use are given on page 4. In each case you should ask the question, "Does the organism or tissue fit the trend stated in the cell theory by consisting of one or more cells?"

Three atypical examples are worth considering:

- Striated muscle is the type of tissue that we use to change the position of our body. The building blocks of this tissue are muscle fibres, which are similar in some ways to cells. They are surrounded by a membrane and are formed by division of pre-existing cells. They have their own genetic material and their own energy release system. However muscle fibres are far from typical. They are much larger than most animal cells.

In humans they have an average length of about 30 mm, whereas other human cells are mostly less than 0.03 mm in length. Instead of having one nucleus they have many, sometimes as many as several hundred.

▲ Figure 12 Striated muscle fibres

- Fungi consist of narrow thread-like structures called hyphae. These hyphae are usually white in colour and have a fluffy appearance. They have a cell membrane and, outside it, a cell wall. In some types of fungi the hyphae are divided up into small cell-like sections by cross walls called septa. However, in aseptate fungi there are no septa. Each hypha is an uninterrupted tube-like structure with many nuclei spread along it.

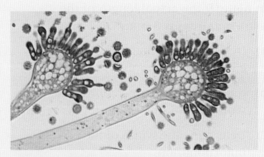

▲ Figure 13 Aseptate hypha

- Algae are organisms that feed themselves by photosynthesis and store their genes inside nuclei, but they are simpler in their structure and organization than plants. Many algae consist of one microscopic cell. There are vast numbers of these unicellular algae in the oceans and they form the basis of most marine food chains. Less common are some algae that grow to a much larger size, yet they still seem to be single cells. They are known as giant algae. *Acetabularia* is one example. It can grow to a length of as much as 100 mm, despite only having one nucleus. If a new organism with a length of 100 mm was discovered, we would certainly expect it to consist of many cells, not just one.

▲ Figure 14 Giant alga

Unicellular organisms

Organisms consisting of only one cell carry out all functions of life in that cell.

The functions of life are things that all organisms must do to stay alive. Some organisms consist of only one cell. This cell therefore has to carry out all the functions of life. Because of this the structure of unicellular organisms is more complex than most cells in multicellular organisms.

Unicellular organisms carry out at least seven functions of life:

- Nutrition – obtaining food, to provide energy and the materials needed for growth.
- Metabolism – chemical reactions inside the cell, including cell respiration to release energy.
- Growth – an irreversible increase in size.
- Response – the ability to react to changes in the environment.
- Excretion – getting rid of the waste products of metabolism.
- Homeostasis – keeping conditions inside the organism within tolerable limits.
- Reproduction – producing offspring either sexually or asexually.

Many unicellular organisms also have a method of movement, but some remain in a fixed position or merely drift in water or air currents.

Limitations on cell size

Surface area to volume ratio is important in the limitation of cell size.

In the cytoplasm of cells, large numbers of chemical reactions take place. These reactions are known collectively as the metabolism of the cell. The rate of these reactions (the metabolic rate of the cell) is proportional to the volume of the cell.

For metabolism to continue, substances used in the reactions must be absorbed by the cell and waste products must be removed. Substances move into and out of cells through the plasma membrane at the surface of the cell. The rate at which substances cross this membrane depends on its surface area.

The surface area to volume ratio of a cell is therefore very important. If the ratio is too small then substances will not enter the cell as quickly as they are required and waste products will accumulate because they are produced more rapidly than they can be excreted.

Surface area to volume ratio is also important in relation to heat production and loss. If the ratio is too small then cells may overheat because the metabolism produces heat faster than it is lost over the cell's surface.

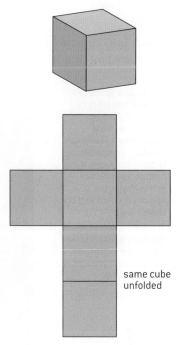

same cube unfolded

▲ Figure 15 Volume and surface area of a cube

🌐 Functions of life in unicellular organisms

Investigation of functions of life in *Paramecium* and one named photosynthetic unicellular organism.

Paramecium is a unicellular organism that can be cultured quite easily in the laboratory. Alternatively collect some pond water and use a centrifuge to concentrate the organisms in it to see if *Paramecium* is present.

Place a drop of culture solution containing *Paramecium* on a microscope slide.

Add a cover slip and examine the slide with a microscope.

The nucleus of the cell can divide to produce the extra nuclei that are needed when the cell reproduces. Often the reproduction is asexual with the parent cell dividing to form two daughter cells.

Food vacuoles contain smaller organisms that the *Paramecium* has consumed. These are gradually digested and the nutrients are absorbed into the cytoplasm where they provide energy and materials needed for growth.

The cell membrane controls what chemicals enter and leave. It allows the entry of oxygen for respiration. Excretion happens simply by waste products diffusing out through the membrane.

The contractile vacuoles at each end of the cell fill up with water and then expel it through the plasma membrane of the cell, to keep the cell's water content within tolerable limits.

Metabolic reactions take place in the cytoplasm, including the reactions that release energy by respiration. Enzymes in the cytoplasm are the catalysts that cause these reactions to happen.

Beating of the cilia moves the *Paramecium* through the water and this can be controlled by the cell so that it moves in a particular direction in response to changes in the environment.

▲ Figure 16 *Paramecium*

Chlamydomonas is a unicellular alga that lives in soil and freshwater habitats. It has been used widely for research into cell and molecular biology. Although it is green in colour and carries out photosynthesis it is not a true plant and its cell wall is not made of cellulose.

The nucleus of the cell can divide to produce genetically identical nuclei for asexual reproduction. Nuclei can also fuse and divide to carry out a sexual form of reproduction. In this image, the nucleus is concealed by chloroplasts.

Metabolic reactions take place in the cytoplasm, with enzymes present to speed them up.

The cell wall is freely permeable and it is the membrane inside it that controls what chemicals enter and leave. Oxygen is a waste product of photosynthesis and is excreted by diffusing out through the membrane.

The contractile vacuoles at the base of the flagella fill up with water and then expel it through the plasma membrane of the cell, to keep the cell's water content within tolerable limits.

Photosynthesis occurs inside chloroplasts in the cytoplasm. Carbon dioxide can be converted into the compounds needed for growth here, but in the dark carbon compounds from other organisms are sometimes absorbed through the cell membrane if they are available.

Beating of the two flagella moves the *Chlamydomonas* through the water. A light-sensitive "eyespot" allows the cell to sense where the brightest light is and respond by swimming towards it.

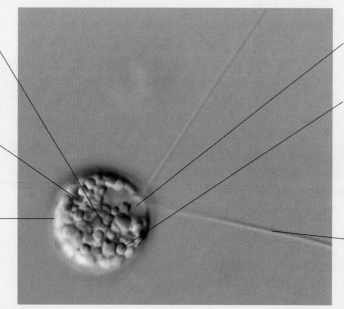

▲ Figure 17 *Chlamydomonas*

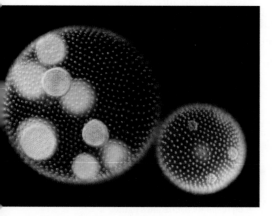

▲ Figure 18 *Volvox* colonies

Multicellular organisms

Multicellular organisms have properties that emerge from the interaction of their cellular components.

Some unicellular organisms live together in colonies, for example a type of alga called *Volvox aureus*. Each colony consists of a ball made of a protein gel, with 500 or more identical cells attached to its surface. Figure 18 shows two colonies, with daughter colonies forming inside them. Although the cells are cooperating, they are not fused to form a single cell mass and so are not a single organism.

Organisms consisting of a single mass of cells, fused together, are multicellular. One of the most intensively researched multicellular organisms is a worm called *Caenorhabditis elegans*. The adult body is about one millimetre long and it is made up of exactly 959 cells. This might seem like a large number, but most multicellular organisms have far more cells. There are about ten million million cells in an adult human body and even more in organisms such as oak trees or whales.

Although very well known to biologists, *Caenorhabditis elegans* has no common name and lives unseen in decomposing organic matter. It feeds on the bacteria that cause decomposition. *C. elegans* has a mouth, pharynx, intestine and anus. It is hermaphrodite so has both male and female reproductive organs. Almost a third of the cells are neurons, or nerve cells. Most of these neurons are located at the front end of the worm in a structure that can be regarded as the animal's brain.

Although the brain in *C. elegans* coordinates responses to the worm's environment, it does not control how individual cells develop. The cells in this and other multicellular organisms can be regarded as cooperative groups, without any cells in the group acting as a leader or supervisor. It is remarkable how individual cells in a group can organize themselves and interact with each other to form a living organism with distinctive overall properties. The characteristics of the whole organism, including the fact that it is alive, are known as *emergent properties*.

Emergent properties arise from the interaction of the component parts of a complex structure. We sometimes sum this up with the phrase: the whole is greater than the sum of its parts. A simple example of an emergent property was described in a Chinese philosophical text written more than 2,500 years ago: *"Pots are fashioned from clay. But it's the hollow that makes the pot work."* So, in biology we can carry out research by studying component parts, but we must remember that some bigger things result from interactions between these components.

Cell differentiation in multicellular organisms

Specialized tissues can develop by cell differentiation in multicellular organisms.

In multicellular organisms different cells perform different functions. This is sometimes called division of labour. In simple terms, a function is a job or a role. For example the function of a red blood cell is to carry oxygen, and the function of a rod cell in the retina of the eye is to absorb light and then transmit impulses to the brain. Often a group of cells specialize in the same way to perform the same function. They are called a tissue.

By becoming specialized, the cells in a tissue can carry out their role more efficiently than if they had many different roles. They can develop the ideal structure, with the enzymes needed to carry out all of the chemical reactions associated with the function. The development of cells in different ways to carry out specific functions is called differentiation. In humans, 220 distinctively different highly specialized cell types have been recognized, all of which develop by differentiation.

Gene expression and cell differentiation

Differentiation involves the expression of some genes and not others in a cell's genome.

There are many different cell types in a multicellular organism but they all have the same set of genes. The 220 cell types in the human body have the same set of genes, despite large differences in their structure and activities. To take an example, rod cells in the retina of the eye produce a pigment that absorbs light. Without it, the rod cell would not be able to do its job of sensing light. A lens cell in the eye produces no pigments and is transparent. If it did contain pigments, less light would

TOK

How can we decide when one model is better than another?

An emergent property of a system is not a property of any one component of the system, but it is a property of the system as a whole. Emergence refers to how complex systems and patterns arise from many small and relatively simple interactions. We cannot therefore necessarily predict emergent properties by studying each part of a system separately (an approach known as reductionism). Molecular biology is an example of the success that a reductionist approach can have. Many processes occurring in living organisms have been explained at a molecular level. However, many argue that reductionism is less useful in the study of emergent properties including intelligence, consciousness and other aspects of psychology. The interconnectivity of the components in cases like these is at least as important as the functioning of each individual component.

One approach that has been used to study interconnectivity and emergent properties is computer modelling. In both animal behaviour and ecology, a programme known as the "Game of Life" has been used. It was devised by John Conway and is available on the Internet. Test the "Game of Life" by creating initial configurations of cells and seeing how they evolve. Research ways in which the model has been applied.

pass through the lens and our vision would be worse. While they are developing, both cell types contain the genes for making the pigment, but these genes are only used in the rod cell.

This is the usual situation – cells do not just have genes with the instructions that they need, they have genes needed to specialize in every possible way. There are approximately 25,000 genes in the human genome, and these genes are all present in a body cell. However, in most cell types less than half of the genes will ever be needed or used.

When a gene is being used in a cell, we say that the gene is being expressed. In simple terms, the gene is switched on and the information in it is used to make a protein or other gene product. The development of a cell involves switching on particular genes and expressing them, but not others. Cell differentiation happens because a different sequence of genes is expressed in different cell types. The control of gene expression is therefore the key to development.

An extreme example of differentiation involves a large family of genes in humans that carry the information for making receptors for odorants – smells. These genes are only expressed in cells in the skin inside the nose, called olfactory receptor cells. Each of these cells expresses just one of the genes and so makes one type of receptor to detect one type of odorant. This is how we can distinguish between so many different smells. Richard Axel and Linda Buck were given the Nobel Prize in 2004 for their work on this system.

Stem cells

The capacity of stem cells to divide and differentiate along different pathways is necessary in embryonic development. It also makes stem cells suitable for therapeutic uses.

A new animal life starts when a sperm fertilizes an egg cell to produce a zygote. An embryo is formed when the zygote divides to give two cells. This two-cell embryo divides again to produce a four-cell embryo, then eight, sixteen and so on. At these early stages in embryonic development the cells are capable of dividing many times to produce large amounts of tissue. They are also extremely versatile and can differentiate along different pathways into any of the cell types found in that particular animal. In the 19th century, the name stem cell was given to the zygote and the cells of the early embryo, meaning that all the tissues of the adult stem from them.

Stem cells have two key properties that have made them one of the most active areas of research in biology and medicine today.

- Stem cells can divide again and again to produce copious quantities of new cells. They are therefore useful for the growth of tissues or the replacement of cells that have been lost or damaged.

- Stem cells are not fully differentiated. They can differentiate in different ways, to produce different cell types.

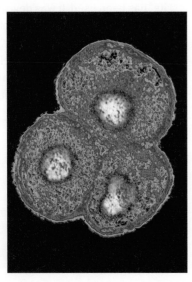

▲ Figure 19 Embryonic stem cells

Embryonic stem cells are therefore potentially very useful. They could be used to produce regenerated tissue, such as skin for people who have suffered burns. They could provide a means of healing diseases such as type 1 diabetes where a particular cell type has been lost or is malfunctioning. They might even be used in the future to grow whole replacement organs – hearts or kidneys, for example. These types of use are called therapeutic, because they provide therapies for diseases or other health problems.

There are also non-therapeutic uses for embryonic stem cells. One possibility is to use them to produce large quantities of striated muscle fibres, or meat, for human consumption. The beef burgers of the future may therefore be produced from stem cells, without the need to rear and slaughter cattle.

It is the early stage embryonic stem cells that are the most versatile. Gradually during embryo development the cells commit themselves to a pattern of differentiation. This involves a series of points at which a cell decides whether to develop along one pathway or another. Eventually each cell becomes committed to develop into one specific cell type. Once committed, a cell may still be able to divide, but all of these cells will differentiate in the same way and they are no longer stem cells.

Small numbers of cells remain as stem cells, however, and they are still present in the adult body. They are present in many human tissues, including bone marrow, skin and liver. They give some human tissues considerable powers of regeneration and repair. The stem cells in other tissues only allow limited repair – brain, kidney and heart for example.

 Therapeutic uses of stem cells

Use of stem cells to treat Stargardt's disease and one other named condition.

There are a few current uses of stem cells to treat diseases, and a huge range of possible future uses, many of which are being actively researched. Two examples are given here: one involving embryonic stem cells and one using adult stem cells.

Stargardt's disease

The full name of this disease is Stargardt's macular dystrophy. It is a genetic disease that develops in children between the ages of six and twelve. Most cases are due to a recessive mutation of a gene called ABCA4. This causes a membrane protein used for active transport in retina cells to malfunction. As a consequence, photoreceptive cells in the retina degenerate. These are the cells that detect light, so vision becomes progressively worse. The loss of vision can be severe enough for the person to be registered as blind.

Researchers have developed methods for making embryonic stem cells develop into retina cells. This was done initially with mouse cells, which were then injected into the eyes of mice that had a condition similar to Stargardt's disease. The injected cells were not rejected, did not develop into tumours or cause any other problems. The cells moved to the retina where they attached themselves and remained. Very encouragingly, they caused an improvement in the vision of the mice.

In November 2010, researchers in the United States got approval for trials in humans. A woman in her 50s with Stargardt's disease was treated by having 50,000 retina cells derived from embryonic stem cells injected into her eyes. Again the cells attached to the retina and remained there during the four-month trial. There was an improvement in her vision, and no harmful side effects.

Further trials with larger numbers of patients are needed, but after these initial trials at least, we can be optimistic about the development of treatments for Stargardt's disease using embryonic stem cells.

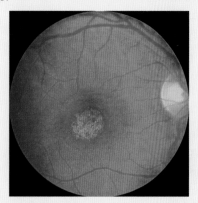

▲ Figure 20 Stargardt's disease

Leukemia

This disease is a type of cancer. All cancers start when mutations occur in genes that control cell division. For a cancer to develop, several specific mutations must occur in these genes in one cell. This is very unlikely to happen, but as there are huge numbers of cells in the body, the overall chance becomes much larger. More than a quarter of a million cases of leukemia are diagnosed each year globally and there are over 200,000 deaths from the disease.

Once the cancer-inducing mutations have occurred in a cell, it grows and divides repeatedly, producing more and more cells. Leukemia involves the production of abnormally large numbers of white blood cells. In most cancers, the cancer cells form a lump or tumour but this does not happen with leukemia. White blood cells are produced in the bone marrow, a soft tissue in the hollow centre of large bones such as the femur. They are then released into the blood, both in normal conditions and when excessive numbers are produced with leukemia. A normal adult white blood cell count is between 4,000 and 11,000 per mm^3 of blood. In a person with leukemia this number rises higher and higher. Counts above 30,000 per mm^3 suggest that a person may have leukemia. If there are more than 100,000 per mm^3 it is likely that the person has acute leukemia.

To cure leukemia, the cancer cells in the bone marrow that are producing excessive numbers of white blood cells must be destroyed. This can be done by treating the patient with chemicals that kill dividing cells. The procedure is known as chemotherapy. However, to remain healthy in the long term the patient must be able to produce the white blood cells needed to fight disease. Stem cells that can produce blood cells must be present, but they are killed by chemotherapy. The following procedure is therefore used:

- A large needle is inserted into a large bone, usually the pelvis, and fluid is removed from the bone marrow.

- Stem cells are extracted from this fluid and are stored by freezing them. They are adult stem cells and only have the potential for producing blood cells.

- A high dose of chemotherapy drugs is given to the patient, to kill all the cancer cells in the bone marrow. The bone marrow loses its ability to produce blood cells.

- The stem cells are then returned to the patient's body. They re-establish themselves in the bone marrow, multiply and start to produce red and white blood cells.

In many cases this procedure cures the leukemia completely.

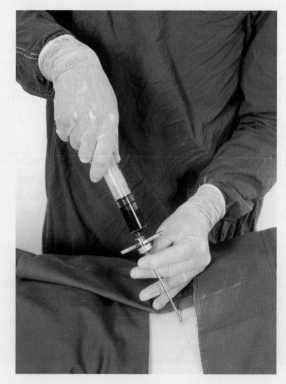

▲ Figure 21 Removal of stem cells from bone marrow

The ethics of stem cell research

Ethical implications of research: research involving stem cells is growing in importance and raises ethical issues.

Stem cell research has been very controversial. Many ethical objections have been raised. Scientists should always consider the ethical implications of their research before doing it. Some of the research that was carried out in the past would not be considered ethically acceptable today, such as medical research carried out on patients without their informed consent.

Decisions about whether research is ethically acceptable must be based on a clear understanding of the science involved. Some people dismiss all stem cell research as unethical, but this shows a misunderstanding of the different possible sources of the stem cells being used. In the next section, three possible sources of stem cells and the ethics of research involving them are discussed.

 # Sources of stem cells and the ethics of using them

Ethics of the therapeutic use of stem cells from specially created embryos, from the umbilical cord blood of a new-born baby and from an adult's own tissues.

Stem cells can be obtained from a variety of sources.

- Embryos can be deliberately created by fertilizing egg cells with sperm and allowing the resulting zygote to develop for a few days until it has between four and sixteen cells. All of the cells are embryonic stem cells.

- Blood can be extracted from the umbilical cord of a new-born baby and stem cells obtained from it. The cells can be frozen and stored for possible use later in the baby's life.

- Stem cells can be obtained from some adult tissues such as bone marrow.

These types of stem cell vary in their properties and therefore in their potential for therapeutic use. The table below gives some properties of the three types, to give the scientific basis for an ethical assessment.

Embryonic stem cells	Cord blood stem cells	Adult stem cells
• Almost unlimited growth potential.	• Easily obtained and stored.	• Difficult to obtain as there are very few of them and they are buried deep in tissues.
• Can differentiate into any type in the body.	• Commercial collection and storage services already available.	• Less growth potential than embryonic stem cells.
• More risk of becoming tumour cells than with adult stem cells, including teratomas that contain different tissue types.	• Fully compatible with the tissues of the adult that grows from the baby, so no rejection problems occur.	• Less chance of malignant tumours developing than from embryonic stem cells.
• Less chance of genetic damage due to the accumulation of mutations than with adult stem cells.	• Limited capacity to differentiate into different cell types – only naturally develop into blood cells, but research may lead to production of other types.	• Limited capacity to differentiate into different cell types.
• Likely to be genetically different from an adult patient receiving the tissue.	• Limited quantities of stem cells from one baby's cord.	• Fully compatible with the adult's tissues, so rejection problems do not occur.
• Removal of cells from the embryo kills it, unless only one or two cells are taken.	• The umbilical cord is discarded whether or not stem cells are taken from it.	• Removal of stem cells does not kill the adult from which the cells are taken.

15

Stem cell research has been very controversial. Many ethical objections have been raised. There are most objections to the use of embryonic stem cells, because current techniques usually involve the death of the embryo when the stem cells are taken. The main question is whether an early stage embryo is as much a human individual as a new-born baby, in which case killing the embryo is undoubtedly unethical.

When does a human life begin? There are different views on this. Some consider that when the sperm fertilizes the egg, a human life has begun. Others say that early stage embryos have not yet developed human characteristics and cannot suffer pain, so they should be thought of simply as groups of stem cells. Some suggest that a human life truly begins when there is a heartbeat, or bone tissue or brain activity. These stages take place after a few weeks of development. Another view is that it is only when the embryo has developed into a fetus that is capable of surviving outside the uterus.

Some scientists argue that if embryos are specially created by **in vitro fertilization (IVF)** in order to obtain stem cells, no human that would otherwise have lived has been denied its chance of living. However, a counterargument is that it is unethical to create human lives solely for the purpose of obtaining stem cells. Also, IVF involves hormone treatment of women, with some associated risk, as well as an invasive surgical procedure for removal of eggs from the ovary. If women are paid for supplying eggs for IVF this could lead to the exploitation of vulnerable groups such as college students.

We must not forget ethical arguments in favour of the use of embryonic stem cells. They have the potential to allow methods of treatment for diseases and disabilities that are currently incurable, so they could greatly reduce the suffering of some individuals.

▲ Figure 22 Harvesting umbilical cord blood

1.2 Ultrastructure of cells

Understanding

→ Prokaryotes have a simple cell structure without compartments.

→ Eukaryotes have a compartmentalized cell structure.

→ Prokaryotes divide by binary fission.

→ Electron microscopes have a much higher resolution than light microscopes.

 ## Applications

→ The structure and function of organelles within exocrine gland cells of the pancreas.

→ The structure and function of organelles within palisade mesophyll cells of the leaf.

 ## Nature of science

→ Developments in scientific research follow improvements in apparatus: the invention of electron microscopes led to greater understanding of cell structure.

 ## Skills

→ Drawing the ultrastructure of prokaryotic cells based on electron micrographs.

→ Drawing the ultrastructure of eukaryotic cells based on electron micrographs.

→ Interpretation of electron micrographs to identify organelles and deduce the function of specialized cells.

The invention of the electron microscope

Developments in scientific research follow improvements in apparatus: the invention of electron microscopes led to greater understanding of cell structure.

Much of the progress in biology over the last 150 years has followed improvements in the design of microscopes. In the second half of the 19th century improved light microscopes allowed the discovery of bacteria and other unicellular organisms. Chromosomes were seen for the first time and the processes of mitosis, meiosis and gamete formation were discovered. The basis of sexual reproduction, which had previously eluded William Harvey and many other biologists, was seen to be the fusion of gametes and subsequent development of embryos. The complexity of organs such as the kidney was revealed and mitochondria, chloroplasts and other structures were discovered within cells.

There was a limit to the discoveries that could be made though. For technical reasons that are explained later in this sub-topic, light microscopes cannot produce clear images of structures smaller than 0.2 micrometres (μm). (A micrometre is a thousandth of a millimetre.) Many biological structures are smaller than this. For example, membranes in cells are about 0.01 μm thick. Progress was hampered until a different type of microscope was invented – the electron microscope.

Electron microscopes were developed in Germany during the 1930s and came into use in research laboratories in the 1940s and 50s. They allowed images to be produced of things as small as 0.001 μm – 200 times smaller than with light microscopes. The structure of eukaryotic cells was found to be far more intricate than most biologists had expected and many previous ideas were shown to be wrong. For example, in the 1890s the light microscope had revealed darker green areas in the chloroplast. They were called grana and interpreted as droplets of chlorophyll. The electron microscope showed that grana are in fact stacks of flattened membrane sacs, with the chlorophyll located in the membranes. Whereas mitochondria appear as tiny structureless rods or spheres under the light microscope, the electron microscope revealed them to have an intricate internal membrane structure.

The electron microscopes revealed what is now called the ultrastructure of cells, including previously unknown features. Ribosomes, lysosomes and the endoplasmic reticulum were all discovered and named in the 1950s, for example. It is unlikely that there are structures as significant as these still to be discovered, but improvements in the design of electron microscopes continue and each improvement allows new discoveries to be made. A recent example, described in sub-topic 8.2, is electron tomography – a method of producing 3-D images by electron microscopy.

The resolution of electron microscopes

Electron microscopes have a much higher resolution than light microscopes.

If we look at a tree with unaided eyes we can see its individual leaves, but we cannot see the cells within its leaves. The unaided eye can see things with a size of 0.1 mm as separate objects, but no smaller. To see the cells within the leaf we need to use a light microscope. This allows us to see things with a size of down to about 0.2 μm as separate objects, so cells can become individually visible – they can be distinguished.

Making the separate parts of an object distinguishable by eye is called **resolution**.

The maximum resolution of a light microscope is 0.2 μm, which is 200 nanometres (nm). However powerful the lenses of a light microscope are, the resolution cannot be higher than this because it is limited by the wavelength of light (400–700 nm). If we try to resolve smaller objects by

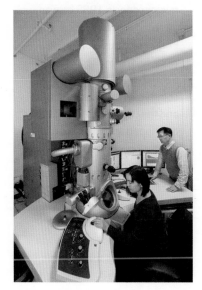

▲ Figure 1 An electron microscope in use

making lenses with greater magnification, we find that it is impossible to focus them properly and get a blurred image. This is why the maximum magnification with light microscopes is usually × 400.

Beams of electrons have a much shorter wavelength, so electron microscopes have a much higher resolution. The resolution of modern electron microscopes is 0.001 μm or 1 nm. Electron microscopes therefore have a resolution that is 200 times greater than light microscopes. This is why light microscopes reveal the structure of cells, but electron microscopes reveal the ultrastructure. It explains why light microscopes were needed to see bacteria with a size of 1 micrometre, but viruses with a diameter of 0.1 micrometres could not be seen until electron microscopes had been invented.

	Resolution		
	Millimetres (mm)	Micrometres (μm)	Nanometres (nm)
Unaided eyes	0.1	100	100,000
Light microscopes	0.0002	0.2	200
Electron microscopes	0.000001	0.001	1

Activity

Commerce and science

While still a young student in Berlin in the late 1920s Ernst Ruska developed magnetic coils that could focus beams of electrons. He worked on the idea of using these lenses to obtain an image as in a light microscope, but with electron beams instead of light. During the 1930s he developed and refined this technology. By 1939 Ruska had designed the first commercial electron microscope. In 1986 he was awarded the Nobel Prize in Physics for this pioneering work. Ruska worked with the German firm Siemens. Other companies in Britain, Canada and the United States also developed and manufactured electron microscopes.

- Scientists in different countries usually cooperate with each other but commercial companies do not. What are the reasons for this difference?

Prokaryotic cell structure
Prokaryotes have a simple cell structure without compartments.

All organisms can be divided into two groups according to their cell structure. Eukaryotes have a compartment within the cell that contains the chromosomes. It is called the nucleus and is bounded by a nuclear envelope consisting of a double layer of membrane. Prokaryotes do not have a nucleus.

Prokaryotes were the first organisms to evolve on Earth and they still have the simplest cell structure. They are mostly small in size and are found almost everywhere – in soil, in water, on our skin, in our intestines and even in pools of hot water in volcanic areas.

All cells have a cell membrane, but some cells, including prokaryotes, also have a cell wall outside the cell membrane. This is a much thicker and stronger structure than the membrane. It protects the cell, maintains its shape and prevents it from bursting. In prokaryotes the cell wall contains peptidoglycan. It is often referred to as being extracellular.

As no nucleus is present in a prokaryotic cell its interior is entirely filled with cytoplasm. The cytoplasm is not divided into compartments by membranes – it is one uninterrupted chamber. The structure is therefore simpler than in eukaryotic cells, though we must remember that it is still very complex in terms of the biochemicals that are present, including many enzymes.

Organelles are present in the cytoplasm of eukaryotic cells that are analogous to the organs of multi-cellular organisms in that they are distinct structures with specialized functions. Prokaryotes do not have cytoplasmic organelles apart from ribosomes. Their size, measured in Svedberg units (S) is 70S, which is smaller than those of eukaryotes.

Part of the cytoplasm appears lighter than the rest in many electron micrographs. This region contains the DNA of the cell, usually in the form of one circular DNA molecule. The DNA is not associated with proteins, which explains the lighter appearance compared with other parts of the cytoplasm that contain enzymes and ribosomes. This lighter area of the cell is called the nucleoid – meaning nucleus-like as it contains DNA but is not a true nucleus.

Cell division in prokaryotes

Prokaryotes divide by binary fission.

All living organisms need to produce new cells. They can only do this by division of pre-existing cells. Cell division in prokaryotic cells is called binary fission and it is used for asexual reproduction. The single circular chromosome is replicated and the two copies of the chromosome move to opposite ends of the cell. Division of the cytoplasm of the cell quickly follows. Each of the daughter cells contains one copy of the chromosome so they are genetically identical.

 ## Drawing prokaryotic cells

Draw the ultrastructure of prokaryotic cells based on electron micrographs.

Because prokaryotes are mostly very small, their internal structure cannot be seen using a light microscope. It is only with much higher magnification in electron micrographs that we can see the details of the structure, called the ultrastructure. Drawings of the ultrastructure of prokaryotes are therefore based on electron micrographs.

Shown below and on the next page are two electron micrographs of *E. coli*, a bacterium found in our intestines. One of them is a thin section and shows the internal structure. The other has been prepared by a different technique and shows the external structure. A drawing of each is also shown. By comparing the drawings with the electron micrographs you can learn how to identify structures within prokaryotic cells.

Electron micrograph of *Escherichia coli* (1–2 μm in length)

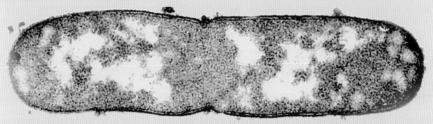

Drawing to help interpret the electron micrograph

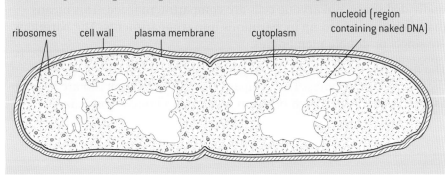

ribosomes   cell wall   plasma membrane   cytoplasm   nucleoid (region containing naked DNA)

19

Electron micrograph of *Escherichia coli* showing surface features

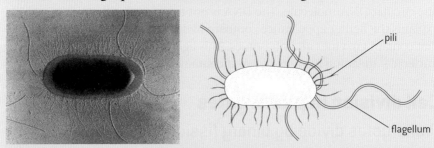

Shown below is another micrograph of a prokaryote. You can use it to practice your skill at drawing the ultrastructure of prokaryotic cells. You can also find other electron micrographs of prokaryotic cells on the internet and try drawing these. There is no need to spend a long time drawing many copies of a particular structure, such as the ribosomes. You can indicate their appearance in one small representative part of the cytoplasm and annotate your drawing to say that they are found elsewhere.

Activity

Garlic cells and compartmentalization

Garlic cells store a harmless sulphur-containing compound called alliin in their vacuoles. They store an enzyme called alliinase in other parts of the cell. Alliinase converts alliin into a compound called allicin, which has a very strong smell and flavour and is toxic to some herbivores. This reaction occurs when herbivores bite into garlic and damage cells, mixing the enzyme and its substrate. Perhaps surprisingly, many humans like the flavour, but to get it garlic must be crushed or cut, not used whole.

● You can test this by smelling a whole garlic bulb, then cutting or crushing it and smelling it again.

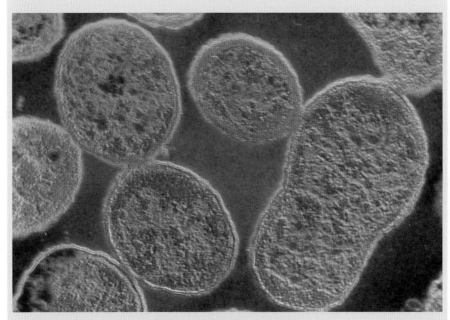

▲ Figure 2 *Brucella abortus* (Bang's bacillus), 2 μm in length

Eukaryotic cell structure

Eukaryotes have a compartmentalized cell structure.

Eukaryotic cells have a much more complicated internal structure than prokaryotic cells. Whereas the cytoplasm of a prokaryotic cell is one undivided space, eukaryotic cells are compartmentalized. This means that they are divided up by partitions into compartments. The partitions are single or double membranes.

The most important of these compartments is the nucleus. It contains the cell's chromosomes. The compartments in the cytoplasm are known as organelles. Just as each organ in an animal's body is specialized to perform a particular role, each organelle in a eukaryotic cell has a distinctive structure and function.

There are several advantages in being compartmentalized:

- Enzymes and substrates for a particular process can be much more concentrated than if they were spread throughout the cytoplasm.

- Substances that could cause damage to the cell can be kept inside the membrane of an organelle. For example, the digestive enzymes of a lysosome could digest and kill a cell, if they were not safely stored inside the lysosome membrane.

- Conditions such as pH can be maintained at an ideal level for a particular process, which may be different to the levels needed for other processes in a cell.

- Organelles with their contents can be moved around within the cell.

 Drawing eukaryotic cells

Draw the ultrastructure of eukaryotic cells based on electron micrographs.

The ultrastructure of eukaryotic cells is very complex and it is often best to draw only part of a cell. Your drawing is an interpretation of the structure, so you need to understand the structure of the organelles that might be present.

The table below contains an electron micrograph of each of the commonly occurring organelles, with a drawing of the structure. Brief notes on recognition features and the function of each organelle are included.

Nucleus 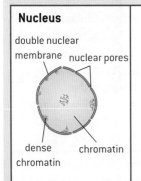	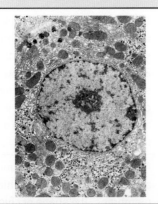	The nuclear membrane is double and has pores through it. The nucleus contains the chromosomes, consisting of DNA associated with histone proteins. Uncoiled chromosomes are spread through the nucleus and are called chromatin. There are often densely staining areas of chromatin around the edge of the nucleus. The nucleus is where DNA is replicated and transcribed to form mRNA, which is exported via the nuclear pores to the cytoplasm.
Rough endoplasmic reticulum 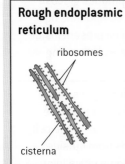	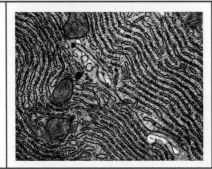	The rER consists of flattened membrane sacs, called cisternae. Attached to the outside of these cisternae are ribosomes. They are larger than in prokaryotes and are classified as 80S. The main function of the rER is to synthesize protein for secretion from the cell. Protein synthesized by the ribosomes of the rER passes into its cisternae and is then carried by vesicles, which bud off and are moved to the Golgi apparatus.

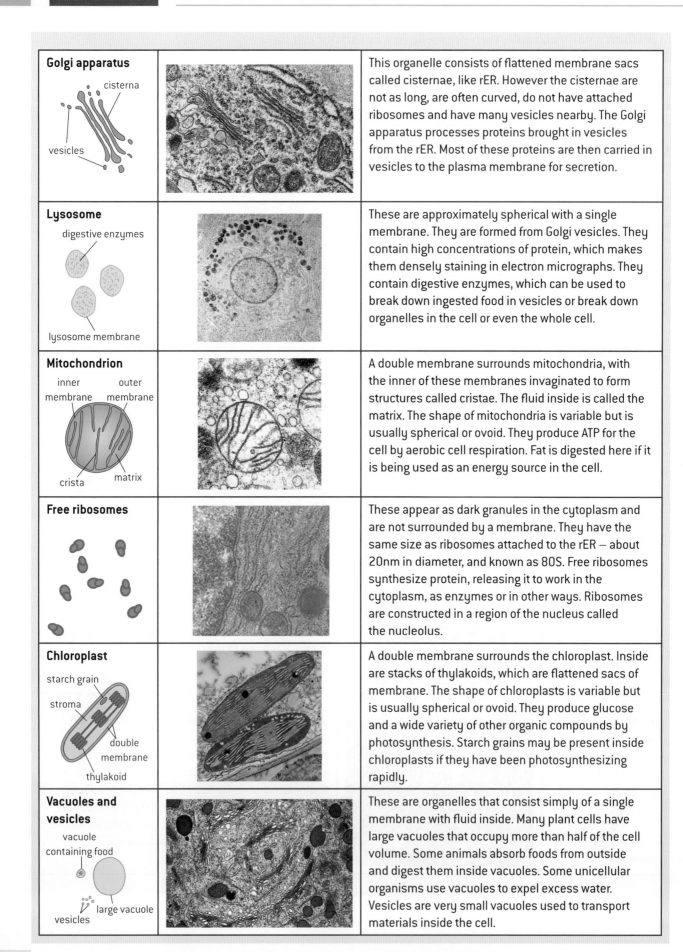

Golgi apparatus		This organelle consists of flattened membrane sacs called cisternae, like rER. However the cisternae are not as long, are often curved, do not have attached ribosomes and have many vesicles nearby. The Golgi apparatus processes proteins brought in vesicles from the rER. Most of these proteins are then carried in vesicles to the plasma membrane for secretion.
Lysosome		These are approximately spherical with a single membrane. They are formed from Golgi vesicles. They contain high concentrations of protein, which makes them densely staining in electron micrographs. They contain digestive enzymes, which can be used to break down ingested food in vesicles or break down organelles in the cell or even the whole cell.
Mitochondrion		A double membrane surrounds mitochondria, with the inner of these membranes invaginated to form structures called cristae. The fluid inside is called the matrix. The shape of mitochondria is variable but is usually spherical or ovoid. They produce ATP for the cell by aerobic cell respiration. Fat is digested here if it is being used as an energy source in the cell.
Free ribosomes		These appear as dark granules in the cytoplasm and are not surrounded by a membrane. They have the same size as ribosomes attached to the rER – about 20nm in diameter, and known as 80S. Free ribosomes synthesize protein, releasing it to work in the cytoplasm, as enzymes or in other ways. Ribosomes are constructed in a region of the nucleus called the nucleolus.
Chloroplast		A double membrane surrounds the chloroplast. Inside are stacks of thylakoids, which are flattened sacs of membrane. The shape of chloroplasts is variable but is usually spherical or ovoid. They produce glucose and a wide variety of other organic compounds by photosynthesis. Starch grains may be present inside chloroplasts if they have been photosynthesizing rapidly.
Vacuoles and vesicles		These are organelles that consist simply of a single membrane with fluid inside. Many plant cells have large vacuoles that occupy more than half of the cell volume. Some animals absorb foods from outside and digest them inside vacuoles. Some unicellular organisms use vacuoles to expel excess water. Vesicles are very small vacuoles used to transport materials inside the cell.

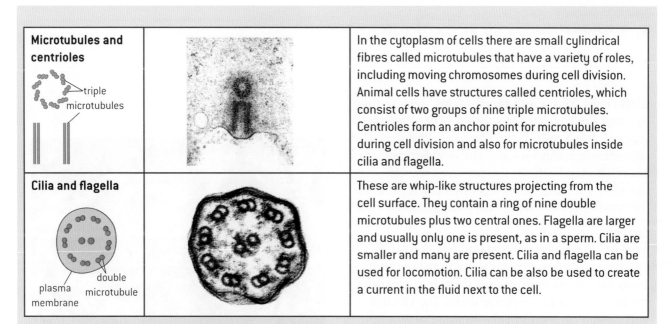

Microtubules and centrioles		In the cytoplasm of cells there are small cylindrical fibres called microtubules that have a variety of roles, including moving chromosomes during cell division. Animal cells have structures called centrioles, which consist of two groups of nine triple microtubules. Centrioles form an anchor point for microtubules during cell division and also for microtubules inside cilia and flagella.
Cilia and flagella		These are whip-like structures projecting from the cell surface. They contain a ring of nine double microtubules plus two central ones. Flagella are larger and usually only one is present, as in a sperm. Cilia are smaller and many are present. Cilia and flagella can be used for locomotion. Cilia can be also be used to create a current in the fluid next to the cell.

The electron micrograph below shows a liver cell with labels to identify some of the organelles that are present.

- Using your understanding of these organelles, draw the whole cell to show its ultrastructure.

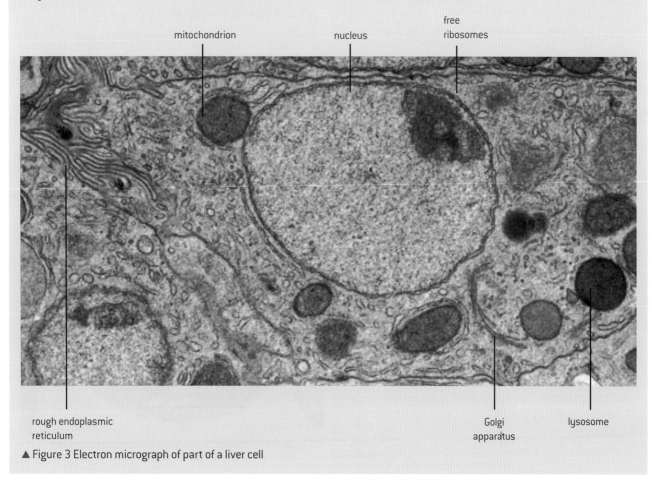

▲ Figure 3 Electron micrograph of part of a liver cell

23

Exocrine gland cells of the pancreas

The structure and function of organelles within exocrine gland cells of the pancreas.

Gland cells secrete substances – they release them through their plasma membrane. There are two types of gland cells in the pancreas. Endocrine cells secrete hormones into the bloodstream. Exocrine gland cells in the pancreas secrete digestive enzymes into a duct that carries them to the small intestine where they digest foods.

Enzymes are proteins, so the exocrine gland cells have organelles needed to synthesize proteins in large quantities, process them to make them ready for secretion, transport them to the plasma membrane and then release them. The electron micrograph on the right shows these organelles:

plasma membrane   Golgi apparatus
mitochondrion   vesicles
nucleus   lysosomes
rough ER

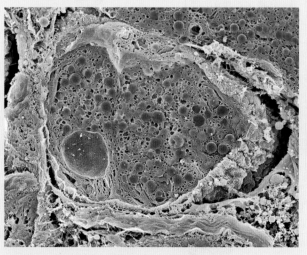

▲ Figure 4 Electron micrograph of pancreas cell

 ## Palisade mesophyll cells

The structure and function of organelles within palisade mesophyll cells of the leaf.

The function of the leaf is photosynthesis – producing organic compounds from carbon dioxide and other simple inorganic compounds, using light energy. The cell type that carries out most photosynthesis in the leaf is palisade mesophyll. The shape of these cells is roughly cylindrical. Like all living plant cells the cell is surrounded by a cell wall, with a plasma membrane inside it. The electron micrograph on the right shows the organelles that a palisade mesophyll cell contains:

cell wall
plasma membrane
chloroplasts
mitochondrion
vacuole
nucleus

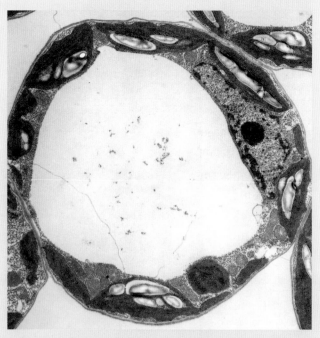

▲ Figure 5 Electron micrograph of palisade mesophyll cell

Interpreting the structure of eukaryotic cells

Interpret electron micrographs to identify organelles and deduce the function of specialized cells.

If the organelles in a eukaryotic cell can be identified and their function is known, it is often possible to deduce the overall function of the cell.

- Study the electron micrographs in figures 6, 7 and 8. Identify the organelles that are present and try to deduce the function of each cell.

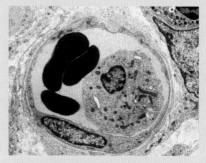

▲ Figure 7

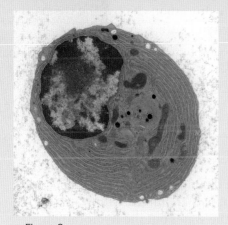

▲ Figure 6

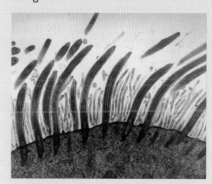

▲ Figure 8

1.3 Membrane structure

Understanding

→ Phospholipids form bilayers in water due to the amphipathic properties of phospholipid molecules.

→ Membrane proteins are diverse in terms of structure, position in the membrane and function.

→ Cholesterol is a component of animal cell membranes.

Applications

→ Cholesterol in mammalian membranes reduces membrane fluidity and permeability to some solutes.

Nature of science

→ Using models as representations of the real world: there are alternative models of membrane structure.

→ Falsification of theories with one theory being superseded by another: evidence falsified the Davson–Danielli model.

Skills

→ Drawing the fluid mosaic model.

→ Analysis of evidence from electron microscopy that led to the proposal of the Davson–Danielli model.

→ Analysis of the falsification of the Davson–Danielli model that led to the Singer–Nicolson model

Figure 1 The molecular structure of a phospholipid. The phosphate often has other hydrophilic groups attached to it, but these are not shown in this diagram

Phospholipid bilayers

Phospholipids form bilayers in water due to the amphipathic properties of phospholipid molecules.

Some substances are attracted to water – they are **hydrophilic**.

Other substances are not attracted to water – they are **hydrophobic**.

Phospholipids are unusual because part of a phospholipid molecule is hydrophilic and part is hydrophobic. Substances with this property are described as **amphipathic**.

The hydrophilic part of a phospholipid is the phosphate group. The hydrophobic part consists of two hydrocarbon chains. The chemical structure of phospholipids is shown in figure 1.

The structure can be represented simply using a circle for the phosphate group and two lines for the hydrocarbon chains.

▲ Figure 2 Simplified diagram of a phospholipid molecule

The two parts of the molecule are often called phosphate heads and hydrocarbon tails. When phospholipids are mixed with water the phosphate heads are attracted to the water but the hydrocarbon tails are attracted to each other, but not to water. Because of this the phospholipids become arranged into double layers, with the hydrophobic hydrocarbon tails facing inwards towards each other and the hydrophilic heads facing the water on either side. These double layers are called phospholipid bilayers. They are stable structures and they form the basis of all cell membranes.

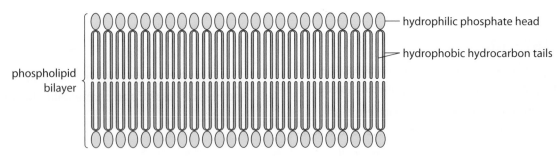

▲ Figure 3 Simplified diagram of a phospholipid bilayer

Models of membrane structure

Using models as representations of the real world: there are alternative models of membrane structure.

In the 1920s, Gorter and Grendel extracted phospholipids from the plasma membrane of red blood cells and calculated that the area that the phospholipids occupied when arranged in a monolayer was twice as large as the area of plasma membrane. They deduced that the membrane contained a **bilayer** of phospholipids. There were several errors in their methods but luckily these cancelled each other out and there is now very strong evidence for cell membranes being based on phospholipid bilayers.

Membranes also contain protein and Gorter and Grendel's model did not explain where this is located. In the 1930s Davson and Danielli proposed layers of protein adjacent to the phospholipid bilayer, on both sides of the membrane. They proposed this sandwich model because they thought it would explain how membranes, despite being very thin, are a very effective barrier to the movement of some substances. High magnification electron micrographs of membranes were made in the 1950s, which showed a railroad track appearance – two dark lines with a lighter band between. Proteins appear dark in electron micrographs and phospholipids appear light, so this appearance fitted the Davson-Danielli model.

Another model of membrane structure was proposed in 1966 by Singer and Nicolson. In this model the proteins occupy a variety of positions in the membrane. Peripheral proteins are attached to the inner or outer surface. Integral proteins are embedded in the phospholipid bilayer, in some cases with parts protruding out from the bilayer on one or both sides. The proteins are likened to the tiles in a mosaic. Because the phospholipid molecules are free to move in each of the two layers of the bilayer, the proteins are also able to move. This gives the model its name – the fluid mosaic model.

Problems with the Davson–Danielli model

Falsification of theories with one theory being superseded by another: evidence falsified the Davson–Danielli model.

The Davson–Danielli model of membrane structure was accepted by most cell biologists for about 30 years. Results of many experiments fitted the model including X-ray diffraction studies and electron microscopy.

In the 1950s and 60s some experimental evidence accumulated that did not fit with the Davson–Danielli model:

- **Freeze-etched electron micrographs.** This technique involves rapid freezing of cells and then fracturing them. The fracture occurs along lines of weakness, including the centre of membranes. Globular structures scattered through freeze-etched images of the centre of membranes were interpreted as transmembrane proteins.

- **Structure of membrane proteins.** Improvements in biochemical techniques allowed proteins to be extracted from membranes. They were found to be very varied in size and globular in shape so were unlike the type of structural protein that would form continuous layers on the

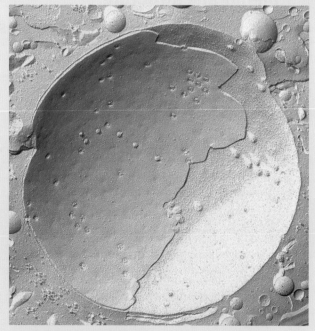

▲ Figure 4 Freeze-etched electron micrograph of nuclear membranes, with nuclear pores visible and vesicles in the surrounding cytoplasm. The diagram on page 28 shows the line of fracture through the centre of the inner and outer nuclear membranes. Transmembrane proteins are visible in both of the membranes

periphery of the membrane. Also the proteins were hydrophobic on at least part of their surface so they would be attracted to the hydrocarbon tails of the phospholipids in the centre of the membrane.

- **Fluorescent antibody tagging.** Red or green fluorescent markers were attached to antibodies that bind to membrane proteins. The membrane proteins of some cells were tagged with red markers and other cells with green markers. The cells were fused together. Within 40 minutes the red and green markers were mixed throughout the membrane of the fused cell. This showed that membrane proteins are free to move within the membrane rather than being fixed in a peripheral layer.

Taken together, this experimental evidence falsified the Davson–Danielli model. A

replacement was needed that fitted the evidence and the model that became widely accepted was the Singer–Nicolson fluid mosaic model. It has been the leading model for over fifty years but it would be unwise to assume that it will never be superseded. There are already some suggested modifications of the model.

An important maxim for scientists is "Think it possible that you might be mistaken." Advances in science happen because scientists reject dogma and instead search continually for better understanding.

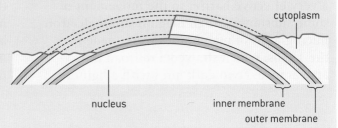

⚗ Evidence for and against the Davson–Danielli model of membrane structure

Analysis of evidence from electron microscopy that led to the proposal of the Davson–Danielli model.

Figure 5 shows the plasma membrane of a red blood cell and some of the cytoplasm near the edge of the cell.

1. Describe the appearance of the plasma membrane. [2]

2. Explain how this appearance suggested that the membrane had a central region of phospholipid with layers of protein on either side. [2]

3. Suggest reasons for the dark grainy appearance of the cytoplasm of the red blood cell. [2]

4. Calculate the magnification of the electron micrograph assuming that the thickness of the membrane is 10 nanometres. [3]

The two sets of data-based questions that follow are based on the types of data that were used to falsify the Davson–Danielli model of membrane structure.

▲ Figure 5 TEM of plasma membrane of a red blood cell

Data-based questions: Membranes in freeze-etched electron micrographs

Figure 6 shows a freeze-etched electron micrograph image of part of a cell. It was prepared by Professor Horst Robenek of Münster University.

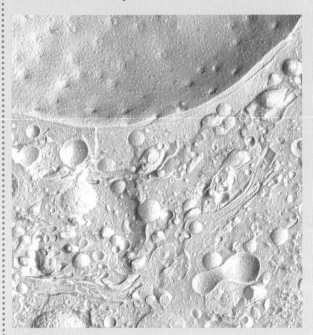

▲ Figure 6

1   In all of the fractured membranes in the micrograph small granules are visible.

    a)   State what these granules are.   [2]

    b)   Explain the significance of these granules in the investigation of membrane structure.   [3]

2   One of the membranes that surround the nucleus is visible at the top of the micrograph. Deduce whether it is the inner or outer nuclear membrane. (Always give your reasons when asked to deduce something.)   [2]

3   Identify three mitochondria visible in the micrograph, either using labels or by describing their positions.   [2]

4   Explain the evidence from the micrograph that this cell was processing proteins in its cytoplasm.   [2]

Extension questions on this topic can be found at www.oxfordsecondary.co.uk/ib-biology

Diffusion of proteins in membranes

Frye and Edidin used an elegant technique to obtain evidence for the fluid nature of membranes. They attached fluorescent markers to membrane proteins – green markers to mouse cells and red markers to human cells. In both cases, spherical cells growing in tissue culture were used. The marked mouse and human cells were then fused together. At first, the fused cells had one green hemisphere and one red one, but over the minutes following fusion, the red and green markers gradually merged, until they were completely mixed throughout the whole of the cell membrane. Blocking of ATP production did not prevent this mixing (ATP supplies energy for active processes in the cell).

Time after fusion / minutes	Cells with markers fully mixed/%				
	Result 1	Result 2	Result 3	Result 4	Mean
5	0	0	–	–	
10	3	0	–	–	
25	40	54	–	–	
40	87	88	93	100	
120	100	–	–	–	

1   Calculate the mean percentage of cells with markers fully mixed for each time after fusion.   [4]

2   Plot a graph of the results, including range bars for times where there was variation in the results. To do this you plot the highest and lowest results with a small bar and join these bars with a ruled line. You should also plot the mean result with a cross. This will lie on the range bar.   [4]

3   Describe the trend shown by the graph.   [1]

4   Explain whether the results fit the Davson–Danielli model or the Singer–Nicolson model more closely.   [2]

5   Explain the benefit of plotting range bars on graphs.   [2]

6   During this experiment the cells were incubated at 37 °C. Suggest a reason for the researchers choosing this temperature.   [1]

7 The experiment was repeated at different temperatures. Figure 7 shows the results. Explain the trends shown in the graph for temperatures between 15 and 35 °C. [2]

8 Explain the trends shown in the graph for temperatures below 15 °C. [2]

9 When ATP synthesis was blocked in the cells, the mixing of the red and green markers still occurred. Explain what conclusion can be drawn from this. [1]

10 Predict, with reasons, the results of the experiment if it was repeated using cells from arctic fish rather than from mice or humans. [1]

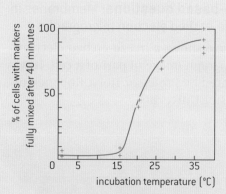

▲ Figure 7 Effect of temperature on the rate of diffusion of fluorescent markers in membranes

Membrane proteins

Membrane proteins are diverse in terms of structure, position in the membrane and function.

Cell membranes have a wide range of functions. The primary function is to form a barrier through which ions and hydrophilic molecules cannot easily pass. This is carried out by the phospholipid bilayer. Almost all other functions are carried out by proteins in the membrane. Six examples are listed in table 1.

Functions of membrane proteins
Hormone binding sites (also called hormone receptors), for example the insulin receptor. Figure 8 shows an example.
Immobilized enzymes with the active site on the outside, for example in the small intestine.
Cell adhesion to form tight junctions between groups of cells in tissues and organs.
Cell-to-cell communication, for example receptors for neurotransmitters at synapses.
Channels for passive transport to allow hydrophilic particles across by facilitated diffusion.
Pumps for active transport which use ATP to move particles across the membrane.

▲ Table 1

Because of these varied functions, membrane proteins are very diverse in structure and in their position in the membrane. They can be divided into two groups.

- Integral proteins are hydrophobic on at least part of their surface and they are therefore embedded in the hydrocarbon chains in the centre of the membrane. Many integral proteins are transmembrane – they extend across the membrane, with hydrophilic parts projecting through the regions of phosphate heads on either side.

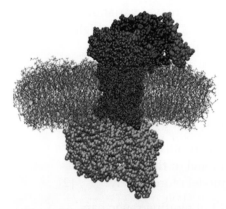

▲ Figure 8 Hormone receptor (purple) embedded in phospholipid bilayer (grey). The hormone (blue/red) is thyroid stimulating hormone. G-protein (brown) conveys the hormone's message to the interior of the cell

- Peripheral proteins are hydrophilic on their surface, so are not embedded in the membrane. Most of them are attached to the surface of integral proteins and this attachment is often reversible. Some have a single hydrocarbon chain attached to them which is inserted into the membrane, anchoring the protein to the membrane surface.

Figure 9 includes examples of both types of membrane protein.

Membranes all have an inner face and an outer face and membrane proteins are orientated so that they can carry out their function correctly. For example, pump proteins in the plasma membranes of root cells in plants are orientated so that they pick up potassium ions from the soil and pump them into the root cell.

The protein content of membranes is very variable, because the function of membranes varies. The more active a membrane, the higher is its protein content. Membranes in the myelin sheath around nerve fibres just act as insulators and have a protein content of only 18%.

The protein content of most plasma membranes on the outside of the cell is about 50%. The highest protein contents are in the membranes of chloroplasts and mitochondria, which are active in photosynthesis and respiration. These have protein contents of about 75%.

 Drawing membrane structure

Draw the fluid mosaic model of membrane structure.

The structure of membranes is far too complicated for us to show all of it in full detail in a drawing, but we can show our understanding of it using symbols to represent the molecules present. A diagram of membrane structure is shown in figure 9.

The diagram shows these components of a membrane:

- phospholipids;
- integral proteins;
- peripheral proteins;
- cholesterol.

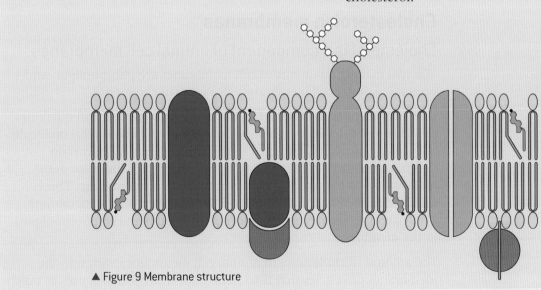

▲ Figure 9 Membrane structure

31

Identify which each component in the diagram is.

Using similar symbols to represent the components draw the structure of a membrane, according to the fluid mosaic model, that contains these proteins: channels for facilitated diffusion, pumps for active transport, immobilized enzymes and receptors for hormones or neurotransmitters.

It is worth thinking about what you have been doing when you draw the fluid mosaic model of membrane structure. Drawings simplify and interpret a structure or process. They are used in science as visual explanations. They show our understanding of a structure or process and not merely what it looks like. Drawings are based on models, hypotheses or theories. For example, when we show an animal tissue as a group of cells with lines to represent the plasma membranes, we are basing our drawing on the cell theory.

A diagram in a book or scientific paper usually starts out as a drawing on paper by the author, which is tidied up to make it suitable for printing. It is now possible to use computer software, but a pencil and paper are perhaps still the best way to draw. No artistic ability is needed for scientific drawing, and all biologists can develop and improve their drawing skills. Of course some biologists produce particularly good drawings. Some examples are shown in figure 10.

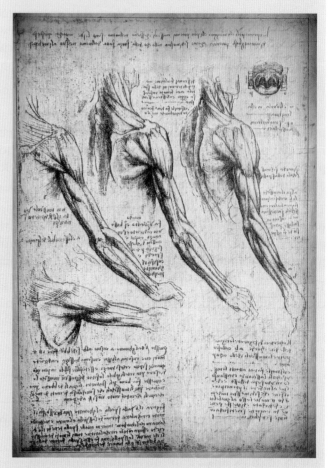

▲ Figure 10 Anatomical drawings by Leonardo da Vinci

Cholesterol in membranes

Cholesterol is a component of animal cell membranes.

The two main components of cell membranes are phospholipids and proteins. Animal cell membranes also contain cholesterol.

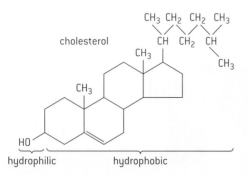

▲ Figure 11 The structure of cholesterol

Cholesterol is a type of lipid, but it is not a fat or oil. Instead it belongs to a group of substances called steroids. Most of a cholesterol molecule is hydrophobic so it is attracted to the hydrophobic hydrocarbon tails in the centre of the membrane, but one end of the cholesterol molecule has a hydroxyl (−OH) group which is hydrophilic. This is attracted to the phosphate heads on the periphery of the membrane. Cholesterol molecules are therefore positioned between phospholipids in the membrane.

The amount of cholesterol in animal cell membranes varies. In the membranes of vesicles that hold neurotransmitters at synapses as much of 30% of the lipid in the membrane is cholesterol.

The role of cholesterol in membranes

Cholesterol in mammalian membranes reduces membrane fluidity and permeability to some solutes.

Cell membranes do not correspond exactly to any of the three states of matter. The hydrophobic hydrocarbon tails usually behave as a liquid, but the hydrophilic phosphate heads act more like a solid. Overall the membrane is fluid as components of the membrane are free to move.

The fluidity of animal cell membranes needs to be carefully controlled. If they were too fluid they would be less able to control what substances pass through, but if they were not fluid enough the movement of the cell and substances within it would be restricted.

Cholesterol disrupts the regular packing of the hydrocarbon tails of phospholipid molecules, so prevents them crystallizing and behaving as a solid. However it also restricts molecular motion and therefore the fluidity of the membrane. It also reduces the permeability to hydrophilic particles such as sodium ions and hydrogen ions. Due to its shape cholesterol can help membranes to curve into a concave shape, which helps in the formation of vesicles during endocytosis.

1.4 Membrane transport

Understanding

→ Particles move across membranes by simple diffusion, facilitated diffusion, osmosis and active transport.
→ The fluidity of membranes allows materials to be taken into cells by endocytosis or released by exocytosis.
→ Vesicles move materials within cells.

 Applications

→ Structure and function of sodium–potassium pumps for active transport and potassium channels for facilitated diffusion in axons.
→ Tissues or organs to be used in medical procedures must be bathed in a solution with the same osmolarity as the cytoplasm to prevent osmosis.

 Nature of science

→ Experimental design: accurate quantitative measurements in osmosis experiments are essential.

 Skills

→ Estimation of osmolarity in tissues by bathing samples in hypotonic and hypertonic solutions. (Practical 2)

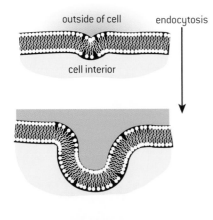

outside of cell   endocytosis

cell interior

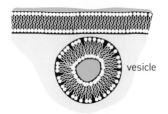

vesicle

▲ Figure 1 Endocytosis

Endocytosis

The fluidity of membranes allows materials to be taken into cells by endocytosis or released by exocytosis.

A vesicle is a small sac of membrane with a droplet of fluid inside. Vesicles are spherical and are normally present in eukaryotic cells. They are a very dynamic feature of cells. They are constructed, moved around and then deconstructed. This can happen because of the fluidity of membranes, which allows structures surrounded by a membrane to change shape and move.

To form a vesicle, a small region of a membrane is pulled from the rest of the membrane and is pinched off. Proteins in the membrane carry out this process, using energy from ATP.

Vesicles can be formed by pinching off a small piece of the plasma membrane of cells. The vesicle is formed on the inside of the plasma membrane. It contains material that was outside the cell, so this is a method of taking materials into the cell. It is called **endocytosis**. Figure 1 shows how the process occurs.

Vesicles taken in by endocytosis contain water and solutes from outside the cell but they also often contain larger molecules needed by the cell that cannot pass across the plasma membrane. For example, in the placenta, proteins from the mother's blood, including antibodies, are absorbed into the fetus by endocytosis. Some cells take in large undigested food particles by endocytosis. This happens in unicellular organisms including *Amoeba* and *Paramecium*. Some types of white blood cells take in pathogens including bacteria and viruses by endocytosis and then kill them, as part of the body's response to infection.

Vesicle movement in cells

Vesicles move materials within cells.

Vesicles can be used to move materials around inside cells. In some cases it is the contents of the vesicle that need to be moved. In other cases it is proteins in the membrane of the vesicle that are the reason for vesicle movement.

An example of moving the vesicle contents occurs in secretory cells. Protein is synthesized by ribosomes on the rough endoplasmic reticulum (rER) and accumulates inside the rER. Vesicles containing the proteins bud off the rER and carry them to the Golgi apparatus. The vesicles fuse with the Golgi apparatus, which processes the protein into its final form. When this has been done, vesicles bud off the Golgi apparatus and move to the plasma membrane, where the protein is secreted.

In a growing cell, the area of the plasma membrane needs to increase. Phospholipids are synthesized next to the rER and become inserted into the rER membrane. Ribosomes on the rER synthesize membrane proteins which also become inserted into the membrane. Vesicles bud off the rER and move to the plasma membrane. They fuse with it, each increasing the area of the plasma membrane by a very small amount. This method can also be used to increase the size of organelles in the cytoplasm such as lysosomes and mitochondria.

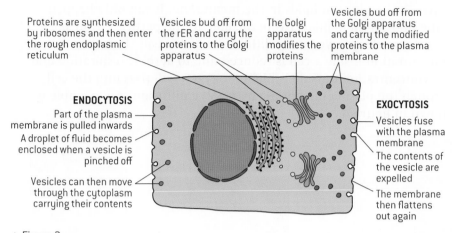

Proteins are synthesized by ribosomes and then enter the rough endoplasmic reticulum

Vesicles bud off from the rER and carry the proteins to the Golgi apparatus

The Golgi apparatus modifies the proteins

Vesicles bud off from the Golgi apparatus and carry the modified proteins to the plasma membrane

ENDOCYTOSIS
Part of the plasma membrane is pulled inwards
A droplet of fluid becomes enclosed when a vesicle is pinched off

Vesicles can then move through the cytoplasm carrying their contents

EXOCYTOSIS
Vesicles fuse with the plasma membrane
The contents of the vesicle are expelled
The membrane then flattens out again

▲ Figure 2

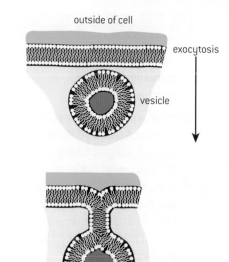
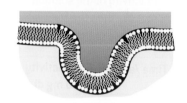

outside of cell
exocytosis
vesicle
cell interior

▲ Figure 3 Exocytosis

Exocytosis

The fluidity of membranes allows materials to be taken into cells by endocytosis or released by exocytosis.

Vesicles can be used to release materials from cells. If a vesicle fuses with the plasma membrane, the contents are then outside the membrane and therefore outside the cell. This process is called **exocytosis**.

Digestive enzymes are released from gland cells by exocytosis. The polypeptides in the enzymes are synthesized by the rER, processed in the Golgi apparatus and then carried to the membrane in vesicles for exocytosis. In this case the release is referred to as secretion, because a useful substance is being released, not a waste product.

Exocytosis can also be used to expel waste products or unwanted materials. An example is the removal of excess water from the cells of unicellular organisms. The water is loaded into a vesicle, sometimes called a contractile vacuole, which is then moved to the plasma membrane for expulsion by exocytosis. This can be seen quite easily in *Paramecium*, using a microscope. Figure 4 shows a drawing of *Paramecium* showing a contractile vesicle at each end of the cell.

Simple diffusion

Particles move across membranes by simple diffusion, facilitated diffusion, osmosis and active transport.

Simple diffusion is one of the four methods of moving particles across membranes.

Diffusion is the spreading out of particles in liquids and gases that happens because the particles are in continuous random motion. More particles move from an area of higher concentration to an area of lower concentration than move in the opposite direction. There is therefore a net movement from the higher to the lower concentration – a movement down the concentration gradient. Living

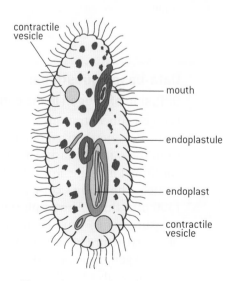

contractile vesicle
mouth
endoplastule
endoplast
contractile vesicle

▲ Figure 4 Drawing of *Paramecium*

TOK

Can the same data justify mutually exclusive conclusions?

In an experiment to test whether NaCl can diffuse through dialysis tubing, a 1% solution of NaCl was placed inside a dialysis tube and the tube was clamped shut. The tube containing the solution was immersed in a beaker containing water. A conductivity meter was inserted into the water surrounding the tubing. If the conductivity of the solution increases, then the NaCl is diffusing out of the tubing.

Time /s $\pm$ 1	Conductivity $\pm$ 10 mg dl^{-1}
0	81.442
30	84.803
60	88.681
90	95.403
120	99.799

Noting the uncertainty of the conductivity probe, discuss whether the data supports the conclusion that NaCl is diffusing out of the dialysis tubing.

organisms do not have to use energy to make diffusion occur so it is a passive process.

Simple diffusion across membranes involves particles passing between the phospholipids in the membrane. It can only happen if the phospholipid bilayer is permeable to the particles. Non-polar particles such as oxygen can diffuse through easily. If the oxygen concentration inside a cell is reduced due to aerobic respiration and the concentration outside is higher, oxygen will pass into the cell through the plasma membrane by passive diffusion. An example is shown in figure 6.

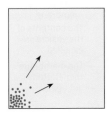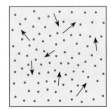

▲ Figure 5 Model of diffusion with dots representing particles

The centre of membranes is hydrophobic, so ions with positive or negative charges cannot easily pass through. Polar molecules, which have partial positive and negative charges over their surface, can diffuse at low rates between the phospholipids of the membrane. Small polar particles such as urea or ethanol pass through more easily than large particles.

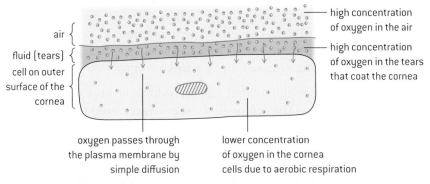

the cornea has no blood supply so its cells obtain oxygen by simple diffusion from the air

air {

fluid (tears) {

cell on outer surface of the cornea {

high concentration of oxygen in the air

high concentration of oxygen in the tears that coat the cornea

lower concentration of oxygen in the cornea cells due to aerobic respiration

oxygen passes through the plasma membrane by simple diffusion

▲ Figure 6 Passive diffusion

Data-based questions:

Diffusion of oxygen in the cornea

Oxygen concentrations were measured in the cornea of anesthetized rabbits at different distances from the outer surface. These measurements were continued into the aqueous humor behind the cornea. The rabbit's cornea is 400 micrometres (400 μm) thick. The graph (figure 7) shows the measurements. You may need to look at a diagram of eye structure before answering the questions. The oxygen concentration in normal air is 20 kilopascals (20 kPa).

1 Calculate the thickness of the rabbit cornea in millimetres. [1]

2 a) Describe the trend in oxygen concentrations in the cornea from the outer to the inner surface. [2]

b) Suggest reasons for the trend in oxygen concentration in the cornea. [2]

3 a) Compare the oxygen concentrations in the aqueous humor with the concentrations in the cornea. [2]

b) Using the data in the graph, deduce whether oxygen diffuses from the cornea to the aqueous humor. [2]

4 Using the data in the graph, evaluate diffusion as a method of moving substances in large multicellular organisms. [2]

5 a) Predict the effect of wearing contact lenses on oxygen concentrations in the cornea. [1]

b) Suggest how this effect could be minimized. [1]

6 The range bars for each data point indicate how much the measurements varied. Explain the reason for showing range bars on the graph. [2]

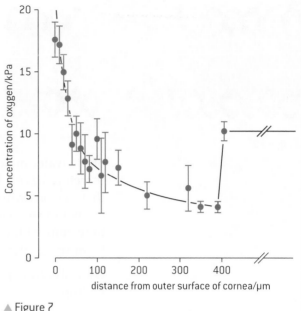

▲ Figure 7

Facilitated diffusion

Particles move across membranes by simple diffusion, facilitated diffusion, osmosis and active transport.

Facilitated diffusion is one of the four methods of moving particles across membranes.

Ions and other particles that cannot diffuse between phospholipids can pass into or out of cells if there are channels for them through the plasma membrane. These channels are holes with a very narrow diameter. The walls of the channel consist of protein. The diameter and chemical properties of the channel ensure that only one type of particle passes through, for example sodium ions, or potassium ions, but not both.

Because these channels help particles to pass through the membrane, from a higher concentration to a lower concentration, the process is called facilitated diffusion. Cells can control which types of channel are synthesized and placed in the plasma membrane and in this way they can control which substances diffuse in and out.

Figure 8 shows the structure of a channel for magnesium ions, viewed from the side and from the outside of the membrane. The structure of the protein making up the channel ensures that only magnesium ions are able to pass through the hole in the centre.

Osmosis

Particles move across membranes by simple diffusion, facilitated diffusion, osmosis and active transport.

Osmosis is one of the four methods of moving particles across membranes.

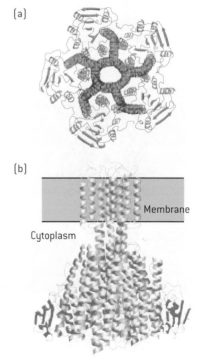

(a)

(b)

Membrane

Cytoplasm

▲ Figure 8 Magnesium channel

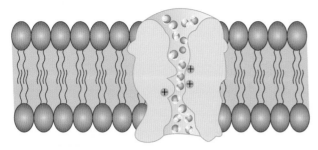

▲ Figure 9

Water is able to move in and out of most cells freely. Sometimes the number of water molecules moving in and out is the same and there is no net movement, but at other times more molecules move in one direction or the other. This net movement is osmosis.

Osmosis is due to differences in the concentration of substances dissolved in water (solutes). Substances dissolve by forming intermolecular bonds with water molecules. These bonds restrict the movement of the water molecules. Regions with a higher solute concentration therefore have a lower concentration of water molecules free to move than regions with a lower solute concentration. Because of this there is a net movement of water from regions of lower solute concentration to regions with higher solute concentration. This movement is passive because no energy has to be expended directly to make it occur.

Osmosis can happen in all cells because water molecules, despite being hydrophilic, are small enough to pass though the phospholipid bilayer. Some cells have water channels called aquaporins, which greatly increase membrane permeability to water. Examples are kidney cells that reabsorb water and root hair cells that absorb water from the soil.

At its narrowest point, the channel in an aquaporin is only slightly wider than water molecules, which therefore pass through in single file. Positive charges at this point in the channel prevent protons (H^+) from passing through.

Active transport

Particles move across membranes by simple diffusion, facilitated diffusion, osmosis and active transport.

Active transport is one of the four methods of moving particles across membranes.

Cells sometimes take in substances, even though there is already a higher concentration inside than outside. The substance is absorbed against the concentration gradient. Less commonly, cells sometimes pump substances out, even though there is already a larger concentration outside.

This type of movement across membranes is not diffusion and energy is needed to carry it out. It is therefore called active transport. Most active transport uses a substance called ATP as the energy supply for this process. Every cell produces its own supply of ATP by cell respiration.

Active transport is carried out by globular proteins in membranes, usually called pump proteins. The membranes of cells contain many different pump proteins allowing the cell to control the content of its cytoplasm precisely.

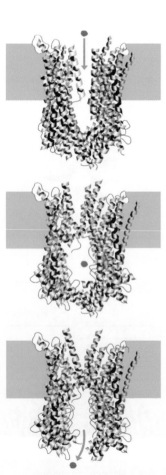

▲ Figure 10 Action of a pump protein

Figure 10 illustrates how a pump protein works. The molecule or ion enters the pump protein and can reach as far as a central chamber. A conformational change to the protein takes place using energy from ATP. After this, the ion or molecule can pass to the opposite side of the membrane and the pump protein returns to its original conformation. The pump protein shown transports Vitamin B_{12} into *E. coli*.

Data-based questions: Phosphate absorption in barley roots

Roots were cut off from barley plants and were used to investigate phosphate absorption. Roots were placed in phosphate solutions and air was bubbled through. The phosphate concentration was the same in each case, but the percentage of oxygen and nitrogen was varied in the air bubbled through. The rate of phosphate absorption was measured. Table 1 shows the results.

Oxygen /%	Nitrogen /%	Phosphate absorption/ $\mu mol\ g^{-1}\ h^{-1}$
0.1	99.9	0.07
0.3	99.7	0.15
0.9	99.1	0.27
2.1	97.1	0.32
21.0	79.0	0.33

▲ Table 1

1 Describe the effect of reducing the oxygen concentration below 21.0% on the rate of phosphate absorption by roots. You should only use information from the table in your answer. [3]

2 Explain the effect of reducing the oxygen percentage from 21.0 to 0.1 on phosphate absorption. In your answer you should use as much biological understanding as possible of how cells absorb mineral ions. [3]

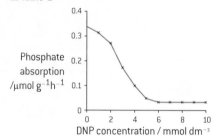

▲ Figure 11 Effect of DNP concentration on phosphate absorption

An experiment was done to test which method of membrane transport was used by the roots to absorb phosphate. Roots were placed in the phosphate solution as before, with 21.0% oxygen bubbling through. Varying concentrations of a substance called DNP were added. DNP blocks the production of ATP by aerobic cell respiration. Figure 11 shows the results of the experiment.

3 Deduce, with a reason, whether the roots absorbed the phosphate by diffusion or active transport. [2]

4 Discuss the conclusions that can be drawn from the data in the graph about the method of membrane transport used by the roots to absorb phosphate. [2]

Active transport of sodium and potassium in axons

Structure and function of sodium–potassium pumps for active transport.

An axon is part of a neuron (nerve cell) and consists of a tubular membrane with cytoplasm inside. Axons can be as narrow as one micrometre in diameter, but as long as one metre. Their function is to convey messages rapidly from one part of the body to another in an electrical form called a nerve impulse.

A nerve impulse involves rapid movements of sodium and then potassium ions across the axon membrane. These movements occur by facilitated diffusion through sodium and potassium channels. They occur because of concentration gradients between the inside and outside of the axon. The concentration gradients are built up by active transport, carried out by a sodium–potassium pump protein.

The sodium–potassium pump follows a repeating cycle of steps that result in three sodium ions being pumped out of the axon and two potassium ions being pumped in. Each time the pump goes round this cycle it uses one ATP. The cycle consists of these steps:

1 The interior of the pump is open to the inside of the axon; three sodium ions enter the pump and attach to their binding sites.

2 ATP transfers a phosphate group from itself to the pump; this causes the pump to change shape and the interior is then closed.

3 The interior of the pump opens to the outside of the axon and the three sodium ions are released.

4 Two potassium ions from outside can then enter and attach to their binding sites.

5 Binding of potassium causes release of the phosphate group; this causes the pump to change shape again so that it is again only open to the inside of the axon.

6 The interior of the pump opens to the inside of the axon and the two potassium ions are released; sodium ions can then enter and bind to the pump again (stage 1).

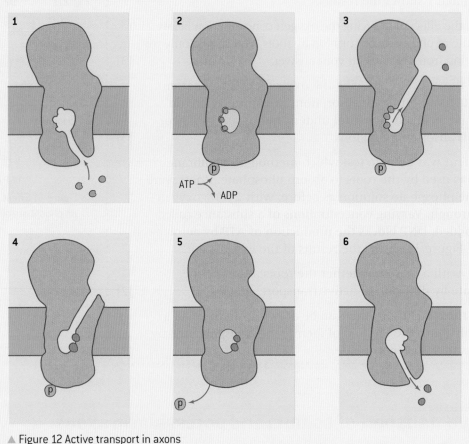

▲ Figure 12 Active transport in axons

Facilitated diffusion of potassium in axons

Structure and function of sodium–potassium pumps for active transport and potassium channels for facilitated diffusion in axons.

A nerve impulse involves rapid movements of sodium and then potassium ions across the axon membrane. These movements occur by facilitated diffusion through sodium and potassium channels. Potassium channels will be described here as a special example of facilitated diffusion. Each potassium channel consists of four protein subunits with a narrow pore between them that allows potassium ions to pass in either direction. The pore is 0.3 nm wide at its narrowest.

Potassium ions are slightly smaller than 0.3 nm, but when they dissolve they become bonded to a shell of water molecules that makes them too large to pass through the pore. To pass through, the bonds between the potassium ion and the surrounding water molecules are broken and bonds form temporarily between the ion and a series of amino acids in the narrowest part of the pore. After the potassium ion has passed through this part of the pore, it can again become associated with a shell of water molecules.

Other positively charged ions that we might expect to pass through the pore are either too large to fit through or are too small to form bonds with the amino acids in the narrowest part of the pore, so they cannot shed their shell of water molecules. This explains the specificity of the pump.

Potassium channels in axons are voltage gated. Voltages across membranes are due to an imbalance of positive and negative charges across the membrane. If an axon has relatively more

positive charges outside than inside, potassium channels are closed. At one stage during a nerve impulse there are relatively more positive charges inside. This causes potassium channels to open, allowing potassium ions to diffuse through. However, the channel rapidly closes again. This seems to be due to an extra globular protein subunit or ball, attached by a flexible chain of amino acids. The ball can fit inside the open pore, which it does within milliseconds of the pore opening. The ball remains in place until the potassium channel returns to its original closed state. This is shown in figure 13.

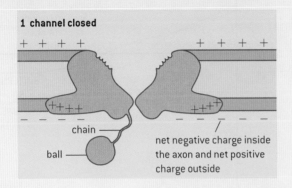

1 channel closed

chain

ball

net negative charge inside the axon and net positive charge outside

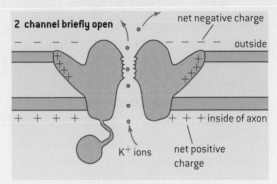

2 channel briefly open

net negative charge

outside

inside of axon

K⁺ ions

net positive charge

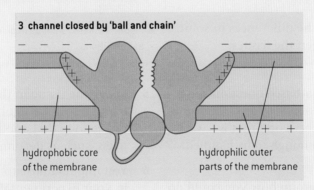

3 channel closed by 'ball and chain'

hydrophobic core of the membrane

hydrophilic outer parts of the membrane

 Figure 13

⚗ Estimation of osmolarity (Practical 2)

Estimation of osmolarity in tissues by bathing samples in hypotonic and hypertonic solutions.

Osmosis is due to solutes that form bonds with water. These solutes are osmotically active. Glucose, sodium ions, potassium ions and chloride ions are all osmotically active and solutions of them are often used in osmosis experiments. Cells contain many different osmotically active solutes.

The osmolarity of a solution is the total concentration of osmotically active solutes. The

units for measuring it are osmoles or milliosmoles (mOsm). The normal osmolarity of human tissue is about 300 mOsm.

An isotonic solution has the same osmolarity as a tissue. A hypertonic solution has a higher osmolarity and a hypotonic solution has a lower osmolarity. If samples of a tissue are bathed in hypertonic and hypotonic solutions, and measurements are taken to find out whether water enters or leaves the tissue, it is possible to deduce what concentration of solution would be isotonic and therefore find out the osmolarity of the tissue. The data-based questions below give the results from an experiment of this type.

Data-based questions: Osmosis in plant tissues

If samples of plant tissue are bathed in salt or sugar solutions for a short time, any increase or decrease in mass is due almost entirely to water entering or leaving the cells by osmosis. Figure 14 shows the percentage mass change of four tissues, when they were bathed in salt solutions of different concentrations.

1 a) State whether water moved into or out of the tissues at 0.0 mol dm^{-3} sodium chloride solution. [1]

   b) State whether water moved into or out of the tissues at 1.0 mol dm^{-3} sodium chloride solution. [2]

2 Deduce which tissue had the lowest solute concentration in its cytoplasm. Include how you reached your conclusion in your answer. [2]

3 Suggest reasons for the differences in solute concentration between the tissues. [3]

4 Explain the reasons for using percentage mass change rather than the actual mass change in grams in this type of experiment. [2]

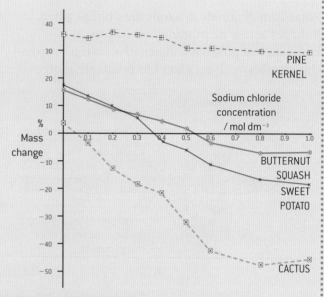

▲ Figure 14 Mass changes in plant tissues bathed in salt solutions

The experiment in the data-based question can be repeated using potato tubers, or any other plant tissue from around the world that is homogeneous and tough enough to be handled without disintegrating.

Discuss with a partner or group how you could do the following things:

1 Dilute a 1 mol dm^{-3} sodium chloride solution to obtain the concentrations shown on the graph.

2 Obtain samples of a plant tissue that are similar enough to each other to give comparable results.

3 Ensure that the surface of the tissue samples is dry when finding their mass, both at the start and end of the experiment.

4 Ensure that all variables are kept constant, apart from salt concentration of the bathing solution.

5 Leave the tissue in the solutions for long enough to get a significant mass change, but not so long that another factor affects the mass, such as decomposition!

6 You might choose to be more inventive in your experimental approach. Figure 15 gives one idea for measuring changes to the turgidity of plant tissue, but other methods could be used.

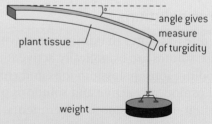

▲ Figure 15 Method of assessing turgidity of plant tissue

Experimental design

Experimental design: accurate quantitative measurements in osmosis experiments are essential.

An ideal experiment gives results that have only one reasonable interpretation. Conclusions can be drawn from the results without any doubts or uncertainties. In most experiments there are some doubts and uncertainties, but if the design of an experiment is rigorous, these can be minimized. The experiment then provides strong evidence for or against a hypothesis.

This checklist can be used when designing an experiment:

- Results should if possible be quantitative as these give stronger evidence than descriptive results.

- Measurements should be as accurate as possible, using the most appropriate and best quality meters or other apparatus.

- Repeats are needed, because however accurately quantitative measurements are taken biological samples are variable.

- All factors that might affect the results of the experiment must be controlled, with only the factors under investigation being allowed to vary and all other factors remaining constant.

After doing an experiment the design can be evaluated using this checklist. The evaluation might lead to improvements to the design that would have made the experiment more rigorous.

If you have done an osmosis experiment in which samples of plant tissue are bathed in solutions of varying solute concentration, you can evaluate its design. If you did repeats for each concentration of solution, and the results were very similar to each other, your results were probably reliable.

▲ Figure 16 Replicates are needed for each treatment in a rigorous experiment

 ## Designing osmosis experiments

Rigorous experimental design is needed to produce reliable results: how can accurate quantitative measurements be obtained in osmosis experiments?

The osmolarity of plant tissues can be investigated in many ways. Figure 17 shows some red onion cells that had been placed in a sodium chloride solution. The following method can be used to observe the consequences of osmosis in red onion cells.

1 Peel off some epidermis from the scale of a red onion bulb.

2 Cut out a sample of it, about 5 × 5mm.

3 Mount the sample in a drop of distilled water on a microscope slide, with a cover slip.

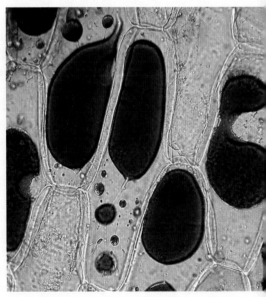

▲ Figure 17 Micrograph of red onion cells placed in salt solution

4 Observe using a microscope. The cytoplasm should fill the space inside the cell wall, with the plasma membrane pushed up against it.

5 Mount another sample of epidermis in sodium chloride solutions with concentration of 0.5mol dm^{-3} or 3%. If water leaves the cells by osmosis and the volume of cytoplasm is reduced, the plasma membrane pulls away from the cell wall, as shown in Figure 17. Plant cells with their membranes pulled away from their cell walls are plasmolysed and the process is plasmolysis.

This method can be used to help design an experiment to find out the osmolarity of onion cells or other cells in which the area occupied by the cytoplasm can easily be seen. The checklist in the previous section can be used to try to ensure that the design is rigorous.

Preventing osmosis in excised tissues and organs

Tissues or organs to be used in medical procedures must be bathed in a solution with the same osmolarity as the cytoplasm to prevent osmosis.

Animal cells can be damaged by osmosis. Figure 18 shows blood cells that have been bathed in solutions with (a) the same osmolarity, (b) higher osmolarity and (c) lower osmolarity.

a)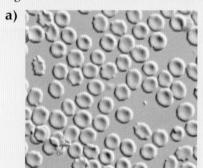

b)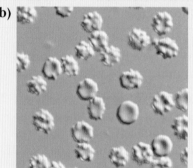

c)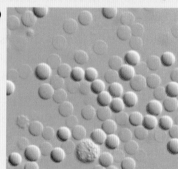

▲ Figure 18 Blood cells bathed in solutions of different solute concentration

In a solution with higher osmolarity (a hypertonic solution), water leaves the cells by osmosis so their cytoplasm shrinks in volume. The area of plasma membrane does not change, so it develops indentations, which are sometimes called crenellations. In a solution with lower osmolarity (hypotonic), the cells take in water by osmosis and swell up. They may eventually burst, leaving ruptured plasma membranes called red cell ghosts.

Both hypertonic and hypotonic solutions therefore damage human cells, but in a solution with same osmolarity as the cells (isotonic), water molecules enter and leave the cells at the same rate so they remain healthy. It is therefore important for any human tissues and organs to be bathed in an isotonic solution during medical procedures. Usually an isotonic sodium chloride solution is used, which is called normal saline. It has an osmolarity of about 300 mOsm (milliOsmoles).

Normal saline is used in many medical procedures. It can be:

- safely introduced to a patient's blood system via an intravenous drip.

- used to rinse wounds and skin abrasions.

- used to keep areas of damaged skin moistened prior to skin grafts.

- used as the basis for eye drops.

- frozen to the consistency of slush for packing hearts, kidneys and other donor organs that have to be transported to the hospital where the transplant operation is to be done.

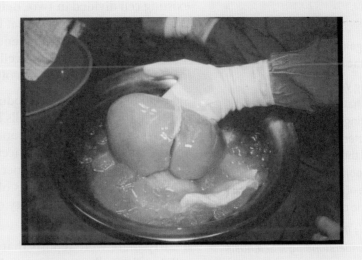

▲ Figure 19 Donor liver packed in an isotonic medium, surrounded by isotonic slush. There is a worldwide shortage of donor organs – in most countries it is possible to register as a possible future donor

1.5 The origin of cells

Understanding
→ Cells can only be formed by division of pre-existing cells.
→ The first cells must have arisen from non-living material.
→ The origin of eukaryotic cells can be explained by the endosymbiotic theory.

Applications
→ Evidence from Pasteur's experiments that spontaneous generation of cells and organisms does not now occur on Earth.

Nature of science
→ Testing the general principles that underlie the natural world: the principle that cells only come from pre-existing cells needs to be verified.

Cell division and the origin of cells

Cells can only be formed by division of pre-existing cells.

Since the 1880s there has been a theory in biology that cells can only be produced by division of a pre-existing cell. The evidence for this hypothesis is very strong and is discussed in the nature of science panel below.

The implications of the hypothesis are remarkable. If we consider the trillions of cells in our bodies, each one was formed when a previously existing cell divided in two. Before that all of the genetic material in the nucleus was copied so that both cells formed by cell division had a nucleus with a full complement of genes. We can trace the origin of cells in the body back to the first cell – the zygote that was the start of our lives, produced by the fusion of a sperm and an egg.

Sperm and egg cells were produced by cell division in our parents. We can trace the origins of all cells in our parents' bodies back to the zygote from which they developed, and then continue this process over the generations of our human ancestors. If we accept that humans evolved from pre-existing ancestral species, we can trace the origins of cells back through hundreds of millions of years to the earliest cells on Earth. There is therefore a continuity of life from its origins on Earth to the cells in our bodies today.

In 2010 there were reports that biologists had created the first artificial cell, but this cell was not entirely new. The base sequence of the DNA of a bacterium (*Mycoplasma mycoides*) was synthesized artificially, with a few deliberate changes. This DNA was transferred to pre-existing cells of a different type of bacterium (*Mycoplasma capricolum*), which was effectively converted into *Mycoplasma mycoides*. This process was therefore an extreme form of genetic modification and the creation of entirely new cells remains an insuperable challenge at the moment.

Activity

The loss of Silphium

The Greek coin in figure 2 depicts a Silphium plant, which grew in a small part of what is now Libya and was highly prized for its medicinal uses, especially as a birth control agent. It seems to have been so widely collected that within a few hundred years of the ancient Greeks colonizing North Africa it had become extinct. Rather than arising again spontaneously, Silphium has remained extinct and we cannot now test its contraceptive properties scientifically. How can we prevent the loss of other plants that could be of use to us?

▲ Figure 2 An ancient Greek coin, showing Silphium

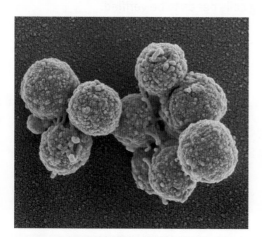

▲ Figure 1 Synthetic *Mycoplasma* bacteria

Spontaneous generation and the origin of cells

Verifying the general principles that underlie the natural world: the principle that cells only come from pre-existing cells needs to be verified.

Spontaneous generation is the formation of living organisms from non-living matter. The Greek philosopher and botanist Theophrastus reported that a plant called Silphium had sprung up from soil where it was not previously present and described this as an example of spontaneous generation. Aristotle wrote about insects being formed from the dew falling on leaves or from the hair, flesh or faeces of animals. In the 16th century the German-Swiss botanist and astrologer Paracelsus quoted observations of spontaneous generation of mice, frogs and eels from water, air or decaying matter.

It is easy to see how ideas of spontaneous generation could have persisted when cells and microorganisms had not been discovered and the nature of sexual reproduction was not understood. From the 17th century onwards biologists carried out experiments to test the theory that life could arise from non-living matter. Francesco Redi showed that maggots only developed in rotting meat if flies were allowed to come into contact with it. Lazzaro Spallanzani boiled soup in eight containers, then sealed four of them and left the others open to the air. Organisms grew in the containers left open but not in the others.

Some biologists remained convinced that spontaneous generation could occur if there was access to the air. Louis Pasteur responded by carrying out carefully designed experiments with swan-necked flasks, which established beyond reasonable doubt that spontaneous generation of life does not now occur. Pasteur's experiments are described in the next section of this sub-topic.

Apart from the evidence from the experiments of Pasteur and others, there are other reasons for biologists universally accepting that cells only come from pre-existing cells:

- A cell is a highly complex structure and no natural mechanism has been suggested for producing cells from simpler subunits.

- No example is known of increases in the number of cells in a population, organism or tissue without cell division occurring.

- Viruses are produced from simpler subunits but they do not consist of cells, and they can only be produced inside the host cells that they have infected.

Spontaneous generation and Pasteur's experiments

Evidence from Pasteur's experiments that spontaneous generation of cells and organisms does not now occur on Earth.

Louis Pasteur made a nutrient broth by boiling water containing yeast and sugar. He showed that if this broth was kept in a sealed flask, it remained unchanged, and no fungi or other organisms appeared. He then passed air though a pad of cotton wool in a tube, to filter out microscopic particles from the air, including bacteria and the spores of fungi. If the pad of cotton wool was placed in broth in a sealed flask, within 36 hours, there were large number of microorganisms in the broth and mould grew over its surface.

The most famous of Pasteur's experiments involved the use of swan-necked flasks. He placed samples of broth in flasks with long necks and then melted the glass of the necks and bent it into a variety of shapes, shown in figure 3.

Pasteur then boiled the broth in some of the flasks to kill any organisms present but left others unboiled as controls. Fungi and other organisms soon appeared in the unboiled flasks but not in the boiled ones, even after long periods of time. The broth in the flasks was in contact with air, which it had been suggested was needed for spontaneous generation, yet no spontaneous generation occurred. Pasteur snapped the necks of some of the flasks to leave a shorter vertical neck. Organisms were soon apparent in these flasks and decomposed the broth.

Pasteur published his results in 1860 and subsequently repeated them with other liquids including urine and milk, with the same results. He concluded that the swan necks prevented organisms from the air getting into the broth or other liquids and that no organisms appeared spontaneously. His experiments convinced most biologists, both at the time of publication and since then.

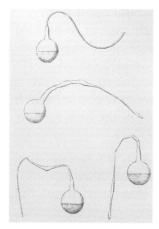

▲ Figure 3 Drawings of Pasteur's swan-necked flasks

Origin of the first cells

The first cells must have arisen from non-living material.

If we trace back the ancestry of cells over billions of years, we must eventually reach the earliest cells to have existed. These were the first living things on Earth. Unless cells arrived on Earth from somewhere else in the universe, they must have arisen from non-living material. This is a logical conclusion, but it gives perhaps the hardest question of all for biologists to answer: how could a structure as complex as the cell have arisen by natural means from non-living material?

It has sometimes been argued that complex structures cannot arise by evolution, but there is evidence that this can happen in a series of stages over long periods of time. Living cells may have evolved over hundreds of millions of years. There are hypotheses for how some of the main stages could have occurred.

1. Production of carbon compounds such as sugars and amino acids

Stanley Miller and Harold Urey passed steam through a mixture of methane, hydrogen and ammonia. The mixture was thought to be representative of the atmosphere of the early Earth. Electrical discharges were used to simulate lightning. They found that amino acids and other carbon compounds needed for life were produced.

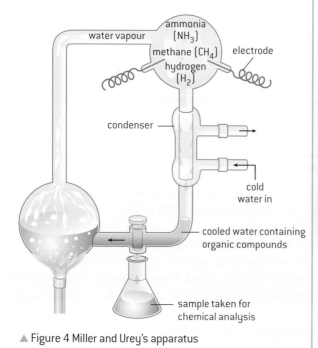

▲ Figure 4 Miller and Urey's apparatus

2. Assembly of carbon compounds into polymers

A possible site for the origin of the first carbon compounds is around deep-sea vents. These are cracks in the Earth's surface, characterized by gushing hot water carrying reduced inorganic chemicals such as iron sulphide. These chemicals represent readily accessible supplies of energy, a source of energy for the assembly of these carbon compounds into polymers.

▲ Figure 5 Deep sea vents

3. Formation of membranes

If phospholipids or other amphipathic carbon compounds were among the first carbon compounds, they would have naturally assembled into bilayers. Experiments have shown that these bilayers readily form vesicles resembling the plasma membrane of a small cell. This would have allowed different internal chemistry from that of the surroundings to develop.

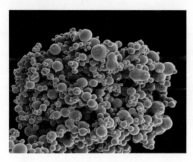

▲ Figure 6 Liposomes

4. Development of a mechanism for inheritance

Living organisms currently have genes made of DNA and use enzymes as catalysts. To replicate DNA and be able to pass genes on to offspring, enzymes are needed. However, for enzymes to be made, genes are needed. The solution to this conundrum may have been an earlier phase in evolution when RNA was the genetic material. It can store information in the same way as DNA but it is both self-replicating and can itself act as a catalyst.

Endosymbiosis and eukaryotic cells

The origin of eukaryotic cells can be explained by the endosymbiotic theory.

The theory of endosymbiosis helps to explain the evolution of eukaryotic cells. It states that mitochondria were once free-living prokaryotic organisms that had developed the process of aerobic cell respiration. Larger prokaryotes that could only respire anaerobically took them in by endocytosis. Instead of killing and digesting the smaller prokaryotes they allowed them to continue to live in their cytoplasm. As long as the smaller prokaryotes grew and divided as fast as the larger ones, they could persist indefinitely inside the larger cells. According to the theory of endosymbiosis they have persisted over hundreds of millions of years of evolution to become the mitochondria inside eukaryotic cells today.

The larger prokaryotes and the smaller aerobically respiring ones were in a symbiotic relationship in which both of them benefited. This is known as a mutualistic relationship. The smaller cell would have been supplied with food by the larger one. The smaller cell would have carried out aerobic respiration to supply energy efficiently to the larger cell. Natural selection therefore favoured cells that had developed this endosymbiotic relationship.

The endosymbiotic theory also explains the origin of chloroplasts. If a prokaryote that had developed photosynthesis was taken in by a larger cell and was allowed to survive, grow and divide, it could have developed into the chloroplasts of photosynthetic eukaryotes. Again, both of the organisms in the endosymbiotic relationship would have benefited.

Activity

Where did life begin?

Erasmus Darwin was Charles Darwin's grandfather. In a poem entitled *The Temple of Nature*, published in 1803, he tells us how and where he believed life to have originated:

> *Organic Life began beneath the waves ...*
> *Hence without parent by spontaneous birth*
> *Rise the first specks of animated earth*

Has Erasmus Darwin's hypothesis that life began in the sea been falsified?

Activity

Bangiomorpha and the origins of sex.

The first known eukaryote and first known multicellular organism is *Bangiomorpha pubescens*. Fossils of this red alga were discovered in 1,200 million year old rocks from northern Canada. It is the first organism known to produce two different types of gamete —a larger sessile female gamete and a smaller motile male gamete. *Bangiomorpha* is therefore the first organism known to reproduce sexually. It seems unlikely that eukaryote cell structure, multicellularity and sexual reproduction evolved simultaneously. What is the most likely sequence for these landmarks in evolution?

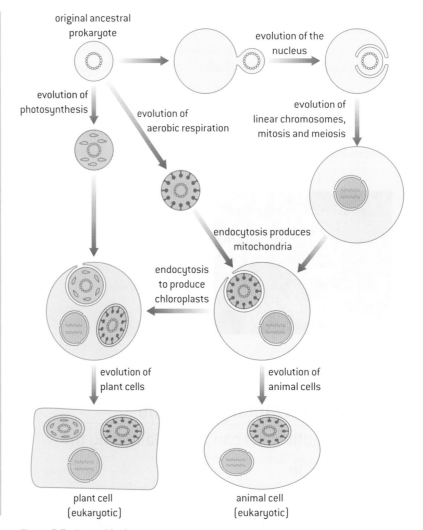

▲ Figure 7 Endosymbiosis

Although no longer capable of living independently, chloroplasts and mitochondria both have features that suggest they evolved from independent prokaryotes:

- They have their own genes, on a circular DNA molecule like that of prokaryotes.

- They have their own 70S ribosomes of a size and shape typical of some prokaryotes.

- They transcribe their DNA and use the mRNA to synthesize some of their own proteins.

- They can only be produced by division of pre-existing mitochondria and chloroplasts.

1.6 Cell division

Understanding

→ Mitosis is division of the nucleus into two genetically identical daughter nuclei.

→ Chromosomes condense by supercoiling during mitosis.

→ Cytokinesis occurs after mitosis and is different in plant and animal cells.

→ Interphase is a very active phase of the cell cycle with many processes occurring in the nucleus and cytoplasm.

→ Cyclins are involved in the control of the cell cycle.

→ Mutagens, oncogenes and metastasis are involved in the development of primary and secondary tumours.

 ## Applications

→ The correlation between smoking and incidence of cancers.

 ## Skills

→ Identification of phases of mitosis in cells viewed with a microscope.

→ Determination of a mitotic index from a micrograph.

 ## Nature of science

→ Serendipity and scientific discoveries: the discovery of cyclins was accidental.

The role of mitosis

Mitosis is division of the nucleus into two genetically identical daughter nuclei.

The nucleus of a eukaryotic cell can divide to form two genetically identical nuclei by a process called mitosis. Mitosis allows the cell to divide into two daughter cells, each with one of the nuclei and therefore genetically identical to the other.

Before mitosis can occur, all of the DNA in the nucleus must be replicated. This happens during interphase, the period before mitosis. Each chromosome is converted from a single DNA molecule into two identical DNA molecules, called chromatids. During mitosis, one of these chromatids passes to each daughter nucleus.

Mitosis is involved whenever cells with genetically identical nuclei are required in eukaryotes: during embryonic development, growth, tissue repair and asexual reproduction.

Although mitosis is a continuous process, cytologists have divided the events into four phases: prophase, metaphase, anaphase and telophase. The events that occur in these phases are described in a later section of this sub-topic.

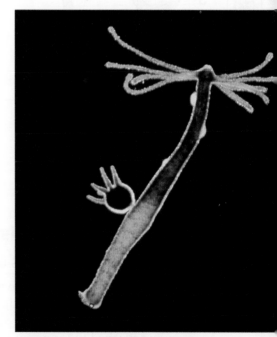

▲ Figure 1 *Hydra viridissima* with a small new polyp attached, produced by asexual reproduction involving mitosis

Interphase

Interphase is a very active phase of the cell cycle with many processes occurring in the nucleus and cytoplasm.

The cell cycle is the sequence of events between one cell division and the next. It has two main phases: interphase and cell division. Interphase is a very active phase in the life of a cell when many metabolic reactions occur. Some of these, such as the reactions of cell respiration, also occur during cell division, but DNA replication in the nucleus and protein synthesis in the cytoplasm only happen during interphase.

During interphase the numbers of mitochondria in the cytoplasm increase. This is due to the growth and division of mitochondria. In plant cells and algae the numbers of chloroplasts increase in the same way. They also synthesize cellulose and use vesicles to add it to their cell walls.

Interphase consists of three phases, the G_1 phase, S phase and G_2 phase. In the S phase the cell replicates all the genetic material in its nucleus, so that after mitosis both the new cells have a complete set of genes. Some do not progress beyond G_1, because they are never going to divide so do not need to prepare for mitosis. They enter a phase called G_0 which may be temporary or permanent.

Supercoiling of chromosomes

Chromosomes condense by supercoiling during mitosis.

During mitosis, the two chromatids that make up each chromosome must be separated and moved to opposite poles of the cell. The DNA molecules in these chromosomes are immensely long. Human nuclei are on average less than 5 μm in diameter but DNA molecules in them are more than 50,000 μm long. It is therefore essential to package chromosomes into much shorter structures. This process is known as condensation of chromosomes and it occurs during the first stage of mitosis.

Condensation occurs by means repeatedly coiling the DNA molecule to make the chromosome shorter and wider. This process is called supercoiling. Proteins called histones that are associated with DNA in eukaryote chromosomes help with supercoiling and enzymes are also involved.

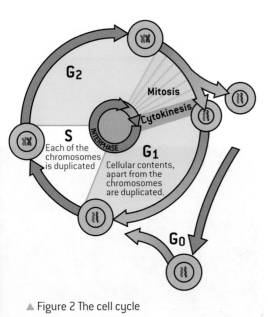

G₂

Mitosis

Cytokinesis

S Each of the chromosomes is duplicated

INTERPHASE

G₁ Cellular contents, apart from the chromosomes are duplicated.

G₀

▲ Figure 2 The cell cycle

🧪 Phases of mitosis

Identification of phases of mitosis in cells viewed with a microscope.

There are large numbers of dividing cells in the tips of growing roots. If root tips are treated chemically to allow the cells to be separated, they can be squashed to form a single layer of cells on a microscope slide. Stains that bind to DNA are used to make the chromosomes visible and stages of mitosis can then be observed using a microscope.

To be able to identify the four stages of mitosis, it is necessary to understand what is happening in them. After studying the information in this section you should be able to observe dividing cells using a microscope or in a micrograph and assign them to one of the phases.

Prophase

The chromosomes become shorter and fatter by coiling. To become short enough they have to coil repeatedly. This is called supercoiling. The nucleolus breaks down. Microtubules grow from structures called microtubule organizing centres (MTOC) to form a spindle-shaped array that links the poles of the cell. At the end of prophase the nuclear membrane breaks down.

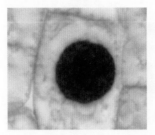

▲ Interphase – chromosomes are visible inside the nuclear membrane

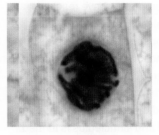

▲ Prophase – nucleoli visible in the nucleus but no individual chromosomes

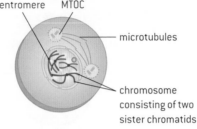

centromere   MTOC

microtubules

chromosome consisting of two sister chromatids

▲ Early prophase

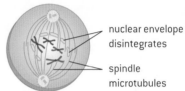

nuclear envelope disintegrates

spindle microtubules

▲ Late prophase

Metaphase

Microtubules continue to grow and attach to the centromeres on each chromosome. The two attachment points on opposite sides of each centromere allow the chromatids of a chromosome to attach to microtubules from different poles. The microtubules are all put under tension to test whether the attachment is correct. This happens by shortening of the microtubules at the centromere. If the attachment is correct, the chromosomes remain on the equator of the cell.

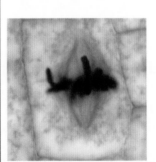

▲ Metaphase – chromosomes aligned on the equator and not inside a nuclear membrane

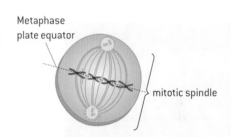

Metaphase plate equator

mitotic spindle

▲ Metaphase

Anaphase

At the start of anaphase, each centromere divides, allowing the pairs of sister chromatids to separate. The spindle microtubules pull them rapidly towards the poles of the cell. Mitosis produces two genetically identical nuclei because sister chromatids are pulled to opposite poles. This is ensured by the way that the spindle microtubules were attached in metaphase.

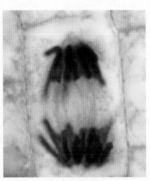

▲ Anaphase – two groups of V-shaped chromatids pointing to the two poles

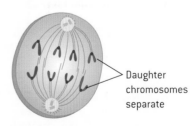

Daughter chromosomes separate

▲ Anaphase

Telophase

The chromatids have reached the poles and are now called chromosomes. At each pole the chromosomes are pulled into a tight group near the MTOC and a nuclear membrane reforms around them. The chromosomes uncoil and a nucleolus is formed. By this stage of mitosis the cell is usually already dividing and the two daughter cells enter interphase again.

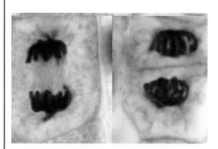

▲ Telophase – tight groups of chromosomes at each pole, new cell wall forming at the equator

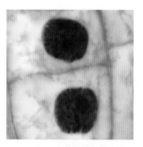

▲ Interphase – nucleoli visible inside the nuclear membranes but not individual chromosomes

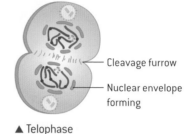

Cleavage furrow

Nuclear envelope forming

▲ Telophase

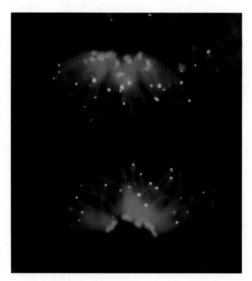

▲ Figure 3 Cell in mitosis

Data-based questions: Centromeres and telomeres

Figure 3 and the other micrographs on the website* show cells undergoing mitosis. In figure 3, DNA has been stained blue. The centromeres have been stained with a red fluorescent dye. At the ends of the chromosomes there are structures called telomeres. These have been stained with a green fluorescent dye.

1 Deduce the stage of mitosis that the cell was in, giving reasons for your answer. [3]

2 The cell has an even number of chromosomes.

   a) State how many chromosomes there are in this cell. [1]

   b) Explain the reason for body cells in plants and animals having an even number of chromosomes. [2]

   c) In the micrograph of a cell in interphase, the centromeres are on one side of the nucleus and the telomeres are on the other side. Suggest reasons for this. [2]

   d) An enzyme called telomerase lengthens the telomeres, by adding many short repeating base sequences of DNA. This enzyme is only active in the germ cells that are used to produce gametes. When DNA is replicated during the cell cycle in body cells, the end of the telomere cannot be replicated, so the telomere becomes shorter. Predict the consequences for a plant or animal of the shortening of telomeres. [2]

*www.oxfordsecondary.co.uk/ib-biology

The mitotic index

Determination of a mitotic index from a micrograph.

The mitotic index is the ratio between the number of cells in mitosis in a tissue and the total number of observed cells. It can be calculated using this equation:

$$\text{Mitotic index} = \frac{\text{number of cells in mitosis}}{\text{total number of cells}}$$

Figure 4 is a micrograph of cells from a tumour that has developed from a Leydig cell in the testis. The mitotic index for this tumour can be calculated if the total number of cells in the micrograph is counted and also the number of cells in meiosis.

To find the mitotic index of the part of a root tip where cells are proliferating rapidly, these instructions can be used:

* Obtain a prepared slide of an onion or garlic root tip. Find and examine the meristematic region, i.e. a region of rapid cell division.

* Create a tally chart. Classify each of about a hundred cells in this region as being either in interphase or in any of the stages of mitosis.

* Use this data to calculate the mitotic index.

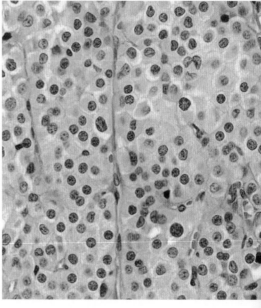

▲ Figure 4 Cells undergoing mitosis in a Leydig cell tumour

Cytokinesis

Cytokinesis occurs after mitosis and is different in plant and animal cells.

Cells can divide after mitosis when two genetically identical nuclei are present in a cell. The process of cell division is called cytokinesis. It usually begins before mitosis has actually been completed and it happens in a different way in plant and animal cells.

In animal cells the plasma membrane is pulled inwards around the equator of the cell to form a cleavage furrow. This is accomplished using a ring of contractile protein immediately inside the plasma membrane at the equator. The proteins are actin and myosin and are similar to proteins that cause contraction in muscle. When the cleavage furrow reaches the centre, the cell is pinched apart into two daughter cells.

In plant cells vesicles are moved to the equator where they fuse to form tubular structures across the equator. With the fusion of more vesicles these tubular structures merge to form two layers of membrane across the whole of the equator, which develop into the plasma membranes of the two daughter cells and are connected to the existing plasma membranes at the sides of the cell, completing the division of the cytoplasm.

The next stage in plants is for pectins and other substances to be brought in vesicles and deposited by exocytosis between the two new membranes. This forms the middle lamella that will link the new cell walls. Both of the daughter cells then bring cellulose to the equator and deposit it by exocytosis adjacent to the middle lamella. As a result, each cell builds its own cell wall adjacent to the equator.

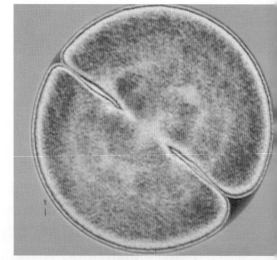

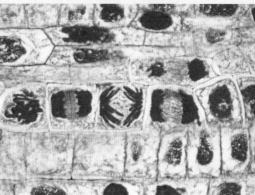

▲ Figure 5 Cytokinesis in (a) fertilized sea urchin egg (b) cell from shoot tip of *Coleus* plant

Cyclins and the control of the cell cycle

Cyclins are involved in the control of the cell cycle.

Each of the phases of the cell cycle involves many important tasks. A group of proteins called cyclins is used to ensure that tasks are performed at the correct time and that the cell only moves on to the next stage of the cycle when it is appropriate.

Cyclins bind to enzymes called cyclin-dependent kinases. These kinases then become active and attach phosphate groups to other proteins in the cell. The attachment of phosphate triggers the other proteins to become active and carry out tasks specific to one of the phases of the cell cycle.

There are four main types of cyclin in human cells. The graph in figure 6 shows how the levels of these cyclins rise and fall. Unless these cyclins reach a threshold concentration, the cell does not progress to the next stage of the cell cycle. Cyclins therefore control the cell cycle and ensure that cells divide when new cells are needed, but not at other times.

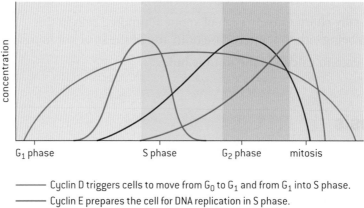

——— Cyclin D triggers cells to move from G_0 to G_1 and from G_1 into S phase.
——— Cyclin E prepares the cell for DNA replication in S phase.
——— Cyclin A activates DNA replication inside the nucleus in S phase.
——— Cyclin B promotes the assembly of the mitotic spindle and other tasks in the cytoplasm to prepare for mitosis.

▲ Figure 6

 Discovery of cyclins

Serendipity and scientific discoveries: the discovery of cyclins was accidental.

During research into the control of protein synthesis in sea urchin eggs, Tim Hunt discovered a protein that increased in concentration after fertilization then decreased in concentration, unlike other proteins which continued to increase. The protein was being synthesized over a period of about 30 minutes and then soon after was being broken down. Further experiments showed that the protein went through repeated increases and decreases in concentration that coincided with the phases of the cell cycle. The breakdown occurred about ten minutes after the start of mitosis. Hunt named the protein cyclin.

Further research revealed other cyclins and confirmed what Hunt had suspected from an early stage – that cyclins are a key factor in the control of the cell cycle. Tim Hunt was awarded a Nobel Prize for Physiology in 2001 to honour his work in the discovery of cyclins. His Nobel Lecture can be downloaded from the internet and viewed. In it he mentions the importance of serendipity several times because he had not set out to discover how the cell cycle is controlled. This discovery is an example of serendipity – a happy and unexpected discovery made by accident.

Tumour formation and cancer

Mutagens, oncogenes and metastasis are involved in the development of primary and secondary tumours.

Tumours are abnormal groups of cells that develop at any stage of life in any part of the body. In some cases the cells adhere to each other and do not invade nearby tissues or move to other parts of the body. These tumours are unlikely to cause much harm and are classified as benign. In other tumours the cells can become detached and move elsewhere in the body and develop into secondary tumours. These tumours are malignant and are very likely to be life-threatening.

Diseases due to malignant tumours are commonly known as cancer and have diverse causes. Chemicals and agents that cause cancer are known as carcinogens, because carcinomas are malignant tumours. There are various types of carcinogens including some viruses. All mutagens are carcinogenic, both chemical mutagens and also high energy radiation such as X-rays and short-wave ultraviolet light. This is because mutagens are agents that cause gene mutations and mutations can cause cancer.

Mutations are random changes to the base sequence of genes. Most genes do not cause cancer if they mutate. The few genes that can become cancer-causing after mutating are known as oncogenes. In a normal cell oncogenes are involved in the control of the cell cycle and cell division. This is why mutations in them can result in uncontrolled cell division and therefore tumour formation.

Several mutations must occur in the same cell for it to become a tumour cell. The chance of this happening is extremely small, but because there are vast numbers of cells in the body, the total chance of tumour formation during a lifetime is significant. When a tumour cell has been formed it divides repeatedly to form two, then four, then eight cells and so on. This group of cells is called a primary tumour. Metastasis is the movement of cells from a primary tumour to set up secondary tumours in other parts of the body.

 Smoking and cancer

The correlation between smoking and incidence of cancers.

A correlation in science is a relationship between two variable factors. The relationship between smoking and cancer is an example of a correlation. There are two types of correlation. With a positive correlation, when one factor increases the other one also increases; they also decrease together. With a negative correlation, when one factor increases the other decreases.

There is a positive correlation between cigarette smoking and the death rate due to cancer. This has been shown repeatedly in surveys. table 1 shows the results of one of the largest surveys, and the longest continuous one. The data shows that the more cigarettes smoked per day, the higher the death rate due to cancer. They also show a higher death rate among those who smoked at one time but had stopped.

The results of the survey also show huge increases in the death rate due to cancers of the mouth, pharynx, larynx and lung. This is expected as smoke from cigarettes comes into contact with each of these parts of the body, but there is also a positive correlation between smoking and cancers of the esophagus, stomach, kidney, bladder, pancreas and cervix. Although the death rate due to other cancers is not significantly different in smokers and non-smokers, table 1 shows smokers are several times more likely to die from all cancers than non-smokers.

It is important in science to distinguish between a correlation and a cause. Finding that there is a positive correlation between smoking and cancer does not prove that smoking causes cancer. However, in this case the causal links are well established. Cigarette smoke contains many different chemical substances. Twenty of these have been shown in experiments to cause tumours in the lungs of laboratory animals or humans. There is evidence that at least forty other chemicals in cigarette smoke are carcinogenic. This leaves little doubt that smoking is a cause of cancer.

Cause of death between 1951 and 2001 (Sample size: 34,439 male doctors in Britain)	Mortality rate per 100,000 men/year				
	Lifelong non-smokers	Former cigarette smokers	Current smokers (cigarettes/day)		
			1–14	15–24	≥25
All cancers	360	466	588	747	1,061
Lung cancer	17	68	131	233	417
Cancer of mouth, pharynx, larynx and esophagus	9	26	36	47	106
All other cancers	334	372	421	467	538

▲ Table 1 from British Medical Journal 328(7455) June 24 2004

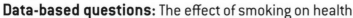

Data-based questions: The effect of smoking on health

One of the largest ever studies of the effect of smoking on health involved 34,439 male British doctors. Information was collected on how much they smoked from 1951 to 2001 and the cause of death was recorded for each of the doctors who died during this period. The table below shows some of the results. The figures given are the number of deaths per hundred thousand men per year.

Type of disease	Non-smokers	1–14 cigarettes per day	15–24 cigarettes per day	>25 cigarettes per day
Respiratory (diseases of the lungs and airways)	107	237	310	471
Circulatory (diseases of the heart and blood vessels)	1,037	1,447	1,671	1,938
Stomach and duodenal ulcers	8	11	33	34
Cirrhosis of the liver	6	13	22	68
Parkinson's disease	20	22	6	18

1   Deduce whether there is a positive correlation between smoking and the mortality rate due to **all** types of disease.   [2]

2   Using the data in the table, discuss whether the threat to health from smoking is greater with respiratory or with circulatory diseases.   [4]

3   Discuss whether the data suggests that smoking a small number of cigarettes is safe.   [3]

4   Discuss whether the data **proves** that smoking is a cause of cirrhosis of the liver.   [3]

5   The table does not include deaths due to cancer. The survey showed that seven types of cancer are linked with smoking. Suggest three cancers that you would expect smoking to cause.   [3]

Questions

1   Figure 7 represents a cell from a multicellular organism.

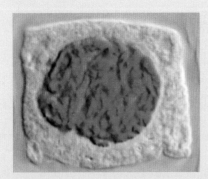

▲ Figure 7

a)   Identify, with a reason, whether the cell is

   (i)   prokaryotic or eukaryotic;   [1]

   (ii)   part of a root tip or a finger tip;   [1]

   (iii)   in a phase of mitosis or in interphase.   [1]

b)   The magnification of the drawing is 2,500 ×.

   (i)   Calculate the actual size of the cell.   [2]

   (ii)   Calculate how long a 5 μm scale bar should be if it was added to the drawing.   [1]

c)   Predict what would happen to the cell if it was placed in a concentrated salt solution for one hour. Include reasons for your answer.   [3]

2   Table 2 shows the area of membranes in a rat liver cell.

Membrane component	Area (μm^2)
Plasma membrane	1,780
Rough endoplasmic reticulum	30,400
Mitochondrial outer membrane	7,470
Mitochondrial inner membrane	39,600
Nucleus	280
Lysosomes	100
Other components	18,500

▲ Table 2

a)   Calculate the total area of membranes in the liver cell.   [2]

b)   Calculate the area of plasma membrane as a percentage of the total area of membranes in the cell. Show your working.   [3]

c)   Explain the difference in area of the inner and outer mitochondrial membranes.   [3]

d)   Using the data in the table, identify two of the main activities of liver cells.   [2]

3   In human secretory cells, for example in the lung and the pancreas, positively charged ions are pumped out, and chloride ions follow passively through chloride channels. Water also moves from the cells into the liquid that has been secreted.

In the genetic disease cystic fibrosis, the chloride channels malfunction and too few ions move out of the cells. The liquid secreted by the cells becomes thick and viscous, with associated health problems.

a)   State the names of the processes that:

   (i)   move positively charged ions out of the secretory cells   [1]

   (ii)   move chloride ions out of the secretory cells.   [1]

   (iii)   move water out of the secretory cells.   [1]

b)   Explain why the fluid secreted by people with cystic fibrosis is thick and viscous.   [4]

4   The amount of DNA present in each cell nucleus was measured in a large number of cells taken from two different cultures of human bone marrow (figure 8).

a)   For each label (I, II and III) in the Sample B graph, deduce which phase of the cell cycle the cells could be in; i.e. G1, G2 or S.   [3]

b)   Estimate the approximate amount of DNA per nucleus that would be expected in the following human cell types:

   (i)   bone marrow at prophase

   (ii)   bone marrow at telophase.   [2]

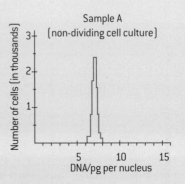
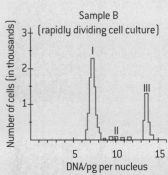

▲ Figure 8

2 MOLECULAR BIOLOGY

Introduction

Water is the medium for life. Living organisms control their composition by a complex web of chemical reactions that occur within this medium. Photosynthesis uses the energy in sunlight to supply the chemical energy needed for life and cell respiration releases this energy when it is needed. Compounds of carbon, hydrogen and oxygen are used to supply and store energy. Many proteins act as enzymes to control the metabolism of the cell and others have a diverse range of biological functions. Genetic information is stored in DNA and can be accurately copied and translated to make the proteins needed by the cell.

2.1 Molecules to metabolism

Understanding

→ Molecular biology explains living processes in terms of the chemical substances involved.

→ Carbon atoms can form four bonds allowing a diversity of compounds to exist.

→ Life is based on carbon compounds including carbohydrates, lipids, proteins and nucleic acids.

→ Metabolism is the web of all the enzyme catalysed reactions in a cell or organism.

→ Anabolism is the synthesis of complex molecules from simpler molecules including the formation of macromolecules from monomers by condensation reactions.

→ Catabolism is the breakdown of complex molecules into simpler molecules including the hydrolysis of macromolecules into monomers.

Applications

→ Urea as an example of a compound that is produced by living organisms but can also be artificially synthesized.

Skills

→ Drawing molecular diagrams of glucose, ribose, a saturated fatty acid and a generalized amino acid.

→ Identification of biochemicals such as carbohydrate, lipid or protein from molecular diagrams.

Nature of science

→ Falsification of theories: the artificial synthesis of urea helped to falsify vitalism.

▲ Figure 1 A molecular biologist at work in the laboratory

Molecular biology

Molecular biology explains living processes in terms of the chemical substances involved.

The discovery of the structure of DNA in 1953 started a revolution in biology that has transformed our understanding of living organisms. It raised the possibility of explaining biological processes from the structure of molecules and how they interact with each other. The structures are diverse and the interactions are very complex, so although molecular biology is more than 50 years old, it is still a relatively young science.

Many molecules are important in living organisms including one as apparently simple as water, but the most varied and complex molecules are nucleic acids and proteins. Nucleic acids comprise DNA and RNA. They are the chemicals used to make genes. Proteins are astonishingly varied in structure and carry out a huge range of tasks within the cell, including controlling chemical reactions of the cell by acting as enzymes. The relationship between genes and proteins is at the heart of molecular biology.

The approach of the molecular biologist is reductionist as it involves considering the various biochemical processes of a living organism and breaking down into its component parts. This approach has been immensely productive in biology and has given us insights into whole organisms that we would not otherwise have. Some biologists argue that the reductionist approach of the molecular biologist cannot explain everything though, and that when component parts are combined there are emergent properties that cannot be studied without looking at the whole system together.

🌐 Synthesis of urea

Urea as an example of a compound that is produced by living organisms but can also be artificially synthesized.

Urea is a nitrogen-containing compound with a relatively simple molecular structure (figure 2). It is a component of urine and this was where it was first discovered. It is produced when there is an excess of amino acids in the body, as a means of excreting the nitrogen from the amino acids. A cycle of reactions, catalysed by enzymes, is used to produce it (figure 3). This happens in the liver. Urea is then transported by the blood stream to the kidneys where it is filtered out and passes out of the body in the urine.

Urea can also be synthesized artificially. The chemical reactions used are different from those in the liver and enzymes are not involved, but the urea that is produced is identical.

ammonia + carbon dioxide → ammonium carbamate
→ urea + water

About 100 million tonnes are produced annually. Most of this is used as a nitrogen fertilizer on crops.

▲ Figure 2 Molecular diagram of urea

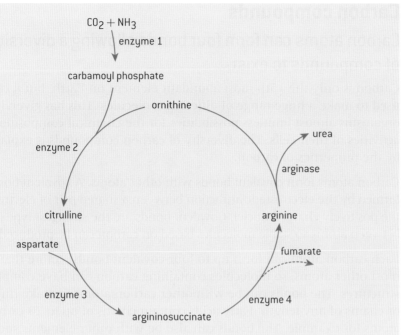

▲ Figure 3 The cycle of reactions occurring in liver cells that is used to synthesize urea

 Urea and the falsification of vitalism

Falsification of theories: the artificial synthesis of urea helped to falsify vitalism.

Urea was discovered in urine in the 1720s and was assumed to be a product of the kidneys. At that time it was widely believed that organic compounds in plants and animals could only be made with the help of a "vital principle". This was part of vitalism – the theory that the origin and phenomena of life are due to a vital principle, which is different from purely chemical or physical forces. Aristotle used the word psyche for the vital principle – a Greek word meaning breath, life or soul.

In 1828 the German chemist Friedrich Wöhler synthesized urea artificially using silver isocyanate and ammonium chloride. This was the first organic compound to be synthesized artificially. It was a very significant step, because no vital principle had been involved in the synthesis. Wöhler wrote this excitedly to the Swedish chemist Jöns Jacob Berzelius:

> In a manner of speaking, I can no longer hold my chemical water. I must tell you that I can make urea without the kidneys of any animal, be it man or dog.

An obvious deduction was that if urea had been synthesized without a vital principle, other organic compounds could be as well. Wöhler's achievement was evidence against the theory of vitalism. It helped to falsify the theory, but it did not cause all biologists to abandon vitalism immediately. It usually requires several pieces of evidence against a theory for most biologists to accept that it has been falsified and sometimes controversies over a theory continue for decades.

Although biologists now accept that processes in living organisms are governed by the same chemical and physical forces as in non-living matter, there remain some organic compounds that have not been synthesized artificially. It is still impossible to make complex proteins such as hemoglobin, for example, without using ribosomes and other components of cells. Four years after his synthesis of urea, Wöhler wrote this to Berzelius:

> Organic chemistry nowadays almost drives one mad. To me it appears like a primeval tropical forest full of the most remarkable things; a dreadful endless jungle into which one dare not enter, for there seems no way out.

Carbon compounds

Carbon atoms can form four bonds allowing a diversity of compounds to exist.

Carbon is only the 15th most abundant element on Earth, but it can be used to make a huge range of different molecules. This has given living organisms almost limitless possibilities for the chemical composition and activities of their cells. The diversity of carbon compounds is explained by the properties of carbon.

Carbon atoms form covalent bonds with other atoms. A covalent bond is formed by the electrostatic attraction between a shared pair of electrons and the positively charged nuclei. Covalent bonds are the strongest type of bond between atoms so stable molecules based on carbon can be produced.

Each carbon atom can form up to four covalent bonds – more than most other atoms, so molecules containing carbon can have complex structures. The bonds can be with other carbon atoms to make rings or chains of any length. Fatty acids contain chains of up to 20 carbon atoms for example. The bonds can also be with other elements such as hydrogen, oxygen, nitrogen or phosphorus.

Carbon atoms can bond with just one other element, such as hydrogen in methane, or they can bond to more than one other element as in ethanol (alcohol found in beer and wine). The four bonds can all be single covalent bonds or there can be two single and one double covalent bond, for example in the carboxyl group of ethanoic acid (the acid in vinegar).

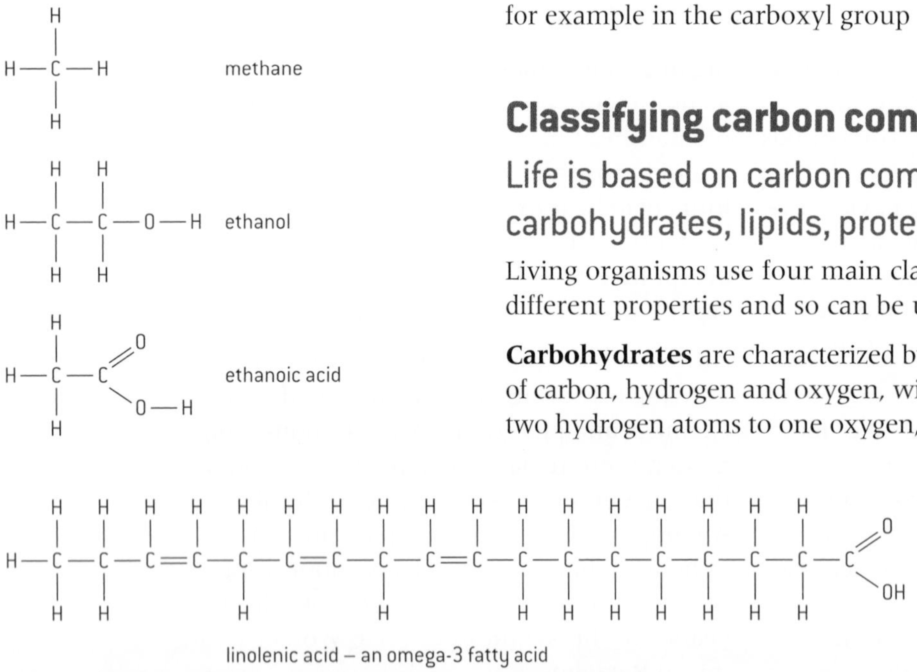

Classifying carbon compounds

Life is based on carbon compounds including carbohydrates, lipids, proteins and nucleic acids.

Living organisms use four main classes of carbon compound. They have different properties and so can be used for different purposes.

Carbohydrates are characterized by their composition. They are composed of carbon, hydrogen and oxygen, with hydrogen and oxygen in the ratio of two hydrogen atoms to one oxygen, hence the name carbo*hydrate*.

Lipids are a broad class of molecules that are insoluble in water, including steroids, waxes, fatty acids and triglycerides. In common language, triglycerides are fats if they are solid at room temperature or oils if they are liquid at room temperature.

Proteins are composed of one or more chains of amino acids. All of the amino acids in these chains contain the elements carbon, hydrogen, oxygen and nitrogen, but two of the twenty amino acids also contain sulphur.

Nucleic acids are chains of subunits called nucleotides, which contain carbon, hydrogen, oxygen, nitrogen and phosphorus. There are two types of nucleic acid: ribonucleic acid (RNA) and deoxyribonucleic acid (DNA).

▲ Figure 4 Some common naturally-occurring carbon compounds

Drawing molecules

Drawing molecular diagrams of glucose, ribose, a saturated fatty acid and a generalized amino acid.

There is no need to memorize the structure of many different molecules but a biologist should be able to draw diagrams of a few of the most important molecules.

Each atom in a molecule is represented using the symbol of the element. For example a carbon atom is represented with C and an oxygen atom with O. Single covalent bonds are shown with a line and double bonds with two lines.

Some chemical groups are shown with the atoms together and bonds not indicated. Table 1 gives examples.

Name of group	Full structure	Simplified notation
hydroxyl	— O — H	–OH
amine	— N (with H above and H below)	–NH$_2$
carboxyl	— C (=O, O — H)	–COOH
methyl	— C (H above, H below, — H)	–CH$_3$

▲ Table 1

Ribose

- The formula for ribose is C$_5$H$_{10}$O$_5$

- The molecule is a five-membered ring with a side chain.

- Four carbon atoms are in the ring and one forms the side chain.

- The carbon atoms can be numbered starting with number 1 on the right.

- The hydroxyl groups (OH) on carbon atoms 1, 2 and 3 point up, down and down respectively.

▲ Ribose

Glucose

- The formula for glucose is C$_6$H$_{12}$O$_6$

- The molecule is a six-membered ring with a side chain.

- Five carbon atoms are in the ring and one forms the side chain.

- The carbon atoms can be numbered starting with number 1 on the right.

- The hydroxyl groups (OH) on carbon atoms 1, 2, 3 and 4 point down, down, up and down respectively, although in a form of glucose used by plants to make cellulose the hydroxyl group on carbon atom 1 points upwards.

▲ Glucose

65

Saturated fatty acids

- The carbon atoms form an unbranched chain.

- In saturated fatty acids they are bonded to each other by single bonds.

- The number of carbon atoms is most commonly between 14 and 20.

- At one end of the chain the carbon atom is part of a carboxyl group

- At the other end the carbon atom is bonded to three hydrogen atoms.

- All other carbon atoms are bonded to two hydrogen atoms.

Amino acids

- A carbon atom in the centre of the molecule is bonded to four different things:

  - an amine group, hence the term amino acid;

  - a carboxyl group which makes the molecule an acid;

  - a hydrogen atom;

  - the R group, which is the variable part of amino acids.

full molecular diagram   simplified molecular diagram

▲ Molecular diagrams of an amino acid

▲ Simplified molecular diagram of a saturated fatty acid

▲ Full molecular diagram of a saturated fatty acid

🧪 Identifying molecules

Identification of biochemicals as carbohydrate, lipid or protein from molecular diagrams.

The molecules of carbohydrates, lipids and proteins are so different from each other that it is usually quite easy to recognize them.

- Proteins contain C, H, O and N whereas carbohydrates and lipids contain C, H and O but not N.

- Many proteins contain sulphur (S) but carbohydrates and lipids do not.

- Carbohydrates contain hydrogen and oxygen atoms in a ratio of 2:1, for example glucose is $C_6H_{12}O_6$ and sucrose (the sugar commonly used in baking) is $C_{12}H_{22}O_{11}$

- Lipids contain relatively less oxygen than carbohydrates, for example oleic acid (an unsaturated fatty acid) is $C_{18}H_{34}O_2$ and the steroid testosterone is $C_{19}H_{28}O_2$

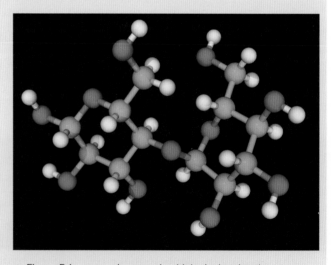

▲ Figure 5 A commonly-occurring biological molecule

Metabolism

Metabolism is the web of all the enzyme catalysed reactions in a cell or organism.

All living organisms carry out large numbers of different chemical reactions. These reactions are catalysed by enzymes. Most of them happen in the cytoplasm of cells but some are extracellular, such as the reactions used to digest food in the small intestine. Metabolism is the sum of all reactions that occur in an organism.

Metabolism consists of pathways by which one type of molecule is transformed into another, in a series of small steps. These pathways are mostly chains of reactions but there are also some cycles. An example is shown in figure 3.

Even in relatively simple prokaryote cells, metabolism consists of over 1,000 different reactions. Global maps showing all reactions are very complex. They are available on the internet, for example in the Kyoto Encyclopedia of Genes and Genomes.

Anabolism

Anabolism is the synthesis of complex molecules from simpler molecules including the formation of macromolecules from monomers by condensation reactions.

Metabolism is often divided into two parts, anabolism and catabolism. Anabolism is reactions that build up larger molecules from smaller ones. An easy way to remember this is by recalling that anabolic steroids are hormones that promote body building. Anabolic reactions require energy, which is usually supplied in the form of ATP.

Anabolism includes these processes:

- Protein synthesis using ribosomes.
- DNA synthesis during replication.
- Photosynthesis, including production of glucose from carbon dioxide and water.
- Synthesis of complex carbohydrates including starch, cellulose and glycogen.

Catabolism

Catabolism is the breakdown of complex molecules into simpler molecules including the hydrolysis of macromolecules into monomers.

Catabolism is the part of metabolism in which larger molecules are broken down into smaller ones. Catabolic reactions release energy and in some cases this energy is captured in the form of ATP, which can then be used in the cell. Catabolism includes these processes:

- Digestion of food in the mouth, stomach and small intestine.
- Cell respiration in which glucose or lipids are oxidized to carbon dioxide and water.
- Digestion of complex carbon compounds in dead organic matter by decomposers.

67

2.2 Water

Understanding

→ Water molecules are polar and hydrogen bonds form between them.

→ Hydrogen bonding and dipolarity explain the adhesive, cohesive, thermal and solvent properties of water.

→ Substances can be hydrophilic or hydrophobic.

 Nature of science

→ Use theories to explain natural phenomena: the theory that hydrogen bonds form between water molecules explains water's properties.

 Applications

→ Comparison of the thermal properties of water with those of methane.

→ Use of water as a coolant in sweat.

→ Methods of transport of glucose, amino acids, cholesterol, fats, oxygen and sodium chloride in blood in relation to their solubility in water.

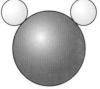

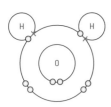

tends to pull the electrons slightly in this direction

small positive charge δ⁺ on each hydrogen atom

Corresponding negative charge $2\delta^-$ on oxygen atom

▲ Figure 1 Water molecules

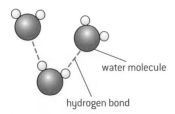

water molecule

hydrogen bond

▲ Figure 2 The dotted line indicates the presence of an intermolecular force between the molecules. This is called a hydrogen bond

Hydrogen bonding in water

Water molecules are polar and hydrogen bonds form between them.

A water molecule is formed by covalent bonds between an oxygen atom and two hydrogen atoms. The bond between hydrogen and oxygen involves unequal sharing of electrons – it is a polar covalent bond. This is because the nucleus of the oxygen atom is more attractive to electrons than the nuclei of the hydrogen atoms (figure 1).

Because of the unequal sharing of electrons in water molecules, the hydrogen atoms have a partial positive charge and oxygen has a partial negative charge. Because water molecules are bent rather than linear, the two hydrogen atoms are on the same side of the molecule and form one pole and the oxygen forms the opposite pole.

Positively charged particles (positive ions) and negatively charged particles (negative ions) attract each other and form an ionic bond. Water molecules only have partial charges, so the attraction is less but it is still enough to have significant effects. The attraction between water molecules is a "hydrogen bond". Strictly speaking it is an intermolecular force rather than a bond. A hydrogen bond is the force that forms when a hydrogen atom in one polar molecule is attracted to a slightly negative atom of another polar covalent molecule.

Although a hydrogen bond is a weak intermolecular force, water molecules are small, so there are many of them per unit volume of water and large numbers of hydrogen bonds (figure 2). Collectively they give water its unique properties and these properties are, in turn, of immense importance to living things.

Hydrogen bonds and the properties of water

Use theories to explain natural phenomena: the theory that hydrogen bonds form between water molecules explains water's properties.

There is strong experimental evidence for hydrogen bonds, but it remains a theory that they form between water molecules. Scientists cannot prove without doubt that they exist as they are not directly visible. However, hydrogen bonds are a very useful way of explaining the properties of water. They explain the cohesive, adhesive, thermal and solvent properties of water. It is these distinctive properties that make water so useful to living organisms.

It might seem unwise to base our understanding of the natural world on something that has not been proven to exist. However this is the way that science works – we can assume that a theory is correct if there is evidence for it, if it helps to predict behaviour, if it has not been falsified and if it helps to explain natural phenomena.

Properties of water

Hydrogen bonding and dipolarity explain the cohesive, adhesive, thermal and solvent properties of water.

Cohesive properties

Cohesion refers to the binding together of two molecules of the same type, for instance two water molecules.

Water molecules are cohesive – they cohere, which means they stick to each other, due to hydrogen bonding, described in the previous section. This property is useful for water transport in plants. Water is sucked through xylem vessels at low pressure. The method can only work if the water molecules are not separated by the suction forces. Due to hydrogen bonding this rarely happens and water can be pulled up to the top of the tallest trees – over a hundred metres.

Adhesive properties

Hydrogen bonds can form between water and other polar molecules, causing water to stick to them. This is called adhesion. This property is useful in leaves, where water adheres to cellulose molecules in cell walls. If water evaporates from the cell walls and is lost from the leaf via the network of air spaces, adhesive forces cause water to be drawn out of the nearest xylem vessel. This keeps the walls moist so they can absorb carbon dioxide needed for photosynthesis.

Thermal properties

Water has several thermal properties that are useful to living organisms:

- **High specific heat capacity.** Hydrogen bonds restrict the motion of water molecules and increases in the temperature of water require hydrogen bonds to be broken. Energy is needed to do this. As a result, the amount of energy needed to raise the temperature of water is relatively large. To cool down, water must lose relatively large amounts of energy. Water's temperature remains relatively stable in comparison to air or land, so it is a thermally stable habitat for aquatic organisms.

- **High latent heat of vaporization.** When a molecule evaporates it separates from other molecules in a liquid and becomes a vapour molecule. The heat needed to do this is called the latent heat of

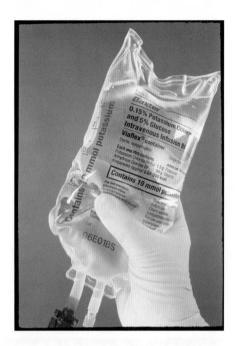

vaporization. Evaporation therefore has a cooling effect. Considerable amounts of heat are needed to evaporate water, because hydrogen bonds have to be broken. This makes it a good evaporative coolant. Sweating is an example of the use of water as a coolant.

- **High boiling point.** The boiling point of a substance is the highest temperature that it can reach in a liquid state. For the same reasons that water has a high latent heat of vaporization, its boiling point is high. Water is therefore liquid over a broad range of temperatures – from 0 °C to 100 °C. This is the temperature range found in most habitats on Earth.

Solvent properties

Water has important solvent properties. The polar nature of the water molecule means that it forms shells around charged and polar molecules, preventing them from clumping together and keeping them in solution. Water forms hydrogen bonds with polar molecules. Its partially negative oxygen pole is attracted to positively charged ions and its partially positive hydrogen pole is attracted to negatively charged ions, so both dissolve. Cytoplasm is a complex mixture of dissolved substances in which the chemical reactions of metabolism occurs.

Hydrophilic and hydrophobic

Substances can be hydrophilic or hydrophobic.

The literal meaning of the word hydrophilic is water-loving. It is used to describe substances that are chemically attracted to water. All substances that dissolve in water are hydrophilic, including polar molecules such as glucose, and particles with positive or negative charges such as sodium and chloride ions. Substances that water adheres to, cellulose for example, are also hydrophilic.

Some substances are insoluble in water although they dissolve in other solvents such as propanone (acetone). The term hydrophobic is used to describe them, though they are not actually water-fearing. Molecules are hydrophobic if they do not have negative or positive charges and are nonpolar. All lipids are hydrophobic, including fats and oils

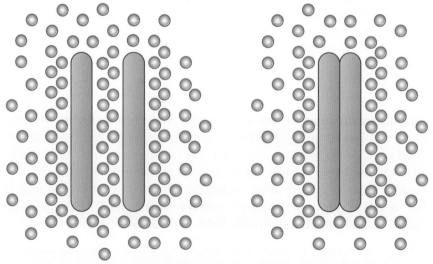

▲ Figure 3 When two nonpolar molecules in water come into contact, weak interactions form between them and more hydrogen bonds form between water molecules

If a nonpolar molecule is surrounded by water molecules, hydrogen bonds form between the water molecules, but not between the nonpolar molecule and the water molecules. If two nonpolar molecules are surrounded by water molecules and random movements bring them together, they behave as though they are attracted to each other. There is a slight attraction between nonpolar molecules, but more significantly, if they are in contact with each other, more hydrogen bonds can form between water molecules. This is not because they are water-fearing: it is simply because water molecules are more attracted to each other than to the nonpolar molecules. As a result, nonpolar molecules tend to join together in water to form larger and larger groups. The forces that cause nonpolar molecules to join together into groups in water are known as hydrophobic interactions.

 ## Comparing water and methane

Comparison of the thermal properties of water with those of methane.

The properties of water have already been described. Methane is a waste product of anaerobic respiration in certain prokaryotes that live in habitats where oxygen is lacking. Methanogenic prokaryotes live in swamps and other wetlands and in the guts of animals, including termites, cattle and sheep. They also live in waste dumps and are deliberately encouraged to produce methane in anaerobic digesters. Methane can be used as a fuel but if allowed to escape into the atmosphere it contributes to the greenhouse effect.

Water and methane are both small molecules with atoms linked by single covalent bonds. However water molecules are polar and can form hydrogen bonds, whereas methane molecules are nonpolar and do not form hydrogen bonds. As a result their physical properties are very different.

The data in table 1 shows some of the physical properties of methane and water. The density and specific heat capacity are given for methane and water in a liquid state. The data shows that water has a higher specific heat capacity, higher latent heat of vaporization, higher melting point and higher boiling point. Whereas methane is liquid over a range of only 22 °C, water is liquid over 100 °C.

Property	Methane	Water
Formula	CH_4	H_2O
Molecular mass	16	18
Density	0.46g per cm³	1g per cm³
Specific heat capacity	2.2 J per g per °C	4.2 J per g per °C
Latent heat of vaporization	760 J/g	2,257 J/g
Melting point	−182 °C	0 °C
Boiling point	−160 °C	100 °C

▲ Table 1 Comparing methane and water

▲ Figure 4 Bubbles of methane gas, produced by prokaryotes decomposing organic matter at the bottom of a pond have been trapped in ice when the pond froze

Cooling the body with sweat

Use of water as a coolant in sweat.

Sweat is secreted by glands in the skin. The sweat is carried along narrow ducts to the surface of the skin where it spreads out. The heat needed for the evaporation of water in sweat is taken from the tissues of the skin, reducing their temperature. Blood flowing through the skin is therefore cooled. This is an effective method of cooling the body because water has a high latent heat of vaporization. Solutes in the sweat, especially ions such as sodium, are left on the skin surface and can sometimes be detected by their salty taste.

Sweat secretion is controlled by the hypothalamus of the brain. It has receptors that monitor blood temperature and also receives sensory inputs from temperature receptors in the skin. If the body is overheated the hypothalamus stimulates the sweat glands to secrete up to two litres of sweat per hour. Usually no sweat is secreted if the body is below the target temperature, though when adrenalin is secreted we sweat even if we are already cold. This is because adrenalin is secreted when our brain anticipates a period of intense activity that will tend to cause the body to overheat.

There are methods of cooling other than sweating, though many of these also rely on heat loss due to evaporation of water. Panting in dogs and birds is an example. Transpiration is evaporative loss of water from plant leaves; it has a cooling effect which is useful in hot environments.

Transport in blood plasma

Methods of transport of glucose, amino acids, cholesterol, fats, oxygen and sodium chloride in blood in relation to their solubility in water.

Blood transports a wide variety of substances, using several methods to avoid possible problems and ensure that each substance is carried in large enough quantities for the body's needs.

Sodium chloride is an ionic compound that is freely soluble in water, dissolving to form sodium ions (Na^+) and chloride ions (Cl^-), which are carried in blood plasma.

Amino acids have both negative and positive charges. Because of this they are soluble in water but their solubility varies depending on the R group, some of which are hydrophilic while others are hydrophobic. All amino acids are soluble enough to be carried dissolved in blood plasma.

Glucose is a polar molecule. It is freely soluble in water and is carried dissolved in blood plasma.

Oxygen is a nonpolar molecule. Because of the small size of the molecule it dissolves in water but only sparingly and water becomes saturated with oxygen at relatively low concentrations. Also, as the temperature of water rises, the solubility of oxygen decreases, so blood plasma at 37 °C can hold much less dissolved oxygen than water at 20 °C or lower. The amount of oxygen that blood plasma can transport around the body is far too little to provide for aerobic cell respiration. This problem is overcome by the use of hemoglobin in red blood cells. Hemoglobin has binding sites for oxygen and greatly increases the capacity of the blood for oxygen transport.

Fats molecules are entirely nonpolar, are larger than oxygen and are insoluble in water. They are carried in blood inside lipoprotein complexes. These are groups of molecules with a single layer of phospholipid on the outside and fats inside. The hydrophilic phosphate heads of the phospholipids face outwards and are in contact with water in the blood plasma. The hydrophobic hydrocarbon tails face inwards and are in contact with the fats. There are also proteins in the phospholipid monolayer, hence the name lipoprotein.

Cholesterol molecules are hydrophobic, apart from a small hydrophilic region at one end. This is not enough to make cholesterol dissolve in water and instead it is transported with fats in lipoprotein complexes. The cholesterol molecules are positioned in the phospholipid monolayers, with the hydrophilic region facing outwards in the region with the phosphate heads of the phospholipids.

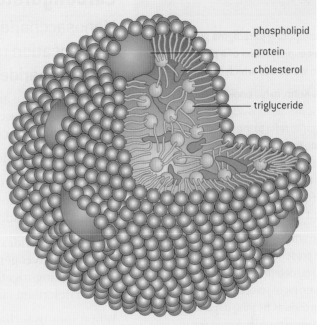

phospholipid
protein
cholesterol
triglyceride

▲ Figure 5 Arrangement of molecules in a lipoprotein complex

2.3 Carbohydrates and lipids

Understanding

→ Monosaccharide monomers are linked together by condensation reactions to form disaccharides and polysaccharide polymers.

→ Fatty acids can be saturated, monounsaturated or polyunsaturated.

→ Unsaturated fatty acids can be cis or trans isomers.

→ Triglycerides are formed by condensation from three fatty acids and one glycerol.

Applications

→ Structure and function of cellulose and starch in plants and glycogen in humans.

→ Scientific evidence for health risks of trans-fats and saturated fats.

→ Lipids are more suitable for long-term energy storage in humans than carbohydrates.

→ Evaluation of evidence and the methods used to obtain evidence for health claims made about lipids.

Nature of science

→ Evaluating claims: health claims made about lipids need to be assessed.

Skills

→ Use of molecular visualization software to compare cellulose, starch and glycogen.

→ Determination of body mass index by calculation or use of a nomogram.

Carbohydrates

Monosaccharide monomers are linked together by condensation reactions to form disaccharides and polysaccharide polymers.

Glucose, fructose and ribose are all examples of monosaccharides. The structure of glucose and ribose molecules was shown in sub-topic 2.1. Monosaccharides can be linked together to make larger molecules.

- Monosaccharides are single sugar units.

- Disaccharides consist of two monosaccharides linked together. For example, maltose is made by linking two glucose molecules together. Sucrose is made by linking a glucose and a fructose.

- Polysaccharides consist of many monosaccharides linked together. Starch, glycogen and cellulose are polysaccharides. They are all made by linking together glucose molecules. The differences between them are described later in this sub-topic.

When monosaccharides combine, they do so by a process called condensation (figure 1). This involves the loss of an –OH from one molecule and an –H from another molecule, which together form H_2O. Thus, condensation involves the combination of subunits and yields water.

Linking together monosaccharides to form disaccharides and polysaccharides is an anabolic process and energy has to be used to do it. ATP supplies energy to the monosaccharides and this energy is then used when the condensation reaction occurs.

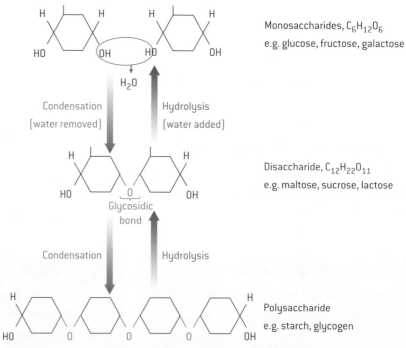

Monosaccharides, $C_6H_{12}O_6$
e.g. glucose, fructose, galactose

H_2O

Condensation (water removed)

Hydrolysis (water added)

Disaccharide, $C_{12}H_{22}O_{11}$
e.g. maltose, sucrose, lactose

Glycosidic bond

Condensation

Hydrolysis

Polysaccharide
e.g. starch, glycogen

▲ Figure 1 Condensation and hydrolysis reactions between monosaccharides and disaccharides

Imaging carbohydrate molecules

Use of molecular visualization software to compare cellulose, starch and glycogen.

The most widely used molecular visualization software is JMol, which can be downloaded free of charge. There are also many websites that use JMol, which are easier to use.

When JMol software is being used, you should be able to make these changes to the image of a molecule that you see on the screen:

- Use the scroll function on the mouse to make the image larger or smaller.

- Left click and move the mouse to rotate the image.

- Right click to display a menu that allows you to change the style of molecular model, label the atoms, make the molecule rotate continuously or change the background colour.

Spend some time developing your skill in molecular visualization and then try these questions to test your skill level and learn more about the structure of polysaccharides.

Questions

1   Select glucose with the ball and stick style with a black background.

- What colours are used to show carbon, hydrogen and oxygen atoms? [2]

2   Select sucrose with sticks style and a blue background.

- What is the difference between the glucose ring and the fructose ring in the sucrose molecule? [1]

3   Select amylose, which is the unbranched form of starch, with the wireframe style and a white background. If possible select a short amylose chain and then a longer one.

- What is the overall shape of an amylose molecule? [1]

- How many glucose molecules in the chain are linked to only one other glucose? [1]

4   Select amylopectin, with the styles and colours that you prefer. Amylopectin is the branched form of starch. Zoom in to look closely at a position where there is a branch. A glucose molecule must be linked to an extra third glucose to make the branch.

- What is different about this linkage, compared to the linkages between glucose molecules in unbranched parts of the molecule? [1]

- How many glucose molecules are linked to only one other glucose in the amylopectin molecule? [1]

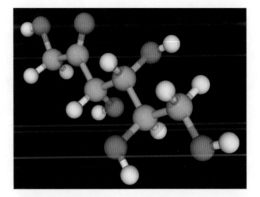

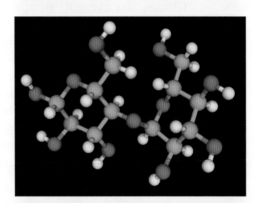

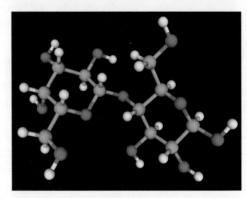

▲ Figure 2 Images of sugars using molecular visualization software – (a) fructose, (b) maltose, (c) lactose

5 Select glycogen. It is similar but not identical to the amylopectin form of starch.

- What is the difference between glycogen and amylopectin? [1]

6 Select cellulose.

- How is it different in shape from the other polysaccharides? [1]

7 Look at the oxygen atom that forms part of the ring in each glucose molecule in the chain.

- What pattern do you notice in the position of these oxygen atoms along the chain?

🌐 Polysaccharides

Structure and function of cellulose and starch in plants and glycogen in humans.

Starch, glycogen and cellulose are all made by linking together glucose molecules, yet their structure and functions are very different. This is due to differences in the type of glucose used to make them and in the type of linkage between glucose molecules.

Glucose has five –OH groups, any of which could be used in condensation reactions, but only three of them are actually used to link to make polysaccharides. The most common link is between the –OH on carbon atom 1 (on the right hand side in molecular diagrams of glucose) and the –OH on carbon atom 4 (shown on the left hand side). The –OH on carbon atom 6 (shown at the top of molecular diagrams) is used to form side branches in some polysaccharides.

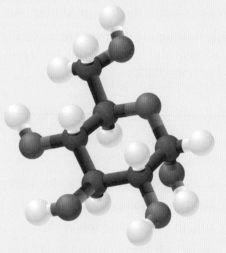

▲ Figure 3 Glucose molecule

Glucose can have the –OH group on carbon atom 1 pointing either upwards or downwards. In alpha glucose (α-glucose) the –OH group points downwards but in beta glucose (β-glucose) it points upwards. This small difference has major consequences for polysaccharides made from glucose.

Cellulose is made by linking together β-glucose molecules. Condensation reactions link carbon atom 1 to carbon atom 4 on the next β-glucose. The –OH groups on carbon atom 1 and 4 point in opposite directions: up on carbon 1 and down on carbon 4. To bring these –OH groups together and allow a condensation reaction to occur, each β-glucose added to the chain has to be positioned at 180° to the previous one. The glucose subunits in the chain are oriented alternately upwards and downwards. The consequence of this is that the cellulose molecule is a straight chain, rather than curved.

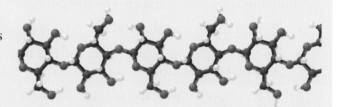

▲ Figure 4 Cellulose

Cellulose molecules are unbranched chains of β-glucose, allowing them to form bundles with hydrogen bonds linking the cellulose molecules. These bundles are called cellulose microfibrils. They have very high tensile strength and are used as the basis of plant cell walls. The tensile strength of cellulose prevents plant cells from bursting, even when very high pressures have developed inside the cell due to entry of water by osmosis.

Starch is made by linking together α-glucose molecules. As in cellulose, the links are made by condensation reactions between the –OH groups on carbon atom 1 of one glucose and carbon atom 4 of the adjacent glucose. These –OH groups both point downwards, so all the glucose molecules in starch can be orientated in the same way. The consequence of this is that the starch molecule is curved, rather than straight. There are two forms of starch. In amylose the chain of α-glucose molecules is unbranched and forms a helix. In amylopectin the chain is branched, so has a more globular shape.

Starch is only made by plant cells. Molecules of both types of starch are hydrophilic but they are too large to be soluble in water. They are therefore useful in cells where large amounts of glucose need to be stored, but a concentrated glucose solution would cause too much water to enter a cell by osmosis. Starch is used as a store of glucose and therefore of energy in seeds and storage organs such as potato cells. Starch is made as a temporary store in leaf cells when glucose is being made faster by photosynthesis than it can be exported to other parts of the plant.

Glycogen is very similar to the branched form of starch, but there is more branching, making the molecule more compact. Glycogen is made by animals and also some fungi. It is stored in the liver and some muscles in humans. Glycogen has the same function as starch in plants: it acts as a store of energy in the form of glucose, in cells where large stores of dissolved glucose would cause osmotic problems. With both starch and

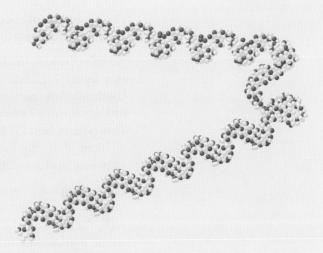

▲ Figure 5 Starch

glycogen it is easy to add extra glucose molecules or remove them. This can be done at both ends of an unbranched molecule or at any of the ends in a branched molecule. Starch and glycogen molecules do not have a fixed size and the number of glucose molecules that they contain can be increased or decreased.

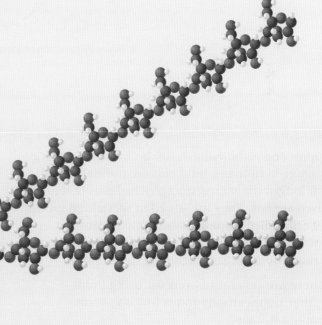

▲ Figure 6 Glycogen

Lipids

Triglycerides are formed by condensation from three fatty acids and one glycerol.

Lipids are a diverse group of carbon compounds that share the property of being insoluble in water. Triglycerides are one of the principal groups of lipid. Examples of triglycerides are the fat in adipose tissue in humans and the oil in sunflower seeds. Fats are liquid at body temperature (37 °C) but solid at room temperature (20 °C) whereas oils are liquid at both body temperature and room temperature.

A triglyceride is made by combining three fatty acids with one glycerol (see figure 7). Each of the fatty acids is linked to the glycerol by a condensation reaction, so three water molecules are produced. The linkage formed between each fatty acid and the glycerol is an ester bond. This type of bond is formed when an acid reacts with the –OH group in an alcohol. In this case the reaction is between the –COOH group on a fatty acid and an –OH on the glycerol.

Triglycerides are used as energy stores. The energy from them can be released by aerobic cell respiration. Because they do not conduct heat well, they are used as heat insulators, for example in the blubber of Arctic marine mammals.

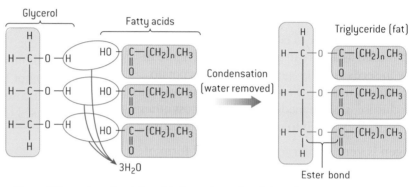

▲ Figure 7 Formation of a triglyceride from glycerol and three fatty acids

🌐 Energy storage

Lipids are more suitable for long term energy storage in humans than carbohydrates.

Lipids and carbohydrates are both used for energy storage in humans, but lipids are normally used for long-term energy storage. The lipids that are used are fats. They are stored in specialized groups of cells called adipose tissue. Adipose tissue is located immediately beneath the skin and also around some organs including the kidneys.

There are several reasons for using lipids rather than carbohydrates for long-term energy storage:

- The amount of energy released in cell respiration per gram of lipids is double the amount released from a gram of carbohydrates. The same amount of energy stored as lipid rather than carbohydrate therefore adds half as much to body mass. In fact the mass advantage of lipids is even

greater because fats form pure droplets in cells with no water associated, whereas each gram of glycogen is associated with about two grams of water, so lipids are actually six times more efficient in the amount of energy that can be stored per gram of body mass. This is important, because we have to carry our energy stores around with us wherever we go. It is even more important for animals such as birds and bats that fly.

- Stored lipids have some secondary roles that could not be performed as well by carbohydrates. Because lipids are poor conductors of heat, they can be used as heat insulators. This is the reason for much of our stored fat being in sub-cutaneous adipose tissue next to the skin. Because fat is liquid at body temperature, it can also act as a shock absorber. This is the reason for adipose tissue around the kidneys and some other organs.

Glycogen is the carbohydrate that is used for energy storage, in the liver and in some muscles. Although lipids are ideal for long-term storage of energy, glycogen is used for short-term storage. This is because glycogen can be broken down to glucose rapidly and then transported easily by the blood to where it is needed. Fats in adipose tissue cannot be mobilized as rapidly. Glucose can be used either in anaerobic or aerobic cell respiration whereas fats and fatty acids can only be used in aerobic respiration. The liver stores up to 150 grams of glycogen and some muscles store up to 2% glycogen by mass.

Data-based questions: Emperor penguins

During the Antarctic winter female Emperor penguins live and feed at sea, but males have to stay on the ice to incubate the single egg the female has laid. Throughout this time the males eat no food. After 16 weeks the eggs hatch and the females return. While the males are incubating the eggs they stand in tightly packed groups of about 3,000 birds. To investigate the reasons for standing in groups, 10 male birds were taken from a colony at Pointe Geologie in Antarctica. They had already survived 4 weeks without food. They were kept for 14 more weeks without food in fenced enclosures where they could not form groups. All other conditions were kept the same as in the wild colony. The mean air temperature was −16.4 °C. The composition of the captive and the wild birds' bodies was measured before and after the 14-week period of the experiment. The results in kilograms are shown in figure 8.

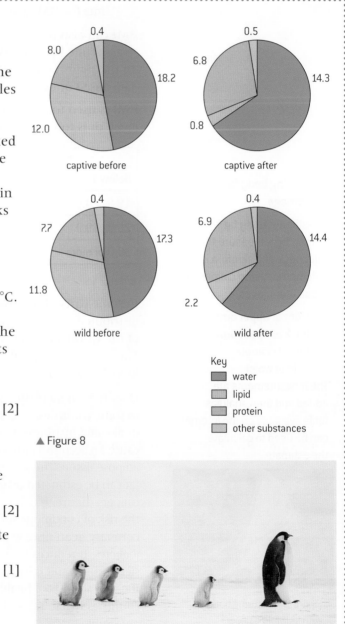

▲ Figure 8

a) Calculate the total mass loss for each group of birds. [2]

  i) wild

  ii) captive

b) Compare the changes in lipid content of the captive birds with those of the birds living free in the colony. [2]

c) Besides being used as an energy source, state another function of lipid which might be important for penguin survival. [1]

Body mass index

Determination of body mass index by calculation or use of a nomogram.

The body mass index, usually abbreviated to BMI, was developed by a Belgian statistician, Adolphe Quetelet. Two measurements are needed to calculate it: the mass of the person in kilograms and their height in metres.

BMI is calculated using this formula:

$$BMI = \frac{\text{mass in kilograms}}{(\text{height in metres})^2}$$

Units for BMI are kg m^{-2}

BMI can also be found using a type of chart called a nomogram. A straight line between the height on the left hand scale and the mass on the right hand scale intersects the BMI on the central scale. The data based questions on page 81 include a BMI nomogram.

BMI is used to assess whether a person's body mass is at a healthy level, or is too high or too low. Table 1 shows how this is done:

BMI	Status
below 18.5	underweight
18.5–24.9	normal weight
25.0–29.9	overweight
30.0 or more	obese

▲ Table 1

In some parts of the world food supplies are insufficient or are unevenly distributed and many people as a result are underweight. In other parts of the world a likelier cause of being underweight is anorexia nervosa. This is a psychological condition that involves voluntary starvation and loss of body mass.

Obesity is an increasing problem in some countries. Excessive food intake and insufficient exercise cause an accumulation of fat in adipose tissue. The amount of body fat can be estimated using skinfold calipers (figure 9). Obesity increases the risk of conditions such as coronary heart disease and type 2 diabetes. It reduces life expectancy significantly and is increasing the overall costs of health care in countries where rates of obesity are rising.

▲ Measuring body mass. What was this person's body mass index if their height was 1.80 metres?

Activity

Estimating body fat percentage

To estimate body fat percentage, measure the thickness of a skinfold in millimetres using calipers in these four places:

- Front of upper arm
- Back of upper arm
- Below scapula
- Side of waist

The measurements are added and then analysis tools available on the internet can be used to calculate the estimate.

▲ Figure 9 Measuring body fat with skinfold callipers

Data based questions: Nomograms and BMI

Use figure 11 to answer these questions.

1 a) State the body mass index of a man who has a mass of 75 kg and a height of 1.45 metres. [1]

b) Deduce the body mass status of this man. [1]

2 a) State the body mass of the person standing on the scales on the previous page. [1]

b) The person has a height of 1.8 metres. Deduce their body mass status. [1]

3 a) A woman has a height of 150 cm and a BMI of 40. Calculate the minimum amount of body mass she must lose to reach normal body mass status. Show all of your working. [3]

b) Suggest two ways in which the woman could reduce her body mass. [2]

4. Outline the relationship between height and BMI for a fixed body mass. [1]

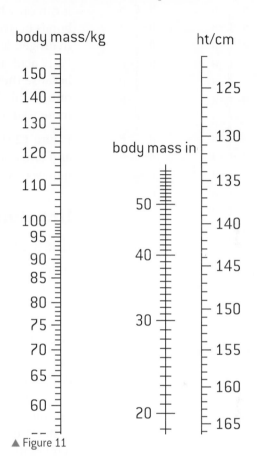
▲ Figure 11

▲ Figure 10 Jogger

Fatty acids

Fatty acids can be saturated, monounsaturated or polyunsaturated.

The basic structure of fatty acids was described in sub-topic 2.1. There is a chain of carbon atoms, with hydrogen atoms linked to them by single covalent bonds. It is therefore a hydrocarbon chain. At one end of the chain is the acid part of the molecule. This is a carboxyl group, which can be represented as –COOH.

The length of the hydrocarbon chain is variable but most of the fatty acids used by living organisms have between 14 and 20 carbon atoms. Another variable feature is the bonding between the carbon atoms. In some fatty

palmitic acid
• saturated
• non-essential

linolenic acid
• polyunsaturated
• all *cis*
• essential
• omega 3

palmitoleic acid
• monounsaturated
• *cis*
• non-essential
• omega 7

▲ Figure 12 Examples of fatty acids

acids all of the carbon atoms are linked by single covalent bonds, but in other fatty acids there are one or more positions in the chain where carbon atoms are linked by double covalent bonds.

If a carbon atom is linked to adjacent carbons in the chain by single bonds, it can also bond to two hydrogen atoms. If a carbon atom is linked by a double bond to an adjacent carbon in the chain, it can only bond to one hydrogen atom. A fatty acid with single bonds between all of its carbon atoms therefore contains as much hydrogen as it possibly could and is called a **saturated fatty acid**. Fatty acids that have one or more double bonds are **unsaturated** because they contain less hydrogen than they could. If there is one double bond, the fatty acid is **monounsaturated** and if it has more than one double bond it is **polyunsaturated**.

Figure 12 shows one saturated fatty acid, one monounsaturated and one polyunsaturated fatty acid. It is not necessary to remember names of specific fatty acids in IB Biology.

Unsaturated fatty acids

Unsaturated fatty acids can be cis or trans isomers.

In unsaturated fatty acids in living organisms, the hydrogen atoms are nearly always on the same side of the two carbon atoms that are double bonded – these are called **cis**-fatty acids. The alternative is for the hydrogens to be on opposite sides – called **trans**-fatty acids. These two conformations are shown in figure 14.

In cis-fatty acids, there is a bend in the hydrocarbon chain at the double bond. This makes triglycerides containing cis-unsaturated fatty acids less good at packing together in regular arrays than saturated fatty acids, so it lowers the melting point. Triglycerides with cis-unsaturated fatty acids are therefore usually liquid at room temperature – they are oils.

Trans-fatty acids do not have a bend in the hydrocarbon chain at the double bond, so they have a higher melting point and are solid at room temperature. Trans-fatty acids are produced artificially by partial hydrogenation of vegetable or fish oils. This is done to produce solid fats for use in margarine and some other processed foods.

cis

trans

▲ Figure 13 Double bonds in fatty acids

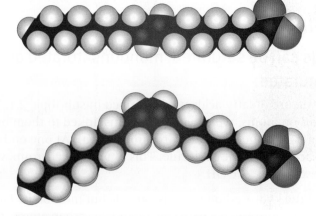

▲ Figure 14 Fatty acid stereochemistry – (a) trans (b) cis

Health risks of fats

Scientific evidence for health risks of trans-fats and saturated fats.

There have been many claims about the effects of different types of fat on human health. The main concern is coronary heart disease (CHD). In this disease the coronary arteries become partially blocked by fatty deposits, leading to blood clot formation and heart attacks.

A positive correlation has been found between saturated fatty acid intake and rates of CHD in many research programs. However, finding a correlation does not prove that saturated fats cause the disease. It could be another factor correlated with saturated fat intake, such as low amounts of dietary fibre, that actually causes CHD.

There are populations that do not fit the correlation. The Maasai of Kenya for example have a diet that is rich in meat, fat, blood and milk. They therefore have a high consumption of saturated fats, yet CHD is almost unknown among the Maasai. Figure 17 shows members of another Kenyan tribe that show this trend.

Diets rich in olive oil, which contains cis-monounsaturated fatty acids, are traditionally eaten in countries around the Mediterranean. The populations of these countries typically have low rates of CHD and it has been claimed that this is due to the intake of cis-monounsaturated fatty acids. However, genetic factors in these populations, or other aspects of the diet such as the use of tomatoes in many dishes could explain the CHD rates.

There is also a positive correlation between amounts of trans-fat consumed and rates of CHD. Other risk factors have been tested, to see if they can account for the correlation, but none did. Trans-fats therefore probably do cause CHD. In patients who had died from CHD, fatty deposits in the diseased arteries have been found to contain high concentrations of trans-fats, which gives more evidence of a causal link.

▲ Figure 15 Triglycerides in olive oil contain cis-unsaturated fatty acids

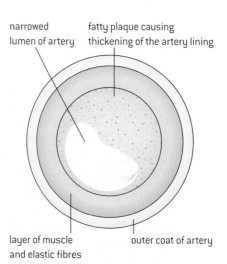

narrowed lumen of artery

fatty plaque causing thickening of the artery lining

layer of muscle and elastic fibres

outer coat of artery

▲ Figure 16 Artery showing fatty plaque

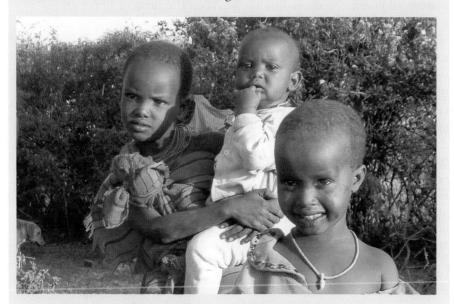

▲ Figure 17 Samburu people of Northern Kenya. Like the Maasai, the Samburu have a diet rich in animal products but rates of heart disease are extremely low

Evaluating the health risks of foods

Evaluating claims: health claims made about lipids need to be assessed.

Many health claims about foods are made. In some cases the claim is that the food has a health benefit and in other cases it is that the food is harmful. Many claims have been found to be false when they are tested scientifically.

It is relatively easy to test claims about the effects of diet on health using laboratory animals. Large numbers of genetically uniform animals can be bred and groups of them with the same age, sex and state of health can be selected for use in experiments. Variables other than diet, such as temperature and amount of exercise, can be controlled so that they do not influence the results of the experiment. Diets can be designed so that only one dietary factor varies and strong evidence can thus be obtained about the effect of this factor on the animal.

Results of animal experiments are often interesting, but they do not tell us with certainty what the health effects are on humans of a factor in the diet. It would be very difficult to carry out similar controlled experiments with humans. It might be possible to select matched groups of experimental subjects in terms of age, sex and health, but unless identical twins were used they would be genetically different. It would also be almost impossible to control other variables such as exercise and few humans would be willing to eat a very strictly controlled diet for a long enough period.

Researchers into the health risks of food must therefore use a different approach. Evidence is obtained by epidemiological studies. These involve finding a large cohort of people, measuring their food intake and following their health over a period of years. Statistical procedures can then be used to find out whether factors in the diet are associated with an increased frequency of a particular disease. The analysis has to eliminate the effects of other factors that could be causing the disease.

Nature of science question: using volunteers in experiments.

During the Second World War, experiments were conducted both in England and in the US using conscientious objectors to military service as volunteers. The volunteers were willing to sacrifice their health to help extend medical knowledge. A vitamin C trial in England involved 20 volunteers. For six weeks they were all given a diet containing 70 mg of vitamin C. Then, for the next eight months, three volunteers were kept on the diet with 70 mg, seven had their dose reduced to 10 mg and ten were given no vitamin C. All of these ten volunteers developed scurvy. Three-centimetre cuts were made in their thighs, with the wounds closed up with five stitches. These wounds failed to heal. There was also bleeding from hair follicles and from the gums. Some of the volunteers developed more serious heart problems. The groups given 10 mg or 70 mg of vitamin C fared equally well and did not develop scurvy.

Experiments on requirements for vitamin C have also been done using real guinea-pigs, which ironically are suitable because guinea-pigs, like humans, cannot synthesize ascorbic acid. During trial periods with various intakes of vitamin C, concentrations in blood plasma and urine were monitored. The guinea-pigs were then killed and collagen in bone and skin was tested. The collagen in guinea-pigs with restricted vitamin C had less cross-linking between the protein fibres and therefore lower strength.

1   Is it ethically acceptable for doctors or scientists to perform experiments on volunteers, where there is a risk that the health of the volunteers will be harmed?

2   Sometimes people are paid to participate in medical experiments, such as drug trials. Is this more or less acceptable than using unpaid volunteers?

3   Is it better to use animals for experiments or are the ethical objections the same as with humans?

4   Is it acceptable to kill animals, so that an experiment can be done?

Analysis of data on health risks of lipids

Evaluation of evidence and the methods used to obtain the evidence for health claims made about lipids.

An evaluation is defined in IB as an assessment of implications and limitations. Evidence for health claims comes from scientific research. There are two questions to ask about this research:

1 Implications – do the results of the research support the health claim strongly, moderately or not at all?

2 Limitations – were the research methods used rigorous, or are there uncertainties about the conclusions because of weaknesses in methodology?

The first question is answered by analysing the results of the research – either experimental results or results of a survey. Analysis is usually easiest if the results are presented as a graph or other type of visual display.

● Is there a correlation between intake of the lipid being investigated and rate of the disease or the health benefit? This might be either a positive or negative correlation.

● How large is the difference between mean (average) rates of the disease with different levels of lipid intake? Small differences may not be significant.

● How widely spread is the data? This is shown by the spread of data points on a scattergraph or the size of error bars on a bar chart. The more widely spread the data, the less likely it is that mean differences are significant.

● If statistical tests have been done on the data, do they show significant differences?

The second question is answered by assessing the methods used. The points below refer to surveys and slightly different questions should be asked to assess controlled experiments.

● How large was the sample size? In surveys it is usually necessary to have thousands of people in a survey to get reliable results.

● How even was the sample in sex, age, state of health and life style? The more even the sample, the less other factors can affect the results.

● If the sample was uneven, were the results adjusted to eliminate the effects of other factors?

● Were the measurements of lipid intake and disease rates reliable? Sometimes people in a survey do not report their intake accurately and diseases are sometimes misdiagnosed.

Data-based questions: Evaluating evidence from a health survey

The Nurses' Health Survey is a highly respected survey into the health consequences of many factors. It began in 1976 with 121,700 female nurses in the USA and Canada, who completed a lengthy questionnaire about their lifestyle factors and medical history. Follow-up questionnaires have been completed every two years since then.

Details of the methods used to assess diet and diagnose coronary heart disease can be found by reading a research paper in the American Journal of Epidemiology, which is freely available on the internet: Oh, K, Hu, FB, Manson, JE, Stampfer, MJ and Willett, WC. (2005) Dietary Fat Intake and Risk of Coronary Heart Disease in Women: 20 Years of Follow-up of the Nurses'

Health Study. *American Journal of Epidemiology*, 161:672–679. doi:10.1093/aje/kwi085

To assess the effects of trans-fats on rates of CHD, the participants in the survey were divided into five groups according to their trans-fat intake. Quintile 1 was the 20% of participants with the lowest intake and quintile 5 was the 20% with the highest intake. The average intake of trans-fats for each quintile was calculated, as a percentage of dietary energy intake. The relative risk of CHD was found for each quintile, with Quintile 1 assigned a risk of 1. The risk was adjusted for differences between the quintiles in age, body mass index, smoking, alcohol intake, parental history of CHD, intake of other foods that affect CHD rates and various other factors. Figure 18 is a graph showing the percentage of energy from trans-fats for each of the five quintiles and the adjusted relative risk of CHD. The effect of trans-fat intake on relative risk of CHD is statistically significant with a confidence level of 99%.

1   Suggest reasons for using only female nurses in this survey.   [3]

2   State the trend shown in the graph.   [1]

3   The mean age of nurses in the five quintiles was not the same. Explain the reasons for adjusting the results to compensate for the effects of age differences.   [2]

4   Calculate the chance, based on the statistical tests, of the differences in CHD risk being due to factors other than trans-fat intake.   [2]

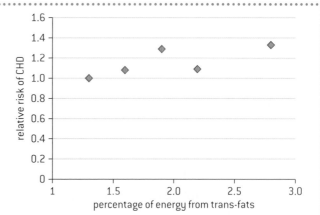

▲ Figure 18

Data for graph

% of energy from trans-fat	1.3	1.6	1.9	2.2	2.8
Relative risk of CHD	1.0	1.08	1.29	1.19	1.33

Data-based questions: Saturated fats and coronary heart disease

Populations ranked by % calories as saturated fat	E. Finland	W. Finland	Zutphen	USA	Slavonia	Belgrade	Crevalcor	Zrenjanin	Dalmatia	Crete	Montegiorgio	Velika	Rome	Corfu	Ushibuka	Tanushimaru
% Calories as saturated fat	22	19	19	18	14	12	10	10	9	9	9	9	8	7	3	3
Death rate/ 100,000 yr^{-1} — CHD	992	351	420	574	214	288	248	152	86	9	150	80	290	144	66	88
Death rate/ 100,000 yr^{-1} — All causes	1727	1318	1175	1088	1477	509	1241	1101	758	543	1080	1078	1027	764	1248	1006

▲ Table 2

1   a) Plot a scattergraph of the data in table 2.   [5]

    b) Outline the trend shown by the scattergraph.   [2]

2   Compare the results for:

    a) East and West Finland;   [2]

    b) Crete and Montegiorgio.   [2]

3   Evaluate the evidence from this survey for saturated fats as a cause of coronary heart disease.   [4]

2.4 Proteins

Understanding

→ Amino acids are linked together by condensation to form polypeptides.

→ There are twenty different amino acids in polypeptides synthesized on ribosomes.

→ Amino acids can be linked together in any sequence giving a huge range of possible polypeptides.

→ The amino acid sequence of polypeptides is coded for by genes.

→ A protein may consist of a single polypeptide or more than one polypeptide linked together.

→ The amino acid sequence determines the three-dimensional conformation of a protein.

→ Living organisms synthesize many different proteins with a wide range of functions.

→ Every individual has a unique proteome.

Applications

→ Rubisco, insulin, immunoglobulins, rhodopsin, collagen and spider silk as examples of the range of protein functions.

→ Denaturation of proteins by heat or deviation of pH from the optimum.

Skills

→ Draw molecular diagrams to show the formation of a peptide bond.

Nature of science

→ Patterns, trends and discrepancies: most but not all organisms assemble polypeptides from the same amino acids.

Amino acids and polypeptides

Amino acids are linked together by condensation to form polypeptides.

Polypeptides are chains of amino acids that are made by linking together amino acids by condensation reactions. This happens on ribosomes by a process called translation, which will be described in sub-topic 2.7. Polypeptides are the main component of proteins and in many proteins they are the only component. Some proteins contain one polypeptide and other proteins contain two or more.

The condensation reaction involves the amine group ($-NH_2$) of one amino acid and the carboxyl group ($-COOH$) of another. Water is eliminated, as

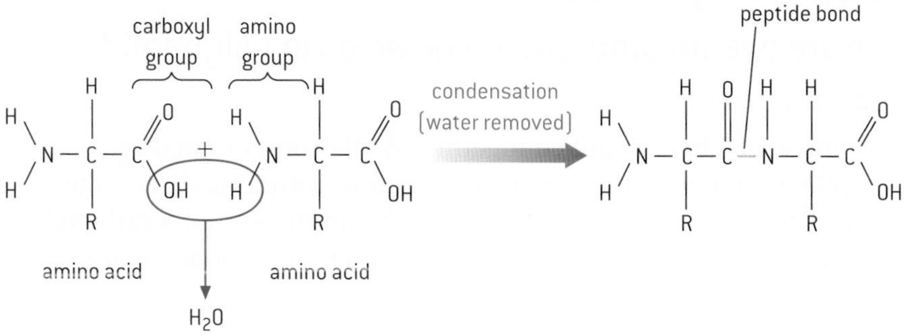

▲ Figure 1 Condensation joins two amino acids with a peptide bond

in all condensation reactions, and a new bond is formed between the two amino acids, called a peptide bond. A dipeptide is a molecule consisting of two amino acids linked by a peptide bond. A polypeptide is a molecule consisting of many amino acids linked by peptide bonds.

Polypeptides can contain any number of amino acids, though chains of fewer than 20 amino acids are usually referred to as oligopeptides rather than polypeptides. Insulin is a small protein that contains two polypeptides, one with 21 amino acids and the other with 30. The largest polypeptide discovered so far is titin, which is part of the structure of muscle. In humans titin is a chain of 34,350 amino acids, but in mice it is even longer with 35,213 amino acids.

⚗ Drawing peptide bonds

Draw molecular diagrams to show the formation of a peptide bond.

To form a dipeptide, two amino acids are linked by a condensation reaction between the amine group of one amino acid and the carboxyl group of the other. This is shown in figure 1.

The peptide bond is the same, whatever R group the amino acid carries. To test your skill at showing how peptide bonds are formed, try showing the formation of a peptide bond between two of the amino acids in figure 2. There are sixteen possible dipeptides that can be produced from these four amino acids.

You could also try to draw an oligopeptide of four amino acids, linked by three peptide bonds. If you do this correctly, you should see these features:

- There is chain of atoms linked by single covalent bonds forming the backbone of the oligopeptide, with a repeating sequence of $-N-C-C-$

- A hydrogen atom is linked by a single bond to each nitrogen atom in the backbone and an oxygen atom is linked by a double bond to one of the two carbon atoms.

- The amine ($-NH_2$) and carboxyl ($-COOH$) groups are used up in forming the peptide bond and only remain at the ends of the chain. These are called the amino and carboxyl terminals of the chain.

- The R groups of each amino acid remain and project outwards from the backbone.

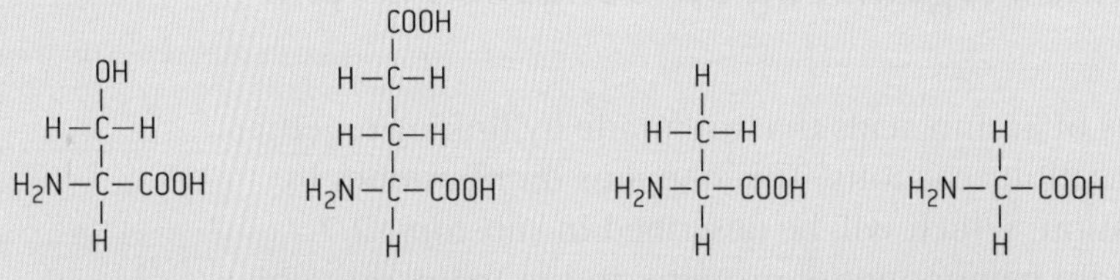

▲ Figure 2 Some common amino acids

The diversity of amino acids

There are twenty different amino acids in polypeptides synthesized on ribosomes.

The amino acids that are linked together by ribosomes to make polypeptides all have some identical structural features: a carbon atom in the centre of the molecule is bonded to an amine group, a carboxyl group and a hydrogen atom. The carbon atom is also bonded to an R group, which is different in each amino acid.

Twenty different amino acids are used by ribosomes to make polypeptides. The amine groups and the carboxyl groups are used up in forming the peptide bond, so it is the R groups of the amino acids that give a polypeptide its character. The repertoire of R groups allows living organisms to make and use an amazingly wide range of proteins. Some of the differences are shown in table 1. It is not necessary to try to learn these specific differences but it is important to remember that because of the differences between their R groups, the twenty amino acids are chemically very diverse.

Some proteins contain amino acids that are not in the basic repertoire of twenty. In most cases this is due to one of the twenty being modified after a polypeptide has been synthesized. There is an example of modification of amino acids in collagen, a structural protein used to provide tensile strength in tendons, ligaments, skin and blood vessel walls. Collagen polypeptides made by ribosomes contain proline at many positions, but at some of these positions it is converted to hydroxyproline, which makes the collagen more stable.

Nine R groups are hydrophobic with between zero and nine carbon atoms		Eleven R groups are hydrophilic			
		Four hydrophilic R groups are polar but never charged	Seven R groups can become charged		
Three R groups contain rings	Six R groups do not contain rings		Four R groups act as an acid by giving up a proton and becoming negatively charged	Three R groups act as a base by accepting a proton and becoming positively charged	

▲ Table 1 Classification of amino acids

 Amino acids and origins

Patterns, trends and discrepancies: most but not all organisms assemble polypeptides from the same amino acids.

It is a remarkable fact that most organisms make proteins using the same 20 amino acids. In some cases amino acids are modified after a polypeptide has been synthesized, but the initial process of linking together amino acids on ribosomes with peptide bonds usually involves the same 20 amino acids.

We can exclude the possibility that this trend is due to chance. There must be one or more reasons for it. Several hypotheses have been proposed:

- These 20 amino acids were the ones produced by chemical processes on Earth before the origin of life, so all organisms used them and have continued to use them. Other amino acids might have been used, if they had been available.

- They are the ideal 20 amino acids for making a wide range of proteins, so natural selection will always favour organisms that use them and do not use other amino acids.

- All life has evolved from a single ancestral species, which used these 20 amino acids. Because of the way that polypeptides are made by ribosomes, it is difficult for any organism to change the repertoire of amino acids, either by removing existing ones or adding new ones.

Biology is a complicated science and discrepancies are commonly encountered. Some species have been found that use one of the three codons that normally signal the end of polypeptide synthesis (stop codons) to encode an extra non-standard amino acid. For example, some species use UGA to code for selenocysteine and some use UAG to code for pyrrolysine.

▲ Figure 3 Kohoutek Comet – 26 different amino acids were found in an artificial comet produced by researchers at the Institut d'Astrophysique Spatiale (CNRS/France), which suggests that amino acids used by the first living organisms on Earth may have come from space

Activity

Calculating polypeptide diversity

Number of amino acids	Number of possible amino acid sequences	
1	20^1	
2	20^2	400
3		8,000
4		
	20^6	64 million
		10.24 trillion

▲ Table 2 Calculate the missing values

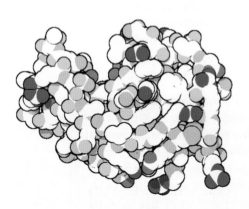

▲ Figure 4 Lysozyme with nitrogen of amine groups shown blue, oxygen red and sulphur yellow. The active site is the cleft upper left

Data-based questions: Commonality of amino acids

1 a) Discuss which of the three hypotheses for use of the same 20 amino acids by most organisms is supported by the evidence.   [3]

   b) Suggest ways of testing one of the hypotheses.   [2]

2 Cell walls of bacteria contain peptidoglycan, a complex carbon compound that contains sugars and short chains of amino acids. Some of these amino acids are different from the usual repertoire of 20. Also, some of them are right-handed forms of amino acids, whereas the 20 amino acids made into polypeptides are always the left-handed forms. Discuss whether this is a significant discrepancy that falsifies the theory that living organisms all make polypeptides using the same 20 amino acids.   [5]

Polypeptide diversity

Amino acids can be linked together in any sequence giving a huge range of possible polypeptides.

Ribosomes link amino acids together one at a time, until a polypeptide is fully formed. The ribosome can make peptide bonds between any pair of amino acids, so any sequence of amino acids is possible.

The number of possible amino acid sequences can be calculated starting with dipeptides (table 2). Both amino acids in a dipeptide can be any of the twenty so there are twenty times twenty possible sequences (20^2). There are $20 \times 20 \times 20$ possible tripeptide sequences (20^3). For a polypeptide of n amino acids there are 20^n possible sequences.

The number of amino acids in a polypeptide can be anything from 20 to tens of thousands. Taking one example, if a polypeptide has 400 amino acids, there are 20^{400} possible amino acid sequences. This is a mind-bogglingly large number and some online calculators simply express it as infinity. If we add all the possible sequences for other numbers of amino acids, the number is effectively infinite.

Genes and polypeptides

The amino acid sequence of polypeptides is coded for by genes.

The number of amino acid sequences that could be produced is immense, but living organisms only actually produce a small fraction of these. Even so, a typical cell produces polypeptides with thousands of different sequences and must store the information needed to do this. The amino acid sequence of each polypeptide is stored in a coded form in the base sequence of a gene.

Some genes have other roles, but most genes in a cell store the amino acid sequence of a polypeptide. They use the genetic code to do this. Three bases of the gene are needed to code for each amino acid in the polypeptide. In theory a polypeptide with 400 amino acids should require a gene with a sequence of 1,200 bases. In practice genes are always longer, with extra base sequences at both ends and sometimes also at certain points in the middle.

The base sequence that actually codes for a polypeptide is known to molecular biologists as the open reading frame. One puzzle is that open reading frames only occupy a small proportion of the total DNA of a species.

Proteins and polypeptides

A protein may consist of a single polypeptide or more than one polypeptide linked together.

Some proteins are single polypeptides, but others are composed of two or more polypeptides linked together.

Integrin is a membrane protein with two polypeptides, each of which has a hydrophobic portion embedded in the membrane. Rather like the blade and handle of a folding knife the two polypeptides can either be adjacent to each other or can unfold and move apart when it is working.

Collagen consists of three long polypeptides wound together to form a rope-like molecule. This structure has greater tensile strength than the three polypeptides would if they were separate. The winding allows a small amount of stretching, reducing the chance of the molecule breaking.

Hemoglobin consists of four polypeptides with associated non-polypeptide structures. The four parts of hemoglobin interact to transport oxygen more effectively to tissues that need it than if they were separate.

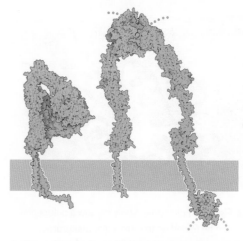

▲ Figure 5 Integrin embedded in a membrane (grey) shown folded and inactive and open with binding sites inside and outside the cell indicated (red and purple)

Number of polypeptides	Example	Background
1	lysozyme	Enzyme in secretions such as nasal mucus and tears; it kills some bacteria by digesting the peptidoglycan in their cell walls.
2	integrin	Membrane protein used to make connections between structures inside and outside a cell.
3	collagen	Structural protein in tendons, ligaments, skin and blood vessel walls; it provides high tensile strength, with limited stretching.
4	hemoglobin	Transport protein in red blood cells; it binds oxygen in the lungs and releases it in tissues with a reduced oxygen concentration.

▲ Table 3 Example of proteins with different numbers of polypeptides

Protein conformations

The amino acid sequence determines the three-dimensional conformation of a protein.

The conformation of a protein is its three-dimensional structure. The conformation is determined by the amino acid sequence of a protein and its constituent polypeptides. Fibrous proteins such as collagen

Activity

Molecular biologists are investigating the numbers of open reading frames in selected species for each of the major groups of living organism. It is still far from certain how many genes in each species code for a polypeptide that the organism actually uses, but we can compare current best estimates:

- *Drosophila melanogaster,* the fruit fly, has base sequences for about 14,000 polypeptides.

- *Caenorhabditis elegans*, a nematode worm with less than a thousand cells, has about 19,000.

- *Homo sapiens* has base sequences for about 23,000 different polypeptides.

- *Arabidopsis thaliana*, a small plant widely used in research, has about 27,000.

Can you find any species with greater or lesser numbers of open reading frames than these?

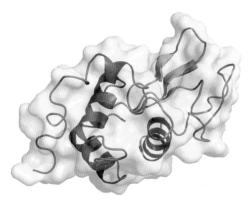

▲ Figure 6 Lysozyme, showing how a polypeptide can be folded up to form a globular protein. Three sections that are wound to form a helix are shown red and a section that forms a sheet is shown yellow. Other parts of the polypeptide including both of its ends are green

are elongated, usually with a repeating structure. Many proteins are globular, with an intricate shape that often includes parts that are helical or sheet-like.

Amino acids are added one by one, to form a polypeptide. They are always added in the same sequence to make a particular polypeptide. In globular proteins the polypeptides gradually fold up as they are made, to develop the final conformation. This is stabilized by bonds between the R groups of the amino acids that have been brought together by the folding.

In globular proteins that are soluble in water, there are hydrophilic R groups on the outside of the molecule and there are usually hydrophobic groups on the inside. In globular membrane proteins there are regions with hydrophobic R groups on the outside of the molecule, which are attracted to the hydrophobic centre of the membrane.

In fibrous proteins the amino acid sequence prevents folding up and ensures that the chain of amino acids remains in an elongated form.

Denaturation of proteins

Denaturation of proteins by heat or pH extremes.

The three-dimensional conformation of proteins is stabilized by bonds or interactions between R groups of amino acids within the molecule. Most of these bonds and interactions are relatively weak and they can be disrupted or broken. This results in a change to the conformation of the protein, which is called denaturation.

A denatured protein does not normally return to its former structure – the denaturation is permanent. Soluble proteins often become insoluble and form a precipitate. This is due to the hydrophobic R groups in the centre of the molecule becoming exposed to the water around by the change in conformation.

Heat can cause denaturation because it causes vibrations within the molecule that can break intermolecular bonds or interactions. Proteins vary in their heat tolerance. Some microorganisms that live in volcanic springs or in hot water near geothermal vents have proteins that are not denatured by temperatures of 80 °C or higher. The best known example is DNA polymerase from *Thermus aquaticus*, a prokaryote that was discovered in hot springs in Yellowstone National Park. It works best at 80 °C and because of this it is widely used in biotechnology. Nevertheless, heat causes denaturation of most proteins at much lower temperatures.

Extremes of pH, both acidic and alkaline, can cause denaturation. This is because charges on R groups are changed, breaking ionic bonds within the protein or causing new ionic bonds to form. As with heat, the three-dimensional structure of the protein is altered and proteins that have been dissolved in water often become insoluble. There are exceptions: the contents of the stomach are normally acidic, with a pH as low as 1.5, but this is the optimum pH for the protein-digesting enzyme pepsin that works in the stomach.

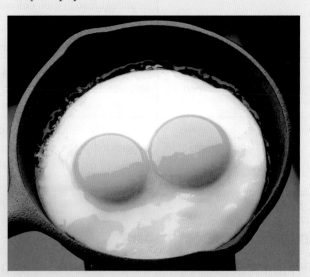

▲ Figure 7 When eggs are heated, proteins that were dissolved in both the white and the yolk are denatured. They become insoluble so both yolk and white solidify

Protein functions

Living organisms synthesize many different proteins with a wide range of functions.

Other groups of carbon compounds have important roles in the cell, but none can compare with the versatility of proteins. They can be compared to the worker bees that perform almost all the tasks in a hive. All of the functions listed here are carried out by proteins.

- **Catalysis** – there are thousands of different enzymes to catalyse specific chemical reactions within the cell or outside it.

- **Muscle contraction** – actin and myosin together cause the muscle contractions used in locomotion and transport around the body.

- **Cytoskeletons** – tubulin is the subunit of microtubules that give animals cells their shape and pull on chromosomes during mitosis.

- **Tensile strengthening** – fibrous proteins give tensile strength needed in skin, tendons, ligaments and blood vessel walls.

- **Blood clotting** – plasma proteins act as clotting factors that cause blood to turn from a liquid to a gel in wounds.

- **Transport of nutrients and gases** – proteins in blood help transport oxygen, carbon dioxide, iron and lipids.

- **Cell adhesion** – membrane proteins cause adjacent animal cells to stick to each other within tissues.

- **Membrane transport** – membrane proteins are used for facilitated diffusion and active transport, and also for electron transport during cell respiration and photosynthesis.

- **Hormones** – some such as insulin, FSH and LH are proteins, but hormones are chemically very diverse.

- **Receptors** – binding sites in membranes and cytoplasm for hormones, neurotransmitters, tastes and smells, and also receptors for light in the eye and in plants.

- **Packing of DNA** – histones are associated with DNA in eukaryotes and help chromosomes to condense during mitosis.

- **Immunity** – this is the most diverse group of proteins, as cells can make huge numbers of different antibodies.

There are many biotechnological uses for proteins including enzymes for removing stains, monoclonal antibodies for pregnancy tests or insulin for treating diabetics. Pharmaceutical companies now produce many different proteins for treating diseases. These tend to be very expensive, as it is still not easy to synthesize proteins artificially. Increasingly, genetically modified organisms are being used as microscopic protein factories.

Activity

Denaturation experiments

A solution of egg albumen in a test tube can be heated in a water bath to find the temperature at which it denatures. The effects of pH can be investigated by adding acids and alkalis to test tubes of egg albumen solution. To quantify the extent of denaturation, a colorimeter can be used as denatured albumen absorbs more light than dissolved albumen.

Activity

Botox

Botox is a neurotoxin obtained from *Clostridium botulinum* bacteria.

1　What are the reasons for injecting it into humans?

2　What is the reason for *Clostridium botulinum* producing it?

3　What are the reasons for injecting it rather than taking it orally?

Examples of proteins

Rubisco, insulin, immunoglobulins, rhodopsin, collagen and spider silk as examples of the range of protein functions.

Six proteins which illustrate some of the functions of proteins are described in table 4.

Rubisco	Insulin
This name is an abbreviation for ribulose bisphosphate carboxylase, which is arguably the most important enzyme in the world. The shape and chemical properties of its active site allow it to catalyse the reaction that fixes carbon dioxide from the atmosphere, which provides the source of carbon from which all carbon compounds needed by living organisms can be produced. It is present at high concentrations in leaves and so is probably the most abundant of all proteins on Earth.	This hormone is produced as a signal to many cells in the body to absorb glucose and help reduce the glucose concentration of the blood. These cells have a receptor for insulin in their cell membrane to which the hormone binds reversibly. The shape and chemical properties of the insulin molecule correspond precisely to the binding site on the receptor, so insulin binds to it, but not other molecules. Insulin is secreted by β cells in the pancreas and is transported by the blood.
Immunoglobulin	**Rhodopsin**
These proteins are also known as antibodies. They have sites at the tips of their two arms that bind to antigens on bacteria or other pathogens. The other parts of the immunoglobulin cause a response, such as acting as a marker to phagocytes that can engulf the pathogen. The binding sites are hypervariable. The body can produce a huge range of immunoglobulins, each with a different type of binding site. This is the basis of specific immunity to disease.	Vision depends on pigments that absorb light. One of these pigments is rhodopsin, a membrane protein of rod cells of the retina. Rhodopsin consists of a light sensitive retinal molecule, not made of amino acids, surrounded by an opsin polypeptide. When the retinal molecule absorbs a single photon of light, it changes shape. This causes a change to the opsin, which leads to the rod cell sending a nerve impulse to the brain. Even very low light intensities can be detected.
Collagen	**Spider silk**
There are a number of different forms of collagen but all are rope-like proteins made of three polypeptides wound together. About a quarter of all protein in the human body is collagen – it is more abundant than any other protein. It forms a mesh of fibres in skin and in blood vessel walls that resists tearing. Bundles of parallel collagen molecules give ligaments and blood vessel walls their immense strength. It forms part of the structure of teeth and bones, helping to prevent cracks and fractures.	Different types of silk with different functions are produced by spiders. Dragline silk is stronger than steel and tougher than Kevlar™. It is used to make the spokes of spiders' webs and the lifelines on which spiders suspend themselves. When first made it contains regions where the polypeptide forms parallel arrays. Other regions seem like a disordered tangle, but when the silk is stretched they gradually extend, making the silk extensible and very resistant to breaking.

Proteomes

Every individual has a unique proteome.

A proteome is all of the proteins produced by a cell, a tissue or an organism. By contrast, the genome is all of the genes of a cell, a tissue or an organism. To find out how many different proteins are being produced, mixtures of proteins are extracted from a sample and are then separated by gel electrophoresis. To identify whether or not a particular protein is present, antibodies to the protein that have been linked to a fluorescent marker can be used. If the cell fluoresces, the protein is present.

Whereas the genome of an organism is fixed, the proteome is variable because different cells in an organism make different proteins. Even in a single cell the proteins that are made vary over time depending on the cell's activities. The proteome therefore reveals what is actually happening in an organism, not what potentially could happen.

Within a species there are strong similarities in the proteome of all individuals, but also differences. The proteome of each individual is unique, partly because of differences of activity but also because of small differences in the amino acid sequence of proteins. With the possible exception of identical twins, none of us have identical proteins, so each of us has a unique proteome. Even the proteome of identical twins can become different with age.

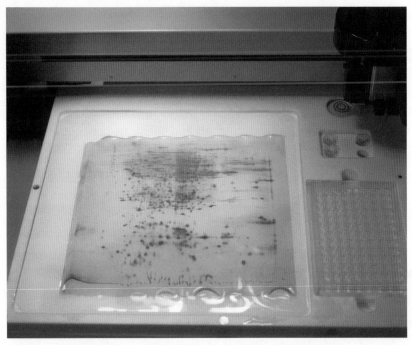

▲ Figure 8 Proteins from a nematode worm have been separated by gel electrophoresis. Each spot on the gel is a different protein

Activity

Active science: genomes and proteomes

We might expect the proteome of an organism to be smaller than its genome, as some genes do not code for polypeptides. In fact the proteome is larger. How could an organism produce more proteins than the number of genes that its genome contains?

2.5 Enzymes

Understanding

→ Enzymes have an active site to which specific substrates bind.

→ Enzyme catalysis involves molecular motion and the collision of substrates with the active site.

→ Temperature, pH and substrate concentration affect the rate of activity of enzymes.

→ Enzymes can be denatured.

→ Immobilized enzymes are widely used in industry.

Applications

→ Methods of production of lactose-free milk and its advantages.

Nature of science

→ Experimental design: accurate quantitative measurements in enzyme experiments require replicates to ensure reliability.

Skills

→ Design of experiments to test the effect of temperature, pH and substrate concentration on the activity of enzymes.

→ Experimental investigation of a factor affecting enzyme activity. (Practical 3)

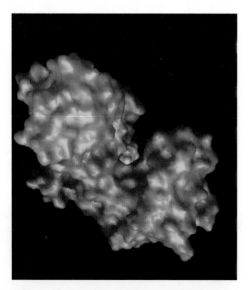

▲ Figure 1 Computer-generated image of the enzyme hexokinase, with a molecule of its substrate glucose bound to the active site. The enzyme bonds a second substrate, phosphate, to the glucose, to make glucose phosphate

Active sites and enzymes

Enzymes have an active site to which specific substrates bind.

Enzymes are globular proteins that work as catalysts – they speed up chemical reactions without being altered themselves. Enzymes are often called biological catalysts because they are made by living cells and speed up biochemical reactions. The substances that enzymes convert into products in these reactions are called **substrates**. A general equation for an enzyme-catalysed reaction is:

$$\text{substrate} \xrightarrow{\text{enzyme}} \text{product}$$

Enzymes are found in all living cells and are also secreted by some cells to work outside. Living organisms produce many different enzymes – literally thousands of them. Many different enzymes are needed, as enzymes only catalyse one biochemical reaction and thousands of reactions take place in cells, nearly all of which need to be catalysed. This property is called **enzyme–substrate specificity**. It is a significant difference between enzymes and non-biological catalysts such as the metals that are used in catalytic converters of vehicles.

To be able to explain enzyme–substrate specificity, we must look at the mechanism by which enzymes speed up reactions. This involves the substrate, or substrates binding to a special region on the surface of the enzyme called the **active site** (see figure 1). The shape and chemical properties of the active site and the substrate match each other. This allows the substrate to bind, but not other substances. Substrates are converted into products while they are bound to the active site and the products are then released, freeing the active site to catalyse another reaction.

Data-based questions: Biosynthesis of glycogen

The Nobel Prize for Medicine was won in 1947 by Gerty Cori and her husband Carl. They isolated two enzymes that convert glucose phosphate into glycogen. Glycogen is a polysaccharide, composed of glucose molecules bonded together in two ways, called 1,4 and 1,6 bonds (see figure 2).

1⟶ 4 bonding

1⟶ 4 bonding plus a
1⟶ 6 bond forming a side-branch

▲ Figure 2 Bonding in glycogen

1 Explain why two different enzymes are needed for the synthesis of glycogen from glucose phosphate. [2]

2 The formation of side-branches increases the rate at which glucose phosphate molecules can be linked on to a growing glycogen molecule. Explain the reason for this. [2]

3 Curve A was obtained using heat-treated enzymes. Explain the shape of curve A. [2]

4 Curve B was obtained using enzymes that had not been heat-treated.

   a) Describe the shape of Curve B. [2]

   b) Explain the shape of Curve B. [2]

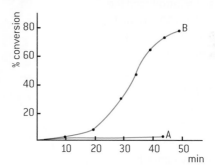

▲ Figure 3 shows the percentage conversion of glucose phosphate to glycogen by the two enzymes, over a 50-minute period

---

Enzyme activity

Enzyme catalysis involves molecular motion and the collision of substrates with the active site.

Enzyme activity is the catalysis of a reaction by an enzyme. There are three stages:

- The substrate binds to the active site of the enzyme. Some enzymes have two substrates that bind to different parts of the active site.

- While the substrates are bound to the active site they change into different chemical substances, which are the products of the reaction.

- The products separate from the active site, leaving it vacant for substrates to bind again.

A substrate molecule can only bind to the active site if it moves very close to it. The coming together of a substrate molecule and an active site is known as a collision. This might suggest a high velocity impact between two vehicles on a road, but that would be a misleading image and we need to think about molecular motion in liquids to understand how substrate–active site collisions occur.

With most reactions the substrates are dissolved in water around the enzyme. Because water is in a liquid state, its molecules and all the particles dissolved in it are in contact with each other and are in continual motion. Each particle can move separately. The direction of movement repeatedly changes and is random, which is the basis of diffusion in liquids. Both substrates and enzymes with active sites are able to move, though most substrate molecules are smaller than the enzyme so their movement is faster.

So, collisions between substrate molecules and the active site occur because of random movements of both substrate and enzyme. The substrate may be at any angle to the active site when the collision occurs. Successful collisions are ones in which the substrate and active site are correctly aligned to allow binding to take place.

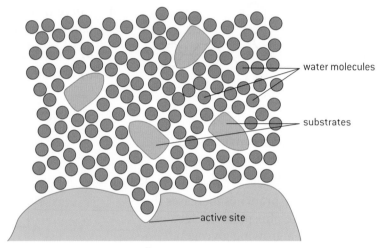

▲ Figure 4 Enzyme-substrate collisions. If random movements bring any of the substrate molecules close to the active site with the correct orientation, the substrate can bind to the active site

Factors affecting enzyme activity

Temperature, pH and substrate concentration affect the rate of activity of enzymes.

Enzyme activity is affected by temperature in two ways

- In liquids, the particles are in continual random motion. When a liquid is heated, the particles in it are given more kinetic energy. Both enzyme and substrate molecules therefore move around faster at higher temperatures and the chance of a substrate molecule colliding with the active site of the enzyme is increased. Enzyme activity therefore increases.

- When enzymes are heated, bonds in the enzyme vibrate more and the chance of the bonds breaking is increased. When bonds in the enzyme break, the structure of the enzyme changes, including the active site. This change is permanent and is called denaturation. When an enzyme molecule has been denatured, it is no longer able to catalyse reactions. As more and more enzyme molecules in a solution become denatured, enzyme activity falls. Eventually it stops altogether, when the enzyme has been completely denatured. So, as temperature rises there are reasons for both increases and decreases in enzyme activity. Figure 5 shows the effects of temperature on a typical enzyme.

Enzymes are sensitive to pH

The pH scale is used to measure the acidity or alkalinity of a solution. The lower the pH, the more acid or the less alkaline a solution is. Acidity is due to the presence of hydrogen ions, so the lower the pH, the higher the hydrogen ion concentration. The pH scale is logarithmic. This means that reducing the pH by one unit makes a solution ten times more acidic. A solution at pH 7 is neutral. A solution at pH 6 is slightly acidic; pH 5 is ten times more acidic than pH 6, pH 4 is one hundred times more acidic than pH 6, and so on.

Most enzymes have an optimum pH at which their activity is highest. If the pH is increased or decreased from the optimum, enzyme activity decreases and eventually stops altogether. When the hydrogen ion concentration is higher or lower than the level at which the enzyme naturally works, the structure of the enzyme is altered, including the active site. Beyond a certain pH the structure of the enzyme is irreversibly altered. This is another example of denaturation.

Enzymes do not all have the same pH optimum – in fact, there is a wide range. This reflects the wide range of pH environments in which enzymes work. For example, the protease secreted by *Bacillus licheniformis* has a pH optimum between 9 and 10. This bacterium is cultured to produce its alkaline-tolerant protease for use in biological laundry detergents, which are alkaline. Figure 6 shows the pH range of some of the places where enzymes work. Figure 7 shows the effects of pH on an enzyme that is adapted to work at neutral pH.

Enzyme activity is affected by substrate concentration

Enzymes cannot catalyse reactions until the substrate binds to the active site. This happens because of the random movements of molecules in liquids that result in collisions between substrates and active sites. If the concentration of substrates is increased, substrate–active site collisions will take place more frequently and the rate at which the enzyme catalyses its reaction increases.

However, there is another trend that needs to be considered. After the binding of a substrate to an active site, the active site is occupied and unavailable to other substrate molecules until products have been formed and released from the active site. As the substrate concentration rises, more and more of the active sites are occupied at any moment. A greater and greater proportion of substrate–active site collisions are therefore blocked. For this reason, the increases in the rate at which enzymes catalyse reactions get smaller and smaller as substrate concentration rises.

If the relationship between substrate concentration and enzyme activity is plotted on a graph, a distinctive curve is seen (figure 8), rising less and less steeply, but never quite reaching a maximum.

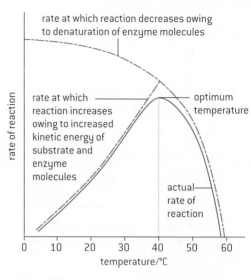

▲ Figure 5 Temperature and enzyme activity

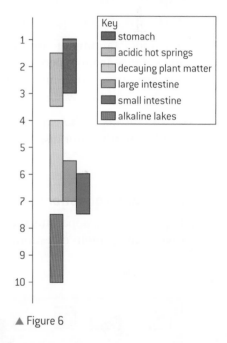

▲ Figure 6

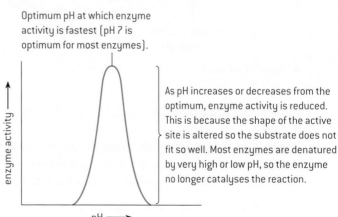

Optimum pH at which enzyme activity is fastest (pH 7 is optimum for most enzymes).

As pH increases or decreases from the optimum, enzyme activity is reduced. This is because the shape of the active site is altered so the substrate does not fit so well. Most enzymes are denatured by very high or low pH, so the enzyme no longer catalyses the reaction.

▲ Figure 7 pH and enzyme activity

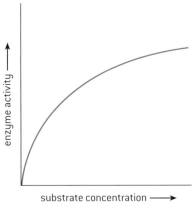

▲ Figure 8 The effect of substrate concentration on enzyme activity

Denaturation

Enzymes can be denatured.

Enzymes are proteins, and like other proteins their structure can be irreversibly altered by certain conditions. This process is denaturation and both high temperatures and either high or low pH can cause it.

When an enzyme has been denatured, the active site is altered so the substrate can no longer bind, or if its binds, the reaction that the enzyme normally catalyses does not occur. In many cases denaturation causes enzymes that were dissolved in water to become insoluble and form a precipitate.

Quantitative experiments

Experimental design: accurate quantitative measurements in enzyme experiments require replicates to ensure reliability.

Our understanding of enzyme activity is based on evidence from experiments. To obtain strong evidence these experiments must be carefully designed and follow some basic principles:

- the results of the experiment should be quantitative, not just descriptive;

- measurements should be accurate, which in science means close to the true value; and

- the experiment should be repeated, so that the replicate results can be compared to assess how reliable they are.

Data-based questions: Digesting jello cubes

Figure 9 shows apparatus that can be used to investigate protein digestion.

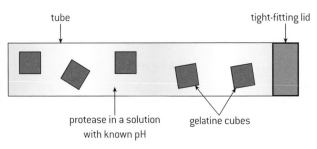

▲ Figure 9 Tube used to investigate the rate of digestion of gelatine

If the cubes are made from sugar-free jello (jelly), the colouring that they contain will gradually be released as the protein is digested by the protease. The questions below assume that strawberry-flavoured jello with red colouring has been used!

1   Explain whether these methods of assessing the rate of protein digestion are acceptable:

a) describing whether the solution around the cubes is colourless or a shade of pink or red

b) taking a sample of the solution and measuring its absorbance in a colorimeter

c) finding the mass of the cubes using an electronic balance. [3]

2   If method (c) was chosen, discuss whether it would be better to find the mass of all of the cubes of jello together, or find the mass of each one separately. [2]

3   If the jello cubes have a mass of 0.5 grams, state whether it is accurate enough to measure their mass to:

a) the nearest gram (g)

b) the nearest milligram (mg)

c) the nearest microgram (μg). [3]

4 To obtain accurate mass measurements of the jello cubes, it is necessary to remove them from the tube and dry their surface to ensure that there are no drips of solution from the tube adhering. Explain the reason for drying the surface of the blocks. [2]

Table 1 gives the results that were obtained using sugar-free jello cubes and a protease called papain, extracted from the flesh of fresh pineapples.

5 Discuss whether the results in table 1 are reliable. [2]

6 Most of the results were obtained using an extract of protease from one pineapple, but after this ran out, a second pineapple was used to obtain more protease for use in the experiment.

a) Deduce which results were obtained using the second extract. [1]

b) Suggest how the use of a second extract could have affected the results. [2]

7 Draw a graph of the results in the table. [5]

8 Describe the relationship between pH and papain activity. [3]

9 Discuss the conclusions that can be drawn from this data about the precise optimum pH of papain. [2]

pH	Mass decrease (mg)		
2	80	87	77
3	122	127	131
4	163	166	164
5	171	182	177
6	215	210	213
7	167	163	84
8	157	157	77
9	142	146	73

▲ Table 1

Designing enzyme experiments

Design of experiments to test the effect of temperature, pH and substrate concentration on the activity of enzymes.

1 The factor that you are going to investigate is the **independent variable**. You need to decide:

- how you are going to vary it, for example with substrate concentration you would obtain a solution with the highest concentration and dilute it to get lower concentrations;

- what units should be used for measuring the independent variable, for example temperature is measured in degrees Celsius;

- what range you need for the independent variable, including the highest and lowest levels and the number of intermediate levels.

2 The variable that you measure to find out how fast the enzyme is catalysing the reaction is the **dependent variable**. You need to decide:

- how you are going to measure it, including the choice of meter or other measuring device, for example an electronic stop

clock could be used to measure the time taken for a colour change;

- what units should be used for measuring the dependent variable, for example seconds rather than minutes or hours would be used for measuring a rapid colour change;

- how many repeats you need to get reliable enough results.

3 Other factors that could affect the dependent are **control variables**. You need to decide:

- what all the control variables are;

- how each of them can be kept constant;

- what level they should be kept at, for example temperature should be kept at the optimum for the enzyme if pH is being investigated, but factors that might inhibit enzymes should be kept at a minimum level.

Enzyme experiments (Practical 3)

Experimental investigation of a factor affecting enzyme activity.

There are many worthwhile enzyme experiments. The method that follows can be used to investigate the effect of substrate concentration on the activity of catalase.

Catalase is one of the most widespread enzymes. It catalyses the conversion of hydrogen peroxide, a toxic by-product of metabolism, into water and oxygen. The apparatus shown in figure 10 can be used to investigate the activity of catalase in yeast.

The experiment could be repeated using the same concentration of yeast, but different hydrogen peroxide concentrations. Another possible investigation would be to assess the catalase concentrations in other cell types, such as liver, kidney or germinating seeds. These tissues would have to be macerated and then mixed with water at the same concentration as the yeast.

1   Describe how the activity of the enzyme catalase could be measured using the apparatus shown in figure 10.   [2]

2   Explain why a yeast suspension must always be thoroughly stirred before a sample of it is taken for use in an experiment.   [2]

3   State two factors, apart from enzyme concentration, that should be kept constant if investigating the effect of substrate concentration.   [2]

4   Predict whether the enzyme activity will change more if substrate concentration is increased by 0.2 mol dm^{-3} or if it is decreased by the same amount.   [2]

5   Explain why tissues such as liver must be macerated before investigating catalase activity in them.   [2]

Safety goggles must be worn if this experiment is performed. Care should be taken not to get hydrogen peroxide on the skin.

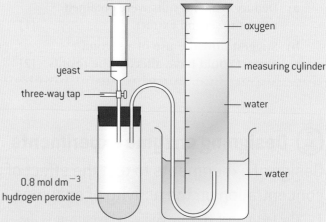

▲ Figure 10 Apparatus for measuring catalase activity

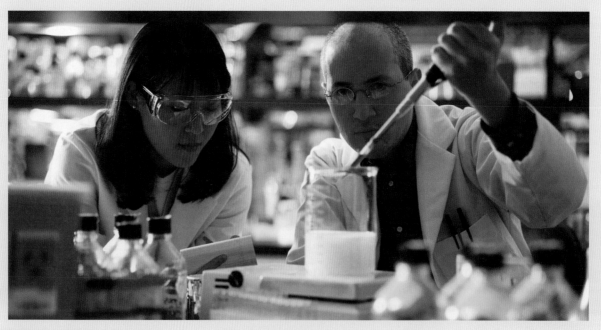

▲ Figure 11 Enzyme experiment

Data-based questions: Designing an experiment to find the effect of temperature on lipase.

Lipase converts fats into fatty acids and glycerol. It therefore causes a decrease in pH. This pH change can be used to measure the activity of lipase. Figure 12 shows suitable apparatus.

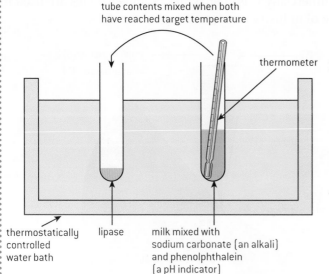

tube contents mixed when both have reached target temperature

thermometer

thermostatically controlled water bath

lipase

milk mixed with sodium carbonate (an alkali) and phenolphthalein (a pH indicator)

▲ Figure 12 Apparatus for investigating the activity of lipase

Phenolphthalein is pink in alkaline conditions, but becomes colourless when the pH drops to 7. The time taken for this colour change can be used to measure the activity of lipase at different temperatures. Alternatively, pH changes could be followed using a pH probe and data-logging software.

1 a) State the independent variable in this experiment and how you would vary it. [2]

   b) State the units for measuring the independent variable. [1]

   c) State an appropriate range for the independent variable. [2]

2 a) Explain how you would measure the dependent variable accurately. [2]

   b) State the units for measuring the dependent variable. [1]

   c) Explain the need for at least three replicate results for each temperature in this experiment. [2]

3 a) List the control factors that must be kept constant in this experiment. [3]

   b) Explain how these control factors can be kept constant. [2]

   c) Suggest a suitable level for each control factor. [3]

4 Suggest reasons for:

   a) milk being used to provide a source of lipids in this experiment rather than vegetable oil. [1]

   b) the thermometer being placed in the tube containing the larger, rather than the smaller, volume of liquid [1]

   c) the substrate being added to the enzyme, rather than the enzyme to the substrate. [1]

5 Sketch the shape of graph that you would expect from this experiment, with a temperature range from 0 °C to 80 °C on the x-axis and time taken for the indicator to change colour on the y-axis. [2]

6 Explain whether lipase from human pancreas or from germinating castor oil seeds would be expected to have the higher optimum temperature. [2]

Immobilized enzymes

Immobilized enzymes are widely used in industry.

In 1897 the Buchner brothers, Hans and Eduard, showed that an extract of yeast, containing no yeast cells, would convert sucrose into alcohol. The door was opened to the use of enzymes to catalyse chemical processes outside living cells.

Louis Pasteur had claimed that fermentation of sugars to alcohol could only occur if living cells were present. This was part of the theory of

What is the difference between dogma and theory?

After the discovery in the 19th century of the conversion of sugar into alcohol by yeast, a dispute developed between two scientists, Justus von Liebig and Louis Pasteur. In 1860 Pasteur argued that this process, called fermentation, could not occur unless live yeast cells were present. Liebig claimed that the process was chemical and that living cells were not needed. Pasteur's view reflected the vitalistic dogma – that the substances in animals and plants could only be made under the influence of a "vital spirit" or "vital force". These contrasting views were as much influenced by political and religious factors as by scientific evidence. The dispute was only resolved after the death of both men. In 1897 the Buchner brothers, Hans and Eduard, showed that an extract of yeast, containing no yeast cells, did indeed convert sucrose into alcohol. The vitalistic dogma was overthrown and the door was opened to the use of enzymes to catalyse chemical processes outside living cells.

vitalism, which stated that substances in animals and plants can only be made under the influence of a "vital spirit" or "vital force". The artificial synthesis of urea, described in sub-topic 2.1, had provided evidence against vitalism, but the Buchners' research provided a clearer falsification of the theory.

More than 500 enzymes now have commercial uses. Figure 13 shows a classification of commercially useful enzymes. Some enzymes are used in more than one type of industry.

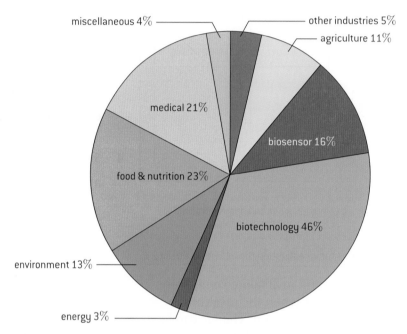

▲ Figure 13

The enzymes used in industry are usually immobilized. This is attachment of the enzymes to another material or into aggregations, so that movement of the enzyme is restricted. There are many ways of doing this, including attaching the enzymes to a glass surface, trapping them in an alginate gel, or bonding them together to form enzyme aggregates of up to 0.1 mm diameter.

Enzyme immobilization has several advantages.

- The enzyme can easily be separated from the products of the reaction, stopping the reaction at the ideal time and preventing contamination of the products.

- After being retrieved from the reaction mixture the enzyme may be recycled, giving useful cost savings, especially as many enzymes are very expensive.

- Immobilization increases the stability of enzymes to changes in temperature and pH, reducing the rate at which they are degraded and have to be replaced.

- Substrates can be exposed to higher enzyme concentrations than with dissolved enzymes, speeding up reaction rates.

Lactose-free milk
Methods of production of lactose-free milk and its advantages.

Lactose is the sugar that is naturally present in milk. It can be converted into glucose and galactose by the enzyme lactase: lactose → glucose + galactose.

Lactase is obtained from *Kluveromyces lactis*, a type of yeast that grows naturally in milk. Biotechnology companies culture the yeast, extract the lactase from the yeast and purify it for sale to food manufacturing companies. There are several reasons for using lactase in food processing:

- Some people are lactose-intolerant and cannot drink more than about 250 ml of milk per day, unless it is lactose-reduced (see figure 14).

- Galactose and glucose are sweeter than lactose, so less sugar needs to be added to sweet foods containing milk, such as milk shakes or fruit yoghurt.

- Lactose tends to crystallize during the production of ice cream, giving a gritty texture. Because glucose and galactose are more soluble than lactose they remain dissolved, giving a smoother texture.

- Bacteria ferment glucose and galactose more quickly than lactose, so the production of yoghurt and cottage cheese is faster.

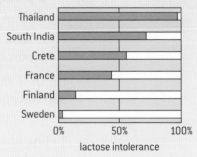
▲ Figure 14 Rates of lactose intolerance

2.6 Structure of DNA and RNA

Understanding
→ The nucleic acids DNA and RNA are polymers of nucleotides.
→ DNA differs from RNA in the number of strands normally present, the base composition and the type of pentose.
→ DNA is a double helix made of two antiparallel strands of nucleotides linked by hydrogen bonding between complementary base pairs.

 ## Applications
→ Crick and Watson's elucidation of the structure of DNA using model-making.

 ## Nature of science
→ Using models as representation of the real world: Crick and Watson used model-making to discover the structure of DNA.

 ## Skills
→ Drawing simple diagrams of the structure of single nucleotides and of DNA and RNA, using circles, pentagons and rectangles to represent phosphates, pentoses and bases.

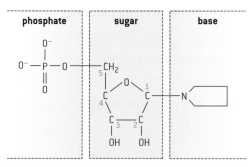

▲ Figure 1 The parts of a nucleotide

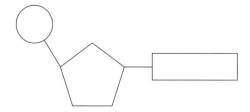

▲ Figure 2 A simpler representation of a nucleotide

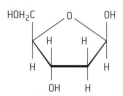

▲ Figure 3 The sugar within DNA is deoxyribose (top) and the sugar in RNA is ribose (bottom)

Nucleic acids and nucleotides

The nucleic acids DNA and RNA are polymers of nucleotides.

Nucleic acids were first discovered in material extracted from the nuclei of cells, hence their name. There are two types of nucleic acid: DNA and RNA. Nucleic acids are very large molecules that are constructed by linking together nucleotides to form a polymer.

Nucleotides consist of three parts:

- a **sugar**, which has five carbon atoms, so is a pentose sugar;
- a **phosphate** group, which is the acidic, negatively-charged part of nucleic acids; and
- a **base** that contains nitrogen and has either one or two rings of atoms in its structure.

Figure 1 shows these parts and how they are linked together. The base and the phosphate are both linked by covalent bonds to the pentose sugar. Figure 2 shows a nucleotide in symbolic form.

To link nucleotides together into a chain or polymer, covalent bonds are formed between the phosphate of one nucleotide and the pentose sugar of the next nucleotide. This creates a strong backbone for the molecule of alternating sugar and phosphate groups, with a base linked to each sugar.

There are four different bases in both DNA and RNA, so there are four different nucleotides. The four different nucleotides can be linked together in any sequence, because the phosphate and sugar used to link them are the same in every nucleotide. Any base sequence is therefore possible along a DNA or RNA molecule. This is the key to nucleic acids acting as a store of genetic information – the base sequence is the store of information and the sugar phosphate backbone ensures that the store is stable and secure.

Differences between DNA and RNA

DNA differs from RNA in the number of strands normally present, the base composition and the type of pentose.

There are three important differences between the two types of nucleic acid:

1 The sugar within DNA is deoxyribose and the sugar in RNA is ribose. Figure 3 shows that deoxyribose has one fewer oxygen atom than ribose. The full names of DNA and RNA are based on the type of sugar in them – deoxyribonucleic acid and ribonucleic acid.

2 There are usually two polymers of nucleotides in DNA but only one in RNA. The polymers are often referred to as strands, so DNA is double-stranded and RNA is single-stranded.

3 The four bases in DNA are adenine, cytosine, guanine and thymine. The four bases in RNA are adenine, cytosine, guanine and uracil, so the difference is that uracil is present instead of thymine in RNA.

Data-based questions: Chargaff's data

DNA samples from a range of species were analysed in terms of their nucleotide composition by Edwin Chargaff, an Austrian biochemist, and by others. The data is presented in table 1.

1   Compare the base composition of *Mycobacterium tuberculosis* (a prokaryote) with the base composition of the eukaryotes shown in the table. [2]

2   Calculate the base ratio A+ G/T + C, for humans and for *Mycobacterium tuberculosis*. Show your working. [2]

3   Evaluate the claim that in the DNA of eukaryotes and prokaryotes the amount of adenine and thymine are equal and the amounts of guanine and cytosine are equal. [2]

4   Explain the ratios between the amounts of bases in eukaryotes and prokaryotes in terms of the structure of DNA. [2]

5   Suggest reasons for the difference in the base composition of bacteriophage T2 and the polio virus. [2]

Source of DNA	Group	Adenine	Guanine	Cytosine	Thymine
Human	Mammal	31.0	19.1	18.4	31.5
Cattle	Mammal	28.7	22.2	22.0	27.2
Salmon	Fish	29.7	20.8	20.4	29.1
Sea urchin	Invertebrate	32.8	17.7	17.4	32.1
Wheat	Plant	27.3	22.7	22.8	27.1
Yeast	Fungus	31.3	18.7	17.1	32.9
Mycobacterium tuberculosis	Bacterium	15.1	34.9	35.4	14.6
Bacteriophage T2	Virus	32.6	18.2	16.6	32.6
Polio virus	Virus	30.4	25.4	19.5	0.0

 Table 1

⚗ Drawing DNA and RNA molecules

Drawing simple diagrams of the structure of single nucleotides and of DNA and RNA, using circles, pentagons and rectangles to represent phosphates, pentoses and bases.

The structure of DNA and RNA molecules can be shown in diagrams using simple symbols for the subunits:

- circles for phosphates;
- pentagons for pentose sugar;
- rectangles for bases.

Figure 2 shows the structure of a nucleotide, using these symbols. The base and the phosphate are linked to the pentose sugar. The base is linked to C_1 – the carbon atom on the right hand side of the pentose sugar. The phosphate is linked to C_5 – the carbon atom on the side

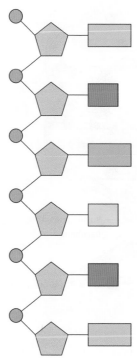

▲ Figure 4 Simplified diagram of RNA

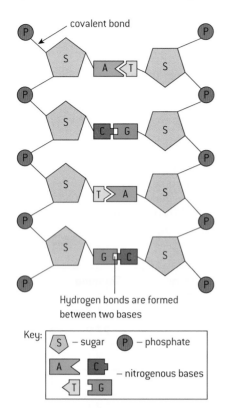

covalent bond

Hydrogen bonds are formed between two bases

Key: S – sugar   P – phosphate

A   C – nitrogenous bases
T   G

▲ Figure 5 Simplified diagram of DNA

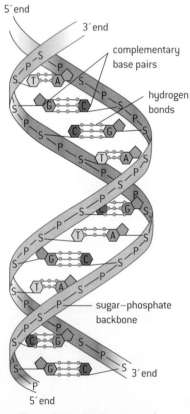

5′ end

3′ end

complementary base pairs

hydrogen bonds

sugar–phosphate backbone

3′ end

5′ end

▲ Figure 6 The double helix

chain on the upper left side of the pentose sugar. The positions of these carbon atoms are shown in figure 1.

To show the structure of RNA, draw a polymer of nucleotides, with a line to show the covalent bond linking the phosphate group of each nucleotide to the pentose in the next nucleotide. The phosphate is linked to C_3 of the pentose – the carbon atom that is on the lower left.

If you have drawn the structure of RNA correctly, the two ends of the polymer will be different. They are referred to as the 3′ and the 5′ terminals.

- The phosphate of another nucleotide could be linked to the C_3 atom of the 3′ terminal.

- The pentose of another nucleotide could be linked to the phosphate of the 5′ terminal.

To show the structure of DNA, draw a strand of nucleotides, as with RNA, then a second strand alongside the first. The second strand should be run in the opposite direction, so that at each end of the DNA molecule, one strand has a C_3 terminal and the other a C_5 terminal. The two strands are linked by hydrogen bonds between the bases. Add letters or names to indicate the bases. Adenine (A) only pairs with thymine (T) and cytosine (C) only pairs with guanine (G).

Structure of DNA

DNA is a double helix made of two antiparallel strands of nucleotides linked by hydrogen bonding between complementary base pairs.

Drawings of the structure of DNA on paper cannot show all features of the three-dimensional structure of the molecule. Figure 6 represents some of these features.

- Each strand consists of a chain of **nucleotides** linked by covalent bonds.

- The two strands are parallel but run in opposite directions so they are said to be antiparallel. One strand is oriented in the direction 5′ to 3′ and the other is oriented in the direction 3′ to 5′.

- The two strands are wound together to form a **double helix**.

- The strands are held together by hydrogen bonds between the nitrogenous bases. Adenine (A) is always paired with thymine (T) and guanine (G) with cytosine (C). This is referred to as **complementary base pairing**, meaning that A and T complement each other by forming base pairs and similarly G and C complement each other by forming base pairs.

Data-based questions: The bases in DNA

Look at the molecular models in figure 7 and answer the following questions.

1 State one difference between adenine and the other bases. [1]

2 Each of the bases in DNA has a nitrogen atom bonded to a hydrogen atom in a similar position, which appears in the lower left in each case in figure 7. Deduce how this nitrogen is used when a nucleotide is being assembled from its subunits. [2]

3 Identify three similarities between adenine and guanine. [3]

4 Compare the structure of cytosine and thymine. [4]

5 Although the bases have some shared features, each one has a distinctive chemical structure and shape. Remembering the function of DNA, explain the importance for the bases each to be distinctive. [5]

Guanine

Adenine

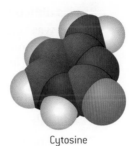
Cytosine

Thymine

▲ Figure 7

 Molecular models

Using models as representation of the real world: Crick and Watson used model-making to discover the structure of DNA.

The word model in English is derived from the Latin word modus, meaning manner or method. Models were originally architects' plans, showing how a new building might be constructed. Three-dimensional models were then developed to give a more realistic impression of what a proposed building would be like.

Molecular models also show a possible structure in three dimensions, but whereas architects' models are used to decide whether a building should become reality in the future, molecular models help us to discover what the structure of a molecule actually is.

Models in science are not always three-dimensional and do not always propose structures. They can be theoretical concepts and they can represent systems or processes. The common feature of models is that they are proposals, which are made to be tested. As with architecture, models in science are often rejected and replaced. Model-making played a critical part in Crick and Watson's discovery of the structure of DNA, but it took two attempts before they were successful.

What is the relative role of competition and cooperation in scientific research?

Three prominent research groups openly competed to elucidate the structure of DNA: Watson and Crick were working at Cambridge; Maurice Wilkins and Rosalind Franklin were working at Kings College of the University of London; and Linus Pauling's research group was operating out of Caltech in the United States.

A stereotype of scientists is that they take a dispassionate approach to investigation. The truth is that science is a social endeavour involving a number of emotion-influenced interactions between science. In addition to the joy of discovery, scientists seek the esteem of their community. Within research groups, collaboration is important, but outside of their research group competition often restricts open communication that might accelerate the pace of scientific discovery. On the other hand, competition may motivate ambitious scientists to work tirelessly.

▲ Figure 8 Crick and Watson and their DNA model

🌐 Crick and Watson's models of DNA structure

Crick and Watson's discovery of the structure of DNA using model-making.

Crick and Watson's success in discovering the structure of DNA was based on using the evidence to develop possible structures for DNA and testing them by model-building. Their first model consisted of a triple helix, with bases on the outside of the molecule and magnesium holding the two strands together with ionic bonds to the phosphate groups on each strand. The helical structure and the spacing between subunits in the helix fitted the X-ray diffraction pattern obtained by Rosalind Franklin.

It was difficult to get all parts of this model to fit together satisfactorily and it was rejected when Franklin pointed out that there would not be enough magnesium available to form the cross links between the strands. Another deficiency of this first model was that is that it did not take account of Chargaff's finding that the amount of adenine equals the thymine and the amount of cytosine equals the amount of guanine.

To investigate the relationship between the bases in DNA pieces of cardboard were cut out to represent their shapes. These showed that A-T and C-G base pairs could be formed, with hydrogen bonds linking the bases. The base pairs were equal in length so would fit between two outer sugar-phosphate backbones.

Another flash of insight was needed to make the parts of the molecule fit together: the two strands in the helix had to run in opposite directions – they must be antiparallel. Crick and Watson were then able to build their second model of the structure of DNA. They used metal rods and sheeting cut to shape and held together with small clamps. Bond lengths were all to scale and bond angles correct. Figure 8 shows Crick and Watson with the newly constructed model.

The model convinced all those who saw it. A typical comment was "It just looked right". The structure immediately suggested a mechanism for copying DNA. It also led quickly to the realization that the genetic code must consist of triplets of bases. In many ways the discovery of DNA structure started the great molecular biology revolution, with effects that are still reverberating in science and in society.

2.7 DNA replication, transcription and translation

Understanding

→ The replication of DNA is semi-conservative and depends on complementary base pairing.

→ Helicase unwinds the double helix and separates the two strands by breaking hydrogen bonds.

→ DNA polymerase links nucleotides together to form a new strand, using the pre-existing strand as a template.

→ Transcription is the synthesis of mRNA copied from the DNA base sequences by RNA polymerase.

→ Translation is synthesis of polypeptides on ribosomes.

→ The amino acid sequence of polypeptides is determined by mRNA according to the genetic code.

→ Codons of three bases on mRNA correspond to one amino acid in a polypeptide.

→ Translation depends on complementary base pairing between codons on mRNA and anticodons on tRNA.

Applications

→ Use of Taq DNA polymerase to produce multiple copies of DNA rapidly by the polymerase chain reaction (PCR).

→ Production of human insulin in bacteria as an example of the universality of the genetic code allowing gene transfer between species.

Skills

→ Use a table of the genetic code to deduce which codon(s) corresponds to which amino acid.

→ Analysis of Meselson and Stahl's results to obtain support for the theory of semi-conservative replication of DNA.

→ Use a table of mRNA codons and their corresponding amino acids to deduce the sequence of amino acids coded by a short mRNA strand of known base sequence.

→ Deducing the DNA base sequence for the mRNA strand.

Nature of science

→ Obtaining evidence for scientific theories: Meselson and Stahl obtained evidence for the semi-conservative replication of DNA.

Semi-conservative replication of DNA

The replication of DNA is semi-conservative and depends on complementary base pairing.

When a cell prepares to divide, the two strands of the double helix separate (see figure 2). Each of these original strands serves as a guide, or template, for the creation of a new strand. The new strands are formed by adding nucleotides, one by one, and linking them together. The result is two DNA molecules, both composed of an original strand and a newly synthesized strand. For this reason, DNA replication is referred to as being **semi-conservative**.

111

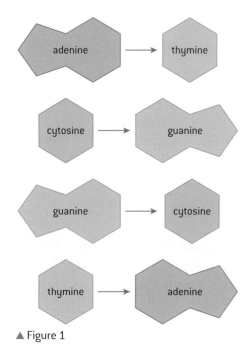

▲ Figure 1

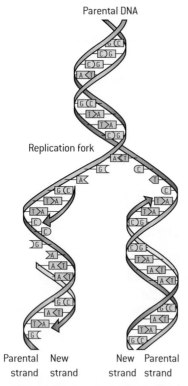

▲ Figure 2 Semi-conservative replication

The base sequence on the template strand determines the base sequence on the new strand. Only a nucleotide carrying a base that is complementary to the next base on the template strand can successfully be added to the new strand (figure 1).

This is because complementary bases form hydrogen bonds with each other, stabilizing the structure. If a nucleotide with the wrong base started to be inserted, hydrogen bonding between bases would not occur and the nucleotide would not be added to the chain. The rule that one base always pairs with another is called **complementary base pairing**. It ensures that the two DNA molecules that result from DNA replication are identical in their base sequences to the parent molecule that was replicated.

Obtaining evidence for the theory of semi-conservative replication

Obtaining evidence for scientific theories: Meselson and Stahl obtained evidence for the semi-conservative replication of DNA.

Semi-conservative replication is an example of a scientific theory that seemed intuitively right, but nonetheless needed to be backed up with evidence. Laboratories around the world attempted to confirm experimentally that replication of DNA is semi-conservative and soon convincing evidence had been obtained.

In 1958 Matthew Meselson and Franklin Stahl published the results of exceedingly elegant experiments that provided very strong evidence for semi-conservative replication. They used ^{15}N, a rare isotope of nitrogen that has one more neutron than the normal ^{14}N isotope, so is denser. In the 1930s Harold Urey had developed methods of purifying stable isotopes that could be used as tracers in biochemical pathways. ^{15}N was one of these.

Meselson and Stahl devised a new method of separating DNA containing ^{15}N in its bases from DNA with ^{14}N. The technique is called caesium chloride density gradient centrifugation. A solution of caesium chloride is spun in an ultracentrifuge at nearly 45,000 revolutions per minute for 20 hours. The dense caesium ions tend to move towards the bottom of the tube but do not sediment fully because of diffusion. A gradient is established, with the greatest caesium concentration, and therefore density, at the bottom and the lowest at the top of the tube. Any substance centrifuged with the caesium chloride solution becomes concentrated at a level corresponding with its density.

Meselson and Stahl cultured the bacterium *E. coli* for fourteen generations in a medium where the only nitrogen source was ^{15}N. Almost all nitrogen atoms in the bases of the DNA in the bacteria were therefore ^{15}N. They then transferred the bacteria abruptly to a medium in which all the nitrogen was ^{14}N. At the temperature used to culture them, the generation time was 50 minutes – the bacteria divided and therefore replicated their DNA once every 50 minutes.

Meselson and Stahl collected samples of DNA from the bacterial culture for several hours from the time when it was transferred to the ^{14}N medium. They extracted the DNA and measured its density by caesium chloride density gradient centrifugation. The DNA could be detected because it absorbs ultraviolet light, and so created a dark band when the tubes were illuminated with ultraviolet. Figure 3 shows the results. In the next part of this sub-topic there is guidance in how to analyse the changes in position of the dark bands.

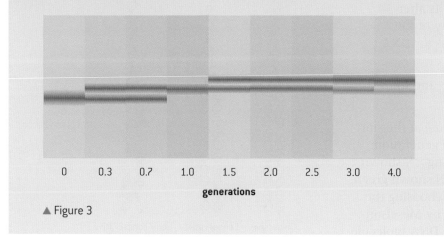

▲ Figure 3

Activity

New experimental techniques

Meselson and Stahl used three techniques in their experiments that that were relatively new. Identify a technique used by them that was developed:

a) by Urey in the 1930s

b) by Pickels in the 1940s

c) by Meselson and Stahl themselves in the 1950s.

Activity

Modelling helicase activity

To model helicase activity you could use some two-stranded rope or string and a split key ring. The strands in the rope are helical and represent the two strands in DNA. Open the key ring and put one strand of the rope inside it. Close the ring so that the other strand is outside. Slide the ring along the string to separate the strands. What problems are revealed by this model of the activity of helicase? Use the internet to find the solution used by living organisms.

⚗ Meselson and Stahl's DNA replication experiments

Analysis of Meselson and Stahl's results to obtain support for the theory of semi-conservative replication of DNA.

The data-based question below will guide you through the analysis of Meselson and Stahl's results and help to build your skills in this aspect of science.

Data-based questions: The Meselson and Stahl experiment

In order for cell division to occur, DNA must be duplicated to ensure that progeny cells have the same genetic information as the parent cells. The process of duplicating DNA is termed replication. The Meselson–Stahl experiment sought to understand the mechanism of replication. Did it occur in a conservative fashion, a semi-conservative fashion or in a dispersive fashion (see figure 4)?

Meselson and Stahl grew *E. coli* in a medium containing "heavy" nitrogen (^{15}N) for a number of generations. They then transferred the bacteria

to a ^{14}N medium. Samples of the bacteria were taken over a period of time and separated by density gradient centrifugation, a method in which heavier molecules settle further down in acentrifuge tube than lighter ones.

1 The single band of DNA at the start (0 generations) had a density of 1.724 g cm^{-3}. The main band of DNA after four generations had a density of 1.710 g cm^{-3}. Explain how DNA with a lower density had been produced by the bacteria. [2]

2 a) Estimate the density of the DNA after one generation. [2]

b) Explain whether the density of DNA after one generation falsifies any of the three possible mechanisms for DNA replication shown in figure 4. [3]

3 a) Describe the results after two generations, including the density of the DNA. [3]

b) Explain whether the results after two generations falsify any of the three possible mechanisms for DNA replication. [3]

4 Explain the results after three and four generations. [2]

5 Figure 4 shows DNA from *E. coli* at the start (0 generations) and after one generation, with strands of DNA containing ^{15}N shown red and strands containing ^{14}N shown green. Redraw either (a), (b) or (c), choosing the mechanism that is supported by Meselson and Stahl's experiment. Each DNA molecule can be shown as two parallel lines rather than a helix and the colours do not have to be red and green. Draw the DNA for two more generations of replication in a medium containing ^{14}N. [3]

6 Predict the results of centrifuging a mixture of DNA from 0 generations and 2 generations. [2]

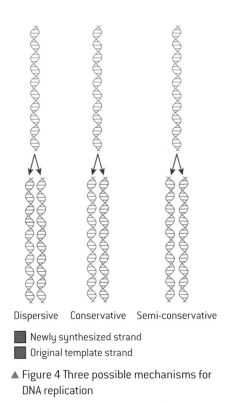

Dispersive   Conservative   Semi-conservative

■ Newly synthesized strand
■ Original template strand

▲ Figure 4 Three possible mechanisms for DNA replication

Helicase

Helicase unwinds the double helix and separates the two strands by breaking hydrogen bonds.

Before DNA replication can occur, the two strands of the molecule must separate so that they can each act as a template for the formation of a new strand. The separation is carried out by helicases, a group of enzymes that use energy from ATP. The energy is required for breaking hydrogen bonds between complementary bases.

One well-studied helicase consists of six globular polypeptides arranged in a donut shape. The polypeptides assemble with one strand of the DNA molecule passing through the centre of the donut and the other outside it. Energy from ATP is used to move the helicase along the DNA molecule, breaking the hydrogen bonds between bases and parting the two stands.

Double-stranded DNA cannot be split into two strands while it is still helical. Helicase therefore causes the unwinding of the helix at the same time as it separates the strands.

DNA polymerase

DNA polymerase links nucleotides together to form a new strand, using the pre-existing strand as a template.

Once helicase has unwound the double helix and split the DNA into two strands, replication can begin. Each of the two strands acts as a template for the formation of a new strand. The assembly of the new strands is carried out by the enzyme DNA polymerase.

DNA polymerase always moves along the template strand in the same direction, adding one nucleotide at a time. Free nucleotides with each of the four possible bases are available in the area where DNA is being replicated. Each time a nucleotide is added to the new strand, only one of the four types of nucleotide has the base that can pair with the base at the position reached on the template strand. DNA polymerase brings nucleotides into the position where hydrogen bonds could form, but unless this happens and a complementary base pair is formed, the nucleotide breaks away again.

Once a nucleotide with the correct base has been brought into position and hydrogen bonds have been formed between the two bases, DNA polymerase links it to the end of the new strand. This is done by making a covalent bond between the phosphate group of the free nucleotide and the sugar of the nucleotide at the existing end of the new strand. The pentose sugar is the 3′ terminal and the phosphate group is the 5′ terminal, so DNA polymerase adds on the 5′ terminal of the free nucleotide to the 3′ terminal of the existing strand.

DNA polymerase gradually moves along the template strand, assembling the new strand with a base sequence complementary to the template strand. It does this with a very high degree of fidelity – very few mistakes are made during DNA replication.

 ## PCR – the polymerase chain reaction

Use of Taq DNA polymerase to produce multiple copies of DNA rapidly by the polymerase chain reaction (PCR).

The polymerase chain reaction (PCR) is a technique used to make many copies of a selected DNA sequence. Only a very small quantity of the DNA is needed at the start. The DNA is loaded into a PCR machine in which a cycle of steps repeatedly doubles the quantity of the selected DNA. This involves double-stranded DNA being separated into two single strands at one stage of the cycle and single strands combining to form double-stranded DNA at another stage.

The two strands in DNA are held together by hydrogen bonds. These are weak interactions, but in a DNA molecule there are large numbers of them so they hold the two strands together successfully at the temperatures normally encountered by most cells. If DNA is heated to a high temperature, the hydrogen bonds eventually break and the two strands separate. If the DNA is then cooled hydrogen bonds can form, so the strands pair up again. This is called re-annealing.

The PCR machine separates DNA strands by heating them to 95 °C for fifteen seconds. It then cools the DNA quickly to 54 °C. This would allow re-annealing of parent strands to form double-stranded DNA. However, a large excess of short sections of single-stranded DNA called primers is present. The primers bind rapidly to target sequences and as a large excess of primers is present, they prevent the re-annealing of the parent strands. Copying of the single parent strands then starts from the primers.

The next stage in PCR is synthesis of double-stranded DNA, using the single strands with primers as templates. The enzyme Taq DNA polymerase is used to do this. It was obtained from a bacterium, *Thermus aquaticus*, found in hot springs, including those of Yellowstone National Park. The temperatures of these springs range from 50 °C to 80 °C. Enzymes in most organisms would rapidly denature at such high temperatures, but those of *Thermus aquaticus*, including its DNA polymerase, are adapted to be very heat-stable to resist denaturation.

Taq DNA polymerase is used because it can resist the brief period at 95 °C used to separate the DNA strands. It would work at the lower temperature of 54 °C that is used to attach the primers, but its optimum temperature is 72 °C. The reaction mixture is therefore heated to this temperature for the period when Taq DNA polymerase is working. At this temperature it adds about 1,000 nucleotides per minute, a very rapid rate of DNA replication.

When enough time has elapsed for replication of the selected base sequence to be complete, the next cycle is started by heating to 95 °C. A cycle of PCR can be completed in less than two minutes. Thirty cycles, which amplify the DNA by a factor of a billion, take less than an hour. With the help of Taq DNA polymerase, PCR allows the production of huge numbers of copies of a selected base sequence in a very short time.

▲ Figure 5

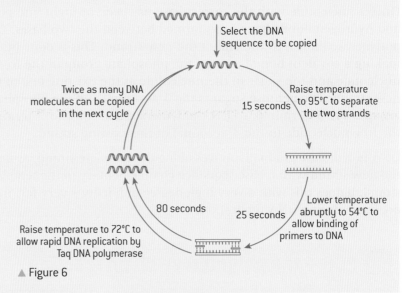

▲ Figure 6

Transcription

Transcription is the synthesis of mRNA copied from the DNA base sequences by RNA polymerase.

This sequence of bases in a gene does not, in itself, give any observable characteristic in an organism. The function of most genes is to specify the sequence of amino acids in a particular polypeptide. It is proteins that often directly or indirectly determine the observable characteristics of an individual. Two processes are needed to produce a specific polypeptide, using the base sequence of a gene. The first of these is **transcription**.

Transcription is the synthesis of RNA, using DNA as a template. Because RNA is single-stranded, transcription only occurs along one of the two strands of DNA. What follows is an outline of transcription:

● The enzyme RNA polymerase binds to a site on the DNA at the start of a gene.

- RNA polymerase moves along the gene separating DNA into single strands and pairing up RNA nucleotides with complementary bases on one strand of the DNA. There is no thymine in RNA, so uracil pairs in a complementary fashion with adenine.

- RNA polymerase forms covalent bonds between the RNA nucleotides.

- The RNA separates from the DNA and the double helix reforms.

- Transcription stops at the end of the gene and the completed RNA molecule is released.

The product of transcription is a molecule of RNA with a base sequence that is complementary to the template strand of DNA. This RNA has a base sequence that is identical to the other strand, with one exception – there is uracil in place of thymine. So, to make an RNA copy of the base sequence of one strand of a DNA molecule, the other strand is transcribed. The DNA strand with the same base sequence as the RNA is called the **sense strand**. The other strand that acts as the template and has a complementary base sequence to both the RNA and the sense strand is called the **antisense strand**.

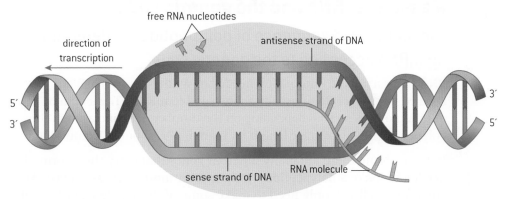

▲ Figure 7

Translation

Translation is synthesis of polypeptides on ribosomes.

The second of the two processes needed to produce a specific polypeptide is **translation**. Translation is the synthesis of a polypeptide, with an amino acid sequence determined by the base sequence of a molecule of RNA. The production of RNA by transcription and how its base sequence is determined by a gene was described in the previous part of this sub-topic.

Translation takes place on cell structures in the cytoplasm known as ribosomes. Ribosomes are complex structures that consist of a small and a large subunit, with binding sites for each of the molecules that take part in the translation. Figure 9 shows the two subunits of a ribosome. Each is composed of RNA molecules (pink and yellow) and proteins (purple). Part of the large subunit (green) is the site that makes peptide bonds between amino acids, to link them together into a polypeptide.

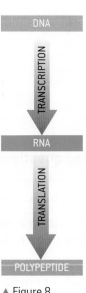

▲ Figure 8

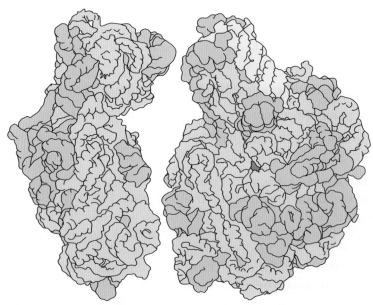

▲ Figure 9 Large and small subunits of the ribosome with proteins shown in purple, ribosomal RNA in pink and yellow and the site that catalyses the formation of peptide bonds green

Messenger RNA and the genetic code

The amino acid sequence of polypeptides is determined by mRNA according to the genetic code.

RNA that carries the information needed to synthesize a polypeptide is called messenger RNA, usually abbreviated to mRNA. The length of mRNA molecules varies depending on the number of amino acids in the polypeptide but an average length for mammals is about 2,000 nucleotides.

In the genome there are many different genes that carry the information needed to make a polypeptide with a specific amino acid sequence. At any time a cell will only need to make some of these polypeptides. Only certain genes are therefore transcribed and only certain types of mRNA will be available for translation in the cytoplasm. Cells that need or secrete large amounts of a particular polypeptide make many copies of the mRNA for that polypeptide. For example, insulin-secreting cells in the pancreas make many copies of the mRNA needed to make insulin.

Although most RNA is mRNA, there are other types; for example, transfer RNA is involved in decoding the base sequence of mRNA into an amino acid sequence during translation and ribosomal RNA is part of the structure of the ribosome. They are usually referred to as tRNA and rRNA.

Data-based questions: Interpreting electron micrographs

The electron micrographs in figure 10 show transcription, translation and DNA replication.

1   Deduce, with reasons, which process is occurring in each electron micrograph.   [5]

2   The colour in the electron micrographs has been added to make the different structures show up more clearly. Identify each of these structures:

a) the red structure in the central micrograph

b) the thin blue molecule near the lower edge of the right-hand micrograph

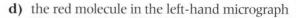

c) the blue molecules of variable length attached to the thin blue molecule in the right-hand micrograph.

d) the red molecule in the left-hand micrograph

e) the green molecules in the left-hand micrograph. [5]

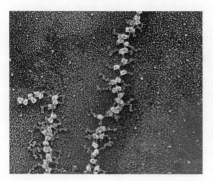
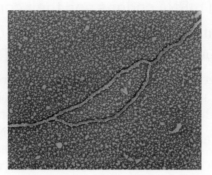
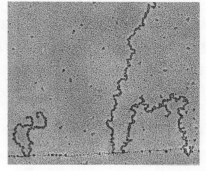

▲ Figure 10

Codons

Codons of three bases on mRNA correspond to one amino acid in a polypeptide.

The "translation dictionary" that enables the cellular machinery to convert the base sequence on the mRNA into an amino acid sequence is called the genetic code. There are four different bases and twenty amino acids, so one base cannot code for one amino acid. There are sixteen combinations of two bases, which is still too few to code for all of the twenty amino acids. Living organisms therefore use a triplet code, with groups of three bases coding for an amino acid.

A sequence of three bases on the mRNA is called a codon. Each codon codes for a specific amino acid to be added to the polypeptide. Table 1 lists all of the 64 possible codons. The three bases of an mRNA codon are designated in the table as first, second and third positions.

Note that different codons can code for the same amino acid. For example the codons GUU and GUC both code for the amino acid valine. For this reason, the code is said to be "degenerate". Note also that three codons are "stop" codons that code for the end of translation.

Amino acids are carried on another kind of RNA, called tRNA. Each amino acid is carried by a specific tRNA, which has a three-base anticodon complementary to the mRNA codon for that particular amino acid.

First position	Second position				Third position
(5' end)	U	C	A	G	(3' end)
U	Phe	Ser	Tyr	Cys	U
	Phe	Ser	Tyr	Cys	C
	Leu	Ser	Stop	Stop	A
	Leu	Ser	Stop	Trp	G
C	Leu	Pro	His	Arg	U
	Leu	Pro	His	Arg	C
	Leu	Pro	Gln	Arg	A
	Leu	Pro	Gln	Arg	G
A	Ile	Thr	Asn	Ser	U
	Ile	Thr	Asn	Ser	C
	Ile	Thr	Lys	Arg	A
	Met	Thr	Lys	Arg	G
G	Val	Ala	Asp	Gly	U
	Val	Ala	Asp	Gly	C
	Val	Ala	Glu	Gly	A
	Val	Ala	Glu	Gly	G

▲ Table 1

Decoding base sequences

Use of a table of the genetic code to deduce which codon(s) corresponds to which amino acid; use of a table of mRNA codons and their corresponding amino acids to deduce the sequence of amino acids coded by a short mRNA strand of known base sequence; deducing the DNA base sequence for the mRNA strand.

There is no need to try to memorize the genetic code, but if a table showing it is available, you should be able to make various deductions.

1 Which codons correspond to an amino acid?

Three letters are used to indicate each amino acid in the table of the genetic code. Each of the 20 amino acids has between one and six codons. Read off the three letters of each codon for the amino acid. For example, the amino acid methionine, shown as Met on the table, has one codon which is AUG.

2 What amino acid sequence would be translated from a sequence of codons in a strand of mRNA?

The first three bases in the mRNA sequence are the codon for the first amino acid, the next three bases are the codon for the second base and so on. Look down the left hand side of the table to find the first base of a codon, across the top of the table to find the second base and down the right hand side to find the third base. For example, GCA codes for the amino acid alanine, which is abbreviated to Ala in the table.

3 What base sequence in DNA would be transcribed to give the base sequence of a strand of mRNA?

A strand of mRNA is produced by transcribing the anti-sense strand of the DNA. This therefore has a base sequence complementary to the mRNA. For example, the codon AUG in mRNA is transcribed from the base sequence TAC on the antisense strand of the DNA. A longer example is that the base sequence GUACGUACG is transcribed from CATGCATGC. Note that adenine pairs with thymine in DNA but with uracil in RNA.

Questions

1 Deduce the codons for

   a) Tryptophan (Trp)

   b) Tyrosine (Tyr)

   c) Arginine (Arg)   [3]

2 Deduce the amino acid sequences that correspond to these mRNA sequences: [3]

   a) ACG **b)** CACGGG **c)** CGCGCGAGG [3]

3 If mRNA contains the base sequence CUCAUCGAAUAACCC

   a) deduce the amino acid sequence of the polypeptide translated from the mRNA   [2]

   b) deduce the base sequence of the antisense strand transcribed to produce the mRNA.   [2]

Codons and anticodons

Translation depends on complementary base pairing between codons on mRNA and anticodons on tRNA.

Three components work together to synthesize polypeptides by translation:

- mRNA has a sequence of codons that specifies the amino acid sequence of the polypeptide;

- tRNA molecules have an anticodon of three bases that binds to a complementary codon on mRNA and they carry the amino acid corresponding to that codon;

- ribosomes act as the binding site for mRNA and tRNAs and also catalyse the assembly of the polypeptide.

A summary of the main events of translation follows:

1   An mRNA binds to the small subunit of the ribosome.

2   A molecule of tRNA with an anticodon complementary to the first codon to be translated on the mRNA binds to the ribosome.

3   A second tRNA with an anticodon complementary to the second codon on the mRNA then binds. A maximum of two tRNAs can be bound at the same time.

4   The ribosome transfers the amino acid carried by the first tRNA to the amino acid on the second tRNA, by making a new peptide bond. The second tRNA is then carrying a chain of two amino acids – a dipeptide.

5   The ribosome moves along the mRNA so the first tRNA is released, the second becomes the first.

6   Another tRNA binds with an anticodon complementary to the next codon on the mRNA.

7   The ribosome transfers the chain of amino acids carried by the first tRNA to the amino acid on the second tRNA, by making a new peptide bond.

Stages 4, 5 and 6 are repeated again and again, with one amino acid added to the chain each time the cycle is repeated. The process continues along the mRNA until a stop codon is reached, when the completed polypeptide is released.

The accuracy of translation depends on complementary base pairing between the anticodon on each tRNA and the codon on mRNA. Mistakes are very rare, so polypeptides with a sequence of hundreds of amino acids are regularly made with every amino acid correct.

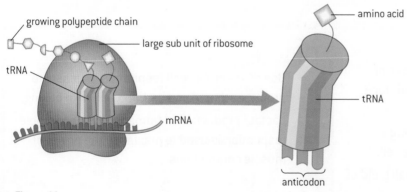

▲ Figure 11

  Production of human insulin in bacteria

Production of human insulin in bacteria as an example of the universality of the genetic code allowing gene transfer between species.

Diabetes in some individuals is due to destruction of cells in the pancreas that secrete the hormone insulin. It can be treated by injecting insulin into the blood. Porcine and bovine insulin, extracted from the pancreases of pigs and cattle, have both been widely used. Porcine insulin has only one difference in amino acid sequence from human insulin and bovine insulin has three differences. Shark insulin, which has been used for treating diabetics in Japan, has seventeen differences.

Despite the differences in the amino acid sequence between animal and human insulin, they all bind to the human insulin receptor and cause lowering of blood glucose concentration. However, some diabetics develop an allergy to animal insulins, so it is preferable to use human insulin. In 1982 human insulin became commercially available for the first time. It was produced using genetically modified *E. coli* bacteria. Since then methods of production have been developed using yeast cells and more recently safflower plants.

Each of these species has been genetically modified by transferring the gene for making human insulin to it. This is done in such a way that the gene is transcribed to produce mRNA and the mRNA is translated to produce harvestable quantities of insulin. The insulin produced has exactly the same amino acid sequence as if the gene was being transcribed and translated in human cells.

This may seem obvious, but it depends on each tRNA with a particular anticodon having the same amino acid attached to it as in humans. In other words, *E. coli*, yeast and safflower (a prokaryote, a fungus and a plant) all use the same genetic code as humans (an animal). It is fortunate for genetic engineers that all organisms, with very few exceptions, use the same genetic code as it makes gene transfer possible between widely differing species.

▲ Figure 12

2.8 Cell respiration

Understanding

→ Cell respiration is the controlled release of energy from organic compounds to produce ATP.

→ ATP from cell respiration is immediately available as a source of energy in the cell.

→ Anaerobic cell respiration gives a small yield of ATP from glucose.

→ Aerobic cell respiration requires oxygen and gives a large yield of ATP from glucose.

Applications

→ Use of anaerobic cell respiration in yeasts to produce ethanol and carbon dioxide in baking.

→ Lactate production in humans when anaerobic respiration is used to maximize the power of muscle contractions.

Nature of science

→ Assessing the ethics of scientific research: the use of invertebrates in respirometer experiments has ethical implications.

Skills

→ Analysis of results from experiments involving measurement of respiration rates in germinating seeds or invertebrates using a respirometer.

Release of energy by cell respiration

Cell respiration is the controlled release of energy from organic compounds to produce ATP.

Cell respiration is one of the functions of life that all living cells perform. Organic compounds are broken down to release energy, which can then be used in the cell. For example, energy is released in muscle fibres by breaking down glucose into carbon dioxide and water. The energy can then be used for muscle contraction.

In humans the source of the organic compounds broken down in cell respiration is the food that we eat. Carbohydrates and lipids are often used, but amino acids from proteins may be used if we eat more protein than needed. Plants use carbohydrates or lipids previously made by photosynthesis.

Cell respiration is carried out using enzymes in a careful and controlled way, so that as much as possible of the energy released is retained in a usable form. This form is a chemical substance called adenosine triphosphate, almost always abbreviated to ATP. To make ATP, a phosphate group is linked to adenosine diphosphate, or ADP. Energy is required to carry out this reaction. The energy comes from the breakdown of organic compounds.

ATP is not transferred from cell to cell and all cells require a continuous supply. This is the reason for cell respiration being an essential function of life in all cells.

ATP is a source of energy

ATP from cell respiration is immediately available as a source of energy in the cell.

Cells require energy for three main types of activity.

- Synthesizing large molecules like DNA, RNA and proteins.

- Pumping molecules or ions across membranes by active transport.

- Moving things around inside the cell, such as chromosomes, vesicles, or in muscle cells the protein fibres that cause muscle contraction.

The energy for all of these processes is supplied by ATP. The advantage of ATP as an energy supply is that the energy is immediately available. It is released simply by splitting ATP into ADP and phosphate. The ADP and phosphate can then be reconverted to ATP by cell respiration.

When energy from ATP is used in cells, it is ultimately all converted to heat. Although heat energy may be useful to keep an organism warm, it cannot be reused for cell activities and is eventually lost to the environment. This is the reason for cells requiring a continual source of ATP for cell activities.

▲ Figure 1 Breaking down 8 grams of glucose in cell respiration provides enough energy to sprint 100 metres

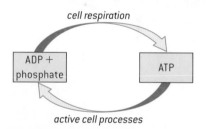

cell respiration

| ADP + phosphate | | ATP |

active cell processes

▲ Figure 2

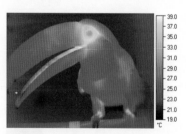

▲ Figure 3 Infra red photo of toucan showing that it is warmer than its surroundings due to heat generated by respiration. Excess heat is dissipated by sending warm blood to the beak

▲ Figure 4 The mud in mangrove swamps is deficient in oxygen. Mangrove trees have evolved vertical roots called pneumatophores which they use to obtain oxygen from the air

Activity

Does bioethanol solve or make more problems?

There has been much debate about bioethanol production. A renewable fuel that cuts down on carbon emissions is obviously desirable. What are the arguments against bioethanol production?

Anaerobic respiration

Anaerobic cell respiration gives a small yield of ATP from glucose.

Glucose is broken down in anaerobic cell respiration without using any oxygen. The yield of ATP is relatively small, but the ATP can be produced quickly. Anaerobic cell respiration is therefore useful in three situations:

- when a short but rapid burst of ATP production is needed;
- when oxygen supplies run out in respiring cells;
- in environments that are deficient in oxygen, for example waterlogged soils.

The products of anaerobic respiration are not the same in all organisms. In humans, glucose is converted to lactic acid, which is usually in a dissolved form known as lactate. In yeast and plants glucose is converted to ethanol and carbon dioxide. Both lactate and ethanol are toxic in excess, so must be removed from the cells that produce them, or be produced in strictly limited quantities.

Summary equations

$$\text{glucose} \longrightarrow \text{lactate}$$
$$\text{ADP} \quad \text{ATP}$$

This occurs in animals including humans.

$$\text{glucose} \longrightarrow \text{ethanol} + \text{carbon dioxide}$$
$$\text{ADP} \quad \text{ATP}$$

This occurs in yeasts and plants.

🌐 Yeast and its uses

Use of anaerobic cell respiration in yeasts to produce ethanol and carbon dioxide in baking.

Yeast is a unicellular fungus that occurs naturally in habitats where glucose or other sugars are available, such as the surface of fruits. It can respire either aerobically or anaerobically. Anaerobic cell respiration in yeast is the basis for production of foods, drinks and renewable energy.

Bread is made by adding water to flour, kneading the mixture to make dough and then baking it. Usually an ingredient is added to the dough to create bubbles of gas, so that the baked bread has a lighter texture. Yeast is often this ingredient. After kneading, the dough is kept warm to encourage the yeast to respire. Any oxygen in the dough is soon used up so the yeast carries out anaerobic cell respiration. The carbon dioxide produced by anaerobic cell respiration cannot escape from the dough and forms bubbles. The swelling of the dough due to

▲ Figure 5

the production of bubbles of carbon dioxide is called rising. Ethanol is also produced by anaerobic cell respiration, but it evaporates during baking.

Bioethanol is ethanol produced by living organisms, for use as a renewable energy source. Although any plant matter can be utilized as a feed stock and various living organisms can be used to convert the plant matter into ethanol, most bioethanol is produced from sugar cane and corn (maize), using yeast. Yeast converts sugars into ethanol in large fermenters by anaerobic respiration. Only sugars can be converted, so starch and cellulose must first be broken down into sugars. This is done using enzymes. The ethanol produced by the yeasts is purified by distillation and various methods are then used to remove water from it to improve its combustion. Most bioethanol is used as a fuel in vehicles, sometimes in a pure state and sometimes mixed with gasoline (petrol).

▲ Figure 6

Data-based questions: Monitoring anaerobic cell respiration in yeast

The apparatus in figure 7 was used to monitor mass changes during the brewing of wine. The flask was placed on an electronic balance, which was connected to a computer for data-logging. The results are shown in figure 8.

1   Calculate the total loss of mass during the experiment and the mean daily loss.   [3]

2   Explain the loss of mass.   [3]

3   Suggest two reasons for the increasing rate of mass loss from the start of the experiment until day 6.   [2]

4   Suggest two reasons for the mass remaining constant from day 11 onwards.   [2]

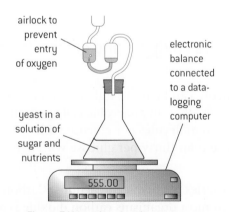

▲ Figure 7 Yeast data-logging apparatus

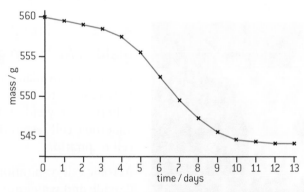

▲ Figure 8 Monitoring anaerobic cell respiration in yeast

🌐 Anaerobic respiration in humans

Lactate production in humans when anaerobic respiration is used to maximize the power of muscle contractions.

The lungs and blood system supply oxygen to most organs of the body rapidly enough for aerobic respiration to be used, but sometimes we resort to anaerobic cell respiration in muscles. The reason is that anaerobic respiration can supply ATP very rapidly for a short period of time. It is therefore used when we need to maximize the power of muscle contractions.

In our ancestors maximally powerful muscle contractions will have been needed for survival by allowing escape from a predator or catching of prey during times of food shortage. These events rarely occur in our lives today. Instead anaerobic respiration is more likely to be used during training or sport. These are examples:

- weight lifters during the lift;

- short-distance runners in races up to 400 metres;

- long-distance runners, cyclists and rowers during a sprint finish.

Anaerobic cell respiration involves the production of lactate, so when it is being used to supply ATP, the concentration of lactate in a muscle increases. There is a limit to the concentration that the body can tolerate and this limits how much anaerobic respiration can be done. This is the reason for the short timescale over which the power of muscle contractions can be maximized. We can only sprint for a short distance – not more than 400 metres.

After vigorous muscle contractions, the lactate must be broken down. This involves the use of oxygen. It can take several minutes for enough oxygen to be absorbed for all lactate to be broken down. The demand for oxygen that builds up during a period of anaerobic respiration is called the oxygen debt.

▲ Figure 9 Short bursts of intense exercise are fuelled by ATP from anaerobic cell respiration

Aerobic respiration

Aerobic cell respiration requires oxygen and gives a large yield of ATP from glucose.

If oxygen is available to a cell, glucose can be more fully broken down to release a greater quantity of energy than in anaerobic cell respiration. Whereas the yield of ATP is only two molecules per glucose with anaerobic cell respiration, it is more than thirty per glucose with aerobic cell respiration.

Aerobic cell respiration involves a series of chemical reactions. Carbon dioxide and water are produced. In most organisms carbon dioxide is a waste product that has to be excreted, but the water is often useful. In humans about half a litre is produced per day.

$$\text{glucose} + \text{oxygen} \xrightarrow[\text{ADP to ATP}]{} \text{carbon dioxide} + \text{water}$$

In eukaryotic cells most of the reactions of aerobic cell respiration, including all of the reactions that produce carbon dioxide, happen inside the mitochondrion.

▲ Figure 10 The desert rat never needs to drink despite only eating dry foods, because aerobic cell respiration supplies its water needs

Respirometers

Analysis of results from experiments involving measurement of respiration rates in germinating seeds or invertebrates using a respirometer.

A respirometer is any device that is used to measure respiration rate. There are many possible designs. Most involve these parts:

- A sealed glass or plastic container in which the organism or tissue is placed.

- An alkali, such as potassium hydroxide, to absorb carbon dioxide.

- A capillary tube containing fluid, connected to the container.

One possible design of respirometer is shown in figure 11, but it is possible to design simpler versions that require only a syringe with a capillary tube attached to it.

If the respirometer is working correctly and the organisms inside are carrying out aerobic cell respiration, the volume of air inside the respirometer will reduce and the fluid in the capillary tube will move towards the container with the organisms. This is because oxygen is used up and carbon dioxide produced by aerobic cell respiration is absorbed by the alkali.

The position of the fluid should be recorded several times. If the rate of movement of the fluid is relatively even, the results are reliable. If the temperature inside the respirometer fluctuates, the results will not be reliable because an increase in air temperature causes an increase in volume. If possible the temperature inside the respirometer should be controlled using a thermostatically controlled water bath.

Respirometers can be used to perform various experiments:

- the respiration rate of different organisms could be compared;

- the effect of temperature on respiration rate could be investigated;

- respiration rates could be compared in active and inactive organisms.

The table below shows the results of an experiment in which the effect of temperature on respiration in germinating pea seeds was investigated.

To analyse these results you should first check to see if the repeats at each temperature are close enough for you to decide that the results are reliable. You should then calculate mean results for each temperature. The next stage is to plot a graph of the mean results, with temperature on the horizontal x-axis and the rate of movement of fluid on the vertical y-axis. Range bars can be added to the graph by plotting the lowest and highest result at each temperature and joining them with a ruled line. The graph will allow you to conclude what the relationship is between the temperature and the respiration rate of the germinating peas.

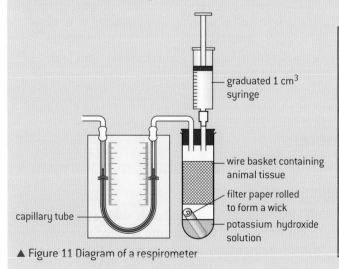

▲ Figure 11 Diagram of a respirometer

Temperature (°C)	Movement of fluid in respirometer (mm min^{-1})		
	1st reading	2nd reading	3rd reading
5	2.0	1.5	2.0
10	2.5	2.5	3.0
15	3.5	4.0	4.0
20	5.5	5.0	6.0
25	6.5	8.0	7.5
30	11.5	11.0	9.5

Data-based questions: Oxygen consumption in tobacco hornworms

Tobacco hornworms are the larvae of *Manduca sexta*. Adults of this species are moths. Larvae emerge from the eggs laid by the adult female moths. There are a series of larval stages called instars. Each instar grows and then changes into the next one by shedding its exoskeleton and developing a new larger one. The exoskeleton includes the tracheal tubes that supply oxygen to the tissues.

The graphs below (figure 12) show measurements made using a simple respirometer of the respiration rate of 3rd, 4th and 5th instar larvae. Details of the methods are given in the paper published by the biologists who carried out the research. The reference to the research is Callier V and Nijhout H F (2011) "Control of body size by oxygen supply reveals size-dependent and size-independent mechanisms of molting and metamorphosis." PNAS;108:14664–14669. This paper is freely available on the internet at http://www.pnas.org/content/108/35/14664.full.pdf+html.

Each data point on the graphs shows the body mass and respiration rate of one larva. For each instar the results have been divided into younger larvae with low to intermediate body mass and older larvae with intermediate to high body mass. The results are plotted on separate graphs. The intermediate body mass is referred to as the critical weight.

1   a) Predict, using the data in the graphs, how the respiration rate of a larva will change as it grows from moulting until it reaches the critical weight.   [1]

    b) Explain the change in respiration rate that you have described.   [2]

2   a) Discuss the trends in respiration rate in larvae above the critical weight.   [2]

    b) Suggest reasons for the difference in the trends between the periods below and above the critical weight.   [2]

The researchers reared some tobacco hornworms in air with reduced oxygen content. They found that the instar larvae moulted at a lower body mass than larvae reared in normal air with 20% oxygen.

3   Suggest a reason for earlier moulting in larvae reared in air with reduced oxygen content.   [2]

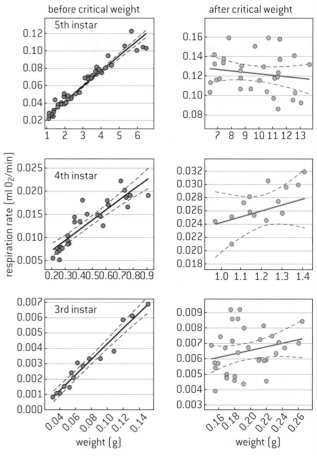

▲ Figure 12 Respiration rates of tobacco hornworms (after Callier and Nijhout, 2011)

 ## Ethics of animal use in respirometers

Assessing the ethics of scientific research: the use of invertebrates in respirometer experiments has ethical implications.

It is important for all scientists to assess the ethics of their research. There has been intense debate about the ethics of using animals in experiments. When discussing ethical issues, do we consider the consequences such as benefits to students who are learning science? Do we consider intentions? For example, if the animals are harmed unintentionally does that change whether the experiment was ethical or not? Are there absolute principles of right and wrong: for example, can we say that animals should never be subject to conditions that are outside what they would encounter in their natural habitat?

Before carrying out respirometer experiments involving animals these questions should be answered to help to decide whether the experiments are ethically acceptable:

1   Is it acceptable to remove animals from their natural habitat for use in an experiment and can they be safely returned to their habitat?

2   Will the animals suffer pain or any other harm during the experiment?

3   Can the risk of accidents that cause pain or suffering to the animals be minimized during the experiment? In particular, can contact with the alkali be prevented?

4   Is the use of animals in the experiment essential or is there an alternative method that avoids using animals?

It is particularly important to consider the ethics of animal use in respirometer experiments because the International Baccalaureate Organization has issued a directive that laboratory or field experiments and investigations need to be undertaken in an ethical way. An important aspect of this is that experiments should not be undertaken in schools that inflict pain or harm on humans or other living animals.

2.9 Photosynthesis

Understanding

→   Photosynthesis is the production of carbon compounds in cells using light energy.

→   Visible light has a range of wavelengths with violet the shortest wavelength and red the longest.

→   Chlorophyll absorbs red and blue light most effectively and reflects green light more than other colours.

→   Oxygen is produced in photosynthesis from photolysis of water.

→   Energy is needed to produce carbohydrates and other carbon compounds from carbon dioxide.

→   Temperature, light intensity and carbon dioxide concentration are possible limiting factors on the rate of photosynthesis.

 Applications

→   Changes to the Earth's atmosphere, oceans and rock deposition due to photosynthesis.

 Skills

→   Design of experiments to investigate limiting factors on photosynthesis.

→   Separation of photosynthetic pigments by chromatography. (Practical 4)

→   Drawing an absorption spectrum for chlorophyll and an action spectrum for photosynthesis.

 Nature of science

→   Experimental design: controlling relevant variables in photosynthesis experiments is essential.

What is photosynthesis?

Photosynthesis is the production of carbon compounds in cells using light energy.

Living organisms require complex carbon compounds to build the structure of their cells and to carry out life processes. Some organisms are able to make all the carbon compounds that they need using only light energy and simple inorganic substances such as carbon dioxide and water. The process that does this is called photosynthesis.

Photosynthesis is an example of energy conversion, as light energy is converted into chemical energy in carbon compounds. The carbon compounds produced include carbohydrates, proteins and lipids.

▲ Figure 1 Leaves absorb carbon dioxide and light and use them in photosynthesis

▲ Figure 2 The trees in one hectare of redwood forest in California can have a biomass of more than 4,000 tonnes, mostly carbon compounds produced by photosynthesis

(⚗) Separating photosynthetic pigments by chromatography

Separation of photosynthetic pigments by chromatography. (Practical 4)

Chloroplasts contain several types of chlorophyll and other pigments called accessory pigments. Because these pigments absorb different ranges of wavelength of light, they look a different colour to us. Pigments can be separated by chromatography. You may be familiar with paper chromatography but thin layer chromatography gives better results. This is done with a plastic strip that has been coated with a thin layer of a porous material. A spot containing pigments extracted from leaf tissue is placed near one end of the strip. A solvent is allowed to run up the strip, to separate the different types of pigment.

1   Tear up a leaf into small pieces and put them in a mortar.

2   Add a small amount of sand for grinding.

▲ Figure 3 Thin layer chromatography (TLC)

3 Add a small volume of propanone (acetone).

4 Use the pestle to grind the leaf tissue and dissolve out the pigments.

5 If the propanone all evaporates, add a little more.

6 When the propanone has turned dark green, allow the sand and other solids to settle, then pour the propanone off into a watch glass.

7 Use a hair drier to evaporate off all the propanone and water from the cells' cytoplasm.

8 When you have just a smear of dry pigments in the watch glass, add 3–4 drops of propanone and use a paint brush to dissolve the pigments.

9 Use the paint brush to transfer a very small amount of the pigment solution to the TLC strip. Your aim is to make a very small spot of pigment in the middle of the strip, 10 millimetres from one end. It should be very dark. This is achieved by repeatedly putting a small drop onto the strip and then allowing it to dry before adding another amount. You can speed up drying by blowing on the spot or by using the hair drier.

10 When the spot is dark enough, slide the other end of the strip into the slot in a cork or bung that fits into a tube that is wider than the TLC strip. The slot should hold the strip firmly.

11 Insert the cork and strip into a specimen tube. The TLC strip should extend nearly to the bottom of the tube, but not quite touch.

Pigment	Colour of pigment	R_f
Carotene	orange	0.98
Chlorophyll a	blue green	0.59
Chlorophyll b	yellow green	0.42
Phaeophytin	olive green	0.81
Xanthophyll 1	yellow	0.28
Xanthophyll 2	yellow	0.15

12 Mark the outside of the tube just below the level of the spot on the TLC strip.

13 Take the strip and cork out of the tube.

14 Pour running solvent into the specimen tube up to the level that you marked.

15 Place the specimen tube on a lab bench where it will not be disturbed. Carefully lower the TLC strip and cork into the tube, so that the tube is sealed and the TLC strip is just dipping into the running solvent. The solvent must NOT touch the pigment spot.

16 Leave the tube completely alone for about five minutes, to allow the solvent to run up through the TLC strip. You can watch the pigments separate, but DO NOT TOUCH THE TUBE.

17 When the solvent has nearly reached the top of the strip, remove it from the tube and separate it from the cork.

18 Rule two pencil lines across the strip, one at the level reached by the solvent and one at the level of the initial pigment spot.

19 Draw a circle around each of the separated pigment spots and a cross in the centre of the circle.

Spot number	Colour	Distance moved (mm)	R_f	Name of pigment
1				
2				
3				
4				
5				
6				
7				
8				

Table of standard R_f values

▲ Figure 4 Chromatogram of leaf pigments

20 Using a ruler with millimetre markings, measure the distance moved by the running solvent (the distance between the two lines) and the distance moved by each pigment (the distance between the lower line and the cross in the centre of the circle).

21 Calculate the R_f for each pigment, where R_f is the distance run by the pigment divided by the distance run by the solvent.

22 Show all your results in the table above, starting with the pigment that had moved least far.

▲ Figure 5 In a rainbow the wavelengths of visible light are separated

Wavelengths of light

Visible light has a range of wavelengths with violet the shortest wavelength and red the longest.

Sunlight or simply light is made up of all the wavelengths of electromagnetic radiation that our eyes can detect. It is therefore visible to us and other wavelengths are invisible. There is a spectrum of electromagnetic radiation from very short to very long wavelengths. Shorter wavelengths such as X-rays and ultraviolet radiation have high energy; longer wavelengths such as infrared radiation and radio waves have lower energy. Visible light has wavelengths longer than ultraviolet and shorter than infrared. The range of wavelengths of visible light is 400 to 700 nanometres.

When droplets of water in the sky split sunlight up and a rainbow is formed, different colours of light are visible. This is because sunlight is a mixture of different wavelengths, which we see as different colours, including violet, blue, green and red. Violet and blue are the shorter wavelengths and red is the longest wavelength.

The wavelengths of light that are detected by the eye are also those used by plants in photosynthesis. A reason for this is that they are emitted by the sun and penetrate the Earth's atmosphere in larger quantities than other wavelengths, so are particularly abundant.

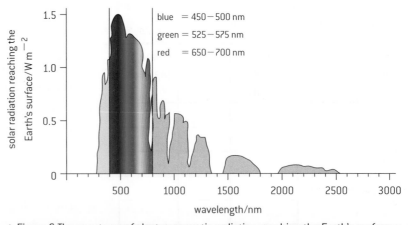

▲ Figure 6 The spectrum of electromagnetic radiation reaching the Earth's surface

Light absorption by chlorophyll

Chlorophyll absorbs red and blue light most effectively and reflects green light more than other colours.

The first stage in photosynthesis is the absorption of sunlight. This involves chemical substances called pigments. A white or transparent substance does not absorb visible light. Pigments are substances that do

absorb light and therefore appear coloured to us. Pigments that absorb all of the colours appear black, because they emit no light.

There are pigments that absorb some wavelengths of visible light but not others. For example, the pigment in a gentian flower absorbs all colours except blue. It appears blue to us, because this part of the sunlight is reflected and can pass into our eye, to be detected by cells in the retina.

Photosynthesizing organisms use a range of pigments, but the main photosynthetic pigment is chlorophyll. There are various forms of chlorophyll but they all appear green to us. This is because they absorb red and blue light very effectively, but the intermediate green light much less effectively. Wavelengths of green light therefore are reflected. This is the reason for the main colour in ecosystems dominated by plants being green.

▲ Figure 7 Gentian flowers contain the pigment delphinidin, which reflects blue light and absorbs all other wavelengths.

 ## Absorption and action spectra

Drawing an absorption spectrum for chlorophyll and an action spectrum for photosynthesis.

An action spectrum is a graph showing the rate of photosynthesis at each wavelength of light. An absorption spectrum is a graph showing the percentage of light absorbed at each wavelength by a pigment or a group of pigments.

- When drawing both action and absorption spectra, the horizontal x-axis should have the legend wavelength, with nanometres shown as the units. The scale should extend from 400 to 700 nanometres.

- On an action spectrum the y-axis should be used for a measure of the relative amount of photosynthesis. This is often given as a percentage of the maximum rate, with a scale from 0 to 100%.

- On an absorption spectrum the y-axis should have the legend "% absorption", with a scale from 0 to 100%.

- Ideally data points for specific wavelengths should be plotted and then a smooth curve be drawn through them. If this is not possible, the curve from a published spectrum could be copied.

It is not difficult to explain why action and absorption spectra are very similar: photosynthesis can only occur in wavelengths of light that chlorophyll or the other photosynthetic pigments can absorb.

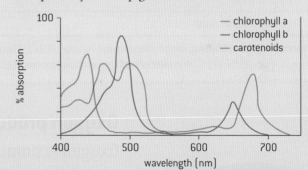

▲ Figure 8 Absorption spectra of plant pigments

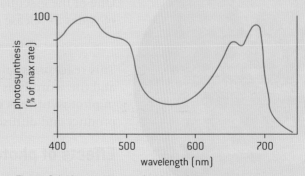

▲ Figure 9 Action spectrum of a plant pigment

133

Data-based questions: Growth of tomato seedlings in red, green and blue light

Tomato seeds were germinated and grown for 30 days in light produced by red, orange, green and blue light emitting diodes. Four different colours of LED were tested and two combinations of colours. In every treatment the tomato plants received the same intensity of photons of light. The peak wavelength of light emitted by each wavelength is shown in the table below, together with the mean leaf area and height of the seedlings. Plants often grow tall, with weak stems and small leaves when they are receiving insufficient light for photosynthesis.

1   Plot a graph to show the relationship between wavelength, leaf area and height. Hint: if you need two different scales on the y-axis you can put one on the left hand side of the graph and the other on the right hand side. Do not attempt to plot the results for combinations of LEDs.   [6]

2   Using your graph, deduce the relationship between the leaf area of the seedlings and their height.   [1]

3   Evaluate the data in the table for a grower of tomato crops in greenhouses who is considering using LEDs to provide light.   [3]

Colours of LEDs	Peak wavelength of light emitted by LED (nm)	Leaf area of seedlings (cm²)	Height of seedlings (mm)
Red	630	5.26	192
Orange	600	4.87	172
Green	510	5.13	161
Blue	450	7.26	128
Red and Blue	–	5.62	99
Red, Green and Blue	–	5.92	85

Source: Xiaoying, Shirong, Taotao, Zhigang and Tezuka (2012). "Regulation of the growth and photosynthesis of cherry tomato seedlings by different light irradiations of light emitting diodes (LED)." *African Journal of Biotechnology* Vol. 11(22), pp. 6169-6177

Oxygen production in photosynthesis

Oxygen is produced in photosynthesis from photolysis of water.

One of the essential steps in photosynthesis is the splitting of molecules of water to release electrons needed in other stages.

$$H_2O \rightarrow 4e^- + 4H^+ + O_2$$

This reaction is called photolysis because it only happens in the light and the word lysis means disintegration. All of the oxygen generated in photosynthesis comes from photolysis of water. Oxygen is a waste product and diffuses away.

Effects of photosynthesis on the Earth

Changes to the Earth's atmosphere, oceans and rock deposition due to photosynthesis.

Prokaryotes were the first organisms to perform photosynthesis, starting about 3,500 million years ago. They were joined millions of years later by algae and plants, which have been carrying out photosynthesis ever since.

▲ Figure 10 Photosynthesizing organisms seem insignificant in relation to the size of the Earth but over billions of years they have changed it significantly

One consequence of photosynthesis is the rise in the oxygen concentration of the atmosphere. This began about 2,400 million years ago (mya), rising to 2% by volume by 2,200 mya. This is known as the Great Oxidation Event.

At the same time the Earth experienced its first glaciation, presumably due to a reduction in the greenhouse effect. This could have been due to the rise in oxygenation causing a decrease in the concentration of methane in the atmosphere and photosynthesis causing a decrease in carbon dioxide concentration. Both methane and carbon dioxide are potent greenhouse gases.

The increase in oxygen concentrations in the oceans between 2,400 and 2,200 mya caused the oxidation of dissolved iron in the water, causing it to precipitate onto the sea bed. A distinctive rock formation was produced called the banded iron formation, with layers of iron oxide alternating with other minerals. The reasons for the banding are not yet fully understood. The banded iron formations are the most important iron ores, so it is thanks to photosynthesis in bacteria billions of years ago that we have abundant supplies of steel today.

The oxygen concentration of the atmosphere remained at about 2% from 2,200 mya until about 750-635 mya. There was then a significant rise to 20% or more. This corresponds with the period when many groups of multicellular organisms were evolving.

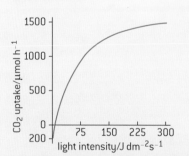

▲ Figure 11

Production of carbohydrates

Energy is needed to produce carbohydrates and other carbon compounds from carbon dioxide.

Plants convert carbon dioxide and water into carbohydrates by photosynthesis. The simple equation below summarizes the process:

$$\text{carbon dioxide} + \text{water} \rightarrow \text{carbohydrate} + \text{oxygen}$$

To carry out this process, energy is required. A chemical reaction that involves putting in energy is described as endothermic. Reactions involving the production of oxygen are usually endothermic in living systems. Reactions involving combining smaller molecules to make larger ones are also often endothermic and molecules of carbohydrate such as glucose are much larger than carbon dioxide or water.

Activity

CO₂ concentration

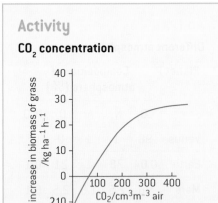

increase in biomass of grass /kg ha⁻¹ h⁻¹

$CO_2/cm^3 m^{-3}$ air

▲ Figure 13 In this graph the rate of photosynthesis was measured indirectly by measuring the change in plant biomass.

1 The maximum carbon dioxide concentration of the atmosphere is 380 $cm^3 m^{-3}$air. Why is the concentration often lower near leaves?

2 In what weather conditions is carbon dioxide concentration likely to be the limiting factor for photosynthesis?

The energy for the conversion of carbon dioxide into carbohydrate is obtained by absorbing light. This is the reason for photosynthesis only occurring in the light. The energy absorbed from light does not disappear – it is converted to chemical energy in the carbohydrates.

Limiting factors

Temperature, light intensity and carbon dioxide concentration are possible limiting factors on the rate of photosynthesis.

The rate of photosynthesis in a plant can be affected by three external factors:

- temperature;
- light intensity;
- carbon dioxide concentration.

Each of these factors can limit the rate if they are below the optimal level. These three factors are therefore called limiting factors. According to the concept of limiting factors, under any combination of light intensity, temperature and carbon dioxide concentration, only one of the factors is actually limiting the rate of photosynthesis. This is the factor that is furthest from its optimum. If the factor is changed to make it closer to the optimum, the rate of photosynthesis increases, but changing the other factors will have no effect, as they are not the limiting factor.

Of course, as the limiting factor is moved closer to its optimum, while keeping the other factors constant, a point will be reached where this factor is no longer the one that is furthest from its optimum and another factor becomes the limiting factor. For example, at night, light intensity is presumably the limiting factor for photosynthesis. When the sun rises and light intensity increases, temperature will usually take over as the limiting factor. As the temperature increases during the morning, carbon dioxide concentration might well become the limiting factor.

 Controlled variables in limiting factor experiments

Experimental design: controlling relevant variables in photosynthesis experiments is essential.

In any experiment, it is important to control all variables other than the independent and dependent variable that you are investigating. The independent variable is the one that you deliberately vary in the experiment with a range of levels that you choose. The dependent variable is what you measure during the experiment, to see if it is affected by the independent variable.

It is essential during this type of experiment to be sure that the independent variable is the only factor that could be affecting the dependent variable. All other variables that might affect the independent variable must therefore be controlled.

These are questions that you need to answer when you are designing an experiment to investigate a limiting factor on photosynthesis:

- Which limiting factor will you investigate? This will be your independent variable.

- How will you measure the rate of photosynthesis? This will be your dependent variable.

- How will you keep the other limiting factors at a constant and optimal level? These will be your controlled variables.

 ## Investigating limiting factors

Design of experiments to investigate limiting factors on photosynthesis.

There are many possible experimental designs. A method that can be used to investigate the effect of carbon dioxide concentration is given below. You could either modify this to investigate a different limiting factor or you could develop an entirely different design.

Investigating the effect of carbon dioxide on photosynthesis

If a stem of pondweed such as *Elodea*, *Cabomba* or *Myriophyllum* is placed upside-down in water and the end of the stem is cut, bubbles of gas may be seen to escape. If these are collected and tested, they are found to be mostly oxygen, produced by photosynthesis. The rate of oxygen production can be measured by counting the bubbles. Factors that might affect the rate of photosynthesis can be varied to find out what effect this has. In the method below carbon dioxide concentration is varied.

1   Enough water to fill a large beaker is boiled and allowed to cool. This removes carbon dioxide and other dissolved gases.

2   The water is poured repeatedly from one beaker to another, to oxygenate the water. Very little carbon dioxide will dissolve.

3   A stem of pondweed is placed upside-down in the water and the end of its stem is cut. No bubbles are expected to emerge, as the water contains almost no carbon dioxide. The temperature of the water should be about 25 °C and the water should be very brightly illuminated. Suitable apparatus is shown in figure 15.

4   Enough sodium hydrogen carbonate is added to the beaker to raise the carbon dioxide concentration by 0.01 mol dm^{-3}. If bubbles emerge, they are counted for 30 seconds, repeating the counts until two or three consistent results are obtained.

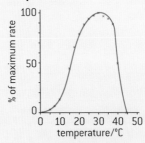

Activity

Temperature

▲ Figure 14 In this graph the rate of photosynthesis was measured indirectly by measuring the change in plant biomass

1   What was the optimum temperature for photosynthesis in this plant?

2   What was the maximum temperature for photosynthesis?

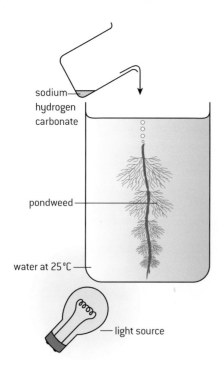

sodium hydrogen carbonate

pondweed

water at 25 °C

light source

▲ Figure 15 Apparatus for measuring photosynthesis rates in different concentrations of carbon dioxide

5   Enough sodium hydrogen carbonate is added to raise the concentration by another 0.01 mol dm^{-3}. Bubble counts are done in the same way.

6   The procedure above is repeated again and again until further increases in carbon dioxide do not affect the rate of bubble production.

Questions

1   Why are the following procedures necessary?

   a)   Boiling and then cooling the water before the experiment.

   b)   Keeping the water at 25 °C and brightly illuminating it.

   c)   Repeating bubble counts until several consistent counts have been obtained.

2   What other factor could be investigated using bubble counts with pondweed and how would you design the experiment?

3   How could you make the measurement of the rate of oxygen production more accurate?

Questions

1 Lipase is a digestive enzyme that accelerates the breakdown of triglycerides in the small intestine. In the laboratory, the rate of activity of lipase can be detected by a decline in pH. Explain what causes the pH to decline. [4]

2 Papain is a protease that can be extracted from pineapple fruits. Figure 16 shows the effect of temperature on the activity of papain. The experiment was performed using papain dissolved in water and then repeated with the same quantity of papain that had been immobilized by attaching it to a solid surface. The results show the percentage of the protein in the reaction mixture that was digested in a fixed time.

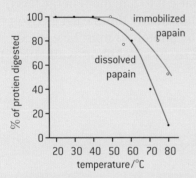

▲ Figure 16

a) (i) Outline the effects of temperature on the activity of dissolved papain. [2]

   (ii) Explain the effects of temperature on the activity of dissolved papain. [2]

b) (i) Compare the effect of temperature on the activity of immobilized papain with the effect on dissolved papain. [2]

   (ii) Suggest a reason for the difference that you have described. [2]

   (iii) In some parts of the human body, enzymes are immobilized in membranes. Suggest one enzyme and a part of the body where it would be useful for it to be immobilized in a membrane. [2]

3 The equation below summarizes the results of metabolic pathways used to produce ATP, using energy from the oxidation of glucose.

glucose + oxygen + (ADP + Pi) →
   180 g   134.4 dm³   18.25 kg

carbon dioxide + water + ATP
   134.4 dm³   108 g   18.25 kg

a) (i) State the volume units that are shown in the equation. [1]

   (ii) State the mass units that are shown in the equation. [2]

b) (i) Calculate the mass of ATP produced per dm³ of oxygen. [2]

   (ii) Calculate the mass of ATP produced per race in table 1. [4]

c) Explain how it is possible to synthesize such large masses of ATP during races. [3]

d) During a 100 m race, 80 g of ATP is needed but only 0.5 dm³ of oxygen is consumed. Deduce how ATP is being produced. [3]

Length of race/m	Volume of oxygen consumed in cell respiration during the race/dm³
1500	36
10,000	150
42,300	700

▲ Table 1

4 Figure 17 shows the effects of varying light intensity on the carbon dioxide absorption by leaves, at different, fixed carbon dioxide concentrations and temperatures.

a) Deduce the limiting factor for photosynthesis at:

   (i) W   (ii) X   (iii) Y   (iv) Z. [4]

b) Explain why curves I and II are the same between 1 and 7 units of light intensity. [3]

c) Explain the negative values for carbon dioxide absorption when the leaves were in low light intensities. [3]

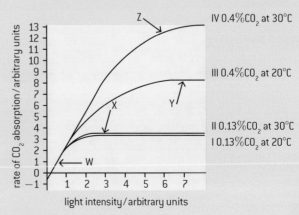

▲ Figure 17

5   Figure 18 shows the results of an experiment in which *Chlorella* cells were given light of wavelengths from 660 nm (red) up to 700 nm (far red). The rate of oxygen production by photosynthesis was measured and the yield of oxygen per photon of light was calculated. This gives a measure of the efficiency of photosynthesis at each wavelength. The experiment was then repeated with supplementary light with a wavelength of 650 nm at the same time as each of the wavelengths from 660 to 700 nm, but with the same overall intensity of light as in the first experiment.

o----o with supplementary light

□——□ without supplementary light

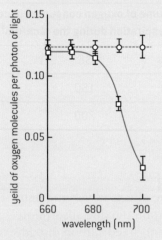

▲ Figure 18 Photon yield of photosynthesis in different light intensities

a)   Describe the relationship between wavelength of light and oxygen yield, when there was no supplementary light. [2]

b)   Describe the effect of the supplementary light. [2]

c)   Explain how the error bars help in drawing conclusions from this experiment. [2]

d)   The probable maximum yield of oxygen was 0.125 molecules per photon of light. Calculate how many photons are needed to produce one oxygen molecule in photosynthesis. [2]

e)   Oxygen production by photolysis involves this reaction:

$$4H_2O \rightarrow O_2 + 2H_2O + 4H^+ + 4e^-$$

Each photon of light is used to excite an electron (raise it to a higher energy level). Calculate how many times each electron produced by photolysis must be excited during the reactions of photosynthesis. [2]

3 GENETICS

Introduction

Every living organism inherits a blueprint for life from its parents. The inheritance of genes follows patterns. Chromosomes carry genes in a linear sequence that is shared by members of a species. Alleles segregate during meiosis allowing new combinations to be formed by the fusion of gametes. Biologists have developed techniques for artificial manipulation of DNA, cells and organisms.

3.1 Genes

Understanding

→ A gene is a heritable factor that consists of a length of DNA and influences a specific characteristic.

→ A gene occupies a specific position on one type of chromosome.

→ The various specific forms of a gene are alleles.

→ Alleles differ from each other by one or a few bases only.

→ New alleles are formed by mutation.

→ The genome is the whole of the genetic information of an organism.

→ The entire base sequence of human genes was sequenced in the Human Genome Project.

Applications

→ The causes of sickle cell anemia, including a base substitution mutation, a change to the base sequence of mRNA transcribed from it and a change to the sequence of a polypeptide in hemoglobin.

→ Comparison of the number of genes in humans with other species.

Skills

→ Use of a database to determine differences in the base sequence of a gene in two species.

Nature of science

→ Developments in scientific research follow improvements in technology: gene sequencers, essentially lasers and optical detectors, are used for the sequencing of genes.

What is a gene?

A gene is a heritable factor that consists of a length of DNA and influences a specific characteristic.

Genetics is the branch of biology concerned with the storage of information in living organisms and how this information can be passed from parents to progeny. The word genetics was used by biologists long before the method of information storage was understood. It came from the word genesis, meaning origins. Biologists were interested in the origins of features such as baldness, blue eyes and much more. Something must be the cause of these features and be passed on to offspring where the features would again develop.

Experiments in the 19th century showed that there were indeed factors in living organisms that influenced specific characteristics and that these factors were heritable. They could be passed on to offspring by pea plants, fruit flies and all other organisms. There was intense research into genetics from the early 20th century onwards and the word gene was invented for the heritable factors.

One obvious question was the chemical composition of genes. By the middle of the 20th century there was strong evidence that genes were made of DNA. There are relatively few DNA molecules in a cell – just 46 in a typical human cell for example – yet there are thousands of *genes*. We can therefore deduce that each gene consists of a much shorter length of DNA than a chromosome and that each chromosome carries many genes.

 ## Comparing numbers of genes

Comparison of the number of genes in humans with other species.

How many genes does it take to make a bacterium, a banana plant or a bat, and how many are needed to make a human? We see ourselves as more complex in structure, physiology and behaviour so we might expect to have more genes. The table shows whether this is true. It gives a range of predicted gene numbers. They are based on evidence from the DNA of these species but are not precise counts of gene numbers as these are not yet known.

Group	Name of species	Brief description	Numbers of genes
Prokaryotes	*Haemophilus influenzae*	Pathogenic bacterium	1,700
	Escherichia coli	Gut bacterium	3,200
Protoctista	*Trichomonas vaginalis*	Unicellular parasite	60,000
Fungi	*Saccharomyces cerevisiae* (Yeast)	Unicellular fungus	6,000
Plants	*Oryza sativa* (Rice)	Crop grown for food	41,000
	Arabidopsis thaliana (Thale cress)	Small annual weed	26,000
	Populus trichocarpa (Black cottonwood)	Large tree	46,000
Animals	*Drosophila melanogaster* (Fruit fly)	Larvae consume ripe fruit	14,000
	Caenorhabditis elegans	Small soil roundworm	19,000
	Homo sapiens (Humans)	Large omnivorous biped	23,000
	Daphnia pulex (Water flea)	Small pond crustacean	31,000

Where are genes located?

A gene occupies a specific position on one type of chromosome.

Experiments in which different varieties of plant or animals are crossed show that genes are linked in groups and each group corresponds to one of the types of chromosome in a species. For example, there are four groups of linked genes in fruit flies and four types of chromosome. Maize has ten groups of linked genes and ten types of chromosome and in humans the number of both is 23.

Each gene occupies a specific position on the type of chromosome where it is located. This position is called the locus of the gene. Maps showing the sequence of genes along chromosomes in fruit flies and other organisms were produced by crossing experiments, but much more detailed maps can now be produced when the genome of a species is sequenced.

Activity

Estimating the number of human genes

In October 1970 *Scientific American* published an estimate that the human genome might consist of as many as 10 million genes. How many times greater than the current predicted number is this? What reasons can you give for such a huge overestimate in 1970?

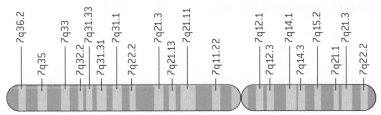

▲ Figure 1 Chromosome 7: an example of a human chromosome. It consists of a single DNA molecule with approximately 170 million base pairs – about 5% of the human genome. The pattern of banding, obtained by staining the chromosome, is different from other human chromosomes. Several thousand genes are located on chromosome 7, mostly in the light bands, each of which has a unique identifying code. The locus of a few of the genes on chromosome 7 is shown

What are alleles?

The various specific forms of a gene are alleles.

Gregor Mendel is usually regarded as the father of genetics. He crossed varieties of pea plants, for example tall pea plants with dwarf peas and white-flowered pea plants with purple-flowered. Mendel deduced that the differences between the varieties that he crossed together were due to different heritable factors. We now know that these pairs of heritable factors are alternative forms of the same gene. For example there are two forms of the gene that influences height, one making pea plants tall and the other making the plants dwarf.

These different forms are called alleles. There can be more than two alleles of a gene. One of the first examples of multiple alleles to be discovered is in mice. A gene that influences coat colour has three alleles, making the mice yellow, grey and black. There are three alleles of the gene in humans that determines ABO blood groups. In some cases there are large numbers of different alleles of a gene, for example the gene that influences eye colour in fruit flies.

As alleles are alternative forms of the same gene, they occupy the same position on one type of chromosome – they have the same locus. Only one allele can occupy the locus of the gene on a chromosome. Most animal and plant cells have two copies of each type of chromosome, so

▲ Figure 2 Different coat colours in mice

143

we can expect two copies of a gene to be present. These could be two of the same allele of the gene or two different alleles.

Differences between alleles

Alleles differ from each other by one or a few bases only.

A gene consists of a length of DNA, with a base sequence that can be hundreds or thousands of bases long. The different alleles of a gene have slight variations in the base sequence. Usually only one or a very small number of bases are different, for example adenine might be present at a particular position in the sequence in one allele and cytosine at that position in another allele.

Positions in a gene where more than one base may be present are called single nucleotide polymorphisms, abbreviated to SNPs and pronounced snips. Several snips can be present in a gene, but even then the alleles of the gene differ by only a few bases.

 Comparing genes

Use of a database to determine differences in the base sequence of a gene in two species

One outcome of the Human Genome Project is that the techniques that were developed have enabled the sequencing of other genomes. This allows gene sequences to be compared. The results of this comparison can be used to determine evolutionary relationships. Also, the identification of conserved sequences allows species to be chosen for exploring the function of that sequence.

- Go to the website called GenBank (http://www.ncbi.nlm.nih.gov/pubmed/)

- Choose 'gene' from the search menu.

- Enter the name of a gene plus the organism, such as cytochrome oxidase 1 (COX1) for pan (chimpanzee).

- Move your mouse over the section 'Genomic regions, transcripts, and products' until 'Nucleotide Links' appears.

- Choose 'Fast A' and the sequence should appear. Copy the sequence and paste it into a .txt file or notepad file.

- Repeat with a number of different species that you want to compare and save the files.

- To have the computer align the sequence for you, download the software called ClustalX and run it.

- In the File menu, choose 'Load Sequences'.

- Select your file. Your sequences should show up in the ClustalX window.

- Under the Alignment menu choose 'Do Complete Alignment.' The example below shows the sequence alignment of 9 different organisms.

▼ Figure 3

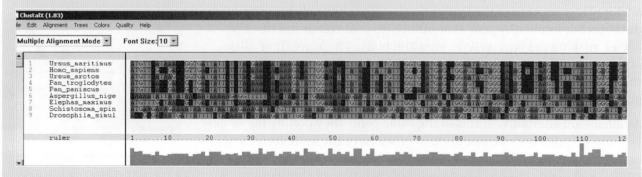

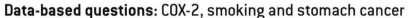

Data-based questions: COX-2, smoking and stomach cancer

COX-2 is a gene that codes for the enzyme cyclooxygenase. The gene consists of over 6,000 nucleotides. Three single nucleotide polymorphisms have been discovered that are associated with gastric adenocarcinoma, a cancer of the stomach. One of these SNPs occurs at nucleotide 1195. The base at this nucleotide can be either adenine or guanine. A large survey in China involved sequencing both copies of the COX-2 gene in 357 patients who had developed gastric adenocarcinoma and in 985 people who did not have the disease. All of these people were asked whether they had ever smoked cigarettes.

Table 1 shows the 357 patients with gastric adenocarcinoma categorized according to whether they were smokers or non-smokers and whether they had two copies of COX-2 with G at nucleotide 1195 (GG) or at least one copy of the gene with A at this position (AG or AA). The results are shown as percentages. Table 2 shows the same categorization for the 985 people who did not have this cancer.

1 Predict, using the data, which of bases G or A is more common at nucleotide 1195 in the non-smokers without cancer. [2]

2 a) Calculate the total percentage of the patients that were smokers and the total percentage of controls that were smokers. [2]

   b) Explain the conclusion that can be drawn from the difference in the percentages. [2]

3 Deduce, with a reason, whether G or A at nucleotide 1195 is associated with an increased risk of gastric adenocarcinoma. [2]

4 Discuss, using the data, whether the risk of gastric adenocarcinoma is increased equally in all smokers. [2]

	GG	AG or AA
Smokers	9.8%	43.7%
Non-smokers	9.5%	40.0%

▲ Table 1 Patients with cancer

	GG	AG or AA
Smokers	9.4%	35.6%
Non-smokers	12.6%	42.4%

▲ Table 2 Individuals without cancer

Mutation

New alleles are formed by mutation.

New alleles are formed from other alleles by gene mutation. Mutations are random changes – there is no mechanism for a particular mutation being carried out. The most significant type of mutation is a base substitution. One base in the sequence of a gene is replaced by a different base. For example, if adenine was present at a particular point in the base sequence it could be substituted by cytosine, guanine or thymine.

A random change to an allele that has developed by evolution over perhaps millions of years is unlikely to be beneficial. Almost all mutations are therefore either neutral or harmful. Some mutations are lethal – they cause the death of the cell in which the mutation occurs. Mutations in body cells are eliminated when the individual dies, but mutations in cells that develop into gametes can be passed on to offspring and cause genetic disease.

Activity

New alleles

Recent research into mutation involved finding the base sequence of all genes in parents and their offspring. It showed that there was one base mutation per 1.2×10^8 bases. Calculate how many new alleles a child is likely to have as a result of mutations in their parents. Assume that there are 25,000 human genes and these genes are 2,000 bases long on average.

Source: Campbell, CD, et al. (2012) "Estimating the human mutation rate using autozygosity in a founder population." *Nature Genetics*, 44: 1277-1281. doi: 10.1038/ng.2418

What criteria can be used to distinguish between correlation and cause and effect?

There is a correlation between high frequencies of the sickle-cell allele in human populations and high rates of infection with *Falciparum* malaria. Where a correlation exists, it may or may not be due to a causal link. Consider the information in figure 4 to decide whether sickle-cell anemia causes infection with malaria.

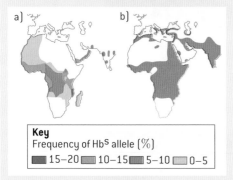

Key
Frequency of Hb^S allele (%)
☐ 15–20 ☐ 10–15 ☐ 5–10 ☐ 0–5

▲ Figure 4 Map (a) shows the frequency of the sickle cell allele and map (b) shows malaria affected areas in Africa and Western Asia

Sickle cell anemia

The causes of sickle cell anemia, including a base substitution mutation, a change to the base sequence of mRNA transcribed from it and a change to the sequence of a polypeptide in hemoglobin.

Sickle-cell anemia is the commonest genetic disease in the world. It is due to a mutation of the gene that codes for the beta-globin polypeptide in hemoglobin. The symbol for this gene is *Hb*. Most humans have the allele *Hb^A*. If a base substitution mutation converts the sixth codon of the gene from GAG to GTG, a new allele is formed, called *Hb^S*. The mutation is only inherited by offspring if it occurs in a cell of the ovary or testis that develops into an egg or sperm.

When the *Hb^S* allele is transcribed, the mRNA produced has GUG as its sixth codon instead of GAG, and when this mRNA is transcribed, the sixth amino acid in the polypeptide is valine instead of glutamic acid. This change causes hemoglobin molecules to stick together in tissues with low oxygen concentrations. The bundles of hemoglobin molecules that are formed are rigid enough to distort the red blood cells into a sickle shape.

These sickle cells cause damage to tissues by becoming trapped in blood capillaries, blocking them and reducing blood flow. When sickle cells return to high oxygen conditions in the lung, the hemoglobin bundles break up and the cells return to their normal shape. These changes occur time after time, as the red blood cells circulate. Both the hemoglobin and the plasma membrane are damaged and the life of a red blood cell can be shortened to as little as 4 days. The body cannot replace red blood cells at a rapid enough rate and anemia therefore develops.

So, a small change to a gene can have very harmful consequences for individuals that inherit the gene. It is not known how often this mutation has occurred but in some parts of the world the *Hb^S* allele is remarkably common. In parts of East Africa up to 5% of newborn babies have two copies of the allele and develop severe anemia. Another 35% have one copy so make both normal hemoglobin and the mutant form. These individuals only suffer mild anemia.

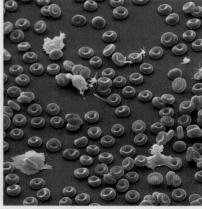

▲ Figure 5 Micrographs of sickle cells and normal red blood cells

What is a genome?

The genome is the whole of the genetic information of an organism.

Among biologists today the word genome means the whole of the genetic information of an organism. Genetic information is contained in DNA, so a living organism's genome is the entire base sequence of each of its DNA molecules.

- In humans the genome consists of the 46 molecules that form the chromosomes in the nucleus plus the DNA molecule in the mitochondrion. This is the pattern in other animals, though the number of chromosomes is usually different.

- In plant species the genome is the DNA molecules of chromosomes in the nucleus plus the DNA molecules in the mitochondrion and the chloroplast.

- The genome of prokaryotes is much smaller and consists of the DNA in the circular chromosome, plus any plasmids that are present.

The Human Genome Project

The entire base sequence of human genes was sequenced in the Human Genome Project.

The Human Genome Project began in 1990. Its aim was to find the base sequence of the entire human genome. This project drove rapid improvements in base sequencing techniques, which allowed a draft sequence to be published much sooner than expected in 2000 and a complete sequence in 2003.

Although knowledge of the entire base sequence has not given us an immediate and total understanding of human genetics, it has given us what can be regarded as a rich mine of data, which will be worked by researchers for many years to come. For example, it is possible to predict which base sequences are protein-coding genes. There are approximately 23,000 of these in the human genome. Originally, estimates for the number of genes were much higher.

Another discovery was that most of the genome is not transcribed. Originally called "junk DNA," it is being increasingly recognized that within these "junk" regions, there are elements that affect gene expression as well as highly repetitive sequences, called satellite DNA.

The genome that was sequenced consists of one set of chromosomes – it is **a** human genome rather than **the** human genome. Work continues to find variations in sequence between different individuals. The vast majority of base sequences are shared by all humans giving us genetic unity, but there are also many single nucleotide polymorphisms which contribute to human diversity.

Since the publication of the human genome, the base sequence of many other species has been determined. Comparisons between these genomes reveal aspects of the evolutionary history of living organisms that were previously unknown. Research into genomes will be a developing theme of biology in the 21st century.

Activity

Ethics of genome research

Ethical questions about genome research are worth discussing.

Is it ethical to take a DNA sample from ethnic groups around the world and sequence it without their permission?

Is it ethical for a biotech company to patent the base sequence of a gene to prevent other companies from using it to conduct research freely?

Who should have access to this genetic information? Should employers, insurance companies and law enforcement agencies know our genetic makeup?

Techniques used for genome sequencing

Developments in scientific research follow improvements in technology: gene sequencers, essentially lasers and optical detectors, are used for the sequencing of genes.

The idea of sequencing the entire human genome seemed impossibly difficult at one time but improvements in technology towards the end of the 20th century made it possible, though still very ambitious. These improvements continued once the project was underway and draft sequences were therefore completed much sooner than expected. Further advances are allowing the genomes of other species to be sequenced at an ever increasing rate.

To sequence a genome, it is first broken up into small lengths of DNA. Each of these is sequenced separately. To find the base sequence of a fragment of DNA, single-stranded copies of it are made using DNA polymerase, but the process is stopped before the whole base sequence has been copied by putting small quantities of a non-standard nucleotide into the reaction mixture. This is done separately with non-standard nucleotides carrying each of the four possible DNA bases. Four samples of DNA copy of varying length are produced, each with one of four DNA bases at the end of each copy. These four samples are separated according to length by gel electrophoresis. For each number of nucleotides in the copy there is a band in just one of the four tracks in the gel, from which the sequence of bases in the DNA can be deduced.

The major advance in technology that speeded up base sequencing by automating it is this:

- Coloured fluorescent markers are used to mark the DNA copies. A different colour of fluorescent marker is used for the copies ending in each of the four bases.

- The samples are mixed together and all the DNA copies are separated in one lane of a gel according to the number of nucleotides.

- A laser scans along the lane to make the fluorescent markers fluoresce.

- An optical detector is used to detect the colours of fluorescence along the lane. There is a series of peaks of fluorescence, corresponding to each number of nucleotides

- A computer deduces the base sequence from the sequence of colours of fluorescence detected.

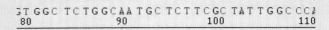

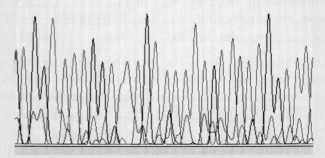

▲ Figure 6 Sequencing read from the DNA of Pinor Noir variety of grape

3.2 Chromosomes

Understanding

→ Prokaryotes have one chromosome consisting of a circular DNA molecule.

→ Some prokaryotes also have plasmids but eukaryotes do not.

→ Eukaryote chromosomes are linear DNA molecules associated with histone proteins.

→ In a eukaryote species there are different chromosomes that carry different genes.

→ Homologous chromosomes carry the same sequence of genes but not necessarily the same alleles of those genes.

→ Diploid nuclei have pairs of homologous chromosomes.

→ Haploid nuclei have one chromosome of each pair.

→ The number of chromosomes is a characteristic feature of members of a species.

→ A karyogram shows the chromosomes of an organism in homologous pairs of decreasing length.

→ Sex is determined by sex chromosomes and autosomes are chromosomes that do not determine sex.

Applications

→ Cairns's technique for measuring the length of DNA molecules by autoradiography.

→ Comparison of genome size in T2 phage, *Escherichia coli*, *Drosophila melanogaster*, *Homo sapiens* and *Paris japonica*.

→ Comparison of diploid chromosome numbers of *Homo sapiens*, *Pan troglodytes*, *Canis familiaris*, *Oryza sativa*, *Parascaris equorum*.

→ Use of karyotypes to deduce sex and diagnose Down syndrome in humans.

Skills

→ Use of online databases to identify the locus of a human gene and its protein product.

Nature of science

→ Developments in scientific research follow improvements in techniques: autoradiography was used to establish the length of DNA molecules in chromosomes.

Bacterial chromosomes

Prokaryotes have one chromosome consisting of a circular DNA molecule.

The structure of prokaryotic cells was described in sub-topic 1.2. In most prokaryotes there is one chromosome, consisting of a circular DNA molecule containing all the genes needed for the basic life processes of the cell. The DNA in bacteria is not associated with proteins, so is sometimes described as naked.

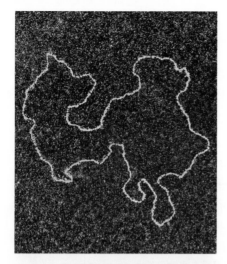

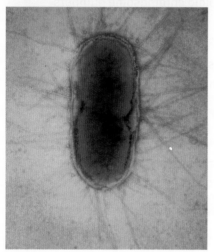

▲ Figure 1 (a) Circular DNA molecule from a bacterium (b) Bacterium preparing to divide

Because only one chromosome is present in a prokaryotic cell, there is usually only a single copy of each gene. Two identical copies are present briefly after the chromosome has been replicated, but this is a preparation for cell division. The two genetically identical chromosomes are moved to opposite poles and the cell then splits in two.

Plasmids

Some prokaryotes also have plasmids but eukaryotes do not.

Plasmids are small extra DNA molecules that are commonly found in prokaryotes but are very unusual in eukaryotes. They are usually small, circular and naked, containing a few genes that may be useful to the cell but not those needed for its basic life processes. For example, genes for antibiotic resistance are often located in plasmids. These genes are beneficial when an antibiotic is present in the environment but are not at other times.

Plasmids are not always replicated at the same time as the chromosome of a prokaryotic cell or at the same rate. Hence there may be multiple copies of plasmids in a cell and a plasmid may not be passed to both cells formed by cell division.

Copies of plasmids can be transferred from one cell to another, allowing spread through a population. It is even possible for plasmids to cross the species barrier. This happens if a plasmid that is released when a prokaryotic cell dies is absorbed by a cell of a different species. It is a natural method of gene transfer between species. Plasmids are also used by biologists to transfer genes between species artificially.

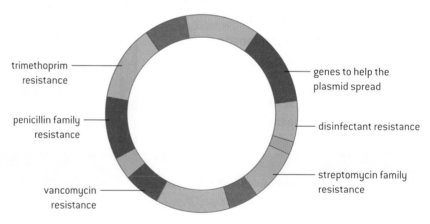

trimethoprim resistance

penicillin family resistance

vancomycin resistance

genes to help the plasmid spread

disinfectant resistance

streptomycin family resistance

▲ Figure 2 The pLW1043 plasmid

Using autoradiography to measure DNA molecules

Developments in scientific research follow improvements in techniques: autoradiography was used to establish the length of DNA molecules in chromosomes.

Quantitative data is usually considered to be the strongest type of evidence for or against a hypothesis, but in biology it is sometimes images that provide the most convincing evidence.

Developments in microscopy have allowed images to be produced of structures that were previously invisible. These sometimes confirm existing ideas but sometimes also change our understanding.

Autoradiography was used by biologists from the 1940s onwards to discover where specific substances were located in cells or tissues. John Cairns used the technique in a different way in the 1960s. He obtained images of whole DNA molecules from *E. coli* bacteria. At the time it was not clear whether the bacterial chromosome was a single DNA molecule or more than one, but the images produced by Cairns answered this question. They also revealed replication forks in DNA for the first time. Cairns's technique was used by others to investigate the structure of eukaryote chromosomes.

 ## Measuring the length of DNA molecules

Cairns's technique for measuring the length of DNA molecules by autoradiography.

John Cairns produced images of DNA molecules from *E.coli* using this technique:

- Cells were grown for two generations in a culture medium containing tritiated thymidine. Thymidine consists of the base thymine linked to deoxyribose and is used by *E. coli* to make nucleotides that it uses in DNA replication. Tritiated thymidine contains tritium, a radioactive isotope of hydrogen, so radioactively labelled DNA was produced by replication in the *E. coli* cells.

- The cells were then placed onto a dialysis membrane and their cell walls were digested using the enzyme lysozyme. The cells were gently burst to release their DNA onto the surface of the dialysis membrane.

- A thin film of photographic emulsion was applied to the surface of the membrane and left in darkness for two months. During that time some of the atoms of tritium in the DNA decayed and emitted high energy electrons, which react with the film.

- At the end of the two-month period the film was developed and examined with a microscope. At each point where a tritium atom decayed there is a dark grain. These indicate the position of the DNA.

The images produced by Cairns showed that the chromosome in *E. coli* is a single circular DNA molecule with a length of 1,100 μm. This is remarkably long given that the length of the *E coli* cells is only 2 μm.

Autoradiography was then used by other researchers to produce images of eukaryotic chromosomes. An image of a chromosome from the fruit fly *Drosophila melanogaster* was produced that was 12,000 μm long. This corresponded with the total amount of DNA known to be in a *D. melanogaster* chromosome, so for this species at least a chromosome contains one very long DNA molecule. In contrast to prokaryotes, the molecule was linear rather than circular.

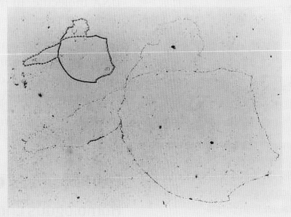

▲ Figure 3

Eukaryote chromosomes

Eukaryote chromosomes are linear DNA molecules associated with histone proteins.

Chromosomes in eukaryotes are composed of DNA and protein. The DNA is a single immensely long linear DNA molecule. It is associated with histone proteins. Histones are globular in shape and are wider than the DNA. There are many histone molecules in a chromosome, with the DNA molecule wound around them. Adjacent histones in the chromosome are separated by short stretches of the DNA molecule that are not in contact with histones. This gives a eukaryotic chromosome the appearance of a string of beads during interphase.

Differences between chromosomes

In a eukaryote species there are different chromosomes that carry different genes.

Eukaryote chromosomes are too narrow to be visible with a light microscope during interphase. During mitosis and meiosis the chromosomes become much shorter and fatter by supercoiling, so are visible if stains that bind either DNA or proteins are used. In the first stage of mitosis the chromosomes can be seen to be double. There are two chromatids, with identical DNA molecules produced by replication.

When the chromosomes are examined during mitosis, different types can be seen. They differ both in length and in the position of the centromere where the two chromatids are held together. The centromere can be positioned anywhere from close to an end to the centre of the chromosome.

There are at least two different types in every eukaryote but in most species there are more than that. In humans for example there are 23 types of chromosome.

Every gene in eukaryotes occupies a specific position on one type of chromosome, called the locus of the gene. Each chromosome type therefore carries a specific sequence of genes arranged along the linear DNA molecule. In many chromosomes this sequence contains over a thousand genes.

Crossing experiments were done in the past to discover the sequence of genes on chromosome types in *Drosophila melanogaster* and other species. The base sequence of whole chromosomes can now be found, allowing more accurate and complete gene sequences to be deduced.

Having the genes arranged in a standard sequence along a type of chromosome allows parts of chromosomes to be swapped during meiosis.

Homologous chromosomes

Homologous chromosomes carry the same sequence of genes but not necessarily the same alleles of those genes.

If two chromosomes have the same sequence of genes they are homologous. Homologous chromosomes are not usually identical to each other because, for at least some of the genes on them, the alleles are different.

If two eukaryotes are members of the same species, we can expect each of the chromosomes in one of them to be homologous with at least one chromosome in the other. This allows members of a species to interbreed.

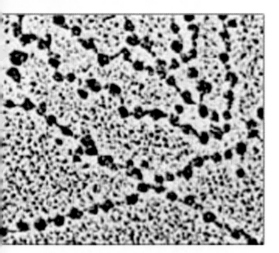

▲ Figure 4 In an electron micrograph the histones give a eukaryotic chromosome the appearance of a string of beads during interphase

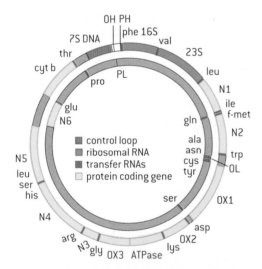

▲ Figure 5 Gene map of the human mitochondrial chromosome. There are genes on both of the two DNA strands. The chromosomes in the nucleus are much longer, carry far more genes and are linear rather than circular

Data-based questions: Comparing the chromosomes of mice and humans

Figure 6 shows all of the types of chromosome in mice and in humans. Numbers and colours are used to indicate sections of mouse chromosomes that are homologous to sections of human chromosomes.

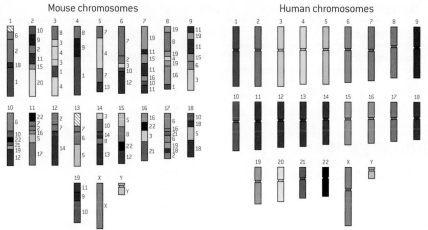

▲ Figure 6 Chromosomes

1. Deduce the number of types of chromosomes in mice and in humans. [2]

2. Identify the two human chromosome types that are most similar to mouse chromosomes. [2]

3. Identify mouse chromosomes which contain sections that are not homologous to human chromosomes. [2]

4. Suggest reasons for the many similarities between the mouse and human genomes. [2]

5. Deduce how chromosomes have mutated during the evolution of animals such as mice and humans. [2]

🌐 Comparing the genome sizes

Comparison of genome size in T2 phage, *Escherichia coli*, *Drosophila melanogaster*, *Homo sapiens* and *Paris japonica*.

The genomes of living organisms vary by a huge amount. The smallest genomes are those of viruses, though they are not usually regarded as living organisms. The table on the next page gives the genome size of one virus and four living organisms.

One of the four living organisms is a prokaryote. It has much the smallest genome. The genome size of eukaryotes depends on the size and number of chromosomes. It is correlated with the complexity of the organism, but is not directly proportional. There are several reasons for this. The proportion of the DNA that acts as functional genes is very variable and also the amount of gene duplication varies.

Activity

Microscope investigation of garlic chromosomes

1. Garlic has large chromosomes so is an ideal choice for looking at chromosomes. Cells in mitosis are needed. Garlic bulbs grow roots if they are kept for 3 or 4 days with their bases in water, at about 25°C. Root tips with cells in mitosis are yellow in colour, not white.

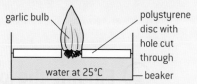

2. Root tips are put in a mixture of a stain that binds to the chromosomes and acid, which loosens the connections between the cell walls. A length of about 5 mm is suitable. Ten parts of aceto-orcein to one part of 1.0 mol dm^{-3} hydrochloric acid gives good results.

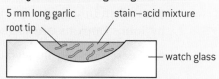

3. The roots are heated in the stain–acid mixture on a hot plate, to 80°C for 5 minutes. One of the root tips is put on a microscope slide, cut in half and the 2.5 mm length furthest from the end of the root is discarded.

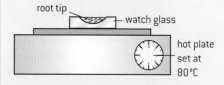

4. A drop of stain and a cover slip is added and the root tip is squashed to spread out the cells to form a layer one cell thick. The chromosomes can then be examined and counted and the various phases of mitosis should also be visible.

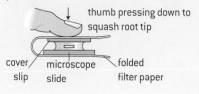

Organism	Genome size (million base pairs)	Description
T2 phage	0.18	Virus that attacks *Escherichia coli*
Escherichia coli	5	Gut bacterium
Drosophila melanogaster	140	Fruit fly
Homo sapiens	3,000	Humans
Paris japonica	150,000	Woodland plant

🧪 Finding the loci of human genes

Use of online databases to identify the locus of a human gene and its protein product.

The locus of a gene is its particular position on homologous chromosomes. Online databases can be used to find the locus of human genes. There is an example of such a database in the Online Mendelian Inheritance in Man website, maintained by Johns Hopkins University.

- Search for the abbreviation OMIM to open the home page.

- Choose Search Gene Map.

- Enter the name of a gene into the Search Gene Map box. This should bring up a table with information about the gene, including its locus, starting with the chromosome on which the gene is located. Suggestions of human genes are shown on the right.

- An alternative to entering the name of a gene is to select a chromosome from 1–22 or one of the sex chromosomes X or Y. A complete sequence of gene loci will be displayed,

together with the total number of gene loci on that chromosome.

Gene name	Description of gene
DRD4	A gene that codes for a dopamine receptor that is implicated in a variety of neurological and psychiatric conditions.
CFTR	A gene that codes for a chloride channel protein. An allele of this gene causes cystic fibrosis.
HBB	The gene that codes for the beta-globin subunit of hemoglobin. An allele of this gene causes sickle cell anemia.
F8	The gene that codes for Factor VIII, one of the proteins needed for the clotting of blood. The classic form of hemophilia is caused by an allele of this gene.
TDF	Testis determining factor – the gene that causes a fetus to develop as a male.

Haploid nuclei

Haploid nuclei have one chromosome of each pair.

A haploid nucleus has one chromosome of each type. It has one full set of the chromosomes that are found in its species. Haploid nuclei in humans contain 23 chromosomes for example.

Gametes are the sex cells that fuse together during sexual reproduction. Gametes have haploid nuclei, so in humans both egg and sperm cells contain 23 chromosomes.

Diploid nuclei

Diploid nuclei have pairs of homologous chromosomes.

A diploid nucleus has two chromosomes of each type. It has two full sets of the chromosomes that are found in its species. Diploid nuclei in humans contain 46 chromosomes for example.

When haploid gametes fuse together during sexual reproduction, a zygote with a diploid nucleus is produced. When this divides by mitosis, more cells with diploid nuclei are produced. Many animals and plants consist entirely of diploid cells, apart from the cells that they are using to produce gametes for sexual reproduction.

Diploid nuclei have two copies of every gene, apart from genes on the sex chromosomes. An advantage of this is that the effects of harmful recessive mutations can be avoided if a dominant allele is also present. Also, organisms are often more vigorous if they have two different alleles of genes instead of just one. This is known as hybrid vigour and is the reason for strong growth of F_1 hybrid crop plants.

▲ Figure 7 Mosses coat the trunks of the laurel trees in this forest in the Canary Islands. Mosses are unusual because their cells are haploid. In most eukaryotes the gametes are haploid but not the parent that produces them

Chromosome numbers

The number of chromosomes is a characteristic feature of members of a species.

One of the most fundamental characteristics of a species is the number of chromosomes. Organisms with a different number of chromosomes are unlikely to be able to interbreed so all the interbreeding members of a species need to have the same number of chromosomes.

The number of chromosomes can change during the evolution of a species. It can decrease if chromosomes become fused together or increase if splits occur. There are also mechanisms that can cause the chromosome number to double. However, these are rare events and chromosome numbers tend to remain unchanged over millions of years of evolution.

▲ Figure 8 *Trillium luteum* cell with a diploid number of 12 chromosomes. Two of each type of chromosome are present

🌐 Comparing chromosome numbers

Comparison of diploid chromosome numbers of Homo sapiens, Pan troglodytes, Canis familiaris, Oryza sativa, Parascaris equorum.

The Oxford English Dictionary consists of twenty large volumes, each containing a large amount of information about the origins and meanings of words. This information could have been published in a smaller number of larger volumes or in a larger number of smaller volumes. There is a parallel with the numbers and sizes of chromosomes in

eukaryotes. Some have a few large chromosomes and others have many small ones.

All eukaryotes have at least two different types of chromosome, so the diploid chromosome number is at least four. In some cases it is over a hundred. The table on the next page shows the diploid chromosome number of selected species.

155

Scientific name of species	English name	Diploid chromosome number
Parascaris equorum	horse threadworm	4
Oryza sativa	rice	24
Homo sapiens	humans	46
Pan troglodytes	chimpanzee	48
Canis familiaris	dog	78

▲ Figure 9 Who has more chromosomes – a dog or its owner?

Data-based questions: Differences in chromosome number

Plants	Chromosome number	Animals
Haplopappus gracilis	4	*Parascaris equorum* (horse threadworm)
Luzula purpurea (woodrush)	6	*Aedes aegypti* (yellow fever mosquito)
Crepis capillaris	8	*Drosophila melanogaster* (fruitfly)
Vicia faba (field bean)	12	*Musca domestica* (house fly)
Brassica oleracea (cabbage)	18	*Chorthippus parallelus* (grasshopper)
Citrullus vulgaris (water melon)	22	*Cricetulus griseus* (Chinese hamster)
Lilium regale (royal lily)	24	*Schistocerca gregaria* (desert locust)
Bromus texensis	28	*Desmodus rotundus* (vampire bat)
Camellia sinesis (Chinese tea)	30	*Mustela vison* (mink)
Magnolia virginiana (sweet bay)	38	*Felis catus* (domestic cat)
Arachis hypogaea (peanut)	40	*Mus musculus* (mouse)
Coffea arabica (coffee)	44	*Mesocricetus auratus* (golden hamster)
Stipa spartea (porcupine grass)	46	*Homo sapiens* (modern humans)
Chrysoplenum alternifolium (saxifrage)	48	*Pan troglodytes* (chimpanzee)
Aster laevis (Michaelmas daisy)	54	*Ovis aries* (domestic sheep)
Glyceria canadensis (manna grass)	60	*Capra hircus* (goat)
Carya tomentosa (hickory)	64	*Dasypus novemcinctus* (armadillo)
Magnolia cordata	76	*Ursus americanus* (American black bear)
Rhododendron keysii	78	*Canis familiaris* (dog)

▲ Table 1

1   There are many different chromosome numbers in the table, but some numbers are missing, for example, 5, 7, 11, 13. Explain why none of the species has 13 chromosomes.   [3]

2   Discuss, using the data in the table, the hypothesis that the more complex an organism is, the more chromosomes it has.   [4]

3   Explain why the size of the genome of a species cannot be deduced from the number of chromosomes.   [1]

4   Suggest, using the data in table 1, a change in chromosome structure that may have occurred during human evolution.   [2]

Sex determination

Sex is determined by sex chromosomes and autosomes are chromosomes that do not determine sex.

There are two chromosomes in humans that determine sex:

- the X chromosome is relatively large and has its centromere near the middle.

- the Y chromosome is much smaller and has its centromere near the end.

Because the X and Y chromosomes determine sex they are called the sex chromosomes. All the other chromosomes are autosomes and do not affect whether a fetus develops as a male or female.

The X chromosome has many genes that are essential in both males and females. All humans must therefore have at least one X chromosome. The Y chromosome only has a small number of genes. A small part of the Y chromosome has the same sequence of genes as a small part of the X chromosome, but the genes on the remainder of the Y chromosome are not found on the X chromosome and are not needed for female development.

One Y chromosome gene in particular causes a fetus to develop as a male. This is called either SRY or TDF. It initiates the development of male features, including testes and testosterone production. Because of this gene a fetus with one X and one Y chromosome develops as a male. A fetus that has two X chromosomes and no Y chromosome does not have the TDF gene so ovaries develop instead of testes and female sex hormones are produced, not testosterone.

Females have two X chromosomes. Females pass on one of their two X chromosomes in each egg cell, so all offspring inherit an X chromosome from their mother. The gender of a human is determined at the moment of fertilization by one chromosome carried in the sperm. This can either be an X or a Y chromosome. When sperm are formed, half contain the X chromosome and half the Y chromosome. Daughters inherit their father's X chromosome and sons inherit his Y chromosome.

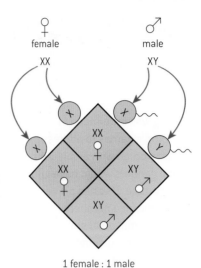

1 female : 1 male

▲ Figure 10 Determination of gender

Karyograms

A karyogram shows the chromosomes of an organism in homologous pairs of decreasing length.

The chromosomes of an organism are visible in cells that are in mitosis, with cells in metaphase giving the clearest view. Stains have to be used to make the chromosomes show up. Some stains give each chromosome type a distinctive banding pattern.

If dividing cells are stained and placed on a microscope slide and are then burst by pressing on the cover slip, the chromosomes become spread. Often they overlap each other, but with careful searching a cell can usually be found with no overlapping chromosomes. A micrograph can be taken of the stained chromosomes.

To what extent is determining gender for sporting competition a scientific question?

Gender testing was introduced at the 1968 Olympic games to address concerns that women with ambiguous physiological genders would have an unfair advantage. This has proven to be problematic for a number of reasons. The chromosomal standard is problematic as non-disjunction can lead to situations where an individual might technically be male, but might not define herself in that way. People with two X chromosomes can develop hormonally as a male and people with an X and a Y can develop hormonally as a female.

The practice of gender testing was discontinued in 1996 in part because of human rights issues including the right to self-expression and the right to identify one's own gender. Rather than being a scientific question, it is more fairly a social question.

Originally analysis involved cutting out all the chromosomes and arranging them manually but this process can now be done digitally. The chromosomes are arranged according to their size and structure. The position of the centromere and the pattern of banding allow chromosomes that are of a different type but similar size to be distinguished.

As most cells are diploid, the chromosomes are usually in homologous pairs. They are arranged by size, starting with the longest pair and ending with the smallest.

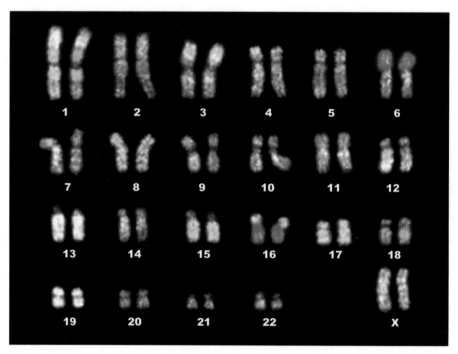

▲ Figure 11 Karyogram of a human female, with fluorescent staining

▲ Figure 12 Child with trisomy 21 or Down syndrome

Karyotypes and Down syndrome
Use of karyotypes to deduce sex and diagnose Down syndrome in humans.

A karyogram is an image of the chromosomes of an organism, arranged in homologous pairs of decreasing length. A karyotype is a property of an organism – it is the number and type of chromosomes that the organism has in its nuclei. Karyotypes are studied by looking at karyograms. They can be used in two ways:

1 To deduce whether an individual is male or female. If two XX chromosomes are present the individual is female whereas one X and one Y indicate a male.

2 To diagnose Down syndrome and other chromosome abnormalities. This is usually done using fetal cells taken from the uterus during pregnancy. If there are three copies of chromosome 21 in the karyotype instead of two, the child has Down syndrome. This is sometimes called trisomy 21. While individuals vary, some of the component features of the syndrome are hearing loss, heart and vision disorders. Mental and growth retardation are also common.

Data-based questions: A human karyotype

The karyogram shows the karyotype of a fetus.

1 State which chromosome type is

    a) longest

    b) shortest. [2]

2 Distinguish between the structure of

    a) human chromosome 2 and chromosome 12

    b) the human X and Y chromosome. [4]

3 Deduce with a reason the sex of the fetus. [2]

4 Explain whether the karyotype shows any abnormalities. [2]

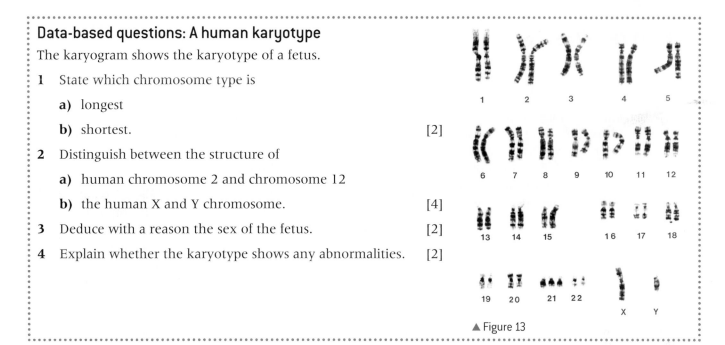

▲ Figure 13

3.3 Meiosis

Understanding

→ One diploid nucleus divides by meiosis to produce four haploid nuclei.

→ The halving of the chromosome number allows a sexual life cycle with fusion of gametes.

→ DNA is replicated before meiosis so that all chromosomes consist of two sister chromatids.

→ The early stages of meiosis involve pairing of homologous chromosomes and crossing over followed by condensation.

→ Orientation of pairs of homologous chromosomes prior to separation is random.

→ Separation of pairs of homologous chromosomes in the first division of meiosis halves the chromosome number.

→ Crossing over and random orientation promotes genetic variation.

→ Fusion of gametes from different parents promotes genetic variation.

Applications

→ Non-disjunction can cause Down syndrome and other chromosome abnormalities. Studies showing age of parents influences chances of non-disjunction.

→ Methods used to obtain cells for karyotype analysis e.g. chorionic villus sampling and amniocentesis and the associated risks.

Skills

→ Drawing diagrams to show the stages of meiosis resulting in the formation of four haploid cells.

Nature of science

→ Making careful observations: meiosis was discovered by microscope examination of dividing germ-line cells.

🧬 The discovery of meiosis

Making careful observations: meiosis was discovered by microscope examination of dividing germ-line cells.

When improved microscopes had been developed in the 19th century that gave detailed images of cell structures, it was discovered that some dyes specifically stained the nucleus of the cell. These dyes revealed thread-like structures in dividing nuclei that were named chromosomes. From the 1880s onwards a group of German biologists carried out careful and detailed observations of dividing nuclei that gradually revealed how mitosis and meiosis occur.

We can appreciate the considerable achievements of these biologists if we try to repeat the observations that they made. The preparation of microscope slides showing meiosis is challenging. Suitable tissue can be obtained from the developing anthers inside a lily bud or from the testis of a dissected locust. The tissue must be fixed, stained and then squashed on a microscope slide. Often no cells in meiosis are visible or the images are not clear enough to show details of the process. Even with prepared slides made by experts it is difficult to understand the images as chromosomes form a variety of bizarre shapes during the stages of meiosis.

A key observation was that in the horse threadworm (*Parascaris equorum*) there are two chromosomes in the nuclei of egg and sperm cells, whereas the fertilized egg contains four. This indicated that the chromosome number is doubled by fertilization. The observation led to the hypothesis that there must be a special nuclear division in every generation that halves the chromosome number.

Nuclear divisions unlike mitosis had already been observed during gamete development in both animals and plants. These divisions were identified as the method used to halve the chromosome number and they were named meiosis. The sequence of events in meiosis was eventually worked out by careful observation of cells taken from the ovaries of rabbits (*Oryctolagus cuniculus*) between 0 and 28 days old. The advantage of this species is that in females meiosis begins at birth and occurs slowly over many days.

▲ Figure 1

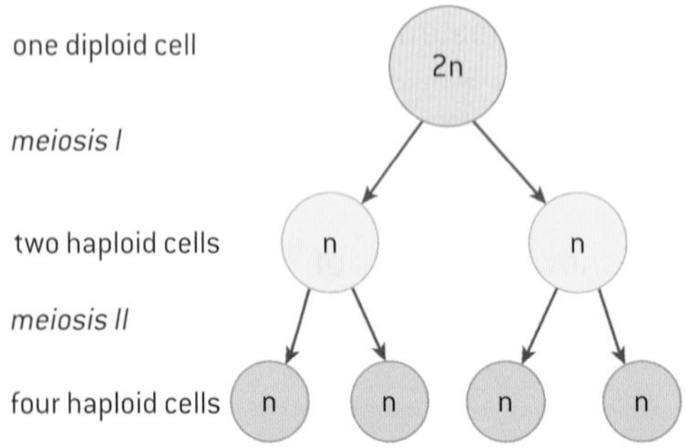

Meiosis in outline

One diploid nucleus divides by meiosis to produce four haploid nuclei.

one diploid cell

meiosis I

two haploid cells

meiosis II

four haploid cells

▲ Figure 2 Overview of meiosis

Meiosis is one of the two ways in which the nucleus of a eukaryotic cell can divide. The other method is mitosis, which was described in sub-topic 1.6. In meiosis the nucleus divides twice. The first division produces two nuclei, each of which divides again to give a total of four nuclei. The two divisions are known as meiosis I and meiosis II.

The nucleus that undergoes the first division of meiosis is diploid – it has two chromosomes of each type. Chromosomes of the same type are known as homologous chromosomes. Each of the four nuclei produced by meiosis has just one chromosome of each type – they are haploid. Meiosis involves a halving of the chromosome number. It is therefore known as a reduction division.

The cells produced by meiosis I have one chromosome of each type, so the halving of the chromosome number happens in the first division,

not the second division. The two nuclei produced by meiosis I have the haploid number of chromosomes, but each chromosome still consists of two chromatids. These chromatids separate during meiosis II, producing four nuclei that have the haploid number of chromosomes, with each chromosome consisting of a single chromatid.

Meiosis and sexual life cycles

The halving of the chromosome number allows a sexual life cycle with fusion of gametes.

The life cycles of living organisms can be sexual or asexual. In an asexual life cycle the offspring have the same chromosomes as the parent so are genetically identical. In a sexual life cycle there are differences between the chromosomes of the offspring and the parents, so there is genetic diversity.

In eukaryotic organisms, sexual reproduction involves the process of fertilization. Fertilization is the union of sex cells, or gametes, usually from two different parents. Fertilization doubles the number of chromosomes each time it occurs. It would therefore cause a doubling of chromosome number every generation, if the number was not also halved at some stage in the life cycle. This halving of chromosome number happens during meiosis.

Meiosis can happen at any stage during a sexual life cycle, but in animals it happens during the process of creating the gametes. Body cells are therefore diploid and have two copies of most genes.

Meiosis is a complex process and it is not at the moment clear how it developed. What is clear is that its evolution was a critical step in the origin of eukaryotes. Without meiosis there cannot be fusion of gametes and the sexual life cycle of eukaryotes could not occur.

▲ Figure 4 Fledgling owls (bottom) produced by a sexual life cycle have diploid body cells but mosses (top) have haploid cells

---

Data-based questions: Life cycles

Figure 3 shows the life cycle of humans and mosses, with n being used to represent the haploid number of chromosomes and 2n to represent the diploid number. Sporophytes of mosses grow on the main moss plant and consist of a stalk and a capsule in which spores are produced.

1   Outline five similarities between the life cycle of a moss and of a human.   [5]

2   Distinguish between the life cycles of a moss and a human by giving five differences.   [5]

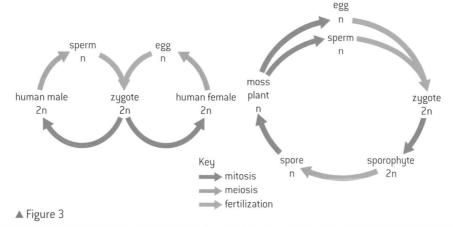

Key
→ mitosis
→ meiosis
→ fertilization

▲ Figure 3

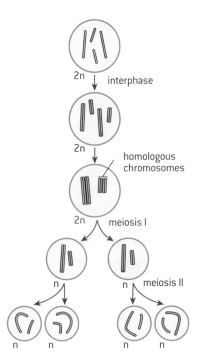

▲ Figure 5 Outline of meiosis

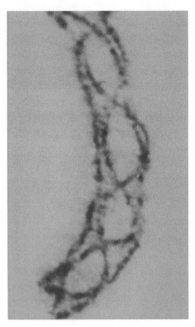

▲ Figure 6 A pair of homologous chromosomes contains four chromatids and is sometimes called a tetrad. Five chiasmata are visible in this tetrad, showing that crossing over can occur more than once

Replication of DNA before meiosis

DNA is replicated before meiosis so that all chromosomes consist of two sister chromatids.

During the early stages of meiosis the chromosomes gradually shorten by supercoiling. As soon as they become visible it is clear that each chromosome consists of two chromatids. This is because all DNA in the nucleus is replicated during the interphase before meiosis, so each chromosome consists of two sister chromatids.

Initially the two chromatids that make up each chromosome are genetically identical. This is because DNA replication is very accurate and the number of mistakes in the copying of the DNA is extremely small.

We might expect the DNA to be replicated again between the first and the second division of meiosis, but it does not happen. This explains how the chromosome number is halved during meiosis. One diploid nucleus, in which each chromosome consists of two chromatids, divides twice to produce four haploid nuclei in which each chromosome consists of one chromatid.

Bivalents formation and crossing over

The early stages of meiosis involve pairing of homologous chromosomes and crossing over followed by condensation.

Some of the most important events of meiosis happen at the start of meiosis I while the chromosomes are still very elongated and cannot be seen with a microscope. Firstly homologous chromosomes pair up with each other. Because DNA replication has already occurred, each chromosome consists of two chromatids and so there are four DNA molecules associated in each pair of homologous chromosomes. A pair of homologous chromosomes is bivalent and the pairing process is sometimes called synapsis.

Soon after synapsis, a process called crossing over takes place. The molecular details of this need not concern us here, but the outcome is very important. A junction is created where one chromatid in each of the homologous chromosomes breaks and rejoins with the other chromatid. Crossing over occurs at random positions anywhere along the chromosomes. At least one crossover occurs in each bivalent and there can be several.

Because a crossover occurs at precisely the same position on the two chromatids involved, there is a mutual exchange of genes between the chromatids. As the chromatids are homologous but not identical, some alleles of the exchanged genes are likely to be different. Chromatids with new combinations of alleles are therefore produced.

Random orientation of bivalents

Orientation of pairs of homologous chromosomes prior to separation is random.

While pairs of homologous chromosomes are condensing inside the nucleus of a cell in the early stages of meiosis, spindle microtubules are growing from the poles of the cell. After the nuclear membrane has

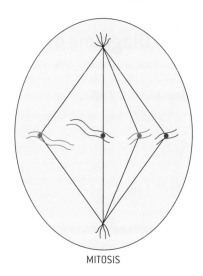

broken down, these spindle microtubules attach to the centromeres of the chromosomes.

The attachment of the spindle microtubules is not the same as in mitosis. The principles are these:

- Each chromosome is attached to one pole only, not to both.

- The two homologous chromosomes in a bivalent are attached to different poles.

- The pole to which each chromosome is attached depends on which way the pair of chromosomes is facing. This is called the orientation.

- The orientation of bivalents is random, so each chromosome has an equal chance of attaching to each pole, and eventually of being pulled to it.

- The orientation of one bivalent does not affect other bivalents. The consequences of the random orientation of bivalents are discussed in the section on genetic diversity later in this topic.

Halving the chromosome number

Separation of pairs of homologous chromosomes in the first division of meiosis halves the chromosome number.

The movement of chromosomes is not the same in the first division of meiosis as in mitosis. Whereas in mitosis the centromere divides and the two chromatids that make up a chromosome move to opposite poles, in meiosis the centromere does not divide and whole chromosomes move to the poles.

Initially the two chromosomes in each bivalent are held together by chiasmata, but these slide to the end of the chromosomes and then the chromosomes can separate. This separation of homologous chromosomes is called disjunction. One chromosome from each bivalent moves to one of the poles and the other chromosome to the other pole.

The separation of pairs of homologous chromosomes to opposite poles of the cell halves the chromosome number of the cell. It is therefore the first division of meiosis that is the reduction division. Because one chromosome of each type moves to each pole, both of the two nuclei formed in the first division of meiosis contain one of each type of chromosome, so they are both haploid.

MITOSIS

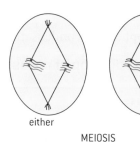

either   or

MEIOSIS

▲ Figure 7 Comparison of attachment of chromosomes to spindle microtubules in mitosis and meiosis

 Obtaining cells from a fetus

Methods used to obtain cells for karyotype analysis e.g. chorionic villus sampling and amniocentesis and the associated risks.

Two procedures are used for obtaining cells containing the fetal chromosomes needed for producing a karyotype. Amniocentesis involves passing a needle through the mother's abdomen wall, using ultrasound to guide the needle. The needle is used to withdraw a sample of amniotic fluid containing fetal cells from the amniotic sac.

The second procedure is chorionic villus sampling. A sampling tool that enters through the vagina is used to obtain cells from the chorion, one of the membranes from which the placenta develops. This can be done earlier in the pregnancy than amniocentesis, but whereas the risk of miscarriage with amniocentesis is 1%, with chorionic villus sampling it is 2%.

Diagrams of the stages of meiosis

Drawing diagrams to show the stages of meiosis resulting in the formation of four haploid cells.

In mitosis four stages are usually recognized: prophase, metaphase, anaphase and telophase. Meiosis can also be divided into these stages, but each stage happens twice: in meiosis I and then a second time in meiosis II. The main events of each stage in mitosis also happen in meiosis:

- prophase: condensation of chromosomes;

- metaphase: attachment of spindle microtubules;

- anaphase: movement of chromosomes to the poles;

- telophase: decondensation of chromosomes.

Usually we draw biological structures from actual specimens, often looking at them down a microscope. Preparation of microscope slides showing meiosis is worth attempting but it is challenging. Permanent slides usually have more cells visible in meiosis than temporary mounts, but even then it is difficult to interpret the structure of bivalents from their appearance. This is why we usually construct diagrams of meiosis rather than draw stages from specimens on microscope slides!

The first division of meiosis

Prophase I - Cell has 2n chromosomes (double chromatid): n is haploid number of chromosomes. - Homologous chromosomes pair (synapsis). - Crossing over occurs.	nuclear membrane spindle microtubules and centriole **Prophase I**	
Metaphase I - Spindle microtubules move homologous pairs to equator of cell. - Orientation of paternal and maternal chromosomes on either side of equator is random and independent of other homologous pairs.	bivalents aligned on the equator **Metaphase I**	
Anaphase I - Homologous pairs are separated. One chromosome of each pair moves to each pole.	homologous chromosomes being pulled to opposite poles **Anaphase I**	
Telophase I - Chromosomes uncoil. During interphase that follows, no replication occurs. - Reduction of chromosome number from diploid to haploid completed. - Cytokinesis occurs.	cell has divided across the equator **Telophase I**	

The second division of meiosis

Prophase II • Chromosomes, which still consist of two chromatids, condense and become visible.	 Prophase II	
Metaphase II	 Metaphase II	
Anaphase II • Centromeres separate and chromatids are moved to opposite poles.	 Anaphase II	
Telophase II • Chromatids reach opposite poles. • Nuclear envelope forms. • Cytokinesis occurs.	 Telophase II	

Meiosis and genetic variation

Crossing over and random orientation promotes genetic variation.

When two parents have a child, they know that it will inherit an unpredictable mixture of characteristics from each of them. Much of the unpredictability is due to meiosis. Every gamete produced by a parent has a new combination of alleles – meiosis is a source of endless genetic variation.

Apart from the genes on the X and Y chromosomes, humans have two copies of each gene. In some cases the two copies are the same allele and there will be one copy of that allele in every gamete produced by the parent. There are likely to be thousands of genes in the parent's genome

Activity

If g is the number of genes in a genome with different alleles, 2^g is the number of combinations of these alleles that can be generated by meiosis. If there were just 69 genes with different alleles (3 in each of the 23 chromosome types in humans) there would be 590,295,810,358,705, 700,000 combinations. Assuming that all humans are genetically different, and that there are 7,000,000 humans, calculate the percentage of all possible genomes that currently exist.

where the two alleles are different. Each of the two alleles has an equal chance of being passed on in a gamete. Let us suppose that there is a gene with the alleles A and a. Half of the gametes produced by the parent will contain A and half will contain a.

Let us now suppose that there is another gene with the alleles B and b. Again half of the gametes will contain B and half b. However, meiosis can result in gametes with different combinations of these genes: AB, Ab, aB and ab. There are two processes in meiosis that generate this diversity.

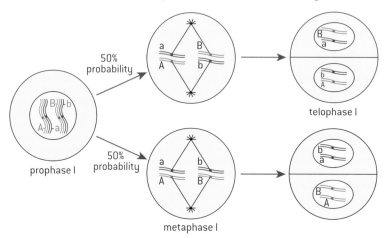

▲ Figure 8 Random orientation in metaphase I

1. Random orientation of bivalents

In metaphase I the orientation of bivalents is random and the orientation of one bivalent does not influence the orientation of any of the others. Random orientation of bivalents is the process that generates genetic variation among genes that are on different chromosome types.

For every additional bivalent, the number of possible chromosome combinations in a cell produced by meiosis doubles. For a haploid number of n, the number of possible combinations is 2^n. For humans with a haploid number of 23 this amounts to 2^{23} or over 8 million combinations.

2. Crossing over

Without crossing over in prophase I, combinations of alleles on chromosomes would be forever linked together. For example, if one chromosome carried the combination CD and another carried cd, only these combinations could occur in gametes. Crossing over allows linked genes to be reshuffled, to produce new combinations such as Cd and cD. It increases the number of allele combinations that can be generated by meiosis so much that it is effectively infinite.

Fertilization and genetic variation

Fusion of gametes from different parents promotes genetic variation.

The fusion of gametes to produce a zygote is a highly significant event both for individuals and for species.

- It is the start of the life of a new individual.

- It allows alleles from two different individuals to be combined in one new individual.

▲ Figure 9

- The combination of alleles is unlikely ever to have existed before.

- Fusion of gametes therefore promotes genetic variation in a species.

- Genetic variation is essential for evolution.

Non-disjunction and Down syndrome

Non-disjunction can cause Down syndrome and other chromosome abnormalities.

Meiosis is sometimes subject to errors. One example of this is when homologous chromosomes fail to separate at anaphase. This is termed non-disjunction. This can happen with any of the pairs of homologous chromosomes. Both of the chromosomes move to one pole and neither to the other pole. The result will be a gamete that either has an extra chromosome or is deficient in a chromosome. If the gamete is involved in human fertilization, the result will be an individual with either 45 or 47 chromosomes.

An abnormal number of chromosomes will often lead to a person possessing a syndrome, i.e. a collection of physical signs or symptoms. For example trisomy 21, also known as Down syndrome, is due to a non-disjunction event that leaves the individual with three of chromosome number 21 instead of two. While individuals vary, some of the component features of the syndrome include hearing loss, heart and vision disorders. Mental and growth retardation are also common.

Most other trisomies in humans are so serious that the offspring do not survive. Babies are sometimes born with trisomy 18 and trisomy 13. Non-disjunction can also result in the birth of babies with abnormal numbers of sex chromosomes. Klinefelter's syndrome is caused by having the sex chromosomes XXY. Turner's syndrome is caused by having only one sex chromosome, an X.

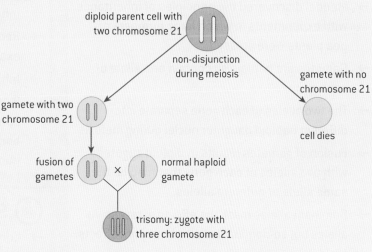

▲ Figure 10 How non-disjunction can give rise to Down syndrome

---

Data-based questions: Parental age and non-disjunction: Studies showing age of parents influences chances of non-disjunction

The data presented in figure 11 shows the relationship between maternal age and the incidence of trisomy 21 and of other chromosomal abnormalities.

1   Outline the relationship between maternal age and the incidence of chromosomal abnormalities in live births.   [2]

2   a)   For mothers 40 years of age, determine the probability that they will give birth to a child with trisomy 21.   [1]

   b)   Using the data in figure 11, calculate the probability that a mother of 40 years of age will give birth to a child with a chromosomal abnormality other than trisomy 21.   [2]

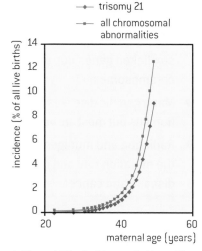

▲ Figure 11 The incidence of trisomy 21 and other chromosomal abnormalities as a function of maternal age

3   Only a small number of possible chromosomal abnormalities are ever found among live births, and trisomy 21 is much the commonest. Suggest reasons for these trends.   [3]

4   Discuss the risks parents face when choosing to postpone having children.   [2]

3.4 Inheritance

Understanding

→ Mendel discovered the principles of inheritance with experiments in which large numbers of pea plants were crossed.

→ Gametes are haploid so contain one allele of each gene.

→ The two alleles of each gene separate into different haploid daughter nuclei during meiosis.

→ Fusion of gametes results in diploid zygotes with two alleles of each gene that may be the same allele or different alleles.

→ Dominant alleles mask the effects of recessive alleles but co-dominant alleles have joint effects.

→ Many genetic diseases in humans are due to recessive alleles of autosomal genes.

→ Some genetic diseases are sex-linked and some are due to dominant or co-dominant alleles.

→ The pattern of inheritance is different with sex-linked genes due to their location on sex chromosomes.

→ Many genetic diseases have been identified in humans but most are very rare.

→ Radiation and mutagenic chemicals increase the mutation rate and can cause genetic disease and cancer.

 Applications

→ Inheritance of ABO blood groups.

→ Red-green colour-blindness and hemophilia as examples of sex-linked inheritance.

→ Inheritance of cystic fibrosis and Huntington's disease.

→ Consequences of radiation after nuclear bombing of Hiroshima and Nagasaki and the nuclear accidents at Chernobyl.

 Skills

→ Construction of Punnett grids for predicting the outcomes of monohybrid genetic crosses.

→ Comparison of predicted and actual outcomes of genetic crosses using real data.

→ Analysis of pedigree charts to deduce the pattern of inheritance of genetic diseases.

 Nature of science

→ Making quantitative measurements with replicates to ensure reliability: Mendel's genetic crosses with pea plants generated numerical data.

Mendel and the principles of inheritance

Mendel discovered the principles of inheritance with experiments in which large numbers of pea plants were crossed.

When living organisms reproduce, they pass on characteristics to their offspring. For example, when blue whales reproduce, the young are also blue whales – they are members of the same species. More than this, variations, such as the markings on the skin of a blue whale, can be passed on. We say that the offspring inherit the parents' characteristics. However, some characteristics cannot be inherited. Scars seen on the tails of some blue whales caused by killer whale attacks and cosmetic surgery in humans are examples of this. According to current theories, acquired characteristics such as these cannot be inherited.

Inheritance has been discussed since the time of Hippocrates and earlier. For example, Aristotle observed that children sometimes resemble their grandparents more than their parents. Many of the early theories involved blending inheritance, in which offspring inherit characters from both parents and so have characters intermediate between those of their parents. Some of the observations that biologists made in the first half of the 19th century could not be explained by blending inheritance, but it was not until Mendel published his paper "Experiments in Plant Hybridization" that an alternative theory was available.

Mendel's experiments were done using varieties of pea plant, each of which reliably had the same characters when grown on its own. Mendel carefully crossed varieties of pea together by transferring the male pollen from one variety to the female parts in flowers of another variety. He collected the pea seeds that were formed as a result and grew them to find out what their characters were. Mendel repeated each cross with many pea plants. He also did this experiment with seven different pairs of characters and so his results reliably demonstrated the principles of inheritance in peas, not just an isolated effect.

In 1866 Mendel published his research. For over thirty years his findings were largely ignored. Various reasons have been suggested for this. One factor was that his experiments used pea plants and there was not great interest in the pattern of inheritance in that species. In 1900 several biologists rediscovered Mendel's work. They quickly did cross-breeding experiments with other plants and with animals. These confirmed that Mendel's theory explained the basis of inheritance in all plants and animals.

▲ Figure 1 Hair styles are acquired characteristics and are fortunately not inherited by offspring

 Replicates and reliability in Mendel's experiments

Making quantitative measurements with replicates to ensure reliability: Mendel's genetic crosses with pea plants generated numerical data.

Gregor Mendel is regarded by most biologists as the father of genetics. His success is sometimes attributed to being the first to use pea plants for research into inheritance. Peas have clear characteristics such as red or white flower colour that can easily be followed from one generation to the next. They can also be crossed to produce hybrids or they can be allowed to self-pollinate.

In fact Mendel was not the first to use pea plants. Thomas Andrew Knight, an English horticulturalist, had conducted research at Downton Castle in Herefordshire in the late 18th century and published his results in the Philosophical Transactions of the Royal Society. Knight made some important discoveries:

- male and female parents contribute equally to the offspring;

- characters such as white flower colour that apparently disappear in offspring can reappear in the next generation, showing that inheritance is discrete rather than blending;

- one character such as red flower colour can show "a stronger tendency" than the alternative character.

Although Mendel was not as pioneering in his experiments as sometimes thought, he deserves credit for another aspect of his research. Mendel was a pioneer in obtaining quantitative results and in having large numbers of replicates. He also did seven different cross experiments, not just one. Table 1 shows the results of his monohybrid crosses.

It is now standard practice in science to include repeats in experiments to demonstrate the reliability of results. Repeats can be compared to see how close they are. Anomalous results can be identified and excluded from analysis. Statistical tests can be done to assess the significance of differences between treatments. It is also standard practice to repeat whole experiments, using a different organism or different treatments, to test a hypothesis in different ways. Mendel should therefore be regarded as one of the fathers of genetics, but even more we should think of him as a pioneer of research methods in biology.

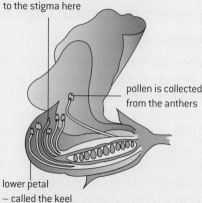

cross pollinating peas:
pollen from another plant is dusted on to the stigma here

pollen is collected from the anthers

lower petal — called the keel

self pollinating peas:
— if the flower is left untouched, the anthers inside the keel pollinate the stigma

▲ Figure 2 Cross and self pollination

(a) Prediction based on blending inheritance

tall plants × dwarf plants

pea plants with an intermediate height

(b) Actual results

tall plants × dwarf plants

pea plants as tall as the tall parent

▲ Figure 3 Example of a monohybrid cross experiment. All the hybrid plants produced by crossing two varieties together had the same character as one of the parents and the character of the other parent was not seen. This is a clear falsification of the theory of blending inheritance

Parental plants	Hybrid plants	Offspring from self-pollinating the hybrids	Ratio
Tall stem × dwarf stem	All tall	787 tall : 277 dwarf	2.84 : 1
Round seed × wrinkled seed	All round	5474 round : 1850 wrinkled	2.96 : 1
Yellow cotyledons × green cotyledons	All yellow	6022 yellow : 2001 green	3.01 : 1
Purple flowers × white flowers	All purple	705 purple : 224 white	3.15 : 1
Full pods × constricted pods	All full	882 full : 299 constricted	2.95 : 1
Green unripe pods × yellow unripe pods	All green	428 green : 152 yellow	2.82 : 1
Flowers along stem × flowers at stem tip	All along stem	651 along stem : 207 at tip	3.14 : 1

▲ Table 1

Gametes

Gametes are haploid so contain one allele of each gene.

Gametes are cells that fuse together to produce the single cell that is the start of a new life. They are sometimes called sex cells, and the single cell produced when male and female gametes fuse is a zygote. Male and female gametes are different in size and motility. The male gamete is generally smaller than the female one. It is usually able to move whereas the female gamete moves less or not at all. In humans, for example, the sperm has a much smaller volume than the egg cell and uses its tail to swim to the egg.

Parents pass genes on to their offspring in gametes. Gametes contain one chromosome of each type so are haploid. The nucleus of a gamete therefore only has one allele of each gene. This is true of both male and female gametes, so male and female parents make an equal genetic contribution to their offspring, despite being very different in overall size.

▲ Figure 4 Pollen on the anthers of a flower contains the male gamete of the plant. The male gametes contain one allele of each of the plant's genes

Zygotes

Fusion of gametes results in diploid zygotes with two alleles of each gene that may be the same allele or different alleles.

When male and female gametes fuse, their nuclei join together, doubling the chromosome number. The nucleus of the zygote contains two chromosomes of each type so is diploid. It contains also two alleles of each gene.

If there were two alleles of a gene, A and a, the zygote could contain two copies of either allele or one of each. The three possible combinations are AA, Aa and aa.

Some genes have more than two alleles. For example, the gene for ABO blood groups in humans has three alleles: I^A, I^B and i. This gives six possible combinations of alleles:

- three with two of the same allele, $I^A I^A$, $I^B I^B$ and ii
- three with two different alleles, $I^A I^B$, $I^A i$ and $I^B i$.

Segregation of alleles

The two alleles of each gene separate into different haploid daughter nuclei during meiosis.

During meiosis a diploid nucleus divides twice to produce four haploid nuclei. The diploid nucleus contains two copies of each gene, but the haploid nuclei contain only one.

- If two copies of one allele of a gene were present, each of the haploid nuclei will receive one copy of this allele. For example, if the two alleles were PP, every gamete will receive one copy of P.

- If two different alleles were present, each haploid nucleus will receive either one of the alleles or the other allele, not both. For example, if the two alleles were Pp, 50% of the haploid nuclei would receive P and 50% would receive p.

▲ Figure 5 Most crop plants are pure-bred strains with two of the same allele of each gene

The separation of alleles into different nuclei is called segregation. It breaks up existing combinations of alleles in a parent and allows new combinations to form in the offspring.

Dominant, recessive and co-dominant alleles

Dominant alleles mask the effects of recessive alleles but co-dominant alleles have joint effects.

In each of Mendel's seven crosses between different varieties of pea plant, all of the offspring showed the character of one of the parents, not the other. For example, in a cross between a tall pea plant and a dwarf pea plant, all the offspring were tall. The difference in height between the parents is due to one gene with two alleles:

- the tall parents have two copies of an allele that makes them tall, TT
- the dwarf parents have two copies of an allele that makes them dwarf, tt
- they each pass on one allele to the offspring, which therefore has one of each allele, Tt
- when the two alleles are combined in one individual, it is the allele for tallness that determines the height because the allele for tallness is dominant
- the other allele, that does not have an effect if the dominant allele is present, is recessive.

In each of Mendel's crosses one of the alleles was dominant and the other was recessive. However, some genes have pairs of alleles where both have an effect when they are present together. They are called co-dominant alleles. A well-known example is the flower colour of *Mirabilis jalapa*. If a red-flowered plant is crossed with a white-flowered plant, the offspring have pink flowers.

- there is an allele for red flowers, C^R
- there is an allele for white flowers, C^W
- these alleles are co-dominant so $C^R C^W$ gives pink flowers.

The usual reason for dominance of one allele is that this allele codes for a protein that is active and carries out a function, whereas the recessive allele codes for a non-functional protein.

▲ Figure 6 There are co-dominant alleles of the gene for coat colour in Icelandic horses.

🧪 Punnett grids

Construction of Punnett grids for predicting the outcomes of monohybrid genetic crosses.

Monohybrid crosses only involve one character, for example the height of a pea plant, so they involve only one gene. Most crosses start with two pure-breeding parents. This means that the parents have two of the same allele, not two different alleles. Each parent therefore produces just one type of gamete, containing one copy of the allele. Their offspring are also identical, although they have two different alleles. The offspring obtained by crossing the parents are called F_1 hybrids or the F_1 generation.

The F_1 hybrids have two different alleles of the gene, so they can each produce two types of gamete. If two F_1 hybrids are crossed together, or if an F_1 plant is allowed to self-pollinate, there are four possible outcomes. This can be shown using a 2×2 table, called a Punnett grid after the geneticist who first used this type of table. The offspring of a cross between two F_1 plants are called the F_2 generation.

To make a Punnett grid as clear as possible the gametes should be labeled and both the alleles and the character of the four possible outcomes should be shown on the grid. It is also useful to give an overall ratio below the Punnett grid.

Figure 7 shows Mendel's cross between tall and dwarf plants. It explains the F_2 ratio of three tall to one dwarf plant.

Figure 8 shows the results of a cross between red and white flowered plants of *Mirabilis jalapa*. It explains the F_2 ratio of one red to two pink to one white flowered plant.

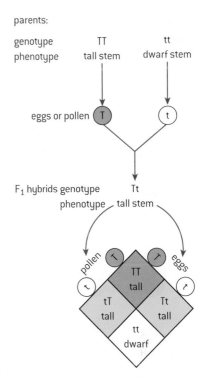

▲ Figure 7 Explanation of Mendel's 3:1 ratio

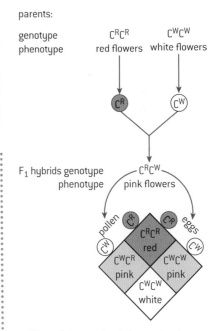

▲ Figure 8 A cross involving co-dominance

Data-based questions: Coat colour in the house mouse

In the early years of the 20th century, many crossing experiments were done in a similar way to those of Mendel. The French geneticist Lucien Cuénot used the house mouse, *Mus musculus*, to see whether the principles that Mendel had discovered also operated in animals. He crossed normal grey-coloured mice with albino mice. The hybrid mice that were produced were all grey. These grey hybrids were crossed together and produced 198 grey and 72 albino offspring.

1   Calculate the ratio between grey and albino offspring, showing your working.   [2]

2   Deduce the colour of coat that is due to a recessive allele, with two reasons for your answer.   [3]

3   Choose suitable symbols for the alleles for grey and albino coat and list the possible combinations of alleles of mice using your symbols, together with the coat colours associated with each combination of alleles.   [3]

typica   *annulata*

▲ Figure 9

▲ Figure 10 F₁ hybrid offspring

▲ Figure 11 F₂ offspring

Activity

ABO blood groups

It is possible for two parents to have an equal chance of having a child with blood group A, B, AB or O. What would be the genotypes of the parents?

4   Using a Punnett grid, explain how the observed ratio of grey and albino mice was produced. [5]

5   The albino mice had red eyes in addition to white coats. Suggest how one gene can determine whether the mice had grey fur and black eyes or white fur and red eyes. [2]

Data-based questions: The two-spot ladybird

Adalia bipunctata is a species of ladybird. In North America ladybirds are called ladybugs. The commonest form of this species is known as *typica*. There is a rarer form called *annulata*. Both forms are shown in figure 9.

1   Compare the *typica* and *annulata* forms of *Adalia bipunctata*. [2]

2   The differences between the two forms are due to a single gene. If male and female *typica* are mated together, all the offspring are *typica*. Similarly, the offspring produced when *annulata* forms are mated are all *annulata*. Explain the conclusions that can be drawn. [2]

3   When *typica* is mated with *annulata*, the F₁ hybrid offspring are not identical to either parent. Examples of these F₁ hybrid offspring are shown in figure 10. Distinguish between the F₁ hybrid offspring and the *typica* and *annulata* parents. [3]

4   If F₁ hybrid offspring are mated with each other, the offspring include both *typica* and *annulata* forms, and also offspring with the same wing case markings as the F₁ hybrid offspring.

    a)   Use a genetic diagram to explain this pattern of inheritance. [6]

    b)   Predict the expected ratio of phenotypes. [2]

⊕ ABO blood groups

Inheritance of ABO blood groups.

The ABO blood group system in humans is an example of co-dominance. It is of great medical importance: before blood is transfused, it is vital to find out the blood group of a patient and ensure that it is matched. Unless this is done, there may be complications due to coagulation of red blood cells. One gene determines the ABO blood group of a person. The genotype I^AI^A gives blood group A and the genotype I^BI^B gives group B. Neither I^A nor I^B is dominant over the other allele and a person with the genotype I^AI^B has a different blood group, called AB. There is a third allele of the ABO blood group gene, usually called *i*. A person with the genotype *ii* is in blood group O. The genotypes I^Ai and I^Bi give blood groups A and B respectively, showing that *i* is

recessive to both I^A and I^B. The reasons for two alleles being co-dominant and the other allele being recessive are as follows:

- All of the three alleles cause the production of a glycoprotein in the membrane of red blood cells.

- I^A alters the glycoprotein by addition of acetyl-galactosamine. This altered glycoprotein is absent from people who do not have the allele I^A so if exposed to it they make anti-A antibodies.

- I^B alters the glycoprotein by addition of galactose. This altered glycoprotein is not present in people who do not have the allele I^B so if exposed to it they make anti-B antibodies.

- The genotype $I^A I^B$ causes the glycoprotein to be altered by addition of acetyl-galactosamine and galactose. As a consequence neither anti-A nor anti-B antibodies are produced. This genotype therefore gives a different phenotype to $I^A I^A$ and $I^B I^B$ so the alleles I^A and I^B are co-dominant.

- The allele i is recessive because it does not cause the basic glycoprotein to be modified, but if either I^A or I^B are also present they do cause modification, so $I^A i$ gives the same phenotype as $I^A I^A$ and $I^B i$ gives the same as $I^B I^B$.

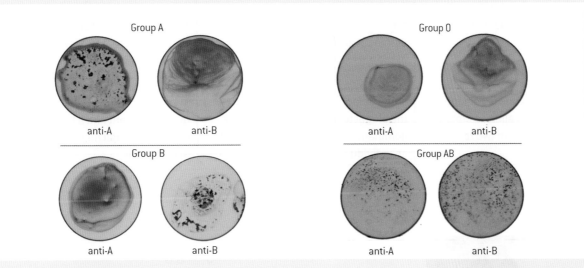

▲ Figure 12 Blood group can easily be determined using test cards

🧪 Testing predictions in cross-breeding experiments

Comparison of predicted and actual outcomes of genetic crosses using real data.

It is in the nature of science to try to find general principles that explain natural phenomena and not just to describe individual examples of a phenomenon. Mendel discovered principles of inheritance that have great predictive power. We can still use them to predict the outcomes of genetic crosses. Table 2 lists possible predictions in monohybrid crosses.

The actual outcomes of genetic crosses do not usually correspond exactly with the predicted outcomes. This is because there is an element of chance involved in the inheritance of genes. The tossing of a coin is a simple analogy. We expect the coin to land 50% of times with each of its two faces uppermost, but if we toss it 1,000 times we do not expect it to land precisely 500 times with one face showing and 500 times with the other face showing.

An important skill in biology is deciding whether the results of an experiment are close enough to the predictions for us to accept that they fit, or whether the differences are too great and either the results or the predictions must be false. An obvious trend is that the greater the difference between observed and expected results, the less likely that the difference is due to chance and the more likely that the predictions do not fit the results.

To assess objectively whether results fit predictions, statistical tests are used. For genetic crosses the chi-squared test can be used. This test is described later in the book in sub-topic 4.1.

Cross	Predicted outcome	Example
Pure-breeding parents one with dominant alleles and one with recessive alleles are crossed.	All of the offspring will have the same character as the parent with dominant alleles.	All offspring of a cross between pure-breeding tall and dwarf pea plants will be tall.
Pure-breeding parents that have different co-dominant alleles are crossed.	All of the offspring will have the same character and the character will be different from either parent.	All offspring of a cross between red and white flowered *Mirabilis jalapa* plants will have pink flowers.
Two parents each with one dominant and one recessive allele are crossed.	Three times as many offspring have the character of the parent with dominant alleles as have the character of the parent with the recessive alleles.	3:1 ratio of tall to dwarf pea plants from a cross between two parents that each have one allele for tall height and one allele for dwarf height.
A parent with one dominant and one recessive allele is crossed with a parent with two recessive alleles.	Equal proportions of offspring with the character of an individual with a dominant allele and the character of an individual with recessive alleles.	1:1 ratio from a cross between a dwarf pea plant and a tall plant with one allele for tall height and one for dwarf height .

▲ Table 2

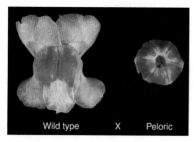

Wild type   X   Peloric

▲ Figure 13 *Antirrhinum* flowers – (a) wild type, (b) peloric

Data-based questions: Analysing genetic crosses

1 Charles Darwin crossed pure breeding wild-type *Antirrhinum majus* plants, which have bilaterally symmetric flowers, with pure breeding plants with peloric flowers that are radially symmetric. All the F_1 offspring produced bilaterally symmetric flowers. Darwin then crossed the F_1 plants together. In the F_2 generation there were 88 plants with bilaterally symmetric flowers and 37 with peloric flowers.

a) Construct a Punnett grid to predict the outcome of the cross between the F_1 plants. [3]

b) Discuss whether the actual results of the cross are close enough to support the predicted outcome. [2]

c) Peloric *Antirrhinum majus* plants are extremely rare in wild populations of this species. Suggest reasons for this. [1]

2 There are three varieties of pheasant with feather coloration called light, ring and buff. When light pheasants were bred together, only light offspring were produced. Similarly, when ring were crossed with ring, all the offspring were ring. When buff pheasants were crossed with buff there were 75 light offspring, 68 ring and 141 buff.

a) Construct a Punnett grid to predict the outcome of breeding together buff pheasants. [3]

b) Discuss whether the actual results of the cross are close enough to support the predicted outcome. [2]

3 Mary and Herschel Mitchell investigated the inheritance of a character called poky in the fungus *Neurospora crassa*. Poky strains of the fungus grow more slowly than the wild-type. The results are shown in table 3.

Male parent	Female parent	Number of wild type offspring	Number of poky offspring
Wild type	Wild type	9,691	90
Poky	Poky	0	10,591
Wild type	Poky	0	7,905
Poky	Wild type	4,816	43

▲ Table 3

a) Discuss whether the data fits any of the Mendelian ratios in table 1 (page 170). [2]

b) Suggest a reason for all the offspring being poky in a cross between wild type and poky strains when a wild type is the male parent. [2]

c) Suggest a reason for a small number of poky offspring in a cross between wild type and poky strains when a wild type is the female parent. [1]

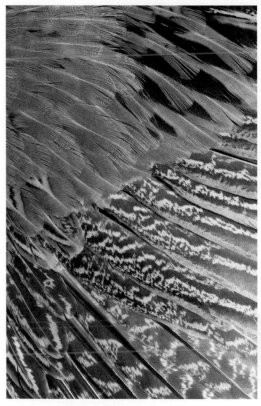

▲ Figure 14 Feather coloration from a buff pheasant

Genetic diseases due to recessive alleles

Many genetic diseases in humans are due to recessive alleles of autosomal genes.

A genetic disease is an illness that is caused by a gene. Most genetic diseases are caused by a recessive allele of a gene. The disease therefore only develops in individuals that do not have the dominant allele of the gene, usually because they have two copies of the recessive allele. If a person has one allele for the genetic disease and one dominant allele, they will not show symptoms of the disease, but they can pass on the recessive allele to their offspring. These individuals are called carriers.

Genetic diseases caused by a recessive allele usually appear unexpectedly. Both parents of a child with the disease must be carriers, but as they do not show symptoms of the disease, they are unaware of this. The probability of these parents having a child with the disease is 25 per cent (see figure 15). Cystic fibrosis is an example of a genetic disease caused by a recessive allele. It is described later in this sub-topic.

Other causes of genetic diseases

Some genetic diseases are sex-linked and some are due to dominant or co-dominant alleles.

A small proportion of genetic diseases are caused by a dominant allele. It is not possible to be a carrier of these diseases. If a person has one dominant allele then they themselves will develop the disease. If one

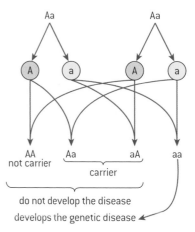

▲ Figure 15 Genetic diseases caused by a recessive allele

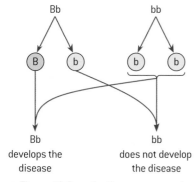

Bb   bb

B   b   b   b

Bb   bb
develops the   does not develop
disease   the disease

▲ Figure 16 Genetic diseases caused by a dominant allele

parent has the allele for the disease, the chance of a child inheriting it is 50 per cent (see figure 16). Huntington's disease is an example of a genetic disease caused by a dominant allele. It is described later in this sub-topic.

A very small proportion of genetic diseases are caused by co-dominant alleles. An example is sickle-cell anemia. The molecular basis of this disease was described in sub-topic 3.1. The normal allele for hemoglobin is Hb^A and the sickle cell allele is Hb^S. Figure 17 shows the three possible combinations of alleles and the characteristics that result. Individuals that have one Hb^A and one Hb^S allele do not have the same characteristics as those who have two copies of either allele, so the alleles are co-dominant.

Most genetic diseases affect males and females in the same way but some show a different pattern of inheritance in males and females. This is called sex linkage. The causes of sex linkage and two examples, red-green colour-blindness and hemophilia, are described later in this sub-topic.

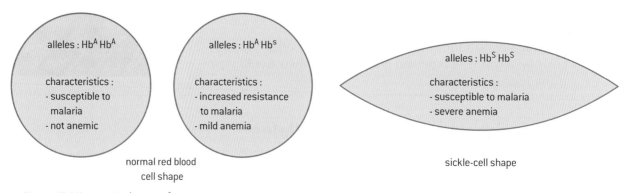

alleles : $Hb^A Hb^A$

characteristics :
- susceptible to malaria
- not anemic

alleles : $Hb^A Hb^S$

characteristics :
- increased resistance to malaria
- mild anemia

alleles : $Hb^S Hb^S$

characteristics :
- susceptible to malaria
- severe anemia

normal red blood cell shape

sickle-cell shape

▲ Figure 17 Effects of Hb^A and Hb^S alleles

Cystic fibrosis and Huntington's disease
Inheritance of cystic fibrosis and Huntington's disease.

Cystic fibrosis is the commonest genetic disease in parts of Europe. It is due to a recessive allele of the CFTR gene. This gene is located on chromosome 7 and the gene product is a chloride ion channel that is involved in secretion of sweat, mucus and digestive juices.

The recessive alleles of this gene result in chloride channels being produced that do not function properly. Sweat containing excessive amounts of sodium chloride is produced, but digestive juices and mucus are secreted with insufficient sodium chloride. As a result not enough water moves by osmosis into the

secretions, making them very viscous. Sticky mucus builds up in the lungs causing infections and the pancreatic duct is usually blocked so digestive enzymes secreted by the pancreas do not reach the small intestine.

In some parts of Europe one in twenty people have an allele for cystic fibrosis. As the allele is recessive, a single copy of the allele does not have any effects. The chance of two parents both being a carrier of the allele is $\frac{1}{20} \times \frac{1}{20}$, which is $\frac{1}{400}$. The chance of such parents having a child with cystic fibrosis can be found using a Punnett grid.

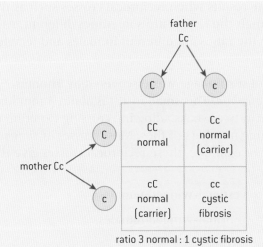

ratio 3 normal : 1 cystic fibrosis

Because of the late onset, many people diagnosed with Huntington's disease have already had children. A genetic test can show before symptoms would develop whether a young person has the dominant allele, but most people at risk choose not to have the test.

About one in 10,000 people have a copy of the Huntington's allele, so it is very unlikely for two parents both to have a copy. A person can nonetheless develop the disease if only one of their parents has the allele because it is dominant.

Huntington's disease is due to a dominant allele of the HTT gene. This gene is located on chromosome 4 and the gene product is a protein named huntingtin. The function of huntingtin is still being researched.

The dominant allele of HTT causes degenerative changes in the brain. Symptoms usually start when a person is between 30 and 50 years old. Changes to behaviour, thinking and emotions become increasingly severe. Life expectancy after the start of symptoms is about 20 years. A person with the disease eventually needs full nursing care and usually succumbs to heart failure, pneumonia or some other infectious disease.

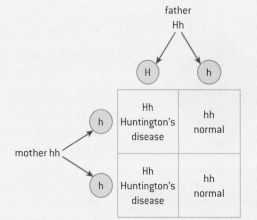

ratio 1 normal : 1 Huntington's disease

Sex-linked genes

The pattern of inheritance is different with sex-linked genes due to their location on sex chromosomes.

Plants such as peas are hermaphrodite – they can produce both male and female gametes. When Thomas Andrew Knight did crossing experiments between pea plants in the late 18th century, he discovered that the results were the same whichever character was in the male gamete and which in the female gamete. For example, these two crosses gave the same results:

- pollen from a plant with green stems placed onto on the stigma of a plant with purple stems;

- pollen from a plant with purple stems placed onto on the stigma of a plant with green stems.

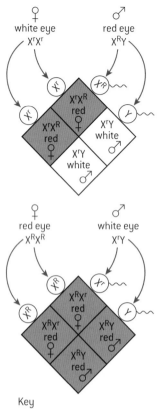

Key
X^R X chromosome with allele
       for red eye (dominant)
X^r X chromosome with allele
       for white eye (recessive)
Y   Y chromosome

▲ Figure 18 Reciprocal sex-linkage
crosses

Plants always give the same results when reciprocal crosses such as these are carried out, but in animals the results are sometimes different. An inheritance pattern where the ratios are different in males and females is called **sex linkage**.

One of the first examples of sex linkage was discovered by Thomas Morgan in the fruit fly, *Drosophila*. This small insect is about 4 mm long and completes its life cycle in two weeks, allowing crossing experiments to be done quickly with large numbers of flies. Most crosses in *Drosophila* do not show sex linkage. For example, these reciprocal crosses give the same results:

- normal-winged males × vestigial-winged females;
- vestigial-winged males × normal-winged females.

These crosses gave different results:

- white-eyed males × red-eyed females gave only red-eyed offspring;
- red-eyed males × white-eyed females gave red-eyed females and white-eyed males.

Geneticists had observed that the inheritance of genes and of chromosomes showed clear parallels and so genes were likely to be located on chromosomes. It was also known that female *Drosophila* have two copies of a chromosome called X and males only have one copy. Morgan deduced that sex linkage of eye colour could therefore be due to the eye colour gene being located on the X chromosome. Male *Drosophila* also have a Y chromosome, but this does not carry the eye-colour gene.

Figure 18 explains the inheritance of eye colour in *Drosophila*. In crosses involving sex linkage, the alleles should always be shown as a superscript letter on a letter X to represent the X chromosome. The Y chromosome should also be shown though it does not carry an allele of the gene.

Red-green colour-blindness and hemophilia

Red-green colour-blindness and hemophilia as examples of sex-linked inheritance.

Many examples of sex linkage have been discovered in humans. They are almost all due to genes located on the X chromosome, as there are very few genes on the Y chromosome. Two examples of sex-linked conditions due to genes on the X chromosomes are described here: red-green colour-blindness and hemophilia.

Red-green colour-blindness is caused by a recessive allele of a gene for one of the photoreceptor proteins. These proteins are made by cone cells in the retina of the eye and detect specific wavelength ranges of visible light.

▲ Figure 19 A person with red-green colour-blindness cannot clearly distinguish between the colours of the flowers and the leaves

▲ Figure 20 Blood should stop quickly flowing from a pricked finger but in hemophiliacs bleeding continues for much longer as blood does not clot properly

proteins involved in the clotting of blood. Life expectancy is only about ten years if hemophilia is untreated. Treatment is by infusing Factor VIII, purified from the blood of donors.

The gene for Factor VIII is located on the X chromosome. The allele that causes hemophilia is recessive. The frequency of the hemophilia allele is about 1 in 10,000. This is therefore the frequency of the disease in boys. Females can be carriers of the recessive hemophilia allele but they only develop the disease if both of their X chromosomes carry the allele. The frequency in girls theoretically is $\left(\frac{1}{10,000}\right)^2 = 1$ in 100,000,000. In practice, there have been even fewer cases of girls with hemophilia due to lack of Factor VIII than this. One reason is that the father would have to be hemophiliac and decide to risk passing on the condition to his children.

Males have only one X chromosome, which they inherit from their mother. If that X chromosome carries the red-green colour-blindness allele then the son will be red-green colour-blind. In parts of northern Europe the percentage of males with this disability is as high as 8%. Girls are red-green colour-blind if their father is red-green colour-blind and they also inherit an X chromosome carrying the recessive gene from their mother. We can predict that the percentage of girls with colour-blindness in the same parts of Europe to be $8\% \times 8\% = 0.64\%$. The actual percentage is about 0.5%, fitting this prediction well.

Whereas red-green colour-blindness is a mild disability, hemophilia is a life-threatening genetic disease. Although there are some rarer forms of the disease, most cases of hemophilia are due to an inability to make Factor VIII, one of the

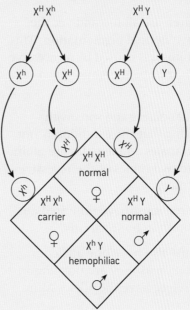

KEY

X^H X chromosome carrying the allele for normal blood clotting

X^h X chromosome carrying the allele for hemophilia.

🧪 Pedigree charts

Analysis of pedigree charts to deduce the pattern of inheritance of genetic diseases.

It isn't possible to investigate the inheritance of genetic diseases in humans by carrying out cross experiments. Pedigree charts can be used instead to deduce the pattern of inheritance. These are the usual conventions for constructing pedigree charts:

- males are shown as squares;

- females are shown as circles;

- squares and circles are shaded or cross-hatched to indicate whether an individual is affected by the disease;

- parents and children are linked using a T, with the top bar of the T between the parents;

- Roman numerals indicate generations;

181

- Arabic numbers are used for individuals in each generation.

Example 1 Albinism in humans

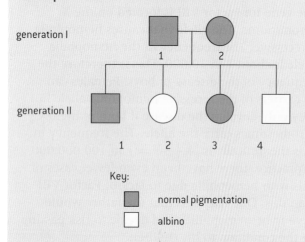

generation I

generation II

Key:

■ normal pigmentation

□ albino

Deductions:

- Two of the children are albino and yet the parents both have normal pigmentation. This suggests that albinism is caused by a recessive allele (m) and normal pigmentation by a dominant allele (M).

- There are both daughters and sons with albinism suggesting that the condition is not sex-linked. Both males and females are albino only if they have two copies of the recessive albinism allele (mm).

- The albino children must have inherited an allele for albinism from both parents.

- Both parents must also have one allele for normal pigmentation as they are not albino. The parents therefore have the alleles Mm.

- The chance of a child of these parents having albinism is $\frac{1}{4}$. Although on average 1 in 4 of

their children will be albino, we could only expect to see that ratio if the parents had very large numbers of children. The actual ratio of 1 in 2 is not unexpected and does not show that our deductions about the inheritance of albinism are incorrect.

Example 2 Vitamin D-resistant rickets

Deductions:

- Two unaffected parents only have unaffected children but two affected parents have an unaffected child, suggesting that this disease is caused by a dominant allele.

- The offspring of the parents in generation I are all affected daughters and unaffected sons. This suggests sex linkage although the number of offspring is too small to be sure of the inheritance pattern.

- If vitamin D-resistant rickets is caused by a dominant X-linked allele, daughters of the father in generation I would inherit his X chromosome carrying the dominant allele, so all of his daughters would have the disease. The data in the pedigree shows this and so supports the theory.

- Similarly if vitamin D-resistant rickets is caused by a dominant X-linked allele, the mother with the disease in generation II would have one X chromosome carrying the dominant allele for the disease and one with the recessive allele. All of her offspring would have a 50% chance of inheriting this X chromosome and of having the disease. The data in the pedigree fits this and so supports the theory.

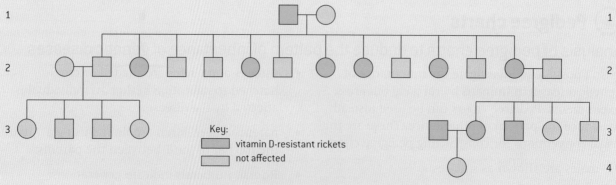

Key:

■ vitamin D-resistant rickets

□ not affected

▲ Figure 21 Pedigree of a family with cases of vitamin D-resistant rickets

Data-based questions: Deducing genotypes from pedigree charts

The pedigree chart in figure 22 shows five generations of a family affected by a genetic disease.

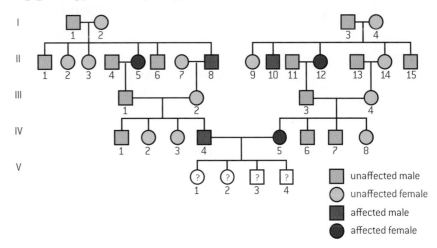

▲ Figure 22 Example of a pedigree chart

1   Explain, using evidence from the pedigree, whether the condition is due to a recessive or a dominant allele.   [3]

2   Explain what the probability is of the individuals in generation V having:

  a) two copies of a recessive allele;

  b) one recessive and one dominant allele;

  c) two copies of the dominant allele.   [3]

3   Deduce, with reasons, the possible alleles of:

  a) 1 in generation III;

  b) 13 in generation II.   [2]

4   Suggest two examples of genetic diseases that would fit this inheritance pattern.   [2]

Legend:
- ☐ unaffected male
- ○ unaffected female
- ■ affected male
- ● affected female

Genetic diseases in humans

Many genetic diseases have been identified in humans but most are very rare.

Several genetic diseases have already been described in this sub-topic, including sickle-cell anemia, cystic fibrosis, hemophilia and Huntington's disease. There are other well-known examples, such as phenylketonuria (PKU), Tay-Sachs disease and Marfan's syndrome.

Medical research has already identified more than 4,000 genetic diseases and more no doubt remain to be found. Given this large number of genetic diseases, it might seem surprising that most of us do not suffer from any of them. The reason for this is that most genetic diseases are caused by very rare recessive alleles which follow Mendelian patterns of inheritance. The chance of inheriting one allele for any specific disease is small but to develop the disease two alleles must be inherited and the chance of this is extremely small.

It is now possible to sequence the genome of an individual human relatively cheaply and quickly and large numbers of humans are being sequenced to allow comparisons. This research is revealing the number of rare recessive alleles that a typical individual is carrying that could cause a genetic disease. Current estimates are that the number is between 75 and 200 alleles among the 25,000 or so genes in the human genome. An individual can only produce a child with a genetic disease due to one of these recessive alleles if the other parent of the child has the same rare allele.

▲ Figure 23 Alleles from two parents come together when they have a child. There is a small chance that two recessive alleles will come together and cause a genetic disease

▲ Figure 24 Abraham Lincoln's features resemble Marfan's syndrome but a more recent theory is that he suffered from MEN2B, another genetic disease

Causes of mutation

Radiation and mutagenic chemicals increase the mutation rate and can cause genetic disease and cancer.

A gene consists of a length of DNA, with a base sequence that can be hundreds or thousands of bases long. The different alleles of a gene have slight variations in the base sequence. Usually only one or a very small number of bases are different. New alleles are formed from other alleles by gene mutation.

A mutation is a random change to the base sequence of a gene. Two types of factor can increase the mutation rate.

- Radiation increases the mutation rate if it has enough energy to cause chemical changes in DNA. Gamma rays and alpha particles from radioactive isotopes, short-wave ultraviolet radiation and X-rays are all mutagenic.

- Some chemical substances cause chemical changes in DNA and so are mutagenic. Examples are benzo[a]pyrene and nitrosamines found in tobacco smoke and mustard gas used as a chemical weapon in the First World War.

Mutations are random changes – there is no mechanism for a particular mutation being carried out. A random change to an allele that has developed by evolution over perhaps millions of years is unlikely to be beneficial. Almost all mutations are therefore either neutral or harmful. Mutations of the genes that control cell division can cause a cell to divide endlessly and develop into a tumour. Mutations are therefore a cause of cancer.

Mutations in body cells, including those that cause cancer, are eliminated when the individual dies, but mutations in cells that develop into gametes can be passed on to offspring. This is the origin of genetic diseases. It is therefore particularly important to minimize the number of mutations in gamete-producing cells in the ovaries and testes. Current estimates are that one or two new mutations occur each generation in humans, adding to the risk of genetic diseases in children.

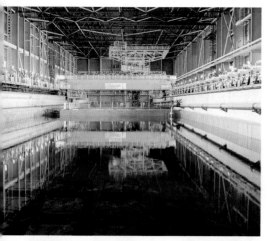

▲ Figure 25 The risk of mutations due to radiation from nuclear waste is minimized by careful storage

 Consequences of nuclear bombing and accidents at nuclear power stations

Consequences of radiation after nuclear bombing of Hiroshima and Nagasaki and the nuclear accidents at Chernobyl.

The common feature of the nuclear bombing of Hiroshima and Nagasaki and the nuclear accidents at Three Mile Island and Chernobyl is that radioactive isotopes were released into the environment and as a result people were exposed to potentially dangerous levels of radiation.

When the atomic bombs were detonated over Hiroshima and Nagasaki 150,000–250,000 people either died directly or within a few months. The health of nearly 100,000 survivors has been followed since then by the Radiation Effects Research Foundation in Japan. Another 26,000 people who were not exposed to radiation have been used as a control group. By 2011 the survivors had developed 17,448 tumours, but only 853 of these could be attributed to the effects of radiation from the atomic bombs.

Apart from cancer the other main effect of the radiation that was predicted was mutations, leading to stillbirths, malformation or death. The health of 10,000 children that were fetuses when the atomic bombs were detonated and 77,000 children that were born later in Hiroshima and Nagasaki has been monitored. No evidence has been found of mutations caused by the radiation. There are likely to have been some mutations, but the number is too small for it to be statistically significant even with the large numbers of children in the study.

Despite the lack of evidence of mutations due to the atomic bombs, survivors have sometimes felt that they were stigmatized. Some found that potential wives or husbands were reluctant to marry them for fear that their children might have genetic diseases.

The accident at Chernobyl, Ukraine, in 1986 involved explosions and a fire in the core of a nuclear reactor. Workers at the plant quickly received fatal doses of radiation. Radioactive isotopes of xenon, krypton, iodine, caesium and tellurium were released and spread over large parts of Europe. About six tonnes of uranium and other radioactive metals in fuel from the reactor was broken up into small particles by the explosions and escaped. An estimated 5,200 million GBq of radioactive material was released

into the atmosphere in total. The effects were widespread and severe:

- 4 km^2 of pine forest downwind of the reactor turned ginger brown and died.

- Horses and cattle near the plant died from damage to their thyroid glands.

- Lynx, eagle owl, wild boar and other wildlife subsequently started to thrive in a zone around Chernobyl from which humans were excluded.

- Bioaccumulation caused high levels of radioactive caesium in fish as far away as Scandinavia and Germany and consumption of lamb contaminated with radioactive caesium was banned for some time as far away as Wales.

- Concentrations of radioactive iodine in the environment rose and resulted in drinking water and milk with unacceptably high levels.

- More than 6,000 cases of thyroid cancer have been reported that can be attributed to radioactive iodine released during the accident.

- According to the report "Chernobyl's Legacy Health, Environmental and Socio-Economic Impacts", produced by The Chernobyl Forum, there is no clearly demonstrated increase in solid cancers or leukemia due to radiation in the most affected populations.

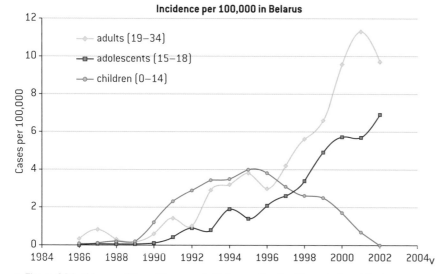

▲ Figure 26 Incidence of thyroid cancer in Belarus after the Chernobyl accident

Activity

Changing rates of thyroid cancer

When would you expect the cases of thyroid cancer in young adults to start to drop, based on the data in figure 26?

▲ Figure 27 Humans have been excluded from a large zone near the Chernobyl reactor. Some plants and animals have shown deformities that may be due to mutations

Data-based questions: The aftermath of Chernobyl

Mutations can cause a cell to become a tumour cell. The release of 6.7 tonnes of radioactive material from the nuclear power station at Chernobyl in 1986 was therefore the cause of large numbers of deaths due to cancer. The UN Chernobyl Forum stated that "up to 4,000 people" may ultimately die as a result of the disaster, but Green Party members of the European Parliament commissioned a report from a radiation scientist, which gave an estimate of 30,000 to 60,000 extra deaths. One way of obtaining an estimate is to use data from previous radiation exposures, such as the detonation of nuclear warheads at Hiroshima and Nagasaki in 1945. The data below is an analysis of deaths due to leukemia and cancer between 1950 and 1990 among those exposed to radiation from these warheads. It was published by the Radiation Effects Research Foundation.

Radiation dose range (Sv)	Number of deaths in people exposed to radiation	Estimate of excess deaths over control groups	Percentage of deaths attributable to radiation exposure
Leukemia			
0.005–0.2	70	10	
0.2–0.5	27	13	48
0.5–1	23	17	74
>1	56	47	
Cancer			
0.005–0.2	3391	63	2
0.2–0.5	646	76	12
0.5–1	342	79	23
>1	308	121	39

1   Calculate the percentage of excess deaths over control groups due to leukemia in people exposed to (a) 0.005-0.2 Sv (sieverts) of radiation (b) >1 Sv of radiation. [4]

2   Construct a suitable type of graph or chart to represent the data in the right-hand column of the table, including the two percentages that you have calculated. There should be two y-axes, for the leukemia deaths and the cancer deaths. [4]

3   Compare the effect of radiation on deaths due to leukemia and deaths due to cancer. [3]

4   Discuss, with reasons, what level of radiation might be acceptable in the environment. [4]

3.5 Genetic modification and biotechnology

Understanding

→ Gel electrophoresis is used to separate proteins or fragments of DNA according to size.

→ PCR can be used to amplify small amounts of DNA.

→ DNA profiling involves comparison of DNA.

→ Genetic modification is carried out by gene transfer between species.

→ Clones are groups of genetically identical organisms, derived from a single original parent cell.

→ Many plant species and some animal species have natural methods of cloning.

→ Animals can be cloned at the embryo stage by breaking up the embryo into more than one group of cells.

→ Methods have been developed for cloning adult animals using differentiated cells.

Applications

→ Use of DNA profiling in paternity and forensic investigations.

→ Gene transfer to bacteria with plasmids using restriction endonucleases and DNA ligase.

→ Assessment of the potential risks and benefits associated with genetic modification of crops.

→ Production of cloned embryos by somatic-cell nuclear transfer.

Skills

→ Design of an experiment to assess one factor affecting the rooting of stem-cuttings.

→ Analysis of examples of DNA profiles.

→ Analysis of data on risks to monarch butterflies of Bt crops.

Nature of science

→ Assessing risks associated with scientific research: scientists attempt to assess the risks associated with genetically modified crops or livestock.

Gel electrophoresis

Gel electrophoresis is used to separate proteins or fragments of DNA according to size.

Gel electrophoresis involves separating charged molecules in an electric field, according to their size and charge. Samples are placed in wells cast in a gel. The gel is immersed in a conducting fluid and an electric field is applied. Molecules in the sample that are charged will move through the gel. Molecules with negative and positive charges move in opposite directions. Proteins may be positively or negatively charged so can be separated according to their charge.

The gel used in gel electrophoresis consists of a mesh of filaments that resists the movement of molecules in a sample. DNA molecules from eukaryotes are too long to move through the gel, so they must be broken up into smaller fragments. All DNA molecules carry negative charges so move in the same direction during gel electrophoresis, but not

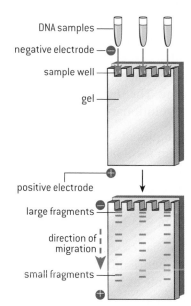

▲ Figure 1 Procedure for gel electrophoresis

187

at the same rate. Small fragments move faster than large ones so they move further in a given time. Gel electrophoresis can therefore be used to separate fragments of DNA according to size.

DNA amplification by PCR

PCR can be used to amplify small amounts of DNA.

The polymerase chain reaction is used to make large numbers of copies of DNA. It is almost always simply called PCR. The details of this technique are described in sub-topic 2.7. Only a very small amount of DNA is needed at the start of the process – in theory just a single molecule. Within an hour or two, millions of copies can be made. This makes it possible to study the DNA further without the risk of using up a limited sample. For example, DNA extracted from fossils can be amplified using PCR. Very small amounts of DNA from blood, semen or hairs can also be amplified for use in forensic investigations.

PCR is not used to copy the entire set of DNA molecules in a sample such as blood or semen. White blood cells contain all chromosomes of the person from whom the blood came, for example, and together the sperm cells in a sample of semen contain a man's entire genome. Instead PCR is used to copy specific DNA sequences. A sequence is selected for copying by using a primer that binds to the start of the desired sequence. The primer binds by complementary base pairing.

The selectivity of PCR allows particular desired sequences to be copied from a whole genome or even greater mixture of DNA. One test for the presence of genetically modified ingredients in foods involves the use of a primer that binds to the genetically modified DNA. Any such DNA present is amplified by the PCR, but if there is none present the PCR has no effect.

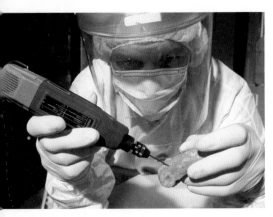

▲ Figure 2 Small samples of DNA being extracted from fossil bones of a Neanderthal for amplification by PCR

Data-based questions: PCR and Neanderthals

The evolution of groups of living organisms can be studied by comparing the base sequences of their DNA. If a species separates into two groups, differences in base sequence between the two species accumulate gradually over long periods of time. The number of differences can be used as an "evolutionary clock".

Samples of DNA were recently obtained from fossil bones of a Neanderthal (*Homo neanderthalensis*). They were amplified using PCR. A section of the Neanderthal mitochondrial DNA was sequenced and compared with sequences from 994 humans and 16 chimpanzees.

The bar chart in figure 3 shows how many base-sequence differences were found within the sample of humans, between the humans and the

Neanderthal and between the humans and the chimpanzees.

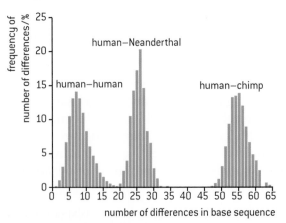

▲ Figure 3 Number of differences in base sequences between humans, chimps and Neanderthals

1 State the most common number of differences in base sequence between pairs of humans. [1]

2 Humans and Neanderthals are both classified in the genus *Homo* and chimpanzees are classified in the genus *Pan*. Discuss whether this classification is supported by the data in the bar chart. [3]

3 Suggest a limitation to drawing any conclusion from the human–Neanderthal comparison. [1]

DNA profiling

DNA profiling involves comparison of DNA.

DNA profiling involves these stages:

- A sample of DNA is obtained, either from a known individual or from another source such as a fossil or a crime scene.

- Sequences in the DNA that vary considerably between individuals are selected and are copied by PCR.

- The copied DNA is split into fragments using restriction endonucleases.

- The fragments are separated using gel electrophoresis.

- This produces a pattern of bands that is always the same with DNA taken from one individual. This is the individual's DNA profile.

- The profiles of different individuals can be compared to see which bands are the same and which are different.

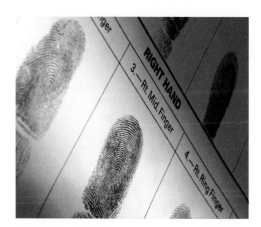

▲ Figure 4 DNA profiles are often referred to as DNA fingerprints as they are used in a similar way to real fingerprints to distinguish one individual from all others

🌐 Paternity and forensic investigations

Use of DNA profiling in paternity and forensic investigations.

DNA profiling is used in forensic investigations.

- Blood stains on a suspect's clothing could be shown to come from the victim.

- Blood stains at the crime scene that are not from the victim could be shown to come from the suspect.

- A single hair at the crime scene could be shown to come from the suspect.

- Semen from a sexual crime could be shown to come from the suspect.

In each example the DNA profile of material from the crime scene is compared with the DNA profile of a sample of DNA taken from the suspect or the victim. If the pattern of bands matches exactly it is highly likely that the two samples of DNA are from the same person. This can provide very strong evidence of who committed the crime. Some countries now have databases of DNA profiles, which have allowed many criminal cases to be solved.

DNA profiling is also used in paternity investigations. These are done to find out whether a man is the father of a child. There are various reasons for paternity investigations being requested.

- Men sometimes claim that they are not the father of a child to avoid having to pay the mother to raise the child.

- Women who have had multiple partners may wish to identify the biological father of a child.

- A child may wish to prove that a deceased man was their father in order to show that they are their heir.

DNA profiles of the mother, the child and the man are needed. DNA profiles of each of the samples are prepared and the patterns of bands are compared. If any bands in the child's profile do not occur in the profile of the mother or man, another person must be the father.

Analysis of DNA profiles

Analysis of examples of DNA profiles.

Analysis of DNA profiles in forensic investigations is straightforward: two DNA samples are very likely to have come from the same person if the pattern of bands on the profile is the same.

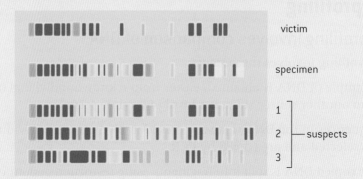

▲ Figure 5 Which of the three suspects' DNA fingerprints matches the specimen recovered from the crime scene?

Analysis of DNA profiles in paternity investigations is more complicated. Each of the bands in the child's DNA profile must be the same as a band in the biological mother or father's profile. Every band in the child's profile must be checked to make sure that it occurs either in the mother's profile or in the profile of the man presumed to be the father. If one or more bands do not, another man must have been the biological father.

Genetic modification

Genetic modification is carried out by gene transfer between species.

Molecular biologists have developed techniques that allow genes to be transferred between species. The transfer of genes from one species to another is known as genetic modification. It is possible because the genetic code is universal, so when genes are transferred between species, the amino acid sequence translated from them is unchanged – the same polypeptide is produced.

Genes have been transferred from eukaryotes to bacteria. One of the early examples was the transfer of the gene for making human insulin to a bacterium. This was done so that large quantities of this hormone can be produced for treating diabetics.

Genetic modification has been used to introduce new characteristics to animal species. For example, goats have been produced that secrete milk containing spider silk protein. Spider silk is immensely strong, but spiders could not be used to produce it commercially.

Genetic modification has also been used to produce many new varieties of crop plant. These are known as genetically modified or GM crops. For example genes from snapdragons have been transferred to tomatoes to produce fruits that are purple rather than red. The production of golden rice involved the transfer of three genes, two from daffodil plants and

▲ Figure 6 Genes have been transferred from daffodil plants to rice, to make the rice produce a yellow pigment in its seeds

one from a bacterium, so that the yellow pigment β-carotene is produced in the rice grains.

Activity

Scientists have an obligation to consider the ethical implications of their research. Discuss the ethics of the development of golden rice. β-carotene is a precursor to vitamin A. The development of golden rice was intended as a solution to the problem of vitamin A deficiency, which is a significant cause of blindness among children globally.

🌐 Techniques for gene transfer to bacteria

Gene transfer to bacteria with plasmids using restriction endonucleases and DNA ligase.

Genes can be transferred from one species to another by a variety of techniques. Together these techniques are known as genetic engineering. Gene transfer to bacteria usually involves plasmids, restriction enzymes and DNA ligase.

- A plasmid is a small extra circle of DNA. The smallest plasmids have about 1,000 base pairs (1 kbp), but they can have over 1,000 kbp. They occur commonly in bacteria. The most abundant plasmids are those with genes that encourage their replication in the cytoplasm and transfer from one bacterium to another. There are therefore some parallels with viruses but plasmids are not pathogenic and natural selection favours plasmids that confer an advantage on a bacterium rather than a disadvantage. Bacteria use plasmids to exchange genes, so naturally absorb them and incorporate them into their main circular DNA molecule. Plasmids are very useful in genetic engineering.

- Restriction enzymes, also known as endonucleases, are enzymes that cut DNA molecules at specific base sequences. They can be used to cut open plasmids and also to cut out desired genes from larger DNA molecules. Some restriction enzymes have the useful property of cutting the two strands of a DNA molecule at different points. This leaves single-stranded sections called sticky ends. The sticky ends created by any one particular restriction enzyme have complementary base sequences so can be used to link together pieces of DNA, by hydrogen bonding between the bases.

- DNA ligase is an enzyme that joins DNA molecules together firmly by making sugar–phosphate bonds between nucleotides. When the desired gene has been inserted into a plasmid using sticky ends there are still nicks in each sugar–phosphate backbone of the DNA but DNA ligase can be used to seal these nicks.

An obvious requirement for gene transfer is a copy of the gene being transferred. It is usually easier to obtain messenger RNA transcripts of genes than the genes themselves. Reverse transcriptase is an enzyme that makes DNA copies of RNA molecules called cDNA. It can be used to make the DNA needed for gene transfer from messenger RNA.

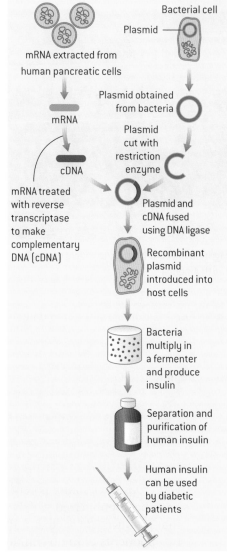

▲ Figure 7 shows the steps involved in one example of gene transfer. It has been used to create genetically modified *E. coli* bacteria that are able to manufacture human insulin, for use in treating diabetes

191

▲ Figure 8 The biohazard symbol indicates any organism or material that poses a threat to the health of living organisms especially humans

▲ Figure 9 GM corn (maize) is widely grown in North America

 Assessing the risks of genetic modification

Assessing risks associated with scientific research: scientists attempt to assess the risks associated with genetically modified crops or livestock.

There have been many fears expressed about the possible dangers of genetic modification. These fears can be traced back to the 1970s when the first experiments in gene transfer were being conducted. Paul Berg planned an experiment in which DNA from the monkey virus SV40 was going to be inserted into the bacterium *E. coli*. Other biologists expressed serious concerns because SV40 was known to cause cancer in mice and *E. coli* lives naturally in the intestines of humans. There was therefore a risk of the genetically engineered bacterium causing cancer in humans.

Since then many other risks associated with genetic modification have been identified. There has been fierce debate both among scientists and between scientists and non-scientists about the safety of the research and the safety of using genetically modified organisms. This has led to bans being imposed in some countries, with potentially useful applications of GM crops or livestock left undeveloped.

Almost everything that we do carries risks and it is not possible to eliminate risk entirely, either in science or in other aspects of our lives. It is natural for humans to assess the risk of an action and decide whether or not go ahead with it. This is what scientists must do – assess the risks associated with their research before carrying it out. The risks can be assessed in two ways:

- What is the chance of an accident or other harmful consequence?

- How harmful would the consequence be?

If there is a high chance of harmful consequences or a significant chance of very harmful consequences then research should not be done.

Risks and benefits of GM crops

Assessment of the potential risks and benefits associated with genetic modification of crops.

GM crops have many potential benefits. These have been publicized widely by the corporations that produce GM seed, but they are questioned by opponents of the technology. Even basic issues such as whether GM crops increase yields and reduce pesticide and herbicide use have been contested. It is not surprising that there is disagreement, because gene transfer to crop plants is a relatively recent procedure, the issues involved are very complex and in science it often takes decades for disputes to be resolved.

Potential benefits can be grouped into environmental benefits, health benefits and agricultural benefits. Economic benefits of GM crops are not included here, because they cannot be assessed on a scientific basis using experimental evidence. It would be impossible in the time available for IB students to assess all claimed benefits for all GM crops. Instead it is better to select one claim from the list given here and assess it for one crop. Much of the evidence relating to potential benefits and also to risks is freely available.

Claims about environmental benefits of GM crops:

- Pest-resistant crop varieties can be produced by transferring a gene for making a toxin to the plants. Less insecticide then has to be sprayed on to the crop so fewer bees and other beneficial insects are harmed.

- Use of GM crop varieties reduces the need for plowing and spraying crops, so less fuel is needed for farm machinery.

- The shelf-life of fruit and vegetables can be improved, reducing wastage and reducing the area of crops that have to be grown.

Claims about the health benefits of GM crops:

- The nutritional value of crops can be improved, for example by increasing the vitamin content.

- Varieties of crops could be produced lacking allergens or toxins that are naturally present in them.

- GM crops could be engineered that produce edible vaccines so by eating the crop a person would be vaccinated against a disease.

Claims about agricultural benefits of GM crops:

- Varieties resistant to drought, cold and salinity can be produced by gene transfer, expending the range over which crops can be produced and increasing total yields.

- A gene for herbicide resistance can be transferred to crop plants allowing all other plants to be killed in the growing crop by spraying with herbicide. With less weed competition crop yields are higher. Herbicides that kill all plants can be used to create weed-free conditions for sowing non-GM crops but they cannot be used once the crop is growing.

- Crop varieties can be produced that are resistant to diseases caused by viruses.

▲ Figure 10 Wild plants growing next to a crop of GM maize

These diseases currently reduce crop yields significantly and the only current method of control is to reduce transmission by killing insect vectors of the viruses with insecticides.

A wide variety of concerns about GM crops have been raised. Some of these, such as the effect on farmer's incomes, cannot be assessed on scientific grounds so are not relevant here. The remaining concerns can be grouped into health risks, environmental risks and agricultural risks. To make overall judgments about the safety of GM crops, each risk needs to be assessed carefully, using all the available experimental evidence. This needs to be done on a case by case basis as it is not possible to assess the risks and benefits of one GM crop from experiments performed on another one.

There is no consensus among all scientists or non-scientists yet about GM crops and it is therefore important for as many of us as possible to look at the evidence for the claims and counter-claims, rather than the publicity. Any of the risks that are included here could be selected for detailed scrutiny.

Claims made about health risks of GM crops:

- Proteins produced by transcription and translation of transferred genes could be toxic or cause allergic reactions in humans or livestock that eat GM crops.

- Antibiotic resistance genes used as markers during gene transfer could spread to pathogenic bacteria.

- Transferred genes could mutate and cause unexpected problems that were not risk-assessed during development of GM crops.

Claims made about environmental risks of GM crops:

- Non-target organisms could be affected by toxins that are intended to control pests in GM crop plants.

- Genes transferred to crop plants to make them herbicide resistant could spread to wild plants, turning them into uncontrollable super-weeds.

- Biodiversity could be reduced if a lower proportion of sunlight energy passes to weed plants, plant-eating insects and organisms that feed on them where GM rather than non-GM crops are being grown.

Claims made about agricultural risks of GM crops:

- Some seed from a crop is always spilt and germinates to become unwanted volunteer plants that must be controlled, but this could become very difficult if the crop contains herbicide resistance genes.

- Widespread use of GM crops containing a toxin that kills insect pests will lead to the spread of resistance to the toxin in the pests that were the initial problem and also to the spread of secondary pests that are resistant to the toxin but were previously scarce.

- Farmers are not permitted by patent law to save and re-sow GM seed from crops they have grown, so strains adapted to local conditions cannot be developed.

 Analysing risks to monarch butterflies of *Bt* corn

Analysis of data on risks to monarch butterflies of *Bt* crops.

Insect pests of crops can be controlled by spraying with insecticides but varieties have been recently been produced by genetic engineering that produce a toxin that kills insects. A gene was transferred from the bacterium *Bacillus thuringiensis* that codes for *Bt* toxin. The toxin is a protein. It kills members of insect orders that contain butterflies, moths, flies, beetles, bees and ants. The genetically engineered corn varieties produce *Bt* toxin in all parts of the plant including pollen.

Bt varieties of many crops have been produced, including *Zea mays*. In North America this crop is called corn, while in Britain it is known as maize, or corn on the cob. The crop is attacked by various insect pests including corn borers, which are the larvae of the moth *Ostrinia nubilalis*. Concerns have been expressed about the effects of *Bt* corn on non-target species of insect. One particular species of concern is the monarch butterfly, *Danaus plexippus*.

The larvae of the monarch butterfly feed on leaves of milkweed, *Asclepias curassavica*. This plant sometimes grows close enough to corn crops to become dusted with the wind-dispersed corn pollen. There is therefore a risk that monarch larvae might be poisoned by *Bt* toxin in pollen from GM corn crops. This risk has been investigated experimentally. Data from these experiments is available for analysis.

Data-based questions: Transgenic pollen and monarch larvae

To investigate the effect of pollen from *Bt* corn on the larvae of monarch butterflies the following procedure was used. Leaves were collected from milkweed plants and were lightly misted with water. A spatula of pollen was gently tapped over the leaves to deposit a fine dusting. The leaves were placed in water-filled tubes. Five three-day-old monarch butterfly larvae were placed on each leaf. The area of leaf eaten by the larvae was monitored over four days. The mass of the larvae was measured after four days. The survival of the larvae was monitored over four days.

Three treatments were included in the experiment, with five repeats of each treatment:

- leaves not dusted with pollen (blue)

- leaves dusted with non-GM pollen (yellow)

- leaves dusted with pollen from *Bt* corn (red)

The results are shown in the table, bar chart and graph on the right.

1 **a)** List the variables that were kept constant in the experiment. [3]

   b) Explain the need to keep these variables constant. [2]

2 **a)** Calculate the total number of larvae used in the experiment. [2]

   b) Explain the need for replicates in experiments. [2]

3 The bar chart and the graph show mean results and error bars. Explain how error bars help in the analysis and evaluation of data. [2]

4 Explain the conclusions that can be drawn from the percentage survival of larvae in the three treatments. [2]

5 Suggest reasons for the differences in leaf consumption between the three treatments. [3]

6 Predict the mean mass of larvae that fed on leaves dusted with non-GM pollen. [2]

7 Outline any differences between the procedures used in this experiment and processes that occur in nature, which might affect whether monarch larvae are actually harmed by *Bt* pollen. [2]

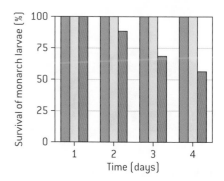

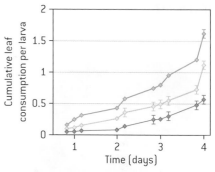

Source: Losey JE, Rayor LS, Carter ME (May 1999). "Transgenic pollen harms monarch larvae". *Nature* 399 (6733): 214.

Treatment	Mean mass of surviving larvae (g)
Leaves not dusted with pollen	0.38
Leaves dusted with non-GM pollen	Not available
Leaves dusted with pollen from Bt corn	0.16

Activity

Estimating the size of a clone

A total of 130,000 hectares of Russet Burbank potatoes were planted in Idaho in 2011. The mean density of planting of potato tubers was 50,000 per hectare. Estimate the size of the clone at the time of planting and at the time of harvest.

Clones

Clones are groups of genetically identical organisms, derived from a single original parent cell.

A zygote, produced by the fusion of a male and female gamete, is the first cell of a new organism. Because zygotes are produced by sexual reproduction, they are all genetically different. A zygote grows and develops into an adult organism. If it reproduces sexually, its

Activity

How many potato clones are there in this photo?

offspring will be genetically different. In some species organisms can also reproduce asexually. When they do this, they produce genetically identical organisms.

The production of genetically identical organisms is called cloning and a group of genetically identical organisms is called a clone.

Although we do not usually think of them in this way, a pair of identical twins is the smallest clone that can exist. They are either the result of a human zygote dividing into two cells, which each develop into separate embryos, or an embryo splitting into two parts which each develop into a separate individual. Identical twins are not identical in all their characteristics and have, for example, different fingerprints. A better term for them is monozygotic. More rarely identical triplets, quadruplets and even quintuplets have been produced.

Sometimes a clone can consist of very large numbers of organisms. For example, commercially grown potato varieties are huge clones. Large clones are formed by cloning happening again and again, but even so all the organisms may be traced back to one original parent cell.

Natural methods of cloning

Many plant species and some animal species have natural methods of cloning.

Although the word clone is now used for any group of genetically identical organisms, it was first used in the early 20th century for plants produced by asexual reproduction. It comes from the Greek word for twig. Many plants have a natural method of cloning. The methods used by plants are very varied and can involve stems, roots, leaves or bulbs. Two examples are given here:

▲ Figure 11 Identical twins are an example of cloning

- A single garlic bulb, when planted, uses its food stores to grow leaves. These leaves produce enough food by photosynthesis to grow a group of bulbs. All the bulbs in the group are genetically identical so they are a clone.

- A strawberry plant grows long horizontal stems with plantlets at the end. These plantlets grow roots into the soil and photosynthesize using their leaves, so can become independent of the parent plant. A healthy strawberry plant can produce ten or more genetically identical new plants in this way during a growing season.

Natural methods of cloning are less common in animals but some species are able to do it.

- *Hydra* clones itself by a process called budding (sub-topic 1.6, figure 1, page 51).

Female aphids can give birth to offspring that have been produced entirely from diploid egg cells that were produced by mitosis rather than meiosis. The offspring are therefore clones of their mother.

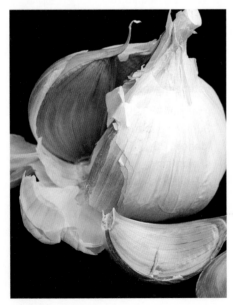

▲ Figure 12 One bulb of garlic clones itself to produce a group of bulbs by the end of the growing season

Investigating factors affecting the rooting of stem-cuttings

Design of an experiment to assess one factor affecting the rooting of stem-cuttings.

Stem-cuttings are short lengths of stem that are used to clone plants artificially. If roots develop from the stem, the cutting can become an independent new plant.

1   Many plants can be cloned from cuttings. *Ocimum basilicum* roots particularly easily.

2   Nodes are positions on the stem where leaves are attached. With most species the stem is cut below a node.

3   Leaves are removed from the lower half of the stem. If there are many large leaves in the upper half they can also be reduced.

4   The lowest third of the cutting is inserted into compost or water. Compost should be sterile and contain plenty of both air and water.

5   A clear plastic bag with a few holes cut in it prevents excessive water loss from cuttings inserted in compost.

6   Rooting normally takes a few weeks. Growth of new leaves usually indicates that the cutting has developed roots.

Not all gardeners have success when trying to clone plants using root cuttings. Successful gardeners are sometimes said to have "green fingers" but a biologist would reject this as the reason for their success. Experiments can give evidence about the factors that determine whether cuttings root or not. You can design and carry out an experiment to investigate one of the factors on the list below, or another factor of your own.

Possible factors to investigate:

- whether the stem is cut above or below a node
- how long the cutting is
- whether the end of the stem is left in the air to callus over
- how many leaves are left on the cutting
- whether a hormone rooting powder is used

- whether the cutting is placed in water or compost
- what type of compost is used
- how warm the cuttings are kept
- whether a plastic bag is placed over the cuttings
- whether holes are cut in the plastic bag.

You should think about these questions when you design your experiment:

1   What is your independent variable?

2   How will you measure the amount of root formation, which is your dependent variable?

3   Which variables should you keep constant?

4   How many different types of plant should you use?

5   How many cuttings should you use for each treatment?

197

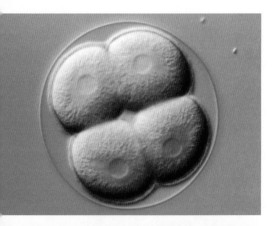

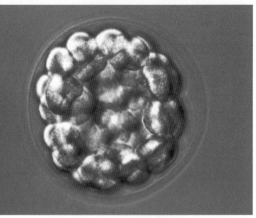

▲ Figure 13 Sea urchin embryo (a) 4-cell stage (b) blastula stage consisting of a hollow ball of cells

Cloning animal embryos

Animals can be cloned at the embryo stage by breaking up the embryo into more than one group of cells.

At an early stage of development all cells in an animal embryo are pluripotent (capable of developing into all types of tissue). It is therefore theoretically possible for the embryo to divide into two or more parts and each part to develop into a separate individual with all body parts. This process is called splitting or fragmentation. Coral embryos have been observed to clone themselves by breaking up into smaller groups of cells or even single cells, presumably because this increases the chance of one embryo surviving.

Formation of identical twins could be regarded as cloning by splitting, but most animal species do not appear to do this naturally. However, it is possible to break up animal embryos artificially and in some cases the separated parts develop into multiple embryos.

In livestock, an egg can be fertilized in vitro and allowed to develop into a multicellular embryo. Individual cells can be separated from the embryo while they are still pluripotent and transplanted into surrogate mothers. Only a limited number of clones can be obtained this way, because after a certain number of divisions the embryo cells are no longer pluripotent. Splitting of embryos is usually most successful at the eight-cell stage.

There has been little interest in this method of artificial cloning because at the embryo stage it is not possible to assess whether a new individual produced by sexual reproduction has desirable characteristics.

Cloning adult animals using differentiated cells

Methods have been developed for cloning adult animals using differentiated cells.

It is relatively easy to clone animal embryos, but at that stage it is impossible to know whether the embryos will have desirable characteristics. Once the embryos have grown into adults it is easy to assess their characteristics, but it is much more difficult to clone them. This is because the cells that make up the body of an adult animal are differentiated. To produce all the tissues in a new animal body undifferentiated pluripotent cells are needed.

The biologist John Gurdon carried out experiments on cloning in the frog *Xenopus* as a postgraduate student in Oxford during the 1950s. He removed nuclei from body cells of *Xenopus* tadpoles and transplanted them into egg cells from which the nucleus had been removed. The egg cells into which the nuclei were transplanted developed as though they were zygotes. They carried out cell division, cell growth and differentiation to form all the tissues of a normal *Xenopus* frog. In 2012 Gurdon was awarded the Nobel Prize for Physiology or Medicine for his pioneering research.

Cloning using differentiated cells proved to be much more difficult in mammals. The first cloned mammal was Dolly the sheep in 1996. Apart from the obvious reproductive uses of this type of cloning, there is also interest in it for therapeutic reasons. If this procedure

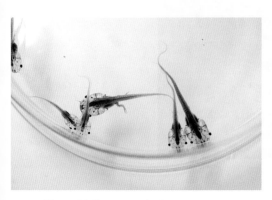

▲ Figure 14 *Xenopus* tadpoles

was done with humans, the embryo would consist of pluripotent stem cells, which could be used to regenerate tissues for the adult. Because the cells would be genetically identical to those of the adult from whom the nucleus was obtained they would not cause rejection problems.

Methods used to produce Dolly
Production of cloned embryos by somatic-cell nuclear transfer.

The production of Dolly was a pioneering development in animal cloning. The method that was used is called somatic-cell nuclear transfer. A somatic cell is a normal body cell with a diploid nucleus. The method has these stages:

- Adult cells were taken from the udder of a Finn Dorset ewe and were grown in the laboratory, using a medium containing a low concentration of nutrients. This made genes in the cells inactive so that the pattern of differentiation was lost.

- Unfertilized eggs were taken from the ovaries of a Scottish Blackface ewe. The nuclei were removed from these eggs. One of the cultured cells from the Finn Dorset was placed next to each egg cell, inside the zona pellucida around the egg, which is a protective coating of gel. A small electric pulse was used to cause the two cells to fuse together. About 10% of the fused cells developed like a zygote into an embryo.

▲ Figure 15 Dolly with Dr Ian Wilmut, the embryologist who led the team that produced her

- The embryos were then injected when about seven days old into the uteri of other ewes that could act as surrogate mothers. This was done in the same way as in IVF. Only one of the 29 embryos implanted successfully and developed through a normal gestation. This was Dolly.

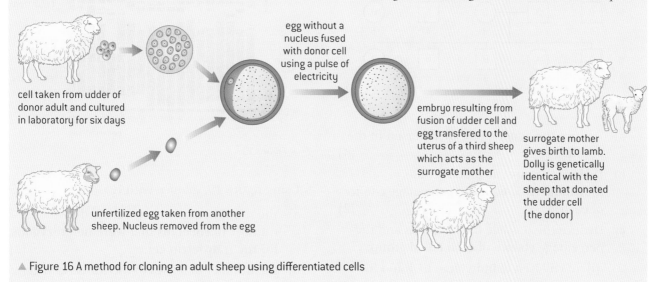

cell taken from udder of donor adult and cultured in laboratory for six days

egg without a nucleus fused with donor cell using a pulse of electricity

unfertilized egg taken from another sheep. Nucleus removed from the egg

embryo resulting from fusion of udder cell and egg transferred to the uterus of a third sheep which acts as the surrogate mother

surrogate mother gives birth to lamb. Dolly is genetically identical with the sheep that donated the udder cell (the donor)

▲ Figure 16 A method for cloning an adult sheep using differentiated cells

Questions

1 Human somatic cells have 46 chromosomes, while our closest primate relatives, the chimpanzee, the gorilla and the orangutan all have 48 chromosomes. One hypothesis is that the human chromosome number 2 was formed from the fusion of two chromosomes in a primate ancestor. The image below shows human chromosome 2 compared to chromosome 12 and 13 from the chimpanzee.

   a) Compare the human chromosome 2 with the two chimpanzee chromosomes (figure 17). [3]

   b) The ends of chromosomes, called telomeres, have many repeats of the same short DNA sequence. If the fusion hypothesis were true, predict what would be found in the region of the chromosome where the fusion is hypothesized to have occurred. [2]

▲ Figure 17

2 The pedigree in figure 18 shows the ABO groups of three generations of a family.

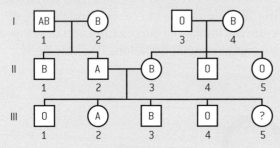

▲ Figure 18

   a) Deduce the genotype of each person in the family. [4]

   b) Deduce the possible blood groups of individual III 5, with the percentage chance of each. [2]

   c) Deduce the possible blood groups and the percentage chance of each blood group:

      (i) of children of individual III 1 and his partner who is also in blood group O [2]

      (ii) of children of III 2 and her partner who has blood group AB. [2]

3 The cheetah (*Acinonyx jubatus*) is an endangered species of large cat found in South and East Africa. A study of the level of variation of the cheetah gene pool was carried out. In one part of this study, blood samples were taken from 19 cheetahs and analysed for the protein transferrin using gel electrophoresis. The results were compared with the electrophoresis patterns for blood samples from 19 domestic cats (*Felis sylvestris*). Gel electrophoresis can be used to separate proteins using the same principles as in DNA profiling. The bands on the gel which represent forms of the protein transferrin are indicated.

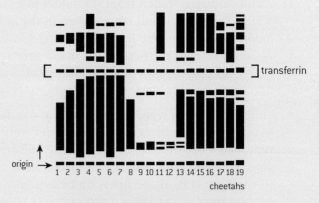

cheetahs

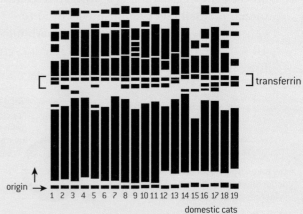

domestic cats

▲ Figure 19

Using figure 19, deduce with reasons:

a) the number of domestic cats and the number of cheetahs that were heterozygous for the transferrin gene; [2]

b) the number of alleles of the transferrin gene in the gene pool of domestic cats; [2]

c) the number of alleles of the transferrin gene in the gene pool of cheetahs. [1]

4 ECOLOGY

Introduction

Ecosystems require a continuous supply of energy to fuel life processes and to replace energy lost as heat. Continued availability of carbon and other chemical elements in ecosystems depends on cycles. The future survival of living organisms including humans depends on sustainable ecological communities. Concentrations of gases in the atmosphere have significant effects on climates experienced at the Earth's surface.

4.1 Species, communities and ecosystems

Understanding

→ Species are groups of organisms that can potentially interbreed to produce fertile offspring.

→ Members of a species may be reproductively isolated in separate populations.

→ Species have either an autotrophic or heterotrophic method of nutrition (a few species have both methods).

→ Consumers are heterotrophs that feed on living organisms by ingestion.

→ Detritivores are heterotrophs that obtain organic nutrients from detritus by internal digestion.

→ Saprotrophs are heterotrophs that obtain organic nutrients from dead organic matter by external digestion.

→ A community is formed by populations of different species living together and interacting with each other.

→ A community forms an ecosystem by its interactions with the abiotic environment.

→ Autotrophs and heterotrophs obtain inorganic nutrients from the abiotic environment.

→ The supply of inorganic nutrients is maintained by nutrient cycling.

→ Ecosystems have the potential to be sustainable over long periods of time.

Skills

→ Classifying species as autotrophs, consumers, detritivores or saprotrophs from a knowledge of their mode of nutrition.

→ Testing for association between two species using the chi-squared test with data obtained by quadrat sampling.

→ Recognizing and interpreting statistical significance.

→ Setting up sealed mesocosms to try to establish sustainability. (Practical 5)

Nature of science

→ Looking for patterns, trends and discrepancies: plants and algae are mostly autotrophic but some are not.

▲ Figure 1 A bird of paradise in Papua New Guinea

Species

Species are groups of organisms that can potentially interbreed to produce fertile offspring.

Birds of paradise inhabit Papua New Guinea and other Australasian islands. In the breeding season the males do elaborate and distinctive courtship dances, repeatedly carrying out a series of movements to display their exotic plumage. One reason for this is to show to a female that they are fit and would be a suitable partner. Another reason is to show that they are the same type of bird of paradise as the female.

There are forty-one different types of bird of paradise. Each of these usually only reproduces with others of its type and hybrids between the different types are rarely produced. For this reason each of the forty-one types of bird of paradise remains distinct, with characters that are different to those of other types. Biologists call types of organism such as these **species**. Although few species have as elaborate courtship rituals as birds of paradise, most species have some method of trying to ensure that they reproduce with other members of their species.

When two members of the same species mate and produce offspring they are interbreeding. Occasionally members of different species breed together. This is called cross-breeding. It happens occasionally with birds of paradise. However, the offspring produced by cross-breeding between species are almost always infertile, which prevents the genes of two species becoming mixed.

The reproductive separation between species is the reason for each species being a recognizable type of organism with characters that distinguish it from even the most closely related other species. In summary, a species is a group of organisms that interbreed to produce fertile offspring.

Populations

Members of a species may be reproductively isolated in separate populations.

A population is a group of organisms of the same species who live in the same area at the same time. If two populations live in different areas they are unlikely to interbreed with each other. This does not mean that they are different species. If they potentially could interbreed, they are still members of the same species.

If two populations of a species never interbreed then they may gradually develop differences in their characters. Even if there are recognizable differences, they are considered to be the same species until they cannot interbreed and produce fertile offspring. In practice it can be very difficult to decide whether two populations have reached this point and biologists sometimes disagree about whether populations are the same or different species.

Autotrophic and heterotrophic nutrition

Species have either an autotrophic or heterotrophic method of nutrition (a few species have both methods).

All organisms need a supply of organic nutrients, such as glucose and amino acids. They are needed for growth and reproduction. Methods of obtaining these carbon compounds can be divided into two types:

- some organisms make their own carbon compounds from carbon dioxide and other simple substances – they are autotrophic, which means self-feeding;

- some organisms obtain their carbon compounds from other organisms – they are heterotrophic, which means feeding on others.

Some unicellular organisms use both methods of nutrition. *Euglena gracilis* for example has chloroplasts and carries out photosynthesis when there is sufficient light, but can also feed on detritus or smaller organisms by endocytosis. Organisms that are not exclusively autotrophic or heterotrophic are mixotrophic.

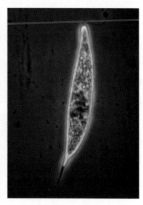

▲ Figure 3 *Arabidopsis thaliana* –the autotroph that molecular biologists use as a model plant

▲ Figure 4 Humming birds are heterotrophic; the plants from which they obtain nectar are autotrophic

▲ Figure 5 *Euglena* – an unusual organism as it can feed both autotrophically and heterotrophically

🔬 Trends in plant and algal nutrition

Looking for patterns, trends and discrepancies: plants and algae are mostly autotrophic but some are not.

Almost all plants and algae are autotrophic – they make their own complex organic compounds using carbon dioxide and other simple substances. A supply of energy is needed to do this, which plants and algae obtain by absorbing light. Their method of autotrophic nutrition is therefore photosynthesis and they carry it out in chloroplasts.

This trend for plants and algae to make their own carbon compounds by photosynthesis in chloroplasts is followed by the majority of species. However there are small numbers of both plants and algae that do not fit the trend, because although they are recognizably plants or algae, they

Activity

Galápagos tortoises

The tortoises that live on the Galápagos islands are the largest in the world. They have sometimes been grouped together into one species, *Chelinoidis nigra*, but more recently have been split into separate species.

Discuss whether each of these observations indicates that populations on the various islands are separate species:

- The Galápagos tortoises are poor swimmers and cannot travel from one island to another so they do not naturally interbreed.

- Tortoises from different islands have recognizable differences in their characters, including shell size and shape.

- Tortoises from different islands have been mated in zoos and hybrid offspring have been produced but they have lower fertility and higher mortality than the offspring of tortoises from the same island.

▲ Figure 2 Galápagos tortoise

do not contain chloroplasts and they do not carry out photosynthesis. These species grow on other plants, obtain carbon compounds from them and cause them harm. They are therefore parasitic.

To decide whether parasitic plants falsify the theory that plants and algae are groups of autotrophic species or whether they are just minor and insignificant discrepancies we need to consider how many species there are and how they evolved.

- The number of parasitic plants and algae is relatively small – only about 1% of all plant and algal species.

- It is almost certain that the original ancestral species of plant and alga were autotrophic and that the parasitic species evolved from them. Chloroplasts can quite easily be lost from cells, but cannot easily be developed. Also, parasitic species are diverse and occur in many different families. This pattern suggests that parasitic plants have evolved repeatedly from photosynthetic species.

Because of this evidence, ecologists regard plants and algae as groups of autotrophs, with a small number of exceptional species that are parasitic.

Data-based questions: Unexpected diets

Although we usually expect plants to be autotrophs and animals to be consumers, living organisms are very varied and do not always conform to our expectations. Figures 6 to 9 show four organisms with diets that are unexpected.

1 Which of the organisms is autotrophic?   [1]

2 Which of the organisms is heterotrophic?   [1]

3 Of the organisms that are heterotrophic, deduce which is a consumer, which a detritivore and which a saprotroph. [4]

▲ Figure 6 Venus fly trap: grows in swamps, with green leaves that carry out photosynthesis and also catch and digest insects, to provide a supply of nitrogen

▲ Figure 7 Ghost orchid: grows underground in woodland, feeding off dead organic matter, occasionally growing a stem with flowers above ground

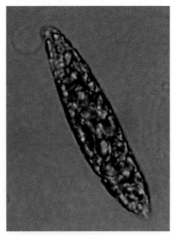

▲ Figure 8 *Euglena*: unicell that lives in ponds, using its chloroplasts for photosynthesis, but also ingesting dead organic matter by endocytosis

▲ Figure 9 Dodder: grows parasitically on gorse bushes, using small root-like structures to obtain sugars, amino acids and other substances it requires, from the gorse

Consumers

Consumers are heterotrophs that feed on living organisms by ingestion.

Heterotrophs are divided into groups by ecologists according to the source of organic molecules that they use and the method of taking them in. One group of heterotrophs is called consumers.

Consumers feed off other organisms. These other organisms are either still alive or have only been dead for a relatively short time. A mosquito sucking blood from a larger animal is a consumer that feeds on an organism that is still alive. A lion feeding off a gazelle that it has killed is a consumer.

Consumers ingest their food. This means that they take in undigested material from other organisms. They digest it and absorb the products of digestion. Unicellular consumers such as *Paramecium* take the food in by endocytosis and digest it inside vacuoles. Multicellular consumers such as lions take food into their digestive system by swallowing it.

Consumers are sometimes divided up into trophic groups according to what other organisms they consume. Primary consumers feed on autotrophs; secondary consumers feed on primary consumers and so on. In practice, most consumers do not fit neatly into any one of these groups because their diet includes material from a variety of trophic groups.

▲ Figure 10 Red kite *(Milvus milvus)* is a consumer that feeds on live prey but also on dead animal remains (carrion)

▲ Figure 11 Yellow-necked mouse *(Apodemus flavicollis)* is a consumer that feeds mostly on living plant matter, especially seeds, but also on living invertebrates

Detritivores

Detritivores are heterotrophs that obtain organic nutrients from detritus by internal digestion.

Organisms discard large quantities of organic matter, for example:

- dead leaves and other parts of plants
- feathers, hairs and other dead parts of animal bodies
- feces from animals.

This dead organic matter rarely accumulates in ecosystems and instead is used as a source of nutrition by two groups of heterotroph – detritivores and saprotrophs.

Detritivores ingest dead organic matter and then digest it internally and absorb the products of digestion. Large multicellular detritivores such as earthworms ingest the dead matter into their gut. Unicellular organisms ingest it into food vacuoles. The larvae of dung beetles feed by ingestion of feces rolled into a ball by their parent.

Saprotrophs

Saprotrophs are heterotrophs that obtain organic nutrients from dead organic matter by external digestion.

Saprotrophs secrete digestive enzymes into the dead organic matter and digest it externally. They then absorb the products of digestion. Many types of bacteria and fungi are saprotrophic. They are also known as decomposers because they break down carbon compounds in dead organic matter and release elements such as nitrogen into the ecosystem so that they can be used again by other organisms.

▲ Figure 12 Saprotrophic fungi growing over the surfaces of dead leaves and decomposing them by secreting digestive enzymes

Activity

Clearcutting

▲ Figure 14

In a classic essay written in 1972, the physicist Philip Anderson stated this:

The ability to reduce everything to simple fundamental laws does not imply the ability to start from those laws and reconstruct the universe. At each level of complexity entirely new properties appear.

Clearcutting is the most common and economically profitable form of logging. It involves clearing every tree in an area so that no canopy remains. With reference to the concept of emergent properties, suggest why the ecological community often fails to recover after clearcutting.

Identifying modes of nutrition

Classifying species as autotrophs, consumers, detritivores or saprotrophs from a knowledge of their mode of nutrition.

By answering a series of simple questions about an organism's mode of nutrition it is usually possible to deduce what trophic group it is in. These questions are presented here as a dichotomous key, which consists of a series of pairs of choices. The key works for unicellular and multicellular organisms but does not work for parasites such as tapeworms or fungi that cause diseases in plants. All multicellular autotrophs are photosynthetic and have chloroplasts containing chlorophyll.

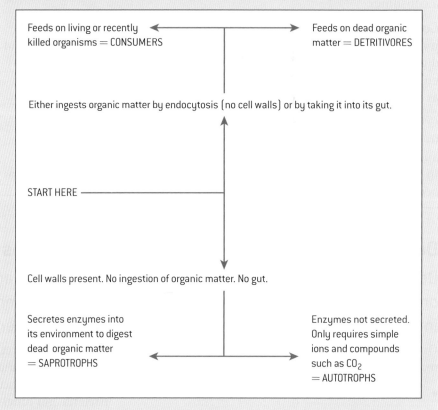

Communities

A community is formed by populations of different species living together and interacting with each other.

An important part of ecology is research into relationships between organisms. These relationships are complex and varied. In some cases the interaction between two species is of benefit to one species and harms the other, for example the relationship between a parasite and its host. In other cases both species benefit, as when a hummingbird feeds on nectar from a flower and helps the plant by pollinating it.

All species are dependent on relationships with other species for their long-term survival. For this reason a population of one species can never live in isolation. Groups of populations live together. A group of populations living together in an area and interacting with each other is known in ecology as a community. Typical communities consist of hundreds or even thousands of species living together in an area.

▲ Figure 13 A coral reef is a complex community with many interactions between the populations. Most corals have photosynthetic unicellular algae called zooxanthellae living inside their cells

 Field work – associations between species

Testing for association between two species using the chi-squared test with data obtained by quadrat sampling.

Quadrats are square sample areas, usually marked out using a quadrat frame. Quadrat sampling involves repeatedly placing a quadrat frame at random positions in a habitat and recording the numbers of organisms present each time.

The usual procedure for randomly positioning quadrats is this:

- A base line is marked out along the edge of the habitat using a measuring tape. It must extend all the way along the edge of the habitat.

- Random numbers are obtained using either a table or a random number generator on a calculator.

- A first random number is used to determine a distance along the measuring tape. All distances along the tape must be equally likely.

- A second random number is used to determine a distance out across the habitat at right angles to the tape. All distances across the habitat must be equally likely.

- The quadrat is placed precisely at the distances determined by the two random numbers.

If this procedure is followed correctly, with a large enough number of replicates, reliable estimates of

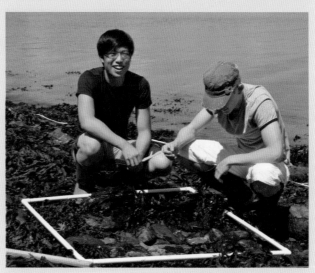

▲ Figure 15 Quadrat sampling of seaweed populations on a rocky shore

population sizes are obtained. The method is only suitable for plants and other organisms that are not motile. Quadrat sampling is not suitable for populations of most animals, for obvious reasons.

If the presence or absence of more than one species is recorded in every quadrat during sampling of a habitat, it is possible to test for an association between species. Populations are often unevenly distributed because some parts of the habitat are more suitable for a species than others. If two species occur in the same parts of a habitat, they will tend to be found in the same quadrats. This is known as a positive association. There can also be negative associations, or the distribution of two species can be independent.

There are two possible hypotheses:

H_0: two species are distributed independently (the null hypothesis).

H_1: two species are associated (either positively so they tend to occur together or negatively so they tend to occur apart).

We can test these hypotheses using a statistical procedure – the chi-squared test.

The chi-squared test is only valid if all the expected frequencies are 5 or larger and the sample was taken at random from the population.

Method for chi-squared test

1 Draw up a contingency table of observed frequencies, which are the numbers of quadrats containing or not containing the two species.

	Species A present	Species A absent	Row totals
Species B present			
Species B absent			
Column totals			

Calculate the row and column totals. Adding the row totals or the column totals should give the same grand total in the lower right cell.

2 Calculate the expected frequencies, assuming independent distribution, for each of the four species combinations. Each expected frequency is calculated from values on the contingency table using this equation:

$$\text{expected frequency} = \frac{\text{row total} \times \text{column total}}{\text{grand total}}$$

3 Calculate the number of degrees of freedom using this equation.

$$\text{degrees of freedom} = (m - 1)(n - 1)$$

where m and n are the number of rows and number of columns in the contingency table.

4 Find the critical region for chi-squared from a table of chi-squared values, using the degrees of freedom that you have calculated and a significance level (p) of 0.05 (5%). The critical region is any value of chi-squared larger than the value in the table.

5 Calculate chi-squared using this equation:

$$X^2 = \Sigma \frac{(f_o - f_e)^2}{f_e}$$

where f_o is the observed frequency

f_e is the expected frequency and

Σ is the sum of.

6 Compare the calculated value of chi-squared with the critical region.

- If the calculated value is in the critical region, there is evidence at the 5% level for an association between the two species. We can reject the hypothesis H_0.

- If the calculated value is not in the critical region, because it is equal or below the value obtained from the table of chi-squared values, H_0 is not rejected. There is no evidence at the 5% level for an association between the two species.

Data-based questions: Chi-squared testing

Figure 16 shows an area on the summit of Caer Caradoc, a hill in Shropshire, England.

The area is grazed by sheep in summer and hill walkers cross it on grassy paths. There are raised hummocks with heather (*Calluna vulgaris*) growing in them. A visual survey of this site suggested that *Rhytidiadelphus squarrosus*, a species of moss growing in this area, was associated with these heather hummocks. The presence or absence of the heather and the moss was recorded in a sample of 100 quadrats, positioned randomly.

Results

Species	Frequency
Heather only	9
Moss only	7
Both species	57
Neither species	27

Questions

1   Construct a contingency table of observed values. [4]

2   Calculate the expected values, assuming no association between the species. [4]

3   Calculate the number of degrees of freedom. [2]

4   Find the critical region for chi-squared at a significance level of 5%. [2]

5   Calculate chi-squared. [4]

6   State the two alternative hypotheses, H_0 and H_1, and evaluate them using the calculated value for chi-squared. [4]

7   Suggest ecological reasons for an association between the heather and the moss. [4]

8   Explain the methods that should have been used to position quadrats randomly in the area of study. [3]

▲ Figure 16 Caer Caradoc, Shropshire

Ⓐ Statistical significance

Recognizing and interpreting statistical significance.

Biologists often use the phrase "statistically significant" when discussing results of an experiment. This refers to the outcome of a statistical hypothesis test. There are two alternative types of hypothesis:

- H_0 is the null hypothesis and is the belief that there is no relationship, for example that two means are equal or that there is no association or correlation between two variables.

- H_1 is the alternative hypothesis and is the belief that there is a relationship, for example that two means are different or that there is an association between two variables.

The usual procedure is to test the null hypothesis, with the expectation of showing that it is false. A statistic is calculated using the results of the research and is compared with a range of possible values called the critical region. If the calculated statistic exceeds the critical region, the null hypothesis is considered to be false and is therefore rejected, though we cannot say that this has been proved with certainty.

When a biologist states that results were statistically significant it means that if the null hypothesis (H_0) was true, the probability of getting results as extreme as the observed results would be very small. A decision has to be made about how small this probability needs to be. This is known as the significance level. It is the cut-off point for the probability of rejecting the null hypothesis when in fact it was true. A level of 5% is often chosen, so the probability is less than one in twenty. That is the minimum acceptable significance level in published research.

- If there is a difference between the mean results for the two treatments in an experiment, a statistical test will show whether the difference is significant at the 5% level. If it is, there is a less than 5% probability of such a large difference between the sample means arising by chance, even when the population means are equal. We say that there is statistically significant evidence that the population means differ.

- In the example of testing for an association between two species, described on previous pages, the chi-squared test shows whether there is a less than 5% probability of the difference between the observed and the expected results being as large as it is without the species being either positively or negatively associated.

When results of biological research are displayed on a bar chart, letters are often used to indicate statistical significance. Two different letters, usually *a* and *b*, indicate mean results with a statistically significant difference. Two of the same letter such as *a* and *a* indicates that any difference is not statistically significant.

Ecosystems

A community forms an ecosystem by its interactions with the abiotic environment.

A community is composed of all organisms living in an area. These organisms could not live in isolation – they depend on their non-living surroundings of air, water, soil or rock. Ecologists refer to these surroundings as the abiotic environment.

In some cases the abiotic environment exerts a powerful influence over the organisms. For example the wave action on a rocky shore creates a very specialized habitat and only organisms adapted to it can survive. On cliffs, the rock type determines whether there are ledges on which birds can nest.

There are also many cases where living organisms influence the abiotic environment. Sand dunes are an example of this. They develop along coasts where sand is blown up the shore and specialized plants grow in the loose wind-blown sand. The roots of these plants stabilize the sand and their leaves break the wind and encourage more sand to be deposited.

So, not only are there complex interactions within communities, there are also many interactions between organisms and the abiotic environment. The community of organisms in an area and their non-living environment can therefore be considered to be a single highly complex interacting system, known as an ecosystem. Ecologists study both the components of ecosystems and the interactions between them.

Inorganic nutrients

Autotrophs and heterotrophs obtain inorganic nutrients from the abiotic environment.

Living organisms need a supply of chemical elements:

- Carbon, hydrogen and oxygen are needed to make carbohydrates, lipids and other carbon compounds on which life is based.

▲ Figure 17 Grasses in an area of developing sand dunes

- Nitrogen and phosphorus are also needed to make many of these compounds.

- Approximately fifteen other elements are needed by living organisms. Some of them are used in minute traces only, but they are nonetheless essential.

Autotrophs obtain all of the elements that they need as inorganic nutrients from the abiotic environment, including carbon and nitrogen. Heterotrophs on the other hand obtain these two elements and several others as part of the carbon compounds in their food. They do however obtain other elements as inorganic nutrients from the abiotic environment, including sodium, potassium and calcium.

Nutrient cycles

The supply of inorganic nutrients is maintained by nutrient cycling.

There are limited supplies on Earth of chemical elements. Although living organisms have been using the supplies for three billion years, they have not run out. This is because chemical elements can be endlessly recycled. Organisms absorb the elements that they require as inorganic nutrients from the abiotic environment, use them and then return them to the environment with the atoms unchanged.

Recycling of chemical elements is rarely as simple as shown in this diagram and often an element is passed from organism to organism before it is released back into the abiotic environment. The details vary from element to element. The carbon cycle is different from the nitrogen cycle for example. Ecologists refer to these schemes collectively as nutrient cycles. The word nutrient is often ambiguous in biology but in this context it simply means an element that an organism needs. The carbon cycle is described as an example of a nutrient cycle in sub-topic 4.2 and the nitrogen cycle in Option C.

Sustainability of ecosystems

Ecosystems have the potential to be sustainable over long periods of time.

The concept of sustainability has risen to prominence recently because it is clear that some current human uses of resources are unsustainable. Something is sustainable if it can continue indefinitely. Human use of fossil fuels is an example of an unsustainable activity. Supplies of fossil fuels are finite, are not currently being renewed and cannot therefore carry on indefinitely.

Natural ecosystems can teach us how to live in a sustainable way, so that our children and grandchildren can live as we do. There are three requirements for sustainability in ecosystems:

- nutrient availability

- detoxification of waste products

- energy availability.

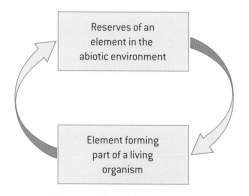

▲ Figure 18 Living organisms have been recycling for billions of years

▲ Figure 19 Sunlight supplies energy to a forest ecosystem and nutrients are recycled

Activity

Cave ecosystems

Organisms have been found living in total darkness in caves, including eyeless fish. Discuss whether ecosystems in dark caves are sustainable.

Figure 20 shows a small ecosystem with photosynthesizing plants near artificial lighting in a cave that is open to visitors in Cheddar Gorge. Discuss whether this is more or less sustainable than ecosystems in dark caves.

▲ Figure 20

Nutrients can be recycled indefinitely and if this is done there should not be a lack of the chemical elements on which life is based. The waste products of one species are usually exploited as a resource by another species. For example, ammonium ions released by decomposers are absorbed and used for an energy source by *Nitrosomonas* bacteria in the soil. Ammonium is potentially toxic but because of the action of these bacteria it does not accumulate.

Energy cannot be recycled, so sustainability depends on continued energy supply to ecosystems. Most energy is supplied to ecosystems as light from the sun. The importance of this supply can be illustrated by the consequences of the eruption of Mount Tambora in 1815. Dust in the atmosphere reduced the intensity of sunlight for some months afterwards, causing crop failures globally and deaths due to starvation. This was only a temporary phenomenon, however, and energy supplies to ecosystems in the form of sunlight will continue for billions of years.

🧪 Mesocosms

Setting up sealed mesocosms to try to establish sustainability. (Practical 5)

Mesocosms are small experimental areas that are set up as ecological experiments. Fenced-off enclosures in grassland or forest could be used as terrestrial mesocosms; tanks set up in the laboratory can be used as aquatic mesocosms. Ecological experiments can be done in replicate mesocosms, to find out the effects of varying one or more conditions. For example, tanks could be set up with and without fish, to investigate the effects of fish on aquatic ecosystems.

Another possible use of mesocosms is to test what types of ecosystems are sustainable. This involves sealing up a community of organisms together with air and soil or water inside a container.

You should consider these questions before setting up either aquatic or terrestrial mesocosms:

- Large glass jars are ideal but transparent plastic containers could also be used. Should the sides of the container be transparent or opaque?

- Which of these groups of organisms must be included to make up a sustainable community: autotrophs, consumers, saprotrophs and detritivores?

- How can we ensure that the oxygen supply is sufficient for all the organisms in the mesocosm as once it is sealed, no more oxygen will be able to enter.

- How can we prevent any organisms suffering as a result of being placed in the mesocosm?

4.2 Energy flow

Understanding

→ Most ecosystems rely on a supply of energy from sunlight.

→ Light energy is converted to chemical energy in carbon compounds by photosynthesis.

→ Chemical energy in carbon compounds flows through food chains by means of feeding.

→ Energy released by respiration is used in living organisms and converted to heat.

→ Living organisms cannot convert heat to other forms of energy.

→ Heat is lost from ecosystems.

→ Energy losses between trophic levels restrict the length of food chains and the biomass of higher trophic levels.

Skills

→ Quantitative representations of energy flow using pyramids of energy.

Nature of science

→ Use theories to explain natural phenomena: the concept of energy flow explains the limited length of food chains.

Sunlight and ecosystems

Most ecosystems rely on a supply of energy from sunlight.

For most biological communities, the initial source of energy is sunlight. Living organisms can harvest this energy by photosynthesis. Three groups of autotroph carry out photosynthesis: plants, eukaryotic algae including seaweeds that grow on rocky shores, and cyanobacteria. These organisms are often referred to by ecologists as producers.

Heterotrophs do not use light energy directly, but they are indirectly dependent on it. There are several groups of heterotroph in ecosystems: consumers, saprotrophs and detritivores. All of them use carbon compounds in their food as a source of energy. In most ecosystems all or almost all energy in the carbon compounds will originally have been harvested by photosynthesis in producers.

The amount of energy supplied to ecosystems in sunlight varies around the world. The percentage of this energy that is harvested by producers and therefore available to other organisms also varies. In the Sahara Desert, for example, the intensity of sunlight is very high but little of it becomes available to organisms because there are very few producers. In the redwood forests of California the intensity of sunlight is less than in the Sahara but much more energy becomes available to organisms because producers are abundant.

213

Activity

Cyanobacteria in caves

Cyanobacteria are photosynthetic bacteria that are often very abundant in marine and freshwater ecosystems. Figure 1 shows an area of green cyanobacteria on an area of wall in a cave that is illuminated by artificial light. The surrounding areas are normally dark. If the artificial light was not present, what other energy sources could be used by bacteria in caves?

▲ Figure 1

Data-based questions: Insolation

Insolation is a measure of solar radiation The two maps in figure 2 show annual mean insolation at the top of the Earth's atmosphere (upper map) and at the Earth's surface (lower map).

Questions

1　State the relationship between distance from the equator and insolation at the top of the Earth's atmosphere. [1]

2　State the mean annual insolation in Watts per square metre for the most northerly part of Australia

   a)　at the top of the atmosphere [1]

   b)　at the Earth's surface. [1]

3　Suggest reasons for differences in insolation at the Earth's surface between places that are at the same distance from the equator. [2]

4　Tropical rainforests are found in equatorial regions of all continents. They have very high rates of photosynthesis. Evaluate the hypothesis that this is due to very high insolation. Include named parts of the world in your answer. [5]

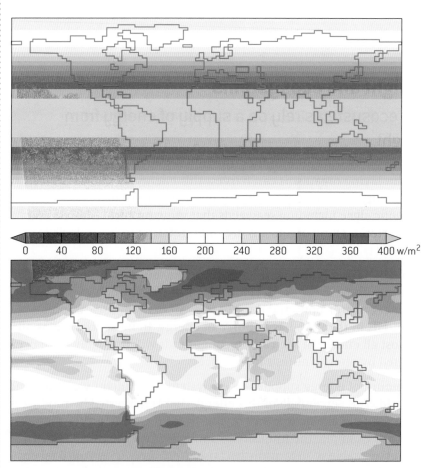

▲ Figure 2

Energy conversion

Light energy is converted to chemical energy in carbon compounds by photosynthesis.

Producers absorb sunlight using chlorophyll and other photosynthetic pigments. This converts the light energy to chemical energy, which is used to make carbohydrates, lipids and all the other carbon compounds in producers.

Producers can release energy from their carbon compounds by cell respiration and then use it for cell activities. Energy released in this way is eventually lost to the environment as waste heat. However, only some of the carbon compounds in producers are used in this way and the largest part remains in the cells and tissues of producers. The energy in these carbon compounds is available to heterotrophs.

Energy in food chains

Chemical energy in carbon compounds flows through food chains by means of feeding.

A food chain is a sequence of organisms, each of which feeds on the previous one. There are usually between two and five organisms in a food chain. It is rare for there to be more organisms in the chain. As they do not obtain food from other organisms, producers are always the first organisms in a food chain. The subsequent organisms are consumers. Primary consumers feed on producers; secondary consumers feed on primary consumers; tertiary consumers feed on secondary consumers, and so on. No consumers feed on the last organism in a food chain. Consumers obtain energy from the carbon compounds in the organisms on which they feed. The arrows in a food chain therefore indicate the direction of energy flow.

Figure 4 is an example of a food chain from the forests around Iguazu falls in northern Argentina.

▲ Figure 4

Respiration and energy release

Energy released by respiration is used in living organisms and converted to heat.

Living organisms need energy for cell activities such as these:

- Synthesizing large molecules like DNA, RNA and proteins.

- Pumping molecules or ions across membranes by active transport.

- Moving things around inside the cell, such as chromosomes or vesicles, or in muscle cells the protein fibres that cause muscle contraction.

ATP supplies energy for these activities. Every cell produces its own ATP supply.

Activity

Bush and forest fires

▲ Figure 3

Figure 3 shows a bush fire in Australia.

What energy conversion is happening in a bush fire?

Bush and forest fires occur naturally in some ecosystems.

Suggest two reasons for this hypothesis: There are fewer heterotrophs in ecosystems where fires are common compared to ecosystems where fires are not common.

All cells can produce ATP by cell respiration. In this process carbon compounds such as carbohydrates and lipids are oxidized. These oxidation reactions are exothermic and the energy released is used in endothermic reactions to make ATP. So cell respiration transfers chemical energy from glucose and other carbon compounds to ATP. The reason for doing this is that the chemical energy in carbon compounds such as glucose is not immediately usable by the cell, but the chemical energy in ATP can be used directly for many different activities.

The second law of thermodynamics states that energy transformations are never 100% efficient. Not all of the energy from the oxidation of carbon compounds in cell respiration is transferred to ATP. The remainder is converted to heat. Some heat is also produced when ATP is used in cell activities. Muscles warm up when they contract for example. Energy from ATP may reside for a time in large molecules when they have been synthesized, such as DNA and proteins, but when these molecules are eventually digested the energy is released as heat.

Data-based questions

Figure 5 shows the results of an experiment in which yellow-billed magpies *(Pica nuttalli)* were put in a cage in which the temperature could be controlled. The birds' rate of respiration was measured at seven different temperatures, from −10 °C to +40 °C. Between −10 °C and 30 °C the magpies maintained constant body temperature, but above 30 °C body temperature increased.

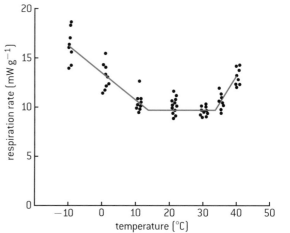

▲ Figure 5 Cell respiration rates at different temperatures in yellow-billed magpies

a) Describe the relationship between external temperature and respiration rate in yellow-billed magpies. [3]

b) Explain the change in respiration rate as temperature drops from +10 °C to −10 °C. [3]

c) Suggest a reason for the change in respiration rate as temperature increased from 30 °C to 40 °C. [2]

d) Suggest two reasons for the variation in respiration rate between the birds at each temperature. [2]

Heat energy in ecosystems

Living organisms cannot convert heat to other forms of energy.

Living organisms can perform various energy conversions:

- Light energy to chemical energy in photosynthesis.

- Chemical energy to kinetic energy in muscle contraction.

- Chemical energy to electrical energy in nerve cells.

- Chemical energy to heat energy in heat-generating adipose tissue.

They cannot convert heat energy into any other form of energy.

Heat losses from ecosystems

Heat is lost from ecosystems.

Heat resulting from cell respiration makes living organisms warmer. This heat can be useful in making cold-blooded animals more active. Birds and mammals increase their rate of heat generation if necessary to maintain their constant body temperatures.

According to the laws of thermodynamics in physics, heat passes from hotter to cooler bodies, so heat produced in living organisms is all eventually lost to the abiotic environment. The heat may remain in the ecosystem for a while, but ultimately is lost, for example when heat is radiated into the atmosphere. Ecologists assume that all energy released by respiration for use in cell activities will ultimately be lost from an ecosystem.

 Explaining the length of food chains

Use theories to explain natural phenomena: the concept of energy flow explains the limited length of food chains.

If we consider the diet of a top carnivore that is at the end of a food chain, we can work out how many stages there are in the food chain leading up to it. For example, if an osprey feeds on fish such as salmon that fed on shrimps, which fed on phytoplankton, there are four stages in the food chain.

There are rarely more than four or five stages in a food chain. We might expect food chains to be limitless, with one species being eaten by another ad infinitum. This does not happen. In ecology, as in all branches of science, we try to explain natural phenomena such as the restricted length of food chains using scientific theories. In this case it is the concept of energy flow along food chains and the energy losses that occur between trophic levels that can provide an explanation.

Energy losses and ecosystems

Energy losses between trophic levels restrict the length of food chains and the biomass of higher trophic levels.

Biomass is the total mass of a group of organisms. It consists of the cells and tissues of those organisms, including the carbohydrates and other carbon compounds that they contain. Because carbon compounds have chemical energy, biomass has energy. Ecologists can measure how much energy is added per year by groups of organisms to their biomass. The results are calculated per square metre of the ecosystem so that different trophic levels can be compared. When this is done, the same trend is always found: the energy added to biomass by each successive trophic level is less. In secondary consumers, for example, the amount of energy is always less per year per square metre of ecosystem than in primary consumers.

The reason for this trend is loss of energy between trophic levels.

- Most of the energy in food that is digested and absorbed by organisms in a trophic level is released by them in respiration for

Activity

Thinking about energy changes

What energy conversions are required to shoot a basketball?

What is the final form of the energy?

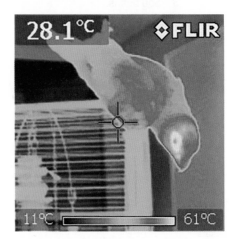

▲ Figure 6 An infrared camera image of an African grey parrot (*Psittacus erithacus*) shows how much heat is being released to the environment by different parts of its body

▲ Figure 7 The osprey (*Pandion halietus*) is a fish-eating top carnivore

217

Activity

Salmon and soy

Most salmon eaten by humans is produced in fish farms. The salmon have traditionally been fed on fish meal, mostly based on anchovies harvested off the coast of South America. These have become scarce and expensive. Feeds based on plant products such as soy beans are increasingly being used. In terms of energy flow, which of these human diets is most and least efficient?

1 Salmon fed on fish meal

2 Salmon fed on soy beans

3 Soy beans.

use in cell activities. It is therefore lost as heat. The only energy available to organisms in the next trophic level is chemical energy in carbohydrates and other carbon compounds that have not been used up in cell respiration.

- The organisms in a trophic level are not usually entirely consumed by organisms in the next trophic level. For example, locusts sometimes consume all the plants in an area but more usually only parts of some plants are eaten. Predators may not eat material from the bodies of their prey such as bones or hair. Energy in uneaten material passes to saprotrophs or detritivores rather than passing to organisms in the next trophic level.

- Not all parts of food ingested by the organisms in a trophic level are digested and absorbed. Some material is indigestible and is egested in feces. Energy in feces does not pass on along the food chain and instead passes to saprotrophs or detritivores.

Because of these losses, only a small proportion of the energy in the biomass of organisms in one trophic level will ever become part of the biomass of organisms in the next trophic level. The figure of 10% is often quoted, but the level of energy loss between trophic levels is variable. As the losses occur at each stage in a food chain, there is less and less energy available to each successive trophic level. After only a few stages in a food chain the amount of energy remaining would not be enough to support another trophic level. For this reason the number of trophic levels in food chains is restricted.

Biomass, measured in grams, also diminishes along food chains, due to loss of carbon dioxide and water from respiration and loss from the food chain of uneaten or undigested parts of organisms. The biomass of higher trophic levels is therefore usually smaller than that of lower levels. There is generally a higher biomass of producers, the lowest trophic level of all, than of any other trophic level.

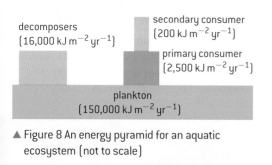

▲ Figure 8 An energy pyramid for an aquatic ecosystem (not to scale)

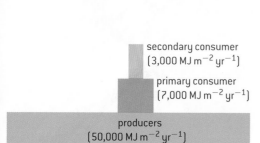

▲ Figure 9 Pyramid of energy for grassland

⚗ Pyramids of energy

Quantitative representations of energy flow using pyramids of energy.

The amount of energy converted to new biomass by each trophic level in an ecological community can be represented with a pyramid of energy. This is a type of bar chart with a horizontal bar for each trophic level. The amounts of energy should be per unit area per year. Often the units are kilojoules per metre squared per year ($kJ\ m^{-2}\ yr^{-1}$). The pyramid should be stepped, not triangular, starting with the producers in the lowest bar. The bars should be labelled producer, first consumer, second consumer and so on. If a suitable scale is chosen, the length of each bar can be proportional to the amount of energy that it shows.

Figure 8 shows an example of a pyramid of energy for an aquatic ecosystem. To be more accurate, the bars should be drawn with relative widths that match the relative energy content at each trophic level. Figure 9 shows a pyramid of energy for grassland, with the bars correctly to scale.

Data-based questions: A simple food web

A sinkhole is a surface feature which forms when an underground cavern collapses. Montezuma Well in the Sonoran desert in Arizona is a sinkhole filled with water. It is an aquatic ecosystem that lacks fish, due in part to the extremely high concentrations of dissolved CO_2. The dominant top predator is *Belostoma bakeri*, a giant water insect that can grow to 70 mm in length.

Figure 10 shows a food web for Montezuma Well.

1 Compare the roles of *Belostoma bakeri* and *Ranatra montezuma* within the food web. [2]

2 Deduce, with a reason, which organism occupies more than one trophic level. [2]

3 Deduce using P values:

a) what would be the most common food chain in this web [2]

b) what is the preferred prey of *B. bakeri*? [1]

4 Construct a pyramid of energy for the first and second trophic levels. [3]

5 Calculate the percentage of energy lost between the first and second trophic levels. [2]

6 Discuss the difficulties of classifying organisms into trophic levels. [2]

7 Outline the additional information that would be required to complete the pyramid of energy for the third and fourth trophic level. [1]

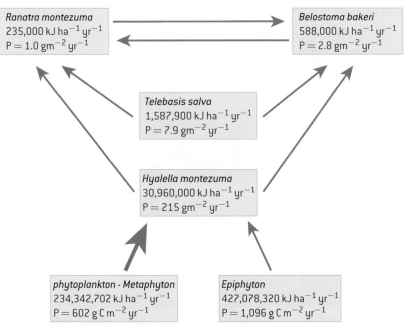

▲ Figure 10 A food web for Montezuma Well. P values represent the biomass stored in the population of that organism each year. Energy values represent the energy equivalent of that biomass. Arrows indicate trophic linkages and arrow thickness indicates the relative amount of energy transferred between trophic levels

219

4.3 Carbon cycling

Understanding

→ Autotrophs convert carbon dioxide into carbohydrates and other carbon compounds.

→ In aquatic habitats carbon dioxide is present as a dissolved gas and hydrogen carbonate ions.

→ Carbon dioxide diffuses from the atmosphere or water into autotrophs.

→ Carbon dioxide is produced by respiration and diffuses out of organisms into water or the atmosphere.

→ Methane is produced from organic matter in anaerobic conditions by methanogenic archaeans and some diffuses into the atmosphere.

→ Methane is oxidized to carbon dioxide and water in the atmosphere.

→ Peat forms when organic matter is not fully decomposed because of anaerobic conditions in waterlogged soils.

→ Partially decomposed organic matter from past geological eras was converted into oil and gas in porous rocks or into coal.

→ Carbon dioxide is produced by the combustion of biomass and fossilized organic matter.

→ Animals such as reef-building corals and molluscs have hard parts that are composed of calcium carbonate and can become fossilized in limestone.

Applications

→ Estimation of carbon fluxes due to processes in the carbon cycle.

→ Analysis of data from atmosphere monitoring stations showing annual fluctuations.

Skills

→ Construct a diagram of the carbon cycle.

Nature of science

→ Making accurate, quantitative measurements: it is important to obtain reliable data on the concentration of carbon dioxide and methane in the atmosphere.

Carbon fixation

Autotrophs convert carbon dioxide into carbohydrates and other carbon compounds.

Autotrophs absorb carbon dioxide from the atmosphere and convert it into carbohydrates, lipids and all the other carbon compounds that they require. This has the effect of reducing the carbon dioxide concentration of the atmosphere. The mean CO_2 concentration of the atmosphere is currently approximately 0.039% or 390 micromoles per mole (μmol/mol) but it is lower above parts of the Earth's surface where photosynthesis rates have been high.

Data-based questions: Carbon dioxide concentration

The two maps in figure 1 were produced by NASA. They show the carbon dioxide concentration of the atmosphere eight kilometres above the surface of the Earth, in May and October 2011.

1   State whether October is in the spring or fall(autumn) in the southern hemisphere. [1]

2   a) Distinguish between carbon dioxide concentrations in May and October in the northern hemisphere. [1]

     b) Suggest reasons for the difference. [2]

3   a) Distinguish between the carbon dioxide concentrations in May between the northern and the southern hemisphere. [1]

     b) Suggest reasons for the difference. [2]

4   a) Deduce the part of the Earth that had the lowest mean carbon dioxide concentration between May and October 2011. [1]

     b) Suggest reasons for the carbon dioxide concentration being lowest in this area. [2]

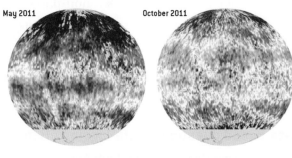

Carbon Dioxide 2011 Mole Fraction [µmol/mol]

388   389   390   391   392   393   394   395

▲ Figure 1

Carbon dioxide in solution

In aquatic habitats carbon dioxide is present as a dissolved gas and hydrogen carbonate ions.

Carbon dioxide is soluble in water. It can either remain in water as a dissolved gas or it can combine with water to form carbonic acid (H_2CO_3). Carbonic acid can dissociate to form hydrogen and hydrogen carbonate ions (H^+ and HCO_3^-). This explains how carbon dioxide can reduce the pH of water.

Both dissolved carbon dioxide and hydrogen carbonate ions are absorbed by aquatic plants and other autotrophs that live in water. They use them to make carbohydrates and other carbon compounds.

Absorption of carbon dioxide

Carbon dioxide diffuses from the atmosphere or water into autotrophs.

Autotrophs use carbon dioxide in the production of carbon compounds by photosynthesis or other processes. This reduces the concentration of carbon dioxide inside autotrophs and sets up a concentration gradient between cells in autotrophs and the air or water around. Carbon dioxide therefore diffuses from the atmosphere or water into autotrophs.

In land plants with leaves this diffusion usually happens through stomata in the underside of the leaves. In aquatic plants the entire surface of the leaves and stems is usually permeable to carbon dioxide, so diffusion can be through any part of these parts of the plant.

Activity

pH changes in rock pools

Ecologists have monitored pH in rock pools on sea shores that contain animals and also photosynthesizing algae. The pH of the water rises and falls in a 24-hour cycle, due to changes in carbon dioxide concentration in the water. The lowest values of about pH 7 have been found during the night, and the highest values of about pH 10 have been found when there was bright sunlight during the day. What are the reasons for these maxima and minima? The pH in natural pools or artificial aquatic mesocosms could be monitored using data loggers.

Release of carbon dioxide from cell respiration

Carbon dioxide is produced by respiration and diffuses out of organisms into water or the atmosphere.

Carbon dioxide is a waste product of aerobic cell respiration. It is produced in all cells that carry out aerobic cell respiration. These can be grouped according to trophic level of the organism:

- non-photosynthetic cells in producers for example root cells in plants
- animal cells
- saprotrophs such as fungi that decompose dead organic matter.

Carbon dioxide produced by respiration diffuses out of cells and passes into the atmosphere or water that surrounds these organisms.

Data-based questions: Data-logging pH in an aquarium

Figure 2 shows the pH and light intensity in an aquarium containing a varied community of organisms including pondweeds, newts and other animals. The data was obtained by data logging using a pH electrode and a light meter. The aquarium was illuminated artificially to give a 24-hour cycle of light and dark using a lamp controlled by a timer.

1 Explain the changes in light intensity during the experiment. [2]

2 Determine how many days the data logging covers. [2]

3 a) Deduce the trend in pH in the light. [1]

   b) Explain this trend. [2]

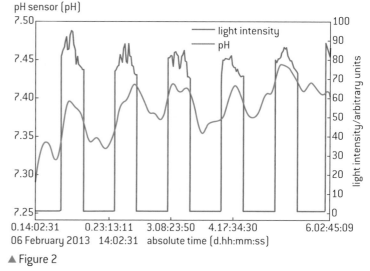

▲ Figure 2

4 a) Deduce the trend in pH in darkness. [1]

   b) Explain this trend. [2]

Methanogenesis

Methane is produced from organic matter in anaerobic conditions by methanogenic archaeans and some diffuses into the atmosphere.

In 1776 Alessandro Volta collected bubbles of gas emerging from mud in a reed bed on the margins of Lake Maggiore in Italy, and found that it was inflammable. He had discovered methane, though Volta did not give it this name. Methane is produced widely in anaerobic environments, as it is a waste product of a type of anaerobic respiration.

Three different groups of anaerobic prokaryotes are involved.

1 Bacteria that convert organic matter into a mixture of organic acids, alcohol, hydrogen and carbon dioxide.

2 Bacteria that use the organic acids and alcohol to produce acetate, carbon dioxide and hydrogen.

3 Archaeans that produce methane from carbon dioxide, hydrogen and acetate. They do this by two chemical reactions:

$$CO_2 + 4H_2 \rightarrow CH_4 + 2H_2O$$

$$CH_3COOH \rightarrow CH_4 + CO_2$$

The archaeans in this third group are therefore methanogenic. They carry out methanogenesis in many anaerobic environments:

- Mud along the shores and in the bed of lakes.

- Swamps, mires, mangrove forests and other wetlands where the soil or peat deposits are waterlogged.

- Guts of termites and of ruminant mammals such as cattle and sheep.

- Landfill sites where organic matter is in wastes that have been buried.

Some of the methane produced by archaeans in these anaerobic environments diffuses into the atmosphere. Currently the concentration in the atmosphere is between 1.7 and 1.85 micromoles per mole. Methane produced from organic waste in anaerobic digesters is not allowed to escape and instead is burned as a fuel.

▲ Figure 3 Waterlogged woodland—a typical habitat for methanogenic prokaryotes

Oxidation of methane

Methane is oxidized to carbon dioxide and water in the atmosphere.

Molecules of methane released into the atmosphere persist there on average for only 12 years, because it is naturally oxidized in the stratosphere. Monatomic oxygen (O) and highly reactive hydroxyl radicals (OH•) are involved in methane oxidation. This explains why atmospheric concentrations are not high, despite large amounts of production of methane by both natural processes and human activities.

Peat formation

Peat forms when organic matter is not fully decomposed because of anaerobic conditions in waterlogged soils.

In many soils all organic matter such as dead leaves from plants is eventually digested by saprotrophic bacteria and fungi. Saprotrophs obtain the oxygen that they need for respiration from air spaces in the soil. In some environments water is unable to drain out of soils so they become waterlogged and anaerobic. Saprotrophs cannot thrive in these conditions so dead organic matter is not fully decomposed. Acidic conditions tend to develop, further inhibiting saprotrophs and also methanogens that might break down the organic matter.

▲ Figure 4 Peat deposits form a blanket on a boggy hill top at Bwlch Groes in North Wales

Data-based questions: Release of carbon from tundra soils

Soils in tundra ecosystems typically contain large amounts of carbon in the form of peat. This accumulates because of low rates of decomposition of dead plant organic matter by saprotrophs. To investigate this, ecologists collected samples of soil from areas of tussock vegetation near Toolik Lake in Alaska. Some of the areas had been fertilized with nitrogen and phosphorus every year for the previous eight years (TF) and some had not (TC). The soils were incubated for 100-day periods at either 7 or 15°C. Some samples were kept moist (M) and others were saturated with water (W). The initial carbon content of the soils was measured and the amount of carbon dioxide given off during the experiment was monitored. The bar chart in figure 5 shows the results.

1 a) State the effect of increasing the temperature of the soils on the rate of release of carbon. [2]

  b) Explain the reasons for this effect. [2]

2 a) Compare the rates of release of carbon in moist soils with those in soils saturated with water. [2]

  b) Suggest reasons for the differences. [2]

3 Outline the effects of fertilizers on rates of release of carbon from the soils. [2]

4 Discuss whether differences in temperature, amount of water in the soil or amount of fertilizer have the greatest impact on the release of carbon. [2]

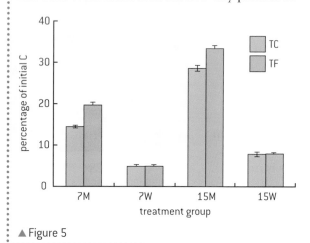

▲ Figure 5

Large quantities of partially decomposed organic matter have accumulated in some ecosystems and become compressed to form a dark brown acidic material called peat. About 3% of the Earth's land surface is covered by peat and as the depth is ten metres or more in some places, the total quantities of this material are immense.

Fossilized organic matter

Partially decomposed organic matter from past geological eras was converted into oil and gas in porous rocks or into coal.

Carbon and some compounds of carbon are chemically very stable and can remain unchanged in rocks for hundreds of millions of years. There are large deposits of carbon from past geological eras. These deposits are the result of incomplete decomposition of organic matter and its burial in sediments that became rock.

- Coal is formed when deposits of peat are buried under other sediments. The peat is compressed and heated, gradually turning into coal. Large coal deposits were formed during the Pennsylvanian sub-period of the Carboniferous. There was a cycle of sea level rises and falls; coastal swamps formed as the level fell and were destroyed and buried when the level rose and the sea spread inland. Each cycle has left a seam of coal.

▲ Figure 6 Coal at a power station

- Oil and natural gas are formed in the mud at the bottom of seas and lakes. Conditions are usually anaerobic and so decomposition is often incomplete. As more mud or other sediments are deposited the partially decomposed matter is compressed and heated. Chemical changes occur, which produce complex mixtures of liquid carbon compounds or gases. We call these mixtures crude oil and natural gas. Methane forms the largest part of natural gas. Deposits are found where there are porous rocks that can hold them such as shales and also impervious rocks above and below the porous rocks that prevent the deposit's escape.

Combustion

Carbon dioxide is produced by the combustion of biomass and fossilized organic matter.

If organic matter is heated to its ignition temperature in the presence of oxygen it will set light and burn. The oxidation reactions that occur are called combustion. The products of complete combustion are carbon dioxide and water.

In some parts of the world it is natural for there to be periodic fires in forests or grassland. Carbon dioxide is released from the combustion of the biomass in the forest or grassland. In these areas the trees and other organisms are often well adapted to fires and communities regenerate rapidly afterwards.

In other areas fires due to natural causes are very unusual, but humans sometimes cause them to occur. Fire is used to clear areas of tropical rainforest for planting oil palms or for cattle ranching. Crops of sugar cane are traditionally burned shortly before they are harvested. The dry leaves burn off, leaving the harvestable stems.

Coal, oil and natural gas are different forms of fossilized organic matter. They are all burned as fuels. The carbon atoms in the carbon dioxide released may have been removed from the atmosphere by photosynthesizing plants hundreds of millions of years ago.

Limestone

Animals such as reef-building corals and molluscs have hard parts that are composed of calcium carbonate and can become fossilized in limestone.

Some animals have hard body parts composed of calcium carbonate ($CaCO_3$):

- mollusc shells contain calcium carbonate;
- hard corals that build reefs produce their exoskeletons by secreting calcium carbonate.

When these animals die, their soft parts are usually decomposed quickly. In acid conditions the calcium carbonate dissolves away but in neutral or alkaline conditions it is stable and deposits of it from hard animal parts can form on the sea bed. In shallow tropical seas calcium

▲ Figure 7 Carbon dioxide is released by combustion of the leaves of sugar cane

▲ Figure 8 *Kodonophyllum*—a Silurian coral, in limestone from Wenlock Edge. The calcium carbonate skeletons of the coral are clearly visible embedded in more calcium carbonate that precipitated 420 million years ago in shallow tropical seas

▲ Figure 9 Chalk cliffs on the south coast of England. Chalk is a form of limestone that consists almost entirely of 90-million-year-old shells of tiny unicellular animals called *foraminifera*

carbonate is also deposited by precipitation in the water. The result is limestone rock, where the deposited hard parts of animals are often visible as fossils.

Approximately 10% of all sedimentary rock on Earth is limestone. About 12% of the mass of the calcium carbonate is carbon, so huge amounts of carbon are locked up in limestone rock on Earth.

🜂 Carbon cycle diagrams

Construct a diagram of the carbon cycle.

Ecologists studying the carbon cycle and the recycling of other elements use the terms pool and flux.

- A pool is a reserve of the element. It can be organic or inorganic. For example the carbon dioxide in the atmosphere is an inorganic pool of carbon. The biomass of producers in an ecosystem is an organic pool.

- A flux is the transfer of the element from one pool to another. An example of carbon flux is the absorption of carbon dioxide from the atmosphere and its conversion by photosynthesis to plant biomass.

Diagrams can be used to represent the carbon cycle. Text boxes can be used for pools and labeled arrows for fluxes. Figure 10 shows an illustrated diagram which can be converted to a diagram of text boxes and arrows.

Figure 10 only shows the carbon cycle for terrestrial ecosystems. A separate diagram could be constructed for marine or aquatic ecosystems, or a combined diagram for all ecosystems. In marine and aquatic ecosystems, the inorganic reserve of carbon is dissolved carbon dioxide and hydrogen carbonate, which is absorbed by producers and by various means is released back into the water.

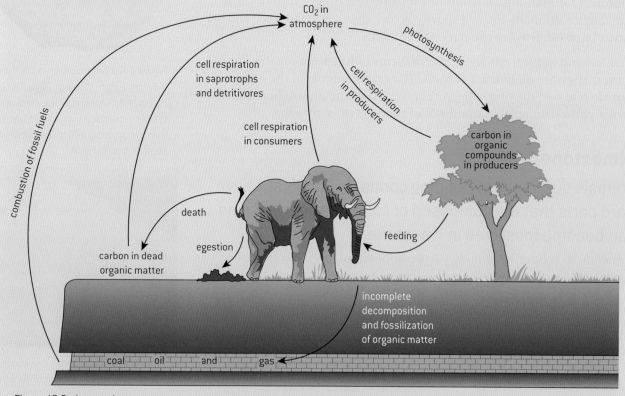

▲ Figure 10 Carbon cycle

Carbon fluxes

Estimation of carbon fluxes due to processes in the carbon cycle.

The carbon cycle diagram in figure 10 shows processes that transfer carbon from one pool to another but it does not show the quantities of these fluxes. It is not possible to measure global carbon fluxes precisely but as these quantities are of great interest, scientists have produced estimates for them. Estimates are based on many measurements in individual natural ecosystems or in mesocosms.

Global carbon fluxes are extremely large so estimates are in gigatonnes (petagrams). One gigatonne is 1×10^{15} grams. Table 1 shows estimates based on *Ocean Biogeochemical Dynamics*, Sarmiento and Gruber, 2006, Princeton University Press.

Process	Flux/gigatonnes year^{-1}
Photosynthesis	120
Cell respiration	119.6
Ocean uptake	92.8
Ocean loss	90.0
Deforestation and land use changes	1.6
Burial in marine sediments	0.2
Combustion of fossil fuels	6.4

▲ Table 1

Data-based questions: Oak woodland and carbon dioxide concentrations

Carbon fluxes have been measured since 1998 in deciduous woodland at Alice Holt Research Forest in England. The trees are mainly oaks, *Quercus robur* and *Quercus petraea*, with some ash, *Fraxinus excelsior*. They were planted in 1935 and are now nearly 20 metres tall.

Carbon dioxide concentrations are measured 20 times a second. From these measurements the net ecosystem production can be deduced. This is the net flux of carbon dioxide between the forest and the atmosphere. Positive values indicate an increase in the carbon pool of the forest and negative values indicate a decrease due to net loss of carbon dioxide. The graph shows the daily average net ecosystem production for several years and also the cumulative net ecosystem production.

1 Calculate whether the carbon pool in the biomass of the forest increases or decreases on more days in the year. [1]

2 Deduce the months in which the carbon pool of biomass in the forest was highest and lowest. [2]

3 Explain the reasons for increases in the carbon pool of biomass in the forest during part of the year and decreases in other parts. [4]

4 State the annual carbon flux to or from the forest. [2]

5 Suggest a reason based on the data for encouraging the planting of more oak forests. [1]

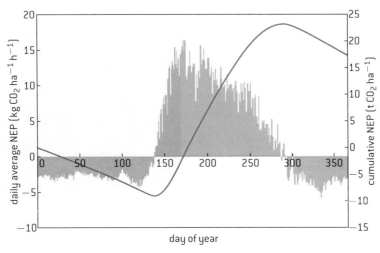

Environmental monitoring

Making accurate, quantitative measurements: it is important to obtain reliable data on the concentration of carbon dioxide and methane in the atmosphere.

Carbon dioxide and methane concentrations in the atmosphere have very important effects. Carbon dioxide concentrations affect photosynthesis rates and the pH of seawater. Both gases influence global temperatures and as a result the extent of ice sheets at the poles. Indirectly they therefore affect sea levels and the position of coast lines. Through their effects on the amount of heat energy in the oceans and the atmosphere they affect ocean currents, the distribution of rainfall and also the frequency and severity of extreme weather events such as hurricanes.

Consider these hypotheses and predictions:

- The carbon dioxide concentration of the atmosphere is currently higher than at any time in the past twenty million years.

- Human activities have increased the carbon dioxide and methane concentrations in the Earth's atmosphere.

- Human activity will cause atmospheric

carbon dioxide concentrations to rise from 397 micromoles per mole in 2014 to a level above 600 by the end of the century.

Reliable data are an essential prerequisite for evaluating hypotheses and predictions such as these. Reliable measurements of atmospheric carbon dioxide and methane concentration are needed over as long a period as possible before we can evaluate the past and possible future consequences of human activity.

Data on concentrations of gases in the atmosphere is collected by the Global Atmosphere Watch programme of the World Meteorological Organization, an agency of the United Nations. Research stations in various parts of the world now monitor the atmosphere, but Mauna Loa Observatory on Hawaii has records from the longest period. Carbon dioxide concentrations have been measured from 1959 onwards and methane from 1984. These and other reliable records are of immense value to scientists.

Trends in atmospheric carbon dioxide

Analysis of data from atmosphere monitoring stations showing annual fluctuations.

Data from atmosphere monitoring stations is freely available allowing any person to analyse it. There are both long-term trends and annual fluctuations in the data. The Mauna Loa Observatory in Hawaii produces vast amounts of data and data from this and other monitoring stations are available for analysis.

▲ Figure 11 Hawaii from space. Mauna Loa is near the centre of the largest island

4.4 Climate change

Understanding

→ Carbon dioxide and water vapour are the most significant greenhouse gases.

→ Other gases including methane and nitrogen oxides have less impact.

→ The impact of a gas depends on its ability to absorb long-wave radiation as well as on its concentration in the atmosphere.

→ The warmed Earth emits longer-wave radiation (heat).

→ Longer-wave radiation is reabsorbed by greenhouse gases which retains the heat in the atmosphere.

→ Global temperatures and climate patterns are influenced by concentrations of greenhouse gases.

→ There is a correlation between rising atmospheric concentrations of carbon dioxide since the start of the industrial revolution two hundred years ago and average global temperatures.

→ Recent increases in atmospheric carbon dioxide are largely due to increases in the combustion of fossilized organic matter.

Applications

→ Correlations between global temperatures and carbon dioxide concentrations on Earth.

→ Evaluating claims that human activities are not causing climate change.

→ Threats to coral reefs from increasing concentrations of dissolved carbon dioxide.

Nature of science

→ Assessing claims: assessment of the claims that human activities are not causing climate change.

Greenhouse gases

Carbon dioxide and water vapour are the most significant greenhouse gases.

The Earth is kept much warmer than it otherwise would be by gases in the atmosphere that retain heat. The effect of these gases has been likened to that of the glass that retains heat in a greenhouse and they are therefore known as greenhouse gases, though the mechanism of heat retention is not the same.

The greenhouse gases that have the largest warming effect on the Earth are carbon dioxide and water vapour.

- Carbon dioxide is released into the atmosphere by cell respiration in living organisms and also by combustion of biomass and fossil

▲ Figure 1 Satellite image of Hurricane Andrew in the Gulf of Mexico. Hurricanes are increasing in frequency and intensity as a result of increases in heat retention by greenhouse gases

fuels. It is removed from the atmosphere by photosynthesis and by dissolving in the oceans.

- Water vapour is formed by evaporation from the oceans and also transpiration in plants. It is removed from the atmosphere by rainfall and snow.

Water continues to retain heat after it condenses to form droplets of liquid water in clouds. The water absorbs heat energy and radiates it back to the Earth's surface and also reflects the heat energy back. This explains why the temperature drops so much more quickly at night in areas with clear skies than in areas with cloud cover.

Other greenhouse gases

Other gases including methane and nitrogen oxides have less impact.

Although carbon dioxide and water vapour are the most significant greenhouse gases there are others that have a smaller but nonetheless significant effect.

- Methane is the third most significant greenhouse gas. It is emitted from marshes and other waterlogged habitats and from landfill sites where organic wastes have been dumped. It is released during extraction of fossil fuels and from melting ice in polar regions.

- Nitrous oxide is another significant greenhouse gas. It is released naturally by bacteria in some habitats and also by agriculture and vehicle exhausts.

The two most abundant gases in the Earth's atmosphere, oxygen and nitrogen, are not greenhouse gases as they do not absorb longer-wave radiation. All of the greenhouse gases together therefore make up less than 1% of the atmosphere.

Assessing the impact of greenhouse gases

The impact of a gas depends on its ability to absorb long-wave radiation as well as on its concentration in the atmosphere.

Two factors together determine the warming impact of a greenhouse gas:

- how readily the gas absorbs long-wave radiation; and

- the concentration of the gas in the atmosphere.

For example, methane causes much more warming per molecule than carbon dioxide, but as it is at a much lower concentration in the atmosphere its impact on global warming is less.

The concentration of a gas depends on the rate at which it is released into the atmosphere and how long on average it remains there. The rate at which water vapour enters the atmosphere is immensely rapid, but it remains there only nine days on average, whereas methane remains in the atmosphere for twelve years and carbon dioxide for even longer.

Long-wavelength emissions from Earth

The warmed Earth emits longer-wave radiation.

The warmed surface of the Earth absorbs short-wave energy from the sun and then re-emits it, but at much longer wavelengths. Most of the re-emitted radiation is infrared, with a peak wavelength of 10,000 nm. The peak wavelength of solar radiation is 400 nm.

Figure 2 shows the range of wavelengths of solar radiation that pass through the atmosphere to reach the Earth's surface and warm it (red) and the range of much longer wavelengths emitted by the Earth that pass out through the atmosphere (blue). The smooth red and blue curves show the range of wavelengths expected to be emitted by bodies of the temperature of the Earth and the sun.

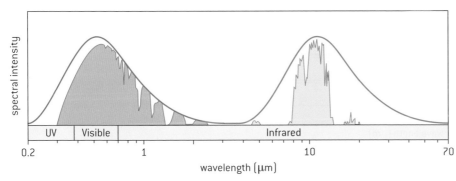

▲ Figure 2

TOK

Questions exist about the reality of scientific phenomenon. What consequences might this have for the public perception and understanding of science?

Much of what science investigates involves entities and concepts beyond everyday experience of the world, such as the nature and behaviour of electromagnetic radiation or the build-up of invisible gases in the atmosphere. This makes it difficult for scientists to convince the general public that such phenomenon actually exist – particularly when the consequences of accepting their existance might run counter to value systems or entrenched beliefs.

Greenhouse gases

Longer-wave radiation is reabsorbed by greenhouse gases which retains the heat in the atmosphere.

25–30% of the short-wavelength radiation from the sun that is passing through the atmosphere is absorbed before it reaches the Earth's surface. Most of the solar radiation absorbed is ultraviolet light, which is absorbed by ozone. 70–75% of solar radiation therefore reaches the Earth's surface and much of this is converted to heat.

A far higher percentage of the longer-wavelength radiation re-emitted by the surface of the Earth is absorbed before it has passed out to space. Between 70% and 85% is captured by greenhouse gases in the atmosphere. This energy is re-emitted, some towards the Earth. The effect is global warming. Without it the mean temperature at the Earth's surface would be about –18°C.

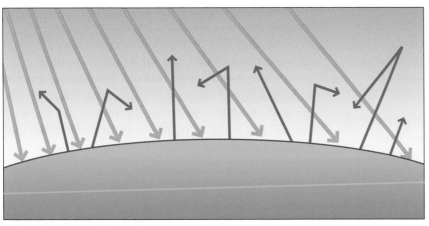

Key

→ short-wave radiation from the sun

→ long-wave radiation from earth

▲ Figure 3 The greenhouse effect

Greenhouse gases in the Earth's atmosphere only absorb energy in specific wavebands. Figure 4 below shows total percentage absorption of radiation by the atmosphere. The graph also shows the bands of wavelengths absorbed by individual gases. The wavelengths re-emitted by the Earth are between 5 and 70 μm. Water vapour, carbon dioxide, methane and nitrous oxide all absorb some of these wavelengths, so each of them is a greenhouse gas.

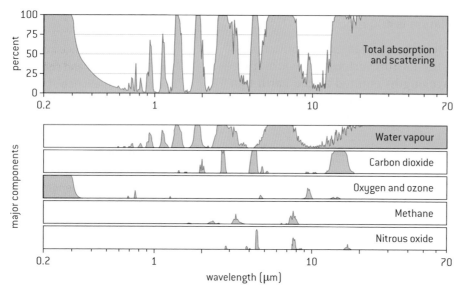

▲ Figure 4

Global temperatures and carbon dioxide concentrations

Correlations between global temperatures and carbon dioxide concentrations on Earth.

If the concentration of any of the greenhouse gases in the atmosphere changes, we can expect the size of its contribution to the greenhouse effect to change and global temperatures to rise or fall. We can test this hypothesis using the carbon dioxide concentration of the atmosphere, because it has changed considerably.

To deduce carbon dioxide concentrations and temperatures in the past, columns of ice have been drilled in the Antarctic. The ice has built up over thousands of years, so ice from deeper down is older than ice near the surface. Bubbles of air trapped in the ice can be extracted and analysed to find the carbon dioxide concentration. Global temperatures can be deduced from ratios of hydrogen isotopes in the water molecules.

Figure 5 shows results for an 800,000 year period before the present. They were obtained from an ice core drilled in Dome C on the Antarctic plateau by the European Project for Ice Coring in Antarctica. During this part of the current Ice Age there has been a repeating pattern of rapid periods of warming followed by much longer periods of gradual cooling. There is a very striking correlation between carbon dioxide concentration and global temperatures – the periods of higher carbon dioxide concentration repeatedly coincide with periods when the Earth was warmer.

The same trend has been found in other ice cores. Data of this type are consistent with the hypothesis that rises in carbon dioxide concentration increase the greenhouse effect. It is important always to remember that correlation does not prove causation, but in this case we know from other research that carbon dioxide is a greenhouse gas. At least some of the temperature variation over the past 800,000 years must therefore have been due to rises and falls in atmospheric carbon dioxide concentrations.

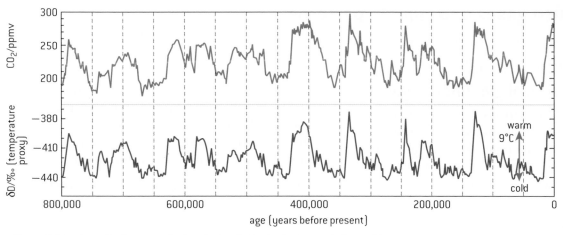

▲ Figure 5 Data from the European Project for Ice Coring in the Antarctic Dome C ice core

Data-based questions: CO₂ concentrations and global temperatures

Figure 6 shows atmospheric carbon dioxide concentrations. The red line shows direct measurements at Mauna Loa Observatory. The points show carbon dioxide concentrations measured from trapped air in polar ice cores.

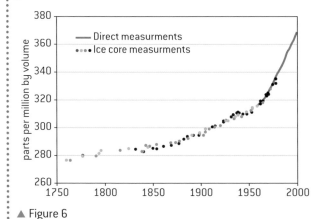

▲ Figure 6

Figure 7 shows a record of global average temperatures compiled by the NASA Goddard Institute for Space Studies. The green points are annual averages and the red curve is a rolling five-year average. The values are given as the deviation from the mean temperature between 1961 and 1990.

1 Discuss whether the measurements of carbon dioxide concentration from ice cores are consistent with direct measurements at Mauna Loa. [2]

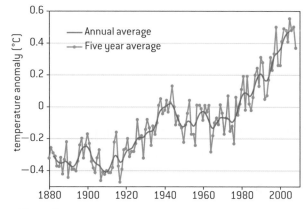

▲ Figure 7

2 Compare the trends in carbon dioxide concentration and global temperatures between 1880 and 2008. [2]

3 Estimate the change in global average temperature between

   a) 1900 and 2000 [1]

   b) 1905 and 2005 [1]

4 a) Suggest reasons for global average temperatures falling for a few years during a period with an overall trend of rising temperatures. [2]

   b) Discuss whether these falls indicate that carbon dioxide concentration does not influence global temperatures. [2]

Greenhouse gases and climate patterns

Global temperatures and climate patterns are influenced by concentrations of greenhouse gases.

The surface of the Earth is warmer than it would be with no greenhouse gases in the atmosphere. Mean temperatures are estimated to be 32°C higher. If the concentration of any of the greenhouse gases rises, more heat will be retained and we should expect an increase in global average temperatures.

This does not mean that global average temperatures are directly proportional to greenhouse gas concentrations. Other factors have an influence, including Milankovitch cycles in the Earth's orbit and variation in sunspot activity. Even so, increases in greenhouse gas concentrations will tend to cause higher global average temperatures and also more frequent and intense heat waves.

Global temperatures influence other aspects of climate. Higher temperatures increase the evaporation of water from the oceans and therefore periods of rain are likely to be more frequent and protracted. The amount of rain delivered during thunderstorms and other intense bursts is likely to increase very significantly. In addition, higher ocean temperatures cause tropical storms and hurricanes to be more frequent and more powerful, with faster wind speeds.

The consequences of any rise in global average temperature are unlikely to be evenly spread. Not all areas would become warmer. The west coast of Ireland and Scotland might become colder if the North Atlantic Current brought less warm water from the Gulf Stream to north-west Europe. The distribution of rainfall would also be likely to change, with some areas becoming more prone to droughts and other areas to intense periods of rainfall and flooding. Predictions about changes to weather patterns are very uncertain, but it is clear that just a few degrees of warming would cause very profound changes to the Earth's climate patterns.

Data-based questions: Phenology

Phenologists are biologists who study the timing of seasonal activities in animals and plants, such as the opening of tree leaves and the laying of eggs by birds. Data such as these can provide evidence of climate changes, including global warming.

The date in the spring when new leaves open on horse chestnut trees (*Aesculus hippocastaneum*) has been recorded in Germany every year since 1951. Figure 8 shows the difference between each year's date of leaf opening and the mean date of leaf opening between 1970 and 2000. Negative values indicate that the date of leaf opening was earlier than the mean. The graph also shows the difference between each year's mean temperature during March and April and the overall mean temperature for these two months. The data for

temperature was obtained from the records of 35 German climate stations.

1   Identify the year in which:

    a) the leaves opened earliest   [1]

    b) mean temperatures in March and April were at their lowest.   [1]

2   Use the data in the graph to deduce the following:

    a) the relationship between temperatures in March and April and the date of opening of leaves on horse chestnut trees.   [1]

    b) whether there is evidence of global warming towards the end of the 20th century.   [2]

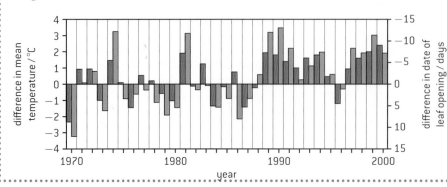

◀ Figure 8 The relationship between temperature and horse chestnut leaf opening in Germany since 1951

Key:
■ temperature
■ leaf opening

Industrialization and climate change

There is a correlation between rising atmospheric concentrations of carbon dioxide since the start of the industrial revolution two hundred years ago and average global temperatures.

The graph of atmospheric carbon dioxide concentrations over the past 800,000 years shown in figure 5 indicates that there have been large fluctuations. During glaciations the concentration dropped to as low as 180 parts per million by volume. During warm interglacial periods they rose as high as 300 ppm. The rise during recent times to concentrations nearing 400 ppm is therefore unprecedented in this period.

Atmospheric carbon dioxide concentrations were between 260 and 280 ppm until the late 18th century. This is when concentrations probably started to rise above the natural levels, but as the rise was initially very slight, it is impossible to say exactly when an unnatural rise in concentrations began. Much of the rise has happened since 1950.

In the late 18th century the industrial revolution was starting in some countries but the main impact of industrialization globally was in the second half of the 20th century. More countries became industrialized, and combustion of coal, oil and natural gas increased ever more rapidly, with consequent increases in atmospheric carbon dioxide concentration.

There is strong evidence for a correlation between atmospheric carbon dioxide concentration and global temperatures, but as already explained, other factors have an effect so temperatures are not directly proportional to carbon dioxide concentration. Nevertheless, since the start of the industrial revolution the correlation between rising atmospheric carbon dioxide concentration and average global temperatures is very marked.

Burning fossil fuels

Recent increases in atmospheric carbon dioxide are largely due to increases in the combustion of fossilized organic matter.

As the industrial revolution spread from the late 18th century onwards, increasing quantities of coal were being mined and burned, causing carbon dioxide emissions. Energy from combustion of the coal provided a source of heat and power. During the 19th century the combustion of oil and natural gas became increasingly widespread in addition to coal.

Increases in the burning of fossil fuels were most rapid from the 1950s onwards and this coincides with the period of steepest rises in atmospheric carbon dioxide. It seems hard to doubt the conclusion that the burning of fossil fuels has been a major contributory factor in the rise of atmospheric carbon dioxide concentrations to higher levels than experienced on Earth for more than 800,000 years.

▲ Figure 9 During the industrial revolution renewable sources of power including wind were replaced with power generated by burning fossil fuels

TOK

What constitutes an unacceptable level of risk?

In situations where the public is at risk, scientists are called upon to advise governments on the setting of policies or restrictions to offset the risk. Because scientific claims are based largely on inductive observation, absolute certainty is difficult to establish. The precautionary principle argues that action to protect the public must precede certainty of risk when the potential consequences for humanity are catastrophic. Principle 15 of the 1992 Rio Declaration on the Environment and Development stated the principle in this way:

Where there are threats of serious or irreversible damage, lack of full scientific certainty shall not be used as a reason for postponing cost-effective measures to prevent environmental degradation.

Data-based questions: Comparing CO₂ emissions

The bar chart in figure 10 shows the cumulative CO_2 emissions from fossil fuels of the European Union and five individual countries between 1950 and 2000. It also shows the total CO_2 emissions including forest clearance and other land use changes.

1 Discuss reasons for higher cumulative CO_2 emissions from combustion of fossil fuels in the United States than in Brazil. [3]

2 Although cumulative emissions between 1950 and 2000 were higher in the United States than any other country, there were four countries in which emissions per capita were higher in the year 2000: Qatar, United Arab Emirates, Kuwait and Bahrain. Suggest reasons for the difference. [3]

3 Although cumulative CO_2 emissions from combustion of fossil fuels in Indonesia and Brazil between 1950 and 2000 were relatively low, total CO_2 emissions were significantly higher. Suggest reasons for this. [3]

4 Australia ranked seventh in the world for emissions of CO_2 in 2000, but fourth when all greenhouse gases are included. Suggest a reason for the difference. [1]

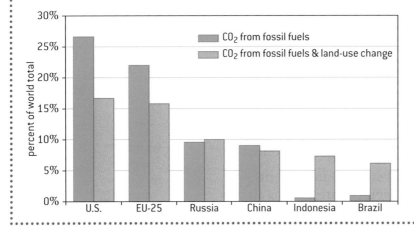

◀ Figure 10 Cummulative CO_2 emmisions from fossil fuels (1950-2000)

Assessing claims and counter-claims

Assessing claims: assessment of the claims that human activities are not causing climate change.

Climate change has been more hotly debated than almost any other area of science. A search of the internet will quickly reveal diametrically opposed views, expressed very vociferously. The author Michael Crichton portrayed climate change scientists as eco-terrorists who were prepared to use mass murder to promote their work in his novel State of Fear. What reasons could there be for such fierce opposition to climate change science and for what reason do climate change scientists defend their findings so vigorously?

These questions are worth discussing. There are many factors that could be having an influence:

- Scientists are trained to be cautious about their claims and to base their ideas on evidence. They are expected to admit when there are uncertainties and this can give the impression that evidence is weaker than it actually is.

- Global climate patterns are very complex and it is difficult to make predictions about the consequences of further increases in greenhouse gas concentrations. There can be tipping points in climate patterns where sudden massive changes occur. This makes prediction even more difficult.

- The consequences of changes in global climate patterns could be very severe for humans and for other species so many feel that there is a need for immediate action even if uncertainties remain in climate change science. Companies make huge profits from coal, oil and natural gas and it is in their interests for fossil fuel combustion to continue to grow. It would not be surprising if they paid for reports to be written that minimized the risks of climate change.

Opposition to the climate change science

Evaluating claims that human activities are not causing climate change.

Many claims that human activities are not causing climate change have been made in newspapers, on television and on the internet. One example of this is:

> *"Global warming stopped in 1998, yet carbon dioxide concentrations have continued to rise, so human carbon dioxide emissions cannot be causing global warming."*

This claim ignores the fact that temperatures on Earth are influenced by many factors, not just greenhouse gas concentrations. Volcanic activity and cycles in ocean currents can cause significant variations from year to year. Because of such factors, 1998 was an unusually warm year and also because of them some recent years have been cooler than they otherwise would have been.

Global warming is continuing but not with equal increases each year. Humans are emitting carbon dioxide by burning fossil fuels and there is strong evidence that carbon dioxide causes warming, so the claim is not supported by the evidence.

Claims that human activities are not causing climate change will continue and these claims need to be evaluated. As always in science, we should base our evaluations on reliable evidence. There is now considerable evidence about emissions of greenhouse gases by humans, about the effects of these gases and about changing climate patterns. Not all sources on the internet are trustworthy and we need to be careful to distinguish between websites with objective assessments based on reliable evidence and others that show bias.

Data-based questions: Uncertainty in temperature rise projections

Figure 11 shows computer-generated forecasts for average global temperatures, based on eight different scenarios for the changes in the emissions of greenhouse gases. The light green band includes the full range of forecasts from research centres around the world, and the dark green band shows the range of most of the forecasts. Figure 12 shows forecasts for arctic temperatures, based on two of the emissions scenarios.

1 Identify the code for the least optimistic emissions scenario. [1]

2 State the minimum and maximum forecasts for average global temperature change. [2]

3 Calculate the difference between the A2 and B2 forecasts of global average temperature rise. [2]

4 Compare the forecasts for arctic temperatures with those for global average temperatures. [2]

5 Suggest uncertainties, apart from greenhouse gas emissions, which affect forecasts for average global temperatures over the next 100 years. [2]

6 Discuss how much more confident we can be in forecasts based on data from a number of different research centres, rather than one. [3]

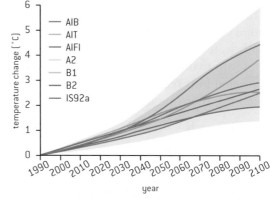

▲ Figure 11 Forecast global average temperatures

7 Discuss whether the uncertainty in temperature forecasts justifies action or inaction. [4]

8 Discuss whether it is possible to balance environmental risks with socio-economic and livelihood risks or whether priorities need to be established. [4]

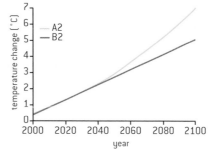

▲ Figure 12 Forecast arctic temperature

Coral reefs and carbon dioxide

Threats to coral reefs from increasing concentrations of dissolved carbon dioxide.

In addition to its contribution to global warming, emissions of carbon dioxide are having effects on the oceans. Over 500 billion tonnes of carbon dioxide released by humans since the start of the industrial revolution have dissolved in the oceans. The pH of surface layers of the Earth's oceans is estimated to have been 8.179 in the late 18th century when there had been little industrialization. Measurements in the mid-1990s showed that it had fallen to 8.104 and current levels are approximately 8.069. This seemingly small change represents a 30% acidification. Ocean acidification will become more severe if the carbon dioxide concentration of the atmosphere continues to rise.

Marine animals such as reef-building corals that deposit calcium carbonate in their skeletons need to absorb carbonate ions from seawater. The concentration of carbonate ions in seawater is low, because they are not very soluble. Dissolved carbon dioxide makes the carbonate concentration even lower as a result of some interrelated chemical reactions. Carbon dioxide reacts with water to form carbonic acid, which dissociates into hydrogen and hydrogen carbonate ions. Hydrogen ions react with dissolved carbonate ions, reducing their concentration.

$$CO_2 + H_2O \rightarrow H_2CO_3 \rightarrow H^+ + HCO_3^-$$

$$H^+ + CO_3^{2-} \rightarrow HCO_3^-$$

If carbonate ion concentrations drop it is more difficult for reef-building corals to absorb them to make their skeletons. Also, if seawater ceases to be a saturated solution of carbonate ions, existing calcium carbonate tends to dissolve, so existing skeletons of reef-building corals are threatened. In 2012 oceanographers from more than 20 countries met in Seattle and agreed to set up a global scheme for monitoring ocean acidification.

There is already evidence for concerns about corals and coral reefs. Volcanic vents near the island of Ischia in the Gulf of Naples have been releasing carbon dioxide into the water for thousands of years, reducing the pH of the seawater. In the area of acidified water there are no corals, sea urchins or other animals that make their skeletons from calcium carbonate. In their place other organisms flourish such as sea grasses and invasive algae. This could be the future of coral reefs around the world if carbon dioxide continues to be emitted from burning fossil fuels.

▲ Figure 13 Skeleton of calcium carbonate from a reef-building coral

Activity

Draw a graph of oceanic pH from the 18th century onwards, using the figures given in the text above, and extrapolate the curve to obtain an estimate of when the pH might drop below 7.

TOK

What are the potential impacts of funding bias?

The costs of scientific research is often met by grant agencies. Scientists submit research proposals to agencies, the application is reviewed and if successful, the research can proceed. Questions arise when the grant agency has a stake in the study's outcome. Further, grant applications might ask scientists to project outcomes or suggest applications of the research before it has even begun. The sponsor may fund several different research groups, suppressing results that run counter to their interests and publishing those that support their industry. For example, a 2006 review of studies examining the health effects of cell phone use revealed that studies funded by the telecommunications industry were statistically least likely to report a significant effect. Pharmaceutical research, nutrition research and climate change research are all areas where claims of funding bias have been prominent in the media.

Questions

1   The total solar energy received by a grassland is 5×10^5 kJ m^{-2} yr^{-1}. The net production of the grassland is 5×10^2 kJ m^{-2} yr^{-1} and its gross production is 6×10^2 kJ m^{-2} yr^{-1}. The total energy passed on to primary consumers is 60 kJ m^{-2} yr^{-1}. Only 10 per cent of this energy is passed on to the secondary consumers.

   a) Calculate the energy lost by plant respiration.   [2]

   b) Construct a pyramid of energy for this grassland.   [3]

2   Figure 14 shows the energy flow through a temperate forest. The energy flow is shown per square metre per year (kJ m^{-2} yr^{-1}).

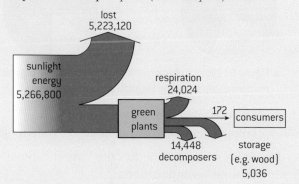

▲ Figure 14

   a) The chart shows that 99.17 per cent of the sunlight energy in the temperate forest is lost. Predict with a reason whether a greater or lesser percentage of sunlight energy would be lost in desert.   [2]

   b) Only a small part of the net production of plants in the temperate forest passes to herbivores. Explain the reasons for this.   [2]

3   Warmer temperatures favour some species of pest, for example the spruce beetle. Since the first major outbreak in 1992, it has killed approximately 400,000 hectares of trees in Alaska and the Canadian Yukon. The beetle normally needs two years to complete its life cycle, but it has recently been able to do it in one year. The graphs in figure 15 show the drought index, a combination of temperatures and precipitation, and the area of spruce trees destroyed annually.

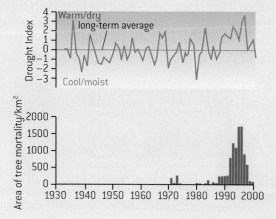

▲ Figure 15 Tree mortality and drought index

   a) Identify the two periods when the drought index remained high for three or more years.   [2]

   b) (i)   Distinguish between the beetle outbreaks in the 1970s and 1990s.   [2]

       (ii) Suggest reasons for the differences between the outbreaks.   [2]

   c) Predict rates of destruction of spruce trees in the future, with reasons for your answer.   [4]

4   Figure 16 shows monthly average carbon dioxide concentrations for Baring Head, New Zealand and Alert, Canada.

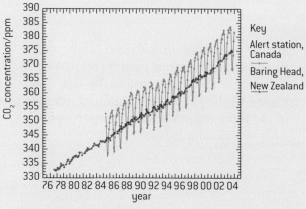

▲ Figure 16

   a) Suggest why scientists have chosen such areas as Mauna Loa, Baring Head and Alert as the locations for monitoring stations.   [1]

   b) Compare and contrast the trends illustrated in both graphs.   [2]

   c) Explain why the graphs show different patterns.   [3]

5 Figure 17 shows the concentration of CO_2 in the atmosphere, measured in parts per million (ppm). In a forest, concentrations of CO_2 change over the course of the day and change with height. The top of the forest is referred to as the canopy.

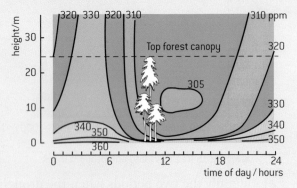

▲ Figure 17

a) (i) State the highest concentration of CO_2 reached in the canopy. [1]

     (ii) Determine the range of concentration found in the canopy. [2]

b) (i) State the time of day (or night) when the highest levels of CO_2 are detected. [1]

     (ii) The highest levels of CO_2 are detected just above the ground. Deduce two reasons why this is the case. [2]

c) Give an example of an hour when CO_2 concentrations are reasonably uniform over the full range of heights. [1]

6 Within an ecosystem, nitrogen can be stored in one of three organic matter compartments: above ground, in roots and in the soil. Figure 18 shows the distribution of nitrogen in the three organic matter compartments for each of six major biomes.

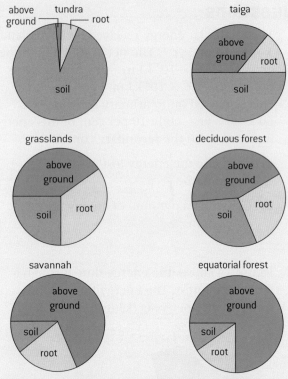

▲ Figure 18 The distribution of nitrogen in the three organic matters compartments for each of six major biomes

a) Deduce what the "above ground" compartment consists of in an ecosystem. [1]

b) State which biome has the largest "above ground" compartment. [1]

c) Explain why it is difficult to grow crops in an area where equatorial forest has been cleared of its vegetation. [2]

d) State the name of the process carried out by decomposers and detritus feeders that releases CO_2 into the atmosphere. [1]

e) Suggest why most of the nitrogen in a tundra ecosystem is in the soil. [1]

f) Explain why warming due to climate change might cause a release of CO_2 from tundra soil. [2]

5 EVOLUTION AND BIODIVERSITY

Introduction

There is overwhelming evidence for the theory that the diversity of life has evolved, and continues to evolve by natural selection. The ancestry of groups of species can be deduced by comparing their base or amino acid sequences. Species are named and classified using an internationally agreed system.

5.1 Evidence for evolution

Understanding

→ Evolution occurs when heritable characteristics of a species change.

→ The fossil record provides evidence for evolution.

→ Selective breeding of domesticated animals shows that artificial selection can cause evolution.

→ Evolution of homologous structures by adaptive radiation explains similarities in structure when there are differences in function.

→ Populations of a species can gradually diverge into separate species by evolution.

→ Continuous variation across the geographical range of related populations matches the concept of gradual divergence.

Applications

→ Comparison of the pentadactyl limb of mammals, birds, amphibians and reptiles with different methods of locomotion.

→ Development of melanistic insects in polluted areas.

Nature of science

→ Looking for patterns, trends and discrepancies: there are common features in the bone structure of vertebrate limbs despite their varied use.

▲ Figure 1 Fossils of dinosaurs show there were animals on Earth in the past that had different characteristics from those alive today

Evolution in summary

Evolution occurs when heritable characteristics of a species change.

There is strong evidence for characteristics of species changing over time. Biologists call this process evolution. It lies at the heart of a scientific understanding of the natural world. An important distinction should be drawn between acquired characteristics that develop during the lifetime of an individual and heritable characteristics that are passed from parent to offspring. Evolution only concerns heritable characteristics.

The mechanism of evolution is now well understood – it is natural selection. Despite the robustness of evidence for evolution by natural selection, there is still widespread disbelief among some religious groups. There are stronger objections to the concept that species can evolve than to the logic of the mechanism that inevitably causes evolution. It is therefore important to look at the evidence for evolution.

Evidence from fossils

The fossil record provides evidence for evolution.

In the first half of the 19th century, the sequence in which layers or strata of rock were deposited was worked out and the geological eras were named. It became obvious that the fossils found in the various layers were different – there was a sequence of fossils. In the 20th century, reliable methods of radioisotope dating revealed the ages of the rock strata and of the fossils in them. There has been a huge amount of research into fossils, which is the branch of science called palaeontology. It has given us strong evidence that evolution has occurred.

▲ Figure 2 Many trilobite species evolved over hundreds of millions of years but the group is now totally extinct

- The sequence in which fossils appear matches the sequence in which they would be expected to evolve, with bacteria and simple algae appearing first, fungi and worms later and land vertebrates later still. Among the vertebrates, bony fish appeared about 420 million years ago (mya), amphibians 340 mya, reptiles 320 mya, birds 250 mya and placental mammals 110 mya.

- The sequence also fits in with the ecology of the groups, with plant fossils appearing before animal, plants on land before animals on land, and plants suitable for insect pollination before insect pollinators.

- Many sequences of fossils are known, which link together existing organisms with their likely ancestors. For example, horses, asses and zebras, members of the genus *Equus*, are most closely related to rhinoceroses and tapirs. An extensive sequence of fossils, extending back over 60 million years, links them to *Hyracotherium*, an animal very similar to a rhinoceros.

Data-based questions: Missing links

An objection to fossil evidence for evolution has been gaps in the record, called missing links, for example a link between reptiles and birds.

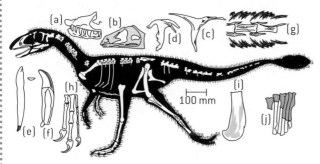

▲ Figure 3 Drawings of fossils recently found in Western China. They show *Dilong paradoxus*, a 130-million-year-old tyrannosauroid dinosaur with protofeathers. a–d: bones of skull; e–f: teeth; g: tail vertebrae with protofeathers; h–j: limb bones

The discovery of fossils that fill in these gaps is particularly exciting for biologists.

1 Calculate the length of *Dilong paradoxus*, from its head to the tip of its tail. [2]

2 Deduce three similarities between *Dilong paradoxus* and reptiles that live on Earth today. [3]

3 Suggest a function for the protofeathers of *Dilong paradoxus*. [1]

4 Suggest two features which *Dilong paradoxus* would have had to evolve to become capable of flight. [2]

5 Explain why it is not possible to be certain whether the protofeathers of *Dilong paradoxus* are homologous with the feathers of birds. [2]

Evidence from selective breeding

Selective breeding of domesticated animals shows that artificial selection can cause evolution.

Humans have deliberately bred and used particular animal species for thousands of years. If modern breeds of livestock are compared with the wild species that they most resemble, the differences are often huge. Consider the differences between modern egg-laying hens and the junglefowl of Southern Asia, or between Belgian Blue cattle and the aurochs of Western Asia. There are also many different breeds of sheep, cattle and other domesticated livestock, with much variation between breeds.

It is clear that domesticated breeds have not always existed in their current form. The only credible explanation is that the change has been achieved simply by repeatedly selecting for and breeding the individuals most suited to human uses. This process is called artificial selection.

The effectiveness of artificial selection is shown by the considerable changes that have occurred in domesticated animals over periods of time that are very short, in comparison to geological time. It shows that selection can cause evolution, but it does not prove that evolution of species has actually occurred naturally, or that the mechanism for evolution is natural selection.

▲ Figure 4 Over the last 15,000 years many breeds of dog have been developed by artificial selection from domesticated wolves

Homology and evolution

Looking for patterns, trends and discrepancies: there are common features in the bone structure of vertebrate limbs despite their varied use.

Vertebrate limbs are used in many different ways, such as walking, running, jumping, flying, swimming, grasping and digging. These varied uses require joints that articulate in different ways, different velocities of movement and also different amounts of force. It would be reasonable to expect them to have very different bone structure, but there are in fact common features of bone structure that are found in all vertebrate limbs.

Patterns like this require explanation. The only reasonable explanation so far proposed in this case is evolution from a common ancestor. As a consequence, the common bone structure of vertebrate limbs has become a classic piece of evidence for evolution.

Data-based questions: Domestication of corn

A wild grass called teosinte that grows in Central America was probably the ancestor of cultivated corn, *Zea mays*. When teosinte is grown as a crop, it gives yields of about 150 kg per hectare. This compares with a world average yield of corn of 4,100 kg per hectare at the start of the 21st century. Table 1 gives the lengths of some cobs. Corn was domesticated at least 7,000 years ago.

1   Calculate the percentage difference in length between teosinte and Silver Queen. [2]

2   Calculate the percentage difference in yield between teosinte and world average yields of corn. [2]

3   Suggest factors apart from cob length, selected for by farmers. [3]

4   Explain why improvement slows down over generations of selection. [3]

Corn variety and origin	Length of cob (mm)
Teosinte – wild relative of corn	14
Early primitive corn from Colombia	45
Peruvian ancient corn from 500 BC	65
Imbricado – primitive corn from Colombia	90
Silver Queen – modern sweetcorn	170

▲ Table 1

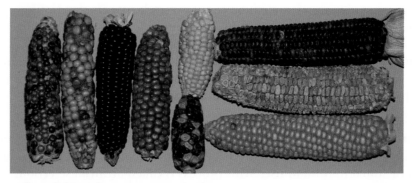

▲ Figure 5 Corn cobs

Evidence from homologous structures

Evolution of homologous structures by adaptive radiation explains similarities in structure when there are differences in function.

Darwin pointed out in *The Origin of Species* that some similarities in structure between organisms are superficial, for example between a dugong and a whale, or between a whale and a fish. Similarities like those between the tail fins of whales and fishes are known as analogous structures. When we study them closely we find that these structures are very different. An evolutionary interpretation is that they have had different origins and have become similar because they perform the same or a similar function. This is called convergent evolution.

Homologous structures are the converse of this. They are structures that may look superficially different and perform a different function, but which have what Darwin called a "unity of type". He gave the example of the forelimbs of a human, mole, horse, porpoise and bat and asked what could be more curious than to find that they "include the same bones, in the same relative positions", despite on the surface appearing completely different. The evolutionary explanation is that they have had the same origin, from an ancestor that had a pentadactyl or five-digit limb, and that they have become different because they perform different functions. This is called adaptive radiation.

There are many examples of homologous structures. They do not prove that organisms have evolved or had common ancestry and do not reveal anything about the mechanism of evolution, but they are difficult to explain without evolution. Particularly interesting are the structures that Darwin called "rudimentary organs" – reduced structures that serve no function. They are now called vestigial organs and examples of them are the beginnings of teeth found in embryo baleen whales, despite adults being toothless, the small pelvis and thigh bone found in the body wall of whales and some snakes, and of course the appendix in humans. These structures are easily explained by evolution as structures that no longer have a function and so are being gradually lost.

 Pentadactyl limbs

Comparison of the pentadactyl limb of mammals, birds, amphibians and reptiles with different methods of locomotion.

The pentadactyl limb consists of these structures:

Bone structure	Forelimb	Hindlimb
single bone in the proximal part	humerus	femur
two bones in the distal part	radius and ulna	tibia and fibula
group of wrist/ankle bones	carpals	tarsals
series of bones in each of five digits	metacarpals and phalanges	metatarsals and phalanges

The pattern of bones or a modification of it is present in all amphibians, reptiles, birds and mammals, whatever the function of their limbs.

The photos in figure 6 show the skeletons of one example of each of the four vertebrates classes that have limbs: amphibians, reptiles, birds and mammals. Each of them has pentadactyl limbs:

- crocodiles walk or crawl on land and use their webbed hind limbs for swimming
- penguins use their hind limbs for walking and their forelimbs as flippers for swimming
- echidnas use all four limbs for walking and also use their forelimbs for digging
- frogs use all four limbs for walking and their hindlimbs for jumping.

Differences can be seen in the relative lengths and thicknesses of the bones. Some metacarpals and phalanges have been lost during the evolution of the penguin's forelimb.

Activity

Pentadactyl limbs in mammals

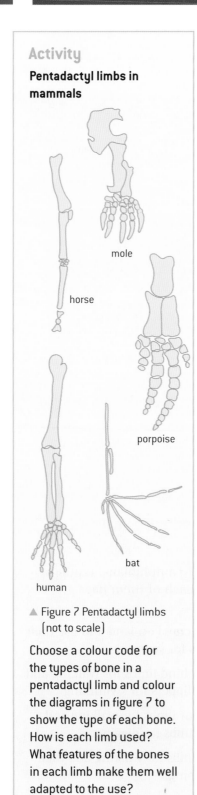

▲ Figure 7 Pentadactyl limbs (not to scale)

Choose a colour code for the types of bone in a pentadactyl limb and colour the diagrams in figure 7 to show the type of each bone. How is each limb used? What features of the bones in each limb make them well adapted to the use?

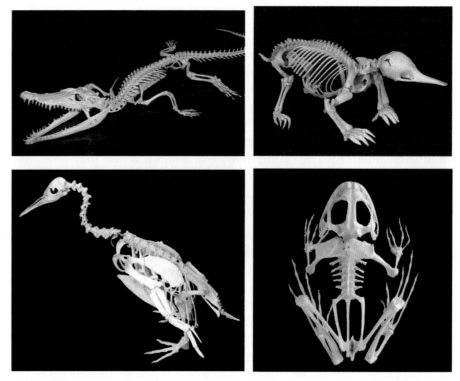

▲ Figure 6

Speciation

Populations of a species can gradually diverge into separate species by evolution.

If two populations of a species become separated so that they do not interbreed and natural selection then acts differently on the two populations, they will evolve in different ways. The characteristics of the two populations will gradually diverge. After a time they will be recognizably different. If the populations subsequently merge and have the chance of interbreeding, but do not actually interbreed, it would be clear that they have evolved into separate species. This process is called speciation.

Speciation often occurs after a population of a species extends its range by migrating to an island. This explains the large numbers of endemic species on islands. An endemic species is one that is found only in a certain geographical area. The lava lizards of the Galápagos Islands are an example of this. One species is present on all the main islands of the archipelago. On six smaller islands there is a closely related but different species, formed by migration to the island and by subsequent divergence.

Evidence from patterns of variation

Continuous variation across the geographical range of related populations matches the concept of gradual divergence.

If populations gradually diverge over time to become separate species, then at any one moment we would expect to be able to find examples of all stages of divergence. This is indeed what we find in nature, as Charles Darwin describes in Chapter II of *The Origin of Species*. He wrote:

> *Many years ago, when comparing, and seeing others compare, the birds from the separate islands of the Galápagos Archipelago, both one with another, and with those from the American mainland, I was much struck how entirely vague and arbitrary is the distinction between species and varieties.*

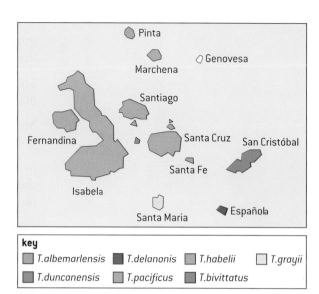

▲ Figure 8 Distribution of lava lizards in the Galápagos Islands

Darwin gave examples of populations that are recognizably different, but not to the extent that they are clearly separate species. One of his examples is the red grouse of Britain and the willow ptarmigan of Norway. They have sometimes been classified as separate species and sometimes as varieties of the species *Lagopus lagopus*. This is a common problem for biologists who name and classify living organisms. Because species can gradually diverge over long periods of time and there is no sudden switch from being two populations of one species to being two separate species, the decision to lump populations together or split them into separate species remains rather arbitrary.

The continuous range in variation between populations does not match either the belief that species were created as distinct types of organism and therefore should be constant across their geographic range or that species are unchanging. Instead it provides evidence for the evolution of species and the origin of new species by evolution.

Industrial melanism

Development of melanistic insects in polluted areas.

Dark varieties of typically light-coloured insects are called melanistic. The most famous example of an insect with a melanistic variety is *Biston betularia*, the peppered moth. It has been widely used as an example of natural selection, as the melanistic variety became commoner in polluted industrial areas where it is better camouflaged than the pale peppered variety. A simple explanation of industrial melanism is this:

- Adult *Biston betularia* moths fly at night to try to find a mate and reproduce.

- During the day they roost on the branches of trees.

- Birds and other animals that hunt in daylight predate moths if they find them.

- In unpolluted areas tree branches are covered in pale-coloured lichens and peppered moths are well camouflaged against them.

- Sulphur dioxide pollution kills lichens. Soot from coal burning blackens tree branches.

- Melanic moths are well camouflaged against dark tree branches in polluted areas.

- In polluted areas the melanic variety of *Biston betularia* replaced the peppered variety over a relatively short time, but not in non-polluted areas.

▲ Figure 9 Museum specimen of the peppered form of *Biston betularia* mounted on tree bark with lichens from an unpolluted area

▲ Figure 10 The ladybug *Adalia bipunctata* has a melanic form which has become common in polluted areas. A melanic male is mating with a normal female here

Biologists have used industrial melanism as a classic example of evolution by natural selection. Perhaps because of this, research findings have been repeatedly attacked. The design of some early experiments into camouflage and predation of the moths has been criticized and this has been used to cast doubt over whether natural selection ever actually occurs.

Michael Majerus gives a careful evaluation of evidence about the development of melanism in *Biston betularia* and other species of moth in his book in the New Naturalist series (*Moths*, Michael Majerus, HarperCollins 2002). His finding is that the evidence for industrial pollution causing melanism in *Biston betularia* and other species of moth is strong, though factors other than camouflage can also influence survival rates of pale and melanic varieties.

Data-based questions: Predation rates in *Biston betularia*

One of the criticisms of the original experiments into predation of *Biston betularia* was that the moths were placed in exposed positions on tree trunks and that this is not normally where they roost. The moths were able to move to more suitable positions but even so the criticisms have persisted on some websites. Experiments done in the 1980s tested the effect of the position in which the moths were placed. Peppered and melanic forms (fifty of each) of *Biston betularia* were placed in exposed positions on tree trunks and 50 millimetres below a joint between a major branch and the tree trunk. This procedure was carried out at two oak woods, one in an unpolluted area of the New Forest in southern England and another in a polluted area near Stoke-on-Trent in the Midlands. The box plots in figure 11 show the percentage of moths eaten and moths surviving.

1 **a)** Deduce, with a reason from the data, whether the moths were more likely to be eaten if they were placed on the exposed trunk or below the junction of a main branch and the trunk. [2]

   b) Suggest a reason for the difference. [1]

2 **a)** Compare and contrast the survival rates of peppered and melanic moths in the New Forest. [3]

   b) Explain the difference in survival rate between the two varieties in the New Forest. [3]

3 Distinguish between the Stoke-on-Trent and New Forest woodlands in relative survival rates of peppered and melanic moths. [2]

4 Pollution due to industry has decreased greatly near Stoke-on-Trent since the 1980s. Predict the consequences of this change for *Biston betularia*. [4]

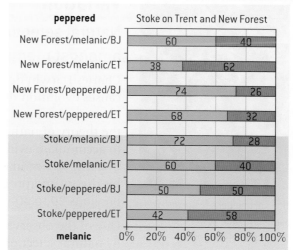

peppered — Stoke on Trent and New Forest

	not eaten	eaten
New Forest/melanic/BJ	60	40
New Forest/melanic/ET	38	62
New Forest/peppered/BJ	74	26
New Forest/peppered/ET	68	32
Stoke/melanic/BJ	72	28
Stoke/melanic/ET	60	40
Stoke/peppered/BJ	50	50
Stoke/peppered/ET	42	58

melanic 0% 20% 40% 60% 80% 100%

key

■ not eaten ■ eaten

ET = exposed trunk BJ = branch junction

▲ Figure 11

Source: Howlett and Majerus (1987) The Understanding of industrial melanism in the peppered moth (*Biston betularia*) *Biol.J.Linn.Soc.* 30, 31–44

5.2 Natural selection

Understanding

→ Natural selection can only occur if there is variation amongst members of the same species.

→ Mutation, meiosis and sexual reproduction cause variation between individuals in a species.

→ Adaptations are characteristics that make an individual suited to its environment and way of life.

→ Species tend to produce more offspring than the environment can support.

→ Individuals that are better adapted tend to survive and produce more offspring while the less well adapted tend to die or produce fewer offspring.

→ Individuals that reproduce pass on characteristics to their offspring.

→ Natural selection increases the frequency of characteristics that make individuals better adapted and decreases the frequency of other characteristics leading to changes within the species.

 ## Applications

→ Changes in beaks of finches on Daphne Major.

→ Evolution of antibiotic resistance in bacteria.

 ## Nature of science

→ Use theories to explain natural phenomena: the theory of evolution by natural selection can explain the development of antibiotic resistance in bacteria.

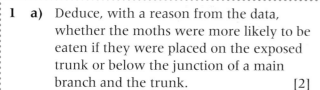

▲ Figure 1 Populations of bluebells (*Hyacinthoides non-scripta*) mostly have blue flowers but white-flowered plants sometimes occur

▲ Figure 2 Dandelions (*Taraxacum officinale*) appear to be reproducing sexually when they disperse their seed but the embryos in the seeds have been produced asexually so are genetically identical

Variation

Natural selection can only occur if there is variation amongst members of the same species.

Charles Darwin developed his understanding of the mechanism that causes evolution over many years, after returning to England from his voyage around the world on HMS Beagle. He probably developed the theory of natural selection in the late 1830s, but then worked to accumulate evidence for it. Darwin published his great work, *The Origin of Species*, in 1859. In this book of nearly 500 pages, he explains his theory and presents the evidence for it that he had found over the previous 20 to 30 years.

One of the observations on which Darwin based the theory of evolution by natural selection is variation. Typical populations vary in many respects. Variation in human populations is obvious – height, skin colour, blood group and many other features. With other species the variation may not be so immediately obvious but careful observation shows that it is there. Natural selection depends on variation within populations – if all individuals in a population were identical, there would be no way of some individuals being favoured more than others.

Sources of variation

Mutation, meiosis and sexual reproduction cause variation between individuals in a species.

The causes of variation in populations are now well understood:

1 Mutation is the original source of variation. New alleles are produced by gene mutation, which enlarges the gene pool of a population.

2 Meiosis produces new combinations of alleles by breaking up the existing combination in a diploid cell. Every cell produced by meiosis in an individual is likely to carry a different combination of alleles, because of crossing over and the independent orientation of bivalents.

3 Sexual reproduction involves the fusion of male and female gametes. The gametes usually come from different parents, so the offspring has a combination of alleles from two individuals. This allows mutations that occurred in different individuals to be brought together.

In species that do not carry out sexual reproduction the only source of variation is mutation. It is generally assumed that such species will not generate enough variation to be able to evolve quickly enough for survival during times of environmental change.

Adaptations

Adaptations are characteristics that make an individual suited to its environment and way of life.

One of the recurring themes in biology is the close relationship between structure and function. For example, the structure of a bird's beak is correlated with its diet and method of feeding. The thick coat of a musk ox is obviously correlated with the low temperatures in its northerly habitats. The water storage tissue in the stem of a cactus is related to infrequent rainfall in desert habitats. In biology characteristics such as these that make an individual suited to its environment or way of life are called adaptations.

The term adaptation implies that characteristics develop over time and thus that species evolve. It is important not to imply purpose in this process. According to evolutionary theory adaptations develop by natural selection, not with the direct purpose of making an individual suited to its environment. They do not develop during the lifetime of one individual. Characteristics that do develop during a lifetime are known as acquired characteristics and a widely accepted theory is that acquired characteristics cannot be inherited.

Overproduction of offspring

Species tend to produce more offspring than the environment can support.

Living organisms vary in the number of offspring they produce.

An example of a species with a relatively slow breeding rate is the southern ground hornbill, *Bucorvus leadbeateri*. It raises one fledgling every three years on average and needs the cooperation of at least two other adults to do this. However they can live for as long as 70 years so in their lifetime a pair could theoretically raise twenty offspring.

Most species have a faster breeding rate. For example, the coconut palm, *Cocos nucifera* usually produces between 20 and 60 coconuts per year. Apart from bacteria, the fastest breeding rate of all may be in the fungus *Calvatia gigantea*. It produces a huge fruiting body called a giant puffball in which there can be as many as 7 trillion spores (7,000,000,000,000).

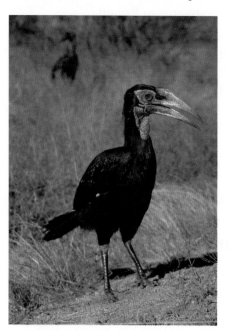

Despite the huge variation in breeding rate, there is an overall trend in living organisms for more offspring to be produced than the environment can support. Darwin pointed out that this will tend to lead to a struggle for existence within a population. There will be competition for resources and not every individual will obtain enough to allow them to survive and reproduce.

◀ Figure 4 The breeding rate of pairs of southern ground hornbills, *Bucorvus leadbeateri*, is as low as 0.3 young per year

Differential survival and reproduction

Individuals that are better adapted tend to survive and produce more offspring while the less well adapted tend to die or produce fewer offspring.

Chance plays a part in deciding which individuals survive and reproduce and which do not, but the characteristics of an individual also have an influence. In the struggle for existence the less well-adapted individuals tend to die or fail to reproduce and the best adapted tend to survive and produce many offspring. This is natural selection.

An example that is often quoted is that of the giraffe. It can graze on grass and herbs but is more adapted to browse on tree leaves. In the wet season its food is abundant but in the dry season there can be periods of food shortage when the only remaining tree leaves are on high branches. Giraffes with longer necks are better adapted to reaching these leaves and surviving periods of food shortage than those with shorter necks.

Inheritance

Individuals that reproduce pass on characteristics to their offspring.

Much of the variation between individuals can be passed on to offspring – it is heritable. Maasai children inherit the dark skin colour of their parents for example and children of light-skinned north European parents inherit a light skin colour. Variation in behaviour can be heritable. The direction of migration to overwintering sites in the blackcap *Sylvia atricapilla* is an example. Due to differences in their genes, some birds of this species migrate southwestwards from Germany to Spain for the winter and others northwestwards to Britain.

Not all features are passed on to offspring. Those acquired during the lifetime of an individual are not usually inherited. An elephant with a broken tusk does not have calves with broken tusks for example. If a person develops darker skin colour through exposure to sunlight, the darker skin is not inherited. Acquired characteristics are therefore not significant in the evolution of a species.

Progressive change

Natural selection increases the frequency of characteristics that make individuals better adapted and decreases the frequency of other characteristics leading to changes within the species.

Because better-adapted individuals survive, they can reproduce and pass on characteristics to their offspring. Individuals that are less well adapted have lower survival rates and less reproductive success. This leads to an increase in the proportion of individuals in a population with characteristics that make them well adapted. Over the generations, the characteristics of the population gradually change – this is evolution by natural selection.

Major evolutionary changes are likely to occur over long time periods and many generations, so we should not expect to be able to observe them during our lifetime, but there are many examples of smaller but significant changes that have been observed. The evolution of dark wing colours in moths has been observed in industrial areas with polluted air. Two examples of evolution are described in the next sections of this book: changes to beaks of finches on the Galapagos Islands and the development of antibiotic resistance in bacteria.

Data-based questions: Evolution in rice plants

The bar charts in figure 6 show the results of an investigation of evolution in rice plants. F_1 hybrid plants were bred by crossing together two rice varieties. These hybrids were then grown at five different sites in Japan. Each year the date of flowering was recorded and seed was collected from the plants, for re-sowing at that site in the following year.

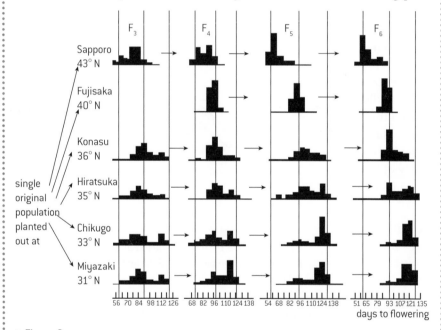

▲ Figure 6

1   Why was the investigation done using hybrids rather than a single pure-bred variety? [2]

2   Describe the changes, shown in the chart, between the F_3 and F_6 generations of rice plants grown at Miyazaki. [2]

3   a)   State the relationship between flowering time and latitude in the F_6 generation. [1]

    b)   Suggest a reason for this relationship. [1]

4   a)   Predict the results if the investigation had been carried on until the F_{10} generation. [1]

    b)   Predict the results of collecting seeds from F_{10} plants grown at Sapporo and from F_{10} plants grown at Miyazaki and sowing them together at Hiratsuka. [3]

▲ Figure 5 A female cheetah's cubs inherit characteristics from her and from one of the several males with whom she mated

🌐 Galápagos finches

Changes in beaks of finches on Daphne Major.

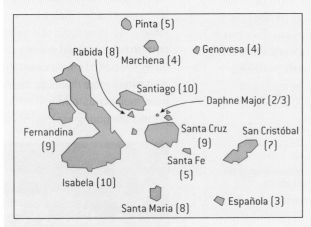

▲ Figure 7 The Galápagos archipelago with the number of species of finch found on each island

(a) *G. fortis* (*large beak*)

(b) *G. fortis* (*small beak*)

(c) *G. magnirostris*

▲ Figure 8 Variation in beak shape in Galápagos finches.
(a) *G. fortis* (large beak). (b) *G. fortis* (small beak).
(c) *G. magnirostris*

Darwin visited the Galápagos Islands in 1835 and collected specimens of small birds, which were subsequently identified as finches. There are 14 species in all. Darwin observed that the sizes and shapes of the beaks of the finches varied, as did their diet. From the overall similarities between the birds and their distribution over the Galapagos islands (see figure 7), Darwin hypothesized that "one might really fancy that from an original paucity of birds in this archipelago, one species had been taken and modified for different ends".

There has since been intense research into what have become known as Darwin's finches. In particular, Peter and Rosemary Grant have shown that beak characters and diet are closely related and when one changes, the other does also. A particular focus of Peter and Rosemary Grant's research has been a population of the medium ground finch, *Geospiza fortis*, on a small island called Daphne Major. On this island, the small ground finch, *Geospiza fuliginosa*, is almost absent. Both species feed on small seeds, though *G. fortis* can also eat larger seeds. In the absence of competition from *G. fuliginosa* for small seeds, *G. fortis* is smaller in body size and beak size on Daphne Major than on other islands.

In 1977, a drought on Daphne Major caused a shortage of small seeds, so *G. fortis* fed instead on larger, harder seeds, which the larger-beaked individuals are able to crack open. Most of the population died in that year, with highest mortality among individuals with shorter beaks. In 1982–83 there was a severe El Niño event, causing eight months of heavy rain and as a result an increased supply of small, soft seeds and fewer large, hard seeds. *G. fortis* bred rapidly, in response to the increase in food availability. With a return to dry weather conditions and greatly reduced supplies of small seeds, breeding stopped until 1987. In that year, only 37 per cent of those alive in 1983 bred and they were not a random sample of the 1983 population. In 1987, *G. fortis* had longer and narrower beaks than the 1983 averages, correlating with the reduction in supply of small seeds.

Variation in the shape and size of the beaks (see figure 8) is mostly due to genes, though the environment has some effect. The proportion of the variation due to genes is called heritability. Using the heritability of beak length and width and data about the birds that had survived to breed, the changes in mean beak length and width between 1983 and 1987 were predicted. The observed results are very close to the predictions. Average beak length was predicted to increase by 10 μm and actually increased by 6 μm. Average beak width was predicted to decrease by 130 μm and actually decreased by 120 μm.

One of the objections to the theory of evolution by natural selection is that significant changes caused by natural selection have not been observed actually occurring. It is unreasonable to expect huge changes to have occurred in a species, even if it had been followed since Darwin's theory was published in 1859, but in the case of *G. fortis*, significant changes have occurred that are clearly linked to natural selection.

Data-based questions: Galápagos finches

When Peter and Rosemary Grant began to study finches on the island of Daphne Major in 1973, there were breeding populations of two species, *Geospiza fortis* and *Geospiza scandens*. *Geospiza magnirostris* established a breeding population on the island in 1982, initially with just two females and three males. Figure 9 shows the numbers of *G. magnirostris* and *G. fortis* on Daphne Major between 1997 and 2006.

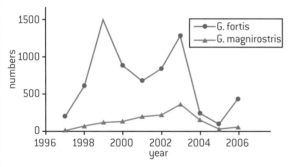

▲ Figure 9 Changes in numbers of *G. fortis* and *G. magnirostris* between 1996 and 2006

1. **a)** Describe the changes in the population of *G. magnirostris* between 1997 and 2006. [2]

   b) Compare and contrast the changes in population of *G. fortis* between 1997 and 2006 with the changes in the population of *G. magnirostris*. [3]

2. Daphne Major has an area of 0.34 km². 1 km² is 100 hectares and 1 hectare is 100 × 100 m. Calculate the maximum and minimum population densities of G. fortis during 1997–2006. [4]

Table 2 shows the percentages of three types of seed in the diets of the three finch species on Daphne Major. Small seeds are produced by 22 plant species, medium seeds by the cactus *Opuntia echios*, and large seeds, which are very hard, by *Tribulus cistoides*.

3. **a)** Outline the diet of each of the species of finch on Daphne Major. [3]

   b) There was a very severe drought on Daphne Major in 2003 and 2004. Deduce how the diet of the finches changed during the drought, using the data in the table. [3]

4. Figure 10 shows an index of beak size of adult *G. fortis* from 1973 to 2006, with the size in 1973 assigned the value zero and the sizes in other years shown in comparison to this.

Species	Geospiza fortis				Geospiza magnirostris			Geospiza scandens			
Year	1977	1985	1989	2004	1985	1989	2004	1977	1985	1989	2004
Small	75	80	77	80	18	5.9	4.5	85	77	23	17
Medium	10	0.0	5.1	11	0.0	12	26	15	22	70	83
Large	17	19	16	8.2	82	82	69	0.0	0.0	0.0	0.0

▲ Table 2

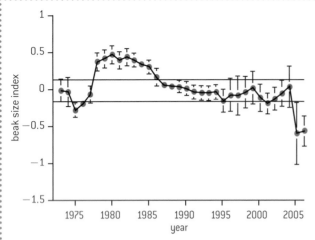

▲ Figure 10 Relative beak size in *G. fortis* between 1973 and 2006

The graph shows two periods of very rapid change in mean beak size, both of which correspond with droughts on Daphne Major.

a) State two periods of most rapid change in mean beak size of *G. fortis*. [2]

b) Suggest two reasons for mean beak size changing most rapidly when there is a drought. [2]

c) In the first severe drought, the mean beak size of *G. fortis* increased, but in the second drought, it decreased. Using the data in this question, explain how natural selection could cause these changes in beak size in the two droughts. [3]

5 The intensity of natural selection on Daphne Major was calculated during the two droughts. The calculated values are called selection differentials. They range from −1.08 for beak length during the second drought, to +0.88 for beak length in the first drought, with similar selection differentials for beak width and depth and overall beak size. These are very large selection differentials, compared to values calculated in other investigations of evolution.
Suggest reasons for natural selection on the beak size of *G. fortis* being unusually intense on the island of Daphne Major. [2]

6 Discuss the advantages of investigations of evolution over long periods and the reasons for few long-term investigations being done. [3]

Natural selection and antibiotic resistance

Use theories to explain natural phenomena: the theory of evolution by natural selection can explain the development of antibiotic resistance in bacteria.

Antibiotics were one of the great triumphs of medicine in the 20th century. When they were first introduced, it was expected that they would offer a permanent method of controlling bacterial diseases, but there have been increasing problems of antibiotic resistance in pathogenic bacteria.

The following trends have become established:

- After an antibiotic is introduced and used on patients, bacteria showing resistance appear within a few years.

- Resistance to the antibiotic spreads to more and more species of pathogenic bacteria.

- In each species the proportion of infections that are caused by a resistant strain increases.

So, during the time over which antibiotics have been used to treat bacterial diseases there have been cumulative changes in the antibiotic resistance properties of populations of bacteria. The development of antibiotic resistance is therefore an example of evolution. It can be explained in terms of the theory of natural selection. A scientific understanding of how antibiotic resistance develops is very useful as it gives an understanding of what should be done to reduce the problem.

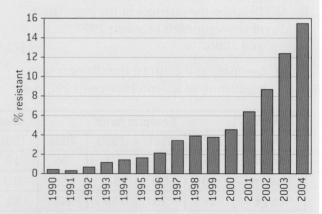

▲ Figure 11 Percentage resistance to ciprofloxacin between 1990 and 2004

Antibiotic resistance

Evolution of antibiotic resistance in bacteria.

Antibiotic resistance is due to genes in bacteria and so it can be inherited. The mechanism that causes antibiotic resistance to become more prevalent or to diminish is summarized in figure 12.

The evolution of multiple antibiotic resistance has occurred in just a few decades. This rapid evolution is due to the following causes:

- There has been very widespread use of antibiotics, both for treating diseases and in animal feeds used on farms.

- Bacteria can reproduce very rapidly, with a generation time of less than an hour.

- Populations of bacteria are often huge, increasing the chance of a gene for antibiotic resistance being formed by mutation.

- Bacteria can pass genes on to other bacteria in several ways, including using plasmids, which allow one species of bacteria to gain antibiotic resistance genes from another species.

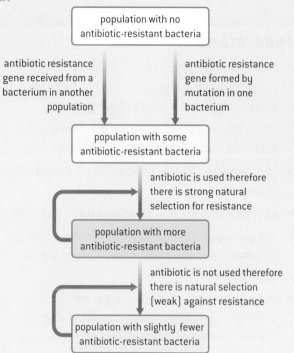

▲ Figure 12 Evolution of antibiotic resistance

Data-based questions: Chlortetracycline resistance in soil bacteria

Bacteria were collected from soil at different distances from a site on a pig farm in Minnesota where manure had been allowed to overflow from an animal pen and accumulate. The feed given to the pigs on this farm contained subtherapeutic low doses of the antibiotic chlortetracycline, in order to promote faster growth rates. The bacteria were tested to find out what percentage of them was resistant to this antibiotic. The results are shown in the bar chart. The yellow bars show the percentage of chlortetracycline resistant bacteria that grew on nutrient-rich medium and the orange bars show the percentage on a nutrient-poor medium that encouraged different types of bacteria to grow.

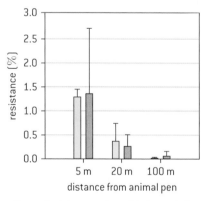

Source: " The effects of subtherapeutic antibiotic use in farm animals on the proliferation and persistence of antibiotic resistance among soil bacteria", Sudeshna Ghosh and Timothy M LaPara, *The International Society for Microbial Ecology Journal* (2007) 1, 191–203

1 a) State the relationship between percentage antibiotic resistance and distance from the animal pen.   [1]

   b) Explain the difference in antibiotic resistance between populations of bacteria near and far from the pen.   [4]

2 Predict whether the percentage antibiotic resistance would have been lower at 200 metres from the pen than at 100 metres.   [3]

3 Discuss the use of subtherapeutic doses of antibiotics in animal feeds.   [2]

5.3 Classification of biodiversity

Understanding

→ The binomial system of names for species is universal among biologists and has been agreed and developed at a series of congresses.

→ When species are discovered they are given scientific names using the binomial system.

→ Taxonomists classify species using a hierarchy of taxa.

→ All organisms are classified into three domains.

→ The principal taxa for classifying eukaryotes are kingdom, phylum, class, order, family, genus and species.

→ In a natural classification the genus and accompanying higher taxa consist of all the species that have evolved from one common ancestral species.

→ Taxonomists sometimes reclassify groups of species when new evidence shows that a previous taxon contains species that have evolved from different ancestral species.

→ Natural classifications help in identification of species and allow the prediction of characteristics shared by species within a group.

Applications

→ Classification of one plant and one animal species from domain to species level.

→ External recognition features of bryophytes, filicinophytes, coniferophytes and angiospermophytes.

→ Recognition features of porifera, cnidaria, platyhelminthes, annelida, mollusca and arthropoda, chordata.

→ Recognition of features of birds, mammals, amphibians, reptiles and fish.

Skills

→ Construction of dichotomous keys for use in identifying specimens.

Nature of science

→ Cooperation and collaboration between groups of scientists: scientists use the binomial system to identify a species rather than the many different local names.

International cooperation and classification

Cooperation and collaboration between groups of scientists: scientists use the binomial system to identify a species rather than the many different local names.

Recognizable groups of organisms are known to biologists as species. The same species can have many different local names, even within one language. For example, in England the species of plant known to scientists as *Arum maculatum* has been called lords-and-ladies, cuckoo-pint, jack in the pulpit, devils and angels, cows and bulls, willy lily and snake's meat. In French there is also a variety of local names: la chandelle, le pied-de-veau, le manteau de la Sainte-Vierge, la pilette or la vachotte. In Spanish there are even more names for this one species of which these are just a few: comida de culebra, alcatrax, barba de arón, dragontia menor, hojas de fuego, vela del diablo and yerba del quemado. The name primaveras is used for *Arum maculatum* in Spanish but for a different plant in other languages.

Local names may be a valuable part of the culture of an area, but science is an international venture so scientific names are needed that are understood throughout the world. The binomial system that has developed is a good example of cooperation and collaboration between scientists.

The credit for devising our modern system of naming species is given to the Swedish biologist Carl Linnaeus who introduced a system of two-part names in the 18th century. This stroke of genius was the basis for the binomial system that is still in use today. In fact Linnaeus was mirroring a style of nomenclature that had been used in many languages before. The style recognizes that there are groups of similar species, so the name for each species in a group consists of a specific name attached to the group name, as in the Ancient Greek αδιαυτου το λευκον and αδιαυτου το μεαυ (used by Threophrastus), Latin *anagallis mas* and *anagallis femina* (used by Pliny), German weiss Seeblumen and geel Seeblumen (used by Fuchs), English wild mynte and water mynte (used by Turner) and Malayan jambu bol and jambu chilli (applied by Malays to different species of *Eugenia*).

▲ Figure 1 *Arum maculatum*

Development of the binomial system

The binomial system of names for species is universal among biologists and has been agreed and developed at a series of congresses.

To ensure that all biologists use the same system of names for living organisms, congresses attended by delegates from around the world are held at regular intervals. There are separate congresses for animals and for plants and fungi.

International Botanical Congresses (IBC) were held every year during the late 19th century. The IBC held in Genoa in 1892 proposed that 1753 be taken as the starting point for both genera and species of plants and fungi as this was the year when Linnaeus published *Species Plantarum*, the book that gave consistent binomials for all species of the plant kingdom then known. The IBC of Vienna in 1905 accepted by 150 votes to 19 the rule that "La nomenclature botanique commence avec Linné, *Species Plantarum* (ann. 1753) pour les groupes de plantes vasculaires." The 19th IBC will be in Shenzhen, China, in 2017.

The first International Zoological Congress was held in Paris in 1889. It was recognized that internationally accepted rules for naming and classifying animal species were needed and these were agreed at this and subsequent congresses. 1758 was chosen as the starting date for valid names of animal species as this was when Linnaeus published *Systema Natura* in which he gave binomials for all species known then. The current International Code for Zoological Nomenclature is the 4th edition and there will no doubt be more editions in the future as scientists refine the methods that they use for naming species.

▲ Figure 2 *Linnaea borealis*. Binomials are often chosen to honour a biologist, or to describe a feature of the organism. *Linnaea borealis* is named in honour of Carl Linnaeus, the Swedish biologist who introduced the binomial system of nomenclature and named many plants and animals using it

The binomial system

When species are discovered they are given scientific names using the binomial system.

The system that biologists use is called binomial nomenclature, because the international name of a species consists of two words. An example is *Linnaea borealis* (figure 2). The first name is the genus name. A genus is a group of species that share certain characteristics. The second name is the species or specific name. There are various rules about binomial nomenclature:

- The genus name begins with an upper-case (capital) letter and the species name with a lower-case (small) letter.

- In typed or printed text, a binomial is shown in italics.

- After a binomial has been used once in a piece of text, it can be abbreviated to the initial letter of the genus name with the full species name, for example: *L. borealis*.

- The earliest published name for a species, from 1753 onwards for plants or 1758 for animals, is the correct one.

The hierarchy of taxa

Taxonomists classify species using a hierarchy of taxa.

The word taxon is Greek and means a group of something. The plural is taxa. In biology, species are arranged or classified into taxa. Every species is classified into a genus. Genera are grouped into families. An example of the genera and species in a family is shown in figure 3. Families are grouped into orders, orders into classes and so on up to the level of kingdom or domain. The taxa form a hierarchy, as each taxon includes taxa from the level below. Going up the hierarchy, the taxa include larger and larger numbers of species, which share fewer and fewer features.

The three domains

All organisms are classified into three domains.

Traditional classification systems have recognized two major categories of organisms based on cell types: eukaryotes and prokaryotes. This classification is now regarded as inappropriate because the prokaryotes have been found to be very diverse. In particular, when the base sequence of ribosomal RNA was determined, it became apparent that there are two distinct groups of prokaryotes. They were given the names Eubacteria and Archaea.

Most classification systems therefore now recognize three major categories of organism, Eubacteria, Archaea and Eukaryota. These categories are called domains, so all organisms are classified into three domains. Table 1 shows some of the features that can be used to distinguish between them. Members of the domains are usually referred to as bacteria, archaeans and eukaryotes. Bacteria and eukaryotes are relatively familiar to most biologists but archaeans are often less well known.

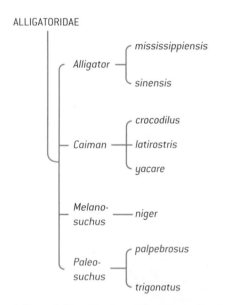

▲ Figure 3 Classification of the alligator family

Feature	Domain		
	Bacteria	**Archaea**	**Eukaryota**
Histones associated with DNA	Absent	Proteins similar to histones bound to DNA	Present
Presence of introns	Rare or absent	Present in some genes	Frequent
Structure of cell walls	Made of chemical called peptidoglycan	Not made of peptidoglycan	Not made of peptidoglycan; not always present
Cell membrane differences	Glycerol-ester lipids; unbranched side chains; d-form of glycerol	Glycerol-ether lipids; unbranched side chains; l-form of glycerol	Glycerol-ester lipids; unbranched side chains; d-form of glycerol

▲ Table 1

Archaeans are found in a broad range of habitats such as the ocean surface, deep ocean sediments and even oil deposits far below the surface of the Earth. They are also found in some fairly extreme habitats such as water with very high salt concentrations or temperatures close to boiling. The methanogens are obligate anaerobes and give off methane as a waste product of their metabolism. Methanogens live in the intestines of cattle and the guts of termites and are responsible for the production of "marsh gas" in marshes.

Viruses are not classified in any of the three domains. Although they have genes coding for proteins using the same genetic code as living organisms they have too few of the characteristics of life to be regarded as living organisms.

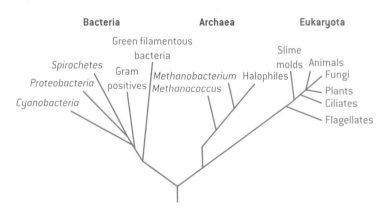

▲ Figure 4 Tree diagram showing relationships between living organisms based on base sequences of ribosomal RNA

Eukaryote classification

The principal taxa for classifying eukaryotes are kingdom, phylum, class, order, family, genus and species.

Eukaryotes are classified into kingdoms. Each kingdom is divided up into phyla, which are divided into classes, then orders, families and genera. The hierarchy of taxa for classifying eukaryotes is thus kingdom, phylum, class, order, family, genus and species.

Most biologists recognize four kingdoms of eukaryote: plants, animals, fungi and protoctista. The last of these is the most controversial as protoctists are very diverse and should be divided up into more kingdoms. At present there is no consensus on how this should be done.

261

Examples of classification

Classification of one plant and one animal species from domain to species level.

Animals and plants are kingdoms of the domain Eukaryota. Table 2 shows the classification of one plant and one animal species from kingdom down to species.

Taxon	Grey wolf	Date palm
Kingdom	Animalia	Plantae
Phylum	Chordata	Angiospermophyta
Class	Mammalia	Monocotyledoneae
Order	Carnivora	Palmales
Family	Canidae	Arecaceae
Genus	*Canis*	*Phoenix*
Species	*lupus*	*dactylifera*

▲ Table 2

Data-based questions: Classifying cartilaginous fish

All the fish shown in figure 6 are in the class Chondrichthyes. They are the most frequently found fish in this class in north-west Europe.

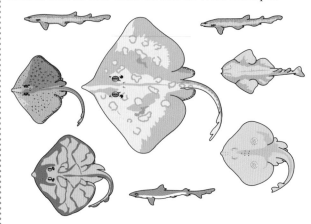

▲ Figure 6 Cartilaginous fish in seas in north-west Europe

1. State the kingdom to which all of the species in figure 6 belong. [1]

2. a) Four of the fish in figure 6 are classified in the same genus. Deduce which these fish are. [1]

   b) Deduce with a reason whether these four fish are in:

      (i) the same or different species [2]

      (ii) the same or different families. [2]

   c) State two characteristics of these four fish that are not possessed by the other four fish. [2]

3. The other four fish are classified into two orders. Deduce, with a reason, how the four fish are split into two orders. [2]

Natural classification

In a natural classification, the genus and accompanying higher taxa consist of all the species that have evolved from one common ancestral species.

Scientific consensus is to classify species in a way that most closely follows the way in which species evolved. Following this convention, all members of a genus or higher taxon should have a common ancestor. This is called a natural classification. Because of the common ancestry we can expect the members of a natural group to share many characteristics.

An example of an unnatural or artificial classification would be one in which birds, bats and insects are grouped together, because they all fly. Flight evolved separately in these groups and as they do not share a common ancestor they differ in many ways. It would not be appropriate to classify them together other than to place them all in the animal kingdom and both birds and bats in the phylum Chordata. Plants and fungi were at one time classified together, presumably because they have cell walls and do not move, but this is an artificial classification as their cell walls evolved separately and molecular research shows that they are no more similar to each other than to animals.

It is not always clear which groups of species do share a common ancestor, so natural classification can be problematic. Convergent evolution can make distantly related organisms appear superficially similar and adaptive radiation can make closely related organisms appear different. In the past, natural classification was attempted by looking at as many visible characteristics as possible, but new molecular methods have been introduced and these have caused significant changes to the classification of some groups. More details of this are given later, in sub-topic 5.4.

TOK

What factors influence the development of a scientific consensus?

Carl Linnaeus's 1753 book *Species Plantarum* introduced consistent two-part names (binomials) for all species of the vegetable kingdom then known. Thus the binomial *Physalis angulata* replaced the obsolete phrase-name, *Physalis annua ramosissima, ramis angulosis glabris, foliis dentato-serratis*. Linnaeus brought the scientific nomenclature of plants back to the simplicity and brevity of the vernacular nomenclature out of which it had grown. Folk-names for species rarely exceed three words. In groups of species alike enough to have a vernacular group-name, the species are often distinguished by a single name attached to the group-name, as in the Ancient Greek αδιαυτου το λενκον and αδιαυτου το μεαυ (used by Threophrastus), Latin anagallis mas and anagallis femina (used by Pliny), German weiss Seeblumen and geel Seeblumen (used by Fuchs), English wild mynte and water mynte (used by Turner) and Malayan jambu bol and jambu chilli (applied by Malays to different species of *Eugenia*).

The International Botanical Congress held in Genoa in 1892 proposed that 1753 be taken as the starting point for both genera and species. This was incorporated in the American "Rochester Code" of 1883 and in the code used at the Berlin Botaniches Museum and supported by British Museum of Natural History, Harvard University botanists and a group of Swiss and Belgian botanists. The International Botanical Congress of Vienna in 1905 accepted by 150 votes to 19 the rule that "La nomenclature botanique commence avec Linné, Species Plantarum (ann. 1753) pour les groupes de plantes vasculaires."

1   Why was Linnaeus's system for naming plants adopted as the international system, rather than any other system?

2   Why do the international rules of nomenclature state that genus and species names must be in Ancient Greek or Latin?

3   Making decisions by voting is rather unusual in science. Why is it done at International Botanical Congresses? What knowledge issues are associated with this method of decision making?

Reviewing classification

Taxonomists sometimes reclassify groups of species when new evidence shows that a previous taxon contains species that have evolved from different ancestral species.

Sometimes new evidence shows that members of a group do not share a common ancestor, so the group should be split up into two or more taxa. Conversely species classified in different taxa are sometimes found to be closely related, so two or more taxa are united, or species are moved from one genus to another or between higher taxa.

The classification of humans has caused more controversy than any other species. Using standard taxonomic procedures, humans are assigned to the order Primates and the family Hominidae. There has been much debate about which, if any, of the great apes to include in this family. Originally all the great apes were placed in another family,

▲ Figure 8 Members of the Hominidae and Pongidae

Activity

Controlling potato blight

Phytophthora infestans, the organism that causes the disease potato blight, has hyphae and was classified as a fungus, but molecular biology has shown that it is not a true fungus and should be classified in a different kingdom, possibly the Protoctista. Potato blight has proved to be a difficult disease to control using fungicides. Discuss reasons for this.

---

the Pongidae, but research has shown that chimpanzees and gorillas are closer to humans than to orang-utans and so should be in the same family. This would just leave orang-utans in the Pongidae. Most evidence suggests that chimpanzees are closer than gorillas to humans, so if humans and chimpanzees are placed in different genera, gorillas should also be in a separate genus. A summary of this scheme for human classification is shown in figure 7.

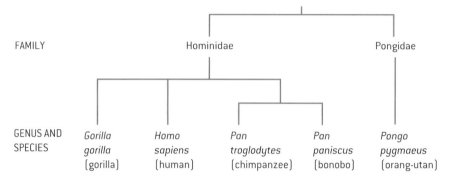

▲ Figure 7 Classification of humans

Advantages of natural classification

Natural classifications help in identification of species and allow the prediction of characteristics shared by species within a group.

There is great interest at the moment in the biodiversity of the world. Groups of biologists are surveying areas where little research has been done before, to find out what species are present. Even in well-known parts of the world new species are sometimes discovered. Natural classification of species is very helpful in research into biodiversity. It has two specific advantages.

1 Identification of species is easier. If a specimen of an organism is found and it is not obvious what species it is, the specimen can be identified by assigning it first to its kingdom, then the phylum within the kingdom, class within the phylum and so on down to species level. Dichotomous keys can be used to help with this process. This process would not work so well with an artificial classification. For example, if flowering plants were classified according to flower colour and a white-flowered bluebell *Hyacinthoides non-scripta* was discovered, it would not be identified correctly as the species normally has blue flowers.

2 Because all of the members of a group in a natural classification have evolved from a common ancestral species, they inherit similar characteristics. This allows prediction of the characteristics of species within a group. For example, if a chemical that is useful as a drug is found in one plant in a genus, this or related chemicals are likely to be found in other species in the genus. If a new species of bat was discovered, we could make many predictions about it with reasonable certainty that they are correct: the bat will have hair, mammary glands, a placenta, a four-chambered heart and many other mammalian features. None of these predictions could be made if bats were classified artificially with all other flying organisms.

Dichotomous keys

Construction of dichotomous keys for use in identifying specimens

Dichotomous keys are often constructed to use for identifying species within a group. A dichotomy is a division into two; a dichotomous key consists of a numbered series of pairs of descriptions. One of these should clearly match the species and the other should clearly be wrong. The features that the designer of the key chooses to use in the descriptions should therefore be reliable and easily visible. Each of the pair of descriptions leads either to another of the numbered pairs of descriptions in the key, or to an identification.

An example of a key is shown in table 3. We can use it to identify the species in figure 9. In the first stage of the key, we must decide if hind limbs are visible. They are not, so we are directed to stage 6 of the key. We must now decide if the species has a blowhole. It does not, so it is a dugong or a manatee. A fuller key would have another stage to separate dugongs and manatees.

1	Fore and hind limbs visible, can emerge on land 2
	Only fore limbs visible, cannot live on land 6
2	Fore and hind limbs have paws 3
	Fore and hind limbs have flippers 4
3	Fur is dark .... sea otters
	Fur is white .... polar bears
4	External ear flap visible .... sea lions and fur seals
	No external ear flap 5
5	Two long tusks .... walruses
	No tusks .... true seals
6	Mouth breathing, no blowhole ... dugongs and manatees
	Breathing through blowholes 7
7	Two blowholes, no teeth .... baleen whales
	One blowhole, teeth .... dolphins, porpoises and whales

▲ Table 3 Key to groups of marine mammals

Activity

Constructing dichotomous keys

Keys are usually designed for use in a particular area. All the groups or species that are found in that area can be identified using the key. There may be a group of organisms in your area for which a key has never been designed.

- You could design a key to the trees in the local forest or on your school campus, using leaf descriptions or bark descriptions.

- You could design a key to birds that visit bird-feeding stations in your area.

- You could design a key to the invertebrates that are associated with one particular plant species.

- You could design a key to the footprints of mammals and birds (figure 10). They are all right front footprints and are not shown to scale.

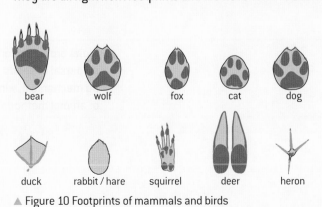

▲ Figure 10 Footprints of mammals and birds

▲ Figure 9 Manatee

Plants

External recognition features of bryophytes, filicinophytes, coniferophytes and angiospermophytes.

All plants are classified together in one kingdom. In the life cycle of every plant, male and female gametes are formed and fuse together. The zygote formed develops into an embryo. The way in which this embryo develops depends on the type of plant it is. The different types of plants are put into phyla.

Most plants are in one of four phyla, but there are other smaller phyla. The *Ginkgo biloba* tree for example is in one of the smaller phyla. The four main plant phyla are:

- Bryophyta – mosses, liverworts and hornworts
- Filicinophyta – ferns
- Coniferophyta – conifers
- Angiospermophyta – flowering plants.

The external recognition features of these phyla are shown in table 4.

	Bryophyta	Filicinophyta	Coniferophyta	Angiospermophyta
Vegetative organs – parts of the plant concerned with growth rather than reproduction	Rhizoids but no true roots. Some with simple stems and leaves; others have only a thallus	Roots, stems and leaves are usually present		
Vascular tissue – tissues with tubular structures used for transport within the plant	No xylem or phloem	Xylem and phloem are both present		
Cambium – cells between xylem and phloem that can produce more of these tissues	No cambium; no true trees and shrubs		Present in conifers and most angiosperms, allowing secondary thickening of stems and roots and development of plants into trees and shrubs	
Pollen – small structures containing male gametes that are dispersed	Pollen is not produced		Pollen is produced in male cones	Pollen is produced by anthers in flowers
Ovules – contains a female gamete and develops into a seed after fertilization	No ovaries or ovules		Ovules are produced in female cones	Ovules are enclosed inside ovaries in flowers
Seeds – dispersible unit consisting of an embryo plant and food reserves, inside a seed coat	No seeds		Seeds are produced and dispersed	
Fruits – seeds together with a fruit wall developed from the ovary wall	No fruits			Fruits produced for dispersal of seeds by mechanical, wind or animal methods

▲ Table 4

Animal phyla

Recognition features of porifera, cnidaria, platyhelminthes, annelida, mollusca and arthropoda, chordata.

Animals are divided up into over 30 phyla, based on their characteristics. Six phyla are featured in table 5. Two examples of each are shown in figure 11.

Phylum	Mouth/anus	Symmetry	Skeleton	Other external recognition features
Porifera – fan sponges, cup sponges, tube sponges, glass sponges	No mouth or anus	None	Internal spicules (sketetal needles)	Many pores over the surface through which water is drawn in for filter feeding. Very varied shapes
Cnidaria – hydras, jellyfish, corals, sea anemones	Mouth only	Radial	Soft, but hard corals secrete $CaCO_3$	Tentacles arranged in rings around the mouth, with stinging cells. Polyps or medusae (jellyfish)
Platyhelminthes – flatworms, flukes, tapeworms	Mouth only	Bilateral	Soft, with no skeleton	Flat and thin bodies in the shape of a ribbon. No blood system or system for gas exchange
Mollusca – bivalves, gastropods, snails, chitons, squid, octopus	Mouth and anus	Bilateral	Most have shell made of $CaCO_3$	A fold in the body wall called the mantle secretes the shell. A hard rasping radula is used for feeding
Annelida – marine bristleworms, oligochaetes, leeches	Mouth and anus	Bilateral	Internal cavity with fluid under pressure	Bodies made up of many ring-shaped segments, often with bristles. Blood vessels often visible
Arthropoda – insects, arachnids, crustaceans, myriapods	Mouth and anus	Bilateral	External skeleton made of plates of chitin	Segmented bodies and legs or other appendages with joints between the sections

▲ Table 5 Characteristics of six animal phyla

1 Study the organisms shown in figure 11 and assign each one to its phylum. [7]

2 List the organisms that are:
   a) bilaterally symmetric
   b) radially symmetric
   c) not symmetrical in their structure. [3]

3 List the organisms that have:
   a) jointed appendages
   b) stinging tentacles
   c) bristles. [3]

4 List the organisms that filter feed by pumping water through tubes inside their bodies. [2]

Adocia cinerea   *Alcyonium glomeratum*

Nymphon gracilis   *Pycnogonum littorale*

Corynactis viridis   *Lepidonotus clara*

Polymastia mammiliaris   *Cyanea capillata*

Procerodes littoralis

Loligo forbesii

Arenicola marina

Prostheceraeus vittatus

Caprella linearis

Gammarus locusta

▲ Figure 11 Invertebrate diversity

Vertebrates

Recognition of features of birds, mammals, amphibians, reptiles and fish.

Most species of chordate belong to one of five major classes, each of which contains more than a thousand species. Although the numbers are not certain and new species are still sometimes discovered, there are about 10,000 bird species, 9,000 reptiles, 6,000 amphibians and 5,700 mammals. All of these classes are outnumbered by the ray-finned bony fish, with more than 30,000 species. The recognition features of the five largest classes of chordate are shown in table 6. All of the organisms are vertebrates, because they have a backbone composed of vertebrae.

Bony ray-finned fish	Amphibians	Reptiles	Birds	Mammals
Scales which are bony plates in the skin	Soft moist skin permeable to water and gases	Impermeable skin covered in scales of keratin	Skin with feathers made of keratin	Skin has follicles with hair made of keratin
Gills covered by an operculum, with one gill slit	Simple lungs with small folds and moist skin for gas exchange	Lungs with extensive folding to increase the surface area	Lungs with para-bronchial tubes, ventilated using air sacs	Lungs with alveoli, ventilated using ribs and a diaphragm
No limbs	Tetrapods with pentadactyl limbs			
Fins supported by rays	Four legs when adult	Four legs (in most species)	Two legs and two wings	Four legs in most (or two legs and two wings/arms)
Eggs and sperm released for external fertilization	Sperm passed into the female for internal fertilization			
Remain in water throughout their life cycle	Larval stage that lives in water and adult that usually lives on land	Female lays eggs with soft shells	Female lays eggs with hard shells	Most give birth to live young and all feed young with milk from mammary glands
Swim bladder containing gas for buoyancy	Eggs coated in protective jelly	Teeth all of one type, with no living parts	Beak but no teeth	Teeth of different types with a living core
Do not maintain constant body temperature			Maintain constant body temperature	

▲ Table 6

5.4 Cladistics

Understanding

→ A clade is a group of organisms that have evolved from a common ancestor.

→ Evidence for which species are part of a clade can be obtained from the base sequences of a gene or the corresponding amino acid sequence of a protein.

→ Sequence differences accumulate gradually so there is a positive correlation between the number of differences between two species and the time since they diverged from a common ancestor.

→ Traits can be analogous or homologous.

→ Cladograms are tree diagrams that show the most probable sequence of divergence in clades.

→ Evidence from cladistics has shown that classifications of some groups based on structure did not correspond with the evolutionary origins of a group of species.

Applications

→ Cladograms including humans and other primates.

→ Reclassification of the figwort family using evidence from cladistics.

Skills

→ Analysis of cladograms to deduce evolutionary relationships.

Nature of science

→ Falsification of theories with one theory being superseded by another: plant families have been reclassified as a result of evidence from cladistics.

Clades

A clade is a group of organisms that have evolved from a common ancestor.

Species can evolve over time and split to form new species. This has happened repeatedly with some highly successful species, so that there are now large groups of species all derived from a common ancestor. These groups of species can be identified by looking for shared characteristics. A group of organisms evolved from a common ancestor is called a clade.

Clades include all the species alive today, together with the common ancestral species and any species that evolved from it and then became extinct. They can be very large and include thousands of species, or very small with just a few. For example, birds form one large clade with about ten thousand living species because they have all evolved from a common ancestral species. The tree *Ginkgo biloba* is the only living member of a clade that evolved about 270 million years ago. There have been other species in this clade but all are now extinct.

Activity

The EDGE of Existence project

The aim of this project is to identify animal species that have few or no close relatives and are therefore members of very small clades. The conservation status of these species is then assessed. Lists are prepared of species that are both Evolutionarily Distinct and Globally Endangered, hence the name of the project. Species on these lists can then be targeted for more intense conservation efforts than other species that are either not threatened or have close relatives. In some cases species are the last members of a clade that has existed for tens or hundreds of millions of years and it would be tragic for them to become extinct as a result of human activities.

What species on EDGE lists are in your part of the world and what can you do to help conserve them?

http://www.edgeofexistence.org/species/

▲ Figure 1 Two species on the EDGE list: *Loris tardigradus tardigradus* (Horton Plains slender loris) from Sri Lanka and *Bradypus pygmaeus* (Pygmy three-toed sloth) from Isla Escudo de Veraguas, a small island off the coast of Panama

Identifying members of a clade

Evidence for which species are part of a clade can be obtained from the base sequences of a gene or the corresponding amino acid sequence of a protein.

It is not always obvious which species have evolved from a common ancestor and should therefore be included in a clade.

The most objective evidence comes from base sequences of genes or amino acid sequences of proteins. Species that have a recent common ancestor can be expected to have few differences in base or amino acid sequence. Conversely, species that might look similar in certain respects but diverged from a common ancestor tens of millions of years ago are likely to have many differences.

Molecular clocks

Sequence differences accumulate gradually so there is a positive correlation between the number of differences between two species and the time since they diverged from a common ancestor.

Differences in the base sequence of DNA and therefore in the amino acid sequence of proteins are the result of mutations. They accumulate gradually over long periods of time. There is evidence that mutations occur at a roughly constant rate so they can be used as a molecular clock. The number of differences in sequence can be used to deduce how long ago different groups split from a common ancestor.

For example, mitochondrial DNA from three humans and four related primates has been completely sequenced. From the differences in base sequence, a hypothetical ancestry has been constructed. It is shown in figure 2. Using differences in base sequence as a molecular clock, these approximate dates for splits between groups have been deduced:

- 70,000 years ago, European–Japanese split

- 140,000 years ago, African–European/Japanese split

- 5,000,000 years ago, human–chimpanzee split

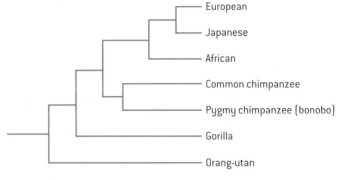

▲ Figure 2

Analogous and homologous traits

Traits can be analogous or homologous.

Similarities between organisms can either be homologous or analogous.

- Homologous structures are similar because of similar ancestry; for example the chicken wing, human arm and other pentadactyl forelimbs.

- Analogous structures are similar because of convergent evolution. The human eye and the octopus eye show similarities in structure and function but they are analogous because they evolved independently.

Problems in distinguishing between homologous and analogous structures have sometimes led to mistakes in classification in the past. For this reason the morphology (form and structure) of organisms is now rarely used for identifying members of a clade and evidence from base or amino acid sequences is trusted more.

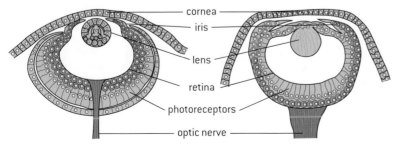

▲ Figure 3 The human eye (left) and the octopus eye (right) are analogous because they are quite similar yet evolved independently

271

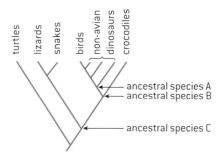

▲ Figure 4 A cladogram showing the hypothesized relationship between birds and the traditional taxonomic group "the reptiles"

Activity

Figure 5 shows an artist's impression of two pterosaurs, which were the first chordates to develop powered flight. They were neither birds nor dinosaurs. Where might pterosaurs have fitted into the cladogram shown in figure 4?

▲ Figure 5 Two pterosaurs in flight

Cladograms

Cladograms are tree diagrams that show the most probable sequence of divergence in clades.

A cladogram is a tree diagram based on similarities and differences between the species in a clade. Cladograms are almost always now based on base or amino acid sequences. Computer programs have been developed that calculate how species in a clade could have evolved with the smallest number of changes of base or amino acid sequence. This is known as the principle of parsimony and although it does not prove how a clade actually evolved, it can indicate the most probable sequence of divergence in clades.

The branching points on cladograms are called nodes. Usually two clades branch off at a node but sometimes there are three or more. The node represents a hypothetical ancestral species that split to form two or more species. Option B includes instructions for constructing cladograms from base sequences using computer software.

Figure 4 is an example of a cladogram for birds and reptiles. It has been based on morphology, so that extinct groups can be included.

- Birds, non-avian dinosaurs and ancestral species A form a clade called dinosauria.

- Birds, non-avian dinosaurs, crocodiles and ancestral species B are part of a clade called archosaurs.

- Lizards, snakes and ancestral species C form a clade called squamates.

This cladogram suggests either that birds should be regarded as reptiles or that reptiles should be divided into two or more groups, as some reptiles are more closely related to birds than to other reptiles.

 Primate cladograms

Cladograms including humans and other primates.

The closest relatives of humans are chimpanzees and bonobos. The entire genome of these three species has been sequenced giving very strong evidence for the construction of a cladogram (figure 6). The numbers on the cladogram are estimates of population sizes and dates when splits occurred. These are based on a molecular clock with a mutation rate of 10^{-9} yr^{-1}.

Figure 7 is a cladogram for primates and the most closely related other groups of mammal. Primates are an order of mammals that have adaptations for climbing trees. Humans, monkeys, baboons, gibbons and lemurs are primates.

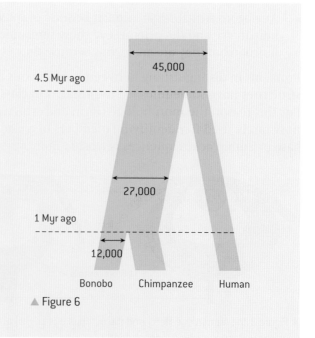

▲ Figure 6

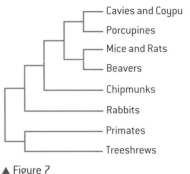

▲ Figure 7

Analysis of cladograms

Analysis of cladograms to deduce evolutionary relationships.

The pattern of branching in a cladogram is assumed to match the evolutionary origins of each species. The sequence of splits at nodes is therefore a hypothetical sequence in which ancestors of existing clades diverged. If two clades on a cladogram are linked at a node, they are relatively closely related. If two species are only connected via a series of nodes, they are less closely related.

Some cladograms include numbers to indicate numbers of differences in base or amino acid sequence or in genes. Because genetic changes are assumed to occur at a relatively constant rate, these numbers can be used to estimate how long ago two clades diverged. This method of estimating times is called a molecular clock. Some cladograms are drawn to scale according to estimates of how long ago each split occurred.

Although cladograms can provide strong evidence for the evolutionary history of a group, they cannot be regarded as proof. Cladograms are constructed on the assumption that the smallest possible number of mutations occurred to account for current base or amino acid sequence differences. Sometimes this assumption is incorrect and pathways of evolution were more convoluted. It is therefore important to be cautious in analysis of cladograms and where possible compare several versions that have been produced independently using different genes.

Activity

A cladogram for the great apes

The great apes are a family of primates. The taxonomic name is Hominidae. There are five species on Earth today, all of which are decreasing in number apart from humans. Figure 6 is a cladogram for three of the species. Use this information to expand the cladogram to include all the great apes: the split between humans and gorillas occurred about 10 million years ago and the split between humans and orang-utans about 15 million years ago.

Data-based questions: Origins of turtles and lizards

Cladograms based on morphology suggest that turtles and lizards are not a clade. To test this hypothesis, microRNA genes have been compared for nine species of chordate. The results were used to construct the cladogram in figure 8. The numbers on the cladogram show which microRNA genes are shared by members of a clade but not members of other clades. For example, there are six microRNA genes found in *humans* and *short-tailed opossums* but not in any of the other chordates on the cladogram.

1   Deduce, using evidence from the cladogram, whether humans are more closely related to the short-tailed opossum or to the duck-billed platypus. [2]

2   Calculate how many microRNA genes are found in the mammal clade on the cladogram but not in the other clades. [2]

3   Discuss whether the evidence in the cladogram supports the hypothesis that turtles and lizards are not a clade. [3]

4   Evaluate the traditional classification of tetrapod chordates into amphibians, reptiles, birds and mammals using evidence from the cladogram. [3]

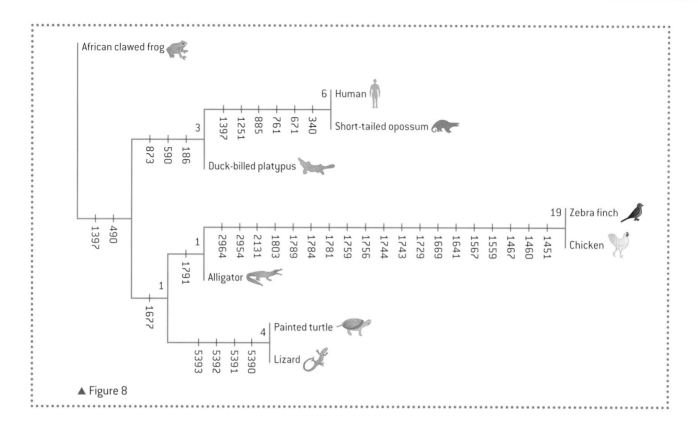

▲ Figure 8

Cladograms and reclassification

Evidence from cladistics has shown that classifications of some groups based on structure did not correspond with the evolutionary origins of a group of species.

The construction of cladograms based on base and amino acid sequences only became possible towards the end of the 20th century. Before that the sequence data was not available and computer software had not been developed to do the analysis. The construction of cladograms and identification of clades is known as cladistics.

Cladistics has caused some revolutions in plant and animal classification. It is now clear from cladograms that traditional classification based on morphology does not always match the evolutionary origins of groups of species. As a result some groups have been reclassified. Some groups have been merged, others have been divided and in some cases species have been transferred from one group to another.

Reclassification of groups of organisms is time-consuming and potentially disruptive for biologists, but it is certainly worthwhile. The new classifications based on cladistics are likely to be much closer to a truly natural classification so their predictive value will be higher. They have revealed some unnoticed similarities between groups and also some significant differences between species previously assumed to be similar.

Cladograms and falsification

Falsification of theories with one theory being superseded by another: plant families have been reclassified as a result of evidence from cladistics.

The reclassification of plants on the basis of discoveries in cladistics is a good example of an important process in science: the testing of theories and of replacement of theories found to be false with new theories. The classification of angiospermophytes into families based on their morphology was begun by the French botanist Antoine Laurent de Jussieu in *Genera plantarum*, published in 1789 and revised repeatedly during the 19th century.

Classification of the figwort family

Reclassification of the figwort family using evidence from cladistics.

There are more than 400 families of angiosperms. Until recently the eighth largest was the Scrophulariaceae, commonly known as the figwort family. It was one of the original families proposed by de Jussieu in 1789. He gave it the name Scrophulariae and included sixteen genera, based on similarities in their morphology. As more plants were discovered, the family grew until there were over 275 genera, with more than 5,000 species.

Taxonomists recently investigated the evolutionary origins of the figwort family using cladistics. One important research project compared the base sequences of three chloroplast genes in a large number of species in genera traditionally assigned to the Scrophulariaceae and genera in closely related families. It was found that species in the figwort family were not a true clade and that five clades had incorrectly been combined into one family.

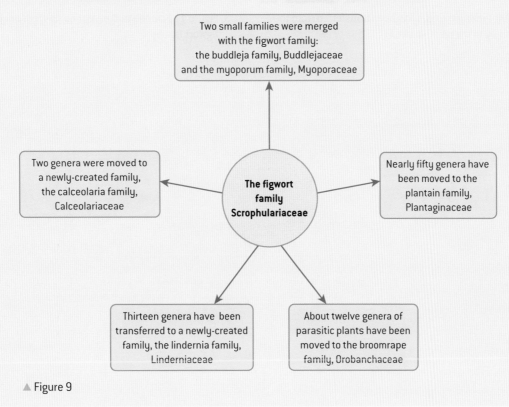

▲ Figure 9

A major reclassification has now been carried out. Less than half of the species have been retained in the family, which is now only the thirty-sixth largest among the angiosperms. A summary of the changes is shown in figure 9. This reclassification has been welcomed as it was widely appreciated before that the Scrophulariaceae had been a rag-bag of species rather than a natural group.

▲ Figure 10 *Antirrhinum majus* has been transferred from the figwort family to the plantain family

▲ Figure 11 *Scrophularia peregrina* has remained in the figwort family

Questions

The bar charts in figure 12 show the growth of three populations of an alga, *Ectocarpus siliculosus*, at different copper concentrations. One population came from an unpolluted environment at Rhosneigr in Wales. The other two came from the undersides of ships that had been painted with a copper-containing anti-fouling paint.

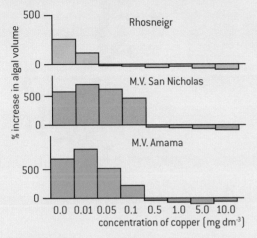

▲ Figure 12

1 How much higher was the maximum copper concentration tolerated by the algae from ships than the algae from an unpolluted environment?

a) 0.09 times higher   **b)** 0.11 times higher

c) 1.0 times higher   **d)** 10 times higher.

2 What is the reason for results lower than zero on the bar charts?

a) The volume of algae decreased.

b) The algae all died.

c) Increases in volume were less than 100%.

d) Results were too small to measure accurately.

3 What was the reason for the difference in copper tolerance between the algae?

a) The algae on the ships absorbed copper.

b) The algae can develop copper tolerance and pass it on to their offspring.

c) The copper in the paint caused mutations.

d) The copper in the paint caused natural selection for higher levels of copper tolerance.

4 Which of the following processes are required for copper tolerance to develop in a population?

(i) variation in copper tolerance

(ii) inheritance of copper tolerance

(iii) failure of algae with lower copper tolerance to survive or reproduce.

a) i) only

b) i) and ii) only

c) i) and iii) only

d) i), ii) and iii).

5 In figure 13, each number represents a species. The closer that two numbers are on the diagram the more similar the two species. The circles represent taxonomic groups. For example, the diagram shows that 2, 3, 4 and 5 are in the same genus.

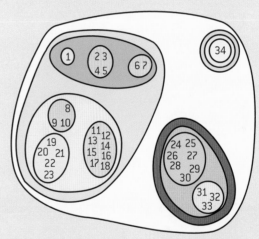

▲ Figure 13

a) State one species that is in a genus with no other species. [1]

b) State the species that are in a family with two genera. [2]

c) State the species that are in an order with two families. [2]

d) State the species that are in a class with three orders. [2]

e) Deduce whether species 8 is more closely related to species 16 or species 6. [2]

f) Explain why three concentric circles have been drawn around species 34 on the diagram. [2]

6   The map in figure 14 shows the distribution
    in the 1950s of two forms of *Biston betularia*
    in Britain and Ireland. *Biston betularia* is a
    species of moth that flies at night. It spends
    the daytime roosting on the bark of trees. The
    non-melanic form has white wings, peppered
    with black spots. The melanic form has black
    wings. Before the industrial revolution, the
    melanic form was very rare. The prevailing
    wind direction is from the Atlantic Ocean, to
    the west.

    a) State the maximum and minimum
         percentages of the melanic form.   [2]

    b) Outline the trends in the distribution of
         the two forms of *Biston betularia*, shown
         in figure 14.   [2]

    c) Explain how natural selection can cause
         moths such as *Biston betularia* to develop
         camouflaged wing markings.   [4]

    d) Suggest reasons for the distribution of
         the two forms.   [2]

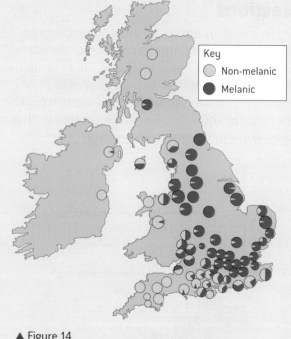

Key
◯ Non-melanic
● Melanic

▲ Figure 14

6 HUMAN PHYSIOLOGY

Introduction

Research into human physiology is the foundation of modern medicine. Body functions are carried out by specialized organ systems. The structure of the wall of the small intestine allows it to move, digest and absorb food. The blood system continuously transports substances to cells and simultaneously collects waste products. The skin and immune system resist the continuous threat of invasion by pathogens. The lungs are actively ventilated to ensure that gas exchange can occur passively. Neurons transmit the message, synapses modulate the message. Hormones are used when signals need to be widely distributed.

6.1 Digestion and absorption

Understanding

→ The contraction of circular and longitudinal muscle layers of the small intestine mixes the food with enzymes and moves it along the gut.

→ The pancreas secretes enzymes into the lumen of the small intestine.

→ Enzymes digest most macromolecules in food into monomers in the small intestine.

→ Villi increase the surface area of epithelium over which absorption is carried out.

→ Villi absorb monomers formed by digestion as well as mineral ions and vitamins.

→ Different methods of membrane transport are required to absorb different nutrients.

Applications

→ Processes occurring in the small intestine that result in the digestion of starch and transport of the products of digestion to the liver.

→ Use of dialysis tubing to model absorption of digested food in the intestine.

Skills

→ Production of an annotated diagram of the digestive system.

→ Identification of tissue layers in transverse sections of the small intestine viewed with a microscope or in a micrograph.

Nature of science

→ Use models as representations of the real world: dialysis tubing can be used to model absorption in the intestine.

Structure of the digestive system

Production of an annotated diagram of the digestive system.

The part of the human body used for digestion can be described in simple terms as a tube through which food passes from the mouth to the anus. The role of the digestive system is to break down the diverse mixture of large carbon compounds in food, to yield ions and smaller compounds that can be absorbed. For proteins, lipids and polysaccharides digestion involves several stages that occur in different parts of the gut.

Digestion requires surfactants to break up lipid droplets and enzymes to catalyse reactions. Glandular cells in the lining of the stomach and intestines produce some of the enzymes.

Surfactants and other enzymes are secreted by accessory glands that have ducts leading to the digestive system. Controlled, selective absorption of the nutrients released by digestion takes place in the small intestine and colon, but some small molecules, notably alcohol, diffuse through the stomach lining before reaching the small intestine.

Figure 1 is a diagram of the human digestive system. The part of the esophagus that passes through the thorax has been omitted. This diagram can be annotated to indicate the functions of different parts. A summary of functions is given in table 1 below.

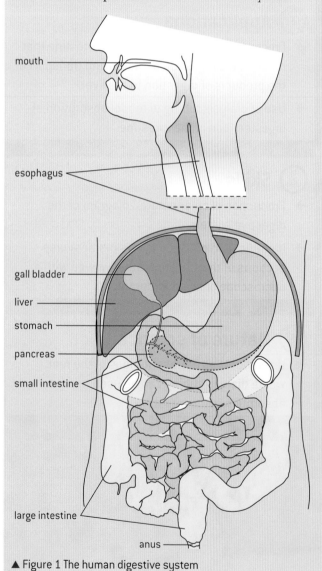

▲ Figure 1 The human digestive system

Structure	Function
Mouth	Voluntary control of eating and swallowing. Mechanical digestion of food by chewing and mixing with saliva, which contains lubricants and enzymes that start starch digestion
Esophagus	Movement of food by peristalsis from the mouth to the stomach
Stomach	Churning and mixing with secreted water and acid which kills foreign bacteria and other pathogens in food, plus initial stages of protein digestion
Small intestine	Final stages of digestion of lipids, carbohydrates, proteins and nucleic acids, neutralizing stomach acid, plus absorption of nutrients
Pancreas	Secretion of lipase, amylase and protease
Liver	Secretion of surfactants in bile to break up lipid droplets
Gall bladder	Storage and regulated release of bile
Large intestine	Re-absorption of water, further digestion especially of carbohydrates by symbiotic bacteria, plus formation and storage of feces

▲ Table 1

Structure of the wall of the small intestine

Identification of tissue layers in transverse sections of the small intestine viewed with a microscope or in a micrograph.

The wall of the small intestine is made of layers of living tissues, which are usually quite easy to distinguish in sections of the wall. From the outside of the wall going inwards there are four layers:

- serosa – an outer coat

- muscle layers – longitudinal muscle and inside it circular muscle

- sub-mucosa – a tissue layer containing blood and lymph vessels

- mucosa – the lining of the small intestine, with the epithelium that absorbs nutrients on its inner surface.

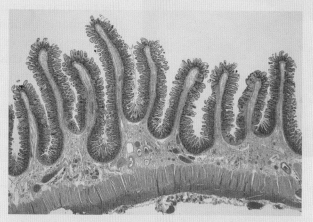

▲ Figure 2 Longitudinal section through the wall of the small intestine. Folds are visible on the inner surface and on these folds are finger-like projections called villi. All of the four main tissue layers are visible, including both circular and longitudinal parts of the muscle layer. The mucosa is stained darker than the sub-mucosa

Peristalsis

The contraction of circular and longitudinal muscle layers of the small intestine mixes the food with enzymes and moves it along the gut.

The circular and longitudinal muscle in the wall of the gut is smooth muscle rather than striated muscle. It consists of relatively short cells, not elongated fibres. It often exerts continuous moderate force, interspersed with short periods of more vigorous contraction, rather than remaining relaxed unless stimulated to contract.

Waves of muscle contraction, called peristalsis, pass along the intestine. Contraction of circular muscles behind the food constricts the gut to prevent it from being pushed back towards the mouth. Contraction of longitudinal muscle where the food is located moves it on along the gut. The contractions are controlled unconsciously not by the brain but by the enteric nervous system, which is extensive and complex.

Swallowed food moves quickly down the esophagus to the stomach in one continuous peristaltic wave. Peristalsis only occurs in one direction, away from the mouth. When food is returned to the mouth from the stomach during vomiting, abdominal muscles are used rather than the circular and longitudinal muscle in the gut wall.

In the intestines the food is moved only a few centimetres at a time so the overall progression through the intestine is much slower, allowing time for digestion. The main function of peristalsis in the intestine is churning of the semi-digested food to mix it with enzymes and thus speed up the process of digestion.

Activity

Tissue plan diagrams of the intestine wall

To practice your skill at identifying tissue layers, draw a plan diagram of the tissues in the longitudinal section of the intestine wall in figure 2. To test your skill further, draw a plan diagram to predict how the tissues of the small intestine would appear in a transverse section.

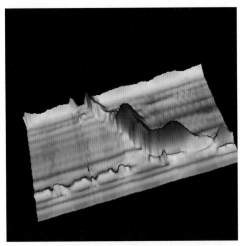

▲ Figure 3 Three-dimensional image showing the wave of muscle contraction (brown) in the esophagus during swallowing. Green indicates when the muscle is exerting less force. Time is shown left to right. At the top the sphincter between the mouth and the esophagus is shown permanently constricted apart from a brief opening when swallowing starts

Pancreatic juice

The pancreas secretes enzymes into the lumen of the small intestine.

The pancreas contains two types of gland tissue. Small groups of cells secrete the hormones insulin and glucagon into the blood. The remainder of the pancreas synthesizes and secretes digestive enzymes into the gut in response to eating a meal. This is mediated by hormones synthesized and secreted by the stomach and also by the enteric nervous system. The structure of the tissue is shown in figure 4. Small groups of gland cells cluster round the ends of tubes called ducts, into which the enzymes are secreted.

The digestive enzymes are synthesized in pancreatic gland cells on ribosomes on the rough endoplasmic reticulum. They are then processed in the Golgi apparatus and secreted by exocytosis. Ducts within the pancreas merge into larger ducts, finally forming one pancreatic duct, through which about a litre of pancreatic juice is secreted per day into the lumen of the small intestine.

Pancreatic juice contains enzymes that digest all the three main types of macromolecule found in food:

- amylase to digest starch
- lipases to digest triglycerides, phospholipids
- proteases to digest proteins and peptides.

secretory vesicles

one acinus

secretory cells

basement membrane

wall of duct

lumen of duct

▲ Figure 4 Arrangement of cells and ducts in a part of the pancreas that secretes digestive enzymes

Digestion in the small intestine

Enzymes digest most macromolecules in food into monomers in the small intestine.

The enzymes secreted by the pancreas into the lumen of the small intestine carry out these hydrolysis reactions:

- starch is digested to maltose by amylase
- triglycerides are digested to fatty acids and glycerol or fatty acids and monoglycerides by lipase
- phospholipids are digested to fatty acids, glycerol and phosphate by phospholipase
- proteins and polypeptides are digested to shorter peptides by protease.

This does not complete the process of digestion into molecules small enough to be absorbed. The wall of the small intestine produces a variety of other enzymes, which digest more substances. Some enzymes produced by gland cells in the intestine wall may be secreted in intestinal juice but most remain immobilized in the plasma membrane of epithelium cells lining the intestine. They are active there and continue to be active when the epithelium cells are abraded off the lining and mixed with the semi-digested food.

- Nucleases digest DNA and RNA into nucleotides.
- Maltase digests maltose into glucose.

- Lactase digests lactose into glucose and galactose.

- Sucrase digests sucrose into glucose and fructose.

- Exopeptidases are proteases that digest peptides by removing single amino acids either from the carboxy or amino terminal of the chain until only a dipeptide is left.

- Dipeptidases digest dipeptides into amino acids.

Because of the great length of the small intestine, food takes hours to pass through, allowing time for digestion of most macromolecules to be completed. Some substances remain largely undigested, because humans cannot synthesize the necessary enzymes. Cellulose for example is not digested and passes on to the large intestine as one of the main components of dietary fibre.

▲ Figure 5 Cystic fibrosis causes the pancreatic duct to become blocked by mucus. Pills containing synthetic enzymes help digestion in the small intestine. The photograph shows one day's supply for a person with cystic fibrosis

Villi and the surface area for digestion

Villi increase the surface area of epithelium over which absorption is carried out.

The process of taking substances into cells and the blood is called absorption. In the human digestive system nutrients are absorbed principally in the small intestine. The rate of absorption depends on the surface area of the epithelium that carries out the process. The small intestine in adults is approximately seven metres long and 25–30 millimetres wide and there are folds on its inner surface, giving a large surface area. This area is increased by the presence of villi.

Villi are small finger-like projections of the mucosa on the inside of the intestine wall. A villus is between 0.5 and 1.5 mm long and there can be as many as 40 of them per square millimetre of small intestine wall. They increase the surface area by a factor of about 10.

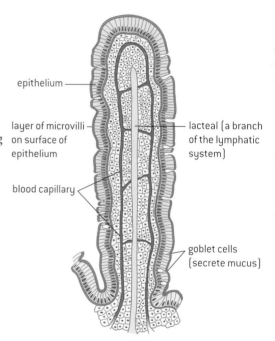

▲ Figure 6 Structure of an intestinal villus

Absorption by villi

Villi absorb monomers formed by digestion as well as mineral ions and vitamins.

The epithelium that covers the villi must form a barrier to harmful substances, while at the same time being permeable enough to allow useful nutrients to pass through.

Villus cells absorb these products of digestion of macromolecules in food:

- glucose, fructose, galactose and other monosaccharides

- any of the twenty amino acids used to make proteins

- fatty acids, monoglycerides and glycerol

- bases from digestion of nucleotides.

They also absorb substances required by the body and present in foods but not needing digestion:

- mineral ions such as calcium, potassium and sodium

- vitamins such as ascorbic acid (vitamin C).

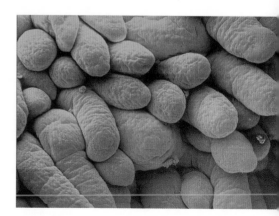

▲ Figure 7 Scanning electron micrograph of villi in the small intestine

283

Some harmful substances pass through the epithelium and are subsequently removed from the blood and detoxified by the liver. Some harmless but unwanted substances are also absorbed, including many of those that give food its colour and flavour. These pass out in urine. Small numbers of bacteria pass through the epithelium but are quickly removed from the blood by phagocytic cells in the liver.

Methods of absorption

Different methods of membrane transport are required to absorb different nutrients.

To be absorbed into the body, nutrients must pass from the lumen of the small intestine to the capillaries or lacteals in the villi. The nutrients must first be absorbed into epithelium cells through the exposed part of the plasma membrane that has its surface area enlarged with microvilli. The nutrients must then pass out of this cell through the plasma membrane where it faces inwards towards the lacteal and blood capillaries of the villus.

Many different mechanisms move nutrients into and out of the villus epithelium cells: simple diffusion, facilitated diffusion, active transport and exocytosis. These methods can be illustrated using two different examples of absorption: triglycerides and glucose.

- Triglycerides must be digested before they can be absorbed. The products of digestion are fatty acids and monoglycerides, which can be absorbed into villus epithelium cells by simple diffusion as they can pass between phospholipids in the plasma membrane.

- Fatty acids are also absorbed by facilitated diffusion as there are fatty acid transporters, which are proteins in the membrane of the microvilli.

- Once inside the epithelium cells, fatty acids are combined with monoglycerides to produce triglycerides, which cannot diffuse back out into the lumen.

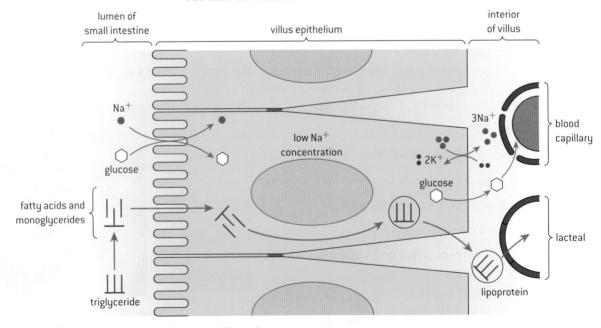

▲ Figure 8 Methods of absorption in the small intestine

- Triglycerides coalesce with cholesterol to form droplets with a diameter of about 0.2 μm, which become coated in phospholipids and protein.

- These lipoprotein particles are released by exocytosis through the plasma membrane on the inner side of the villus epithelium cells. They then either enter the lacteal and are carried away in the lymph, or enter the blood capillaries in the villi.

- Glucose cannot pass through the plasma membrane by simple diffusion because it is polar and therefore hydrophilic.

- Sodium–potassium pumps in the inwards-facing part of the plasma membrane pump sodium ions by active transport from the cytoplasm to the interstitial spaces inside the villus and potassium ions in the opposite direction. This creates a low concentration of sodium ions inside villus epithelium cells.

- Sodium–glucose co-transporter proteins in the microvilli transfer a sodium ion and a glucose molecule together from the intestinal lumen to the cytoplasm of the epithelium cells. This type of facilitated diffusion is passive but it depends on the concentration gradient of sodium ions created by active transport.

- Glucose channels allow the glucose to move by facilitated diffusion from the cytoplasm to the interstitial spaces inside the villus and on into blood capillaries in the villus.

🌐 Starch digestion in the small intestine

Processes occurring in the small intestine that result in the digestion of starch and transport of the products of digestion to the liver.

Starch digestion illustrates some important processes including catalysis, enzyme specificity and membrane permeability. Starch is a macromolecule, composed of many α-glucose monomers linked together in plants by condensation reactions. It is a major constituent of plant-based foods such as bread, potatoes and pasta. Starch molecules cannot pass through membranes so must be digested in the small intestine to allow absorption.

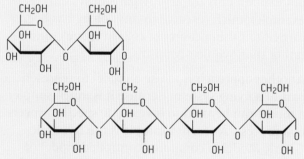

▲ Figure 9 Small portion of an amylopectin molecule showing six α-glucose molecules, all linked by 1,4 bonds apart from one 1,6 bond that creates a branch

All of the reactions involved in the digestion of starch are exothermic, but without a catalyst they happen at very slow rates. There are two types of molecule in starch:

- amylose has unbranched chains of α-glucose linked by 1,4 bonds;

- amylopectin has chains of α-glucose linked by 1,4 bonds, with some 1,6 bonds that make the molecule branched.

The enzyme that begins the digestion of both forms of starch is amylase. Saliva contains amylase but most starch digestion occurs in the small intestine, catalysed by pancreatic amylase. Any 1,4 bond in starch molecules can be broken by this enzyme, as long as there is a chain of at least four glucose monomers. Amylose is therefore digested into a mixture of two- and three-glucose fragments called maltose and maltotriose.

Because of the specificity of its active site, amylase cannot break 1,6 bonds in amylopectin. Fragments of the amylopectin molecule containing a 1,6 bond that amylase cannot digest are called dextrins. Digestion of starch is completed by three enzymes in the membranes of microvilli on villus epithelium cells. Maltase, glucosidase and dextrinase digest maltose, maltotriose and dextrins into glucose.

Glucose is absorbed into villus epithelium cells by co-transport with sodium ions. It then moves by facilitated diffusion into the fluid in interstitial spaces inside the villus. The dense network of

capillaries close to the epithelium ensures that glucose only has to travel a short distance to enter the blood system. Capillary walls consist of a single layer of thin cells, with pores between adjacent cells, but these capillaries have larger pores than usual, aiding the entry of glucose.

Blood carrying glucose and other products of digestion flows though villus capillaries to venules in the sub-mucosa of the wall of the small intestine. The blood in these venules is carried via the hepatic portal vein to the liver, where excess glucose can be absorbed by liver cells and converted to glycogen for storage. Glycogen is similar in structure to amylopectin, but with more 1,6 bonds and therefore more extensive branching.

 ## Modelling physiological processes

Use models as representations of the real world: dialysis tubing can be used to model absorption in the intestine.

Living systems are complex and when experiments are done on them, many factors can influence the results. It can be very difficult to control all of the variables and analysis of results becomes difficult. Sometimes it is better to carry out experiments using only parts of systems. For example, much research in physiology has been carried out using clones of cells in tissue culture rather than whole organisms.

Another approach is to use a model to represent part of a living system. Because it is much simpler, a model can be used to investigate specific aspects of a process. A recent example is the Dynamic Gastric Model, a computer-controlled model of the human stomach that carries out mechanical and chemical digestion of real food samples. It can be used to investigate the effects of diet, drugs, alcohol and other factors on digestion.

A simpler example is the use of dialysis tubing made from cellulose. Pores in the tubing allow water and small molecules or ions to pass through freely, but not large molecules. These properties

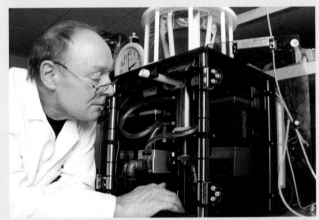

▲ Figure 10 The Dynamic Gastric Model with its inventor, Richard Faulks, adjusting the antrum mechanism

mimic the wall of the gut, which is also more permeable to small rather than large particles. Dialysis tubing can be used to model absorption by passive diffusion and by osmosis. It cannot model active transport and other processes that occur in living cells

Modelling the small intestine

Use of dialysis tubing to model absorption of digested food in the intestine.

To make a model of the small intestine, cut a length of dialysis tubing and seal one end by tying a knot in the tubing or tying with a piece of cotton thread. Pour in a suitable mixture of foods and seal the open end by tying with a piece of cotton thread. Two experiments using model intestines made in this way are suggested here:

1 Investigating the need for digestion using a model of the small intestine

Set up the apparatus shown in figure 11 and leave it for one hour.

Results

To obtain the results for the experiment, take the bags out of each tube, open them and pour the solutions from them into separate test tubes from the liquids in the tubes. You should now have four samples of fluid. Divide each of these samples into two halves and test one half for starch and the other half for sugars.

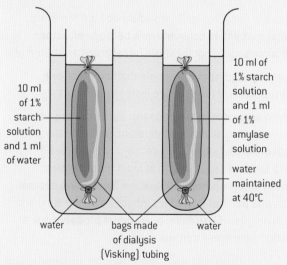

10 ml of 1% starch solution and 1 ml of water

10 ml of 1% starch solution and 1 ml of 1% amylase solution

water maintained at 40°C

water

bags made of dialysis (Visking) tubing

water

▲ Figure 11 Apparatus for showing the need for digestion

Record all the results in the way that you think is most appropriate.

Conclusions and evaluation

State carefully all the conclusions that you can make from your results.

Discuss the strengths and weaknesses of this method of investigating the need for digestion.

Suggest improvements to the method, or suggest an entirely different method of investigating the need for digestion.

2 Investigating membrane permeability using a model of the small intestine

Cola drinks contain a mixture of substances with different particle sizes. They can be used to represent food in the small intestine. Dialysis tubing is semi-permeable so can be used to model the wall of the small intestine.

Predictions

Cola contains glucose, phosphoric acid and caramel, a complex carbohydrate added to produce a brown colour. Predict which of these substances will diffuse out of the bag, with reasons for your predictions. Predict whether the bag will gain or lose mass during the experiment.

Instructions

1   Make the model intestine with cola inside.

2   Rinse the outside of the bag to wash off any traces of cola and then dry the bag.

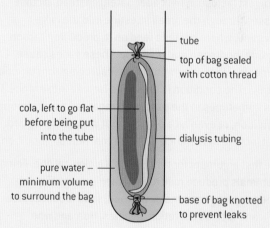

tube

top of bag sealed with cotton thread

cola, left to go flat before being put into the tube

dialysis tubing

pure water – minimum volume to surround the bag

base of bag knotted to prevent leaks

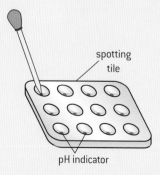

spotting tile

pH indicator

▲ Figure 12 Apparatus for membrane permeability experiment

3   Find the mass of the bag using an electronic balance.

4   When you are ready to start the experiment, place the bag in pure water in a test tube.

5   Test the water around the bag at suitable time intervals. A suggested range is 1, 2, 4, 8 and 16 minutes. At each time lift the bag up and down a few times to mix the water in the tube, then do these tests:

   • Look carefully at the water to see whether it is still clear or has become brown.

   • Use a dropping pipette to remove a few drops of the water and test them in a spotting tile with a narrow-range pH indicator. Use a colour chart to work out the pH.

   • Dip a glucose test strip into the water and record the colour that it turns. Instructions vary for these test strips. Follow the instructions and work out the glucose concentration of the water.

6   After testing the water for the last time, remove the bag, dry it and find its mass again with the electronic balance.

Conclusions

a) Explain the conclusions that you can draw about the permeability of the dialysis tubing from the tests of the water and from the change in mass of the bag.   [5]

b) Compare and contrast the dialysis tubing and the plasma membranes that carry out absorption in villus epithelium cells in the wall of the intestine.   [5]

c) Use the results of your experiment to predict the direction of movement of water by osmosis across villus epithelium cells.   [5]

TOK

What are some of the variables that affect perspectives as to what is "normal"?

In some adult humans, levels of lactase are too low to digest lactose in milk adequately. Instead, lactose passes through the small intestine into the large intestine, where bacteria feed on it, producing carbon dioxide, hydrogen and methane. These gases cause some unpleasant symptoms, discouraging consumption of milk. The condition is known as lactose intolerance. It has sometimes in the past been regarded as an abnormal condition, or even as a disease, but it could be argued that lactose intolerance is the normal human condition.

The first argument for this view is a biological one. Female mammals produce milk to feed their young offspring. When a young mammal is weaned, solid foods replace milk and lactase secretion declines. Humans who continue to consume milk into adulthood are therefore unusual. Inability to consume milk because of lactose intolerance should not therefore be regarded as abnormal.

The second argument is a simple mathematical one: a high proportion of humans are lactose intolerant.

The third argument is evolutionary. Our ancestors were almost certainly all lactose intolerant, so this is the natural or normal state. Lactose tolerance appears to have evolved separately in at least three centres: Northern Europe, parts of Arabia, the Sahara and eastern Sudan, and parts of East Africa inhabited by the Tutsi and Maasai peoples. Elsewhere, tolerance is probably due to migration from these centres.

6.2 The blood system

Understanding

→ Arteries convey blood at high pressure from the ventricles to the tissues of the body.

→ Arteries have muscle and elastic fibres in their walls.

→ The muscle and elastic fibres assist in maintaining blood pressure between pump cycles.

→ Blood flows through tissues in capillaries with permeable walls that allow exchange of materials between cells in the tissue and the blood in the capillary.

→ Veins collect blood at low pressure from the tissues of the body and return it to the atria of the heart.

→ Valves in veins and the heart ensure circulation of blood by preventing backflow.

→ There is a separate circulation for the lungs.

→ The heartbeat is initiated by a group of specialized muscle cells in the right atrium called the sinoatrial node.

→ The sinoatrial node acts as a pacemaker.

→ The sinoatrial node sends out an electrical signal that stimulates contraction as it is propagated through the walls of the atria and then the walls of the ventricles.

→ The heart rate can be increased or decreased by impulses brought to the heart through two nerves from the medulla of the brain.

→ Epinephrine increases the heart rate to prepare for vigorous physical activity.

Applications

→ William Harvey's discovery of the circulation of the blood with the heart acting as the pump.

→ Causes and consequences of occlusion of the coronary arteries.

→ Pressure changes in the left atrium, left ventricle and aorta during the cardiac cycle.

Skills

→ Identification of blood vessels as arteries, capillaries or veins from the structure of their walls.

→ Recognition of the chambers and valves of the heart and the blood vessels connected to it in dissected hearts or in diagrams of heart structure.

Nature of science

→ Theories are regarded as uncertain: William Harvey overturned theories developed by the ancient Greek philosopher Galen on movement of blood in the body.

William Harvey and the circulation of blood

William Harvey's discovery of the circulation of the blood with the heart acting as the pump.

William Harvey is usually credited with the discovery of the circulation of the blood as he combined earlier discoveries with his own research findings to produce a convincing overall theory for blood flow in the body. He overcame widespread opposition by publishing his results and also by touring Europe to demonstrate experiments that falsified previous theories and provided evidence for his theory. As a result his theory became generally accepted.

Harvey demonstrated that blood flow through the larger vessels is unidirectional, with valves to prevent backflow. He also showed that the rate of flow through major vessels was far too high for blood to be consumed in the body after being pumped out by the heart, as earlier theories proposed. It must therefore return to the heart and be recycled. Harvey showed that the heart pumps blood out in the arteries and it returns in veins. He predicted the presence of numerous fine vessels too small to be seen with contemporary equipment that linked arteries to veins in the tissues of the body.

Blood capillaries are too narrow to be seen with the naked eye or with a hand lens. Microscopes had not been invented by the time that Harvey published his theory about the circulation of blood in 1628. It was not until 1660, after his death, that blood was seen flowing from arteries to veins though capillaries as he had predicted.

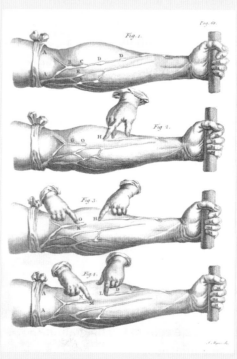

▲ Figure 1 Harvey's experiment to demonstrate that blood flow in the veins of the arm is unidirectional

 ## Overturning ancient theories in science

Theories are regarded as uncertain: William Harvey overturned theories developed by the ancient Greek philosopher Galen on movement of blood in the body.

During the Renaissance, interest was reawakened in the classical writings of Greece and Rome. This stimulated literature and the arts, but in some ways it hampered progress in science. It became almost impossible to question the doctrines of such writers as Aristotle, Hippocrates, Ptolemy and Galen.

According to Galen, blood is formed in the liver and is pumped to and fro between the liver and the right ventricle of the heart. A little blood passes into the left ventricle, where it meets air from the lungs and becomes "vital spirits". The vital spirits are distributed to the body by the arteries. Some of the vital spirits flow to the brain, to be converted into "animal spirits", which are then distributed by the nerves to the body.

William Harvey was unwilling to accept these doctrines without evidence. He made careful observations and did experiments, from which he deduced that blood circulates through the pulmonary and systemic circulations. He predicted the existence of capillaries, linking arteries and veins, even though the lenses of the time were not powerful enough for him to see them.

The following extract is from Harvey's book *On the Generation of Animals*, published in 1651 when he was 73.

And hence it is that without the due admonition of the senses, without frequent observation and reiterated experiment, our mind goes astray after phantoms and appearances. Diligent observation is therefore requisite in every science, and the senses are frequently to be appealed to. We are, I say, to strive after personal experience, not to rely of the experience of others: without which no one can properly become a student of any branch of natural science. I would not have you therefore, gentle reader, to take anything on trust from me concerning the Generation of Animals: I appeal to your own eyes as my witness and judge. The method of pursuing truth commonly pursued at this time therefore is to be held erroneous and almost foolish, in which so many enquire what things others have said, and omit to ask whether the things themselves be actually so or not.

Arteries

Arteries convey blood at high pressure from the ventricles to the tissues of the body.

Arteries are vessels that convey blood from the heart to the tissues of the body. The main pumping chambers of the heart are the ventricles. They have thick strong muscle in their walls that pumps blood into the arteries, reaching a high pressure at the peak of each pumping cycle. The artery walls work with the heart to facilitate and control blood flow. Elastic and muscle tissue in the walls are used to do this.

Elastic tissue contains elastin fibres, which store the energy that stretches them at the peak of each pumping cycle. Their recoil helps propel the blood on down the artery. Contraction of smooth muscle in the artery wall determines the diameter of the lumen and to some extent the rigidity of the arteries, thus controlling the overall flow through them.

Both the elastic and muscular tissues contribute to the toughness of the walls, which have to be strong to withstand the constantly changing and intermittently high blood pressure without bulging outwards (aneurysm) or bursting. The blood's progress along major arteries is thus pulsatile, not continuous. The pulse reflects each heartbeat and can easily be felt in arteries that pass near the body surface, including those in the wrist and the neck.

Each organ of the body is supplied with blood by one or more arteries. For example, each kidney is supplied by a renal artery and the liver by the hepatic artery. The powerful, continuously active muscles of the heart itself are supplied with blood by coronary arteries.

Artery walls

Arteries have muscle and elastic fibres in their walls.

The wall of the artery is composed of several layers:

- tunica externa – a tough outer layer of connective tissue

- tunica media – a thick layer containing smooth muscle and elastic fibres made of the protein elastin

- tunica intima – a smooth endothelium forming the lining of the artery.

Activity

Discussion questions on William Harvey's methods

1 William Harvey refused to accept doctrines without evidence. Are there academic contexts where it is reasonable to accept doctrines on the basis of authority rather than evidence gathered from primary sources?

2 Harvey welcomed questions and criticisms of his theories when teaching anatomy classes. Suggest why he might have done this.

3 Can you think of examples of the "phantoms and appearances" that Harvey refers to?

4 Why does Harvey recommend "reiteration" of experiments?

5 Harvey practised as a doctor, but after the publication in 1628 of his work on the circulation of the blood, far fewer patients consulted him. Why might this have been?

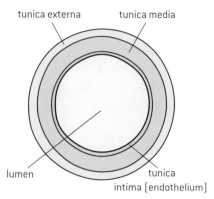

tunica externa   tunica media

lumen   tunica intima (endothelium)

▲ Figure 3 Structure of an artery

Activity

Measuring blood pressures

Because arteries are distensible, blood pressure in those that pass near the body surface can be measured relatively easily. A common method is to inflate an arm cuff until it squeezes the tissues (skin, superficial fat as well as the vessels themselves) enough to stop blood flow. The pressure is then released slowly until flow resumes and the operator or instrument can hear the pulse again. The pressures at which blood flow stops and resumes are the systolic and diastolic pressures. They are measured with a pressure monitor. According to the American Heart Association the desired blood pressures for adults of 18 years or older measured in this way are:

systolic 90-119 mmHg

diastolic 60-79 mmHg

▲ Figure 4 Blood pressure monitor

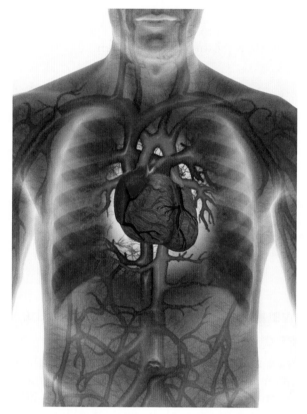

▲ Figure 2 The cardiovascular system. The main artery that supplies oxygenated blood to the tissues of the body is the aorta, shown as the red vessel that emerges from the heart and forms an arch with branches carrying blood to the arms and head. The aorta continues through the thorax and abdomen, with branches serving the liver, kidneys, intestines and other organs

Arterial blood pressure

The muscle and elastic fibres assist in maintaining blood pressure between pump cycles.

The blood entering an artery from the heart is at high pressure. The peak pressure reached in an artery is called the systolic pressure. It pushes the wall of the artery outwards, widening the lumen and stretching elastic fibres in the wall, thus storing potential energy.

At the end of each heartbeat the pressure in the arteries falls sufficiently for the stretched elastic fibres to squeeze the blood in the lumen. This mechanism saves energy and prevents the minimum pressure inside the artery, called the diastolic pressure, from becoming too low. Because it is relatively high, blood flow in the arteries is relatively steady and continuous although driven by a pulsating heart.

The circular muscles in the wall of the artery form a ring so when they contract, in a process called vasoconstriction, the circumference is reduced and the lumen is narrowed. Vasoconstriction increases blood pressure in the arteries. Branches of arteries called arterioles have a particularly high density of muscle cells that respond to various hormone and neural signals to control blood flow to downstream tissues. Vasoconstriction of arterioles restricts blood flow to the part of the body that they supply and the opposite process, called vasodilation, increases it.

Capillaries

Blood flows through tissues in capillaries with permeable walls that allow exchange of materials between cells in the tissue and the blood in the capillary.

Capillaries are the narrowest blood vessels with diameter of about 10 µm. They branch and rejoin repeatedly to form a capillary network with a huge total length. Capillaries transport blood through almost all tissues in the body. Two exceptions are the tissues of the lens and the cornea in the eye which must be transparent so cannot contain any blood vessels. The density of capillary networks varies in other tissues but all active cells in the body are close to a capillary.

The capillary wall consists of one layer of very thin endothelium cells, coated by a filter-like protein gel, with pores between the cells. The wall is thus very permeable and allows part of the plasma to leak out and form tissue fluid. Plasma is the fluid in which the blood cells are suspended. Tissue fluid contains oxygen, glucose and all other substances in blood plasma apart from large protein molecules, which cannot pass through the capillary wall. The fluid flows between the cells in a tissue, allowing the cells to absorb useful substances and excrete waste products. The tissue fluid then re-enters the capillary network.

The permeabilities of capillary walls differ between tissues, enabling particular proteins and other large particles to reach certain tissues but not others. Permeabilities can also change over time and capillaries repair and remodel themselves continually in response to the needs of tissues that they perfuse.

Veins

Veins collect blood at low pressure from the tissues of the body and return it to the atria of the heart.

Veins transport blood from capillary networks back to the atria of the heart. By now the blood is at much lower pressure than it was in the arteries. Veins do not therefore need to have as thick a wall as arteries and the wall contains far fewer muscle and elastic fibres. They can therefore dilate to become much wider and thus hold more blood than arteries. Around 80% of a sedentary person's blood is in the veins though this proportion falls during vigorous exercise.

Blood flow in veins is assisted by gravity and by pressures exerted on them by other tissues especially skeletal muscles. Contraction makes a muscle shorter and wider so it squeezes on adjacent veins like a pump. Walking, sitting or even just fidgeting greatly improves venous blood flow.

Each part of the body is served by one or more veins. For example blood is carried from the arms in the subclavian veins and from the head in the jugular veins. The hepatic portal vein is unusual because it does not carry blood back to the heart. It carries blood from the stomach and intestines to the liver. It is regarded as a portal vein rather than an artery because the blood it carries is at low pressure so it is relatively thin.

Activity

Bruises

Bruises are caused by damage to capillary walls and leakage of plasma and blood cells into spaces between cells in a tissue. The capillaries are quickly repaired, hemoglobin is broken down to green and yellow bile pigments which are transported away and phagocytes remove the remains of the blood cells by endocytosis. When you next have a bruise, make observations over the days after the injury to follow the healing process and the rate at which hemoglobin is removed.

▲ Figure 5 Which veins in this gymnast will need valves to help with venous return?

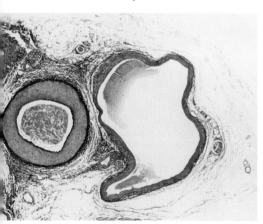

▲ Figure 6 Artery and vein in transverse section. The tunica externa and tunica intima are stained more darkly than the tunica media. Clotted blood is visible in both vessels

Valves in veins

Valves in veins and the heart ensure circulation of blood by preventing backflow.

Blood pressure in veins is sometimes so low that there is a danger of backflow towards the capillaries and insufficient return of blood to the heart. To maintain circulation, veins contain pocket valves, consisting of three cup-shaped flaps of tissue.

- If blood starts to flow backwards, it gets caught in the flaps of the pocket valve, which fill with blood, blocking the lumen of the vein.

- When blood flows towards the heart, it pushes the flaps to the sides of the vein. The pocket valve therefore opens and blood can flow freely.

These valves allow blood to flow in one direction only and make efficient use of the intermittent and often transient pressures provided by muscular and postural changes. They ensure that blood circulates in the body rather than flowing to and fro.

 Identifying blood vessels

Identification of blood vessels as arteries, capillaries or veins from the structure of their walls.

Blood vessels can be identified as arteries, capillaries or veins by looking at their structure. Table 1 below gives differences that may be useful.

	Artery	Capillary	Vein
Diameter	Larger than 10 μm	Around 10 μm	Variable but much larger than 10 μm
Relative thickness of wall and diameter of lumen	Relatively thick wall and narrow lumen	Extremely thin wall	Relatively thin wall with variable but often wide lumen
Number of layers in wall	Three layers, tunica externa, media and intima. These layers may be sub-divided to form more layers	Only one layer – the tunica intima which is an endothelium consisting of a single layer of very thin cells	Three layers – tunica externa, media and intima
Muscle and elastic fibres in the wall	Abundant	None	Small amounts
Valves	None	None	Present in many veins

▲ Table 1

The double circulation

There is a separate circulation for the lungs.

There are valves in the veins and heart that ensure a one-way flow, so blood circulates through arteries, capillaries and veins. Fish have a single circulation. Blood is pumped at high pressure to their gills to be oxygenated. After flowing through the gills the blood still has enough pressure to flow directly, but relatively slowly, to other organs of the body and then back to the heart. In contrast, the lungs used by mammals for gas exchange are supplied with blood by a separate circulation.

Blood capillaries in lungs cannot withstand high pressures so blood is pumped to them at relatively low pressure. After passing through the capillaries of the lungs the pressure of the blood is low, so it must return to the heart to be pumped again before it goes to other organs. Humans therefore have two separate circulations:

- the pulmonary circulation, to and from the lungs

- the systemic circulation, to and from all other organs, including the heart muscles.

Figure 7 shows the double circulation in a simplified form. The pulmonary circulation receives deoxygenated blood that has returned from the systemic circulation, and the systemic circulation receives blood that has been oxygenated by the pulmonary circulation. It is therefore essential that blood flowing to and from these two circulations is not mixed. The heart is therefore a double pump, delivering blood under different pressures separately to the two circulations.

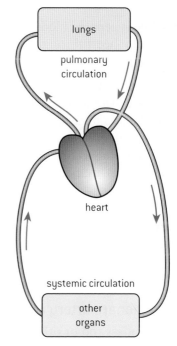

▲ Figure 7 The double circulation

⚗ Heart structure

Recognition of the chambers and valves of the heart and the blood vessels connected to it in dissected hearts or in diagrams of heart structure.

- The heart has two sides, left and right, that pump blood to the systemic and pulmonary circulations.

- Each side of the heart has two chambers, a ventricle that pumps blood out into the arteries and an atrium that collects blood from the veins and passes it to the ventricle.

- Each side of the heart has two valves, an atrioventricular valve between the atrium and the ventricle and a semilunar valve between the ventricle and the artery.

- Oxygenated blood flows into the left side of the heart through the pulmonary veins from the lungs and out through the aorta.

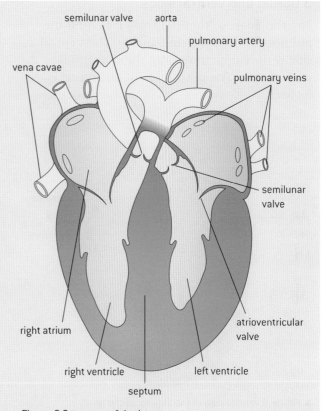

▲ Figure 8 Structure of the heart

- Deoxygenated blood flows into the right side of the heart through the vena cava and out in the pulmonary arteries.

The heart is a complicated three-dimensional structure. The best way to learn about its structure is by doing a dissection. A fresh specimen of a mammalian heart, with blood vessels still attached, a dissecting dish or board and dissecting instruments are needed.

1 Arteries and veins

Tidy up the blood vessels attached to the heart by removing membranes and other tissue from around them. Identify the thick-walled arteries and the thin-walled veins.

2 Pulmonary artery and aorta

Push a glass rod or other blunt-ended instrument into the heart through the arteries and feel through the wall of the heart to where the end of the rod has reached. Identify the pulmonary artery, through which you will reach the thinner-walled right ventricle, and the aorta, through which you will reach the thicker-walled left ventricle.

3 Dorsal and ventral sides

Lay the heart so that the aorta is behind the pulmonary artery, as in figure 9. The ventral side is now uppermost and the dorsal side underneath. The dorsal side of an animal is its back.

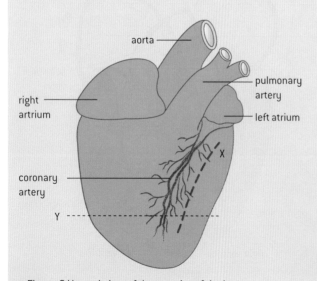

aorta

pulmonary artery

right artrium

left atrium

coronary artery

Y

X

▲ Figure 9 Ventral view of the exterior of the heart

4 Left ventricle

Identify the left ventricle. It has a smooth wall, with a tree-like pattern of blood vessels. Using a sharp scalpel, make an incision as shown by the dashed line X in figure 9. This should open up the left ventricle. Look at the thick muscular wall that you have cut through.

5 Atrioventricular valve

Extend the incision further towards the atrium if necessary until you can see the two thin flaps of the atrioventricular valve. Tendons attached to the sides of the left ventricle prevent the valve inverting into the atrium.

6 Left atrium and pulmonary vein

Identify the left atrium. It will look surprisingly small as there is no blood inside it. The outer surface of its wall has a wrinkled appearance. Extend the incision that you have already made, either with the scalpel or with scissors, to cut through the wall of the left atrium as far as the pulmonary vein. Look at the thin wall of the atrium and the opening of the pulmonary vein or veins (there may be two).

7 Aorta

Find the aorta again and measure the diameter of its lumen, in millimetres. Using scissors, cut through the wall of the aorta, starting at its end and working towards the left ventricle. Look at the smooth inner surface of the aorta and try stretching the wall to see how tough it is.

8 Semilunar valve

Where the aorta exits the left ventricle, there will be three cup-shaped flaps in the wall. These form the semilunar valve. Try pushing a blunt instrument into the flaps to see how blood flowing backwards pushes the flaps together, closing the valve.

9 Coronary artery

Look carefully at the inner surface of the aorta, near the semilunar valve. A small hole should be visible, which is the opening to the coronary arteries. Measure the diameter of the lumen of this artery. The coronary arteries supply the wall of the heart with oxygen and nutrients.

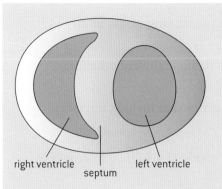

10 Septum

Make a transverse section through the heart near the base of the ventricles, along the dotted line marked Y in figure 9. Measure the thickness in millimetres of the walls of the left and right ventricles and of the septum between them (figure 10). The septum contains conducting fibres, which help to stimulate the ventricles to contract.

▲ Figure 10 Transverse section through the ventricles

Atherosclerosis

Causes and consequences of occlusion of the coronary arteries.

One of the commonest current health problems is atherosclerosis, the development of fatty tissue called atheroma in the artery wall adjacent to the endothelium. Low density lipoproteins (LDL) containing fats and cholesterol accumulate and phagocytes are then attracted by signals from endothelium cells and smooth muscle. The phagocytes engulf the fats and cholesterol by endocytosis and grow very large. Smooth muscle cells migrate to form a tough cap over the atheroma. The artery wall bulges into the lumen narrowing it and thus impeding blood flow.

Small traces of atheroma are normally visible in children's arteries by the age of ten, but do not affect health. In some older people atherosclerosis becomes much more advanced but often goes unnoticed until a major artery becomes so blocked that the tissues it supplies become compromised.

Coronary occlusion is a narrowing of the arteries that supply blood containing oxygen and nutrients to the heart muscle. Lack of oxygen (anoxia) causes pain, known as angina, and impairs the muscle's ability to contract, so the heart beats faster as it tries to maintain blood circulation with some of its muscle out of action. The fibrous cap covering atheromas sometimes ruptures, which stimulates the formation of blood clots that can block arteries supplying blood to the heart and cause acute heart problems. This is described in sub-topic 6.3.

The causes of atherosclerosis are not yet fully understood. Various factors have been shown to be associated with an increased risk of atheroma but are not the sole causes of the condition:

* high blood concentrations of LDL (low density lipoprotein)

* chronic high blood glucose concentrations, due to overeating, obesity or diabetes

Activity

Structure and function of the heart

Discuss the answers to these questions.

1 Why are the walls of the atria thinner than the walls of the ventricles?

2 What prevents the atrioventricular valve from being pushed into the atrium when the ventricle contracts?

3 Why is the left ventricle wall thicker than the right ventricle wall?

4 Does the left side of the heart pump oxygenated or deoxygenated blood?

5 Why does the wall of the heart need its own supply of blood, brought by the coronary arteries?

6 Does the right side of the heart pump a greater volume of blood per minute, a smaller volume, or the same volume as the left?

Activity

Carnitine and coronary occusion

A chemical called carnitine that is found in certain foods is converted into TMAO by bacteria in the gut. Find out what foods contain the highest concentrations of carnitine and discuss whether this finding should influence dietary advice.

- chronic high blood pressure due to smoking, stress or any other cause

- consumption of *trans* fats, which damage the endothelium of the artery.

There are also some more recent theories that include microbes:

- infection of the artery wall with *Chlamydia pneumoniae*

- production of trimethylamine N-oxide (TMAO) by microbes in the intestine.

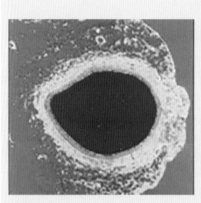
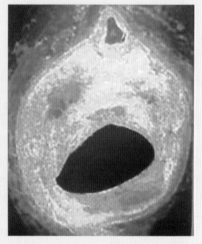

▲ Figure 11 A normal artery (left) has a much wider lumen than an artery that is occluded by atheroma (right)

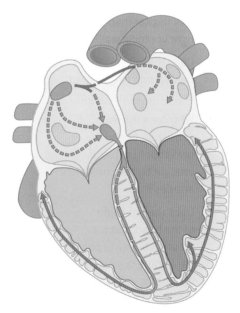

▲ Figure 12 The sinoatrial node

The sinoatrial node

The heartbeat is initiated by a group of specialized muscle cells in the right atrium called the sinoatrial node.

The heart is unique in the body as its muscles can contract without stimulation from motor neurons. The contraction is called myogenic, meaning that it is generated in the muscle itself. The membrane of a heart muscle cell depolarizes when the cell contracts and this activates adjacent cells, so they also contract. A group of cells therefore contracts almost simultaneously at the rate of the fastest.

The region of the heart with the fastest rate of spontaneous beating is a small group of special muscle cells in the wall of the right atrium, called the sinoatrial node. These cells have few of the proteins that cause contraction in other muscle cells, but they have extensive membranes. The sinoatrial node therefore initiates each heartbeat, because the membranes of its cells are the first to depolarize in each cardiac cycle.

Initiating the heartbeat

The sinoatrial node acts as a pacemaker.

Because the sinoatrial node initiates each heartbeat, it sets the pace for the beating of the heart and is often called the pacemaker. If it becomes defective, its output may be regulated or even replaced entirely by an artificial pacemaker. This is an electronic device, placed under the skin with electrodes implanted in the wall of the heart that initiate each heartbeat in place of the sinoatrial node.

Atrial and ventricular contraction

The sinoatrial node sends out an electrical signal that stimulates contraction as it is propagated through the walls of the atria and then the walls of the ventricles.

The sinoatrial node initiates a heartbeat by contracting and simultaneously sends out an electrical signal that spreads throughout the walls of the atria. This can happen because there are interconnections between adjacent fibres across which the electrical signal can be propagated. Also the fibres are branched so each fibre passes the signal on to several others. It takes less than a tenth of a second for all cells in the atria to receive the signal. This propagation of the electrical signal causes the whole of both left and right atria to contract.

After a time delay of about 0.1 seconds, the electrical signal is conveyed to the ventricles. The time delay allows time for the atria to pump the blood that they are holding into the ventricles. The signal is then propagated throughout the walls of the ventricles, stimulating them to contract and pump blood out into the arteries. Details of the electrical stimulation of the heartbeat are included in Option D.

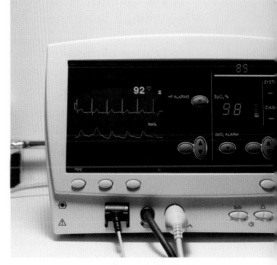

▲ Figure 13 Heart monitor displaying the heart rate, the electrical activity of the heart and the percentage saturation with oxygen of the blood

TOK

What matters more in ethical decision making: intent or consequences?

There are some circumstances in which prolonging the life of an individual who is suffering brings in to question the role of the physician. Sometimes, an active pacemaker may be involved in prolonging the life of a patient and the physician receives a request to deactivate the device. This will accelerate the pace of the patient's death. Euthanasia involves taking active steps to end the life of a patient and it is illegal in many jurisdictions. However, there is a widely accepted practice of withdrawing life-sustaining interventions such as dialysis, mechanical ventilation, or tube feeding from terminally ill patients. This is often a decision of the family of the patient. The withdrawal of life support is seen as distinct from euthanasia because the patient dies of their condition rather than the active steps to end the patient's life in the case of euthanasia. However, the distinction can be subtle. The consequence is the same: the death of the patient. The intent can be the same: to end the patient's suffering. Yet in many jurisdictions, one action is illegal and the other is not.

The cardiac cycle

Pressure changes in the left atrium, left ventricle and aorta during the cardiac cycle.

The pressure changes in the atrium and ventricle of the heart and the aorta during a cardiac cycle are shown in figure 15. To understand them it is necessary to appreciate what occurs at each stage of the cycle. Figure 14 below summarizes the events, with timings assuming a heart rate of 75 beats per minute. Typical volumes of blood are shown and also an indication of the direction of blood flow to or from a chamber of the heart.

0.0 – 0.1 seconds

- The atria contract causing a rapid but relatively small pressure increase, which pumps blood from the atria to the ventricles, through the open atrioventricular valves.

- The semilunar valves are closed and blood pressure in the arteries gradually drops to its minimum as blood continues to flow along them but no more is pumped in.

0.1 – 0.15 seconds

- The ventricles contract, with a rapid pressure build up that causes the atrioventricular valves to close.

- The semilunar valves remain closed.

0.15 – 0.4 seconds

- The pressure in the ventricles rises above the pressure in the arteries so the semilunar valves open and blood is pumped from the ventricles into the arteries, transiently maximizing the arterial blood pressure.

- Pressure slowly rises in the atria as blood drains into them from the veins and they fill.

0.4 – 0.45 seconds

- The contraction of the ventricular muscles wanes and pressure inside the ventricles rapidly drops below the pressure in the arteries, causing the semilunar valves to close.

- The atrioventricular valves remain closed.

0.45 – 0.8 seconds

- Pressure in the ventricles drops below the pressure in the atria so the atrioventricular valves open.

- Blood from the veins drains into the atria and from there into the ventricles, causing a slow increase in pressure.

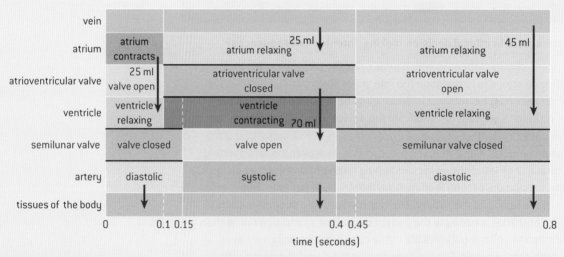

▲ Figure 14 One cardiac cycle is represented on the diagram, starting on the left with contraction of the atrium. Vertical arrows show flows of blood to and from the atrium and ventricle

Data-based questions: Heart action and blood pressures

Figure 15 shows the pressures in the atrium, ventricle and artery on one side of the heart, during one second in the life of the heart.

1 Deduce when blood is being pumped from the atrium to the ventricle. Give both the start and the end times. [2]

2 Deduce when the ventricle starts to contract. [1]

3 The atrioventricular valve is the valve between the atrium and the ventricle. State when the atrioventricular valve closes. [1]

4 The semilunar valve is the valve between the ventricle and the artery. State when the semilunar valve opens. [1]

5 Deduce when the semilunar valve closes. [1]

6 Deduce when blood is being pumped from the ventricle to the artery. Give both the start and the end times. [2]

7 Deduce when the volume of blood in the ventricle is:

   a) at a maximum [1]

   b) at a minimum. [1]

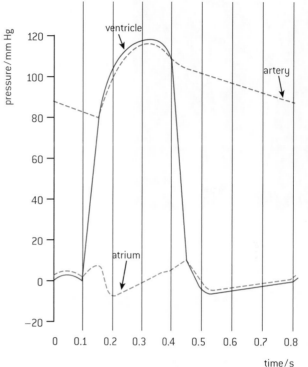

▲ Figure 15 Pressure changes during the cardiac cycle

Changing the heart rate

The heart rate can be increased or decreased by impulses brought to the heart through two nerves from the medulla of the brain.

The sinoatrial node that sets the rhythm for the beating of the heart responds to signals from outside the heart. These include signals from branches of two nerves originating in a region in the medulla of the brain called the cardiovascular centre. Signals from one of the nerves cause the pacemaker to increase the frequency of heartbeats. In healthy young people the rate can increase to three times the resting rate. Signals from the other nerve decrease the rate. These two nerve branches act rather like the throttle and brake of a car.

The cardiovascular centre receives inputs from receptors that monitor blood pressure and its pH and oxygen concentration. The pH of the blood reflects its carbon dioxide concentration.

- Low blood pressure, low oxygen concentration and low pH all suggest that the heart rate needs to speed up, to increase the flow rate of blood to the tissues, deliver more oxygen and remove more carbon dioxide.

- High blood pressure, high oxygen concentration and high pH are all indicators that the heart rate may need to slow down.

Activity

Listening to heart sounds

Sounds produced by blood flow can be heard with a simple tube or stethoscope placed on the chest near the heart. The consequences of this whole cardiac cycle for the flow of blood out of the heart can be felt as the pulse in a peripheral artery.

(a)

(b)

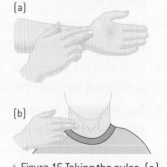

▲ Figure 16 Taking the pulse: (a) radial pulse (b) carotid pulse

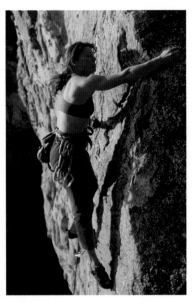

▲ Figure 17 Adventure sports such as rock climbing cause epinephrine secretion

Epinephrine

Epinephrine increases the heart rate to prepare for vigorous physical activity.

The sinoatrial node also responds to epinephrine in the blood, by increasing the heart rate. This hormone is also sometimes called adrenalin and is produced by the adrenal glands. The secretion of epinephrine is controlled by the brain and rises when vigorous physical activity may be necessary because of a threat or opportunity. So epinephrine has the nickname "fight or flight hormone".

In the past when humans were hunter-gatherers rather than farmers, epinephrine would have been secreted when humans were hunting for prey or when threatened by a predator. In the modern world athletes often use pre-race routines to stimulate adrenalin secretion so that their heart rate is already increased when vigorous physical activity begins.

6.3 Defence against infectious disease

Understanding

→ The skin and mucous membranes form a primary defence against pathogens that cause infectious disease.

→ Cuts in the skin are sealed by blood clotting.

→ Clotting factors are released from platelets.

→ The cascade results in the rapid conversion of fibrinogen to fibrin by thrombin.

→ Ingestion of pathogens by phagocytic white blood cells gives non-specific immunity to diseases.

→ Production of antibodies by lymphocytes in response to particular pathogens gives specific immunity.

→ Antibiotics block processes that occur in prokaryotic cells but not in eukaryotic cells.

→ Viral diseases cannot be treated using antibiotics because they lack a metabolism.

→ Some strains of bacteria have evolved with genes which confer resistance to antibiotics and some strains of bacteria have multiple resistance.

Applications

→ Causes and consequences of blood clot formation in coronary arteries.

→ Effects of HIV on the immune system and methods of transmission.

→ Florey and Chain's experiments to test penicillin on bacterial infections in mice.

Nature of science

→ Risks associated with scientific research: Florey and Chain's tests on the safety of penicillin would not be compliant with current protocols on testing.

Skin as a barrier to infection

The skin and mucous membranes form a primary defence against pathogens that cause infectious disease.

There are many different microbes in the environment that can grow inside the human body and cause a disease. Some microorganisms are opportunistic and although they can invade the body they also commonly live outside it. Others are specialized and can only survive inside a human body. Microbes that cause disease are called pathogens.

The primary defence of the body against pathogens is the skin. Its outermost layer is tough and provides a physical barrier against the entry of pathogens and protection against physical and chemical damage. Sebaceous glands are associated with hair follicles and they secrete a chemical called sebum, which maintains skin moisture and slightly lowers skin pH. The lower pH inhibits the growth of bacteria and fungi.

Mucous membranes are a thinner and softer type of skin that is found in areas such as the nasal passages and other airways, the head of the penis and foreskin and the vagina. The mucus that these areas of skin secrete is a sticky solution of glycoproteins. Mucus acts as a physical barrier; pathogens and harmful particles are trapped in it and either swallowed or expelled. It also has antiseptic properties because of the presence of the anti-bacterial enzyme lysozyme.

Cuts and clots

Cuts in the skin are sealed by blood clotting.

When the skin is cut, blood vessels in it are severed and start to bleed. The bleeding usually stops after a short time because of a process called clotting. The blood emerging from a cut changes from being a liquid to a semi-solid gel. This seals up the wound and prevents further loss of blood and blood pressure. Clotting is also important because cuts breach the barrier to infection provided by the skin. Clots prevent entry of pathogens until new tissue has grown to heal the cut.

Platelets and blood clotting

Clotting factors are released from platelets.

Blood clotting involves a cascade of reactions, each of which produces a catalyst for the next reaction. As a result blood clots very rapidly. It is important that clotting is under strict control, because if it occurs inside blood vessels the resulting clots can cause blockages.

The process of clotting only occurs if platelets release clotting factors. Platelets are cellular fragments that circulate in the blood. They are smaller than either red or white blood cells. When a cut or other injury involving damage to blood vessels occurs, platelets aggregate at the site forming a temporary plug. They then release the clotting factors that trigger off the clotting process.

▲ Figure 1 Scanning electron micrograph of bacteria on the surface of teeth. Mucous membranes in the mouth prevent these and other microbes from invading body tissues

Activity

Imaging human skin

A digital microscope can be used to produce images of the different types of skin covering the human body. Figure 2 shows four images produced in this way.

▲ Figure 2

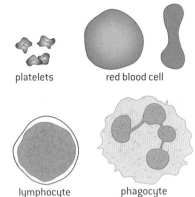

platelets   red blood cell

lymphocyte   phagocyte

▲ Figure 3 Cells and cell fragments from blood. Lymphocytes and phagocytes are types of white blood cell

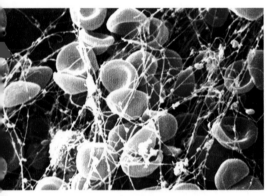

▲ Figure 4 Scanning electron micrograph of clotted blood with fibrin and trapped blood cells

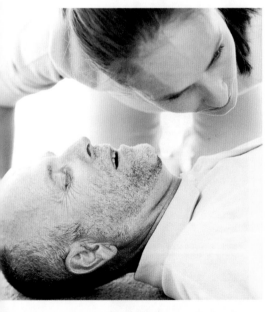

▲ Figure 5 Early intervention during a heart attack can save the patient's life so it is important to know what to do by being trained

Fibrin production

The cascade results in the rapid conversion of fibrinogen to fibrin by thrombin.

The cascade of reactions that occurs after the release of clotting factors from platelets quickly results in the production of an enzyme called thrombin. Thrombin in turn converts the soluble protein fibrinogen into the insoluble fibrin. The fibrin forms a mesh in cuts that traps more platelets and also blood cells. The resulting clot is initially a gel, but if exposed to the air it dries to form a hard scab.

Figure 4 shows red blood cells trapped in this fibrous mesh.

🌐 Coronary thrombosis

Causes and consequences of blood clot formation in coronary arteries.

In patients with coronary heart disease, blood clots sometimes form in the coronary arteries. These arteries branch off from the aorta close to the semilunar valve. They carry blood to the wall of the heart, supplying the oxygen and glucose needed by cardiac muscle fibres for cell respiration. The medical name for a blood clot is a thrombus. Coronary thrombosis is the formation of blood clots in the coronary arteries.

If the coronary arteries become blocked by a blood clot, part of the heart is deprived of oxygen and nutrients. Cardiac muscle cells are then unable to produce sufficient ATP by aerobic respiration and their contractions become irregular and uncoordinated. The wall of the heart makes quivering movements called fibrillation that do not pump blood effectively. This condition can prove fatal unless it resolves naturally or through medical intervention.

Atherosclerosis causes occlusion in the coronary arteries. Where atheroma develops the endothelium of the arteries tends to become damaged and roughened; especially, the artery wall is hardened by deposition of calcium salts. Patches of atheroma sometimes rupture causing a lesion. Coronary occlusion, damage to the capillary epithelium, hardening of arteries and rupture of atheroma all increase the risk of coronary thrombosis.

There are some well-known factors that are correlated with an increased risk of coronary thrombosis and heart attacks:

- smoking
- high blood cholesterol concentration
- high blood pressure
- diabetes
- obesity
- lack of exercise.

Of course correlation does not prove causation, but doctors nonetheless advise patients to avoid these risk factors if possible.

Phagocytes

Ingestion of pathogens by phagocytic white blood cells gives non-specific immunity to diseases.

If microorganisms get past the physical barriers of skin and mucous membranes and enter the body, white blood cells provide the next line of defence. There are many different types of white blood cell. Some are phagocytes that squeeze out through pores in the walls of capillaries and move to sites of infection. There they engulf pathogens by endocytosis and digest them with enzymes from lysosomes. When wounds become infected, large numbers of phagocytes are attracted, resulting in the formation of a white liquid called pus.

Antibody production

Production of antibodies by lymphocytes in response to particular pathogens gives specific immunity.

If microorganisms get past the physical barriers of the skin and invade the body, proteins and other molecules on the surface of pathogens are recognized as foreign by the body and they stimulate a specific immune response. Any chemical that stimulates an immune response is referred to as an antigen. The specific immune response is the production of antibodies in response to a particular pathogen. The antibodies bind to an antigen on that pathogen.

Antibodies are produced by types of white blood cell called lymphocytes. Each lymphocyte produces just one type of antibody, but our bodies can produce a vast array of different antibodies. This is because we have small numbers of lymphocytes for producing each of the many types of antibody. There are therefore too few lymphocytes initially to produce enough antibodies to control a pathogen that has not previously infected the body. However, antigens on the pathogen stimulate cell division of the small group of lymphocytes that produce the appropriate type of antibody. A large clone of lymphocytes called plasma cells are produced within a few days and they secrete large enough quantities of the antibody to control the pathogen and clear the infection.

Antibodies are large proteins that have two functional regions: a hyper-variable region that binds to a specific antigen and another region that helps the body to fight the pathogen in one of a number of ways, including these:

- making a pathogen more recognizable to phagocytes so they are more readily engulfed

- preventing viruses from docking to host cells so that they cannot enter the cells.

Antibodies only persist in the body for a few weeks or months and the plasma cells that produce them are also gradually lost after the infection has been overcome and the antigens associated with it are no longer present. However, some of the lymphocytes produced during an infection are not active plasma cells but instead become memory cells

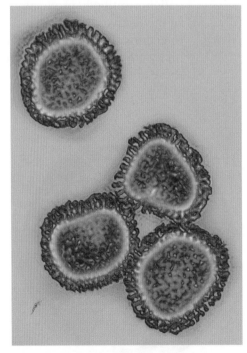

▲ Figure 6 Avian influenza viruses. In this electron micrograph of a virus in transverse section, false colour has been used to distinguish the protein coat that is recognized as antigens by the immune system (purple) from the RNA of the virus (green)

that are very long-lived. These memory cells remain inactive unless the same pathogen infects the body again, in which case they become active and divide to produce plasma cells very rapidly. Immunity to an infectious disease involves either having antibodies against the pathogen, or memory cells that allow rapid production of the antibody.

Human immunodeficiency virus

Effects of HIV on the immune system and methods of transmission.

The production of antibodies by the immune system is a complex process and includes different types of lymphocyte, including helper T-cells. The human immunodeficiency virus (HIV) invades and destroys helper T-cells. The consequence is a progressive loss of the capacity to produce antibodies. In the early stages of infection, the immune system makes antibodies against HIV. If these can be detected in a person's body, they are said to be HIV-positive.

HIV is a retrovirus that has genes made of RNA and uses reverse transcriptase to make DNA copies of its genes once it has entered a host cell. The rate at which helper T-cells are destroyed by HIV varies considerably and can be slowed down by using anti-retroviral drugs. In most HIV-positive patients antibody production eventually becomes so ineffective that a group of opportunistic infections strike, which would be easily fought off by a healthy immune system. Several of these are normally so rare that they are marker diseases for the latter stages of HIV infection, for example Kaposi's sarcoma. A collection of several diseases or conditions existing together is called a syndrome. When the syndrome of conditions due to HIV is present, the person is said to have acquired immune deficiency syndrome (AIDS).

AIDS spreads by HIV infection. The virus only survives outside the body for a short time and infection normally only occurs if there is blood to blood contact between infected and uninfected people. There are various ways in which this can occur:

* sexual intercourse, during which abrasions to the mucous membranes of the penis and vagina can cause minor bleeding

* transfusion of infected blood, or blood products such as Factor VIII

* sharing of hypodermic needles by intravenous drug users.

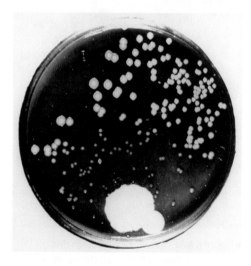

▲ Figure 7 Fleming's petri dish which first showed the inhibition of bacterial growth by penicillin from a mycelium of *Penicillium*

Antibiotics

Antibiotics block processes that occur in prokaryotic cells but not in eukaryotic cells.

An antibiotic is a chemical that inhibits the growth of microorganisms. Most antibiotics are antibacterial. They block processes that occur in prokaryotes but not in eukaryotes and can therefore be used to kill bacteria inside the body without causing harm to human cells. The processes targeted by antibiotics are bacterial DNA replication, transcription, translation, ribosome function and cell wall formation.

Many antibacterial antibiotics were discovered in saprotrophic fungi. These fungi compete with saprotrophic bacteria for the dead organic matter on which they both feed. By secreting antibacterial antibiotics, saprotrophic fungi inhibit the growth of their bacterial competitors. An example is penicillin. It is produced by some strains of the *Penicillium* fungus, but only when nutrients are scarce and competition with bacteria would be harmful.

Testing penicillin

Florey and Chain's experiments to test penicillin on bacterial infections in mice.

Howard Florey and Ernst Chain formed a research team in Oxford in the late 1930s that investigated the use of chemical substances to control bacterial infections. The most promising of these was penicillin, discovered by Alexander Fleming in 1928. Florey and Chain's team developed a method of growing the fungus *Penicillium* in liquid culture in conditions that stimulated it to secrete penicillin. They also developed methods for producing reasonably pure samples of penicillin from the cultures.

The penicillin killed bacteria on agar plates, but they needed to test whether it would control bacterial infections in humans. They first tested it on mice. Eight mice were deliberately infected with *Streptococcus* bacteria that cause death from pneumonia. Four of the infected mice were given injections with penicillin. Within 24 hours all the untreated mice were dead but the four given penicillin were healthy. Florey and Chain decided that they should next do tests on human patients, which required much larger quantities.

When enough penicillin had been produced, a 43-year-old policeman was chosen for the first human test. He had an acute and life-threatening bacterial infection caused by a scratch on the face from a thorn on a rose bush. He was given penicillin for four days and his condition improved considerably, but supplies of penicillin ran out and he suffered a relapse and died from the infection.

Larger quantities of penicillin were produced and five more patients with acute infections were tested. All were cured of their infections, but sadly one of them died. He was a small child who had an infection behind the eye. This had weakened the wall of the artery carrying blood to the brain and although cured of the infection, the child died suddenly of brain hemorrhage when the artery burst.

Pharmaceutical companies in the United States then began to produce penicillin in much larger quantities, allowing more extensive testing, which confirmed that it was a highly effective treatment for many previously incurable bacterial infections.

Activity

World AIDS Day

The red AIDS awareness ribbon is an international symbol of awareness and support for those living with HIV. It is worn on World AIDS Day each year – December 1st.

Are you aware how many people in your area are affected and what can be done to support them?

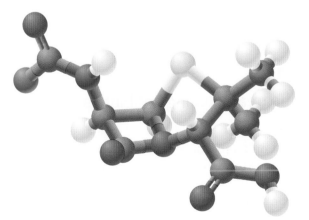

▲ Figure 8 Penicillin – the green ball represents a variable part of the molecule

Penicillin and drug testing

Risks associated with scientific research: Florey and Chain's tests on the safety of penicillin would not be compliant with current protocols on testing.

When any new drug is introduced there are risks that it will prove to be ineffective in some or all patients or that it will cause harmful side effects. These risks are minimized by strict protocols that pharmaceutical companies must follow. Initial tests are performed on animals and then on small numbers of healthy humans. Only if a drug passes these tests is it tested on patients with the disease that the drug is intended to treat. The last tests involve very large numbers of patients to test whether the drug is effective in all patients and to check that there are no severe or common side effects.

There are some famous cases of drugs causing problems during testing or after release.

- Thalidomide was introduced in the 1950s as a treatment for various mild conditions but when it was found to relieve morning sickness in pregnant women it was prescribed for that purpose. The side effects of the drug on the fetus had not been tested and more than 10,000 children were born with birth deformities before the problem was recognized.

- In 2006 six healthy volunteers were given TGN1412, a new protein developed for treatment of autoimmune diseases and leukemia. All six rapidly became very ill and suffered multiple organ failure. Although the volunteers recovered, they may have suffered long-term damage to their immune systems.

It is very unlikely that Florey and Chain would have been allowed to carry out tests on a new drug today with the methods that they used for penicillin. They tested the drug on human patients after only a very brief period of animal testing. Penicillin was a new type of drug and there could easily have been severe side effects. Also the samples that they were using were not pure and there could have been side effects from the impurities.

On the other hand, the patients that they used were all on the point of death and several were cured of their infections as a result of the experimental treatment. Because of expeditious testing with greater risk-taking than would now be allowed, penicillin was introduced far more quickly than would be possible today. During the D-day landings in June 1944 penicillin was used to treat wounded soldiers and the number of deaths from bacterial infection was greatly reduced.

▲ Figure 9 Wounded US troops on Omaha beach 6 June 1944

Viruses and antibiotics

Viral diseases cannot be treated using antibiotics because they lack a metabolism.

Viruses are non-living and can only reproduce when they are inside living cells. They use the chemical processes of a living host cell, instead of having a metabolism of their own. They do not have their own means of transcription or protein synthesis and they rely on the host cell's enzymes for ATP synthesis and other metabolic pathways. These processes cannot be targeted by drugs as the host cell would also be damaged.

All of the commonly used antibiotics such as penicillin, streptomycin, chloramphenicol and tetracycline control bacterial infections and are not effective against viruses. Not only is it inappropriate for doctors to prescribe them for a viral infection, but it contributes to the overuse of antibiotics and increases in antibiotic resistance in bacteria.

There are a few viral enzymes which can be used as targets for drugs to control viruses without harming the host cell. Only a few drugs have been discovered or developed to control viruses in this way. These are known as antivirals rather than antibiotics.

Resistance to antibiotics

Some strains of bacteria have evolved with genes which confer resistance to antibiotics and some strains of bacteria have multiple resistance.

In 2013 the government's chief medical officer for England, Sally Davies, said this:

> *The danger posed by growing resistance to antibiotics should be ranked along with terrorism on a list of threats to the nation. If we don't take action, then we may all be back in an almost 19th-century environment where infections kill us as a result of routine operations. We won't be able to do a lot of our cancer treatments or organ transplants.*

The development of resistance to antibiotics by natural selection is described in sub-topic 5.2. Strains of bacteria with resistance are usually discovered soon after the introduction of an antibiotic. This is not of huge concern unless a strain develops multiple resistance, for example methicillin-resistant *Staphylococcus aureus* (MRSA) which has infected the blood or surgical wounds of hospital patients and resists all commonly used antibiotics. Another example of this problem is multidrug-resistant tuberculosis (MDR-TB). The WHO has reported more than 300,000 cases worldwide per year with the disease reaching epidemic proportions in some areas.

Antibiotic resistance is an avoidable problem. These measures are required:

- doctors prescribing antibiotics only for serious bacterial infections
- patients completing courses of antibiotics to eliminate infections completely
- hospital staff maintaining high standards of hygiene to prevent cross-infection
- farmers not using antibiotics in animal feeds to stimulate growth
- pharmaceutical companies developing new types of antibiotic – no new types have been introduced since the 1980s.

Data-based questions: Antibiotic resistance

Bacterial resistance to antibiotics is a direct consequence of the overuse of these drugs. In the USA, currently more than half of the doctor visits for upper respiratory tract infections (URIs) are prescribed antibiotics, despite knowledge that most URIs are caused by viruses.

In the early 1990s, Finnish public health authorities began discouraging the use of the antibiotic erythromycin for URIs in response to rising bacterial resistance to the antibiotic, and the national erythromycin consumption per capita dropped by 43 per cent.

The data in figure 11 shows the incidence in Finland, over a 10-year period, of *Streptococcus pyogenes* strains that are resistant to the antibiotic erythromycin. *S. pyogenes* is responsible for the condition known as "strep throat".

1 a) Describe the pattern of erythromycin resistance over the period from 1992 to 2002. [3]

b) Suggest a reason for the pattern shown. [2]

2 Calculate the percentage difference in antibiotic resistance between 2002 and 1992. [2]

3 Evaluate the claim that reduction in the use of erythromycin has led to a reduction in the incidence of antibiotic resistance in *S. pyogenes*. [3]

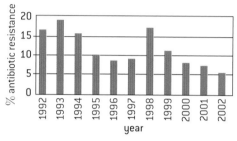

▲ Figure 11 The incidence of *Streptococcus pyogenes* strains that are resistant to the antibiotic erythromycin over a 10-year period in Finland

6.4 Gas exchange

Understanding

→ Ventilation maintains concentration gradients of oxygen and carbon dioxide between air in alveoli and blood flowing in adjacent capillaries.

→ Type I pneumocytes are extremely thin alveolar cells that are adapted to carry out gas exchange.

→ Type II pneumocytes secrete a solution containing surfactant that creates a moist surface inside the alveoli to prevent the sides of the alveolus adhering to each other by reducing surface tension.

→ Air is carried to the lungs in the trachea and bronchi and then to the alveoli in bronchioles.

→ Muscle contractions cause the pressure changes inside the thorax that force air in and out of the lungs to ventilate them.

→ Different muscles are required for inspiration and expiration because muscles only do work when they contract.

 Applications

→ External and internal intercostal muscles, and diaphragm and abdominal muscles as examples of antagonistic muscle action.

→ Causes and consequences of lung cancer.

→ Causes and consequences of emphysema.

 Skills

→ Monitoring of ventilation in humans at rest and after mild and vigorous exercise. (Practical 6)

 Nature of science

→ Obtain evidence for theories: epidemiological studies have contributed to our understanding of the causes of lung cancer.

Ventilation

Ventilation maintains concentration gradients of oxygen and carbon dioxide between air in alveoli and blood flowing in adjacent capillaries.

All organisms absorb one gas from the environment and release a different one. This process is called gas exchange. Leaves absorb carbon dioxide to use in photosynthesis and release the oxygen produced by this process. Humans absorb oxygen for use in cell respiration and release the carbon dioxide produced by this process. Terrestrial organisms exchange gases with the air. In humans gas exchange occurs in small air sacs called alveoli inside the lungs (figure 1).

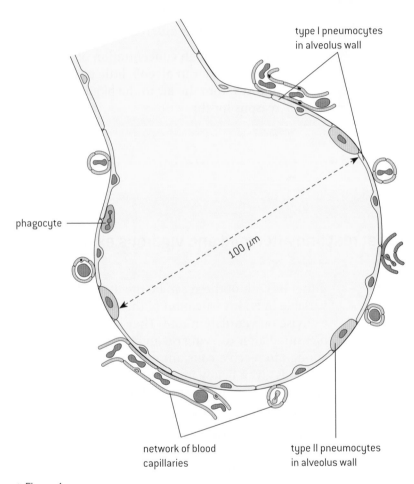

type I pneumocytes
in alveolus wall

phagocyte

100 µm

network of blood
capillaries

type II pneumocytes
in alveolus wall

▲ Figure 1

Gas exchange happens by diffusion between air in the alveoli and blood flowing in the adjacent capillaries. The gases only diffuse because there is a concentration gradient: the air in the alveolus has a higher concentration of oxygen and a lower concentration of carbon dioxide than the blood in the capillary. To maintain these concentration gradients fresh air must be pumped into the alveoli and stale air must be removed. This process is called **ventilation**.

Data-based questions: Concentration gradients

Figure 2 shows the typical composition of atmospheric air, air in the alveoli and gases dissolved in air returning to the lungs in the pulmonary arteries.

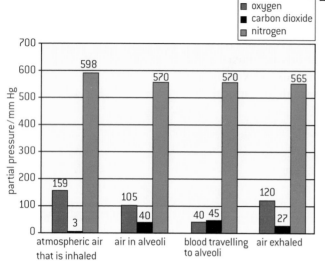

▲ Figure 2 Partial pressures of gases in the pulmonary system

1. Explain why the oxygen concentration in the alveoli is not as high as in fresh air that is inhaled. [2]

2. a) Calculate the difference in oxygen concentration between air in the alveolus and blood arriving at the alveolus. [1]

    b) Deduce the process caused by this concentration difference. [1]

    c) (i) Calculate the difference in carbon dioxide concentration between air inhaled and air exhaled. [1]

    (ii) Explain this difference. [2]

    d) Despite the high concentration of nitrogen in air in alveoli, little or none diffuses from the air to the blood. Suggest reasons for this. [2]

⚗ Ventilation experiments

Monitoring of ventilation in humans at rest and after mild and vigorous exercise. (Practical 6)

In an investigation of the effect of exercise on ventilation, the type or intensity of exercise is the independent variable and the ventilation parameter that is measured is the dependent variable.

- A simple approach for the independent variable is to choose levels of activity ranging from inactive to very active, such as lying down, sitting and standing, walking, jogging and sprinting. A more quantitative approach is to do the same activity at different measured rates, for example running at different speeds on a treadmill. This allows the ventilation parameters to be correlated with work rate in joules per minute during exercise.

Ventilation of the lungs is carried out by drawing some fresh air into the lungs and then expelling some of the stale air from the lungs. The volume of air drawn in and expelled is the tidal volume. The number of times that air is drawn in or expelled per minute is the ventilation rate.

Either or both of these can be the dependent variable in an investigation of the effect of exercise on ventilation rate. They should be measured after carrying on an activity for long enough to reach a constant rate. The example methods given below include a simple and a more advanced technique that could be used for the investigation.

1 **Ventilation rate**
- The most straightforward way to measure ventilation rate is by simple observation. Count the number of times air is exhaled or inhaled in a minute. Breathing should be maintained at a natural rate, which is as slow as possible without getting out of breath.

- Ventilation rate can also be measured by data logging. An inflatable chest belt is placed around the thorax and air is pumped in with a bladder. A differential pressure sensor is then used to measure pressure variations inside the belt due to chest expansions. The rate of ventilations can be deduced and the relative size of ventilations may also be recorded.

2 Tidal volume

- Simple apparatus is shown in figure 3. One normal breath is exhaled through the delivery tube into a vessel and the volume is measured. It is not safe to use this apparatus for repeatedly inhaling and exhaling air as the CO_2 concentration will rise too high.

- Specially designed spirometers are available for use with data logging. They measure flow rate into and out of the lungs and from these measurements lung volumes can be deduced.

To ensure that the experimental design is rigorous, all variables apart from the independent and dependent variables should be kept constant. Ventilation parameters should be measured several times at all levels of exercise with each person in the trial. As many different people as possible should be tested.

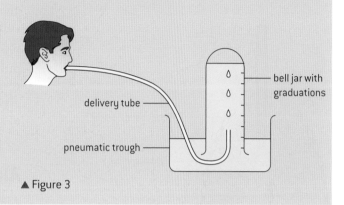

▲ Figure 3

Type I pneumocytes

Type I pneumocytes are extremely thin alveolar cells that are adapted to carry out gas exchange.

The lungs contain huge numbers of alveoli with a very large total surface area for diffusion. The wall of each alveolus consists of a single layer of cells, called the epithelium. Most of the cells in this epithelium are Type I pneumocytes. They are flattened cells, with the thickness of only about 0.15 μm of cytoplasm.

The wall of the adjacent capillaries also consists of a single layer of very thin cells. The air in the alveolus and the blood in the alveolar capillaries are therefore less than 0.5 μm apart. The distance over which oxygen and carbon dioxide has to diffuse is therefore very small, which is an adaptation to increase the rate of gas exchange.

Type II pneumocytes

Type II pneumocytes secrete a solution containing surfactant that creates a moist surface inside the alveoli to prevent the sides of the alveolus adhering to each other by reducing surface tension.

Type II pneumocytes are rounded cells that occupy about 5% of the alveolar surface area. They secrete a fluid which coats the inner surface of the alveoli. This film of moisture allows oxygen in the alveolus to dissolve and then diffuse to the blood in the alveolar capillaries. It also provides an area from which carbon dioxide can evaporate into the air and be exhaled.

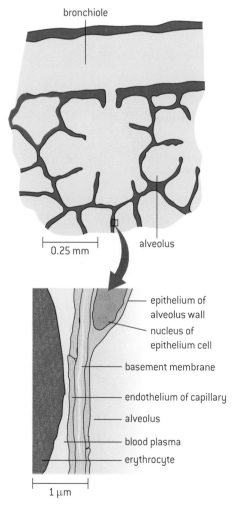

▲ Figure 4 Structure of alveoli

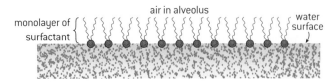

▲ Figure 5 Pulmonary surfactant molecules on the surface of the film of moisture lining the alveoli

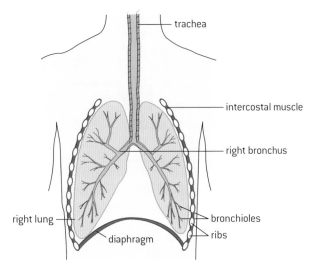

▲ Figure 6 The ventilation system

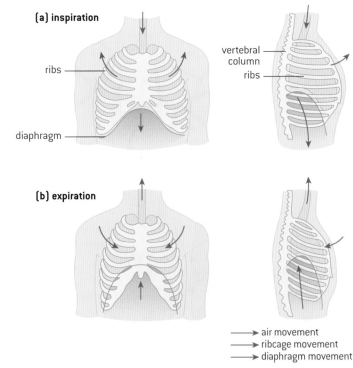

▲ Figure 7 Ventilation of the lungs

The fluid secreted by the Type II pneumocytes contains a pulmonary surfactant. Its molecules have a structure similar to that of phospholipids in cell membranes. They form a monolayer on the surface of the moisture lining the alveoli, with the hydrophilic heads facing the water and the hydrophobic tails facing the air. This reduces the surface tension and prevents the water from causing the sides of the alveoli to adhere when air is exhaled from the lungs. This helps to prevent collapse of the lung.

Premature babies are often born with insufficient pulmonary surfactant and can suffer from infant respiratory distress syndrome. Treatment involves giving the baby oxygen and also one or more doses of surfactant, extracted from animal lungs.

Airways for ventilation

Air is carried to the lungs in the trachea and bronchi and then to the alveoli in bronchioles.

Air enters the ventilation system through the nose or mouth and then passes down the trachea. This has rings of cartilage in its wall to keep it open even when air pressure inside is low or pressure in surrounding tissues is high. The trachea divides to form two bronchi, also with walls strengthened with cartilage. One bronchus leads to each lung.

Inside the lungs the bronchi divide repeatedly to form a tree-like structure of narrower airways, called bronchioles. The bronchioles have smooth muscle fibres in their walls, allowing the width of these airways to vary. At the end of the narrowest bronchioles are groups of alveoli, where gas exchange occurs.

Pressure changes during ventilation

Muscle contractions cause the pressure changes inside the thorax that force air in and out of the lungs to ventilate them.

Ventilation of the lungs involves some basic physics. If particles of gas spread out to occupy a larger volume, the pressure of the gas becomes lower. Conversely, if a gas is compressed to occupy a smaller volume, the pressure rises. If gas is free to move, it will always flow from regions of higher pressure to regions of lower pressure.

During ventilation, muscle contractions cause the pressure inside the thorax to drop below atmospheric pressure. As a consequence, air is drawn into the lungs from the atmosphere (inspiration) until the lung pressure has risen to atmospheric pressure. Muscle contractions then cause pressure inside the thorax to rise above atmospheric, so air is forced out from the lungs to the atmosphere (expiration).

Antagonistic muscles

Different muscles are required for inspiration and expiration because muscles only do work when they contract.

Muscles can be in two states: contracting and relaxing.

- Muscles do work when they contract by exerting a pulling force (tension) that causes a particular movement. They become shorter when they do this.

- Muscles lengthen while they are relaxing, but this happens passively – they do not lengthen themselves. Most muscles are pulled into an elongated state by the contraction of another muscle. They do not exert a pushing force (compression) while relaxing so do no work at this time.

▲ Figure 8 Different muscles are used for bending the leg at the knee and for the opposite movement of straightening it

Muscles therefore can only cause movement in one direction. If movement in opposite directions is needed at different times, at least two muscles will be required. When one muscle contracts and causes a movement, the second muscle relaxes and is elongated by the first. The opposite movement is caused by the second muscle contracting while the first relaxes. When muscles work together in this way they are known as an antagonistic pair of muscles.

Inspiration and expiration involve opposite movements, so different muscles are required, working as antagonistic pairs.

 Antagonistic muscle action in ventilation

External and internal intercostal muscles, and diaphragm and abdominal muscles as examples of antagonistic muscle action.

Ventilation involves two pairs of opposite movements that change the volume and therefore the pressure inside the thorax:

	Inspiration	**Expiration**
Diaphragm	Moves downwards and flattens	Moves upwards and becomes more domed
Ribcage	Moves upwards and outwards	Moves downwards and inwards

Antagonistic pairs of muscles are needed to cause these movements.

	Inspiration	**Expiration**
Volume and pressure changes	The volume inside the thorax increases and consequently the pressure decreases	The volume inside the thorax decreases and consequently the pressure increases

Movement of the diaphragm	Diaphragm	The diaphragm contracts and so it moves downwards and pushes the abdomen wall out	The diaphragm relaxes so it can be pushed upwards into a more domed shape
	Abdomen wall muscles	Muscles in the abdomen wall relax allowing pressure from the diaphragm to push it out	Muscles in the abdomen wall contract pushing the abdominal organs and diaphragm upwards
Movement of the ribcage	External intercostal muscles	The external intercostal muscles contract, pulling the ribcage upwards and outwards	The external intercostal muscles relax and are pulled back into their elongated state.
	Internal intercostal muscles	The internal intercostal muscles relax and are pulled back into their elongated state	The internal intercostal muscles contract, pulling the ribcage inwards and downwards

 ## Epidemiology

Obtain evidence for theories: epidemiological studies have contributed to our understanding of the causes of lung cancer.

Epidemiology is the study of the incidence and causes of disease. Most epidemiological studies are observational rather than experimental because it is rarely possible to investigate the causes of disease in human populations by carrying out experiments.

As in other fields of scientific research, theories about the causes of a disease are proposed. To obtain evidence for or against a theory, survey data is collected that allows the association between the disease and its theoretical cause to be tested. For example, to test the theory that smoking causes lung cancer, the smoking habits of people who have developed lung cancer and people who have not are needed. Examples of very large epidemiological surveys that provided strong evidence for a link between smoking and lung cancer are included in sub-topic 1.6.

A correlation between a risk factor and a disease does not prove that the factor causes the disease. There are usually confounding factors which also have an effect on the incidence. They can cause spurious associations between a disease and a factor that does not cause it. For example, an association has repeatedly been found by epidemiologists between leanness and an increased risk of lung cancer. Careful analysis showed that among smokers leanness is not significantly associated with an increased risk. Smoking reduces appetite and so is associated with leanness and of course smoking is a cause of lung cancer. This explains the spurious association between leanness and lung cancer.

To try to compensate for confounding factors it is usually necessary to collect data on many factors apart from the one being investigated. This allows statistical procedures to be carried out to take account of confounding factors and try to isolate the effect of single factors. Age and sex are almost always recorded and sometimes epidemiological surveys include only males or females or only people in a specific age range.

Causes of lung cancer

Causes and consequences of lung cancer.

Lung cancer is the most common cancer in the world, both in terms of the number of cases and the number of deaths due to the disease. The general causes of cancer are described in sub-topic 1.6. The specific causes of lung cancer are considered here.

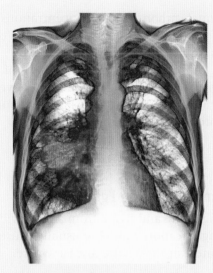

▲ Figure 9 A large tumour (red) is visible in the right lung. The tumour is a bronchial carcinoma

- Smoking causes about 87% of cases. Tobacco smoke contains many mutagenic chemicals. As every cigarette carries a risk, the incidence of lung cancer increases with the number smoked per day and the number of years of smoking.

- Passive smoking causes about 3% of cases. This happens when non-smokers inhale tobacco smoke exhaled by smokers. The number of cases will decline in countries where smoking is banned indoors and in public places.

- Air pollution probably causes about 5% of lung cancers. The sources of air pollution that are most significant are diesel exhaust fumes, nitrogen oxides from all vehicle exhaust fumes and smoke from burning coal, wood or other organic matter.

- Radon gas causes significant numbers of cases in some parts of the world. It is a radioactive gas that leaks out of certain rocks such as granite. It accumulates in badly ventilated buildings and people then inhale it.

- Asbestos, silica and some other solids can cause lung cancer if dust or other particles of them are inhaled. This usually happens on construction sites or in quarries, mines or factories.

The consequences of lung cancer are often very severe. Some of them can be used to help diagnose the disease: difficulties with breathing, persistent coughing, coughing up blood, chest pain, loss of appetite, weight loss and general fatigue.

In many patients the tumour is already large when it is discovered and may also have metastasized, with secondary tumours in the brain or elsewhere. Mortality rates are high. Only 15% of patients with lung cancer survive for more than 5 years. If a tumour is discovered early enough, all or part of the affected lung may be removed surgically. This is usually combined with one or more courses of chemotherapy. Other patients are treated with radiotherapy.

The minority of patients who are cured of lung cancer, but have lost some of their lung tissue, are likely to continue to have pain, breathing difficulties, fatigue and also anxiety about the possible return of the disease.

Emphysema

Causes and consequences of emphysema.

In healthy lung tissue each bronchiole leads to a group of small thin-walled alveoli. In a patient with emphysema these are replaced by a smaller number of larger air sacs with much thicker walls. The total surface area for gas exchange is considerably reduced and the distance over which diffusion of gases occurs is increased, and so gas exchange is therefore much less effective. The lungs also become less elastic, so ventilation is more difficult.

The molecular mechanisms involved are not fully understood, though there is some evidence for these theories:

- Phagocytes inside alveoli normally prevent lung infections by engulfing bacteria and produce elastase, a protein-digesting enzyme, to kill them inside the vesicles formed by endocytosis.

- An enzyme inhibitor called alpha 1-antitrypsin (A1AT) usually prevents elastase and other proteases from digesting lung tissue. In smokers, the number of phagocytes in the lungs increases and they produce more elastase.

- Genetic factors affect the quantity and effectiveness of A1AT produced in the lungs.

In about 30% of smokers digestion of proteins in the alveolus wall by the increased quantity of proteases is not prevented and alveolus walls are weakened and eventually destroyed.

Emphysema is a chronic disease because the damage to alveoli is usually irreversible. It causes low oxygen saturation in the blood and higher than normal carbon dioxide concentrations. As a result the patient lacks energy and may eventually find even tasks such as climbing stairs too onerous. In mild cases there is shortness of breath during vigorous exercise but eventually even mild activity causes it. Ventilation is laboured and tends to be more rapid than normal.

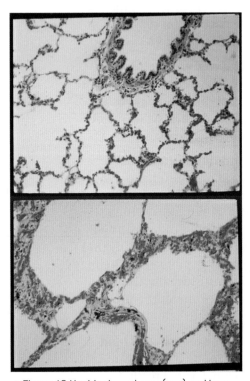

▲ Figure 10 Healthy lung tissue (top) and lung tissue showing emphysema (bottom)

Data-based questions: Emphysema and gas exchange

Figure 10 shows healthy lung tissue and tissue from a lung with emphysema, at the same magnification. Smoking usually causes emphysema. Breathing polluted air makes the disease worse.

1   a) Place a ruler across each micrograph and count how many times the edge of the ruler crosses a gas exchange surface. Repeat this several times for each micrograph, in such a way that the results are comparable. State your results using suitable units. [3]

   b) Explain the conclusions that you draw from the results. [3]

2   Explain why people who have emphysema feel tired all the time. [3]

3   Suggest why people with emphysema often have an enlarged and strained right side of the heart. [1]

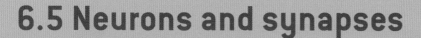

6.5 Neurons and synapses

Understanding

→ Neurons transmit electrical impulses.

→ The myelination of nerve fibres allows for saltatory conduction.

→ Neurons pump sodium and potassium ions across their membranes to generate a resting potential.

→ An action potential consists of depolarization and repolarization of the neuron.

→ Nerve impulses are action potentials propagated along the axons of neurons.

→ Propagation of nerve impulses is the result of local currents that cause each successive part of the axon to reach the threshold potential.

→ Synapses are junctions between neurons and between neurons and receptor or effector cells.

→ When pre-synaptic neurons are depolarized they release a neurotransmitter into the synapse.

→ A nerve impulse is only initiated if the threshold potential is reached.

 ## Applications

→ Secretion and reabsorption of acetylcholine by neurons at synapses.

→ Blocking of synaptic transmission at cholinergic synapses in insects by binding of neonicotinoid pesticides to acetylcholine receptors.

 ## Skills

→ Analysis of oscilloscope traces showing resting potentials and action potentials.

 ## Nature of science

→ Cooperation and collaboration between groups of scientists: biologists are contributing to research into memory and learning.

Neurons

Neurons transmit electrical impulses.

Two systems of the body are used for internal communication: the endocrine system and the nervous system. The endocrine system consists of glands that release hormones. The nervous system consists of nerve cells called neurons. There are about 85 billion neurons in the human nervous system. Neurons help with internal communication by transmitting nerve impulses. A nerve impulse is an electrical signal.

Neurons have a cell body with cytoplasm and a nucleus but they also have narrow outgrowths called nerve fibres along which nerve impulses travel.

- Dendrites are short branched nerve fibres, for examples those used to transmit impulses between neurons in one part of the brain or spinal cord.

- Axons are very elongated nerve fibres, for example those that transmit impulses from the tips of the toes or the fingers to the spinal cord.

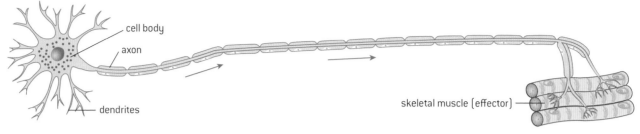

▲ Figure 1 Neuron with dendrites that transmit impulses to the cell body and an axon that transmits impulses a considerable distance to muscle fibres

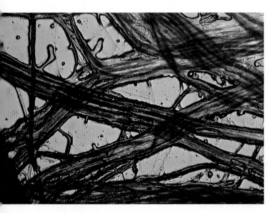

▲ Figure 2 Nerve fibres (axons) transmitting electrical impulses to and from the central nervous system are grouped into bundles

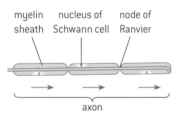

▲ Figure 3 Detail of a myelinated nerve fibre showing the gaps between adjacent Schwann cells (nodes of Ranvier)

Myelinated nerve fibres

The myelination of nerve fibres allows for saltatory conduction.

The basic structure of a nerve fibre along which a nerve impulse is transmitted is very simple: the fibre is cylindrical in shape, with a plasma membrane enclosing a narrow region of cytoplasm. The diameter in most cases is about 1 µm, though some nerve fibres are wider than this. A nerve fibre with this simple structure conducts nerve impulses at a speed of about 1 metre per second.

Some nerve fibres are coated along most of their length by a material called myelin. It consists of many layers of phospholipid bilayer. Special cells called Schwann cells deposit the myelin by growing round and round the nerve fibre. Each time they grow around the nerve fibre a double layer of phospholipid bilayer is deposited. There may be 20 or more layers when the Schwann cell stops growing.

There is a gap between the myelin deposited by adjacent Schwann cells. The gap is called a node of Ranvier. In myelinated nerve fibres the nerve impulse can jump from one node of Ranvier to the next. This is called saltatory conduction. It is much quicker than continuous transmission along a nerve fibre so myelinated nerve fibres transmit nerve impulses much more rapidly than unmyelinated nerve fibres. The speed can be as much as 100 metres per second.

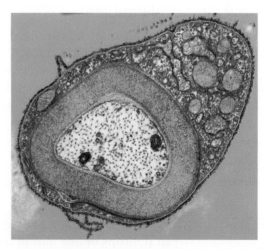

▲ Figure 4 Transverse section of axon showing the myelin sheath formed by the Schwann cell's membrane wrapped round the axon many times (red)

Resting potentials

Neurons pump sodium and potassium ions across their membranes to generate a resting potential.

A neuron that is not transmitting a signal has a potential difference or voltage across its membrane that is called the resting potential. This potential is due to an imbalance of positive and negative charges across the membrane.

- Sodium–potassium pumps transfer sodium (Na^+) and potassium (K^+) ions across the membrane. Na^+ ions are pumped out and K^+ ions are pumped in. The numbers of ions pumped is unequal – when three Na^+ ions are pumped out, only two K^+ ions are pumped in, creating concentration gradients for both ions.

- Also the membrane is about 50 times more permeable to K^+ ions than Na^+ ions, so K^+ ions leak back across the membrane faster than Na^+ ions. As a result, the Na^+ concentration gradient across the membrane is steeper than the K^+ gradient, creating a charge imbalance.

- In addition to this, there are proteins inside the nerve fibre that are negatively charged (organic anions), which increases the charge imbalance.

These factors together give the neuron a resting membrane potential of about −70 mV.

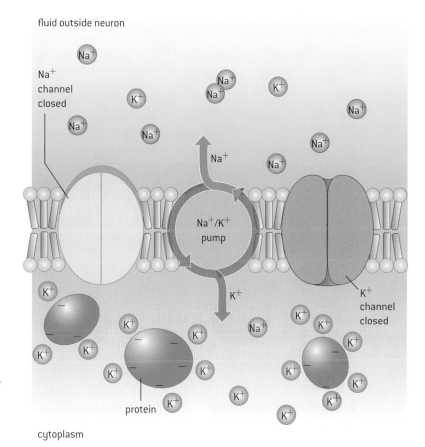

fluid outside neuron

▲ Figure 5 The resting potential is generated by the sodium–potassium pump

Action potentials

An action potential consists of depolarization and repolarization of the neuron.

An action potential is a rapid change in membrane potential, consisting of two phases:

- depolarization – a change from negative to positive

- repolarization – a change back from positive to negative.

Depolarization is due to the opening of sodium channels in the membrane, allowing Na^+ ions to diffuse into the neuron down the concentration gradient. The entry of Na^+ ions reverses the charge imbalance across the membrane, so the inside is positive relative to the outside. This raises the membrane potential to a positive value of about +30 mV.

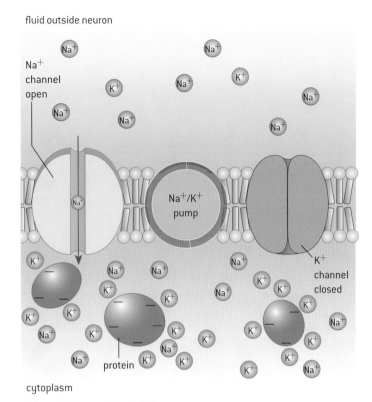

fluid outside neuron

Na⁺ channel open

Na⁺/K⁺ pump

K⁺ channel closed

protein

cytoplasm

▲ Figure 6 Neuron depolarizing

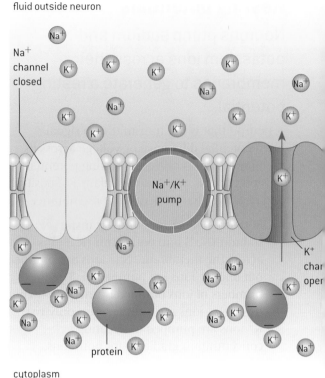

fluid outside neuron

Na⁺ channel closed

Na⁺/K⁺ pump

K⁺ chan oper

protein

cytoplasm

▲ Figure 7 Neuron repolarizing

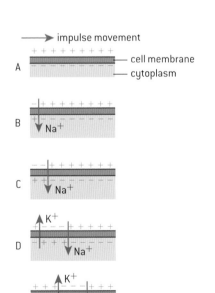

▲ Figure 8 Action potentials are propagated along axons

Repolarization happens rapidly after depolarization and is due to the closing of the sodium channels and opening of potassium channels in the membrane. This allows potassium ions to diffuse out of the neuron, down their concentration gradient, which makes the inside of the cell negative again relative to the outside. The potassium channels remain open until the membrane has fallen to a potential close to −70 mV. The diffusion of potassium repolarizes the neuron, but it does not restore the resting potential as the concentration gradients of sodium and potassium ions have not yet been re-established. This takes a few milliseconds and the neuron can then transmit another nerve impulse.

Propagation of action potentials

Nerve impulses are action potentials propagated along the axons of neurons.

A nerve impulse is an action potential that starts at one end of a neuron and is then propagated along the axon to the other end of the neuron. The propagation of the action potential happens because the ion movements that depolarize one part of the neuron trigger depolarization in the neighbouring part of the neuron.

Nerve impulses always move in one direction along neurons in humans and other vertebrates. This is because an impulse can only be initiated at one terminal of a neuron and can only be passed on to other neurons or different cell types at the other terminal. Also, there is a refractive period after a depolarization that prevents propagation of an action potential backwards along an axon.

Local currents

Propagation of nerve impulses is the result of local currents that cause each successive part of the axon to reach the threshold potential.

The propagation of an action potential along an axon is due to movements of sodium ions. Depolarization of part of the axon is due to diffusion of sodium ions into the axon through sodium channels. This reduces the concentration of sodium ions outside the axon and increases it inside. The depolarized part of the axon therefore has different sodium ion concentrations to the neighbouring part of the axon that has not yet depolarized. As a result, sodium ions diffuse between these regions both inside and outside the axon.

Inside the axon there is a higher sodium ion concentration in the depolarized part of the axon so sodium ions diffuse along inside the axon to the neighbouring part that is still polarized. Outside the axon the concentration gradient is in the opposite direction so sodium ions diffuse from the polarized part back to the part that has just depolarized. These movements are shown in figure 10. They are called local currents.

Local currents reduce the concentration gradient in the part of the neuron that has not yet depolarized. This makes the membrane potential rise from the resting potential of −70mV to about −50 mV. Sodium channels in the axon membrane are voltage-gated and open when a membrane potential of −50mV is reached. This is therefore known as the threshold potential. Opening of the sodium channels causes depolarization.

Thus local currents cause a wave of depolarization and then repolarization to be propagated along the axon at a rate of between one and a hundred (or more) metres per second.

Activity

Neurons in a sea anemone and an anemonefish

Anemonefish have a nervous system similar to ours, with a central nervous system and neurons that transmit nerve impulses in one direction only. Sea anemones have no central nervous system. Their neurons form a simple network and will transmit impulses in either direction along their nerve fibres. They both protect each other from predators more effectively than they can themselves. Explain how they do this.

▲ Figure 9 Anemonefish among the tentacles of a sea anemone

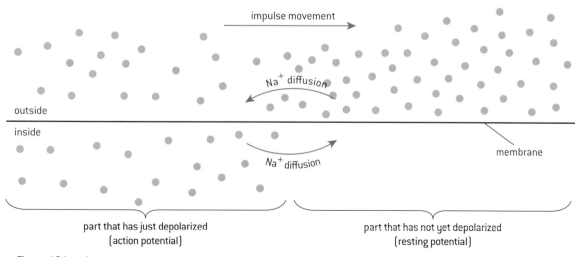

▲ Figure 10 Local currents

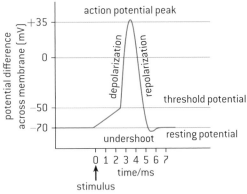

▲ Figure 11 Changes in membrane polarity during an action potential

Analysing oscilloscope traces

Analysis of oscilloscope traces showing resting potentials and action potentials.

Membrane potentials in neurons can be measured by placing electrodes on each side of the membrane. The potentials can be displayed using an oscilloscope. The display is similar to a graph with time on the x-axis and the membrane potential on the y-axis. If there is a resting potential, a horizontal line appears on the oscilloscope screen at a level of −70 mV, assuming that this is the resting potential of the neuron.

If an action potential occurs, a narrow spike is seen, with the rising and falling phases showing the depolarization and repolarization. The oscilloscope trace may also show the potential rising before the depolarization until the threshold potential is reached. The repolarization does not usually return the membrane potential to −70 mV immediately and there is a phase in which the potential changes gradually until the resting potential is reached.

Data-based questions: Analysing an oscilloscope trace

The oscilloscope trace in figure 12 was taken from a digital oscilloscope. It shows an action potential in a mouse hippocampal pyramidal neuron that happened after the neuron was stimulated with a pulse of current.

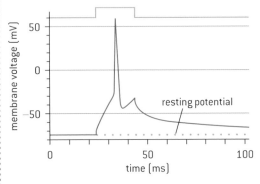

▲ Figure 12

1. State the resting potential of the mouse hippocampal pyramidal neuron. [1]

2. Deduce with a reason the threshold potential needed to open voltage-gated sodium channels in this neuron. [2]

3. Estimate the time taken for the depolarization, and the repolarization. [2]

4. Predict the time taken from the end of the depolarization for the resting potential to be regained. [2]

5. Discuss how many action potentials could be stimulated per second in this neuron. [2]

6. Suggest a reason for the membrane potential rising briefly at the end of the repolarization. [1]

Synapses

Synapses are junctions between neurons and between neurons and receptor or effector cells.

Synapses are junctions between cells in the nervous system. In sense organs there are synapses between sensory receptor cells and neurons. In both the brain and spinal cord there are immense numbers of synapses between neurons. In muscles and glands there are synapses between neurons and

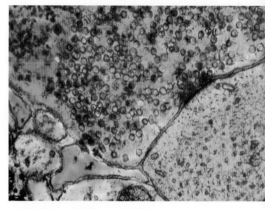

muscle fibres or secretory cells. Muscles and glands are sometimes called effectors, because they effect (carry out) a response to a stimulus.

Chemicals called neurotransmitters are used to send signals across synapses. This system is used at all synapses where the pre-synaptic and post-synaptic cells are separated by a fluid-filled gap, so electrical impulses cannot pass across. This gap is called the synaptic cleft and is only about 20 nm wide.

▲ Figure 13 Electron micrograph of a synapse. False colour has been used to indicate the pre-synaptic neuron (purple) with vesicles of neurotransmitter (blue) and the post-synaptic neuron (pink). The narrowness of the synaptic cleft is visible

Synaptic transmission

When pre-synaptic neurons are depolarized they release a neurotransmitter into the synapse.

Synaptic transmission occurs very rapidly as a result of these events:

- A nerve impulse is propagated along the pre-synaptic neuron until it reaches the end of the neuron and the pre-synaptic membrane.

- Depolarization of the pre-synaptic membrane causes calcium ions (Ca^{2+}) to diffuse through channels in the membrane into the neuron.

- Influx of calcium causes vesicles containing neurotransmitter to move to the pre-synaptic membrane and fuse with it.

- Neurotransmitter is released into the synaptic cleft by exocytosis.

- The neurotransmitter diffuses across the synaptic cleft and binds to receptors on the post-synaptic membrane.

- The binding of the neurotransmitter to the receptors causes adjacent sodium ion channels to open.

- Sodium ions diffuse down their concentration gradient into the post-synaptic neuron, causing the post-synaptic membrane to reach the threshold potential.

- An action potential is triggered in the post-synaptic membrane and is propagated on along the neuron.

- The neurotransmitter is rapidly broken down and removed from the synaptic cleft.

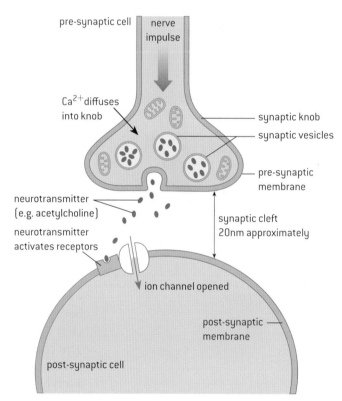

▲ Figure 14 A nerve impulse is propagated across a synapse by the release, diffusion and post-synaptic binding of neurotransmitter

---

Data-based questions: Parkinson's disease

Dopamine is one of the many neurotransmitters that are used at synapses in the brain. In Parkinson's disease, there is a loss of dopamine-secreting neurons, which causes slowness in initiating movement, muscular rigidity and in many cases shaking. Figure 15 shows the metabolic pathways involved in the formation and breakdown of dopamine.

1 Explain how symptoms of Parkinson's disease are relieved by giving the following drugs:

a) L-DOPA   [1]

b) selegeline, which is an inhibitor of monoamine oxidase-B (MAO-B) [1]

c) tolcapone, which is an inhibitor of catechol-O-methyl transferase (COMT) [1]

d) ropinirole, which is an agonist of dopamine [1]

e) safinamide, which inhibits reuptake of dopamine by pre-synaptic neurons. [1]

2 Discuss how a cure for Parkinson's disease might in the future be developed by:

a) stem cell therapy [3]

b) gene therapy. [2]

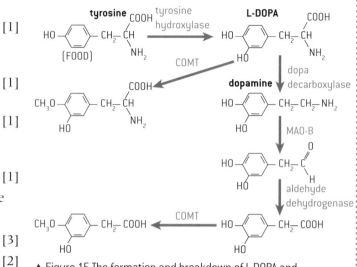

▲ Figure 15 The formation and breakdown of L-DOPA and dopamine. The enzymes catalysing each step are shown in red

Acetylcholine

Secretion and reabsorption of acetylcholine by neurons at synapses.

Acetylcholine is used as the neurotransmitter in many synapses, including synapses between neurons and muscle fibres. It is produced in the pre-synaptic neuron by combining choline, absorbed from the diet, with an acetyl group produced during aerobic respiration. The acetylcholine is loaded into vesicles and then released into the synaptic cleft during synaptic transmission.

The receptors for acetylcholine in the post-synaptic membrane have a binding site to which acetylcholine will bind. The acetylcholine only remains bound to the receptor for a short time, during which only one action potential is initiated in the post-synaptic neuron. This is because the enzyme acetylcholinesterase is present in the synaptic cleft and rapidly breaks acetylcholine down into choline and acetate. The choline is reabsorbed into the pre-synaptic neuron, where it is converted back into active neurotransmitter by recombining it with an acetyl group.

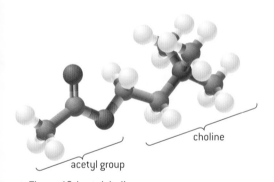

▲ Figure 16 Acetylcholine

🌐 Neonicotinoids

Blocking of synaptic transmission at cholinergic synapses in insects by binding of neonicotinoid pesticides to acetylcholine receptors.

Neonicotinoids are synthetic compounds similar to nicotine. They bind to the acetylcholine receptor in cholinergic synapses in the central nervous system of insects. Acetylcholinesterase does not break down neonicotinoids, so the binding is irreversible. The receptors are blocked, so acetylcholine is unable to bind and synaptic transmission is prevented. The consequence in insects is paralysis and death. Neonicotinoids are therefore very effective insecticides.

One of the advantages of neonicotinoids as pesticides is that they are not highly toxic to humans and other mammals. This is because a much greater proportion of synapses in the central nervous system are cholinergic in insects than in mammals and also because neonicotinoids bind much less strongly to acetylcholine receptors in mammals than insects.

Neonicotinoid pesticides are now used on huge areas of crops. In particular one neonicotinoid, imidacloprid, is the most widely used insecticide in the world. However, concerns have been raised about the effects of these insecticides on honeybees and other beneficial insects. There has been considerable controversy over this and the evidence of harm is disputed by the manufacturers and some government agencies.

Threshold potentials

A nerve impulse is only initiated if the threshold potential is reached.

Nerve impulses follow an all-or-nothing principle. An action potential is only initiated if the threshold potential is reached, because only at this potential do voltage-gated sodium channels start to open, causing depolarization. The opening of some sodium channels and the inward diffusion of sodium ions increases the membrane potential causing more sodium channels to open – there is a positive feedback effect. If the threshold potential is reached there will therefore always be a full depolarization.

At a synapse, the amount of neurotransmitter secreted following depolarization of the pre-synaptic membrane may not be enough to cause the threshold potential to be reached in the post-synaptic membrane. The post-synaptic membrane does not then depolarize. The sodium ions that have entered the post-synaptic neuron are pumped out by sodium–potassium pumps and the post-synaptic membrane returns to the resting potential.

A typical post-synaptic neuron in the brain or spinal cord has synapses not just with one but with many pre-synaptic neurons. It may be necessary for several of these to release neurotransmitter at the same time for the threshold potential to be reached and a nerve impulse to be initiated in the post-synaptic neuron. This type of mechanism can be used to process information from different sources in the body to help in decision-making.

Activity

Research updates on neonicotinoids

There are currently intense research efforts to try to discover whether neonicotinoids are to blame for collapses in honeybee colonies. What are the most recent research findings and do they suggest that these insecticides should be banned?

▲ Figure 17 Research has shown that the neonicotinoid pesticide imidacloprid reduces growth of bumblebee colonies

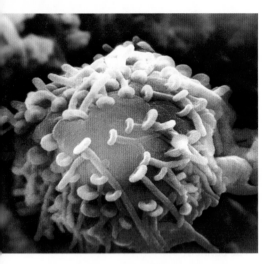

▲ Figure 18 Many synapses are visible in this scanning electron micrograph between the cell body of one post-synaptic neuron and a large number of different pre-synaptic neurons (blue)

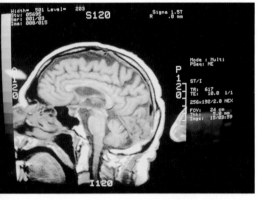

▲ Figure 19 Memory and learning are functions of the cerebrum—the folded upper part of the brain

Research into memory and learning

Cooperation and collaboration between groups of scientists: biologists are contributing to research into memory and learning.

Higher functions of the brain including memory and learning are only partly understood at present and are being researched very actively. They have traditionally been investigated by psychologists but increasingly the techniques of molecular biology and biochemistry are being used to unravel the mechanisms at work. Other branches of science are also making important contributions, including biophysics, medicine, pharmacology and computer science.

The Centre for Neural Circuits and Behaviour at Oxford University is an excellent example of collaboration between scientists with different areas of expertise. The four group leaders of the research team and the area of science that they originally studied are:

- Professor Gero Miesenböck – medicine and physiology
- Dr Martin Booth – engineering and optical microscopy
- Dr Korneel Hens – chemistry and biochemistry
- Professor Scott Waddell – genetics, molecular biology and neurobiology.

The centre specializes in research techniques known as optogenetics. Neurons are genetically engineered to emit light during synaptic transmission or an action potential, making activity in specific neurons in brain tissue visible. They are also engineered so specific neurons in brain tissue respond to a light signal with an action potential. This allows patterns of activity in the neurons of living brain tissue to be studied.

There are many research groups in universities throughout the world that are investigating memory, learning and other brain functions. Although there is sometimes competition between scientists to be the first group to make a discovery, there is also a strongly collaborative element to scientific research. This extends across scientific disciplines and national boundaries. Success in understanding how the brain works will undoubtedly be the achievement of many groups of scientists in many countries throughout the world.

6.6 Hormones, homeostasis and reproduction

Understanding

→ Insulin and glucagon are secreted by α and β cells in the pancreas to control blood glucose concentration.

→ Thyroxin is secreted by the thyroid gland to regulate the metabolic rate and help control body temperature.

→ Leptin is secreted by cells in adipose tissue and acts on the hypothalamus of the brain to inhibit appetite.

→ Melatonin is secreted by the pineal gland to control circadian rhythms.

→ A gene on the Y chromosome causes embryonic gonads to develop as testes and secrete testosterone.

→ Testosterone causes prenatal development of male genitalia and both sperm production and development of male secondary sexual characteristics during puberty.

→ Estrogen and progesterone cause prenatal development of female reproductive organs and female secondary sexual characteristics during puberty.

→ The menstrual cycle is controlled by negative and positive feedback mechanisms involving ovarian and pituitary hormones.

Applications

→ Causes and treatment of type I and type II diabetes.

→ Testing of leptin on patients with clinical obesity and reasons for the failure to control the disease.

→ Causes of jet lag and use of melatonin to alleviate it.

→ The use in IVF of drugs to suspend the normal secretion of hormones, followed by the use of artificial doses of hormones to induce superovulation and establish a pregnancy.

→ William Harvey's investigation of sexual reproduction in deer.

Skills

→ Annotate diagrams of the male and female reproductive system to show names of structures and their functions.

Nature of science

→ Developments in scientific research follow improvements in apparatus: William Harvey was hampered in his observational research into reproduction by lack of equipment. The microscope was invented 17 years after his death.

Control of blood glucose concentration

Insulin and glucagon are secreted by α and β cells in the pancreas to control blood glucose concentration.

Cells in the pancreas respond to changes in blood glucose levels. If the glucose concentration deviates substantially from the set point of about 5 mmol L^{-1}, homeostatic mechanisms mediated by the pancreatic hormones insulin and glucagon are initiated.

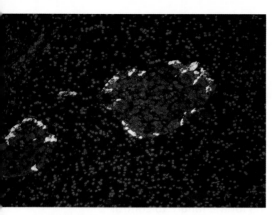

▲ Figure 1 Fluorescent light micrograph of the pancreas showing two islets of Langerhans surrounded by exocrine gland tissue. Alpha cells in the islets are stained yellow and beta cells are stained red

The pancreas is effectively two glands in one organ. Most of the pancreas is exocrine glandular tissue that secretes digestive enzymes into ducts leading to the small intestine. There are small regions of endocrine tissue called islets of Langerhans dotted through the pancreas that secrete hormones directly into the blood stream. The two cell types in the islets of Langerhans secrete different hormones.

- Alpha cells (α cells) synthesize and secrete glucagon if the blood glucose level falls below the set point. This hormone stimulates breakdown of glycogen into glucose in liver cells and its release into the blood, increasing the concentration.

- Beta cells (β cells) synthesize insulin and secrete it when the blood glucose concentration rises above the set point. This hormone stimulates uptake of glucose by various tissues, particularly skeletal muscle and liver, in which it also stimulates the conversion of glucose to glycogen. Insulin therefore reduces blood glucose concentration. Like most hormones, insulin is broken down by the cells it acts upon, so its secretion must be ongoing. Secretion begins within minutes of eating and may continue for several hours after a meal.

🌐 Diabetes

Causes and treatment of type I and type II diabetes.

Diabetes is the condition where a person has consistently elevated blood glucose levels even during prolonged fasting, leading to the presence of glucose in the urine. Continuously elevated glucose damages tissues, particularly their proteins. It also impairs water reabsorption from urine while it is forming in the kidney, resulting in an increase in the volume of urine and body dehydration. If a person needs to urinate more frequently, is constantly thirsty, feels tired and craves sugary drinks, they should test for glucose in the urine to check whether they have developed diabetes.

There are two main types of this disease:

- Type I diabetes, or early-onset diabetes, is characterized by an inability to produce sufficient quantities of insulin. It is an autoimmune disease arising from the destruction of beta cells in the islets of Langerhans by the body's own immune system. In children and young people the more severe and obvious symptoms of the disease usually start rather suddenly. The causes of this and other autoimmune diseases are still being researched.

- Type II diabetes, sometimes called late-onset diabetes, is characterized by an inability to process or respond to insulin because of a deficiency of insulin receptors or glucose transporters on target cells. Onset is slow and the disease may go unnoticed for many years. Until the last few decades, this form of diabetes was very rare in people under 50 and common only in the over 65s. The causes of this form of diabetes are not well understood but the main risk factors are sugary, fatty diets, prolonged obesity due to habitual overeating and lack of exercise, together with genetic factors that affect energy metabolism.

The treatment of the two types of diabetes is different:

- Type I diabetes is treated by testing the blood glucose concentration regularly and injecting insulin when it is too high or likely to become too high. Injections are often done before a meal to prevent a peak of blood glucose as the food is digested and absorbed. Timing is very important because insulin molecules do not last long in the blood. Better treatments are being developed using implanted devices that can release exogenous insulin into the blood as and when it is necessary. A permanent cure may be achievable by coaxing stem cells to become fully functional replacement beta cells.

- Type II diabetes is treated by adjusting the diet to reduce the peaks and troughs of blood glucose. Small amounts of food should be eaten frequently rather than infrequent large meals. Foods with high sugar content should be avoided. Starchy food should only be eaten if it has a low glycemic index, indicating that it is digested slowly. High-fibre foods should be included to slow the digestion of other foods. Strenuous exercise and weight loss are beneficial as they improve insulin uptake and action.

Data-based questions: The glucose tolerance test

The glucose tolerance test is a method used to diagnose diabetes. In this test, the patient drinks a concentrated glucose solution. The blood glucose concentration is monitored to determine the length of time required for excess glucose to be cleared from the blood.

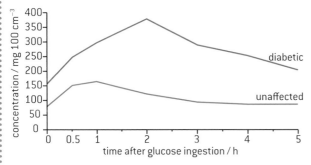

▲ Figure 3 A person with diabetes and an unaffected person give very different responses to the glucose tolerance test

With reference to figure 3, compare the person with normal glucose metabolism to the person with diabetes with respect to:

a) The concentration of glucose at time zero, i.e. before the consumption of the glucose drink.

b) The length of time required to return to the level at time zero.

c) The maximum glucose level reached.

d) The time before glucose levels start to fall.

Activity

Foods for type II diabetics

Discuss which of the foods in figure 2 are suitable for a person with type II diabetes. They should be foods with a low glycemic index.

▲ Figure 2

Thyroxin

Thyroxin is secreted by the thyroid gland to regulate the metabolic rate and help control body temperature.

The hormone thyroxin is secreted by the thyroid gland in the neck. Its chemical structure is unusual as the thyroxin molecule contains four atoms of iodine. Prolonged deficiency of iodine in the diet therefore prevents the synthesis of thyroxin. This hormone is also unusual as almost all cells in the body are targets. Thyroxin regulates the body's metabolic rate, so all cells need to respond but the most metabolically active, such as liver, muscle and brain are the main targets.

Higher metabolic rate supports more protein synthesis and growth and it increases the generation of body heat. In a person with normal physiology, cooling triggers increased thyroxin secretion by the thyroid gland, which stimulates heat production so body temperature rises.

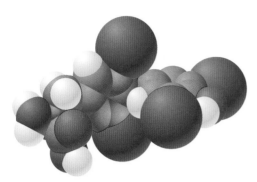

▲ Figure 4 Structure of thyroxin with atoms of iodine shown purple

Thyroxin thus regulates the metabolic rate and also helps to control body temperature.

The importance of thyroxin is revealed by the effects of thyroxin deficiency (hypothyroidism):

- lack of energy and feeling tired all the time

- forgetfulness and depression

- weight gain despite loss of appetite as less glucose and fat are being broken down to release energy by cell respiration

- feeling cold all the time because less heat is being generated

- constipation because contractions of muscle in the wall of the gut slow down.

- impaired brain development in children.

Leptin

Leptin is secreted by cells in adipose tissue and acts on the hypothalamus of the brain to inhibit appetite.

Leptin is a protein hormone secreted by adipose cells (fat storage cells). The concentration of leptin in the blood is controlled by food intake and the amount of adipose tissue in the body. The target of this hormone is groups of cells in the hypothalamus of the brain that contribute to the control of appetite. Leptin binds to receptors in the membrane of these cells. If adipose tissue increases, blood leptin concentrations rise, causing long-term appetite inhibition and reduced food intake.

The importance of this system was demonstrated by research with a strain of mice discovered in the 1950s that feed ravenously, become inactive and gain body weight, mainly through increased adipose tissue. They grow to a body weight of about 100 grams, compared with wild type mice of 20–25 grams. Breeding experiments showed that the obese mice had two copies of a recessive allele, *ob*. In the early 1990s it was shown that the wild-type allele of this gene supported the synthesis of a new hormone that was named leptin. Adipose cells in mice that have two recessive *ob* alleles cannot produce leptin. When *ob/ob* mice were injected with leptin their appetite declined, energy expenditure increased and body mass dropped by 30% in a month.

▲ Figure 5 Mouse with obesity due to lack of leptin and a mouse with normal body mass

🌐 Leptin and obesity

Testing of leptin on patients with clinical obesity and reasons for the failure to control the disease.

The discovery that obesity in mice could be caused by a lack of leptin and cured by leptin injections soon led to attempts to treat obesity in humans in this way. Amgen, a biotechnology company based in California, paid $20 million for the commercial rights to leptin and a large clinical trial was carried out. Seventy-three obese volunteers injected themselves either with one of several leptin doses or with a placebo. A double blind procedure was used, so neither the researchers nor the volunteers knew who was injecting leptin until the results were analysed.

The leptin injections induced skin irritation and swelling and only 47 patients completed the trial. The eight patients receiving the highest dose lost 7.1 kg of body mass on average compared with a loss of 1.3 kg in the 12 volunteers who were injecting the placebo. However, in the group receiving the highest dose the results varied very widely from a loss of 15 kg to a gain of 5 kg. Also any body mass lost during the trial was usually regained rapidly afterwards. Such disappointing outcomes are frequent in drug research – the physiology of humans is different in many ways from mice and other rodents.

In contrast to *ob/ob* mice, most obese humans have exceptionally high blood leptin concentrations. The target cells in the hypothalamus may have become resistant to leptin so fail to respond to it, even at high concentrations. Appetite is therefore not inhibited and food intake is excessive. More adipose tissue develops, causing a rise in blood leptin concentration but the leptin resistance prevents inhibition of appetite. Injection of extra leptin inevitably fails to control obesity if the cause is leptin resistance, just as insulin injections alone are ineffective with early-stage type II diabetes.

A very small proportion of cases of obesity in humans are due to mutations in the genes for leptin synthesis or its various receptors on target cells. Trials in people with such obesity have shown significant weight loss while the leptin injections are continuing. However leptin is a short-lived protein and has to be injected several times a day and consequently most of those offered this treatment have refused it. Also leptin has been shown to affect the development and functioning of the reproductive system, so injections are not suitable in children and young adults. All in all leptin has not fulfilled its early promise as a means of solving the human obesity problem.

Melatonin

Melatonin is secreted by the pineal gland to control circadian rhythms.

Humans are adapted to live in a 24-hour cycle and have rhythms in behaviour that fit this cycle. These are known as circadian rhythms. They can continue even if a person is placed experimentally in continuous light or darkness because an internal system is used to control the rhythm.

Circadian rhythms in humans depend on two groups of cells in the hypothalamus called the suprachiasmatic nuclei (SCN). These cells set a daily rhythm even if grown in culture with no external cues about the time of day. In the brain they control the secretion of the hormone melatonin by the pineal gland. Melatonin secretion increases in the evening and drops to a low level at dawn and as the hormone is rapidly removed from the blood by the liver, blood concentrations rise and fall rapidly in response to these changes in secretion.

The most obvious effect of melatonin is the sleep-wake cycle. High melatonin levels cause feelings of drowsiness and promote sleep through the night. Falling melatonin levels encourage waking at the end of the night. Experiments have shown that melatonin contributes to the night-time drop in core body temperature, as blocking the rise in melatonin levels reduces it and giving melatonin artificially during the day causes a drop in core temperature. Melatonin receptors have been discovered in the kidney, suggesting that decreased urine production at night may be another effect of this hormone.

When humans are placed experimentally in an environment without light cues indicating the time of day, the SCN and pineal gland usually

▲ Figure 6 Until a baby is about three months old it does not develop a regular day-night rhythm of melatonin secretion so sleep patterns do not fit those of the baby's parents

maintain a rhythm of slightly longer than 24 hours. This indicates that timing of the rhythm is normally adjusted by a few minutes or so each day. A special type of ganglion cell in the retina of the eye detects light of wavelength 460–480 nm and passes impulses to cells in the SCN. This indicates to the SCN the timing of dusk and dawn and allows it to adjust melatonin secretion so that it corresponds to the day-night cycle.

🌐 Jet lag and melatonin

Causes of jet lag and use of melatonin to alleviate it.

Jet lag is a common experience for someone who has crossed three or more time zones during air travel. The symptoms are difficulty in remaining awake during daylight hours and difficulty sleeping through the night, fatigue, irritability, headaches and indigestion. The causes are easy to understand: the SCN and pineal gland are continuing to set a circadian rhythm to suit the timing of day and night at the point of departure rather than the destination.

Jet lag only lasts for a few days, during which impulses sent by ganglion cells in the retina to the SCN when they detect light help the body to adjust to the new regime. Melatonin is sometimes used to try to prevent or reduce jet lag. It is taken orally at the time when sleep should ideally be commencing. Most trials of melatonin have shown that it is effective at promoting sleep and helping to reduce jet lag, especially if flying eastwards and crossing five or more time zones.

Sex determination in males

A gene on the Y chromosome causes embryonic gonads to develop as testes and secrete testosterone.

Human reproduction involves the fusion of a sperm from a male with an egg from a female. Initially the development of the embryo is the same in all embryos and embryonic gonads develop that could either become ovaries or testes. The developmental pathway of the embryonic gonads and thereby the whole baby depends on the presence or absence of one gene.

- If the gene SRY is present, the embryonic gonads develop into testes. This gene is located on the Y chromosome, so is only present in 50% of embryos. SRY codes for a DNA-binding protein called TDF (testis determining factor). TDF stimulates the expression of other genes that cause testis development.

- 50% of embryos have two X chromosomes and no Y so they do not have a copy of the SRY gene. TDF is therefore not produced and the embryonic gonads develop as ovaries.

Testosterone

Testosterone causes prenatal development of male genitalia and both sperm production and development of male secondary sexual characteristics during puberty.

The testes develop from the embryonic gonads in about the eighth week of pregnancy, at the time when the embryo is becoming a fetus and is about 30mm long. The testes develop testosterone-secreting cells at an early stage and these produce testosterone until about the fifteenth week

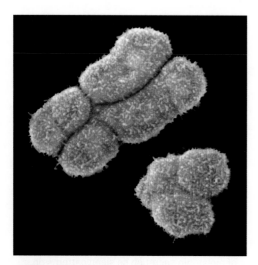
▲ Figure 7 X and Y chromosomes

of pregnancy. During the weeks of secretion, testosterone causes male genitalia to develop, which are shown in figure 8.

At puberty the secretion of testosterone increases. This stimulates sperm production in the testes, which is the primary sexual characteristic of males. Testosterone also causes the development of secondary sexual characteristics during puberty such as enlargement of the penis, growth of pubic hair and deepening of the voice due to growth of the larynx.

Sex determination in females

Estrogen and progesterone cause prenatal development of female reproductive organs and female secondary sexual characteristics during puberty.

If the gene SRY is not present in an embryo because there is no Y chromosome, the embryonic gonads develop as ovaries. Testosterone is therefore not secreted, but the two female hormones, estrogen and progesterone, are always present in pregnancy. At first they are secreted by the mother's ovaries and later by the placenta. In the absence of fetal testosterone and the presence of maternal estrogen and progesterone, female reproductive organs develop which are shown in figure 9.

During puberty the secretion of estrogen and progesterone increases, causing the development of female secondary sexual characteristics. These include enlargement of the breasts and growth of pubic and underarm hair.

 Male and female reproductive systems

Annotate diagrams of the male and female reproductive system to show names of structures and their functions.

The tables on the next page indicate functions that should be included when diagrams of male and female reproductive systems are annotated.

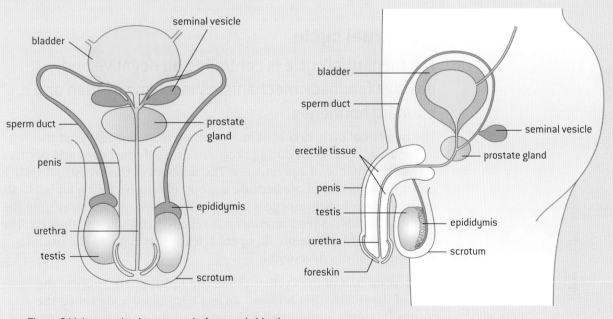

▲ Figure 8 Male reproductive system in front and side view

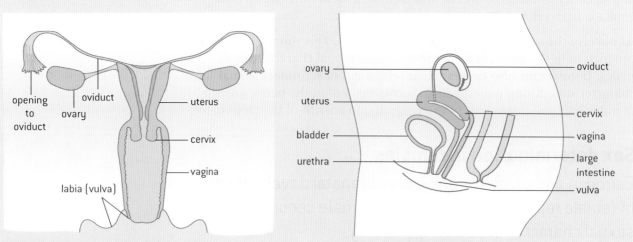

▲ Figure 9 Female reproductive system in front and side view

Male reproductive system

Testis	Produce sperm and testosterone
Scrotum	Hold testes at lower than core body temperature
Epididymis	Store sperm until ejaculation
Sperm duct	Transfer sperm during ejaculation
Seminal vesicle and prostate gland	Secrete fluid containing alkali, proteins and fructose that is added to sperm to make semen
Urethra	Transfer semen during ejaculation and urine during urination
Penis	Penetrate the vagina for ejaculation of semen near the cervix

Female reproductive system

Ovary	Produce eggs, estrogen and progesterone
Oviduct	Collect eggs at ovulation, provide a site for fertilization then move the embryo to the uterus
Uterus	Provide for the needs of the embryo and then fetus during pregnancy
Cervix	Protect the fetus during pregnancy and then dilate to provide a birth canal
Vagina	Stimulate penis to cause ejaculation and provide a birth canal
Vulva	Protect internal parts of the female reproductive system

Menstrual cycle

The menstrual cycle is controlled by negative and positive feedback mechanisms involving ovarian and pituitary hormones.

The menstrual cycle occurs in most women from puberty until the menopause, apart from during pregnancies. Each time the cycle occurs it gives the chance of a pregnancy. The first half of the menstrual cycle is called the follicular phase because a group of follicles is developing in the ovary. In each follicle an egg is stimulated to grow. At the same time the lining of the uterus (endometrium) is repaired and starts to thicken. The most developed follicle breaks open, releasing its egg into the oviduct. The other follicles degenerate.

The second half of the cycle is called the luteal phase because the wall of the follicle that released an egg becomes a body called the corpus luteum. Continued development of the endometrium prepares it for the implantation of an embryo. If fertilization does not occur the corpus luteum in the ovary breaks down. The thickening of the endometrium in the uterus also breaks down and is shed during menstruation.

Figure 10 shows hormone levels in a woman over a 36-day period, including one complete menstrual cycle. The pattern of changes is typical for a woman who is not pregnant. The hormone levels are measured in mass per millilitre. The actual masses are very small, so progesterone, FSH and LH are measured in nanograms (ng) and estrogen is measured in picograms (pg). Figure 10 also shows the state of the ovary and of the endometrium.

The four hormones in figure 10 all help to control the menstrual cycle by both negative and positive feedback. FSH and LH are protein hormones produced by the pituitary gland that bind to FSH and LH receptors in the membranes of follicle cells. Estrogen and progesterone are ovarian hormones, produced by the wall of the follicle and corpus

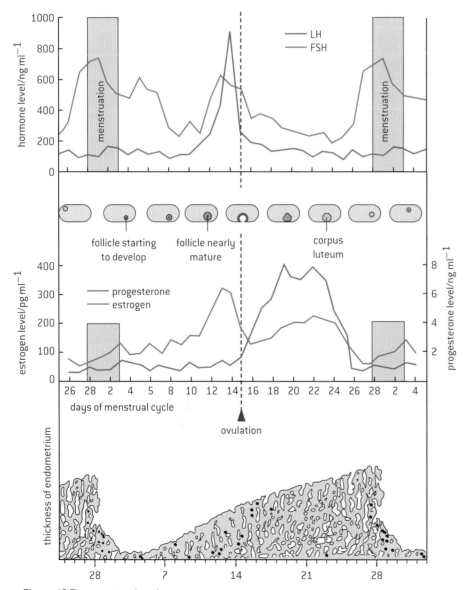

▲ Figure 10 The menstrual cycle

luteum. They are absorbed by many cells in the female body, where they influence gene expression and therefore development.

- FSH rises to a peak towards the end of the menstrual cycle and stimulates the development of follicles, each containing an oocyte and follicular fluid. FSH also stimulates secretion of estrogen by the follicle wall.

- Estrogen rises to a peak towards the end of the follicular phase. It stimulates the repair and thickening of the endometrium after menstruation and an increase in FSH receptors that make the follicles more receptive to FSH, boosting estrogen production (positive feedback). When it reaches high levels estrogen inhibits the secretion of FSH (negative feedback) and stimulates LH secretion.

- LH rises to a sudden and sharp peak towards the end of the follicular phase. It stimulates the completion of meiosis in the oocyte and partial digestion of the follicle wall allowing it to burst open at ovulation. LH also promotes the development of the wall of the follicle after ovulation into the corpus luteum which secretes estrogen (positive feedback) and progesterone.

- Progesterone levels rise at the start of the luteal phase, reach a peak and then drop back to a low level by the end of this phase. Progesterone promotes the thickening and maintenance of the endometrium. It also inhibits FSH and LH secretion by the pituitary gland (negative feedback).

Data-based questions: The female athlete triad

The female athlete triad is a syndrome consisting of three interrelated disorders that can affect female athletes: osteoporosis, disordered eating and menstrual disorders. Osteoporosis is reduced bone mineral density. It can be caused by a diet low in calcium, vitamin D or energy, or by low estrogen levels. Figure 11 shows the bone mineral density in two parts of the femur for female runners who had different numbers of menstrual cycles per year. The t-score is the number of standard deviations above or below mean peak bone mass for young women.

1 **a)** Outline the relationship between number of menstrual cycles per year and bone density. [3]

   b) Compare and contrast the results for the neck of the femur with the results for the trochanter. [3]

2 Explain the reasons for some of the runners having:

   a) higher bone density than the mean [2]

   b) lower bone density than the mean. [4]

3 **a)** Suggest reasons for female athletes having few or no menstrual cycles. [2]

   b) Suggest one reason for eating disorders and low body weight in female athletes. [1]

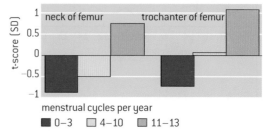

▲ Figure 11 Bone mass in women grouped by number of menstrual cycles

In vitro fertilization

The use in IVF of drugs to suspend the normal secretion of hormones, followed by the use of artificial doses of hormones to induce superovulation and establish a pregnancy.

The natural method of fertilization in humans is *in vivo*, meaning that it occurs inside the living tissues of the body. Fertilization can also happen outside the body in carefully controlled laboratory conditions. This is called *in vitro* fertilization, almost always abbreviated to IVF. This procedure has been used extensively to overcome fertility problems in either the male or female parent.

There are several different protocols for IVF, but the first stage is usually down-regulation. The woman takes a drug each day, usually as a nasal spray, to stop her pituitary gland secreting FSH or LH. Secretion of estrogen and progesterone therefore also stops. This suspends the normal menstrual cycle and allows doctors to control the timing and amount of egg production in the woman's ovaries.

Intramuscular injections of FSH and LH are then given daily for about ten days, to stimulate follicles to develop. The FSH injections give a much higher concentration of this hormone than during a normal menstrual cycle and as

a consequence far more follicles develop than usual. Twelve is not unusual and there can be as many twenty follicles. This stage of IVF is therefore called superovulation.

When the follicles are 18 mm in diameter they are stimulated to mature by an injection of HCG, another hormone that is normally secreted by the embryo. A micropipette mounted on an ultrasound scanner is passed through the uterus wall to wash eggs out of the follicles. Each egg is mixed with 50,000 to 100,000 sperm cells in sterile conditions in a shallow dish, which is then incubated at 37 °C until the next day.

If fertilization is successful then one or more embryos are placed in the uterus when they are about 48 hours old. Because the woman has not gone through a normal menstrual cycle extra progesterone is usually given as a tablet placed in the vagina, to ensure that the uterus lining is maintained. If the embryos implant and continue to grow then the pregnancy that follows is no different from a pregnancy that began by natural conception.

William Harvey and sexual reproduction

William Harvey's investigation of sexual reproduction in deer.

William Harvey is chiefly remembered for his discovery of the circulation of the blood, but he also had a lifelong obsession with how life is transmitted from generation to generation and pioneered research into sexual reproduction. He was taught the "seed and soil" theory of Aristotle, according to which the male produces a seed, which forms an egg when it mixes with menstrual blood. The egg develops into a fetus inside the mother.

William Harvey tested Aristotle's theory using a natural experiment. Deer are seasonal breeders and only become sexually active during the autumn. Harvey examined the uterus of female deer during the mating season by slaughtering and dissecting them. He expected to find eggs developing in the uterus immediately after mating, but only found signs of anything developing in females two or more months after the start of the mating season.

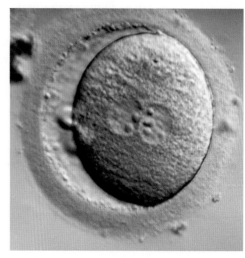

▲ Figure 12 IVF allows the earliest stages in a human life to be seen. This micrograph shows a zygote formed by fertilization. The nuclei of the egg and sperm are visible in the centre of the zygote. There is a protective layer of gel around the zygote called the fertilization membrane

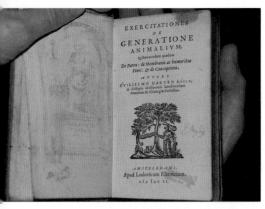

▲ Figure 13 William Harvey's book on the reproduction of animals *Exercitationes de Generatione Animalium* published in 1651

He regarded his experiments with deer as proof that Aristotle's theory of reproduction was false and concluded "the fetus doth neither proceed from the seed of male or female in coition, nor yet from any commixture of that seed". Although Aristotle's "seed and soil" theory was false, Harvey's conclusion that the fetus did not result from events during coitus (sexual intercourse) was also false.

Harvey was well aware that he had not discovered the basis of sexual reproduction: "neither the philosophers nor the physicians of yesterday or today have satisfactorily explained, or solved the problem of Aristotle."

Improvements in apparatus and research breakthroughs

Developments in scientific research follow improvements in apparatus: William Harvey was hampered in his observational research into reproduction by lack of equipment. The microscope was invented seventeen years after his death.

Harvey was understandably reluctant to publish his research into sexual reproduction, but he did eventually do so in 1651 when he was 73 years old in his work *Exercitationes de Generatione Animalium*. He knew that he had not solved the mystery of sexual reproduction:

When I plainly see nothing at all doth remain in the uterus after coition, ... no more than remains in the braine after sensation, ... I have invented this Fable. Let the learned and ingenious flock of men consider of it; let the supercilious reject it: and for the scoffing ticklish generation, let them laugh their swinge. Because I say, there is no sensible thing in the uterus after coition; and yet there is a necessity, that something should be there, which may render the animal fruitful.

William Harvey failed to solve the mystery because effective microscopes were not available when he was working, so fusion of gametes and subsequent embryo development remained undiscovered. He was unlucky with his choice of experimental animal because embryos in the deer that he used remain microscopically small for an unusually long period. Microscopes were invented seventeen years after Harvey's death, allowing the discovery of sperm, eggs and early stage embryos.

Scientific research has often been hampered for a time by deficiencies in apparatus, with discoveries only being made following improvements. This will continue into the future and we can look forward to further transformations in our understanding of the natural world as new techniques and technology are invented.

Questions

1 Using the data in table 1:

a) outline the relationship between the age of the mother and the success rate of IVF [3]

b) outline the relationship between the number of embryos transferred and the chance of having a baby as a result of IVF [3]

c) discuss how many embryos fertility centres should be allowed to transfer. [4]

Age of mother	Percentage of pregnancies per IVF cycle according to the number of embryos transferred					
	1	2		3		
	single	single	twins	single	twins	triplets
< 30	10.4	20.1	9.0	17.5	3.6	0.4
30–34	13.4	21.8	7.9	18.2	7.8	0.6
35–39	19.1	19.1	5.0	17.4	5.6	0.6
> 39	4.1	12.5	3.5	12.7	1.7	0.1

▲ Table 1

2 Figure 14 shows variations in liver glycogen over the course of one day.

a) Explain the variation in liver glycogen. [3]

b) Evaluate the contribution of glycogen to blood sugar homeostasis. [2]

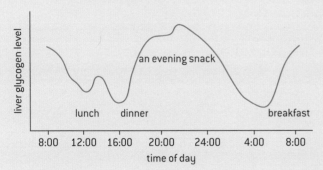

▲ Figure 14

3 Sometimes the ventilation of the lungs stops. This is called apnea. One possible cause is the blockage of the airways by the soft palate during sleep. This is called obstructive sleep apnea. It has some potentially harmful consequences, including an increased risk of

accidents during the daytime as a result of disrupted sleep and tiredness. Figure 15 shows the percentage oxygen saturation of arterial blood during a night of sleep in a patient with severe obstructive sleep apnea.

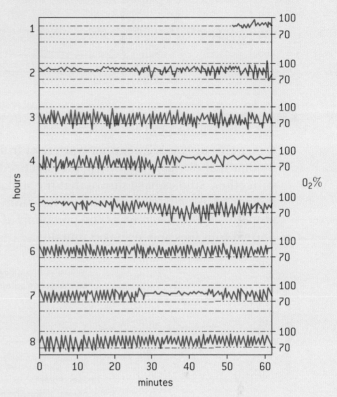

▲ Figure 15

a) Hour 8 shows a typical pattern due to obstructive sleep apnea.

(i) Explain the causes of falls in saturation. [2]

(ii) Explain the causes of rises in saturation. [2]

(iii) Calculate how long each cycle of falling and rising saturation takes. [2]

b) Estimate the minimum oxygen saturation that the patient experienced during the night, and when it occurred. [2]

c) Deduce the sleep patterns of the patient during the night when the trace was taken. [2]

4 The action potential of a squid axon was recorded, with the axon in normal sea water. The axon was then placed in water with a Na^+ concentration of one-third of that of sea water.

The action potential was recorded again. Figure 16 shows these recordings.

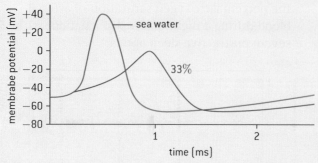

▲ Figure 16

Geneticists discovered a mutant variety of fruit fly that shakes vigorously when anaesthetized with ether. Studies have shown that the shaker mutant has K⁺ channels that do not function properly. Figure 17 shows action potentials in normal fruit flies and in shaker mutants.

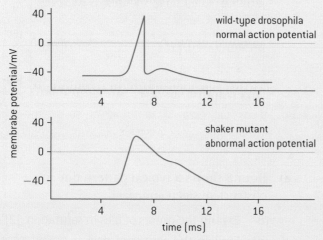

▲ Figure 17

a) Using only the data in figure 16, outline the effect of reduced Na^+ concentration on:

  (i) the magnitude of depolarization   [2]

  (ii) the duration of the action potential.   [2]

b) Explain the effects of reduced Na^+ concentration on the action potential.   [3]

c) Discuss the effect of reduced Na^+ concentration on the time taken to return to the resting potential.   [2]

d) Compare the action potentials of shaker and normal fruit flies.   [3]

e) Explain the differences between the action potentials.   [2]

Introduction

The discovery of the structure of DNA revolutionized biology. Information stored in a coded form in DNA is copied onto mRNA. The structure of DNA is ideally suited to its function. Information transferred from DNA to mRNA is translated into an amino acid sequence.

7.1 DNA structure and replication

Understanding

→ DNA structure suggested a mechanism for DNA replication.

→ Nucleosomes help to supercoil the DNA.

→ DNA replication is continuous on the leading strand and discontinuous on the lagging strand.

→ DNA replication is carried out by a complex system of enzymes.

→ DNA polymerases can only add nucleotides to the 3' end of a primer.

→ Some regions of DNA do not code for proteins but have other important functions.

Nature of science

→ Making careful observations: Rosalind Franklin's X-ray diffraction provided crucial evidence that DNA is a double helix.

Applications

→ Rosalind Franklin's and Maurice Wilkins' investigation of DNA structure by X-ray diffraction.

→ Tandem repeats are used in DNA profiling.

→ Use of nucleotides containing dideoxyribonucleic acid to stop DNA replication in preparation of samples for base sequencing.

Skills

→ Analysis of results of the Hershey and Chase experiment providing evidence that DNA is the genetic material.

→ Utilization of molecular visualization software to analyse the association between protein and DNA within a nucleosome.

The Hershey–Chase experiment

Analysis of the results of the Hershey–Chase experiment providing evidence that DNA is the genetic material.

From the late 1800s, scientists were convinced that chromosomes played a role in heredity and that the hereditary material had a chemical nature. Aware that chromosomes were composed of both protein and nucleic acid, both molecules were contenders to be the genetic material. Until the 1940s, the view that protein was the hereditary material was favoured, as it was a class of macromolecules that had great variety due to twenty naturally occurring sub-units as opposed to four nucleotide sub-units. Further, many specific functions had been identified for proteins. Variety and specificity of function were two properties that were expected to be essential requirements for the hereditary material.

Alfred Hershey and Martha Chase wanted to ascertain whether the genetic material of viruses was protein or DNA. In the 1950s, it was known that viruses are infectious particles which transform cells into virus-producing factories by becoming bound to host cells and injecting their genetic material. The non-genetic portion of the virus remains outside the cell. An infected cell then manufactures large numbers of new viruses and bursts, releasing them to the environment (see figure 1). Viruses are often specific to a certain cell type. The virus they chose to work with was the T2 bacteriophage because of its very simple structure. It has a coat composed entirely of protein while DNA is found inside the coat.

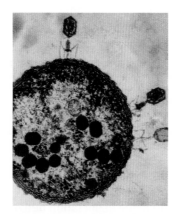

▲ Figure 1 Coloured transmission electron micrograph (TEM) of T2 viruses (blue) bound to an *Escherichia coli* bacterium. Each virus consists of a large DNA-containing head and a tail composed of a central sheath with several fibres. The fibres attach to the host cell surface, and the virus DNA is injected into the cell through the sheath. It instructs the host to build copies of the virus (blue, in cell)

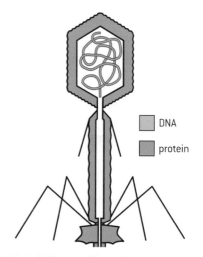

DNA

protein

▲ Figure 2 Diagram illustrating the structure of the T2 virus

Data-based questions: The Hershey–Chase experiment

Alfred Hershey and Martha Chase were two scientists who worked to resolve the debate over the chemical nature of the genetic material. In their experiment, they took advantage of the fact that DNA contains phosphorus but not sulphur while proteins contain sulphur but not phosphorus. They cultured viruses that contained proteins with radioactive (^{35}S) sulphur and they separately cultured viruses that contained DNA with radioactive (^{32}P)

phosphorus. They infected bacteria separately with the two types of viruses. They used a blender to separate the non-genetic component of the virus from the cell and then centrifuged the culture solution to concentrate the cells in a pellet. The cells were expected to have the radioactive genetic component of the virus in them. They measured the radioactivity in the pellet and the supernatant. Figure 3 represents the process and results of the experiment.

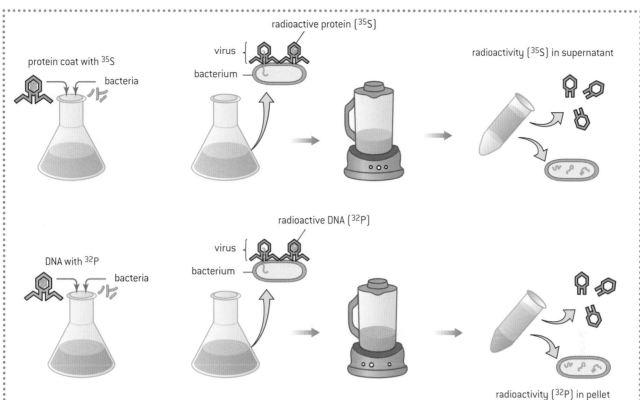

radioactive protein (^{35}S)

virus
bacterium

radioactivity (^{35}S) in supernatant

protein coat with ^{35}S
bacteria

radioactive DNA (^{32}P)

virus
bacterium

DNA with ^{32}P
bacteria

radioactivity (^{32}P) in pellet

Questions

a) Explain what a supernatant is.

b) Explain why the genetic material should be found in the pellet and not the supernatant.

c) Determine the percentage of the ^{32}P that remains in the supernatant.

d) Determine the percentage of ^{35}S that remains in the supernatant.

e) Discuss the evidence that DNA is the chemical which transforms the bacteria into infected cells.

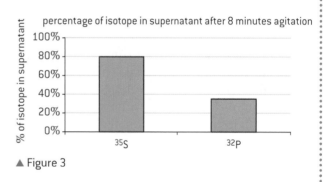

percentage of isotope in supernatant after 8 minutes agitation

▲ Figure 3

X-ray diffraction patterns as evidence of molecular structures

Making careful observations: Rosalind Franklin's X-ray diffraction provided crucial evidence that DNA was a helix.

Two names are usually remembered in connection with the discovery of DNA, Crick and Watson. Flashes of insight led to their success, but they could not have achieved it without skilled experimental work and careful observations by other scientists. One of these was Erwin Chargaff. His research into the percentage base composition of DNA is described in the data-based question in sub-topic 2.6 (page 107).

Another key figure in the discovery of DNA was Rosalind Franklin. In 1950, she became a research associate in the biophysics unit at King's College, London. The unit was already investigating the structure of DNA by X-ray diffraction. Franklin had already become skilled in techniques of crystallography and X-ray diffraction while researching other carbon compounds at an institute in Paris.

At King's College she improved the resolution of a camera, so she could make more detailed measurements of the X-ray diffraction patterns than had previously been possible. She also produced high quality samples of DNA with the molecules aligned in narrow fibres. By careful control of humidity two types of pure sample could be produced and as Franklin was unsure which represented the normal structure of DNA, she investigated both.

Soon after starting work at King's College, Franklin had obtained the sharpest X-ray diffraction images of DNA in existence. They have been described as "amongst the most beautiful X-ray photographs of any substance ever taken". Their implications are described in the next section. She was unwilling to publish her findings until there was strong evidence. She therefore embarked on a rigorous analysis of the diffraction patterns that allowed her to calculate the dimensions of the DNA helix.

Without Franklin's knowledge or permission, James Watson was shown the best diffraction pattern and the calculations based on it. Before Franklin could publish her results Crick and Watson had used them to build their model of DNA structure. It is widely accepted that Rosalind Franklin deserved a Nobel Prize for her research, but this never happened. Crick and Watson were awarded prizes in 1962, but she died of cancer in 1958, aged thirty-seven. Nobel Prizes cannot be awarded posthumously, but Rosalind Franklin is remembered more than many prize winners. What we can remember from her life is that discoveries may sometimes be made through serendipity or flashes of insight, but the real foundations of science are rigorous experimental techniques and diligent observation.

🌐 Rosalind Franklin's investigation of DNA structure

Rosalind Franklin and Maurice Wilkins' investigation of DNA structure by X-ray diffraction.

If a beam of X-rays is directed at a material, most of it passes through but some is scattered by the particles in the material. This scattering is called diffraction. The wavelength of X-rays makes them particularly sensitive to diffraction by the particles in biological molecules including DNA.

In a crystal the particles are arranged in a regular repeating pattern, so the diffraction occurs in a regular way. DNA cannot be crystallized but the molecules were arranged in an orderly enough array in Franklin's samples for a diffraction pattern to be obtained, rather than random scattering.

An X-ray detector is placed close to the sample to collect the scattered rays. The sample can be rotated in three different dimensions to investigate the pattern of scattering. Diffraction patterns can be recorded using X-ray film. Franklin developed a high resolution camera containing X-ray film to obtain very clear images of diffraction patterns from DNA. Figure 4 shows the most famous of these diffraction patterns.

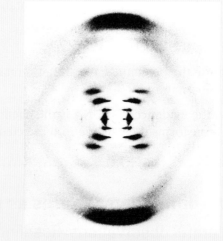

▲ Figure 4 Rosalind Franklin's X-ray diffraction photograph of DNA

From the diffraction pattern in figure 4 Franklin was able to make a series of deductions about the structure of DNA:

- The cross in the centre of the pattern indicated that the molecule was helical in shape.

- The angle of the cross shape showed the pitch (steepness of angle) of the helix.

- The distance between the horizontal bars showed turns of the helix to be 3.4 nm apart.

- The distance between the middle of the diffraction pattern and the top showed that there was a repeating structure within the molecule, with a distance of 0.34 nm between the repeats. This turned out to be the vertical distance between adjacent base pairs in the helix.

These deductions that were made from the X-ray diffraction pattern of DNA were critically important in the discovery of the structure of DNA.

The Watson and Crick model suggested semi-conservative replication

DNA structure suggested a mechanism for DNA replication.

Several lines of experimental evidence came together to lead to the knowledge of the structure of DNA: molecular modelling pioneered by the Nobel prize winner Linus Pauling, X-ray diffraction patterns discerned from the careful photographs of Rosalind Franklin and the base composition studies of Erwin Chargaff. But insight and imagination played a role as well.

One of Watson and Crick's first models had the sugar-phosphate strands wrapped around one another with the nitrogen bases facing outwards. Rosalind Franklin countered this model with the knowledge that the nitrogen bases were relatively hydrophobic in comparison to the sugar-phosphate backbone and would likely point in to the centre of the helix.

Franklin's X-ray diffraction studies showed that the DNA helix was tightly packed so when Watson and Crick built their models, their choices required the bases to fit together such that the strands were not too far apart. As they trialled various models, Watson and Crick found the tight packing they were looking for would occur if a pyrimidine was paired with a purine and if the bases were "upside down" in relation to one another. In addition to being structurally similar, adenine has a surplus negative charge and thymine has a surplus positive charge so that pairing was electrically compatible. Pairing cytosine with guanine allows for the formation of three hydrogen bonds which enhances stability.

Once the model was proposed, the complementary base pairing immediately suggested a mechanism by which DNA replication could occur – one of the key requirements that any structural model would have to address. The Watson–Crick model led to the hypothesis of semi-conservative replication.

The role of nucleosomes in DNA packing

Nucleosomes help to supercoil DNA.

One difference between eukaryotic DNA and bacterial DNA is that eukaryotic DNA is associated with proteins called histones. Most groups of prokaryotes have DNA that is not associated with histones, or proteins like histones. For this reason, prokaryotic DNA is referred to as being naked.

Histones are used by the cell to package the DNA into structures called nucleosomes. A nucleosome consists of a central core of eight histone

TOK

What options do scientists have when theories and predictions don't fully match experimental evidence?

Chargaff wrote about his observations:

the results serve to disprove the tetranucleotide hypothesis. It is, however, noteworthy - whether this is more than accidental, cannot yet be said - that in all deoxypentose nucleic acids examined thus far the molar ratios of total purines to total pyrimidines and also of adenine to thymine and of guanine to cytosine were not far from 1

H. H. Bauer, author of the book *Scientific Literacy and the Myth of the Scientific Method*, argues that Chargaff needed to:

stick his neck out beyond the actual results and say that they mean exact equality and hence some sort of pairing in the molecular structure .... Watson and Crick, on the other hand were speculating and theorizing about the molecular nature and biological functions of DNA and they postulated a structure in which the equalities are exactly one and the deviation form this in the data could be regarded as experimental error. Ideas and theory turned out to be a better guide than raw data.

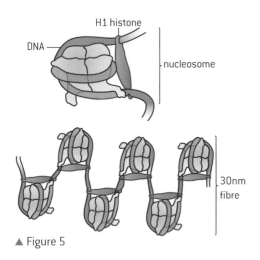

DNA — H1 histone

nucleosome

30nm fibre

▲ Figure 5

proteins with DNA coiled around the proteins. The eight proteins, or octamer, consist of two copies of four different types of histones. A short section of "linker" DNA connects one nucleosome to the next. An additional histone protein molecule, called H1, serves to bind the DNA to the core particle (figure 5).

The association of histones with the DNA contributes to a pattern known as supercoiling. An analogy is if you twist an elastic band repeatedly eventually it forms an additional pattern of coils. Supercoiling allows a great length of DNA to be packed into a much smaller space within the nucleus. The nucleosome is an adaptation that facilitates the packing of the large genomes that eukaryotes possess. The H1 histone binds in such a way to form a structure called the 30 nm fibre that facilitates further packing.

Activity

Determining packing ratio

Packing ratio is defined as the length of DNA divided by the length into which it is packaged. Use the information below to estimate the packing ratio of:

(a) a nucleosome; and

(b) chromosome 22 (one of the smallest human chromosomes).

- The distance between base pairs is 0.34 nm.

- There is approximately 200 bp of DNA coiled around a nucleosome.

- A nucleosome is approximately 10 nm long.

- There is an estimated 5.0×10^7 total base pairs (bp) present in the shortest human autosome (chromosome 22).

- Chromosome 22 in its most condensed form is approximately 2 μm long.

Ⓐ Visualizing nucleosomes

Utilization of molecular visualization software to analyse the association between protein and DNA within a nucleosome.

Visit the protein data bank at http://www.rcsb.org/pdb/home/home.do.

1 Rotate the molecule to see the two copies of each histone protein. In figure 6, they are identified by the tails that extend from the core. Each protein has such a tail that extends out from the core.

2 Note also the approximately 150 bp of DNA wrapped nearly twice around the octamer core.

3 Note the N-terminal tail that projects from the histone core for each protein. Chemical modification of this tail is involved in regulating gene expression.

4 Visualize the positively charged amino acids on the nucleosome core. Suggest how they play a role in the association of the protein core with the negatively charged DNA.

▲ Figure 6

Data-based questions: Apoptosis and the length of DNA between nucleosomes

Under natural conditions, programmed cell death sometimes occurs. This is known as apoptosis and it plays an important role in such processes as metamorphosis and embryological development. One mechanism involved in this auto-destruction is the digestion of DNA by enzymes called DNAases. The DNA associated with the nucleosome is normally not as accessible to the DNAase as the linking sections. DNA gets digested into fragments of lengths equal to multiples of the distance between nucleosomes.

The left hand column of figure 7 shows the results of separation by gel electrophoresis of the DNA released by the action of DNAase on rat liver cells. The right column represents fragments used as a reference called a ladder.

Once the DNA had been cut, nucleosomes were digested by protease.

1   Identify on the diagram the fragment that represents:

(i) the length of DNA between the two sections of linker DNA on either side of one nucleosome;

(ii) the length of DNA between two linker DNA regions with two nucleosomes between them;

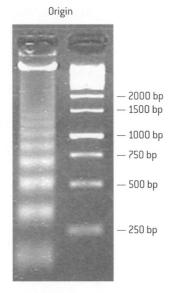

Origin

— 2000 bp
— 1500 bp

— 1000 bp

— 750 bp

— 500 bp

— 250 bp

▲ Figure 7

(iii) the length of DNA between two linker DNA regions with three nucleosomes between them.

2   Deduce the length of DNA associated with a nucleosome.

3   Suggest how the pattern in the left-hand column would change if very high concentrations of DNAase were applied to the cells.

The leading strand and the lagging strand

DNA replication is continuous on the leading strand and discontinuous on the lagging strand.

Because the two strands of the DNA double helix are arranged in an anti-parallel fashion, synthesis on the two strands occurs in very different ways. One strand, the leading strand, is made continuously following the fork as it opens. The other strand, known as the lagging strand, is made in fragments moving away from the replication fork. New fragments are created on the lagging strand as the replication fork exposes more of the template strand. These fragments are called Okazaki fragments.

Proteins involved in replication

DNA replication is carried out by a complex system of enzymes.

Replication involves the formation and movement of the replication fork and synthesis of the leading and lagging strands. Proteins are involved as enzymes at each stage but also serve a number of other functions.

The enzyme helicase unwinds the DNA at the replication fork and the enzyme topoisomerase releases the strain that develops ahead of the helicase. Single-stranded binding proteins keep the strands apart long enough to allow the template strand to be copied.

Starting replication requires an RNA primer. Note that on the lagging strand there are a number of primers but there is just one on the leading strand. The enzyme DNA primase creates one RNA primer on the leading strand and many RNA primers on the lagging strand. The RNA primer is necessary to initiate the activity of DNA polymerase.

DNA polymerase is responsible for covalently linking the deoxyribonucleotide monophosphate to the 3' end of the growing strand. Different organisms have different kinds of DNA polymerases, each with different functions such as proof-reading, polymerization and removal of RNA primers once they are no longer needed.

DNA ligase connects the gaps between fragments.

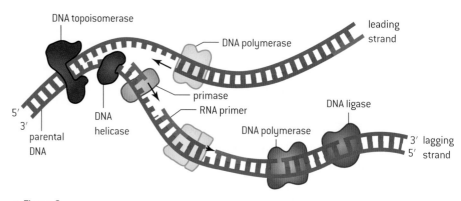

▲ Figure 8

The direction of replication

DNA polymerases can only add nucleotides to the 3' end of a primer

Within DNA molecules, DNA replication begins at sites called origins of replication. In prokaryotes there is one site and in eukaryotes there are many. Replication occurs in both directions away from the origin. The result appears as a replication bubble in electron micrographs.

The five carbons of the deoxyribose sugar have a number (see figure 9).

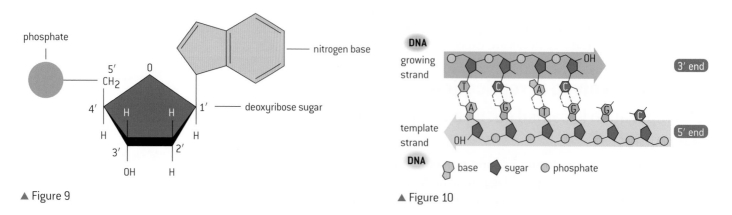

▲ Figure 9

▲ Figure 10

The phosphate group of new DNA nucleotides is added to the 3' carbon of the deoxyribose of the nucleotide at the end of the chain. Replication therefore occurs in the 5' to 3' direction.

Non-coding regions of DNA have important functions

Some regions of DNA do not code for proteins but have other important functions.

The cellular machinery operates according to a genetic code. DNA is used as a guide for the production of polypeptides using the genetic code. However, only some DNA sequences code for the production of polypeptides. These are called coding sequences. There are a number of non-coding sequences found in genomes. Some of them have functions, such as those sequences that are used as a guide to produce tRNA and rRNA. Some non-coding regions play a role in the regulation of gene expression such as enhancers and silencers. In sub-topic 7.2 we will explore non-coding sequences called introns.

Most of the eukaryotic genome is non-coding.

Within the genome, especially in eukaryotes, repetitive sequences can be common. There are two types of repetitive sequences: moderately repetitive sequences and highly repetitive sequences (satellite DNA). Together they can form between 5 and 60 per cent of the genome. In humans, nearly 60% of the DNA consists of repetitive sequences.

One such area of repetitive sequences occurs on the ends of eukaryotic chromosomes called telomeres. The telomere serves a protective function. During interphase, the enzymes that replicate DNA cannot continue replication all the way to the end of the chromosome. If cells went through the cell cycle without telomeres, they would lose the genes at the end of the chromosomes. Sacrificing the repetitive sequences found in telomeres serves a protective function.

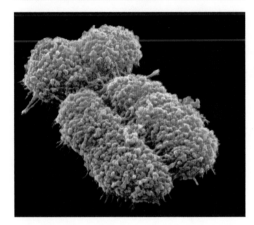

▲ Figure 11 False colour scanning electron micrograph with telomeres coloured pink. The grey region in the centre is the centromere which also consists of non-coding repetitive sequences

🌐 DNA profiling

Tandem repeats are used in DNA profiling.

A variable number tandem repeat (VNTR) is a short nucleotide sequence that shows variations between individuals in terms of the number of times the sequence is repeated. Each variety can be inherited as an allele. Analysis of VNTR allele combinations in individuals is the basis behind DNA profiling for use in such applications as genealogical investigations.

A locus is the physical location of a heritable element on the chromosome. In the hypothetical example shown in figure 12, locus A has a VNTR of the sequence "AT" and locus B has a VNTR of the sequence "TCG". In the two individuals shown, there are two different alleles (varieties) of locus A, two repeats (allele A2) and four repeats (allele A4). In the same individuals, there are three alleles for locus B, three repeats (allele B3), four repeats (allele B4) and five repeats (allele B5). The asterisk mark indicates where the restriction enzyme would cut.

The DNA profile that would result is shown in the lower part of figure 12. Note that the two individuals have some bands in common and some unique bands.

Genealogists deduce paternal lineage by analysing short tandem repeats from the Y-chromosome, and deduce maternal lineage by analysing mitochondrial DNA variations in single nucleotides at specific locations called hyper-variable regions.

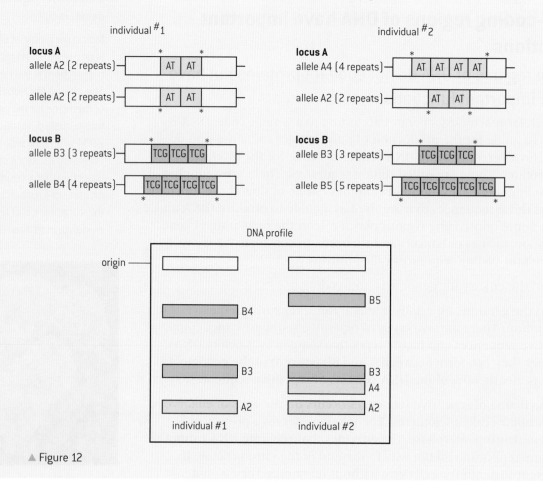

▲ Figure 12

Activity

Analysis of a DNA profile involving alleles of short tandem repeats of DNA

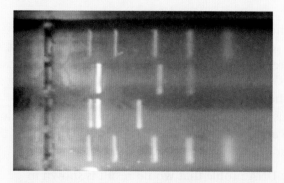

▲ Figure 13 Gel electrophoresis. The outside columns represent ladders of known length. The two inside columns represent samples of unknown length

A logarithm is an alternative way to express an exponent. For example,

$$\log 1{,}000 = \log 10^3 \qquad \log 100 = \log 10^2$$
$$= 3 \qquad\qquad = 2$$

In biology, very large changes in a variable are easier to represent graphically if logarithms are used.

In the example (figure 13), DNA fragments were separated using gel electrophoresis. The fragments vary in size from 100 bp (base pairs) up to 5,000 bp. The two outside columns of the gel represent ladders, i.e. mixtures of DNA fragments of known size. These were used to obtain the data in table 1 and create the plot shown in figure 14. The other inner columns shown in figure 13 are unknowns.

Known ladder fragment size (bp)	Distance moved (mm)
5,000	58
2,000	96
850	150
400	200
100	250

▲ Table 1

1 Using figure 14 determine the size of DNA fragments in the two centre digests:

Fragment size (bp) (column 2)	Distance moved (mm) (column 2)	Fragment size (bp) (column 3)	Distance moved (mm) (column 3)
	60		70
	70		160
	130		200

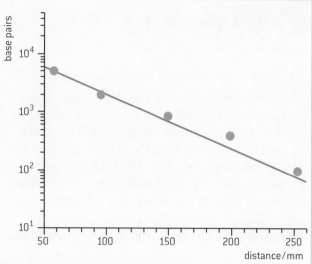

▲ Figure 14 Distance moved as a function of fragment size in gel electrophoresis. Notice that the y-axis scale on this graph goes up in powers of 10. This is a logarithmic scale

Data-based questions: Analysis of DNA profiles using D1S80

One commonly studied DNA locus is a VNTR named D1S80. D1S80 is located on human chromosome 1. This locus is composed of repeating units of 16-nucleotide-long segments of DNA. The number of repeats varies from one individual to the next with 29 known alleles ranging from 15 repeats to 41.

In the image of a DNA profile (figure 15) the outside lanes show ladders representing multiples of 123 bp.

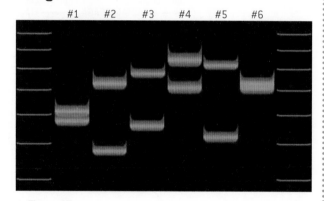

▲ Figure 15

a) Identify the lengths of the fragments represented by each of the bands in the ladder.

b) Using a ruler measure the distance between the origin and the bands in the ladder. Use the length and distance data, to create a standard curve using a logarithmic graph.

c) Measure the distance travelled by each band from the origin, for individuals #1-#6.

d) Using the standard curve, estimate the lengths of the bands in each individual.

e) Estimate the number of repeats represented by each band.

f) It is unclear whether individual #6 has two different copies of the same allele or different alleles. Suggest what could be done to further resolve the genotype of the final individual.

DNA sequencing

Use of nucleotides containing dideoxyribonucleic acid to stop DNA replication in preparation of samples for base sequencing.

The determination of the sequence of bases in a genome is carried out most commonly using a method that employs fluoresence. Many copies of the unknown DNA that is to be sequenced are placed into test tubes with all of the raw materials including deoxyribonucleotides and the enzymes necessary to carry out replication. In addition very small quantities of dideoxyribonucleotides that have been labelled with different fluorescent markers are added. The dideoxyribonucleotides will be incorporated into some of the new DNA, but when they are incorporated, they will stop the replication at precisely the point where they were added. The fragments are separated by length using electrophoresis. The sequence of bases can be automatically analysed by comparing the colour of the fluorescence with the length of the fragment.

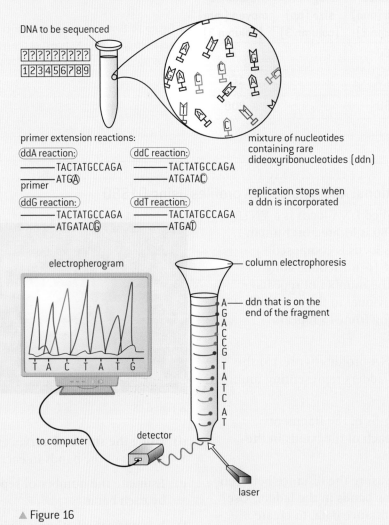

▲ Figure 16

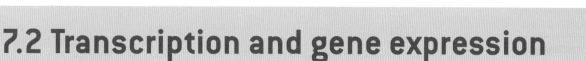

7.2 Transcription and gene expression

Understanding

→ Gene expression is regulated by proteins that bind to specific base sequences in DNA.

→ The environment of a cell and of an organism has an impact on gene expression.

→ Nucleosomes help to regulate transcription in eukaryotes.

→ Transcription occurs in a 5′ to 3′ direction.

→ Eukaryotic cells modify mRNA after transcription.

→ Splicing of mRNA increases the number of different proteins an organism can produce.

 Applications

→ The promoter as an example of non-coding DNA with a function.

 Skills

→ Analysis of changes in DNA methylation patterns.

 Nature of science

→ Looking for patterns, trends and discrepancies: there is mounting evidence that the environment can trigger heritable changes in epigenetic factors.

The function of the promoter

The promoter as an example of non-coding DNA with a function.

Only some DNA sequences code for the production of polypeptides. These are called coding sequences. There are a number of non-coding sequences found in genomes. Some of them have functions, such as those sequences that produce tRNA and rRNA.

Some non-coding regions play a role in the regulation of gene expression such as enhancers and silencers.

The promoter is a sequence that is located near a gene. It is the binding site of RNA polymerase, the enzyme that catalyses the formation of the covalent bond between nucleotides during the synthesis of RNA. The promoter is not transcribed but plays a role in transcription.

Regulation of gene expression by proteins

Gene expression is regulated by proteins that bind to specific base sequences in DNA.

Some proteins are always necessary for the survival of the organism and are therefore expressed in an unregulated fashion. Other proteins need to be produced at certain times and in certain amounts; i.e., their expression must be regulated.

Gene expression is regulated in prokaryotes as a consequence of variations in environmental factors. For example, the genes responsible for the absorption and metabolism of lactose by *E.coli* are expressed in the presence of lactose and are not expressed in the absence of lactose. In this case, the breakdown of lactose results in regulation of gene expression by negative feedback. In the presence of lactose a repressor protein is deactivated (figure 1). Once the lactose has been broken

355

lactose not in the environment;
repressor blocks transcription

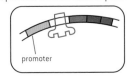

lactose present in the environment;
repressor deactivated; genes involved in lactose
use are transcribed

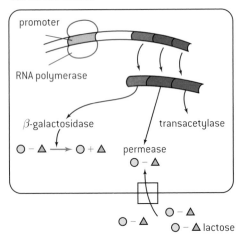

▲ Figure 1

down, the repressor protein is no longer deactivated and proceeds to block the expression of lactose metabolism genes.

As in prokaryotes, eukaryotic genes are regulated in response to variations in environmental conditions. Each cell of a multicellular eukaryotic organism expresses only a fraction of its genes.

The regulation of eukaryotic gene expression is also a critical part of cellular differentiation as well as the process of development. This is seen in the passage of an insect through its life cycle stages or in human embryological development.

There are a number of proteins whose binding to DNA regulates transcription. These include enhancers, silencers and promoter-proximal elements. Unlike the promoter sequence, the sequences linked to regulatory transcription factors are unique to the gene.

Regulatory sequences on the DNA which increase the rate of transcription when proteins bind to them are called enhancers. Those sequences on the DNA which decrease the rate of transcription when proteins bind to them are called silencers. While enhancers and silencers can be distant from the promoter, another series of sequences called "promoter-proximal elements" are nearer to the promoter and binding of proteins to them is also necessary to initiate transcription.

The impact of the environment on gene expression

The environment of a cell and of an organism has an impact on gene expression.

In the history of Western thought, much debate has gone in to the "nature–nurture" debate. This is a debate centred on the extent to which a particular human behaviour or phenotype should be attributed to the environment or to heredity. Much effort has gone into twin studies especially for twins raised apart.

Data-based questions: Identical twin studies

Twin studies have been used to identify the relative influence of genetic factors and environmental factors in the onset of disease (figure 2). Identical twins have 100% of the same DNA while fraternal twins have approximately 50% of the same DNA.

Questions

1 Determine the percentage of identical twins where both have diabetes. [2]

2 Explain why a higher percentage of identical twins sharing a trait suggests that a genetic component contributes to the onset of the trait. [3]

3 With reference to any four conditions, discuss the relative role of the environment and genetics in the onset of the condition. [3]

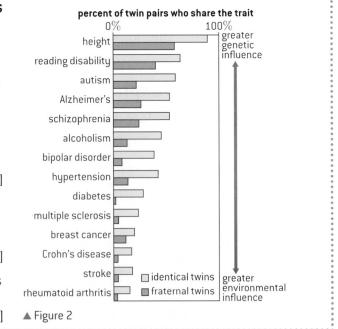

▲ Figure 2

The influence of the environment on gene expression for some traits is unequivocal. Environmental factors can affect gene expression such as the production of skin pigmentation during exposure to sunlight in humans.

In embryonic development, the embryo contains an uneven distribution of chemicals called morphogens. Concentrations of the morphogens affect gene expression contributing to different patterns of gene expression and thus different fates of the embryonic cells depending on their position in the embryo.

In coat colour in cats, the "C" gene codes for the production of the enzyme tyrosinase, the first step in the production of pigment. A mutant allele of the gene, "c^s" allows normal pigment production only at temperatures below body temperature. This mutant allele has been selected for in the selective breeding of Siamese cats. At higher temperatures, the protein product is inactive or less active, resulting in less pigment.

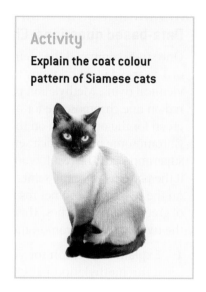

Activity

Explain the coat colour pattern of Siamese cats

Nucleosomes regulate transcription

Nucleosomes help to regulate transcription in eukaryotes.

Eukaryotic DNA is associated with proteins called histones. Chemical modification of the tails of histones is an important factor in determining whether a gene will be expressed or not.

A number of different types of modification can occur to the tails of histones including the addition of an acetyl group, the addition of a methyl group or the addition of a phosphate group.

$$CH_3C- \text{(with } O \text{ double bonded)} \quad \text{Acetyl group} \quad CH_3- \text{ Methyl group}$$

For example, residues of the amino acid lysine on histone tails can have acetyl groups either removed or added. Normally the lysine residues on histone tails bear a positive charge that can bind to the negatively charged DNA to form a condensed structure that inhibits transcription. Histone acetylation neutralizes these positive charges allowing a less condensed structure with higher levels of transcription.

Chemical modification of histone tails can either activate or deactivate genes by decreasing or increasing the accessibility of the gene to transcription factors.

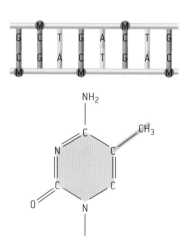

Analysing methylation patterns

Analysis of changes in DNA methylation patterns

The addition of methyl groups directly to DNA is thought to play a role in gene expression. Whereas methylation of histones can promote or inhibit transcription, direct methylation of DNA tends to decrease gene expression. The amount of DNA methylation varies during a lifetime and is affected by environmental factors.

▲ Figure 3 DNA methylation is the addition of a methyl group (green M) to the DNA base cytosine

Data-based questions: Changes in methylation pattern with age in identical twins

One study compared the methylation patterns of 3-year-old identical twins with 50-year-old identical twins. Methylation patterns were dyed red on one chromosome for one twin and dyed green for the other twin on the same chromosome. Chromosome pairs in each set of twins were digitally superimposed. The result would be a yellow colour if the patterns were the same. Differences in patterns on the two chromosomes results in mixed patterns of green and red patches. This was done for four of the twenty-three chromosome pairs in the genome.

1   Explain the reason for yellow coloration if the methylation pattern is the same in the two twins. [3]

2   Identify the chromosome with the least changes as twins age. [1]

3   Identify the chromosomes with the most changes as twins age. [1]

4   Explain how these differences could arise. [3]

5   Predict with a reason whether identical twins will become more or less similar to

each other in their characteristics as they grow older. [2]

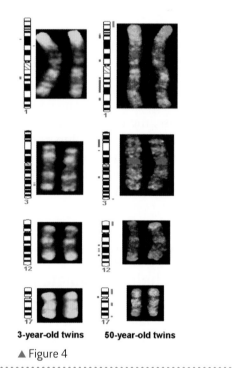

3-year-old twins   **50-year-old twins**

▲ Figure 4

Epigenetics

Looking for patterns, trends and discrepancies: there is mounting evidence that the environment can trigger heritable changes in epigenetic factors.

The chemical modifications of chromatin that impact gene expression, including acetylation, methylation and phosphorylation of amino acid tails of histones (figure 5) as well as methylation of DNA (figure 6), all have an impact on gene expression and thus impact the visible characteristics of an individual (figure 7). These chemical modifications are called epigenetic tags. There is mounting evidence that the chemical modifications that occur to the hereditary material in one generation might, in certain circumstances, be passed on to the next generation both at the cellular as well as whole organism level. The sum of all the epigenetic tags constitutes the epigenome.

Different cells have their own methylation pattern so that a unique set of proteins will be produced in order for that cell to perform its function. During cell division, the methylation pattern will be passed over to the daughter cell. In other words, the environment is affecting inheritance.

Sperm and eggs develop from cells with epigenetic tags. When two reproductive cells meet, the epigenome is erased through a process called "reprogramming".

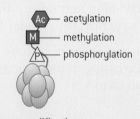

Ac — acetylation
M — methylation
P — phosphorylation

▲ Figure 5 Histone modifications

▲ Figure 6 DNA methylation

About 1% of the epigenome is not erased and survives yielding a result called "imprinting".

For example, when a mammalian mother has gestational diabetes, the high levels of glucose in the fetal circulation trigger epigenetic changes in the daughter's DNA such that she is predisposed to develop gestational diabetes herself.

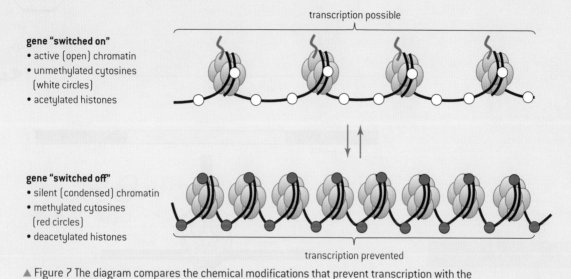

gene "switched on"
• active (open) chromatin
• unmethylated cytosines (white circles)
• acetylated histones

transcription possible

gene "switched off"
• silent (condensed) chromatin
• methylated cytosines (red circles)
• deacetylated histones

transcription prevented

▲ Figure 7 The diagram compares the chemical modifications that prevent transcription with the chemical modifications that allow transcription

The direction of transcription

Transcription occurs in a 5' to 3' direction.

The synthesis of mRNA occurs in three stages: initiation, elongation and termination. Transcription begins near a site in the DNA called the promoter. Once binding of the RNA polymerase occurs, the DNA is unwound by the RNA polymerase forming an open complex. The RNA polymerase slides along the DNA, synthesizing a single strand of RNA.

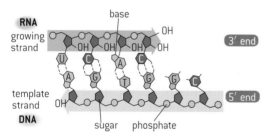

▲ Figure 8

Post-transcriptional modification

Eukaryotic cells modify mRNA after transcription.

The regulation of gene expression can occur at several points. Transcription, translation and post-translational regulation occur in both eukaryotes and prokaryotes. However, most regulation of prokaryotic gene expression occurs at transcription. In addition, post-transcriptional modification of RNA is a method of gene regulation that does not occur in prokaryotes.

▲ Figure 9 Coloured transmission electron micrograph of DNA transcription coupled with translation in the bacterium *Escherichia coli*. During transcription, complementary messenger ribonucleic acid (mRNA) strands (green) are synthesized using DNA (pink) as a template and immediately translated by ribosomes (blue)

a)

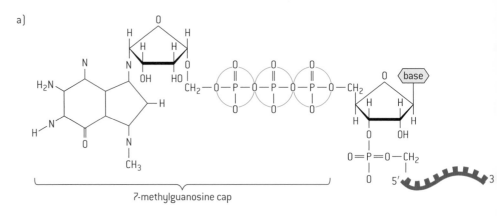

7-methylguanosine cap

b)

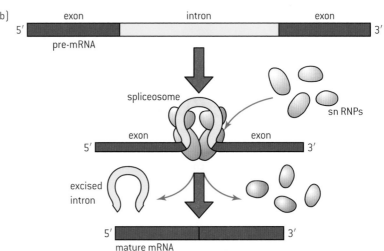

mature mRNA

c)

A poly A tail consisting of 100–200 adenine nucleotides is added after transcription.

poly A tail

▲ Figure 10

One of the most significant differences between eukaryotes and prokaryotes is the absence of a nuclear membrane surrounding the genetic material in prokaryotes. The absence of a compartment in prokaryotes means that transcription and translation can be coupled.

The separation of the location of transcription and translation into separate compartments in eukaryotes allows for significant post-transcriptional modification to occur before the mature transcript exits the nucleus. An example would be the removal of intervening sequences, or introns, from the RNA transcript. Prokaryotic DNA does not contain introns.

In eukaryotes, the immediate product of mRNA transcription is referred to as pre-mRNA, as it must go through several stages of post-transcriptional modification to become mature mRNA.

One of these stages is called RNA splicing, shown in figure 10b. Interspersed throughout the mRNA are sequences that will not contribute to the formation of the polypeptide. They are referred to as intervening sequences, or introns. These introns must be removed. The remaining coding portions of the mRNA are called exons. These will be spliced together to form the mature mRNA.

Post-transcriptional modification also includes the addition of a 5′ cap that usually occurs before transcription has been completed (see figure 10a). A poly-A tail is added after the transcript has been made (see figure 10c).

mRNA splicing

Splicing of mRNA increases the number of different proteins an organism can produce.

Alternative splicing is a process during gene expression whereby a single gene codes for multiple proteins. This occurs in genes with multiple exons. A particular exon may or may not be included in the final messenger RNA. As a result, the proteins translated from alternatively spliced mRNAs will differ in their amino acid sequence and possibly in their biological functions.

In mammals, the protein tropomyosin is encoded by a gene that has eleven exons. Tropomyosin pre-mRNA is spliced differently in different tissues resulting in five different forms of the protein. For example, in skeletal muscle, exon "2" is missing from the mRNA and in smooth muscle, exons "3" and "10" are not present.

In fruit flies, the *Dscam* protein is involved in guiding growing nerve cells to their targets. Research has shown that there are potentially 38,000 different mRNAs possible based on the number of different introns in the gene that could be spliced alternatively.

7.3 Translation

Understanding

→ Initiation of translation involves assembly of the components that carry out the process.

→ Synthesis of the polypeptide involves a repeated cycle of events.

→ Disassembly of the components follows termination of translation.

→ Free ribosomes synthesize proteins for use primarily within the cell.

→ Bound ribosomes synthesize proteins primarily for secretion or for use in lysosomes.

→ Translation can occur immediately after transcription in prokaryotes due to the absence of a nuclear membrane.

→ The sequence and number of amino acids in the polypeptide is the primary structure.

→ The secondary structure is the formation of alpha helices and beta pleated sheets stabilized by hydrogen bonding.

→ The tertiary structure is the further folding of the polypeptide stabilized by interactions between R groups.

→ The quaternary structure exists in proteins with more than one polypeptide chain.

Applications

→ tRNA-activating enzymes illustrate enzyme-substrate specificity and the role of phosphorylation.

Skills

→ The use of molecular visualization software to analyse the structure of eukaryotic ribosomes and a tRNA molecule.

→ Identification of polysomes in an electron micrograph.

Nature of science

→ Developments in scientific research follow improvements in computing: the use of computers has enabled scientists to make advances in bioinformatics applications such as locating genes within genomes and identifying conserved sequences.

The structure of the ribosome

The use of molecular visualization software to analyse the structure of eukaryotic ribosomes and a tRNA molecule.

Ribosome structure includes:

• Proteins and ribosomal RNA molecules (rRNA).

• Two sub-units, one large and one small.

• Three binding sites for tRNA on the surface of the ribosome. Two tRNA molecules can bind at the same time to the ribosome.

• There is a binding site for mRNA on the surface of the ribosome.

Each ribosome has three tRNA binding sites – the "E" or exit site, the "P" or peptidyl site and the "A" or aminoacyl site (see figure 1).

The protein data bank (PDB) is a public database containing data regarding the three-dimensional structure for a large number of biological molecules. In 2000, structural biologists Venkatraman Ramakrishnan, Thomas A. Steitz and Ada E. Yonath made the first data about

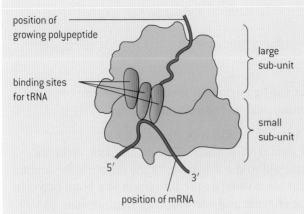

▲ Figure 1

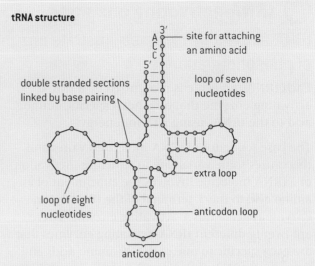

tRNA structure

▲ Figure 3

ribosome subunits available through the PDB. In 2009, they received a Nobel Prize for their work on the structure of ribosomes.

Visit the protein databank to obtain images of the *Thermus thermophilus* ribosome (images 1jgo and 1giy), or download these images from the companion website to the textbook. Using Jmol, rotate the image to visualize the small sub-unit and the large sub-unit. In the image in figure 2, an mRNA molecule is coloured yellow. The pink, purple and blue areas in the image represent the three tRNA binding sites with tRNA molecules bound.

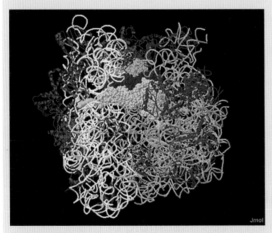

▲ Figure 2

The generalized structure of a tRNA molecule is shown in figure 3.

All tRNA molecules have:

- sections that become double-stranded by base pairing, creating loops

- a triplet of bases called the anticodon which is part of a loop of seven unpaired bases

- two other loops

- the base sequence CCA at the 3′ end which forms a site for attaching an amino acid.

Visit the PDB to obtain an image of a tRNA molecule to explore the structure in a programme such as Jmol. Figure 4 shows such an image. The parts marked green represent the amino acid binding site and the anti-codon. The part in purple shows a region of the molecule where a triplet of bases are hydrogen bonded. This is shown in the second image.

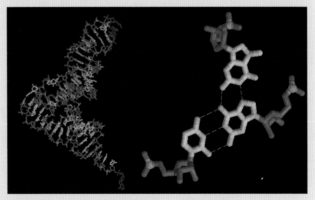

▲ Figure 4 Whole view of a tRNA molecule with a close-up of a triplet of bases connected by hydrogen bonds

tRNA-activating enzymes

tRNA-activating enzymes illustrate enzyme-substrate specificity and the role of phosphorylation.

Each tRNA molecule is recognized by a tRNA-activating enzyme that attaches a specific amino acid to the tRNA, using ATP for energy.

The base sequence of tRNA molecules varies and this causes some variability in structure. Activation of a tRNA molecule involves the attachment of an amino acid to the 3′ terminal of the tRNA by an enzyme called a tRNA-activating enzyme. There are twenty different tRNA-activating enzymes that are each specific to one of the 20 amino acids and the correct tRNA molecule. The active site of the

activating enzyme is specific to both the correct amino acid and the correct tRNA.

Energy from ATP is needed for the attachment of amino acids. Once ATP and an amino acid are attached to the active site of the enzyme, the amino acid is activated by the formation of a bond between the enzyme and adenosine monophosphate (AMP). Then the activated amino acid is covalently attached to the tRNA. Energy from this bond is later used to link the amino acid to the growing polypeptide chain during translation.

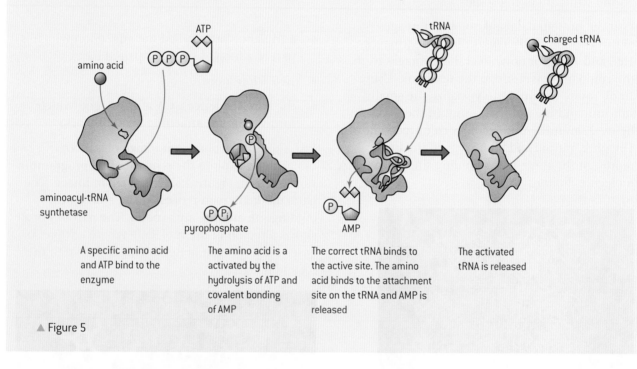

A specific amino acid and ATP bind to the enzyme

The amino acid is activated by the hydrolysis of ATP and covalent bonding of AMP

The correct tRNA binds to the active site. The amino acid binds to the attachment site on the tRNA and AMP is released

The activated tRNA is released

▲ Figure 5

Initiation of translation

Initiation of translation involves assembly of the components that carry out the process.

To begin the process of translation, an mRNA molecule binds to the small ribosomal subunit at an mRNA binding site. An initiator tRNA molecule carrying methionine then binds at the start codon "AUG".

The large ribosomal subunit then binds to the small one.

The initiator tRNA is in the P site. The next codon signals another tRNA to bind. It occupies the A site. A peptide bond is formed between the amino acids in the P and A site.

▲ Figure 6

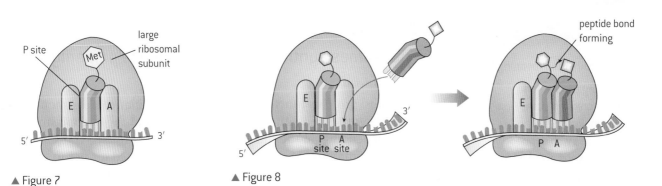

▲ Figure 7

▲ Figure 8

Elongation of the polypeptide

Synthesis of the polypeptide involves a repeated cycle of events.

Following initiation, elongation occurs through a series of repeated steps. The ribosome translocates three bases along the mRNA, moving the tRNA in the P site to the E site, freeing it and allowing a tRNA with the appropriate anticodon to bind to the next codon and occupy the vacant A site.

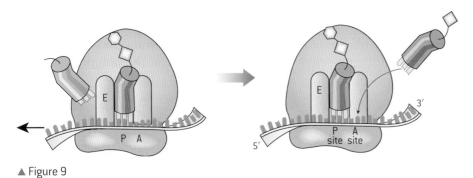

▲ Figure 9

Termination of translation

Disassembly of the components follows termination of translation.

The process continues until a stop codon is reached when the free polypeptide is released. Note the direction of movement along the mRNA is from the 5'end to the 3' end.

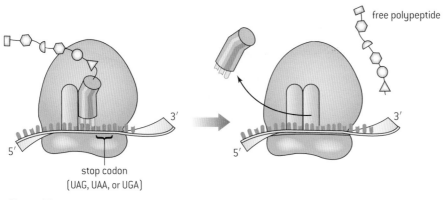

▲ Figure 10

Free ribosomes

Free ribosomes synthesize proteins for use primarily within the cell.

In eukaryotes, proteins function in a particular cellular compartment. Proteins are synthesized either in the cytoplasm or at the endoplasmic reticulum depending on the final destination of the protein. Translation occurs more commonly in the cytosol. Proteins destined for use in the cytoplasm, mitochondria and chloroplasts are synthesized by ribosomes free in the cytoplasm.

Bound ribosomes

Bound ribosomes synthesize proteins primarily for secretion or for use in lysosomes.

In eukaryotic cells, thousands of proteins are made. In many cases, proteins perform a function within a specific compartment of the cell or they are secreted. Proteins must therefore be sorted so that they end up in their correct location. Proteins that are destined for use in the ER, the Golgi apparatus, lysosomes, the plasma membrane or outside the cell are synthesized by ribosomes bound to the ER.

Whether the ribosome is free in the cytosol or bound to the ER depends on the presence of a signal sequence on the polypeptide

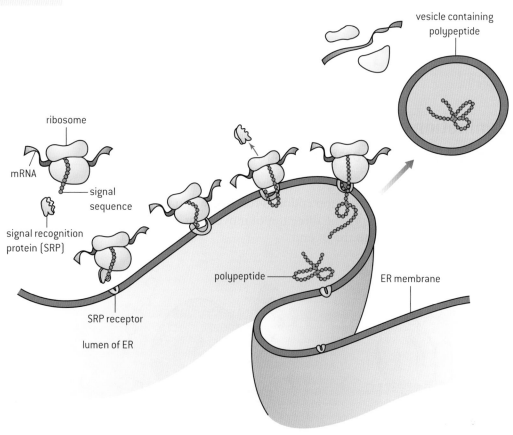

▲ Figure 11

being translated. It is the first part of the polypeptide translated. As the signal sequence is created it becomes bound to a signal recognition protein that stops the translation until it can bind to a receptor on the surface of the ER. Once this happens, translation begins again with the polypeptide moving into the lumen of the ER as it is created.

The coupling of transcription and translation in prokaryotes

Translation can occur immediately after transcription in prokaryotes due to the absence of a nuclear membrane.

In eukaryotes, cellular functions are compartmentalized whereas in prokaryotes they are not. Once transcription is complete in eukaryotes, the transcript is modified in several ways before exiting the nucleus. Thus there is a delay between transcription and translation due to compartmentalization. In prokaryotes, as soon as the mRNA is transcribed, translation begins.

 Identification of polysomes

Identification of polysomes in an electron micrograph.

Polysomes are structures visible in an electron microscope. They appear as beads on a string. They represent multiple ribosomes attached to a single mRNA molecule. Because translation and transcription occur in the same compartment in prokaryotes, as soon as the mRNA is transcribed, translation begins. Thus, multiple polysomes are visible associated with one gene. In eukaryotes, polysomes occur in both the cytoplasm and next to the ER.

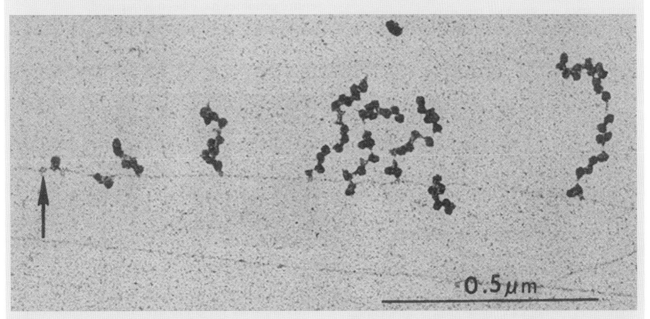

▲ Figure 12 Strings of polysomes attached to a DNA molecule in a prokaryote. The arrow designates where investigators believe RNA polymerase is sitting at, or near, the initiation site for a gene

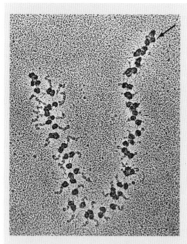

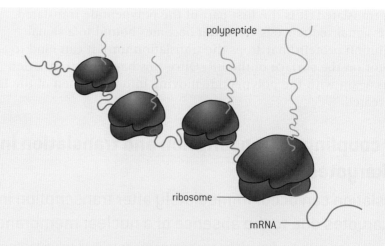

▲ Figure 13 The image shows multiple ribosomes translating a single mRNA molecule within the cytoplasm at the same time. The beginning of the mRNA is to the right (at the arrow). The polypeptides being synthesized get longer and longer, the closer the end of the mRNA the ribosomes get

Bioinformatics

Developments in scientific research follow improvements in computing: the use of computers has enabled scientists to make advances in bioinformatics applications such as locating genes within genomes and identifying conserved sequences.

Bioinformatics involves the use of computers to store and analyse the huge amounts of data being generated by the sequencing of genomes and the identification of gene and protein sequences.

Such information is often amassed in databases, for example, GenBank (a US-based database), the DDBJ (DNA databank of Japan) or the nucleotide sequence database maintained by the EMBL (the European Molecular Biology Laboratory), which then become accessible to the global community including scientists and the general public.

A scientist studying a particular genetic disorder in humans might identify sequence similarities that exist in people with the disorder. They might then search for homologous sequences in other organisms. These sequences might have a common ancestral origin but have accumulated differences over time due to random mutation.

To carry out the search for a homologous nucleotide or amino acid sequence, the scientist would conduct a BLAST search. The acronym stands for <u>b</u>asic <u>l</u>ocal <u>a</u>lignment <u>s</u>earch <u>t</u>ool.

Sometimes the homologous sequences are identical or nearly identical across species. These are called conserved sequences. The fact that they are conserved across species suggests they play a functional role.

The functions of conserved sequences are often investigated in model organisms such as *E. coli*, yeast (*S. cerevisiae*), fruit flies (*D. melanogaster*), a soil roundworm *C. elegans*, thale cress *A. thalania* and mice *M. musculus*. These particular organisms are often used because, along with humans, their entire genomes have been sequenced.

Functions are often discovered by knockout studies where the conserved gene is disrupted or altered and the impact on the organism's phenotype is observed.

▲ Figure 14 Examples of model organisms

In addition to the BLAST program, there are other software programs available. ClustalW can be used to align homologous sequences to search for changes. PhyloWin can be used to construct evolutionary trees based on sequence similarities.

Primary structure

The sequence and number of amino acids in the polypeptide is the primary structure.

A chain of amino acids is called a polypeptide. Given that the 20 commonly occurring amino acids can be combined in any sequence, it should not be surprising that there is a huge diversity of proteins.

The sequence of amino acids in a polypeptide is termed its primary structure.

Data-based questions

The hemoglobin molecule transports oxygen in the blood. It consists of 4 polypeptide chains. In human adults the molecule has two kinds of chains, alpha chains and beta chains, and there are two each. The alpha chain has 141 amino acid residues and the beta chain has 146 amino acid residues. The primary sequence of both polypeptides is shown below. The single residue in the beta chain marked in blue is the site of a mutation in sickle cell anemia. In the mutation, the glutamic acid is replaced by valine.

alpha chain:

> 1 val * leu ser pro ala asp lys thr asn
> val lys ala ala trp gly lys val gly ala
> his ala gly glu tyr gly ala glu ala leu
> glu arg met phe leu ser phe pro thr
> thr lys thr tyr phe pro his phe * asp
> leu ser his gly ser ala * * * * * gln val
> lys gly his gly lys lys val ala asp ala
> leu thr asn ala val ala his val asp asp
> met pro asn ala leu ser ala leu ser asp
> leu his ala his lys leu arg val asp pro
> val asp phe lys leu leu ser his cys leu
> leu val thr leu ala ala his leu pro ala

> glu phe thr pro ala val his ala ser leu
> asp lys phe leu ala ser val ser thr val
> leu thr ser lys tyr arg 141

beta chain:

> 1 val his leu thr pro **glu** glu lys ser ala
> val thr ala leu trp gly lys val asn * * val
> asp glu val gly gly glu ala leu gly arg
> leu leu val val tyr pro trp thr gln arg
> phe phe glu ser phe gly asp leu ser thr
> pro asp ala val met gly asn pro lys val
> lys ala his gly lys lys val leu gly ala phe
> ser asp gly leu ala his leu asp asn leu
> lys gly thr phe ala thr leu ser glu leu
> his cys asp lys leu his val asp pro glu
> asn phe arg leu leu gly asn val leu val
> cys val leu ala his his phe gly lys glu
> phe thr pro pro val gln ala ala tyr gln
> lys val val ala gly val ala asp ala leu ala
> his lys tyr his 146

Compare the primary structure of the two polypeptides. The asterix (*) symbols indicates locations where sections of the amino acid sequence are missing to facilitate comparison. [4]

Secondary structure

The secondary structure is the formation of alpha helices and beta pleated sheets stabilized by hydrogen bonding.

Because the chain of amino acids in a polypeptide has polar covalent bonds within its backbone, it tends to fold in such a way that hydrogen bonds form between the carboxyl (C=O) group of one residue and the amino group (N—H) group of an amino acid in another part of the chain. This results in the formation of patterns within the polypeptide called secondary structures. The α-helix and the β-pleated sheet are examples of secondary structures.

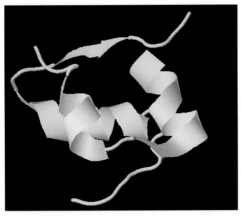

▲ Figure 15 The structure of insulin showing three areas where the α-helix can be seen. It also shows the quaternary structure of insulin, i.e. the relative positions of the two polypeptides

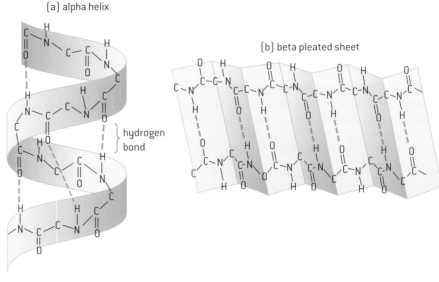

▲ Figure 16 Two examples of protein secondary structure

Tertiary structure

The tertiary structure is the further folding of the polypeptide stabilized by interactions between R groups.

Tertiary structure refers to the overall three-dimensional shape of the protein (figure 18). This shape is a consequence of the interaction of R-groups with one another and with the surrounding water medium. There are several different types of interaction.

- Positively charged R-groups will interact with negatively charged R-groups.
- Hydrophobic amino acids will orientate themselves toward the centre of the polypeptide to avoid contact with water, while hydrophilic amino acids will orientate themselves outward.
- Polar R-groups will form hydrogen bonds with other polar R-groups.
- The R-group of the amino acid cysteine can form a covalent bond with the R-group of another cysteine forming what is called a disulphide bridge.

▲ Figure 17 Collagen—the quaternary structure consists of three polypeptides wound together to fom a tough, rope-like protein

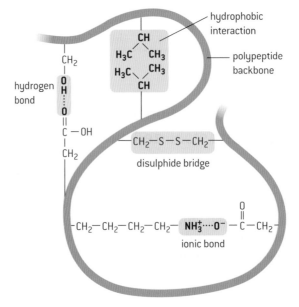

▲ Figure 18 R-group interactions contribute to tertiary structure

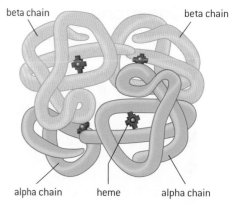

Quartenary structure

The quaternary structure exists in proteins with more than one polypeptide chain.

Proteins can be formed from a single polypeptide chain or from more than one polypeptide chain. Lysozyme is composed of a single chain, so lysozyme is both a polypeptide and a protein. Insulin is formed from two polypeptides, and hemoglobin is made up of four chains. Quaternary structure refers to the way polypeptides fit together when there is more than one chain. It also refers to the addition of non-polypeptide components. The quaternary structure of the hemoglobin molecule consists of four polypeptide chains and four heme groups.

The biological activity of a protein is related to its primary, secondary, tertiary and quaternary structure. Certain treatments such as exposure to high temperatures, or changes in pH can cause alterations in the structure of a protein and therefore disrupt its biological activity. When a protein has permanently lost its structure it is said to be denatured.

▲ Figure 19 The quaternary structure of hemoglobin in adults consists of four chains: two α-chains and two β-chains. Each subunit contains a molecule called a heme group

Data-based questions

Hemoglobin is a protein composed of two pairs of globin subunits. During the process of development from conception through to 6 months after birth, human hemoglobin changes in composition. Adult hemoglobin consists of two alpha- and two beta-globin subunits. Four other polypeptides are found during development: zeta, delta, epsilon and gamma.

Figure 20 illustrates the changes in hemoglobin composition during gestation and after birth in a human.

a) State which two subunits are present in highest amounts early in gestation. [1]

b) Compare changes in the amount of the gamma-globin gene with beta-globin. [3]

c) Determine the composition of the hemoglobin at 10 weeks of gestation and at 6 months of age. [2]

d) State the source of oxygen for the fetus. [1]

e) The different types of hemoglobin have different affinities for oxygen. Suggest reasons

for the changes in hemoglobin type during development and after birth. [3]

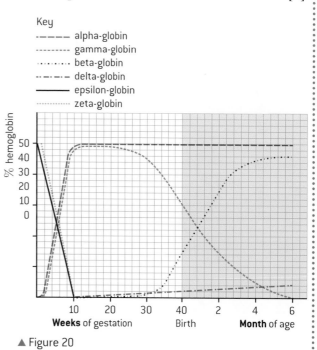

Key
------ alpha-globin
- - - - gamma-globin
········· beta-globin
-·-·- delta-globin
——— epsilon-globin
·········· zeta-globin

▲ Figure 20

Questions

1 Different samples of bacteria were supplied with radioactive nucleoside triphosphates for a series of times (10, 30 or 60 seconds). This was the "pulse" period. This was followed by adding a large excess of non-radioactive nucleoside triphosphates for a longer period of time. This is called the "chase" period. The appearance of radioactive nucleotides (incorporated during the pulse) in parts of the product DNA give an indication of the process of converting intermediates to final products.

DNA was isolated from the bacterial cells, denatured (separated into two strands by heat) and centrifuged to separate molecules by size. The closer to the top of the centrifuge tube, the smaller the molecule.

a) Compare the sample that was pulsed for 10 seconds with the sample that was pulsed for 30 seconds. [2]

b) Explain why the sample that was pulsed for 30 seconds provides evidence for the presence of both a leading strand and many lagging strands. [3]

c) Explain why the sample that was pulsed for 60 seconds provides evidence for the activity of DNA ligase. [3]

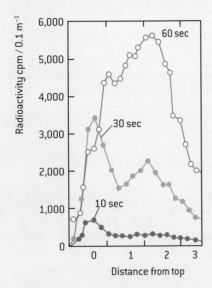
▲ Figure 21

2 With reference to Figure 22, answer the following questions.

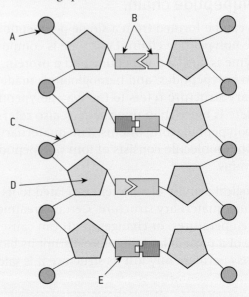
▲ Figure 22

a) What part of the nucleotide is labelled A? [1]

b) What kind of bond forms between the structures labelled B? [1]

c) What kind of bond is indicated by label C? [1]

d) What sub-unit is indicated by label D? [1]

e) What sub-unit is indicated by label E? [1]

3 Refer to figure 23 when answering the following questions.

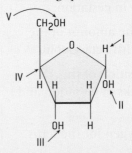
▲ Figure 23

a) State what molecule is represented. [1]

b) State whether the molecule would be found in DNA or RNA. [1]

c) State the part of the molecule to which phosphates bind. [1]

d) Identify the part of the molecule that refers to the 3′ end. [1]

8 METABOLISM, CELL RESPIRATION AND PHOTOSYNTHESIS (AHL)

Introduction

Life is sustained by a complex web of chemical reactions inside cells. These metabolic reactions are regulated in response to the needs of the cell and the organism. Energy is converted to a usable form in cell respiration. In photosynthesis light energy is converted into chemical energy and a huge diversity of carbon compounds is produced.

8.1 Metabolism

Understanding

→ Metabolic pathways consist of chains and cycles of enzyme-catalysed reactions.

→ Enzymes lower the activation energy of the chemical reactions that they catalyse.

→ Enzyme inhibitors can be competitive or non-competitive.

→ Metabolic pathways can be controlled by end-product inhibition.

Applications

→ End-product inhibition of the pathway that converts threonine to isoleucine.

→ Use of databases to identify potential new anti-malarial drugs.

Skills

→ Distinguishing different types of inhibition from graphs at specified substrate concentration.

→ Calculating and plotting rates of reaction from raw experimental results.

Nature of science

→ Developments in scientific research follow improvements in computing: developments in bioinformatics, such as the interrogation of databases, have facilitated research into metabolic pathways.

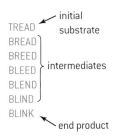

▲ Figure 1 Word game analogy for metabolic pathways

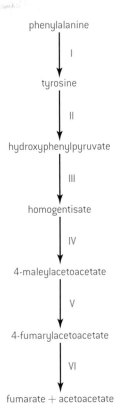

▲ Figure 2 Example of a metabolic pathway

Metabolic pathways

Metabolic pathways consist of chains and cycles of enzyme-catalysed reactions.

The word "metabolism" was introduced in the 19th century by the German cytologist and physiologist Theodor Schwann, to refer to the chemical changes that take place in living cells. It is now known that a huge range of chemical reactions occur in cells, catalysed by over 5,000 different types of enzyme. Although metabolism is very complex, there are some common patterns.

1 Most chemical changes happen not in one large jump, but in a sequence of small steps, together forming what is called a metabolic pathway. The word game in figure 1 is an analogy.

2 Most metabolic pathways involve a *chain* of reactions. Figure 2 shows a reaction chain that is used by cells to convert phenylalanine into fumarate and acetoacetate, which can be used as energy sources in respiration. Phenylalanine causes severe health problems if there is an excess of it in the blood.

3 Some metabolic pathways form a *cycle* rather than a chain. In this type of pathway, the end product of one reaction is the reactant that starts the rest of the pathway.

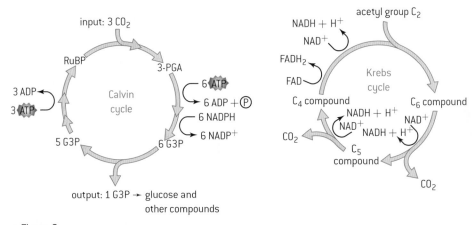

▲ Figure 3

Enzymes and activation energy

Enzymes lower the activation energy of the chemical reactions that they catalyse.

Chemical reactions are not single-step processes. Substrates have to pass through a transition state before they are converted into products. Energy is required to reach the transition state, and although energy is released in going from the transition state to the product, some energy must be put in to reach the transition state. This is called the activation energy. The activation energy is used to break or weaken bonds in the substrates. Figure 4 shows these energy

changes for an exergonic (energy releasing) reaction that is and is not catalysed by an enzyme.

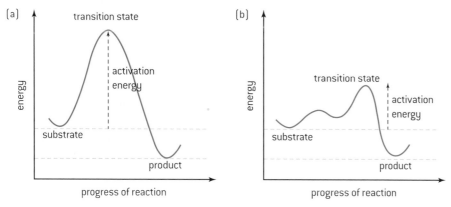

▲ Figure 4 Graphs showing activation energy (a) without an enzyme and (b) with an enzyme

When an enzyme catalyses a reaction, the substrate binds to the active site and is altered to reach the transition state. It is then converted into the products, which separate from the active site. This binding lowers the overall energy level of the transition state. The activation energy of the reaction is therefore reduced. The net amount of energy released by the reaction is unchanged by the involvement of the enzyme. However as the activation energy is reduced, the rate of the reaction is greatly increased, typically by a factor of a million or more.

Types of enzyme inhibitors

Enzyme inhibitors can be competitive or non-competitive.

Some chemical substances bind to enzymes and reduce the activity of the enzyme. They are therefore known as inhibitors. The two main types are competitive and non-competitive inhibitors.

Competitive inhibitors interfere with the active site so that the substrate cannot bind. Non-competitive inhibitors bind at a location other than the active site. This results in a change of shape in the enzyme so that the enzyme cannot bind to the substrate. Table 1 shows examples of each type.

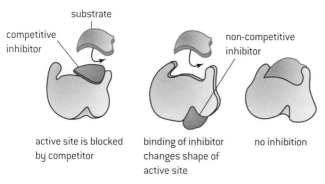

▲ Figure 6

▲ Figure 5 A molecular model of the restriction enzyme EcoRV (purple and pink) bound to a DNA molecule (deoxyribonucleic acid, yellow and orange). Restriction enzymes, also known as restriction endonucleases, recognize specific nucleotide sequences and cut the DNA at these sites. They are found in bacteria and archaea and are thought to have evolved as a defence against viral infection

TOK

To what extent should ethics constrain the development of knowledge in science?

Sarin was a chemical developed as an insectide before being applied is a chemical weapon. It is a competitive inhibitor of the neurotransmitter acetylcholinesterase. Chemical weapons would not exist without the activities of scientists. In fact, the name Sarin is an acronym of the surnames of the scientists who first synthesized it.

Fritz Haber received the 1918 Nobel Prize for Chemistry for his work in developing the chemistry behind the industrial production of ammonia fertilizer. Some scientists boycotted the award ceremony because Haber had been instrumental in encouraging and developing the use of chlorine gas in the First World War. Haber is quoted as saying: "During peace time a scientist belongs to the World, but during war time he belongs to his country."

Enzyme	Substrate	Inhibitor	Binding
dihydropteroate synthetase	para-aminobenzoate	sulfadiazine	The inhibitor binds reversibly to the enzyme's active site. While it remains bound, substrates cannot bind. This is competitive inhibition.
phosphofructokinase	fructose-6-phosphate	xylitol-5-phosphate	The inhibitor binds reversibly to a site away from the active site. While it remains bound, the active site is distorted and substrate cannot bind. This is non-competitive inhibition.

▲ Table 1 Examples of each type of inhibitor

 ## Effects of enzyme inhibitors

Distinguishing different types of inhibition from graphs at specified substrate concentration.

Figure 7 represents the effect of substrate concentration on the rate of an enzyme controlled reaction.

The orange line represents the effect of substrate concentration on enzyme activity in the absence of an inhibitor.

The red line shows the effect of substrate concentration on the rate of reaction when a competitive inhibitor is present. When the concentration of substrate begins to exceed the amount of inhibitor, the maximum rate of the uninhibited enzyme can be achieved; however, it takes a much higher concentration of substrate to achieve this maximum rate.

The blue line shows the effect of substrate concentration on the rate of reaction when a non-competitive inhibitor is present. In the presence of a non-competitive inhibitor, the enzyme does not reach the same maximum rate because the binding of the non-competitive inhibitor prevents some of the enzymes from being able to react regardless of substrate concentration. Those enzymes that do not bind inhibitors follow the same pattern as the normal enzyme. It takes approximately the same concentration of substrate to reach the maximum rate, but the maximum rate is lower than the uninhibited enzyme.

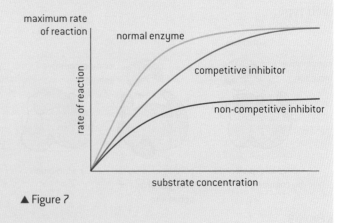

▲ Figure 7

End-product inhibition

Metabolic pathways can be controlled by end-product inhibition.

Many enzymes are regulated by chemical substances that bind to special sites on the enzyme away from the active site. These are called allosteric interactions and the binding site is called an allosteric site. In many cases, the enzyme that is regulated catalyses one of the first reactions in a metabolic pathway and the substance that binds to the allosteric site is the end product of the pathway. The end product acts as an inhibitor. The pathway works rapidly in cells with a shortage of end product but can be switched off completely in cells where there is an excess.

To see why this is such an economical way to control metabolic pathways, we need to understand how the concentration of the product of a reaction can influence the rate of reaction. Reactions often do not go to completion – instead an equilibrium position is reached with a characteristic ratio of substrates and products. So, if the concentration of products increases, a reaction will eventually slow down and stop. This effect reverberates back through a metabolic pathway when the end product accumulates, with all the intermediates accumulating. End-product inhibition prevents this build-up of intermediate products.

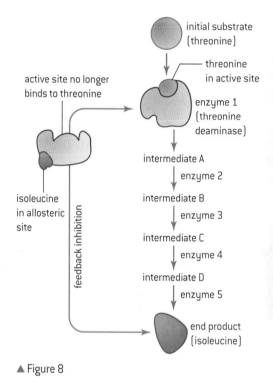

▲ Figure 8

 An example of end-product inhibition

End-product inhibition of the pathway that converts threonine to isoleucine.

Through a series of five reactions, the amino acid threonine is converted to isoleucine. As the concentration of isoleucine builds up, it binds to the allosteric site of the first enzyme in the chain, threonine deaminase, thus acting as a non-competitive inhibitor (figure 8).

 Investigating metabolism through bioinformatics

Developments in scientific research follow improvements in computing: developments in bioinformatics, such as the interrogation of databases, have facilitated research into metabolic pathways.

Computers have increased the capacity of scientists to organize, store, retrieve and analyse biological data. Bioinformatics is an approach whereby multiple research groups can add information to a database enabling other groups to query the database.

One promising bioinformatics technique that has facilitated research into metabolic pathways is referred to as chemogenomics. Sometimes when a chemical binds to a target site, it can

significantly alter metabolic activity. Scientists looking to develop new drugs test massive libraries of chemicals individually on a range of related organisms. For each organism a range of target sites are identified and a range of chemicals which are known to work on those sites are tested. One researcher called chemogenomics "the chemical universe tested against the target universe".

Chemogenomics applied to malaria drugs
Use of databases to identify potential new anti-malarial drugs.

Malaria is a disease caused by the pathogen *Plasmodium falciparum*. The increasing resistance of *P. falciparum* to anti-malarial drugs such as chloroquine, the dependence of all new drug combinations on a narrow range of medicines and increasing global efforts to eradicate malaria all drive the need to develop new anti-malarial drugs.

Plasmodium falciparum strain 3D7 is a variety of the malarial parasite for which the genome has been sequenced. In one study, approximately 310,000 chemicals were screened against a chloroquine-sensitive 3D7 strain and the chloroquine-resistant K1 strain to see if these chemicals inhibited metabolism. Other related and unrelated organisms, including human cell lines, were also screened. One promising outcome was the identification of 19 new chemicals that inhibit the enzymes normally targeted by anti-malarial drugs and 15 chemicals that bind to a total of 61 different malarial proteins. This provides other scientists with possible lines of investigation in the search for new anti-malarials.

 ## Calculating rates of reaction
Calculating and plotting rates of reaction from raw experimental results.

A large number of different protocols are available for investigating enzyme activity. Determining the rate of an enzyme-controlled reaction involves measuring either the rate of disappearance of a substrate or the rate of appearance of a product. Sometimes this will require conversion of units to yield a rate unit which should include s^{-1}.

Data-based questions: The effectiveness of enzymes

The degree to which enzymes increase the rate of reactions varies greatly. By calculating the ratio between the rate of reactions with and without an enzyme catalyst, the affinity between an enzyme and its substrate can be estimated. Table 2 shows the rates of four reactions with and without an enzyme. The ratio between these rates has been calculated for one of the reactions.

1   State which reaction has the slowest rate in the absence of an enzyme. [1]

2   State which enzyme catalyses its reaction at the most rapid rate. [1]

3   Calculate the ratios between the rate of reaction with and without an enzyme for ketosteroid isomerase, nuclease and OMP decarboxylase. [3]

4   Discuss which of the enzymes is the more effective catalyst. [3]

5   Explain how the enzymes increase the rate of the reactions that they catalyse. [2]

Enzyme	Rate without enzyme/s^{-1}	Rate with enzyme/s^{-1}	Ratio between rate with and without enzyme
Carbonic anhydrase	1.3×10^{-1}	1.0×10^{6}	7.7×10^{6}
Ketosteroid isomerase	1.7×10^{-7}	6.4×10^{4}	
Nuclease	1.7×10^{-13}	9.5×10^{6}	
OMP decarboxylase	2.8×10^{-16}	3.9×10^{8}	

▲ Table 2

Data-based questions: Calculating rates of reaction

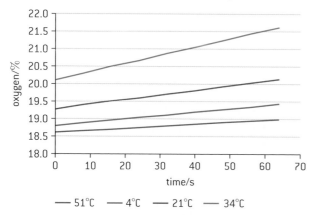

▲ Figure 9 Percentage of oxygen concentration over time at various temperatures after adding catalase to a 1.5% hydrogen peroxide solution

Ten drops of a commercial catalase solution were added to four reaction vessels containing a 1.5% hydrogen peroxide solution. Each of the solutions had been kept at a different temperature. The % oxygen in the reaction vessel was determined using a data logger in a set-up similar to figure 10.

▲ Figure 10

1 Explain the variation in the % oxygen at time zero.

2 Determine the rate of reaction at each temperature using the graph.

3 Construct a scatter plot of reaction rate versus temperature.

Activity

For each of the following enzyme experiments, describe how the rate of reaction can be determined:

a) Paper discs soaked in the enzyme catalase are added to different concentrations of hydrogen peroxide. The reaction produces oxygen bubbles.

b) Lipase catalyses the breakdown of triglycerides to fatty acids and water. The pH of the reaction solution will lower as the reaction proceeds.

c) Papain is a protease that can be extracted from pineapple fruits. Gelatin cubes will be digested by papain.

d) Catechol oxidase converts catechol to a yellow pigment in cut fruit. It can be extracted from bananas. The yellow pigment reacts with oxygen in the air to turn brown.

8.2 Cell respiration

Understanding

→ Cell respiration involves the oxidation and reduction of compounds.

→ Phosphorylation of molecules makes them less stable.

→ In glycolysis, glucose is converted to pyruvate in the cytoplasm.

→ Glycolysis gives a small net gain of ATP without the use of oxygen.

→ In aerobic cell respiration pyruvate is decarboxylated and oxidized.

→ In the link reaction pyruvate is converted into acetyl coenzyme A.

→ In the Krebs cycle, the oxidation of acetyl groups is coupled to the reduction of hydrogen carriers, liberating carbon dioxide.

→ Energy released by oxidation reactions is carried to the cristae of the inner mitochondrial membrane by reduced NAD and FAD.

→ Transfer of electrons between carriers in the electron transport chain is coupled to proton pumping.

→ In chemiosmosis protons diffuse through ATP synthase to generate ATP.

→ Oxygen is needed to bind with the free protons to form water to maintain the hydrogen gradient.

→ The structure of the mitochondrion is adapted to the function it performs.

Applications

→ Electron tomography used to produce images of active mitochondria.

Skills

→ Analysis of diagrams of the pathways of aerobic respiration to deduce where decarboxylation and oxidation reactions occur.

→ Annotation of a diagram to indicate the adaptations of a mitochondrion to its function.

Nature of science

→ Paradigm shifts: the chemiosmotic theory led to a paradigm shift in the field of bioenergetics.

Oxidation and reduction

Cell respiration involves the oxidation and reduction of compounds.

Oxidation and reduction are chemical processes that always occur together. This happens because they involve transfer of electrons from one substance to another. Oxidation is the loss of electrons from a substance and reduction is the gain of electrons.

A useful example to help visualize this in the laboratory is in the Benedict's test, a test for certain types of sugar. The test involves the use of copper sulphate solution, containing copper ions with a charge of two positive (Cu^{2+}). Cu^{2+} often imparts a blue or green colour to solutions. These copper ions are reduced and become atoms of copper by being given electrons. Copper atoms are insoluble and form a red or orange precipitate. The electrons come from sugar molecules, which are therefore oxidized.

Electron carriers are substances that can accept and give up electrons as required. They often link oxidations and reductions in cells. The main electron carrier in respiration is NAD (nicotinamide adenine dinucleotide). In photosynthesis a phosphorylated version of NAD is used, NADP (nicotinamide adenine dinucleotide phosphate). The structure of the NAD molecule is shown in figure 1.

The equation below shows the basic reaction.

$$NAD + 2\ electrons \rightarrow reduced\ NAD$$

The chemical details are a little more complicated. NAD initially has one positive charge and exists as NAD^+. It accepts two electrons in the following way: two hydrogen atoms are removed from the substance that is being reduced. One of the hydrogen atoms is split into a proton and an electron. The NAD^+ accepts the electron, and the proton (H^+) is released. The NAD accepts both the electron and proton of the other hydrogen atom. The reaction can be shown in two ways:

$$NAD^+ + 2H^+ + 2\ electrons\ (2e^-) \rightarrow NADH + H^+$$

$$NAD^+ + 2H \rightarrow NADH + H^+$$

This reaction demonstrates that reduction can be achieved by accepting atoms of hydrogen, because they have an electron. Oxidation can therefore be achieved by losing hydrogen atoms.

Oxidation and reduction can also occur through loss or gain of atoms of oxygen. There are fewer examples of this in biochemical processes, perhaps because in the early evolution of life oxygen was absent from the atmosphere. A few types of bacteria can oxidize hydrocarbons using oxygen:

$$C_7H_{15}-CH_3 + \frac{1}{2}O_2 \rightarrow C_7H_{15}-CH_2OH$$
$$\phantom{C_7H_{15}-CH_3}\text{n-octane}\phantom{+ \frac{1}{2}O_2 \rightarrow}\text{n-octanol}$$

Nitrifying bacteria oxidize nitrite ions to nitrate.

$$NO_2^- + \frac{1}{2}O_2 \rightarrow NO_3^-$$

Adding oxygen atoms to a molecule or ion is oxidation, because the oxygen atoms have a high affinity for electrons and so tend to draw them away from other parts of the molecule or ion. In a similar way, losing oxygen atoms is reduction.

Phosphorylation

Phosphorylation of molecules makes them less stable.

Phosphorylation is the addition of a phosphate molecule (PO_4^{3-}) to an organic molecule. Biochemists indicate that certain amino acid sequences tend to act as binding sites for the phosphate molecule on proteins. For many reactions, the purpose of phosphorylation is to make

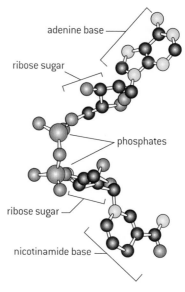

adenine base

ribose sugar

phosphates

ribose sugar

nicotinamide base

▲ Figure 1 Structure of NAD

the phosphorylated molecule more unstable; i.e., more likely to react. Phosphorylation can be said to activate the molecule.

The hydrolysis of ATP releases energy to the environment and is therefore termed an exergonic reaction. Many chemical reactions in the body are endergonic (energy absorbing) and therefore do not proceed spontaneously unless coupled with an exergonic reaction that releases more energy.

For example, depicted below is the first reaction in the series of reactions known as glycolysis.

Glucose $\longrightarrow$ Glucose-6-phosphate
ATP   ADP

The conversion of glucose to glucose-6-phosphate is endergonic and the hydrolysis of ATP is exergonic. Because the reactions are coupled, the combined reaction proceeds spontaneously. Many metabolic reactions are coupled to the hydrolysis of ATP.

Glycolysis and ATP

Glycolysis gives a small net gain of ATP without the use of oxygen.

The most significant consequence of glycolysis is the production of a small yield of ATP without the use of any oxygen, by converting sugar into pyruvate. This cannot be done as a single-step process and instead is an example of a metabolic pathway, composed of many small steps. The first of these may seem rather perverse: ATP is used up in phosphorylating sugar.

Glucose $\longrightarrow$ Glucose-6-phosphate $\rightarrow$ Fructose–6–phosphate $\longrightarrow$ Fructose-1,6-bisphosphate
ATP   ADP   ATP   ADP

However, these phosphorylation reactions reduce the activation energy required for the reactions that follow and so make them much more likely to occur.

Pyruvate is a product of glycolysis

In glycolysis, glucose is converted to pyruvate.

In the next step, the fructose bisphosphate is split to form two molecules of triose phosphate. Each of these triose phosphates is then oxidized to glycerate-3-phosphate in a reaction that yields enough energy to make ATP. This oxidation is carried out by removing hydrogen. Note that it is hydrogen atoms that are removed. If only hydrogen ions were removed (H^+), no electrons would be removed and it would not be an oxidation. The hydrogen is accepted by NAD^+, which becomes $NADH + H^+$. In the final stages of glycolysis, the phosphate group is transferred to ADP to produce more ATP and also pyruvate. These stages are summarized in the equation below, which occurs twice per glucose.

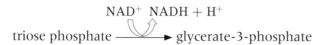

$$\text{triose phosphate} \xrightarrow[\text{NAD}^+ \quad \text{NADH} + \text{H}^+]{} \text{glycerate-3-phosphate}$$

The fate of pyruvate

In aerobic cell respiration pyruvate is decarboxylated and oxidized.

Two molecules of pyruvate are produced in glycolysis per molecule of glucose. If oxygen is available, this pyruvate is absorbed into the mitochondrion, where it is fully oxidized.

$$2CH_3\text{–}CO\text{–}COOH + 5O_2 \rightarrow 6CO_2 + 4H_2O$$
$$\text{pyruvate}$$

As with glycolysis, this is not a single-step process. Carbon and oxygen are removed in the form of carbon dioxide, in reactions called decarboxylations. The oxidation of pyruvate is achieved by the removal of pairs of hydrogen atoms. The hydrogen carrier NAD^+, and a related compound called FAD, accept these hydrogen atoms and pass them on to the electron transport chain where oxidative phosphorylation will occur. These reactions are summarized in figure 2.

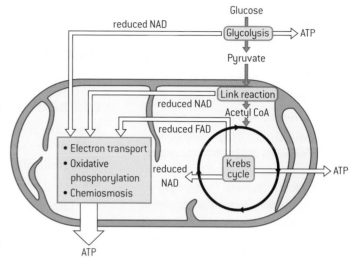

▲ Figure 2 A summary of aerobic respiration

The link reaction

In the link reaction pyruvate is converted into acetyl coenzyme A.

The first step, represented by figure 3, occurs after the pyruvate, which has been produced in the cytoplasm, is shuttled into the mitochondrial matrix. Once there, the pyruvate is decarboxylated and oxidized to form an acetyl group. Two high energy electrons are removed from pyruvate. These react with NAD^+ to produce reduced NAD. This is called the link reaction, because it links glycolysis with the cycle of reactions that follow.

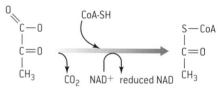

▲ Figure 3 The link reaction

The Krebs cycle

In the Krebs cycle, the oxidation of acetyl groups is coupled to the reduction of hydrogen carriers.

This cycle has several names but is often called the Krebs cycle, in honour of the biochemist who was awarded the Nobel Prize for its discovery. The link reaction involves one decarboxylation and one oxidation. There are two more decarboxylations and four more oxidations in the Krebs cycle.

If glucose is oxidized by burning in air, energy would be released as heat. Most of the energy released in the oxidations of the link reaction and the Krebs cycle is used to reduce hydrogen carriers (NAD^+ and FAD).

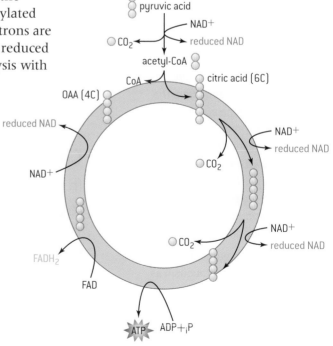

▲ Figure 4 Summary of the Krebs cycle

383

What kinds of explanations do scientists offer, and how do these explanations compare with those offered in other areas of knowledge?

Hans Krebs was awarded the Nobel Prize in 1953. The two final paragraphs of the lecture that he gave on this occasion are reproduced here.

The reactions of the cycle have been found to occur in representatives of all forms of life, from unicellular bacteria and protozoa to the highest mammals. The study of intermediary metabolism shows that the basic metabolic processes, in particular those providing energy and those leading to the synthesis of cell constituents, are also shared by all forms of life.

The existence of common features in different forms of life indicates some relationship between the different organisms, and according to the concept of evolution these relations stem from the circumstance that the higher organisms, in the course of millions of years, have gradually evolved from simpler ones. The concept of evolution postulates that living organisms have common roots, and in turn the existence of common features is powerful support for the concept of evolution. The presence of the same mechanism of energy production in all forms of life suggests two other inferences: firstly that the mechanism of energy production has arisen very early in the evolutionary process; and secondly that life, in its present forms, has arisen only once.

1 Outline the argument for similarities of metabolism as evidence for evolution.

2 Are there any alternative explanations for the similarities?

The energy therefore remains in chemical form and can be passed on to the final part of aerobic cell respiration: oxidative phosphorylation.

For every turn of the cycle, the production of reduced NAD occurs three times, decarboxylation occurs twice and the reduction of FAD occurs once. One molecule of ATP is also generated.

Oxidative phosphorylation

Energy released by oxidation reactions is carried to the cristae of the mitochondria by reduced NAD and FAD.

In aerobic respiration, there are several points where energy released by oxidation reactions is coupled to the reduction of mainly NAD but also FAD. Reduced NAD is produced during glycolysis, the link reaction and the Krebs cycle. $FADH_2$ is produced during the Krebs cycle.

The final part of aerobic respiration is called oxidative phosphorylation, because ADP is phosphorylated to produce ATP, using energy released by oxidation. The substances oxidized include the $FADH_2$ generated in the Krebs cycle and the reduced NAD generated in glycolysis, the link reaction and the Krebs cycle. Thus these molecules are used to carry the energy released in these stages to the mitochondrial cristae.

The electron transport chain

Transfer of electrons between carriers in the electron transport chain is coupled to proton pumping.

The final part of aerobic respiration is called oxidative phosphorylation, because ADP is phosphorylated to produce ATP, using energy released by oxidation. The main substance oxidized is reduced NAD.

The energy is not released in a single large step, but in a series of small steps, carried out by a chain of electron carriers. Reduced NAD and $FADH_2$ donate their electrons to electron carriers. As the electrons are passed from carrier to carrier, energy is utilized to transfer protons across the inner membrane from the matrix into the intermembrane space. The protons then flow through ATP synthase down their concentration gradient providing the energy needed to make ATP.

Chemiosmosis

In chemiosmosis protons diffuse through ATP synthase to generate ATP.

The mechanism used to couple the release of energy by oxidation to ATP production remained a mystery for many years, but is now known to be chemiosmosis. This happens in the inner mitochondrion membrane. It is called chemiosmosis because a chemical substance (H^+) moves across a membrane, down the concentration gradient. This releases the energy needed for the enzyme ATP synthase to make ATP. The main steps in the process are as follows (also see figure 5).

- NADH + H⁺ supplies pairs of hydrogen atoms to the first carrier in the chain, with the NAD⁺ returning to the matrix.

- The hydrogen atoms are split, to release two electrons, which pass from carrier to carrier in the chain.

- Energy is released as the electrons pass from carrier to carrier, and three of these use this energy to transfer protons (H⁺) across the inner mitochondrial membrane, from the matrix to the intermembrane space.

- As electrons continue to flow along the chain and more and more protons are pumped across the inner mitochondrial membrane, a concentration gradient of protons builds up. This proton gradient is a store of potential energy.

- To allow electrons to continue to flow, they must be transferred to a terminal electron acceptor at the end of the chain. In aerobic respiration this is oxygen, which briefly becomes •O_2^-, but then combines with two H⁺ ions from the matrix to become water.

- Protons pass back from the intermembrane space to the matrix through ATP synthase. As they are moving down the concentration gradient, energy is released and this is used by ATP synthase to phosphorylate ADP.

The role of oxygen

Oxygen is needed to bind with the free protons to form water to maintain the hydrogen gradient.

Oxygen is the final electron acceptor in the mitochondrial electron transport chain. The reduction of the oxygen molecule involves both accepting electrons and forming a covalent bond with hydrogen.

By using up hydrogen, the proton gradient across the inner mitochondrial membrane is maintained so that chemiosmosis can continue.

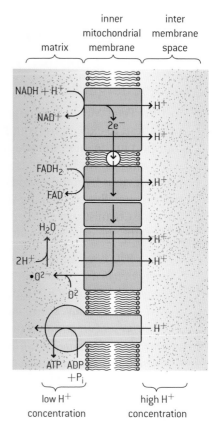

▲ Figure 5 Summary of oxidative phosphorylation

Data-based questions: Oxygen consumption by mitochondria

Figure 6 shows the results of an experiment in which mitochondria were extracted from liver cells and were kept in a fluid medium, in which oxygen levels were monitored. Pyruvate was added at point I on the graph, and ADP was added at points II, III and IV.

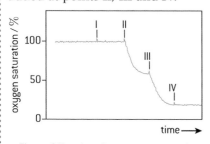

▲ Figure 6 Results of oxygen consumption experiment

1 Explain why oxygen consumption by the mitochondria could not begin unless pyruvate had been added. [3]

2 Deduce what prevented oxygen consumption between points I and II. [2]

3 Predict, with reasons, what would have happened if ADP had not been added at point III. [2]

4 Discuss the possible reasons for oxygen consumption not being resumed after ADP was added at point IV. [3]

The chemiosmotic theory

Paradigm shifts: the chemiosmotic theory produced a paradigm shift in the field of bioenergetics.

In 1961 Peter Mitchell proposed the chemiosmotic hypothesis to explain the coupling of electron transport in the inner mitochondrial membrane to ATP synthesis. His hypothesis was a radical departure from previous hypotheses and only after many years was it generally accepted. He was awarded the Nobel Prize for Chemistry in 1978 and part of the Banquet Speech that he gave is reproduced here:

Emile Zola described a work of art as a corner of nature seen through a temperament. The philosopher Karl Popper, the economist F.A. Hayek and the art historian K.H. Gombrich have shown that the creative process in science and art consists of two main activities: an imaginative jumping forward to a new abstraction or simplified representation, followed by a critical looking back to see how nature appears in the light of the new vision. The imaginative leap forward is a hazardous, unreasonable activity. Reason can be used only when looking critically back. Moreover, in the experimental sciences, the scientific fraternity must test a new theory

to destruction, if possible. Meanwhile, the creator of a theory may have a very lonely time, especially if his colleagues find his views of nature unfamiliar and difficult to appreciate.

The final outcome cannot be known, either to the originator of a new theory, or to his colleagues and critics, who are bent on falsifying it. Thus, the scientific innovator may feel all the more lonely and uncertain.

On the other hand, faced with a new theory, the members of the scientific establishment are often more vulnerable than the lonely innovator. For, if the innovator should happen to be right, the ensuing upheaval of the established order may be very painful and uncongenial to those who have long committed themselves to develop and serve it. Such, I believe, has been the case in the field of knowledge with which my work has been involved. Naturally I have been deeply moved, and not a little astonished, by the accidents of fortune that have brought me to this point.

Structure and function in the mitochondrion

The structure of the mitochondrion is adapted to the function it performs.

There is often a clear relationship between the structures of the parts of living organisms and the functions they perform. This can be explained in terms of natural selection and evolution. The mitochondrion can be used as an example. If mitochondrial structure varied, those organisms with the mitochondria that produced ATP most efficiently would have an advantage. They would have an increased chance of survival and would tend to produce more offspring. These offspring would inherit the type of mitochondria that produce ATP more efficiently. If this trend continued, the structure of mitochondria would gradually evolve to become more and more efficient. This is called adaptation – a change in structure so that something carries out its function more efficiently.

Examine figure 7 showing an electron micrograph of a mitochondrion and a drawing representing that mitochondrion.

The mitochondrion is a semi-autonomous organelle in that it can grow and reproduce itself but it still depends on the rest of the cell for resources and is otherwise part of the cellular system. 70S ribosomes and a naked loop of DNA are found within the mitochondrial matrix.

The mitochondrion is the site of aerobic respiration. The outer mitochondrial membrane separates the contents of the mitochondrion from the rest of the cell creating a compartment specialized for the biochemical reactions of aerobic respiration.

The inner mitochondrial membrane is the site of oxidative phosphorylation. It contains electron transport chains and ATP synthase, which carry out oxidative phosphorylation. Cristae are tubular projections of the inner membrane which increase the surface area available for oxidative phosphorylation.

The intermembrane space is the location where protons build up as a consequence of the electron transport chain. The proton build-up is used to produce ATP via the ATP synthase. The volume of the space is small, so a concentration gradient across the inner membrane can be built up rapidly.

The matrix is the site of the Krebs cycle and the link reaction. The matrix fluid contains the enzymes necessary to support these reaction systems.

🧪 Annotating a diagram of a mitochondrion

Annotation of a diagram to indicate the adaptations of a mitochondrion to its function.

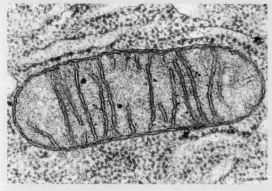

Outer mitochondrial membrane separates the contents of the mitochondrion from the rest of the cell, creating a cellular compartment with ideal conditions for aerobic respiration

Inner mitochondrial membrane contains electron transport chanins and ATP synthase

Cristae are projections of the inner membrane which increase the surface area available for oxidative phosphorylation

Matrix contains enzymes for the Krebs cycle and the link reaction

Intermembrane space Proteins are pumped into this space by the electron transport chain. The space is small so the concentration builds up quickly

Ribosomes and DNA for expression of mitochondrial genes

▲ Figure 7

Activity

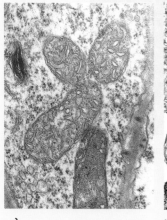

a)

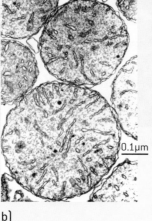

b)

0.1µm

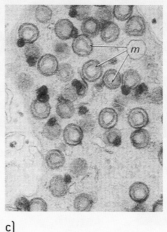

c)

m

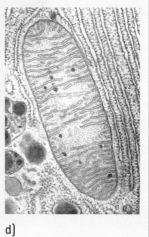

d)

▲ Figure 8 Electron micrographs of mitochondria: (a) from a bean plant (b) from mouse liver (c) from axolotl sperm (d) from bat pancreas

Study the electron micrographs in figure 8 and then answer the multiple-choice questions.

1   The fluid-filled centre of the mitochondrion is called the matrix. What separates the matrix from the cytoplasm around the mitochondrion?

   a) One wall.   **c)** Two membranes.

   b) One membrane.   **d)** One wall and one membrane.

2   The mitochondrion matrix contains 70S ribosomes, whereas the cytoplasm of eukaryotic cells contains

80S ribosomes. Which of these hypotheses is consistent with this observation?

   (i)   Protein is synthesized in the mitochondrion.

   (ii) Ribosomes in mitochondria have evolved from ribosomes in bacteria.

   (iii) Ribosomes are produced by aerobic cell respiration.

   a) (i) only   **c)** (i) and (ii)

   b) (ii) only   **d)** (i), (ii) and (iii)

Mitochondrial membranes are dynamic

Electron tomography used to produce images of active mitochondria.

Ideas in science sometimes change gradually. But sometimes they remain stable for years or even decades and then undergo a sudden change. This can be due to the insight or enthusiasm of a particular scientist, or team.

The development of new techniques can sometimes be the stimulus. The technique of electron tomography has recently allowed three-dimensional images of the interior of mitochondria to be made. One of the leaders in this field is Dr. Carmen Mannella, former Director, Division of Molecular Medicine, Wadsworth Center, Albany NY: Resource for Visualization of Biological Complexity. He recently gave this brief comment on developments in our understanding of mitochondrial structure and function.

The new take-home message about the mitochondrial inner membrane is that the cristae

are not simple infoldings but are invaginations, defining micro-compartments in the organelle. The cristae originate at narrow openings (crista junctions) that likely restrict diffusion of proteins and metabolites between the compartments. The membranes are not only very flexible but also dynamic, undergoing fusion and fission in response to changes in metabolism and physiological stimuli.

The working hypothesis is that the observed changes in membrane shape (topology) are not random and passive but rather a specific mechanism by which mitochondrial function is regulated by changes in internal diffusion pathways, e.g., allowing more efficient utilization of ADP. It appears that there are specific proteins and lipids that actively regulate the topology of the inner membrane. This is a bit speculative at the time but it gives a sense of where things are headed in the field.

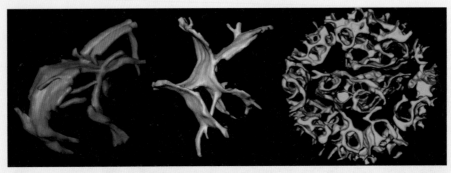

▲ Figure 9 Three images of the inner mitochondrial membrane of mitochondria from liver cells show the dynamic nature of this membrane

TOK

There are some scientific fields that depend entirely upon technology for their existence, for example, spectroscopy, radio or X-ray astronomy. What are the knowledge implications of this? Could there be problems of knowledge that are unknown now, because the technology needed to reveal them does not exist yet?

Activity

Answer the following questions with respect to the three images in figure 9.

a) The diameter of the mitochondrion was 700 nm. Calculate the magnification of the image. [3]

b) Electron tomography has shown that cristae are dynamic structures and that the volume of the intracristal compartment increases when the mitochondrion is active in electron transport. Suggest how electron transport could cause an increase in the volume of fluid inside the cristae. [2]

c) Junctions between the cristae and boundary region of the inner mitochondrial membrane can have the shape of slots or tubes and can be narrow or wide. Suggest how narrow tubular connections could help in ATP synthesis by one of the cristae in a mitochondrion. [2]

8.3 Photosynthesis

Understanding

→ Light-dependent reactions take place in the intermembrane space of the thylakoids.

→ Reduced NADP and ATP are produced in the light-dependent reactions.

→ Light-independent reactions take place in the stroma.

→ Absorption of light by photosystems generates excited electrons.

→ Photolysis of water generates electrons for use in the light-dependent reactions.

→ Transfer of excited electrons occurs between carriers in thylakoid membranes.

→ Excited electrons from Photosystem II are used to generate a proton gradient.

→ ATP synthase in thylakoids generates ATP using the proton gradient.

→ Excited electrons from Photosystem I are used to reduce NADP.

→ In the light-independent reactions a carboxylase catalyses the carboxylation of ribulose bisphosphate.

→ Glycerate 3-phosphate is reduced to triose phosphate using reduced NADP and ATP.

→ Triose phosphate is used to regenerate RuBP and produce carbohydrates.

→ Ribulose bisphosphate is reformed using ATP.

→ The structure of the chloroplast is adapted to its function in photosynthesis.

Applications

→ Calvin's experiment to elucidate the carboxylation of RuBP.

Skills

→ Annotation of a diagram to indicate the adaptations of a chloroplast to its function.

Nature of science

→ Developments in scientific research follow improvements in apparatus: sources of ^{14}C and autoradiography enabled Calvin to elucidate the pathways of carbon fixation.

Location of light-dependent reactions

Light-dependent reactions take place in the intermembrane space of the thylakoids.

Research into photosynthesis has shown that it consists of two very different parts, one of which uses light directly (light-dependent reactions) and the other does not use light directly (light-independent reactions). The light-independent reactions can only carry on in darkness for a few seconds because they depend on substances produced by the light-dependent reactions which rapidly run out.

The chloroplast has an outer membrane and an inner membrane. The inner membrane encloses a third system of interconnected membranes called the thylakoid membranes. Within the thylakoid is a compartment called the thylakoid space.

The light-dependent reactions take place in the thylakoid space and across the thylakoid membranes.

Data-based questions: Freeze-fracture images of chloroplasts

If chloroplasts are frozen rapidly in liquid nitrogen and then split, they fracture across planes of weakness. These planes of weakness are usually the centres of membranes, between the two layers of phospholipid, where there are no hydrogen bonds attracting water molecules to each other. Structures within the membrane such as the photosystems are then visible in electron micrographs (see figure 1).

1   Describe the evidence, visible in the electron micrograph, for chloroplasts having many layers of membrane.   [2]

2   Explain how photosystems become visible as lumps in freeze-fracture electron micrographs of chloroplasts.   [2]

3   Some membranes contain large particles arranged in rectangular arrays. These are Photosystem II. They have a diameter of 18 nm. Calculate the magnification of the electron micrograph.   [3]

4   Other membranes visible in the electron micrograph contain a variety of other structures. Use the information on the following pages to deduce what these are.   [3]

▲ Figure 1 Freeze-fracture electron micrograph of spinach chloroplast

---

The products of the light-dependent reactions

Reduced NADP and ATP are produced in the light-dependent reactions.

Light energy is converted into chemical energy in the form of ATP and reduced NADP in the light reacations. The ATP and reduced NADP serve as energy sources for the light-independent reactions.

The location of the light-independent reactions

Light-independent reactions take place in the stroma.

The inner membrane of the chloroplast encloses a compartment called the stroma. This is a thick protein-rich medium containing enzymes for use in the light-independent reactions, also known as the Calvin cycle. In the light-independent reactions the Calvin cycle is an anabolic pathway that requires endergonic reactions to be coupled to the hydrolysis of ATP and the oxidation of reduced NADP.

Figure 2 summarizes the processes of both the light-dependent and light-independent reactions.

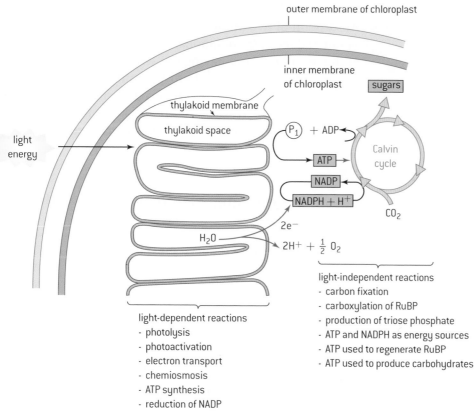

▲ Figure 2

Photoactivation

Absorption of light by photosystems generates excited electrons.

Chlorophyll and the accessory pigments are grouped together in large light-harvesting arrays called photosystems. These photosystems are located in the thylakoids, an arrangement of membranes inside the chloroplast. There are two types of light-harvesting arrays, called Photosystems I and II. In addition to light-harvesting arrays, the photosystems have reaction centres (figure 3).

Both types of photosystem contain many chlorophyll molecules, which absorb light energy and pass it to two special chlorophyll molecules in the reaction centre of the photosystem. Like other chlorophylls, when these special chlorophyll molecules absorb the energy from a photon of light an electron within the molecule becomes excited. The chlorophyll is then **photoactivated**. The chlorophylls at the reaction centre have the special property of being able to donate excited electrons to an electron acceptor.

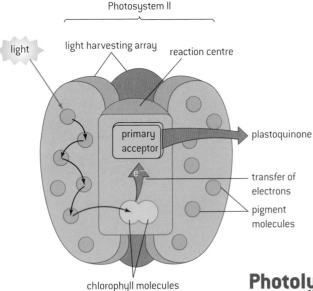

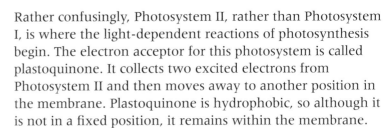

▲ Figure 3 Diagram showing the relationship between the light-harvesting array, the reaction centre and plastoquinone

Rather confusingly, Photosystem II, rather than Photosystem I, is where the light-dependent reactions of photosynthesis begin. The electron acceptor for this photosystem is called plastoquinone. It collects two excited electrons from Photosystem II and then moves away to another position in the membrane. Plastoquinone is hydrophobic, so although it is not in a fixed position, it remains within the membrane.

Absorption of two photons of light causes the production of one reduced plastoquinone, with one of the chlorophylls at the reaction centre having lost two electrons to a plastoquinone molecule. Photosystem II can repeat this process, to produce a second reduced plastoquinone, so the chlorophyll at the reaction centre has lost four electrons and two plastoquinone molecules have been reduced.

Photolysis

Photolysis of water generates electrons for use in the light-dependent reactions.

Once the plastoquinone becomes reduced, the chlorophyll in the reaction centre is then a powerful oxidizing agent and causes the water molecules nearest to it to split and give up electrons, to replace those that it has lost:

$$2H_2O \rightarrow O_2 + 4H^+ + 4e^-$$

The splitting of water, called photolysis, is how oxygen is generated in photosynthesis. Oxygen is a waste product and diffuses away. The useful product of Photosystem II is the reduced plastoquinone, which not only carries a pair of electrons, but also much of the energy absorbed from light. This energy drives all the subsequent reactions of photosynthesis.

The electron transport chain

Transfer of excited electrons occurs between carriers in thylakoid membranes.

The production of ATP, using energy derived from light is called **photophosphorylation**. It is carried out by the **thylakoids**. These are regular "stacks" of membranes, with very small fluid-filled spaces inside (see figure 4). The thylakoid membranes contain the following structures:

- Photosystem II

- ATP synthase

- a chain of electron carriers

- Photosystem I.

Reduced plastoquinone is needed, carrying the pair of excited electrons from the reaction centre of Photosystem II. Plastoquinone carries the electrons to the start of the chain of electron carriers.

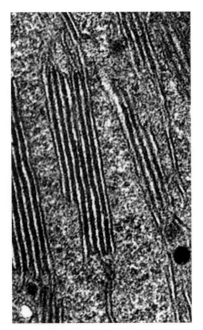

▲ Figure 4 Electron micrograph of thylakoids × 75,000

The proton gradient

Excited electrons from Photosystem II are used to generate a proton gradient.

Once plastoquinone transfers its electrons, the electrons are then passed from carrier to carrier in this chain. As the electrons pass, energy is released, which is used to pump protons across the thylakoid membrane, into the space inside the thylakoids.

A concentration gradient of protons develops across the thylakoid membrane, which is a store of potential energy. Photolysis, which takes place in the fluid inside the thylakoids, also contributes to the proton gradient.

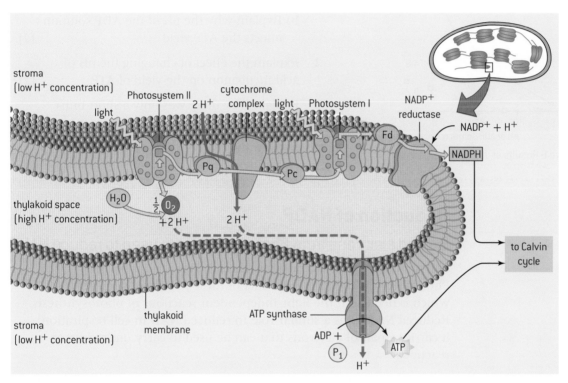

▲ Figure 5

Chemiosmosis

ATP synthase in thylakoids generates ATP using the proton gradient.

The protons can travel back across the membrane, down the concentration gradient, by passing through the enzyme ATP synthase. The energy released by the passage of protons down their concentration gradient is used to make ATP from ADP and inorganic phosphate. This method of producing ATP is strikingly similar to the process that occurs inside the mitochondrion and is given the same name: **chemiosmosis**.

When the electrons reach the end of the chain of carriers they are passed to plastocyanin, a water-soluble electron acceptor in the fluid inside the thylakoids. Reduced plastocyanin is needed in the next stage of photosynthesis.

Data-based questions: Evidence for chemiosmosis

One of the first experiments to give evidence for ATP production by chemiosmosis was performed in the summer of 1966 by André Jagendorf. Thylakoids were incubated for several hours in darkness, in acids with a pH ranging from 3.8 to 5.2. The lower the pH of an acid, the higher its concentration of protons. During the incubation, protons diffused into the space inside the thylakoids, until the concentrations inside and outside were equal. The thylakoids were then transferred, still in darkness, into a solution of ADP and phosphate that was more alkaline. There was a brief burst of ATP production by the thylakoids. The graph shows the yield of ATP at three acid incubation pHs and a range of pHs of the ADP solution.

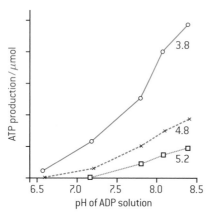

▲ Figure 6 Results of Jagendorf experiment

1 a) Describe the relationship between pH of ADP solution and ATP yield, when acid incubation was at pH 3.8. [2]

   b) Explain why the pH of the ADP solution affects the ATP yield. [2]

2 Explain the effect of changing the pH of acid incubation on the yield of ATP. [2]

3 Explain why there was only a short burst of ATP production. [2]

4 Explain the reason for performing the experiment in darkness. [2]

Reduction of NADP

Excited electrons from Photosystem I are used to reduce NADP.

The remaining parts of the light-dependent reactions involve Photosystem I. The useful product of these reactions is reduced NADP, which is needed in the light-independent reactions of photosynthesis. Reduced NADP has a similar role to reduced NAD in cell respiration: it carries a pair of electrons that can be used to carry out reduction reactions.

Chlorophyll molecules within Photosystem I absorb light energy and pass it to the special two chlorophyll molecules in the reaction centre. This raises an electron in one of the chlorophylls to a high energy level. As with Photosystem II, this is called photoactivation. The excited electron passes along a chain of carriers in Photosystem I, at the end of which it is passed to ferredoxin, a protein in the fluid outside the thylakoid. Two molecules of reduced ferredoxin are then used to reduce NADP, to form reduced NADP.

The electron that Photosystem I donated to the chain of electron carriers is replaced by an electron carried by plastocyanin. Photosystems I and II are therefore linked: electrons excited in Photosystem II are passed along the chain of carriers to plastocyanin, which transfers them to Photosystem I. The electrons are re-excited with light energy and are eventually used to reduce NADP.

The supply of NADP sometimes runs out. When this happens the electrons return to the electron transport chain that links the two photosystems, rather than being passed to NADP. As the electrons flow

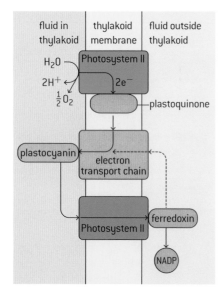

▲ Figure 7 Summary of the light-dependent reactions of photosynthesis

back along the electron transport chain to Photosystem I, they cause pumping of protons, which allows ATP production. This process is **cyclic photophosphorylation**.

Carbon fixation

In the light-independent reactions a carboxylase catalyses the carboxylation of ribulose bisphosphate.

Carbon dioxide is the carbon source for all organisms that carry out photosynthesis. The carbon fixation reaction in which it is converted into another carbon compound is arguably the most important in all living organisms. In plants and algae it occurs in the stroma – the fluid that surrounds the thylakoids in the chloroplast. The product of this carbon fixation reaction is a three-carbon compound: glycerate 3-phosphate. As so often occurs in biological research, the details of the reaction were a surprise when they were discovered. Carbon dioxide does not react with a two-carbon compound to produce glycerate 3-phosphate. Instead, it reacts with a five-carbon compound called ribulose bisphosphate (RuBP), to produce two molecules of glycerate 3-phosphate. The enzyme that catalyses this reaction is called ribulose bisphosphate carboxylase, usually abbreviated to rubisco. The stroma contains large amounts of rubisco to maximize carbon fixation.

The role of reduced NADP and ATP in the Calvin cycle

Glycerate 3-phosphate is reduced to triose phosphate using reduced NADP and ATP.

RuBP is a 5-carbon sugar derivative, but when it is converted to glycerate 3-phosphate by adding carbon and oxygen, the amount of hydrogen in relation to oxygen is reduced. In sugars and other carbohydrates, the ratio of hydrogen to oxygen is 2:1. Hydrogen has to be added to glycerate 3-phosphate by a reduction reaction to produce carbohydrate. This involves both ATP and reduced NADP, produced by the light-dependent reactions of photosynthesis. ATP provides the energy needed to perform the reduction and reduced NADP provides the hydrogen atoms. The product is a three-carbon sugar derivative, triose phosphate.

The fate of triose phosphate

Triose phosphate is used to regenerate RuBP and produce carbohydrates.

The first carbohydrate produced by the light-independent reactions of photosynthesis is triose phosphate. Two triose phosphate molecules can be combined to form hexose phosphate and hexose phosphate can be combined by condensation reactions to form starch. However, if all of the triose phosphate produced by photosynthesis was converted to hexose or starch, the supplies of RuBP in the chloroplast would soon be used up. Some triose phosphate in the chloroplast therefore has to be

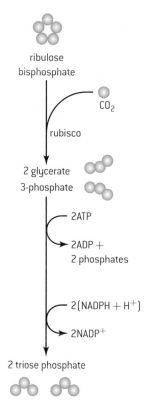

ribulose
bisphosphate

CO_2

rubisco

2 glycerate
3-phosphate

2ATP

2ADP +
2 phosphates

2(NADPH + H$^+$)

2NADP$^+$

2 triose phosphate

▲ Figure 8 Summary of carbon fixation reactions

To what extent is it acceptable to adjust empirical evidence to conform to theoretical prediction?

One of the most famous experiments in the history of biology is that of the Flemish scientist Johannes Baptista van Helmont, published in 1648. It is regarded as the first quantitative biology experiment and also changed our understanding of the growth of plants. At this time, plants were thought to be "soil-eaters". To test this idea, van Helmont put 200 pounds (90 kg) of dry soil in large pot and in it planted a willow tree, which had a mass of 5 pounds (2.2 kg). He attempted to keep dust out of the pot by covering it with a perforated metal plate. He watered the tree with rainwater or distilled water over a period of five years. When the willow was reweighed at the end of this time it had increased to 169 pounds (76 kg). After drying the soil from the pot he found that it had remained almost unchanged in mass, having lost only one eighth of a pound (about 50g). Removal of soil from willow roots is very difficult as soil particles inevitably get stuck to the roots. van Helmont's masses for the soil before and after the five-year period are therefore surprisingly close. Some have questioned whether van Helmont made his data fit pre-decided conclusions.

1 What evidence against the hypothesis that plants are soil eaters does van Helmont's experiment provide?

2 van Helmont concluded from his results that, "164 pounds of Wood, Barks, and Roots, arose out of water only." (164 pounds is 73 kg.) This was not a new idea - 2000 years earlier the Greek philosopher Thales had stated that all matter arose from water. To what extent was van Helmont's conclusion correct?

used to regenerate RuBP. This process is a conversion of 3-carbon sugars into 5-carbon sugars and it cannot be done in a single step. Instead a series of reactions take place.

As RuBP is both consumed and produced in the light-independent reactions of photosynthesis, these reactions form a cycle. It is called the Calvin cycle to honour Melvin Calvin, who was given the Nobel Prize for Chemistry in 1961 for his work in elucidating this process. For the Calvin cycle to continue indefinitely, as much RuBP must be produced as consumed. If three RUBP molecules are used, six triose phosphates are produced. Five of these are needed to regenerate the three RuBP molecules. This leaves just one triose phosphate for conversion to hexose, starch or other products of photosynthesis. To produce one molecule of glucose for example, six turns of the Calvin cycle are needed, each of which contributes one of the fixed carbon atoms in the glucose.

Data-based questions: The effect of light and dark on carbon dioxide fixation

One of the pioneers of photosynthesis research was James Bassham. The results of one of his experiments are shown in figure 9. Concentrations of ribulose bisphosphate and glycerate 3-phosphate were monitored in a culture of cells of the alga, *Scenedesmus*. The algae were kept in bright light and then in the dark.

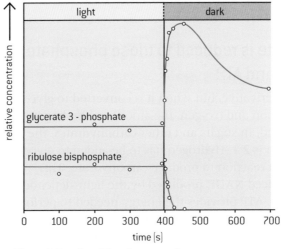

▲ Figure 9 Results of Bassham experiment

1 Compare the effects of the dark period on the concentrations of ribulose bisphosphate and glycerate 3-phosphate. [2]

2 Explain the change that took place in the 25 seconds after the start of darkness, to the concentration of:
   a) glycerate 3-phosphate [3]
   b) ribulose bisphosphate. [1]

3 Predict what the effect would be of turning the light back on after the period of darkness. [2]

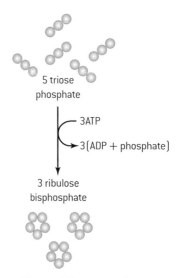

4 Predict the effect of reducing the carbon dioxide concentration from 1.0% to 0.003%, instead of changing from light to darkness:

a) on glycerate 3-phosphate concentration [2]

b) on ribulose bisphosphate concentration. [2]

5 triose phosphate

3ATP

3(ADP + phosphate)

3 ribulose bisphosphate

▲ Figure 10 Summary of RuBP regeneration

RuBP regeneration

Ribulose bisphosphate is reformed using ATP.

In the last phase of the Calvin cycle, a series of enzyme-catalysed reactions convert triose phosphate molecules into RuBP. After the RuBP is regenerated, it can serve to fix CO_2 and begin the cycle again. Figure 10 summarizes the regeneration process.

 Calvin's lollipop apparatus

Developments in scientific research follow improvements in apparatus: sources of ^{14}C and autoradiography enabled Calvin to elucidate the pathways of carbon fixation.

Sometimes progress in biological research suddenly becomes possible because of other discoveries. Martin Kamen and Samuel Ruben discovered ^{14}C in 1945. The half-life of this radioactive isotope of carbon makes it ideal for use in tracing the pathways of photosynthesis. Figure 11 shows apparatus used by Melvin Calvin and his team. At the start of their experiment, they replaced the $^{12}CO_2$ supplied to algae with $^{14}CO_2$. They took samples of the algae at very short time intervals and found what carbon compounds in the algae contained radioactive ^{14}C. The results are shown in figure 12. The amount of radioactivity of each carbon compound is shown as a percentage of the total amount of radioactivity.

1 Explain the evidence from the graph that convinced Calvin that glycerate 3-phosphate is the first product of carbon dioxide fixation. [4]

2 Explain the evidence from the graph for the conversion of glycerate 3-phosphate to triose phosphate and other sugar phosphates. [4]

3 Using the data in the graph, estimate how rapidly carbon dioxide can diffuse into cells and be converted with RuBP to glycerate 3-phosphate. [2]

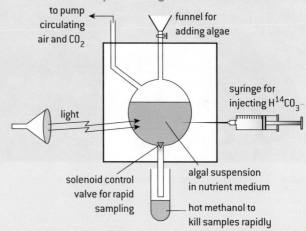

▲ Figure 11 Calvin's lollipop apparatus

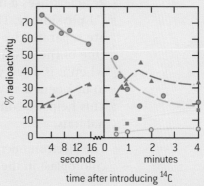

● glycerate-3-phosphate
▲ triose phosphate and other sugar phosphates
■ malate and aspartate
○ alanine

▲ Figure 12 Graph showing Calvin's results

Chloroplast structure and function

The structure of the chloroplast is adapted to its function in photosynthesis.

Chloroplasts are quite variable in structure but share certain features:

- a double membrane forming the outer **chloroplast envelope**

- an extensive system of internal membranes called **thylakoids**, which are an intense green colour

- small fluid-filled spaces inside the thylakoids

- a colourless fluid around the thylakoids called **stroma** that contains many different enzymes.

- In most chloroplasts there are stacks of thylakoids, called **grana**. If a chloroplast has been photosynthesizing rapidly then there may be **starch grains** or **lipid droplets** in the stroma.

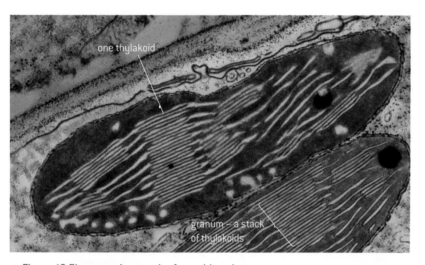

▲ Figure 13 Electron micrograph of pea chloroplast

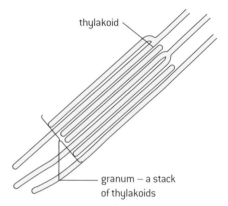

▲ Figure 14 Drawing of part of the pea chloroplast to show the arrangement of the thylakoid membranes

Data-based questions: Photosynthesis in *Zea mays*

Zea mays uses a modified version of photosynthesis, referred to as C4 physiology. The processes of photolysis and the Calvin cycle are separated by being carried out in different types of chloroplast. One of the advantages is that carbon dioxide can be fixed even when it is at very low concentrations, so the stomata do not need to be opened as widely as in plants that do not have C_4 physiology. This helps to conserve water in the plant, so is useful in dry habitats.

The electron micrograph (figure 15) shows the two types of chloroplast in the leaves of *Zea mays*.

One type (Chloroplast X) is from mesophyll tissue and the other (Chloroplast Y) is from the sheath of cells around the vascular tissue that transports materials to and from the leaf.

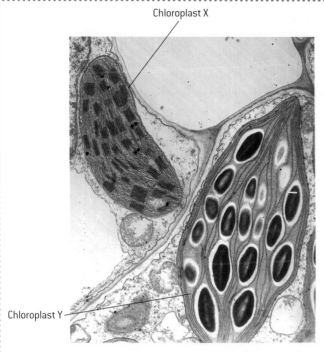

Chloroplast X

Chloroplast Y

▲ Figure 15 Two types of chloroplast in *Zea mays* leaf

1 Draw a small portion of each chloroplast to show its structure. [5]

2 Compare the structure of the two types of chloroplast. [4]

3 Deduce, with a reason:
   a) which type of chloroplast has the greater light absorption capacity [2]
   b) which is the only type of chloroplast to carry out the reactions of the Calvin cycle [2]
   c) which is the only type of chloroplast to produce oxygen. [2]

⚗ Diagram showing chloroplast structure– function relationship

Annotation of a diagram to indicate the adaptations of a chloroplast to its function.

There is a clear relationship between the structure of the chloroplast and its function.

1 **Chloroplasts absorb light**. Pigment molecules, arranged in photosystems in the thylakoid membranes, carry out light absorption. The large area of thylakoid membranes ensures that the chloroplast has a large light-absorbing capacity. The thylakoids are often arranged in stacks called grana. Leaves that are brightly illuminated typically have chloroplasts with deep grana, which allow more light to be absorbed.

2 **Chloroplasts produce ATP by photophosphorylation**. A proton gradient is needed. This develops between the inside and

outside of the thylakoids. The volume of fluid inside the thylakoids is very small, so when protons are pumped in, a proton gradient develops after relatively few photons of light have been absorbed, allowing ATP synthesis to begin.

3   **Chloroplasts carry out the many chemical reactions of the Calvin cycle**. The stroma is a compartment of the plant cell in which the enzymes needed for the Calvin cycle are kept together with their substrates and products. This concentration of enzymes and substrates speeds up the whole Calvin cycle. ATP and reduced NADP, needed for the Calvin cycle, are easily available because the thylakoids, where they are produced, are distributed throughout the stroma.

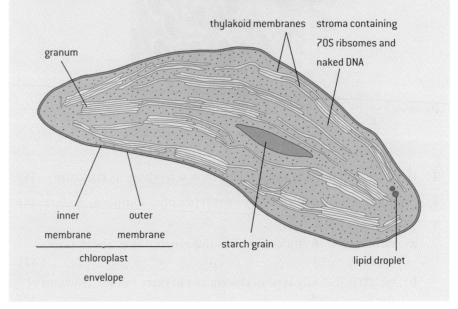

Questions

1 a) State the meaning of the term "metabolic pathway". [2]

Glucose phosphate (G6P) is converted to pyruvate in one of the metabolic pathways of cell respiration. This process happens whether oxygen is available or not.

Figure 16 shows the concentrations of the intermediates of this pathway in rat heart tissue. The concentrations are shown as a percentage of the concentrations in the heart when it has been starved of oxygen.

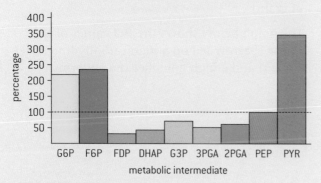

▲ Figure 16

b) Compared with concentrations during oxygen starvation, state which metabolic intermediate:

(i) increased in concentration most [1]

(ii) decreased in concentration most [1]

(iii) did not change in concentration. [1]

c) (i) The concentrations shown in Figure 16 suggest that the rate of this metabolic pathway has been greater than is needed by the heart cells. Explain how the data in the bar chart shows this. [2]

(ii) Because rate of the pathway has been greater than necessary, the enzyme catalysing one of the reactions in the pathway has been inhibited. Deduce which reaction this enzyme catalyses, giving reasons for your answer. [3]

2 Water with mineral nutrients dissolved in it was sterilized and then placed in a 2 dm³ fermenter. The temperature was kept at 25°C. The fermenter was kept in natural sunlight, but a lamp was also used to increase the light intensity. The lamp was controlled by an electronic timer, which switched it off at night. A light meter was placed against the side of the fermenter, near the base, to measure the intensity of light passing through the liquid in the fermenter. The maximum reading it could give was 1,200 lux. At the start of the experiment, a small quantity of *Chlorella*, a type of algae, was added to the fluid in the fermenter. Figure 17 shows the light intensity measured over the 45 days of the experiment.

a) The light intensity followed a similar pattern, every day from Day 12 onwards.

(i) Outline the daily changes in light intensity over a typical day after Day 12. [2]

(ii) Explain these daily changes in light intensity. [2]

b) Each day there is a maximum light intensity. Outline the trends in maximum light intensity.

(i) from Day 1 to Day 12 [1]

(ii) from Day 13 to Day 38 [1]

(iii) from Day 39 to Day 45. [1]

c) Explain why the light intensity when the light was switched on was lower at the end of the experiment than at the start. [3]

d) Suggest reasons for the trend in maximum daily light intensity between Day 39 and Day 45. [3]

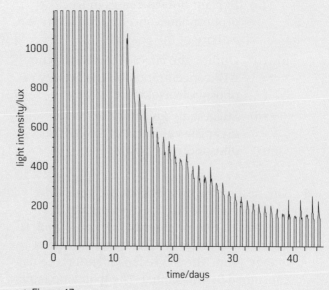

▲ Figure 17

401

3   At the start of glycolysis, glucose is phosphorylated to produce glucose 6-phosphate, which is converted into fructose 6-phosphate. A second phosphorylation reaction is then carried out, in which fructose 6-phosphate is converted into fructose 1,6-bisphosphate. This reaction is catalysed by the enzyme phosphofructokinase. Biochemists measured the enzyme activity of phosphofructokinase (the rate at which it catalysed the reaction) at different concentrations of fructose 6-phosphate. The enzyme activity was measured with a low concentration of ATP and a high concentration of ATP in the reaction mixture. The graph below shows the results.

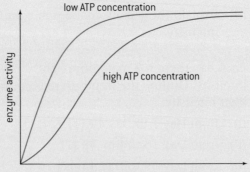

a) (i) Using **only** the data in the above graph, outline the effect of increasing fructose 6-phosphate concentration on the activity of phosphofructokinase, at a low ATP concentration.   [2]

   (ii) Explain how increases in fructose 6-phosphate concentration affect the activity of the enzyme.   [2]

b) (i) Outline the effect of increasing the ATP concentration on the activity of phosphofructokinase.   [2]

   (ii) Suggest an advantage to living organisms of the effect of ATP on phosphofructokinase.   [1]

4   The respiratory quotient (RQ) is a measure of the metabolic activity of an animal. It is the ratio of CO_2 produced to O_2 consumed. In general, the lower the RQ value the higher the energy yield. The RQ is dependent on the diet consumed by the animal. The following table lists the typical RQ values for specified diets.

Diet	RQ
Lipid	0.71
Carbohydrate	1.00
Protein	0.74

Source: Walsberg and Wolf, *Journal of Experimental Biology*, (1995), **198**, pages 213–219. Reproduced by permission of The Company of Biologists Ltd.

In an experiment to assess RQ values for house sparrows, the birds were fed a diet of pure mealworms (beetle larvae) or millet (a type of grain).

The graph below shows the RQ values of a house sparrow fed on a high carbohydrate diet (millet) and a high lipid diet (mealworms).

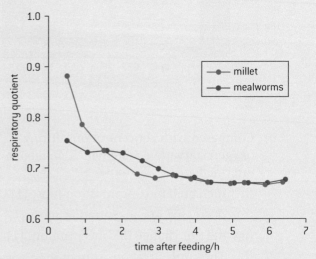

Source: Walsberg and Wolf, *Journal of Experimental Biology*, (1995), **198**, pages 213–219. Reproduced by permission of The Company of Biologists Ltd.

a) Compare the RQ values for millet and mealworms between 1 hour and 6 hours after feeding.   [2]

   The expected RQ value for house sparrows metabolizing millet is 0.93. The expected value when metabolizing mealworms is 0.75.

b) Explain why the expected RQ values for millet and mealworms are different.   [2]

c) Suggest reasons for

   (i) the high initial RQ values for house sparrows fed on millet;   [1]

   (ii) the rapid fall in RQ values for house sparrows fed on millet.   [1]

9 PLANT BIOLOGY (AHL)

Introduction

Plants are highly diverse in structure and physiology. They act as the producers in almost all terrestrial ecosystems. Structure and function are correlated in the xylem and phloem of plants.

Plants have sophisticated methods of adapting their growth to environmental conditions. Reproduction in flowering plants is influenced by both the biotic and abiotic environment.

9.1 Transport in the xylem of plants

Understanding

→ Transpiration is the inevitable consequence of gas exchange in the leaf.

→ Plants transport water from the roots to the leaves to replace losses from transpiration.

→ The cohesive property of water and the structure of the xylem vessels allow transport under tension.

→ The adhesive property of water and evaporation generate tension forces in leaf cell walls.

→ Active uptake of mineral ions in the roots causes absorption of water by osmosis.

Applications

→ Adaptations of plants in deserts and in saline soils for water conservation.

→ Models of water transport in xylem using simple apparatus including blotting or filter paper, porous pots and capillary tubing.

Skills

→ Drawing the structure of primary xylem vessels in sections of stems based on microscope images.

→ Measurement of transpiration rates using potometers. (Practical 7)

→ Design of an experiment to test hypotheses about the effect of temperature or humidity on transpiration rates.

Nature of science

→ Use models as representations of the real world: mechanisms involved in water transport in the xylem can be investigated using apparatus and materials that show similarities in structure to plant tissues.

Transpiration

Transpiration is the inevitable consequence of gas exchange in the leaf.

Plant leaves are the primary organ of photosynthesis. Photosynthesis involves the synthesis of carbohydrates using light energy. Carbon dioxide is used as a raw material. Oxygen is produced as a waste product. Exchange of these two gases must take place to sustain photosynthesis.

Absorption of carbon dioxide is essential for photosynthesis and the waxy cuticle has very low permeability to it, so pores through the epidermis are needed. These pores are called stomata. Figure 1 shows that the problem for plants is that if stomata allow carbon dioxide to be absorbed, they will usually also allow water vapour to escape.

This is an intractable problem for plants and other organisms: having gas exchange without water loss. The loss of water vapour from the leaves and stems of plants is called transpiration.

Plants minimize water losses through stomata using **guard cells**. These are the cells that are found in pairs, one on either side of a stoma. Guard cells control the aperture of the stoma and can adjust from wide open to fully closed. Stomata are found in nearly all groups of land plants for at least part of the plant's life cycle. The exception is a group called the liverworts.

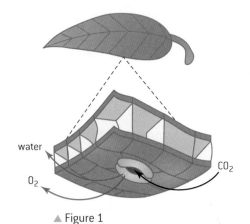

water

O_2

CO_2

▲ Figure 1

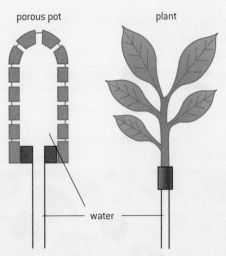

🌐 Modelling water transport

Models of water transport in xylem using simple apparatus including blotting or filter paper, porous pots and capillary tubing.

porous pot

plant

water

▲ Figure 2 Porous pots can be used to model evaporation from leaves. Water fills pores within the pot demonstrating adhesion to the clay molecules within the pot. As the water is drawn into the pot, cohesion causes water molecules to be drawn up the glass tubing

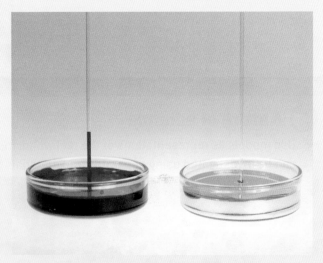

▲ Figure 3 Capillary tubes dipped into water with dye and mercury. Unlike water, there is no adhesion of mercury to the glass nor is there cohesion between mercury atoms, so the mercury does not climb the glass

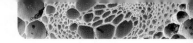

▲ Figure 4 The ability of adhesive forces to result in the movement of water is demonstrated in this image. A folded paper towel with one end immersed in water will transport water into an empty container by capillary action

 Using a potometer

Measurement of transpiration rates using potometers. (Practical 7)

Mechanisms involved in water transport in the xylem can be investigated using apparatus and materials that show similarities in structure to plant tissues.

Figure 5 shows a potometer. This is a device used to measure water uptake in plants. The apparatus consists of a leafy shoot in a tube (right), a reservoir (left of shoot), and a graduated capillary tube (horizontal). A bubble in the capillary tube marks the zero point. As the plant takes up water through its roots, the bubble will move along the capillary tube. The progress of the bubble is being timed here, along with noting the distance travelled. The tap below the reservoir allows the bubble to be reset to carry out new measurements.

▲ Figure 5

 Effect of humidity on transpiration

Design of an experiment to test hypotheses about the effect of temperature or humidity on transpiration rates.

The rate of transpiration is difficult to measure directly. Instead, the rate of water uptake is usually measured, using a potometer. Figure 6 shows one type of potometer.

To design an investigation you will need to discuss the following questions.

1   How will you measure the rate of transpiration in your investigation?

2 What biotic or abiotic factor will you investigate?

3 How will you vary the level of this factor?

4 How many results do you need, at each level of the factor that you are varying?

5 How will you keep other factors constant, so that they do not affect the rate of transpiration?

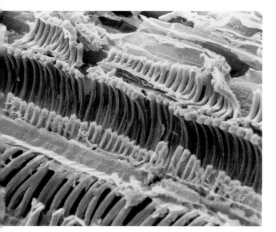

▲ Figure 7 Longitudinal section through a rhubarb stem, *Rheum rhaponticum*. Cut xylem vessels are coloured brown. Xylem vessels are reinforced and strengthened with spiral bands of lignin. Spiral bands allow xylem vessels to elongate and grow lengthwise

fresh shoot, cut under water and transferred to apparatus under water to avoid introducing air bubbles

reservoir from which water can be let into thecapillary tube, pushing the air bubble back to the start of the tube

tap

air tight seal

capillary tube

scale calibrated in mm³

air bubble moves along tube as water is absorbed by shoot

▲ Figure 6 Diagram of a potometer

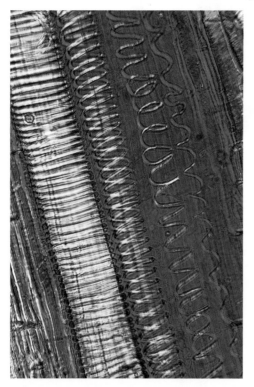

▲ Figure 8 Light micrograph of a vertical section of the primary wood or xylem of a tree showing wood vessels with lignified supporting thickenings

Xylem structure helps withstand low pressure

The cohesive property of water and the structure of the xylem vessels allow transport under tension.

The structure of xylem vessels allows them to transport water inside plants very efficiently. Xylem vessels are long continuous tubes. Their walls are thickened, and the thickenings are impregnated with a polymer called lignin. This strengthens the walls, so that they can withstand very low pressures without collapsing.

Xylem vessels are formed from files of cells, arranged end-to-end. In flowering plants, the cell wall material in some areas between adjacent cells in the file is largely removed and the plasma membranes and contents of the cells break down (see figures 7 and 8). When mature, these xylem cells are nonliving, so the flow of water along them must be a passive process. The pressure inside xylem vessels is usually much lower than atmospheric pressure but the rigid structure prevents the xylem vessels from collapsing.

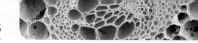

Water molecules are polar and the partial negative charge on the oxygen atom in one water molecule attracts the hydrogen atom in a neighbouring water molecule. This is termed cohesion. Water is also attracted to hydrophilic parts of the cell walls of xylem. This is termed adhesion. As a result of the connections between the molecules, water can be pulled up from the xylem in a continuous stream.

Data-based questions: The Renner experiment

Figure 9 shows the results of an experiment by the German plant physiologist Otto Renner in 1912. A transpiring woody shoot was placed in a potometer and the rate of water uptake was measured. A clamp was attached to the stem to restrict the flow of water up to the leaves. Later on, the top of the shoot, with all of its leaves, was removed. A vacuum pump was then attached to the top of the shoot.

Questions

1 Describe the effect of clamping the stem on the rate of water uptake. [3]

2 Explain the effect of cutting off the top of the shoot on the rate of water uptake. [3]

3 Calculate the difference between the rate of water uptake caused by the vacuum pump and the rate caused by the leaves immediately before the shoot top was cut off. [2]

4 The water in the potometer was at atmospheric pressure. The vacuum pump generated a pressure of zero. Discuss what the results of the experiment showed about the pressures generated in the xylem by the leaves of the shoot. [2]

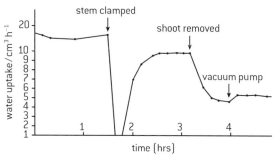

▲ Figure 9 Results of the Renner experiment

Tension in leaf cell walls maintains the transpiration stream

The adhesive property of water and evaporation generate tension forces in leaf cell walls.

When water evaporates from the surface of the wall in a leaf, adhesion causes water to be drawn through the cell wall from the nearest available supply to replace the water lost by evaporation. The nearest available supply is the xylem vessels in the veins of the leaf.

Even if the pressure in the xylem is already low, the force of adhesion between water and cell walls in the leaf is strong enough to suck water out of the xylem, further reducing its pressure.

The low pressure generates a pulling force that is transmitted through the water in the xylem vessels down the stem and to the ends of the xylem in the roots. This is called transpiration-pull and is strong enough to move water upwards, against the force of gravity, to the top of the tallest tree. For the plant, it is a passive process, with all the energy needed for it coming from the thermal energy (heat) that causes transpiration. The pulling of water upwards in xylem vessels depends on the cohesion that exists between water molecules. Many liquids would be unable to resist the very low pressures in xylem vessels and the column of liquid would break. This is called cavitation and it does occasionally happen even with water, but it is unusual. Even though water is a liquid, it can transmit pulling forces in the same way as a solid length of rope does.

Active transport of minerals in the roots

Active uptake of mineral ions in the roots causes absorption of water by osmosis.

Water is absorbed into root cells by osmosis. This happens because the solute concentration inside the root cells is greater than that in the water in the soil. Most of the solutes in both the root cells and the soil are mineral ions. The concentrations of mineral ions in the root can be 100 or more times higher than those in the soil. These concentration gradients are established by active transport, using protein pumps in the plasma membranes of root cells. There are separate pumps for each type of ion that the plant requires. Mineral ions can only be absorbed by active transport if they make contact with an appropriate pump protein. This can occur by diffusion, or by mass flow when water carrying the ions drains through the soil.

Some ions move through the soil very slowly because the ions bind to the surface of soil particles. To overcome this problem, certain plants have developed a relationship with a fungus. The fungus grows on the surface of the roots and sometimes even into the cells of the root. The thread-like hyphae of the fungus grow out into the soil and absorb mineral ions such as phosphate from the surface of soil particles. These ions are supplied to the roots, allowing the plant to grow successfully in mineral-deficient soils. This relationship is found in many trees, in members of the heather

---

Data-based questions: Fungal hyphae and mineral ion absorption

Figure 10 shows the results of an experiment in which seedlings of Sitka spruce, *Picea sitchensis*, were grown for 6 months in sterilized soil either with or without fungi added: C was the control with no fungi added. The species of fungi added were:

I = *Laccaria laccata*; II = *Laccaria ameythestea*;
III = *Thelophora terrestris* from a tree nursery;
IV = *Thelophora terrestris* from a forest;
V = *Paxillus involutus*; VI = *Pisolithus tinctorius*.

1 a) Discuss the effects of the five species of fungi on the growth of the roots and shoots of the tree seedlings. [4]

   b) Explain the effects of the fungi on the growth of tree seedlings. [2]

2 a) State the relationship between root growth and shoot growth in the tree seedlings. [1]

   b) Suggest a reason for the relationship. [1]

   c) Using the data in Figure 10, deduce whether the effects of closely related fungi on tree growth are the same. [2]

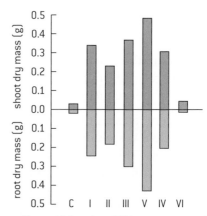

▲ Figure 10 Results of Sitka spruce experiment

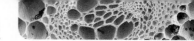

family and in orchids. Most, but not all, of these plants supply sugars and other nutrients to the fungus, so both the fungus and the plant benefit. This is an example of a mutualistic relationship.

Replacing losses from transpiration

Plants transport water from roots to leaves to replace losses from transpiration.

The movement of water from roots to leaves is summarized in figure 11. Water leaving through stomata by transpiration is replaced by water from xylem. Water in the xylem climbs the stem through the pull of transpiration combined with the forces of adhesion and cohesion. Water moves from soil into roots by osmosis due to the active transport of minerals into the roots. Once the water is in the root it travels to the xylem through cell walls (the apoplast pathway) and through cytoplasm (the symplast pathway).

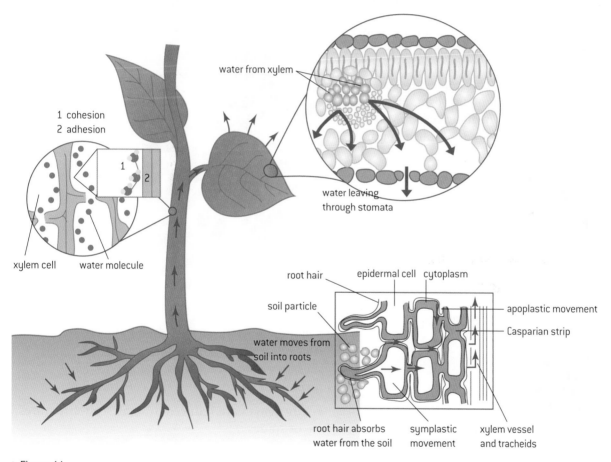

▲ Figure 11

Adaptations for water conservation

Adaptation of plants in deserts and in saline soils for water conservation.

Xerophytes are plants adapted to growing in deserts and other dry habitats. There are various strategies that plants can use to survive in these habitats, including increasing the rate of water uptake from the soil and reducing the rate of water loss by transpiration. Some xerophytes are ephemeral, with a very short life cycle that is completed in the brief period when water is available after rainfall. They then remain dormant as embryos inside seeds until the next rains, sometimes years later. Other plants are perennial and rely on storage of water in specialized leaves, stems or roots.

Most cacti are xerophytes, with leaves that are so reduced in size that they usually only consist of spines. The stems contain water storage tissue and become swollen after rainfall. Pleats allow the stem to expand and contract in volume rapidly. The epidermis of cactus stems has a thick waxy cuticle and unlike most plant stems there are stomata, though they are spaced more widely than in leaves. The stomata usually open at night rather than in the day, when it is much cooler and transpiration occurs more slowly. Carbon dioxide is absorbed at night and stored in the form of a four-carbon compound, malic acid. Carbon dioxide is released from the malic acid during the day, allowing photosynthesis even with the stomata closed. This is called Crassulacean acid metabolism. Plants such as cacti that use this system are called CAM plants. C_4 physiology also helps to reduce transpiration.

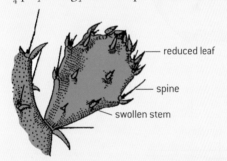

Gymnocalycium baldianum (cactus) viewed from above

10 mm

Euphorbia obesa viewed from above

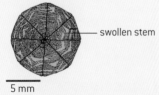

5 mm

▲ Figure 12 Xerophytes

Cacti are native plants of North and South America. Xerophytes in other parts of the world belong to different plant families. The adaptations

in these xerophytes are often very similar to those of cacti. Some African species of *Euphorbia* for example, are difficult to distinguish from cacti until they produce flowers.

Marram Grass (*Ammophila arenaria*) is a xerophyte, i.e. it is a plant adapted for dry conditions. It has a rolled leaf. This creates a localized environment of water vapour which helps to prevent losses of water. The stomata sit in small pits within the curls of the structure, which make them less likely to open and to lose water. The folded leaves have hairs on the inside to slow or stop air movement, much like many other xerophytes. This slowing of air movement once again reduces the amount of water vapour being lost.

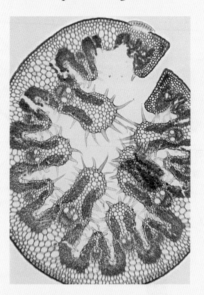

▲ Figure 13

Saline soils are those that contain high concentrations of salts. Plants that live in saline soils are called halophytes. Halophytes have several adaptations for water conservation:

- the leaves are reduced to small scaly structures or spines

- the leaves are shed when water is scarce and the stem becomes green and takes over the function of photosynthesis when the leaves are absent

- water storage structures develop in the leaves

- they have a thick cuticle and a multiple layered epidermis

- they have sunken stomata

- they have long roots, which go in search of water

- they have structures for removing salt build-up.

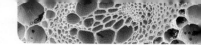

Drawing xylem vessels

Drawing the structure of primary xylem vessels in sections of stems based on microscope images.

Primary xylem vessels are visible in cross sections of young stems such as in young *Helianthus*. Figure 16 shows a longitudinal section through a stem illustrating the structure of xylem. Primary xylem has a thin primary wall that is unlignified and freely permeable, plus lignified secondary thickening of the wall that is usually annular or helical. The thickening allows the xylem vessel to continue growing in length because the rings of annular thickening can move further apart or helical thickening can be stretched so the pitch of the helix is greater.

Once extension growth of a root or stem is complete the plant produces secondary xylem which is much more extensively lignified. Secondary thickening of its cell wall provides more strength but does not allow growth in length.

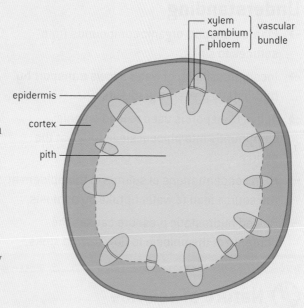

▲ Figure 14

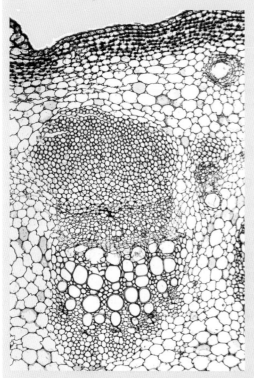

▲ Figure 15 Light micrograph of a section through a young stem from a sunflower (*Helianthus annuus*), showing one of the many vascular bundles. The vascular bundles have an outer layer of sclerenchyma tissue (crimson). Next is the phloem (dark blue) with phloem tubes, parenchyma and companion cells. Then the xylem (red) and at the end of the xylem are patches of fibres (red). In between the phloem and xylem is the cambium (light blue)

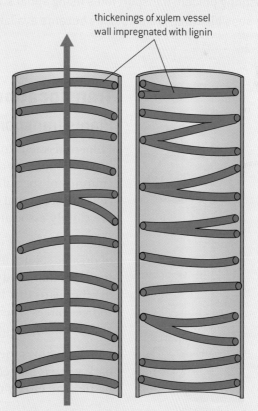

thickenings of xylem vessel wall impregnated with lignin

continuous tubular structure

▲ Figure 16 Structure of xylem vessels

9.2 Transport in the phloem of plants

Understanding

→ Plants transport organic compounds from sources to sinks.

→ Incompressibility of water allows transport by hydrostatic pressure gradients.

→ Active transport is used to load organic compounds into phloem sieve tubes at the source.

→ High concentrations of solutes in the phloem at the source lead to water uptake by osmosis.

→ Raised hydrostatic pressure causes the contents of the phloem to flow toward sinks.

Applications

→ Structure–function relationships of phloem sieve tubes.

Skills

→ Analysis of data from experiments measuring phloem transport rates using aphid stylets and radioactively-labelled carbon dioxide.

→ Identification of xylem and phloem in microscope images of stem and root.

Nature of science

→ Developments in scientific research follow improvements in apparatus: experimental methods for measuring phloem transport rates using aphid stylets and radioactively-labelled carbon dioxide were only possible when radioisotopes became available.

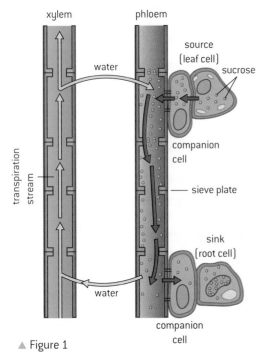

▲ Figure 1

Translocation occurs from source to sink

Plants transport organic compounds from sources to sinks.

Phloem tissue is found throughout plants, including the stems, roots and leaves. Phloem is composed of sieve tubes. Sieve tubes are composed of columns of specialized cells called sieve tube cells. Individual sieve tube cells are separated by perforated walls called sieve plates. Sieve tube cells are closely associated with companion cells (figure 1).

Phloem transports organic compounds throughout the plant. The transport of organic solutes in a plant is called **translocation**. Phloem links parts of the plant that need a supply of sugars and other solutes such as amino acids to other parts that have a surplus. Table 1 classifies parts of the plant into sources (areas where sugars and amino acids are loaded into the phloem) and sinks (where the sugars and amino acids are unloaded and used).

Figure 2 shows the results of a simple experiment in which two rings of bark were removed from an apple tree. The bark contains the phloem tissue. The effects on apple growth are clearly visible.

Sometimes sinks turn into sources, or vice versa. For this reason the tubes in phloem must be able to transport biochemicals in either direction and, unlike the blood system of animals, there are no valves or central pump in phloem. However there are similarities between transport in phloem and blood vessels: in both systems a fluid flows inside tubes because of pressure gradients. Energy is needed to generate the pressures, so the flow of blood and the movement of phloem sap are both active processes.

Sources	Sinks
Photosynthetic tissues: • mature green leaves • green stems. Storage organs that are unloading their stores: • storage tissues in germinating seeds • tap roots or tubers at the start of the growth season.	Roots that are growing or absorbing mineral ions using energy from cell respiration. Parts of the plant that are growing or developing food stores: • developing fruits • developing seeds • growing leaves • developing tap roots or tubers.

▲ Table 1

Phloem loading

Active transport is used to load organic compounds into phloem sieve tubes at the source.

The data in table 2 indicates that sucrose is transported in the phloem. Sucrose is the most prevalent solute in phloem sap. Sucrose is not as readily available for plant tissues to metabolize directly in respiration and therefore makes a good transport form of carbohydrate as it will not be metabolized during transport.

Plants differ in the mechanism by which they bring sugars into the phloem, a process called phloem loading. In some species, a significant amount travels through cell walls from mesophyll cells to the cell walls of companion cells, and sometimes sieve cells, where a sucrose transport protein then actively transports the sugar in. This is referred to as the apoplast route.

In this case, a concentration gradient of sucrose is established by active transport. Figure 3 shows that this is achieved by a mechanism whereby H^+ ions are actively transported out of the companion cell from surrounding tissues using ATP as an energy source. The build-up of H^+ then flows down its concentration gradient through a co-transport protein. The energy released is used to carry sucrose into the companion cell-sieve tube complex.

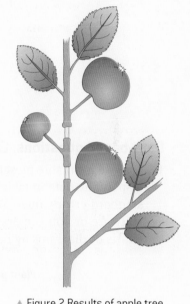

▲ Figure 2 Results of apple tree ringing experiment

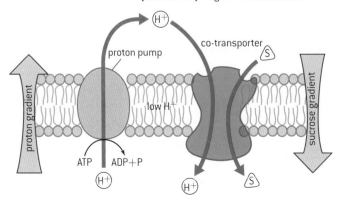

[outside cell] - high H^+ concentration

proton pump

co-transporter

proton gradient

low H^+

sucrose gradient

ATP   ADP+P

[inside cell] - low H^+ concentration

▲ Figure 3 Movement of sucrose (S) across a sieve tube membrane

In other species, much of the sucrose travels between cells through connections between cells called plasmodesmata (singular plasmodesma). This is referred to as the symplast route. Once the sucrose reaches the companion cell it is converted to an oligosaccharide to maintain the sucrose concentration gradient.

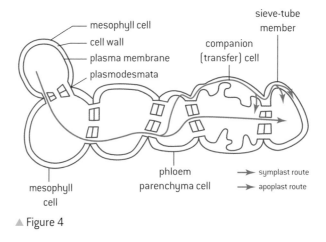

▲ Figure 4

Data-based questions: Carbohydrates in cyclamen

1   Choose a suitable presentation format to display the data in table 2, including the standard error values. You can use graphing software or you can draw graphs, tables, charts or diagrams by hand.

2   Describe the trends in the data and suggest reasons for them based on your knowledge of photosynthesis, the structure of disaccharides and polysaccharides and the transport and storage of carbohydrates in plants.

Plant part	Mean carbohydrate content (μg g^{-1} fresh mass $\pm$ standard error of mean)			
	sucrose	glucose	fructose	starch
Leaf blade	1,312 $\pm$ 212	210 $\pm$ 88	494 $\pm$ 653	62 $\pm$ 25
Vascular bundle in the leaf stalk, consisting of xylem and phloem	5,757 $\pm$ 1,190	479 $\pm$ 280	1,303 $\pm$ 879	<18
Tissue surrounding the vascular bundle in the leaf stalk	417 $\pm$ 96	624 $\pm$ 714	1,236 $\pm$ 1,015	<18
Buds, roots and tubers (underground storage organs)	2,260 $\pm$ 926	120 $\pm$ 41	370 $\pm$ 242	152 $\pm$ 242

▲ Table 2

Pressure and water potential differences play a role in translocation

Incompressibility of water allows transport by hydrostatic pressure gradients.

The build up of sucrose and other carbohydrates draws water into the companion cell through osmosis. The rigid cell walls combined with the incompressibility of water result in a build-up of pressure. Water will flow from this area of high pressure to an area of low pressure.

At the sink end, sucrose is withdrawn from the phloem and either utilized as an energy source for such processes as growth or converted to starch. In either case, the loss of solute causes a reduction in osmotic pressure and the water that carried the solute to the sink is then drawn back in to the transpiration stream in the xylem.

Data-based questions: Explaining water movement

Water potential is a measure of the tendency of water to move from one area to another. It is represented by the variable Ψ_w. It is defined as the sum of solute potential and pressure potential.

Pure water has a solute potential, Ψ_s, of zero. Once solute is added, the value of the solute potential becomes more negative. The more negative the solute potential, the more likely water will be drawn from another area with higher solute potential; i.e., lower solute concentration.

Pressure potential (Ψ_p) in a plant cell is the pressure exerted by the rigid cell wall that limits further intake of water. In a plant cell, pressure exerted by the rigid cell wall limits further water uptake despite solute potential differences.

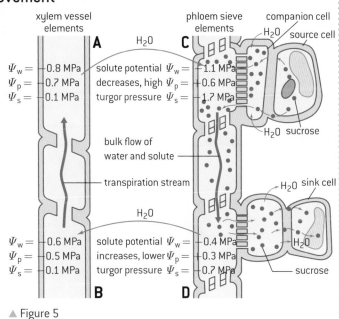

▲ Figure 5

1 Explain the movement of water from point A to point C. [3]

2 Explain the movement of water from point C to point D. [3]

3 Explain the movement of water from point D to point B. [3]

4 Explain the movement of water from point B to point A. [3]

🌐 Phloem sieve tubes

Structure–function relationship in phloem sieve tubes.

The functions of phloem include loading of carbohydrates; transport of the carbohydrates sometimes over long distances; and unloading of the carbohydrates at sinks.

Phloem is composed of sieve tubes. Sieve tubes are composed of columns of specialized cells called sieve tube cells. Unlike the vascular elements of xylem, sieve tube elements are living, though they do have reduced quantities of cytoplasm and no nucleus. One reason that sieve cells need to be living is that they depend on the membrane to help maintain the sucrose and organic molecule concentration that has been established by active transport.

Sieve tube cells are closely associated with companion cells. This is due in part to the fact that the sieve tube cell and its companion cell share the same parent cell. The companion cells perform many of the genetic and metabolic functions of the sieve tube cell and maintain the viability of the sieve tube cell. Note the abundant mitochondria in the companion cell shown in figure 6 to support active transport of sucrose. The infolding of the plasma membrane seen in the companion cell image increases the phloem loading capacity using the apoplastic route. Plasmodesmata connect the cytoplasm of companion cells with the sieve tube cells and have a larger diameter than plasmodesmata found in other parts of the plant to accommodate the movement of oligosaccharides and genetic elements between the two cells.

The accumulation of sucrose in the sieve tube element-companion cell pair requires the presence of active transport proteins or enzyme activity in the companion cells to produce the oligosaccharides.

The rigid cell walls of the sieve tube cell allow for the establishment of the pressure necessary to achieve the flow of phloem in the sieve tube cell.

Individual sieve tube cells are separated by perforated walls called sieve plates, shown in figure 7. These are the remnants of cell walls that separated the cells. The perforated walls in combination with the reduced cytoplasm means that the resistance to the flow of phloem sap will be lower.

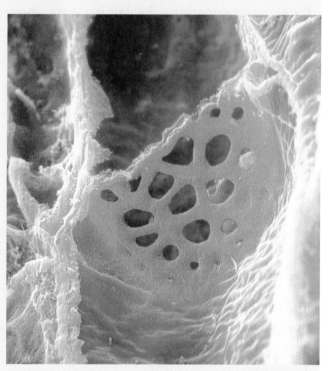
▲ Figure 7

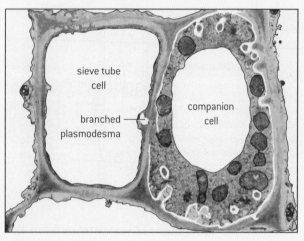
▲ Figure 6

Activity

Analysis of electron micrograph of phloem tissue

1 In the electron micrograph in figure 8, identify the following:

   (i) sieve cells

   (ii) the sieve plate

   (iii) the companion cell

   (iv) plastids with starch granules within the sieve cell

   (v) plasmodesma

   (vi) sieve cell cytoplasm

   (vii) mitochondria within the companion cell.

2 If the scale bar in image represents 5μm, then estimate the width of the sieve tube cell in the region of the sieve plate.

3 Suggest the evidence seen in the micrograph that suggests that the sieve tube cell is living.

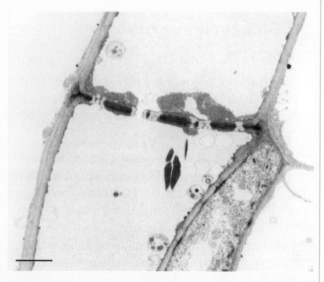
▲ Figure 8

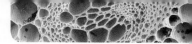

 Experiments using aphid stylets

Analysis of data from experiments measuring phloem transport rates using aphid stylets and radioactively-labelled carbon dioxide.

Phloem sap is nutrient-rich compared with many other plant products and the nutrients in it are small soluble molecules that do not need to be digested. Despite this, the only animals to consume it as the main part of their diet are insects belonging to a group called the Hemiptera which includes aphids, whitefly, mealybugs and psyllids.

Aphids penetrate plant tissues to reach the phloem (p in the first picture in figure 9) using mouth parts called stylets (st in the first picture). If the aphid is anaesthetized and the stylet severed (process about to occur in the middle picture), phloem will continue to flow out of the stylet (final picture) and both the rate of flow and the composition of the sap can be analysed. The closer the stylet is to the sink, the slower the rate at which the phloem sap will come out.

▲ Figure 9

Data-based questions

1 **a)** The only animals to consume phloem sap as the main part of their diet are insects belonging to a group called the Hemiptera. The data in this question comes from research into aphids.

The sugar content of phloem sap is very high – often greater than 1 mol dm^{-3}.

(i) Explain how plants increase the sugar concentration of phloem sap to such high levels. [1]

(ii) Explain how high sugar concentrations cause a high pressure to develop in the phloem. [2]

b) Aphids only ingest a small proportion of the sugar in phloem sap. The remainder passes out in the faeces, which is a liquid called honeydew. Because of the high sugar concentrations, phloem sap has a much higher solute concentration than aphid cells. Enzymes secreted into the aphid gut reduce the solute concentration of phloem sap by converting sugars into oligosaccharides. Figure 11 shows the relationship between the sucrose concentrations of phloem sap ingested by aphids and the oligosaccharide content of the honeydew.

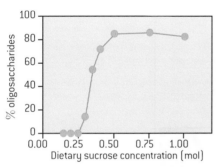

▲ Figure 10

(i) Describe the relationship between the sucrose concentration of phloem sap ingested by aphids and the percentage of oligosaccharides in the honeydew. [3]

(ii) Suggest reasons for aphids secreting enzymes to reduce the solute concentration of the fluid in the gut. [2]

c) Aphids ingest larger volumes of phloem sap than they need, to obtain sufficient sugar for cell respiration. This is because they also need to obtain amino acids and the concentration of amino acids in phloem sap is low. Figure 11 shows the percentages of individual amino acids in phloem sap and the percentages in aphid protein. Nine of the amino acids cannot be synthesized in aphid cells and so are called essential amino acids. The other amino acids can be synthesized from other amino acids and so are non-essential.

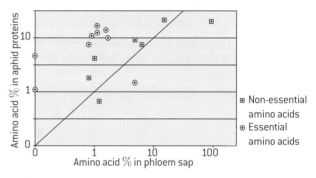

▲ Figure 11

(i) Evaluate phloem sap as a source of amino acids for aphids. [3]

(ii) Suggest reasons for the differences in amino acid content between phloem sap and aphid protein. [2]

d) Specialized cells have been discovered in aphids called bacteriocytes. These organisms contain bacteria called *Buchnera*, which synthesize essential amino acids from aspartic acid and sucrose. Aspartic acid is a non-essential amino acid that is found in much higher concentrations in phloem sap than any other amino acid. When aphids reproduce, they pass on *Buchnera* bacteria to their offspring.

(i) Explain how antibiotics could be used to obtain evidence for the role of *Buchnera* in aphids. [2]

(ii) Using the data in this question, discuss the reasons for few animals using phloem sap as the main part of their diet. [3]

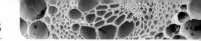

Radioisotopes as important tools in studying translocation

Developments in scientific research follow improvements in apparatus: experimental methods for measuring phloem transport rates using aphid stylets and radioactively-labelled carbon dioxide were only possible when radioisotopes became available.

Carbon-14 is an isotope of carbon that is radioactive. Radioactively-labelled carbon within carbon dioxide can be fixed by plants during photosynthesis. It will release radiation that can be detected either using film or radiation detectors. As the carbon is metabolized, it will be found in different molecules within the plant. In other words, both the formation and movement of radioactive molecules can be traced. Figure 12 shows a device known as a Geiger counter measuring radiation levels in a crop of sunflowers. The sunflowers in the picture are being used for bioremediation of soil contaminated with radiation.

▲ Figure 12

Data-based questions: Radioactive labelling (1)

Source leaves were supplied with a pulse of radioactively-labelled carbon and the time taken for the radioactive carbon to be found in sink leaves was measured by radiophotography. The photosynthetic rate was varied, primarily, by altering the concentration of unlabelled carbon dioxide. The experiment was carried out at three different intensities of light (green squares are 20,000 lux; orange diamonds are 40,000 lux; purple circles are 80,000 lux).

a) Outline the relationship between photosynthesis rate and translocation rate. [1]

b) (i) Deduce the relationship between light intensity and translocation. [2]

   (ii) Suggest whether this is a correlation or a cause and effect relationship. [3]

c) Determine the ratio of translocation to net photosynthesis at two different points on the graph. [2]

d) Deduce, with a reason, whether the source leaf is a growing or mature leaf. [2]

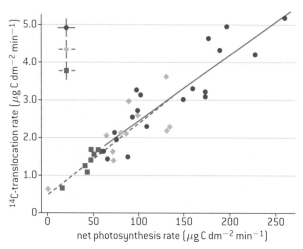

▲ Figure 13

Data-based questions: Radioactive labelling (2)

The distribution of radioactivity in leaves from a sugar beet plant (*Beta vulgaris*) was determined 1 week after $^{14}CO_2$ was supplied for 4 hours to a single source leaf (labelled with an arrow in figure 14). The degree of radioactive labelling is indicated by the intensity of shading of the leaves. Leaves are numbered according to their age; the youngest, newly emerged leaf is designated 1.

The purpose of the experiment was to determine the position of sink leaves in relation to the position of source leaves. The hypothesis was that leaves directly above and below the source leaf are most likely to receive photosynthate (the products of photosynthesis) and that pruning causes a rerouting of translocation pathways to include lateral leaves. Figure 14A shows the distribution of photosynthate in an intact plant. Figure 14B shows the pattern after several leaves have been removed.

(i) In figure 14A, identify the two leaves that received the most photosynthate. [2]

(ii) Using figure 14A, describe the location of the sink leaves receiving the most photosynthate in relation to the source leaf. [2]

(iii) Evaluate the hypothesis that leaves directly above and below the source leaf are most likely to receive photosynthate and that pruning causes a rerouting of translocation pathways to include lateral leaves. [3]

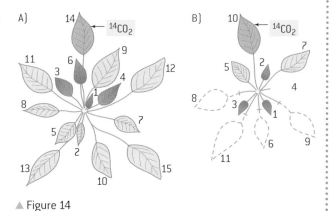

▲ Figure 14

⚗ Identifying xylem and phloem in light micrographs

Identification of xylem and phloem in microscope images of stem and root.

Xylem cells are generally larger than phloem cells. Within one vascular bundle, phloem cells tend to be closer to the outside of the plant in stems and roots.

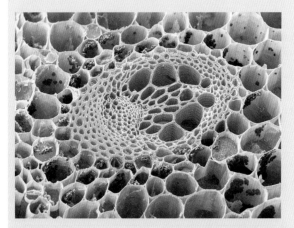

▲ Figure 15 Buttercup stem. Coloured scanning electron micrograph (SEM) of a transverse (cross) section through part of a stem of a buttercup, *Ranunculus repens*, showing a vascular bundle. This is a typical dicotyledon stem. At the centre is an oval vascular bundle embedded in the cortex cells of the stem. Some cells contain chloroplasts (green). The vascular bundle contains large xylem vessels (centre right) which serve to conduct water; the nutrient-conducting phloem is orange

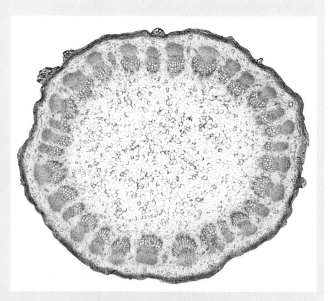

▲ Figure 16 Light micrograph of a transverse section through the stem of a sunflower (*Helianthus annuus*)

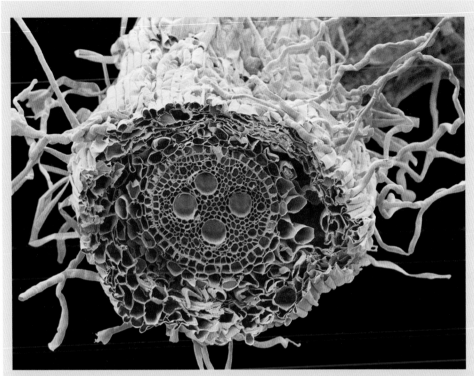

▲ Figure 17 Coloured scanning electron micrograph (SEM) of a section through a rootlet of a dicotyledon plant. The vascular bundle consists of xylem (four yellow circles, centre) and phloem (beige) tissue. Xylem transports water and mineral nutrients from the roots throughout the plant and phloem transports carbohydrates and plant hormones around the plant. Surrounding the vascular bundle is a single layer of endodermis (orange), then cortex (brown), which consists of parenchyma cells. The outermost layer (cream) is the epidermis

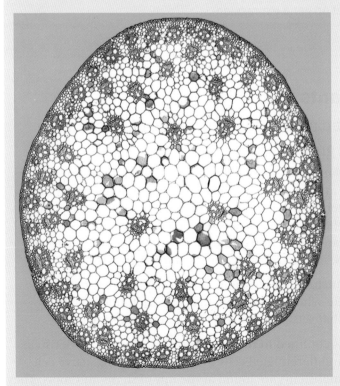

▲ Figure 18 Light micrograph of a section through the stem of a maize plant (*Zea mays*). Vascular bundles (coloured clusters) can be seen containing xylem (larger openings, red/black) and phloem (smaller openings, light blue) tissues

9.3 Growth in plants

Understanding

→ Undifferentiated cells in the meristems of plants allow indeterminate growth.

→ Mitosis and cell division in the shoot apex provide cells needed for extension of the stem and development of leaves.

→ Plant hormones control growth in the shoot apex.

→ Plants respond to the environment by tropisms.

→ Auxin influences cell growth rates by changing the pattern of gene expression.

→ Auxin efflux pumps can set up concentration gradients of auxin in plant tissues.

Applications

→ Micropropagation of plants using tissue from the shoot apex, nutrient agar gels and growth hormones.

→ Use of micropropagation for rapid bulking up of new varieties, production of virus-free strains of existing varieties and propagation of orchids and other rare species.

Nature of science

→ Developments in scientific research follow improvements in analysis and deduction: improvements in analytical techniques allowing the detection of trace amounts of substances have led to advances in the understanding of plant hormones and their effect on gene expression.

Growth in plants

Undifferentiated cells in the meristems of plants allow indeterminate growth.

The growth of a plant is an everyday phenomenon, but it is nonetheless remarkable. Most animals and some plant organs undergo determinate growth; that is, there is either a defined juvenile or embryonic period or growth stops when a certain size is reached or a structure is fully formed. Growth can also be indeterminate when cells continue to divide indefinitely. Plants, in general, have indeterminate growth.

Many plant cells, including some fully differentiated types, have the capacity to generate whole plants; i.e., the cells are totipotent. This phenomenon is what sets plant cells apart from most animals.

Growth in plants is confined to regions known as meristems. Meristems are composed of undifferentiated cells that are undergoing active cell division. Primary meristems are found at the tips of stems and roots. They are called **apical meristems**. The root apical meristem is responsible for the growth of the root. The shoot apical meristem is at the tip of the stem. Many **dicotyledenous** plants also develop lateral meristems.

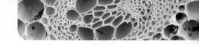

Role of mitosis in stem extension and leaf development

Mitosis and cell division in the shoot apex provide cells needed for extension of the stem and development of leaves.

Cells in meristems are small and go through the cell cycle repeatedly to produce more cells, by mitosis and cytokinesis. These new cells absorb nutrients and water and so increase in volume and mass.

The root apical meristem is responsible for the growth of the root. The shoot apical meristem is more complex. It throws off the cells that are needed for the growth of the stem and also produces the groups of cells that grow and develop into leaves and flowers. With each division, one cell remains in the meristem while the other increases in size and differentiates as it is pushed away from the meristem region. Figure 1 shows the shoot apical meristem of a **dicotyledonous** plant.

Each apical meristem can give rise to additional meristems including protoderm, procambium and ground meristem. In general these give rise to different tissues. For example, protoderm gives rise to epidermis, procambium usually gives rise to vascular tissue and ground meristem can give rise to pith. The position of these tissues and some of the tissues they give rise to are shown in figure 4. Chemical influences also play a large role in determining which type of specialized tissue arises from unspecialized plant cells. Young leaves are produced at the sides of the shoot apical meristem. They appear as small bumps known as leaf primordia.

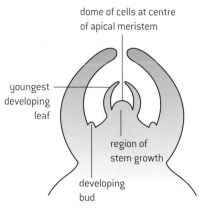

▲ Figure 1 Structure of a shoot apical meristem

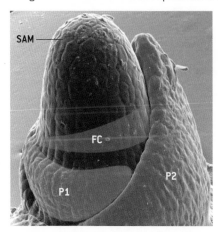

▲ Figure 2 In this image, the shoot apical meristem is labeled SAM. P1 and P2 represent newly forming leaves (primordial leaves) and FC refers to the founding cells of a new leaf that has yet to differentiate

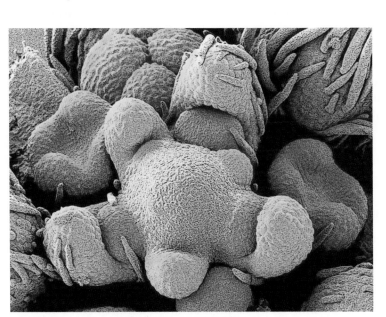

▲ Figure 3 This is a developing flower bud on a shoot apical meristem of a variety of the gunsight clarkia (*Clarkia xantiana*) plant. A shoot apical meristem is where new growth takes place in a flowering plant. Floral buds (red) are developing between leaf axils (green), surrounding the floral meristem dome (blue)

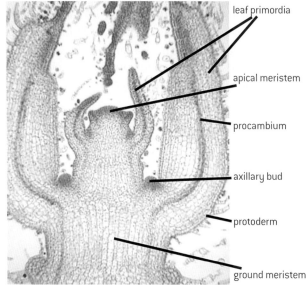

▲ Figure 4

Plant hormones affect shoot growth

Plant hormones control growth in the shoot apex.

A hormone is a chemical message that is produced and released in one part of an organism to have an effect in another part of the organism. Auxins are hormones that have a broad range of functions including initiating the growth of roots, influencing the development of fruits and regulating leaf development. The most abundant auxin is indole-3-acetic acid (IAA). IAA has a role in the control of growth in the shoot apex. Among other effects, IAA promotes the elongation of cells in stems. IAA is synthesized in the apical meristem of the shoot and is transported down the stem to stimulate growth. At very high concentrations, it can inhibit growth.

Axillary buds are shoots that form at the junction, or node, of the stem and the base of a leaf. As the shoot apical meristem grows and forms leaves, regions of meristem are left behind at the node. Growth at these nodes is inhibited by auxin produced by the shoot apical meristem. This is termed apical dominance. The further distant a node is from the shoot apical meristem, the lower the concentration of auxin and the less likely that growth in the axillary bud will be inhibited by auxin. In addition, cytokinins, hormones produced in the root, promote axillary bud growth. The relative ratio of cytokinins and auxins determine whether the axillary bud will develop. Gibberellins are another category of hormones that contribute to stem elongation.

Data-based questions: The acid growth hypothesis

The acid growth hypothesis of auxin effect is that auxin stimulates the action of a proton (H^+) pump. The pump moves protons out of the cell, leading to an increase in the acidity of the cell wall. This leads to the activation of the protein called expansin. Expansin plays a role in breaking and reforming the connections between cellulose fibres and the polysaccharides that cross-link the cellulose. As the cell wall becomes weak, turgor pressure from within the cell pushes the wall outward, causing elongation.

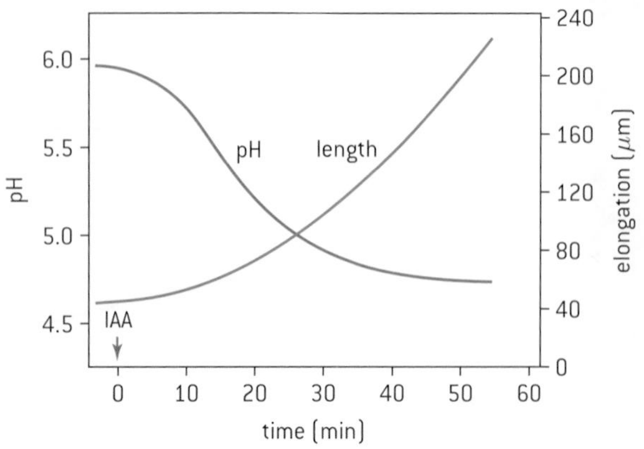

When a shoot first emerges, it has a protective sheath called a coleoptile. Oat coleoptiles were bathed in a solution containing IAA. The pH of the solution surrounding the coleoptiles was determined (see figure 5).

a) Suggest the effect of the application of IAA on the pH of the solution surrounding the coleoptiles. [3]

b) Estimate the time where the change in length of the coleoptiles was the greatest. [1]

c) Outline the relationship between pH and change in length. [2]

In a further experiment, coleoptiles were immersed in a pH 3 solution at time zero. The first arrow in figure 6 indicates the point where the coleoptiles were transferred to a pH 7 solution. The second arrow indicates the point at which IAA was added.

▲ Figure 5

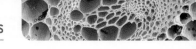

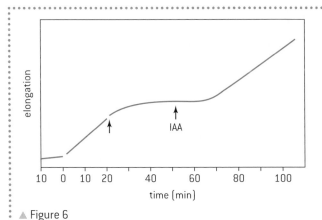

▲ Figure 6

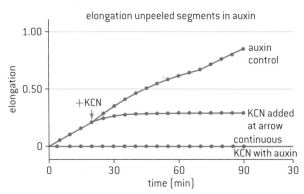

▲ Figure 7

d) Compare the effect of pH 3 on elongation with the effect of pH 7 on elongation.

e) State the effect of the addition of IAA on elongation.

To test the hypothesis that active transport plays a role in mechanism of action of auxin, a respiratory inhibitor (potassium cyanide, KCN) was applied continuously to one treatment group, and to a second treatment group at the arrow. A

third treatment group (the control) did not have KCN applied.

f) State the effect of the addition of KCN on elongation.

g) Based on the data, to what extent is there support for the conclusion that auxin stimulates the active transport of protons out of the shoot and that these protons promote elongation.

Plant tropisms

Plants respond to the environment by tropisms.

Plants use hormones to control the growth of stems and roots. Both the rate and the direction of growth are controlled. The direction in which stems grow can be influenced by two external stimuli: light and gravity. Stems grow towards the source of the brightest light or in the absence of light they grow upwards, in the opposite direction to gravity. These directional growth responses to directional external stimuli are called **tropisms**. Growth towards the light is called **phototropism** and growth in response to gravitational force is called **gravitropism**.

Auxin influences gene expression

Auxin influences cell growth rates by changing the pattern of gene expression.

The first stage in phototropism is the absorption of light by photoreceptors. Proteins called phototropins have this role. When they absorb light of an appropriate wavelength, their conformation changes. They can then bind to receptors within the cell, which control the transcription of specific genes. Although much research is still needed in this field, it seems likely that the genes involved are those coding for a group of glycoproteins located in the plasma membrane of cells in the stem that transport the plant hormone **auxin** from cell to cell, called PIN3 proteins.

▲ Figure 8 Fuchsia plant (*Fuchsia* sp.) growing to the left towards a light source. This kind of directional plant growth in response to light is known as phototropism

▲ Figure 9 A seed of *Brassica napus* showing geotropism

425

Intracellular pumps

Auxin efflux pumps can set up concentration gradients of auxin in plant tissue.

The position and type of PIN3 proteins can be varied to transport auxin to where growth is needed. If phototropins in the tip detect a greater intensity of light on one side of the stem than the other, auxin is transported laterally from the side with brighter light to the more shaded side. Higher concentrations of auxin on the shadier side of the stem cause greater growth on this side, so the stem grows in a curve towards the source of the brighter light. The leaves attached to the stem will therefore receive more light and be able to photosynthesize at a greater rate.

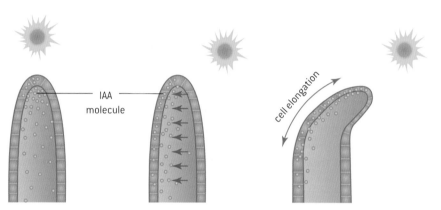

▲ Figure 10

Gravitropism is also auxin dependent. The upward growth of shoots and the downward growth of roots occurs in response to gravity. If a root is placed on its side, gravity causes cellular organelles called statoliths to accumulate on the lower side of cells. This leads to the distribution of PIN3 transporter proteins that direct auxin transport to the bottom of the cells. High concentrations of auxin inhibit root cell elongation so the top cells elongate at a higher rate than the bottom cells causing the root to bend downward. Note that the pattern of auxin effect is opposite to what happens in the shoot. In the shoot, auxin promotes elongation but in the root auxin inhibits shoot elongation.

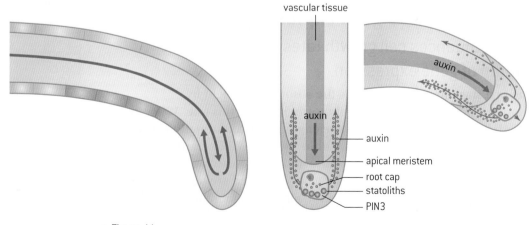

▲ Figure 11

Activity

Design experiments to determine if potato "eyes" positively or negatively gravitropic. Are they positively or negatively phototropic? Does phototropism occur if the apical meristems are pinched off?

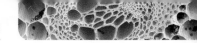

Micropropagation of plants

Micropropagation of plants using tissue from the shoot apex, nutrient agar gels and growth hormones.

Micropropagation is an *in vitro* procedure that produces large numbers of identical plants.

A stock plant is identified that often has some desirable feature. Micropropagation depends on the totipotency of plant tissues; i.e., their ability to differentiate into any functional plant part.

Tissues from the stock plant are sterilized and cut into pieces called explants. For most applications, the least differentiated tissue serves as the source tissue such as a meristem. The explant is placed into sterilized growth media that includes plant hormones. Inclusion of equal proportions of auxin and cytokinin into the media leads to the formation of an undifferentiated mass called a callus. If the growth media contains a ratio of auxin that is greater than ten times the amount of cytokinin, then this is called rooting media and roots develop. If the ratio of auxin to cytokinin is less than 10:1, then this is called shoot media and shoots develop. Once roots and shoots are developed, the cloned plant can be transferred to soil.

▲ Figure 12

Micropropagation is used for rapid bulking up

Use of micropropagation for rapid bulking up of new varieties, production of virus-free strains of existing varieties and propagation of orchids and other rare species.

The international exchange of plant materials carries with it the risk of transmission of pathogens. Micropropagation techniques can be used to produce virus-free strains of plants. Viruses are transported within a plant from cell to cell through vascular tissue and via plasmodesmata. The apical meristem is therefore often free of viruses.

Micropropagation can be used in the production of plants with desirable characteristics, producing many identical copies of an individual. The process is also much faster and takes up less space than traditional methods of production. For example, it is being used in the preservation of species such as orchids. Often the target of collection in the wild, the bulk production of endangered varieties of orchids allows for wild replanting as well as a method for commercial production. Further, the seeds of orchids are difficult to germinate. Asexual reproduction is often more successful. Micropropagated plantlets can be stored in liquid nitrogen – a technique known as cryopreservation. This is equivalent in function to a seed bank.

▲ Figure 13 *Ophrys lutea*

The loss of habitat for the orchid species *Ophrys lutea* (figure 13) in Malta, combined with their normally low seed production and low rates of successful germination identified them as a target for conservation. The material needed to start the cultures was collected from open fields. Once the process of plantlet production is complete, the intent is to both replant the orchid back into the wild habitat as well as to maintain a stock of the threatened species.

Genomics has improved understanding of the role of plant hormones

Developments in scientific research follow improvements in analysis and deduction: improvements in analytical techniques allowing the detection of trace amounts of substances have led to advances in the understanding of plant hormones and their effect on gene expression.

Many of the classical experiments on the action of auxin such as those by Darwin and Went involved experiments with coleoptiles. Modern genomics has opened up opportunities to understand mechanisms and pathways in a way that would not have been previously possible.

Microarrays allow researchers to detect gene expression. If a gene is being expressed, then when the tissue is tested on the microarray, it will cause fluorescence.

In one such study, researchers found that seven genes are expressed at higher levels on the bottom cells in gravitrophically stimulated cells and on the shady side of phototrophically stimulated cells.

The analysis of gene expression takes advantage of the knowledge of model plants like *Arabidopsis thaliana* and its close relative *Brassica oleracea*.

Brassica plant cells are relatively large and so cellular activity is readily observable.

Encoded protein	Level of increase "shaded" vs "lit" flank	Level of increase "bottom" vs "top" flank
α-expansin	3.9 ± 2.2	3.9 ± 2.9
putative oxidase	5.2 ± 0.5	1.4 ± 0.4
IAA-amido synthetase (asp)	1.6 ± 0.3	1.7 ± 0.3
SAUR protein	1.3 ± 0.5	1.4 ± 0.2
BHLH transcription factor	1.7 ± 0.2	2.0 ± 0.9
HD-zip transcription factor	1.9 ± 0.3	2.3 ± 0.4
IAA-amido synthetase (ala)	4.6 ± 1.9	1.9 ± 0.4

▲ Figure 14 The effect of light and gravity on the expression of seven genes

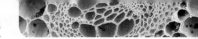

9.4 Reproduction in plants

Understanding

- → Flowering involves a change in gene expression in the shoot apex.
- → The switch to flowering is a response to the length of light and dark periods in many plants.
- → Most flowering plants use mutualistic relationships with pollinators in sexual reproduction.
- → Success in plant reproduction depends on pollination, fertilization and seed dispersal.

Applications

- → Methods used to induce short-day plants to flower out of season.

Skills
- → Drawing internal structure of seeds.
- → Drawing of half-views of animal-pollinated flowers.
- → Design of experiments to test hypotheses about factors affecting germination.

Nature of science
- → Paradigm shifts; more than 85% of the world's 250,000 species of flowering plants depend on pollinators for reproduction. This knowledge has led to protecting entire ecosystems rather than individual species.

Flowering and gene expression

Flowering involves a change in gene expression in the shoot apex.

When a seed germinates, a young plant is formed that grows roots, stems and leaves. These are called **vegetative** structures and the plant is in the vegetative phase. This can last for weeks, months or years, until a trigger causes the plant to change into the reproductive phase and produce flowers. The change from the vegetative to the reproductive phase happens when meristems in the shoot start to produce parts of flowers instead of leaves.

Flowers are structures that allow for sexual reproduction, thereby increasing variety. They are produced by the shoot apical meristem and are therefore a reproductive shoot.

Temperature can play a role in transforming a leaf-producing shoot into a flower-producing shoot, but day length is the main trigger, or more precisely the length of the dark period. Some plants such as the poinsettia (*Euphorbia pulcherrima*) are categorized as short-day plants because they flower when the dark period becomes longer than a critical length, for example in the autumn. Other plants such as red clover (*Trifolium pratense*) are long-day plants because they flower during the long days of early summer when nights are short.

Light plays a role in the production of either inhibitors or activators of genes that control flowering. For example in long-day plants, the active form of the pigment phytochrome leads to the transcription of

▲ Figure 1 The poinsettia is a short-day plant

▲ Figure 2 Red clover is a long-day plant

a flowering time (FT gene). The FT mRNA is then transported in the phloem to the shoot apical meristem where it is translated into FT protein. The FT protein binds to a transcription factor. This interaction leads to the activation of many flowering genes which transform the leaf-producing apical meristem into a reproductive meristem.

Photoperiods and flowering

The switch to flowering is a response to the length of light and dark periods in many plants.

Long-day plants flower in summer when the nights have become short enough.

Short-day plants flower in the autumn (fall), when the nights have become long enough.

Observations of flowering suggested that the trigger for this in some plants might be a particular day length, but experiments have shown that it is the length of darkness that matters, not the length of daylight.

A pigment was discovered in leaves that plants use to measure the length of dark periods. It is called phytochrome and is unusual as it can switch between two forms, P_R and P_{FR}.

- When P_R absorbs red light of wavelength 660 nm it is converted into P_{FR}.

- When P_{FR} absorbs far-red light, of wavelength 730 nm, it is converted to P_R. This conversion is not of great importance as sunlight contains more light of wavelength 660 nm than 730 nm, so in normal sunlight phytochrome is rapidly converted to P_{FR}.

- However, P_R is more stable than P_{FR}, so in darkness P_{FR} very gradually changes into P_R.

Further experiments have shown that P_{FR} is the active form of phytochrome and that receptor proteins are present in the cytoplasm to which P_{FR} but not P_R binds.

- In long-day plants, large enough amounts of P_{FR} remain at the end of short nights to bind to the receptor, which then promotes transcription of genes needed for flowering.

- In short-day plants, the receptor inhibits the transcription of the genes needed for flowering when P_{FR} binds to it. However, at the end of long nights, very little P_{FR} remains, so the inhibition fails and the plant flowers.

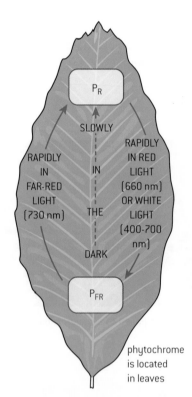

▲ Figure 3 Interconversions of phytochrome

PR

SLOWLY

RAPIDLY IN FAR-RED LIGHT (730 nm)

IN THE

RAPIDLY IN RED LIGHT (660 nm) OR WHITE LIGHT (400-700 nm)

DARK

P FR

phytochrome is located in leaves

Data-based questions: Sowing times for soybeans

Soybeans are rich in protein and are eaten both by humans and livestock. After germination, soybean plants grow a series of sections of stem, with nodes between them. Leaves are produced at the nodes. The stem sections are called internodes. Flowers are produced at each node and pods containing beans develop from them. When they start to flower, soybean plants stop growing more nodes and internodes.

Figure 4 shows the mean numbers of nodes of soybean plants sown on different dates in Nebraska.

1   Compare the growth of the soybean plants sown on the different dates. [5]

2   **a)**   Deduce when the soybeans started to flower. [2]

     b)   Deduce with reasons, the factor that triggers flowering in soybeans. [3]

3   **a)**   Explain the advantage, in terms of soybean yields, of sowing the crop as early as possible. [3]

     b)   Suggest two possible disadvantages of sowing soybeans earlier than the dates used in the trial. [2]

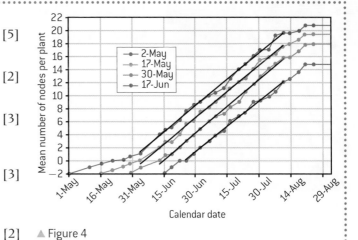

▲ Figure 4

Inducing plants to flower out of season

Methods used to induce short-day plants to flower out of season.

Flower forcing is a procedure designed to get flowers to bloom out of season or at a specific time such as during holiday time. Growers can manipulate the length of the days and nights to force flowering.

The Siam tulip *Curcuma alismatifolia* is sold as cut flowers. It normally produces flowers seasonally during the rainy season where long-day conditions apply. Providing additional light in the middle of the night leads to flowering in the off-season provided that enough humidity and nutrients are provided.

Draw an animal pollinated flower

Drawing of half-views of animal-pollinated flowers.

Figure 5 shows a flower of *Prunus domestica*. In the base of the flower are nectar-secreting glands, which attract insects, especially bees. The petals are large and white, helping insects to find the flower. The sepals protect the flower bud during its development and at night when buds close. The anthers produce pollen, containing the male gametes. The filaments hold the anthers in a position where they are likely to brush pollen onto visiting insects. The female part of the flower is called a carpel. It consists of a stigma, style and ovary. The stigma is sticky and will capture pollen from the visiting insect. The stigma is held up by the style. The ovary is located inside a small rounded structure called an ovule.

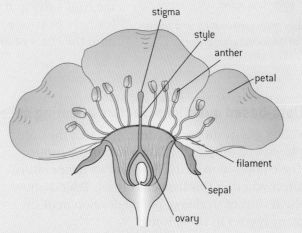

▲ Figure 5 Structure of a plum flower

▲ Figure 6 Honeybee pollinating common mallow flower

▲ Figure 7 Purple-throated carib hummingbird

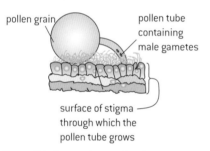

pollen grain

pollen tube containing male gametes

surface of stigma through which the pollen tube grows

▲ Figure 8 Pollen grain germinating on a stigma at the start of the fertilization process

Mutualism between flowers and pollinators

Most flowering plants use mutualistic relationships with pollinators in sexual reproduction.

Sexual reproduction in flowering plants depends on the transfer of pollen from the stamen to a stigma of another plant. Pollen is transferred between plants via a number of strategies including wind and less commonly water but most commonly by animals known as pollinators. Examples of pollinators include birds, bats and insects such as butterflies and bees.

Mutualism is a close association between two organisms where both organisms benefit from the relationship. Pollinators gain food in the form of nectar and the plant gains a means to transfer pollen to another plant. Figure 6 shows a honeybee (*Apis mellifera*) covered in pollen after visiting a mallow flower (*Malva sylvestris*). Figure 7 shows a purple-throated carib hummingbird (*Eulampis jugularis*) that has a curved bill which is an adaptation for extracting nectar from the elongated flower of *Heliconia bihai*.

Pollination, fertilization and seed dispersal

Success in plant reproduction depends on pollination, fertilization and seed dispersal.

The next process after pollination is **fertilization**. From each pollen grain on the stigma a tube grows down the style to the ovary. The pollen tube carries male gametes to fertilize the ovary. The ovary is located inside a small rounded structure called an ovule.

The fertilized ovule develops into a seed and the ovary develops into a fruit.

Seeds cannot move themselves, but nonetheless they often travel long distances from the parent plant. This is called **seed dispersal** and it reduces competition between offspring and parent and helps to spread the species. The type of seed dispersal depends on the structure of the fruit – dry and explosive, fleshy and attractive for animals to eat, feathery or winged to catch the wind, or covered in hooks that catch onto the coats of animals.

Data-based questions: Factors affecting pollen development

Pollen grains sometimes develop when they are placed in a drop of fluid on a microscope slide. The composition of the fluid and its temperature affect whether this happens or not. Table 1 show the results of studies of pollen development in plant species in Hong Kong.

1 The data in table 1 is difficult to analyse in its current form. Choose suitable presentation formats to display the data clearly and allow you to identify any significant trends. You can use ICT or you can draw graphs, tables, charts or diagrams by hand.

2 Describe clearly any trends that you have found in the data. Try to explain each trend that you describe, using your biological knowledge.

3 Identify any weaknesses in the data obtained. Suggest how the investigation could have been improved.

Plant species	Diameter of pollen grain (μm)	Mean growth of pollen tube (μm h^{-1})	Optimal sucrose conc. (mmol dm^{-3})
Bougainvillea glabra	44.00	41.8	0.75
Delonix regia	70.30	4.9	0.45
Leucaena leucocephala	64.60	111.0	0.75
Bauhinia purpurea	71.50	69.9	0.45
Lilium bulbiferum	91.60	11.1	0.30
Gladiolus gandavensis	86.82	50.6	0.45

Sucrose concentration (mmol dm^{-3})	Percentage of *Camellia japonica* pollen grains that developed
0.30	22.5
0.46	23.0
0.60	13.0
0.75	0.0
0.90	0.0

Copper ion concentration (ppm)	Mean growth of pollen tubes of *Bougainvillea glabra* (μm h^{-1})
0.0	33.6
1.0	25.1
2.5	15.5
5.0	10.8
25.0	0.0

▲ Table 1

Preserving habitats as a conservation measure

Paradigm shifts: more than 85% of the world's 250,000 species of flowering plants depend on pollinators for reproduction. This knowledge has led to protecting entire ecosystems rather than individual species.

The growth in number and nature of threats to biodiversity in combination with scarce resources being devoted to conservation means that traditional conservation measures have to be re-evaluated. Traditionally, the focus of conservation efforts was on populations and species of particular concern. The close association between such organisms as pollinators and flowering plants suggests that it is the ecosystem and biological processes which must be protected.

The Saguaro cactus (*Carnegiea gigantea*) is a keystone species of the Sonoran desert. They provide important perching and nesting sites for birds such as red-tailed hawks and nesting cavities for gilded flickers, gila woodpeckers, elf owls, purple martins and other birds. Once the Saguaro fruit ripens, lesser long-nosed bats (*Leptonycteris yerbabuenae*), white-winged doves, gila woodpeckers and other birds consume the

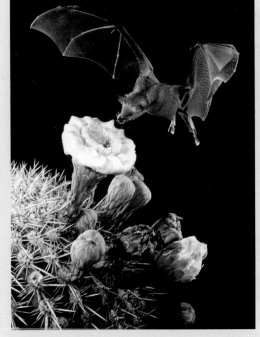

▲ Figure 9 Bat approaching a Saguaro cactus flower.

fruits and disperse the seeds, which pass through their guts intact.

The flowers of the cactus bloom for just one evening each year. They attract lesser long-nosed bats and mexican long-tongued bats (*Choeronycteris mexicana*) to their nectar. The bats use their elongated snouts to reach into the flowers for nectar, covering their heads with pollen that they then transfer from flower to flower as they fly from cactus to cactus throughout the night.

The lesser long-nosed bat is listed as an endangered species under the US Endangered Species Act. However, invasive grasses, development of the desert for human habitation and changes in natural fire cycles threaten the cactus. The survival of both bats and their desert food plants are threatened by loss of habitat. Ensuring the future of the Sonoran desert ecosystem depends on protecting the roles played by the bat, the cactus and seed-dispersing animals.

 ## The structure of seeds

Drawing internal structure of seeds.

A seed is a package containing an embryo plant and food reserves, all inside a protective seed coat. The embryo plant consists of an **embryo root**, **embryo shoot** and one or two **cotyledons**, depending on whether the plant is monocotyledonous or dicotyledonous. The cotyledons are the embryo leaves and in many plants they contain the food reserves of the seed. In other seeds there is a special food storage tissue called endosperm. The scientific name for the seed coat is the **testa**. There is a small hole through the testa, called the **micropyle**. It is located next to a scar where the seed was attached to the parent plant. Figure 10 shows the external and internal structure of a bean seed (*Phaseolus vulgaris*). Figure 11 shows an annotated diagram of the same seed.

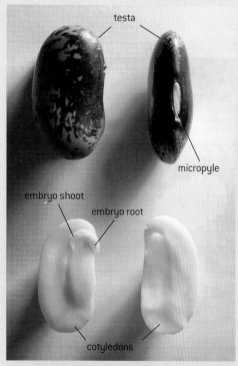

▲ Figure 10 Structure of bean seed (*Phaseolus vulgaris*); external structure (above); internal structure (below)

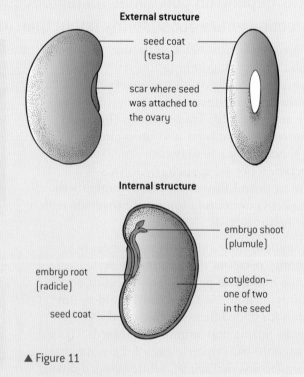

▲ Figure 11

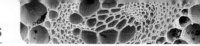

Germination experiment design

Design of experiments to test hypotheses about factors affecting germination.

The early growth of a seed is called germination. Some seeds do not immediately germinate, even if given the conditions normally required. This is called dormancy and it allows time for seeds to be dispersed. It may also help to avoid germination at an unfavourable time. All seeds need **water** for germination. Many seeds are dry and need to rehydrate their cells. Some seeds contain a hormone that inhibits germination and water is needed to wash it out of the seed. Germination involves growth of the embryo root and shoot and this also requires water.

The metabolic rate of a dry and dormant seed is close to zero, but after absorption of water, metabolic processes begin again, including energy release by aerobic cell respiration. Another requirement for germination is therefore a supply of **oxygen**. Because germination involves enzyme-catalysed metabolic reactions, **warmth** is required and germination often fails at low temperatures.

Another metabolic process occurring at the start of germination is synthesis of **gibberellin**, a plant hormone. Several genes have to be expressed to produce the various enzymes of the metabolic pathway leading to gibberellin. This hormone stimulates mitosis and cell division in the embryo. In starchy seeds it also stimulates the production of amylase. This enzyme is needed to break down starch in the food reserves into maltose. Other enzymes convert the maltose into sucrose or glucose. Whereas starch is insoluble and immobile, sucrose and glucose can be transported from the food reserves to where they are needed in the germinating seed. The embryo root and shoot need sugars for growth, together with amino acids and other substances released from the food stores. All parts of the embryo need glucose for aerobic cell respiration.

Most vegetable crop varieties have been bred to germinate quickly – they do not usually have long periods of seed dormancy. Nevertheless, growers of vegetable crops sometimes have difficulty in getting crops to germinate after sowing.

Choose one of the possible causes of crop failure shown in the mind-map, to investigate.

Design an experiment and see whether you obtain evidence for or against your cause.

You will need to decide:

- which seed type to use

- how to vary the factor that you are investigating

- how to keep other factors constant

- how to collect your results, including how to assess whether germination has occurred.

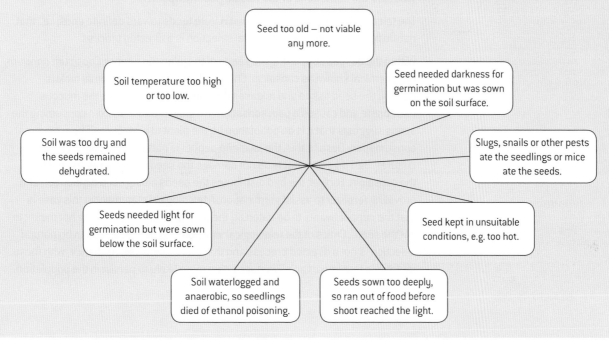

Data-based questions: Fire and seed dormancy in a plant of the chaparral

Emmenanthe penduliflora grows in chaparral (shrubland) in California. It is rarely seen in unburnt chaparral, but appears after fires, growing to about 250 mm, flowering, forming seed and dying in a few months. The electron micrographs below show the results of an experiment in which seeds of the plant were treated with smoke for 3 minutes and then soaked in a solution of lanthanum nitrate hexahydrate.

1   The scale bars in the electron micrographs represent 1 μm. Calculate the thickness of waxy cuticle between the testa and the embryo and food stores inside the control seed. [2]

2   The lanthanum solution appears as dark staining in the electron micrographs and shows how far water was able to penetrate. Deduce how far water could penetrate into the control seeds. [2]

3   a)   Compare the staining of the waxy cuticle in the smoke-treated seeds with the staining of the cuticle in the control seeds. [2]

b)   Suggest a hypothesis for the germination of plants of *Emmenanthe penduliflora* after fires, based on the differences in staining that you have described. [2]

4   Suggest two advantages to *Emmenanthe penduliflora* of dormancy ending after fires in the chaparral. [2]

A = control seed

testa

waxy cuticle

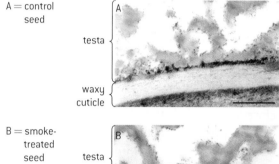

B = smoke-treated seed

testa

waxy cuticle

embryo

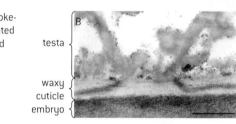

▲ Figure 12 Two electron micrographs of *Emmenanthe penduliflora*. (A) control seed (above); (B) smoke-treated seed (below)

TOK

What are the limitations of the teleological viewpoint?

The teleological viewpoint argues that nature tends toward definite ends; i.e., that nature has intention and that natural selection is a directed process.

Within the tissues supporting the seeds of the jalapeño pepper, *Capsicum annuum*, is a chemical known as capsaicin. Chewing by human and mammal molars destroy the seed tissue and releases the chemical. This irritates the mucous membrane and causes a pain sensation. The pain production from consuming the seeds suggests that it is an adaptation by the plant to protect itself from mammal consumption. Despite this, the jalapeño pepper is part of the cuisine of several cultures. Bird digestive tracts do not damage the seeds and are unaffected by the capsaicin. Further, the bird distributes the seeds aiding seed distribution and providing fertilizer to assist germination. A teleological statement in this case is that the pepper "wants to be eaten by the bird" and that humans "are not meant to eat the seed". Critics of the teleological viewpoint argue that evolution by natural selection is not a directed process and that mutations arise by chance, with those mutations that impart an advantage being more likely to persist in the population.

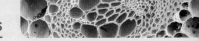

Questions

1 The graphs in figure 13 show the results of investigations into the rate at which water is able to diffuse though the waxy cuticle of plants, which is called the water permeance of the cuticle. Figure 13a shows the relationship between temperature and water permeance of four species of plant. Figure 13b shows the relationship between the thickness of cuticular wax and water permeance. The results of the experiment show how important it is to test hypotheses, even when it may seem that this is not necessary.

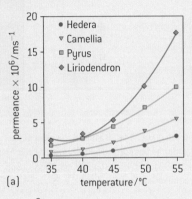

(a)

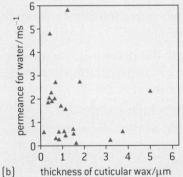

(b)

▲ Figure 13 Factors affecting water permeance of waxy cuticle

a) Using the data in figure 13a, describe the relationship between temperature and water permeance. [2]

b) Discuss the consequences for plants of the effect of temperature on cuticular water permeance. [3]

c) Using the data in figure 13b, state the thickness of cuticular wax with:

(i) the highest water permeance

(ii) the lowest water permeance. [2]

d) Evaluate the hypothesis that the water permeance of the cuticle is positively correlated with its thickness, using the data in figure 13b. [3]

2 In order to prevent transfer of pollen from an anther of one plant to the stigma of the same plant (self-pollination), the sunflower (*Helianthus spp.*) anther sheds its pollen before the stigma is mature enough to receive it. Early in the morning the anther is exposed by elongation of the filaments. The anthers open at this time to release their pollen (anthesis). The stigma appears above the anthers by late afternoon, and by the following morning it is fully receptive.

To see how the filament (F) and the style (S) are affected by light, their lengths were measured at time intervals starting 12 hours before anthesis (−12). Some plants were grown in continuous white light (L24) and some plants grown under cycles of 16 hours white light followed by 8 hours dark (L16/D8). The results are shown in the graph.

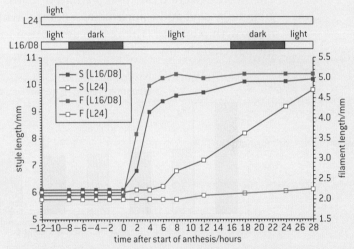

Source: Lobello et al, *Journal of Experimental Botany*, (2000), **51**, pages 1403–1412

a) Filaments of the plants grown in continuous white light increased in length by 0.25 mm in the 28 hours after anthesis. Calculate how much the filaments of the plants grown in alternating white light and dark increased during the same period. [1]

b) Compare the increase in the length of the style in the plants grown in continuous white light with those grown in alternating white light and dark. [2]

The table compares the percentage of ovules that have been fertilized and developed into seeds in sunflower plants grown under continuous white light with those grown under alternating light and dark. The numbers represent the mean ± one standard deviation.

Light treatments	Percentage of fertilized ovules
Continuous white light (L24)	11.40 ± 7.76
Alternating light and dark (L16/D8)	58.26 ± 4.06

c) Explain the differences in the percentages of ovules fertilized using the data in the graph about the growth of filaments and styles. [3]

d) Explain how standard deviation (SD) shown in this table can be used to help in comparing the effect of light treatments on the fertilization of ovules. [3]

To analyse the effect of growth regulators on filament elongation, further experiments were performed in the dark, white light and red light. The flowers were treated with auxin or with gibberellic acid and compared to a control with no growth regulator. The results are shown in the bar chart below.

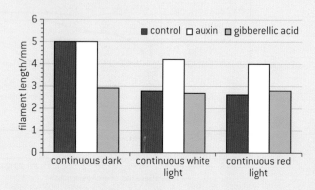

e) Identify, with reasons, which factors promote and which factors inhibit the elongation of filaments. [3]

f) Explain the disadvantages to a plant of self-pollination. [2]

3 Sweet pepper (*Capsicum annuum*) is an important widespread agricultural crop. Scientists studied the transport and distribution of sodium in sweet pepper by growing plants in sodium chloride solutions.

The graph below shows the sodium ion concentration in plant parts of sweet pepper grown in 15 mM sodium chloride for three weeks.

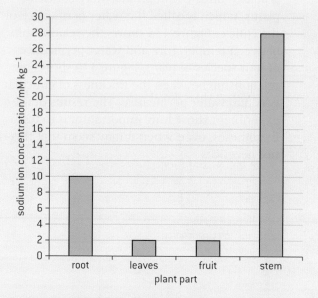

Source: M Blom-Zandstra *et. al.*, "Sodium fluxes in sweet pepper exposed to varying sodium concentrations", *Journal of Experimental Botany* (1 November 1998), vol. 49, issue 328, pp. 1863–1868, by permission of Oxford University Press

a) (i) State the concentration of sodium ions in fruits. [1]

(ii) Calculate the percentage increase in sodium ion concentration between root and stem. [1]

b) Suggest why a high sodium ion concentration in the cells of the stem is important in providing support to this type of plant. [1]

c) State **one** possible use of sodium in plants. [1]

d) Scientists also found that the concentrations of sodium ion in cells of the stem and in xylem sap were the same. Explain why this led the scientists to believe there was no active transport between xylem and stem. [2]

e) Suggest **one** possible method of transport of sodium ions between xylem and stem. [1]

10 GENETICS AND EVOLUTION (AHL)

Introduction

Inheritance follows principles that have been discovered by research from the 19th century onwards. Genes may be linked or unlinked and are inherited accordingly. Meiosis leads to independent assortment of chromosomes and the unique composition of alleles in daughter cells. Gene pools change over time.

10.1 Meiosis

Understanding

→ Chromosomes replicate in interphase before meiosis.

→ Crossing over is the exchange of DNA material between non-sister homologous chromatids.

→ Chiasmata formation between non-sister chromatids in a bivalent can result in an exchange of alleles.

→ Crossing over produces new combinations of alleles on the chromosomes of the haploid cells.

→ Homologous chromosomes separate in meiosis.

→ Independent assortment of genes is due to the random orientation of pairs of homologous chromosomes in meiosis I.

→ Sister chromatids separate in meiosis II.

Skills

→ Drawing diagrams to show chiasmata formed by crossing over.

Nature of science

→ Making careful observations: careful observation and record keeping turned up anomalous data that Mendel's law of independent assortment could not account for. Thomas Hunt Morgan developed the notion of linked genes to account for the anomalies.

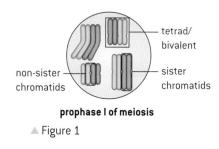

tetrad/bivalent

non-sister chromatids

sister chromatids

prophase I of meiosis

▲ Figure 1

Chromosome replication

Chromosomes replicate in interphase before meiosis.

Like mitosis, meiosis follows a period of interphase with the cell cycle phases of G_1, S and G_2. In the S phase, DNA is replicated so that each chromosome consists of two chromatids. At the start of meiosis, the chromosomes condense and are visible as two chromatids, called sister chromatids. Unlike mitosis, pairing, or synapsis, occurs where homologous chromosomes come to align beside each other. The combination is referred to as a tetrad as it is composed of four chromatids. It is also referred to as a bivalent as it is composed of a homologous pair. In many eukaryotic cells, a protein-based structure forms between the homologous chromosomes called the synaptonemal complex.

Exchange of genetic material

Crossing over is the exchange of DNA material between non-sister homologous chromatids.

During prophase I of meiosis breaks in the DNA occur. Following these chromosome breaks, non-sister chromatids "invade" a homologous sequence on a non-sister chromatid and bind in the region of the break. Once crossing over is complete, the non-sister chromatids continue to adhere at the site where crossing over occurred. These connection points are called chiasmata (plural) or chiasma (singular). Evidence suggests that connections via chiasmata are essential for the successful completion of meiosis.

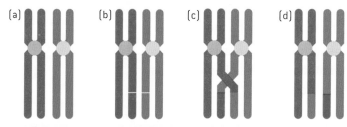

▲ Figure 2 The process of crossing over

Chiasmata formation

Chiasmata formation between non-sister chromatids in a bivalent can result in an exchange of alleles.

Two consequences of chiasmata formation are the increased stability of bivalents at chiasmata as well as an increase genetic variability if crossing over occurs. The process of crossing over results in the exchange of DNA between the maternal and paternal chromosomes. Crossing over can decouple linked combinations of alleles and therefore lead to independent assortment. Further, crossing over can occur multiple times and between different chromatids within the same homologous pair.

Explaining discrepancies in Mendelian ratios

Making careful observations: careful observation and record keeping turned up anomalous data that Mendel's law of independent assortment could not account for. Thomas Hunt Morgan developed the notion of linked genes to account for anomalies.

Mendel's paper was published in 1866. Initially it had little impact and was cited about three times over the next thirty-five years. Starting with the turn of the century, there was increasing recognition of his work. At the same time, some discrepancies arose between observations and Mendel's principle of independent assortment. William Bateson and Reginald Punnett conducted crosses with sweet peas. One of the parent plants had long pollen (LL) and purple flowers (PP). The other had round pollen (ll) and red flowers (pp). As expected all of the F_1 had long pollen and purple flowers (LlPp). The surprising result came in the F_2 generation of a dihybrid cross. Instead of the expected ratio of 9:3:3:1, there were far more of the individuals with the parental phenotypes seen in the P generation and much smaller numbers of the non-parental phenotypes, known as recombinants.

Even though Bateson and Punnett realized their results did not conform to Mendel's principle of independent assortment, they did not develop a clear explanation for the discrepancy. Thomas Hunt Morgan observed similar discrepancies in fruit flies. His discovery of sex linkage led him to develop a theory of gene linkage that accounted for the higher than expected number of parental phenotypes and the notion of crossing over to explain the presence of the recombinants.

New combinations of alleles

Crossing over produces new combinations of alleles on the chromosomes of the haploid cells.

Figure 3 summarizes how crossing over can produce new combinations of alleles on the chromosomes of the haploid cells. The red and the blue diagrams represent homologous chromosomes. As such they have the same length, the same centromere position and the same gene content, but they will have different combinations of alleles. In the diagram, the chromosomes are the locus for genes A, B and D. The individual is heterozygous for all three alleles; i.e, the individual has the genotype AaBbDd. Because the genes are linked, the individual can produce gametes with the combinations AbD and aBd.

The diagram illustrates how crossing over can produce additional combinations of alleles.

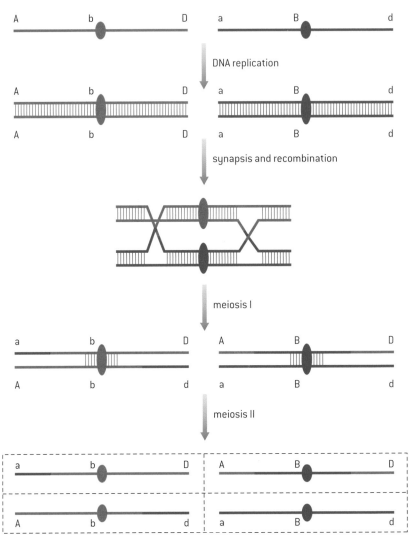

▲ Figure 3 Crossing over occurs multiple times and between different chromatids within the same homologous pair

⚗️ Diagrams of crossing over

Drawing diagrams to show chiasmata formed by crossing over.

A chiasma is an X-shaped knot-like structure that forms where crossing over has occurred. To draw a bivalent with one or more chiasmata it is necessary to use two coloured pens or pencils, so that the two homologous chromosomes, i.e., the maternal and paternal chromosomes can be distinguished. A series of drawings can be used to show different stages in the process. Remember to start with the chromosomes very elongated.

The position at which the crossing over is going to occur can be shown with breaks at the same point in two chromatids, one in each chromosome. As the position of the crossover is random you can show this anywhere along the bivalent. You can have more than one crossover if you want.

It is hard to show the crossover while the chromosomes are still tightly paired as part of it will be hidden, but one of the new connections between the chromatids can be shown clearly.

After crossing over has occurred the chromatids condense by supercoiling. The tight pairing between the homologous chromosomes ends, but they are still held together at each point where crossing over has occurred. This is because the two chromatids of each chromosome remain closely aligned, but chromatids in different chromosomes are now linked to each other. The result is an X-shaped, knot-like structure called a chiasma.

Chiasmata hold homologous chromosomes together for a while, but then slide to the end of the bivalent, allowing the chromosomes to move to opposite poles of the cell.

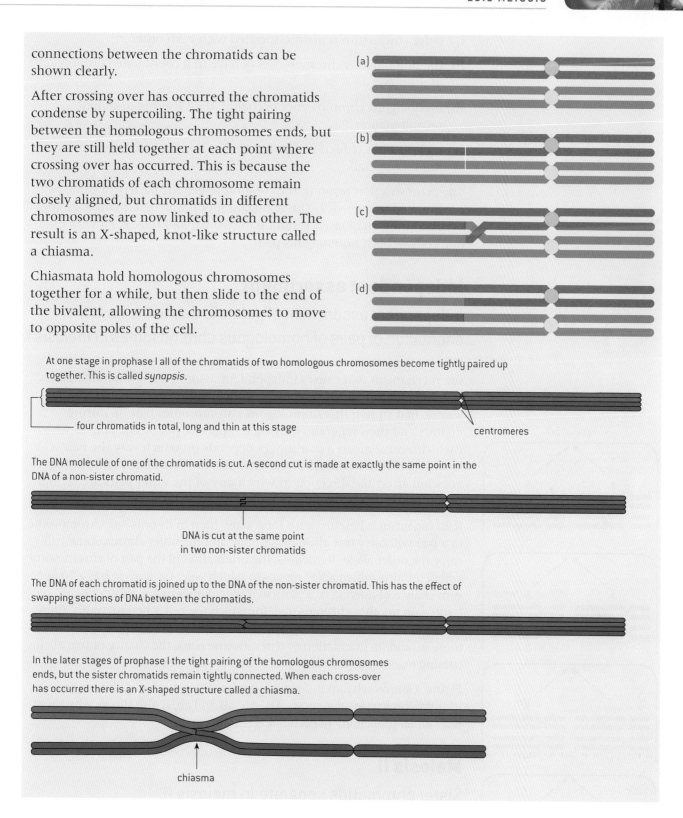

At one stage in prophase I all of the chromatids of two homologous chromosomes become tightly paired up together. This is called *synapsis*.

four chromatids in total, long and thin at this stage

centromeres

The DNA molecule of one of the chromatids is cut. A second cut is made at exactly the same point in the DNA of a non-sister chromatid.

DNA is cut at the same point in two non-sister chromatids

The DNA of each chromatid is joined up to the DNA of the non-sister chromatid. This has the effect of swapping sections of DNA between the chromatids.

In the later stages of prophase I the tight pairing of the homologous chromosomes ends, but the sister chromatids remain tightly connected. When each cross-over has occurred there is an X-shaped structure called a chiasma.

chiasma

Meiosis I

Homologous chromosomes separate in meiosis.

The first meiotic division is unique while the second round resembles mitosis. There are a number of ways in which meiosis I differs from mitosis and meiosis II:

i) sister chromatids remain associated with each other

ii) the homologous chromosomes behave in a coordinated fashion in prophase

iii) homologous chromosomes exchange DNA leading to genetic recombination

iv) meiosis I is a reduction division in that it reduces the chromosome number by half.

The processes that result in the creation of genetic variety of gametes are initiated in meiosis I. The segregation of homologous chromosomes occurs during anaphase I resulting in two haploid cells, each with only one copy of each homologous pair.

Independent assortment

Independent assortment of genes is due to the random orientation of pairs of homologous chromosomes in meiosis I.

When Mendel's work was rediscovered at the start of the 20th century, the mechanism that causes independent assortment of unlinked genes was soon identified. Observations of meiosis in a grasshopper, *Brachystola magna*, had shown that homologous chromosomes pair up during meiosis and then separate, moving to opposite poles. The pole to which each chromosome in a pair moves depends on which way the pair is facing. This is random. Also, the direction in which one pair is facing does not affect the direction in which any of the other pairs are facing. This is called independent orientation.

If an organism is heterozygous for a gene, then in its cells one chromosome in a pair will carry one allele of the gene and the other chromosome will carry the other allele. In meiosis, the orientation of the pair of chromosomes will determine which allele moves to which pole. Each allele has a 50 per cent chance of moving to a particular pole. Similarly for another gene, located on another chromosome, for which the cell is heterozygous, there is a 50 per cent chance of an allele moving to a particular pole. Because there is random orientation of chromosome pairs, the chance of two alleles coming together to the same pole is 25 per cent (see figure 4).

Figure 4 shows why an individual that has the genotype AaBb can produce four different types of gamete: AB, Ab, aB and ab. It also shows why there is an equal probability of each being produced.

Meiosis II

Sister chromatids separate in meiosis II.

After meiosis I, the daughter cells enter meiosis II without passing through interphase. Meiosis II is similar to mitosis in that the replicated chromosome is separated into chromatids. Sister chromatids are separated but they are likely to be non-identical sister chromatids due to the occurrence of crossing over.

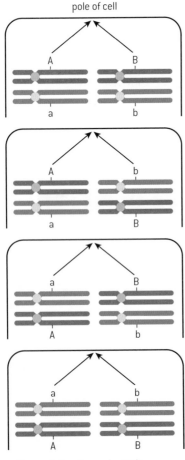

pole of cell

▲ Figure 4 Random orientation

10.2 Inheritance

Understanding

→ Unlinked genes segregate independently as a result of meiosis.

→ Gene loci are said to be linked if on the same chromosome.

→ Variation can be discrete or continuous.

→ The phenotypes of polygenic characteristics tend to show continuous variation.

→ Chi-squared tests are used to determine whether the difference between an observed and expected frequency distribution is statistically significant.

Applications

→ Completion and analysis of Punnett squares for dihybrid traits.

→ Morgan's discovery of non-Mendelian ratios in *Drosophila*.

→ Polygenic traits such as human height may also be influenced by environmental factors.

Nature of science

→ Looking for patterns, trends and discrepancies: Mendel used observations of the natural world to find and explain patterns and trends. Since then, scientists have looked for discrepancies and asked questions based on further observations to show exceptions to the rules. For example, Morgan discovered non-Mendelian ratios in his experiments with *Drosophila*.

Skills

→ Calculation of the predicted genotypic and phenotypic ratio of offspring of dihybrid crosses involving unlinked autosomal genes.

→ Identification of recombinants in crosses involving two linked genes.

→ Use of a chi-squared test on data from dihybrid crosses.

Segregation and independent assortment

Unlinked genes segregate independently as a result of meiosis.

Segregation is the separation of the two alleles of every gene that occurs during meiosis. Independent assortment is the observation that the alleles of one gene segregate independently of the alleles of other genes.

Genes found on different chromosomes are unlinked and do segregate independently as a result of meiosis. However, genes which are on the same chromosome are linked and therefore do not segregate independently. The exception is for linked genes that are far apart on the chromosome. Crossing over between genes occurs more frequently the further the separation of genes and can make it appear that the genes are unlinked.

The examples discussed below are based on the assumption that different alleles segregate independently.

🌐 Punnett squares for dihybrid traits

Completion and analysis of Punnett squares for dihybrid traits.

In a dihybrid cross, the inheritance of two genes is investigated together. Mendel performed dihybrid crosses. As an example, he crossed pure-breeding peas that had round yellow seeds with pure-breeding peas that had wrinkled green seeds.

All the F_1 (first-generation) hybrids had round yellow seeds. This is not surprising, as these characters are due to dominant alleles. When Mendel allowed the F_1 plants to self-pollinate, he found that four different phenotypes appeared in the F_2 generation:

> round yellow – one of the original parental phenotypes
>
> round green – a new phenotype
>
> wrinkled yellow – another new phenotype
>
> wrinkled green – the other original parental phenotype.

If the genotype of the F_1 hybrids is SsYy, the gametes produced by these hybrids could contain either S or s with either Y or y. The four possible gametes are SY, Sy, sY and sy. If the inheritance of these two genes is independent, then the chance of a gamete containing S or s will not affect its chance of containing either Y or y. The chance of a gamete containing each allele is $\frac{1}{2}$, so the combined chance of containing two specific alleles is $\frac{1}{2} \times \frac{1}{2} = \frac{1}{4}$. This theory that the alleles of two genes pass into gametes without influencing each other is called **independent assortment**.

A Punnett square is a diagram that is used to help form predictions about the outcome of a particular breeding event where independent assortment of alleles is occurring. It is used to directly determine the probability of a particular genotype but can also be used to determine the probability of a particular phenotype. It is a table that is used to systematically combine every possible combination of maternal allele and paternal allele.

To create a Punnett square:

Step 1: determine the genotypes of the parents.

Step 2: identify the different varieties of gametes the parents can produce. Note by Mendel's principle of segregation, **one** copy of each gene is present in the gamete. A common mistake is for students to include two copies or no copies.

Step 3: Set up a Punnett grid for your cross, with as many rows as there are unique male gametes (sperm) and as many columns as there are unique female gametes (eggs).

Step 4: Fill in the offspring genotypes inside the table by matching the egg allele at the top of the column with the sperm allele from the row.

Step 5: Determine the genotype ratio for the predicted offspring.

Step 6: Determine the phenotype ratio for the predicted offspring.

The Punnett grid (figure 1) shows how a ratio of F_2 phenotypes is predicted, on the basis of independent assortment. Create a tally chart to verify that the predicted phenotypic ratio is:

9 yellow round: 3 green round: 3 yellow wrinkled: 1 green wrinkled

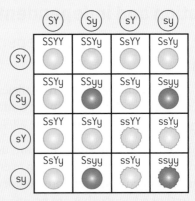

▲ Figure 1 Punnett grid for a dihybrid cross

Making predictions using Punnett squares

Calculation of the predicted genotypic and phenotypic ratio of offspring of dihybrid crosses involving unlinked autosomal genes.

Use the following questions to develop your skill in dihybrid cross calculations.

1   A farmer has rabbits with two particular traits, each controlled by a separate gene. Coat colour brown is completely dominant to white. Tailed is completely dominant to tail-less. A brown, tailed male rabbit that is heterozygous at both loci is crossed with a white, tail-less female rabbit. A large number of offspring is produced with only two phenotypes: brown and tailed, white and tail-less, and the two types are in equal numbers.

   (i) State both parents' genotypes and the gametes that are produced by each during the process of meiosis.

   Male genotype: ....

   Female genotype: ....

   Male gametes: ....

   Female gametes: ....

   (ii) Predict the genotypic and phenotypic ratios of the F_2 generation. Show your working.

2   In peas the allele for smooth seed (S) is dominant over the allele for wrinkled seed (s). The allele for tall plants (T) is dominant over the allele for short plants (t).

   A pure breeding tall plant with smooth seeds was crossed with a pure breeding short plant with wrinkled seeds. All the F_1 plants were tall with smooth seeds. Two of these F_1 plants were crossed and four different phenotypes were obtained in the 320 plants produced.

   How many tall plants with wrinkled seeds would you expect to find?

3   In *Drosophila* the allele for normal wings (W) is dominant over the allele for vestigial wings (w) and the allele for normal body (G) is dominant over the allele for ebony body (g). If two *Drosophila* with the genotypes Wwgg and wwGg are crossed together, what ratio of phenotypes is expected in the offspring?

 Exceptions to Mendel's rules

Looking for patterns, trends and discrepancies: Mendel used observations of the natural world to find and explain patterns and trends. Since then, scientists have looked for discrepancies and asked questions based on further observations to show exceptions to the rules. For example, Morgan discovered non-Mendelian ratios in his experiments with *Drosophila*.

Thomas Hunt Morgan discovered non-Mendelian ratios in his experiments with the fruit fly, *Drosophila melanogaster*. Morgan wasn't the first scientist to use the fruit fly as a research organism, but his success popularized its use. Many students of biology will have some experience of breeding fruit flies – either actually or virtually.

At the start of his investigations, Morgan was critical of Mendel's theory of inheritance and wasn't convinced by aspects of the emerging chromosomal theory of inheritance. He believed that the variation that he saw in organisms was better explained by environmental influence. However, his own subsequent observations of the pattern of inheritance of white eyes led him to reconsider his own perspective. At the same time as his results reinforced aspects of Mendel's conclusions, Morgan's studies identified exceptions to Mendel's principle of independent assortment.

Activity

How can the presence of the three white-eyed flies among 1,200 in Morgan's experiment be explained in the F_1 generation?

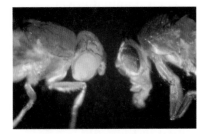

▲ Figure 2 The fly on the right, with the red compound eye, is the common, or wild type. The fly on the left is a mutant type known as White Miniature Forked. It has white eyes, shorter wings than the normal fly, and the bristles on its face and body are distorted and forked. *D. melanogaster* has been used for many years in genetic studies because it is easy to raise in large numbers, reproduces rapidly, and many of its mutations are easy to spot under a low-powered light microscope

▲ Figure 3 Coloured scanning electron micrograph of a fruit fly (*Drosophila melanogaster*) four wing mutant. Two of the mutant's four wings are visible (blue) on one side of its body (brown). The multi-faceted right eye of the fly (red) is also visible. The wild-type fly has two wings

 ## Implications of Morgan's discovery of sex linkage

Morgan's discovery of non-Mendelian ratios in *Drosophila*.

After breeding thousands of *Drosophila* in his "fruit fly room" at Columbia University, Morgan noticed a single fruit fly with white eyes instead of the normal red colour. He mated this white-eyed specimen to an ordinary red-eyed fly. Although the first generation involving over 1,200 offspring was all red-eyed except for three flies, white-eyed flies appeared in much larger numbers in the second generation. In the second generation, approximately three red-eyed flies appeared for every white-eyed fly as predicted by Mendel's principle of dominance and recessiveness. What surprised Morgan was that all of the white-eyed flies in the second generation were male. Mendel's principle of dominance and recessiveness would predict a three to one ratio of red to white in both males and females, but all of the females had red eyes.

Morgan began to reverse his earlier position to entertain the possibility that association of eye colour and sex in fruit flies had a physical basis in the chromosomes. One of *Drosophila's* four chromosome pairs was thought by other researchers to be used for sex determination. Morgan's own idea was that gender was determined by quantity of chromatin. Males possess the XY chromosome pair while flies with the XX chromosome are female. Since the Y chromosome is smaller, Morgan was technically correct. However, if the factor for eye colour was located exclusively on the X chromosome, Morgan could use Mendelian rules for inheritance of dominant and recessive traits to explain his observations. Further, the chromosomal theory could explain why sex and eye colour did not assort independently.

Linked genes

Gene loci are said to be linked if they are on the same chromosome.

Further investigations led Morgan to discover more mutant traits in *Drosophila* – about two dozen between 1911 and 1914. One of Morgan's students showed that a yellow-bodied mutant characteristic was inherited in the same way as white eyes. Further, these two mutations were not inherited independently. They were able to establish the notion of gene linkage through their experiments.

Morgan, and other geneticists working in the early part of the 20th century, went on to discover a group of genes that were all located on the X chromosome of *Drosophila*. By careful crossing experiments they were able to show that these genes were arranged in a linear sequence along the X chromosome. Groups of genes were then assigned to the other chromosomes in *Drosophila*, again arranged in a specific sequence. The same pattern has been found in other species – each particular gene is found in a specific position on one chromosome type. This is called the

locus of a gene. If two chromosomes have the same sequence of genes they are **homologous**. Homologous chromosomes are not usually identical to each other because, for at least some of the genes on them, the alleles will be different.

Since Morgan's time it has been discovered that all the genes on a chromosome are part of one DNA molecule. In *Drosophila* there are eight chromosomes in diploid nuclei. In males one of these is an X and another is a Y chromosome. In females two of them are X chromosomes. The other six chromosomes are common to males and females – they are called **autosomes**.

Diploid nuclei have two of each type of autosome, so in *Drosophila* there are three types of autosome. Geneticists working early in the 20th century found four groups of linked genes in *Drosophila*, corresponding to the three types of autosome and to the X chromosome.

There are two types of linkage – **autosomal gene linkage**, when the genes are on the same autosome, and **sex linkage**, when the genes are located on the X chromosome.

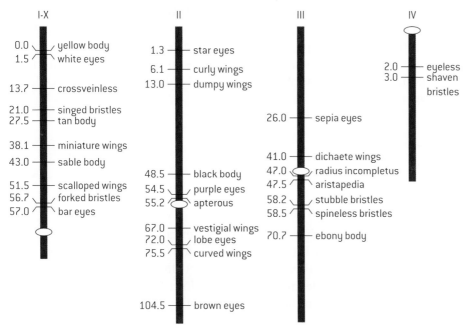

▲ Figure 4 Map showing linkage groups of the four fruit fly chromosomes. Chromosome 1 is the X chromosome

Types of variation

Variation can be discrete or continuous.

The differences between individual organisms are referred to as variation. Where individuals fall into a number of distinct categories, the variation is discrete or discontinuous. Blood types are an example of discrete variation. While there are several blood types, there are no in-between categories. Figure 5 shows the frequency of each of the blood type phenotypes in a population sample from Iceland.

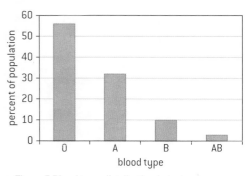

▲ Figure 5 Blood type distribution in Iceland

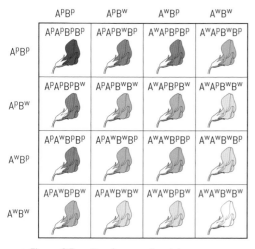

▲ Figure 6 Results of a cross involving polygenic inheritance

$$1:2:1$$

$$1:4:6:4:1$$

$$1:6:15:20:15:6:1$$

$$1:8:28:56:70:56:28:8:1$$

▲ Figure 7 Pascal's triangle

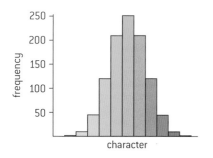

▲ Figure 8 Variation due to polygenic inheritance

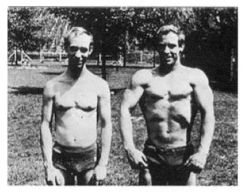

Otto   Ewald

▲ Figure 9

Continuous variation

The phenotypes of polygenic characteristics tend to show continuous variation.

There are examples of inheritance in which two or more genes affect the same character. The genes have an additive effect. Mendel discovered an example of this in beans, where a cross between a purple-flowered plant and a white-flowered plant gave purple-flowered plants in the F_1 generation, but when these were self-pollinated, the expected 3:1 ratio did not occur and instead a range of flower colours was seen. This can be explained if there are two unlinked genes, with co-dominant alleles (see figure 6). Self-pollination of the F_1 should give five different shades of flower colour in a ratio of 1:4:6:4:1. If the number of unlinked genes, with co-dominant alleles, were larger then there would be more phenotypic variants. The number and frequency of variants can be predicted using alternate rows of Pascal's triangle (figure 7). A frequency distribution is shown (figure 8) for a character affected by five genes with co-dominant alleles. As the number of genes increases, the distribution becomes increasingly close to the **normal distribution**. Many characters in humans and other organisms are close to the normal distribution, for example the mass of bean seeds, height in humans and intelligence in humans. The closeness to a normal distribution suggests that more than one gene is involved. This situation is called polygenic inheritance.

🌐 Environmental influence

Polygenic traits such as human height may also be influenced by environmental factors.

When the variation due to polygenic inheritance is examined carefully, it is usually found to be continuous – there is a complete range of variation, rather than the distinct classes that might be predicted from Mendelian inheritance. This is because the differences in phenotype between the classes are subtle and the effects of the environment blur these differences so much that they are undetectable.

Skin colour in humans is an example of continuous variation. It is partly due to the environment – sunlight stimulates the production of the black pigment melanin in the skin. It is also due to the influence of several genes, so is an example of polygenic inheritance.

Figure 9 shows an image of identical twins who were both competitive athletes with long-term differences in their dietary and exercise regimes. A number of traits which show continuous variation are visible in the photo including height. Note the differences in their height in the second image, despite having identical genomes.

Figure 10 shows genetically identical mice that differ in terms of the nutrition received by their mothers.

The pups of mice on the standard diet generally had golden fur. But a high proportion of those born to mice on the enriched diet had dark brown fur.

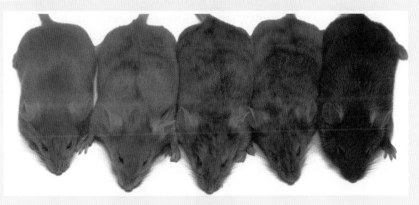

▲ Figure 10 Genetically identical mice showing variation in size and coat colour as a consequence of environmental differences in utero

Identifying recombinants

Identification of recombinants in crosses involving two linked genes.

William Bateson, Edith Saunders and Reginald Punnett discovered the first exception to the law of independent assortment in 1903. When they crossed sweet pea plants with purple flowers and long pollen grains with plants with red flowers and round pollen grains, all the F_1 hybrids had purple flowers and long pollen grains. When these F_1 plants were self-pollinated, four phenotypes were observed in the F_2 generation, but not in the familiar 9:3:3:1 ratio. The cross was repeated with larger numbers and the F_2 contained the numbers of plants shown in table 1.

Phenotype	Observed frequency	Observed %	9:3:3:1 %
purple long	4,831	69.5	56.25
purple round	390	5.6	18.75
red long	393	5.6	18.75
red round	1,338	19.3	6.25

▲ Table 1

Although the observed percentages do not fit the 9:3:3:1 ratio, results like this were not unexpected. Genes were thought to be part of chromosomes by a number of scientists at the time, and as there are far more genes than chromosomes, some genes must be found together on the same chromosome. Alleles of these genes would therefore not follow the law of independent assortment and would pass together into a gamete.

This is seen in the results for the sweet pea cross – there were more of the purple long and red round plants than expected. This is because these were the original parental combinations of alleles. This pattern of inheritance is called gene linkage. Since 1903, many more examples have been found, always with a higher frequency of the parental combinations than predicted from Mendelian ratios.

A genetic diagram explaining the cross in sweet peas is shown in figure 11, using lines to symbolize the chromosomes on which the linked genes are located.

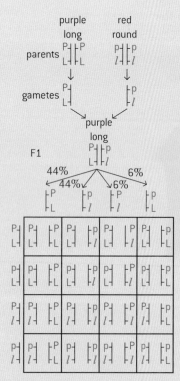

▲ Figure 11 Cross involving gene linkage

The linkage between pairs of genes in a linkage group is not usually complete and new combinations of alleles are sometimes formed. This happens as a result of crossing over, which was described as a part of meiosis in sub-topics 3.3 and 10.1. Figure 12 shows how crossing over gives new allele combinations. The formation of a chromosome or DNA with a new combination of alleles is **recombination**. An individual that has this recombinant chromosome and therefore has a different combination of characters from either of the original parents is called a **recombinant**.

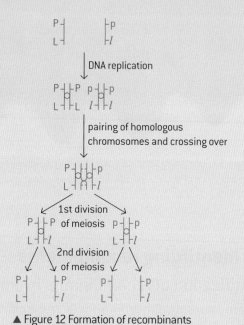

▲ Figure 12 Formation of recombinants

Data-based questions: Gene linkage in *Zea mays*

Corncobs are often used for showing inheritance patterns. All the grains on a cob have the same female parent, and with careful pollination they can also have the same male parent. A variety with coloured and starchy grains was crossed with a variety with white and waxy grains. The F_1 grains were all coloured and starchy. The F_1 plants grown from these grains were crossed ($F_1 \times F_1$).

1   Calculate the expected ratio of F_2 plants, assuming that the genes for coloured/white and starchy/waxy grains are unlinked. Use a genetic diagram to show how you reached your answer. [3]

2   The actual frequencies were:

coloured starchy	1,774
coloured waxy	263
white starchy	279
white waxy	420

Using this data, deduce whether the genes for coloured/white and starchy/waxy are linked. [2]

A variety with coloured and shrunken grains was crossed with a variety with white and non-shrunken grains. The F_1 grains were all coloured and non-shrunken. The F_1 plants grown from these grains were test crossed using pollen from a homozygous recessive variety with white shrunken grains.

3   Calculate the expected ratio of F_2 plants, assuming that the genes are unlinked, using a diagram to show how you reached your answer. [2]

4 The actual frequencies were:

coloured non-shrunken	638
coloured shrunken	21,379
white non-shrunken	21,096
white shrunken	672

Using this data, deduce whether the genes for coloured/white and non-shrunken/shrunken are linked. [2]

5 Deduce whether the genes for starchy/waxy and non-shrunken/shrunken are linked. [1]

Chi-squared tests are used to determine whether the difference between an observed and expected frequency distribution is statistically significant

Use of a chi-squared test on data from dihybrid crosses.

In 1901, Bateson reported one of the first post-Mendelian studies of a cross involving two traits. White leghorn chickens with large "single" combs, were crossed to Indian game fowl with dark feathers and small "pea" combs. All of the F_1 were white with pea combs, and the ratio of F_2 phenotypes involving 190 offspring was: 111 white pea, 37 white single, 34 dark pea and 8 dark single. The expected ratio is 9:3:3:1. The observed ratio was different. Were the differences between observed and expected due to sampling error or were the differences statistically significant, suggesting that the traits do not assort independently? This can be tested using the chi-squared test.

There are two possible hypotheses:

H_0 : the traits assort independently

H_1 : the traits do not assort independently

We can test these hypotheses using a statistical procedure – the chi-squared test.

Method for chi-squared test

1 Draw up a contingency table of observed frequencies, which are the numbers of individuals of each phenotype.

2 Calculate the expected frequencies, assuming independent assortment, for each of the four phenotypes. Each expected frequency is calculated from values on the contingency table using the expected probability from the Punnett grid multiplied by the actual total.

	White pea	White single	Dark pea	Dark single	Total
Observed	111	37	34	8	190
Expected	$\left(\frac{9}{16}\right) \times 190$ $= 106.9$	$\left(\frac{3}{16}\right) \times 190$ $= 35.6$	$\left(\frac{3}{16}\right) \times 190$ $= 35.6$	$\left(\frac{1}{16}\right) \times 190$ $= 11.9$	190

3 Determine the number of degrees of freedom, which is one less than the total number of classes $(4 - 1) = 3$ degrees of freedom.

TOK

When might a persuasive statistical representation be praised as "effective", or, in contrast, condemned as "manipulative"?

In popular discourse, there is a distrust of knowledge claims supported by statistics. One aphorism popularized by Mark Twain is that there are three kinds of lies: lies, damned lies and statistics. The misuse of statistics can be inadvertent or it can be intentional. Some examples include:

- conclusions can be based on statistical analysis of samples that are selected with bias and are therefore not representative of the population

- rejection of the alternate hypothesis can be mistakenly interpreted as proof of the null hypothesis

- a sample size that is too small is likely to be poorly representative of the population even if it is selected without bias

- experimenters may discount data that they believe does not conform to theory.

The effects of such issues can be minimized through a diligent and honest approach meaning that it is the user rather than the tool that needs to be the subject of scrutiny.

4   Find the critical region for chi-squared from a table of chi-squared values, using the degrees of freedom that you have calculated and a significance level (p) of 0.05 (5%). The critical region is any value of chi-squared larger than the value in the table.

				Critical values of the χ^2 distribution					
	p								
df	0.995	0.975	0.9	0.5	0.1	0.05	0.025	0.01	0.005	df
1	0.000	0.000	0.016	0.455	2.706	3.841	5.024	6.635	7.879	1
2	0.010	0.051	0.211	1.386	4.605	5.991	7.378	9.210	10.597	2
3	0.072	0.216	0.584	2.366	6.251	7.815	9.348	11.345	12.838	3

At the 0.05 level of significance, the critical value is 7.815.

5   Calculate chi-squared using this equation:

$$\chi^2 = \sum \frac{(obs - exp)^2}{exp}$$
$$= \frac{(111 - 106.9)^2}{106.9} + \frac{(37 - 35.6)^2}{35.6} + \frac{(34 - 35.6)^2}{35.6} + \frac{(8 - 11.9)^2}{11.9}$$
$$= 1.56$$

6   Compare the calculated value of chi-squared with the critical region.

- If the calculated value is in the critical region, there is evidence at the 5% level for an association between the two traits; i.e., the traits are linked. We can reject the hypothesis H_0.

- If the calculated value is not in the critical region, because it is equal to or below the value obtained from the table of chi-squared values, H_0 is not rejected. There is no evidence at the 5% level for an association between the two traits.

The probability value is outside of the critical region (0.9 >p>0.5) so we reject the alternate hypothesis and accept the null hypothesis.

Data-based questions: Using the chi-squared test

Warren and Hutt (1936) test-crossed a double heterozygote for two pairs of alleles in hens: one for the presence (Cr) or absence (cr) of a crest and one for white (I) or non-white (i) plumage.

For their F_2 cross, there was a total of 754 offspring.

337 were white, crested;

337 were non-white, non-crested;

34 were non-white crested; and

46 were white, non-crested.

1   Construct a contingency table of observed values.   [4]

2   Calculate the expected values, assuming independent assortment.   [4]

3   Determine the number of degrees of freedom.   [2]

4   Find the critical region for chi-squared at a significance level of 5%.   [2]

5   Calculate chi-squared.   [4]

6   State the two alternative hypotheses, H_0 and H_1 and evaluate them using the calculated value for chi-squared.   [4]

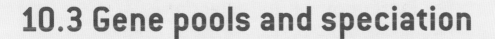

10.3 Gene pools and speciation

Understanding

→ A gene pool consists of all the genes and their different alleles, present in an interbreeding population.

→ Evolution requires that allele frequencies change with time in populations.

→ Reproductive isolation of populations can be temporal, behavioural or geographic.

→ Speciation due to divergence of isolated populations can be gradual.

→ Speciation can occur abruptly.

Applications

→ Identifying examples of directional, stabilizing and disruptive.

→ Speciation in the genus *Allium* by polyploidy.

Skills

→ Comparison of allele frequencies of geographically isolated populations.

Nature of science

→ Looking for patterns, trends and discrepancies: patterns of chromosome number in some genera can be explained by speciation due to polyploidy.

Gene pools

A gene pool consists of all the genes and their different alleles, present in an interbreeding population.

The most commonly accepted definition of a species is the biological species concept. This defines a species as a group of potentially interbreeding populations, with a common gene pool that is reproductively isolated from other species. Some populations of the same species are geographically isolated so it is possible for multiple gene pools to exist for the same species.

Individuals that reproduce contribute to the gene pool of the next generation. Genetic equilibrium exists when all members of a population have an equal chance of contributing to the future gene pool.

Allele frequency and evolution

Evolution requires that allele frequencies change with time in populations.

Evolution is defined as the cumulative change in the heritable characteristics of a population over time. Evolution can occur due to a number of reasons such as mutations introducing new alleles, selection pressures favouring the reproduction of some varieties over others and barriers to gene flow emerging between different populations. If a population is small, random events can also have a significant effect on allele frequency.

Activity

In the cross depicted in figure 1, the frequency of flower colour phenotypes in Japanese four o'clocks is shown over three generations. The genotype C^RC^R yields red flowers, the genotype C^WC^W yields white flowers and because the alleles are co-dominant, the genotype C^RC^W yields pink flowers:

- in the first generation, 50% of the population is red and 50% is white

- in the second generation, 100% of the flowers are pink

- in the third generation, there are 50% pink, 25% white and 25% red.

Show that the allele frequency is 50% C^R and 50% C^W in each of the three generations. While phenotype frequencies can change between generations, it is possible that allele frequency is not changing. This population is not evolving because allele frequencies are not changing.

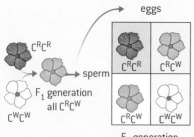

▲ Figure 1 A change in phenotypic frequency between generations does not necessarily indicate that evolution is occurring

🌐 Patterns of natural selection

Identifying examples of directional, stabilizing and disruptive selection.

Fitness of a genotype or phenotype is the likelihood that it will be found in the next generation. Selection pressures are environmental factors that act selectively on certain phenotypes resulting in natural selection. There are three patterns of natural selection: stabilizing selection, disruptive selection, and directional selection.

In stabilizing selection, selection pressures act to remove extreme varieties. For example, average birth weights of human babies are favoured over low birth weight or high birth weight. A clutch is the number of eggs a female lays in a particular reproductive event. Small clutch sizes may mean that none of the offspring survive into the next generation. Very large clutch sizes may mean higher mortality as the parent cannot provide adequate nutrition and resources and may impact their own survival to the next season. This means that a medium clutch size is favoured.

In disruptive natural selection, selection pressures act to remove intermediate varieties, favouring the extremes. One example is in the red crossbill *Loxia curvirostra*. The asymmetric lower part of the bill of red crossbills is an adaptation to extract seeds from conifer cones. An ancestor with a "straight" bill could have experienced disruptive selection, given that a lower part of the bill crossed to either side enables a more efficient exploitation of conifer cones. Both left over right and right over left individuals exist within the same population allowing them to access seeds from cones hanging in different positions.

In directional selection, the population changes as one extreme of a range of variation is better adapted.

Data-based questions: Stabilizing selection

A population of bighorn sheep (*Ovis canadensis*) on Ram Mountain in Alberta, Canada, has been monitored since the 1970s. Hunters can buy a licence to shoot male bighorn sheep on the mountain. The large horns of this species are very attractive to hunters, who display them as hunting trophies.

Most horn growth takes place between the second and the fourth year of life in male bighorn sheep. They use their horns for fighting other males during the breeding season to try to defend groups of females and then mate with them. Figure 2 shows the mean horn length of four-year-old males on Ram Mountain, between 1975 and 2002.

a) Outline the trend in horn length over the study period.

b) Explain the concept of directional selection referring to this example.

c) Discuss the trade-off between short and long horns as an adaptation in this case.

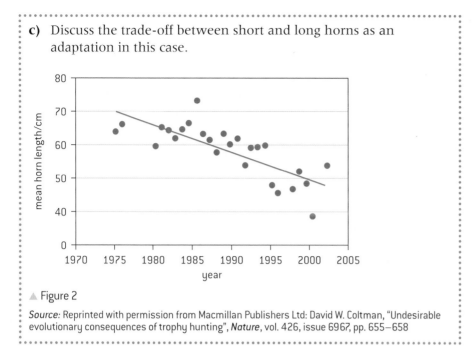

▲ Figure 2

Source: Reprinted with permission from Macmillan Publishers Ltd: David W. Coltman, "Undesirable evolutionary consequences of trophy hunting", *Nature*, vol. 426, issue 6967, pp. 655–658

Data-based questions

Researchers carried out a study on 3,760 children born in a London hospital over a period of 12 years. Data was collected on the children's mass at birth and their mortality rate. The purpose of the study was to determine how natural selection acts on mass at birth. The chart in figure 3 shows the frequency of babies of each mass at birth. The line superimposed on the bar chart indicates the percentage mortality rate (the children that did not survive for more than 4 weeks).

a) Identify the mode value for mass at birth.

b) Identify the optimum mass at birth for survival.

c) Outline the relationship between mass at birth and mortality.

d) Explain how this example illustrates the pattern of natural selection called directional selection.

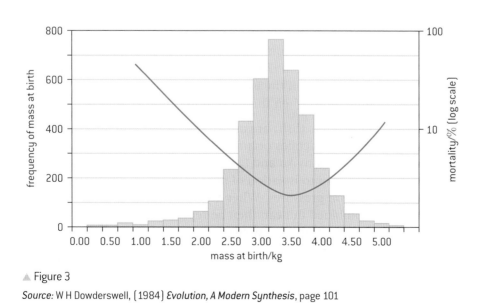

▲ Figure 3

Source: W H Dowderswell, (1984) *Evolution, A Modern Synthesis*, page 101

457

Data-based questions

In coho salmon (*Oncorhynchus kisutch),* some males reach maturity as much as 50% earlier and as small as 30% of the body size of other males in the population. Success in spawning (breeding) depends on the male releasing sperm in close proximity to the egg-laying female. Small and large males employ different strategies to gain access to females. The small-sized males called jacks are specialized at "sneaking". The large-sized males are specialized at fighting and coercing females to spawn. In contrast, intermediate-sized males are at a competitive disadvantage to both jacks and large males as they are more targeted for fights which they lose and are more likely to be prevented from sneaking. The graph in figure 4 shows the average proximity to females achieved by the two strategies.

a) Determine the mean proximity to females achieved by 35–39 cm males by:

  (i) sneaking

  (ii) fighting.

b) Determine the size range that gets nearest to the females by:

  (i) fighting

  (ii) sneaking.

c) Identify a size of male fish that never gets within 100 cm (1 m) by following either strategy.

d) Explain how this example illustrates the pattern of natural selection known as disruptive selection.

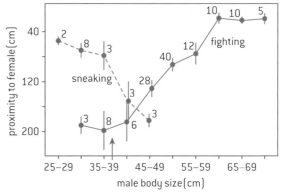

Nature, Vol. 313, No. 5997, pp. 47–48, 3 January 1985

▲ Figure 4 Effect of body size and courting strategy on proximity to females

There are different categories of reproductive isolation

Reproductive isolation of populations can be temporal, behavioural or geographic.

Speciation is the formation of a new species by the splitting of an existing population. Various barriers can isolate the gene pool of one population from that of another population. Speciation may occur when this happens. If the isolation occurs because of geographic separation of populations, then the speciation is termed allopatric speciation.

The cichlids (fish) are one of the largest families of vertebrates. Most species of cichlids occur in three East African lakes, Lake Victoria, Lake Tanganyika and Lake Malawi. Annual fluctuations in water levels lead to isolation of populations that are then subject to different selection pressures. When the rainy season comes, the populations are recombined but can then be reproductively isolated. This can result in the formation of new species.

Sometimes isolation of gene pools occurs within the same geographic area. If speciation occurs, then the process is termed sympatric speciation. For example, isolation can be behavioural. When closely

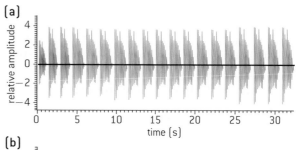

Data-based questions: Lacewing songs

Songs are part of the process of mate selection in members of different species within the genus *Chrysoperla* (lacewings). Males and females of the same species have precisely the same "song" and during the pre-mating period take turns making the songs. The oscillograph for two species of lacewings are shown in figure 5.

1   Compare the songs of the two species of lacewings.   [3]

2   Explain why differences in mating songs might lead to speciation.   [3]

3   The ranges of the two species currently overlap. Suggest how differences in song could have developed:

    a)   by allopatric speciation

    b)   by sympatric speciation.   [4]

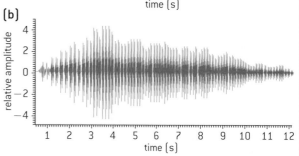

▲ Figure 5 Pre-mating songs of lacewings: (a) *C. lucasina* and (b) *C. mediterranea*. *C. lucasina* ranges across most of Europe and eastward into western Asia, as well as across the northern quarter of Africa. *C. mediterranea* ranges across southern to central Europe and across the north African Mediterranean

related individuals differ in their courtship behaviour, they are often only successful in attracting members of their own population.

There can be temporal isolation of gene pools in the same area. Populations may mate or flower at different seasons or different times of day. For example, three tropical orchid species of the genus *Dendrobium* each flower for a single day. Flowering occurs in response to sudden drops in temperature in all three species. However, the lapse between the stimulus and flowering is 8 days in one species, 9 in another, and 10 to 11 in the third. Isolation of the gene pools occurs because, at the time the flowers of one species are open, those of the other species have already withered or have not yet matured.

Different populations have different allele frequencies

Comparison of allele frequencies of geographically isolated populations.

Online databases such as the Allele Frequency Database (AlFreD) hosted by Yale University contains the frequencies of a variety of human populations. Most human populations are no longer in geographic isolation because of the ease of travel and the significant culture to culture contact that exists due to globalization. Nonetheless, patterns of variation do exist, especially when comparing remote island populations with mainland populations.

*Pan*I is a gene in cod fish that codes for an integral membrane protein called pantophysin. Two alleles of the gene, *Pan*I^A and *Pan*I^B, code for versions of pantophysin that differ by four amino acids in one region of the protein. Samples of cod fish were collected from 23 locations in the north Atlantic and were tested to find the proportions of *Pan*I^A and *Pan*I^B alleles in each population. The results are shown in pie charts, numbered 1–23, on the map in figure 6. The proportions of alleles in a population are called the allele frequencies. The frequency of an allele can vary from 0.0 to 1.0. The light grey sectors of the pie charts show the allele frequency of $PanI^A$ and the black sectors show the allele frequency of $PanI^B$.

1 State the **two** populations with the highest $PanI^B$ allele frequencies. [2]

2 Deduce the allele frequencies of a population in which half of the cod fish had the genotype $PanI^A\ PanI^A$, and half had the genotype $PanI^A\ PanI^B$. [2]

3 Suggest two populations which are likely geographically isolated. [2]

4 Suggest two possible reasons why the $PanI^B$ allele is more common in population 14 than population 21. [2]

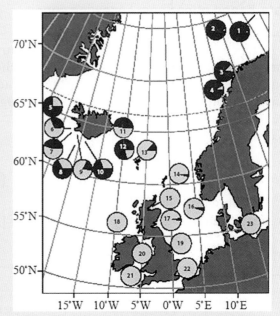

▲ Figure 6

Source: R A J Case, *et al.*, (2005), *Marine Ecology Progress Series*, 201, pages 267–278

Gradualism in speciation

Speciation due to divergence of isolated populations can be gradual.

There are two theories about the pace of evolutionary change. Gradualism, as depicted in figure 7, is the idea that species slowly change through a series of intermediate forms. The axis label "structure" might refer to such things as beak length in birds or cranial capacity in hominids.

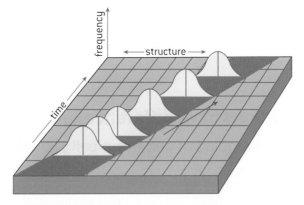

▲ Figure 7 In the gradualist framework, new species emerge from a long sequence of intermediate forms

Gradualism was, for a long time, the dominant framework in palaeontology. However, it was confronted by gaps in the fossil record, i.e. an absence of intermediate forms. Gradualism predicted that evolution occurred by a long sequence of continuous intermediate forms. The absence of these intermediate forms was explained as imperfections in the fossil record.

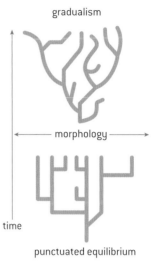

Punctuated equilibrium

Speciation can occur abruptly.

Punctuated equilibrium holds that long periods of relative stability in a species are "punctuated" by periods of rapid evolution. According to the theory of punctuated equilibrium, gaps in the fossil record might not be gaps at all, as there was no long sequence of intermediate forms. Events such as geographic isolation (allopatric speciation) and the opening of new niches within a shared geographic range can lead to rapid speciation.

Rapid change is much more common in organisms with short generation times like prokaryotes and insects.

Figure 8 compares the two models. The top model shows the gradualist model with slow change over geological time. The punctuated equilibrium model on the bottom consists of relatively rapid changes over a short period of time followed by periods of stability.

▲ Figure 8

 Polyploidy can lead to speciation

Looking for patterns, trends and discrepancies: patterns of chromosome number in some genera can be explained by speciation due to polyploidy.

A polyploid organism is one that has more than two sets of homologous chromosomes. Polyploidy can result from hybridization events between different species. There are also polyploids whose chromosomes originate from the same ancestral species. This can occur when chromosomes duplicate in preparation for meiosis but then meiosis doesn't occur. The result is a diploid gamete that when fused with a haploid gamete produces a fertile offspring. In other words, the polyploid has now become reproductively isolated from the original population. The polyploid plant can self-pollinate or it can mate with other polyploid plants. Polyploidy can lead to sympatric speciation.

Polyploidy occurs most commonly in plants, though it does also occur in less complex animals. The red viscacha (*Tympanoctomys barrerae)*, a rodent from Argentina, has the highest chromosome number of any mammal and it has been hypothesized that this is the result of polyploidy. Its chromosome number is 102 and its cells are roughly twice normal size. Its closest living relative is *Octomys mimax*, the Andean viscacha-rat of the same family, whose 2n = 56. Researchers propose that an *Octomys*-like ancestor produced tetraploid offspring (i.e. 4n = 112) that were reproductively isolated from their parent species, eventually shedding some of the additional chromosomes gained at this doubling. Recent scholarship has tested this hypothesis but results are

ambiguous: probes detected only two copies of each autosome pair but it has also been observed that there are several genes that exist in four copies.

▲ Figure 9 *Tympanoctomys barrerae*

▲ Figure 10 *Octomys mimax*

Polyploidy has occurred frequently in *Allium*

Speciation in the genus *Allium* by polyploidy.

Estimates of the number of species of angiosperms that have experienced a polyploidy event range between 50 to 70%.

The *Allium* genus includes onions, leeks, garlic and chives, and as such has played an important role in the food of multiple cultures. Determining the number of species in the genus presents a challenge to taxonomists as polyploidy events are common within the genus. These result in a number of reproductively isolated but otherwise similar populations.

Many species of *Allium* reproduce asexually and polyploidy may confer an advantage over diploidy under certain selection pressures.

Wild onion *(Allium canadense)* is a native of North America. The diploid number for the plant is 14. However, there are variants such as *A. c. ecristatum* (2n = 28) and *A. c. lavendulae* (2n = 28).

Allium angulosum and *Allium oleraceum* are two species that occur in Lithuania. One is a diploid plant with 16 chromosomes and one is a tetraploid plant with 32 chromosomes.

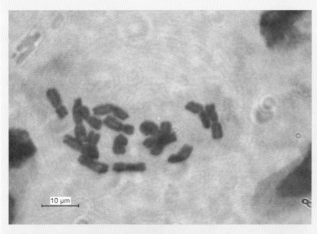

▲ Figure 11 Metaphase chromosomes of *Allium angulosum*, 2n=16

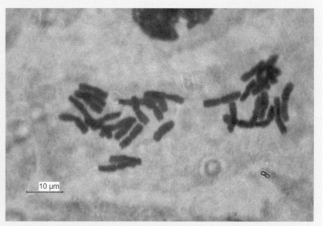

▲ Figure 12 Metaphase chromosomes of *Allium oleraceum*, 2n=32

Questions

1 Identify the stages of meiosis shown in figures 13 and 14.

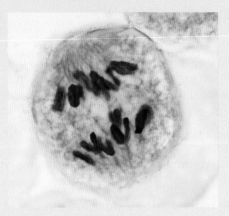

▲ Figure 13

▲ Figure 14

2 The DNA content of cells can be estimated using a stain that binds specifically to DNA. A narrow beam of light is then passed through a stained nucleus and the amount of light absorbed by the stain is measured, to give an estimate of the quantity of DNA. The results in table 1 are for leaf cells in a species of bog moss (*Sphagnum*) from the Svalbard islands.

a) Compare the DNA content of the bog mosses. [2]

b) Suggest a reason for six of the species of bog moss on the Svalbard islands all having the same number of chromosomes in their nuclei. [2]

c) *S. arcticum* and *S. olafii* probably arose as new species when meiosis failed to occur in one their ancestors.

(i) Deduce the chromosome number of nuclei in their leaf cells. Give two reasons for your answer. [3]

(ii) Suggest a disadvantage to *S. arcticum* and *S. olafii* of having more DNA than other bog mosses. [1]

d) It is unusual for plants and animals to have an odd number of chromosomes in their nuclei. Explain how mosses can have odd numbers of chromosomes in their leaf cells. [2]

Sphagnum species	Mass of DNA/pg	Number of chromosomes
S. aongstroemii	0.47	19
S. arcticum	0.95	
S. balticum	0.45	19
S. fimbriatum	0.48	19
S. olafii	0.92	
S. teres	0.42	19
S. tundrae	0.44	19
S. warnstorfii	0.48	19

▲ Table 1

3 The mechanisms of speciation in ferns have been studied in temperate and tropical habitats. One group of three species from the genus *Polypodium* lives in rocky areas in temperate forests in North America. Members of this group have similar morphology (form and structure). Another group of four species from the genus *Pleopeltis* live at different altitudes in tropical mountains in Mexico and Central America. Members of this group are morphologically distinct.

Data from the different species within each group was compared in order to study the mechanisms of speciation.

Genetic identity was determined by comparing the similarities of certain proteins and genes in each species. Values between 0 and 1 were assigned to pairs of species to indicate the degree of similarity in genetic identity. A value of 1 would mean that all the genetic factors studied were identical between the species being compared.

a) Compare the geographic distributions of the two groups. [1]

b) (i) Identify, giving a reason, which group, *Polypodium* or *Pleopeltis*, is most genetically diverse. [1]

   (ii) Identify the **two** species that are most similar genetically. [1]

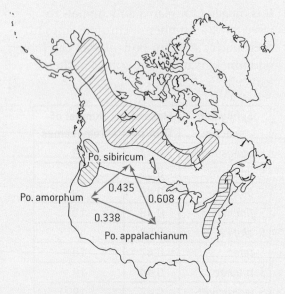

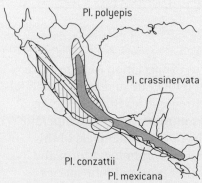

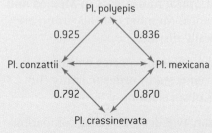

▲ Figure 15 The approximate distribution in North America of the three species of *Polypodium (Po.)* and a summary of genetic identity

Source: C Haufler, E Hooper and J Therrien, (2000), *Plant Species Biology,* **15**, pages 223–236

c) Suggest how the process of speciation could have occurred in *Polypodium*. [1]

d) Explain which of the two groups has most probably been genetically isolated for the longest period of time. [2]

4 In *Zea mays*, the allele for coloured seed (C) is dominant over the allele for colourless seed (c). The allele for starchy endosperm (W) is dominant over the allele for waxy endosperm (w). Pure breeding plants with coloured seeds and starchy endosperm were crossed with pure breeding plants with colourless seeds and waxy endosperm.

a) State the genotype and the phenotype of the F_1 individuals produced as a result of this cross.

   genotype ....

   phenotype ....[2]

b) The F_1 plants were crossed with plants that had the genotype c c w w. Calculate the expected ratio of phenotypes in the F_2 generation, assuming that there is independent assortment.

   Expected ratio .... [3]

   The observed percentages of phenotypes in the F_2 generation are shown below.

coloured starchy	37%
colourless starchy	14%
coloured waxy	16%
colourless waxy	33%

   The observed results differ significantly from the results expected on the basis of independent assortment.

c) State the name of a statistical test that could be used to show that the observed and the expected results are significantly different. [1]

d) Explain the reasons for the observed results of the cross differing significantly from the expected results. [2]

Introduction

Immunity is based on recognition of self and destruction of foreign material. The roles of the musculoskeletal system are movement, support and protection. All animals excrete nitrogenous waste products and some animals also balance water and solute concentrations. Sexual reproduction involves the development and fusion of haploid gametes.

11.1 Antibody production and vaccination

Understanding

→ Every organism has unique molecules on the surface of their cells.

→ B lymphocytes are activated by T lymphocytes in mammals.

→ Plasma cells secrete antibodies.

→ Activated B cells multiply to form a clone of plasma cells and memory cells.

→ Antibodies aid the destruction of pathogens.

→ Immunity depends upon the persistence of memory cells.

→ Vaccines contain antigens that trigger immunity but do not cause the disease.

→ Pathogens can be species-specific although others can cross species barriers.

→ White cells release histamine in response to allergens.

→ Histamines cause allergic symptoms.

→ Fusion of a tumour cell with an antibody-producing plasma cell creates a hybridoma cell.

→ Monoclonal antibodies are produced by hybridoma cells.

 ## Applications

→ Antigens on the surface of red blood cells stimulate antibody production in a person with a different blood group.

→ Smallpox was the first infectious disease of humans to have been eradicated by vaccination.

→ Monoclonal antibodies to hCG are used in pregnancy test kits.

 ## Skills

→ Analysis of epidemiological data related to vaccination programmes.

 ## Nature of science

→ Consider ethical implications of research: Jenner tested his vaccine for smallpox on a child.

Antigens in blood transfusion

Every organism has unique molecules on the surface of their cells.

Any foreign molecule that can trigger an immune response is referred to as an antigen. The most common antigens are proteins and very large polysaccharides. Such molecules are found on the surface of cancer cells, parasites and bacteria, on pollen grains and on the envelopes of viruses.

As an example, figure 1 shows a representation of an influenza virus. Hemagglutinin and neuraminidase are two antigens found on the surface of the virus. Hemagglutinin allows the virus to stick to host cells. Neuraminidase helps with the release of newly-formed virus particles.

The surface of our own cells contains proteins and polypeptides. Immune systems function based on recognizing the distinction between "foreign" antigens and "self". Figure 2 shows a mixture of pollen grains from several species. The antigens on the surface of these grains are responsible for triggering immune responses that are called "allergies" or "hay fever" in common language.

▲ Figure 2 Pollen grains

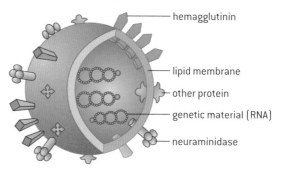

▲ Figure 1 Influenza virus

🌐 Antigens in blood transfusion

Antigens on the surface of red blood cells stimulate antibody production in a person with a different blood group.

Blood groups are based on the presence or absence of certain types of antigens on the surface of red blood cells. Knowledge of this is important in the medical procedure called transfusion where a patient is given blood from a donor. The ABO blood group and the Rhesus (Rh) blood group are the two most important antigen systems in blood transfusions as mismatches between donor and recipient can lead to an immune response.

In figure 3, the differences between the three A, B and O phenotypes are displayed. All three alleles involve a basic antigen sequence called antigen H. In blood type A and B, this antigen H is modified by the addition of an additional molecule. If the additional molecule is galactose, antigen B results. If the additional molecule is N-acetylgalactosamine,

antigen A results. Blood type AB involves the presence of both types of antigens.

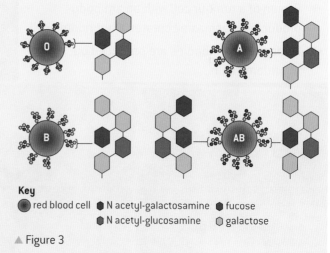

Key
- 🔵 red blood cell
- ⬣ N acetyl-galactosamine
- ⬣ fucose
- ⬣ N acetyl-glucosamine
- ⬡ galactose

▲ Figure 3

If a recipient is given a transfusion involving the wrong type of blood, the result is an immune response called agglutination followed by hemolysis where red blood cells are destroyed and blood may coagulate in the vessels (figure 4).

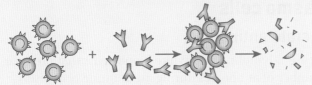

red blood cells with surface antigens from an incompatable donor | antibodies from recipient | agglutination (clumping) | hemolysis

▲ Figure 4

Blood typing involves mixing samples of blood with antibodies. Figure 5 shows the result of a blood group test showing reactions between blood types (rows) and antibody serums (columns). The first column shows the blood's appearance prior to the tests. There are four human blood types: A, B, AB and O. Type A blood has type A antigens (surface proteins) on its blood cells. Type B blood has type B antigens. Mixing type A blood with anti-A+B serum causes an agglutination reaction, producing dense red dots that are different from the control in the first column. Type B blood undergoes the same reaction with anti-B serum and anti A+B serum. AB blood agglutinates in all three anti-serums. Type O blood has neither the A or B antigen, so it does not react to the serums.

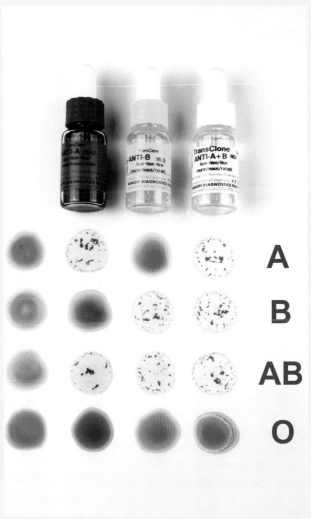

▲ Figure 5

The specific immune response

B lymphocytes are activated by T lymphocytes in mammals.

The principle of "challenge and response" has been used to explain how the immune system produces the large amounts of the specific antibodies that are needed to fight an infection, and avoid producing any of the hundreds of thousands of other types of antibodies that could be produced. Antigens on the surface of pathogens that have invaded the body are the "challenge". The "response" involves the following stages.

Pathogens are ingested by macrophages, and antigens from them are displayed in the plasma membrane of the macrophages. Lymphocytes called helper T cells each have an antibody-like receptor protein in their plasma membranes, which can bind to antigens displayed by macrophages. Of the many types of helper T cell, only a few have receptor proteins that fit the antigen. These helper T cells bind and are activated by the macrophage.

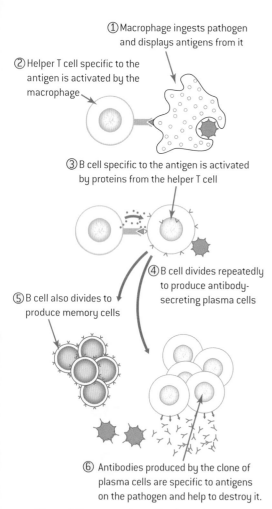

① Macrophage ingests pathogen and displays antigens from it

② Helper T cell specific to the antigen is activated by the macrophage

③ B cell specific to the antigen is activated by proteins from the helper T cell

④ B cell divides repeatedly to produce antibody-secreting plasma cells

⑤ B cell also divides to produce memory cells

⑥ Antibodies produced by the clone of plasma cells are specific to antigens on the pathogen and help to destroy it.

▲ Figure 6 The stages in antibody production

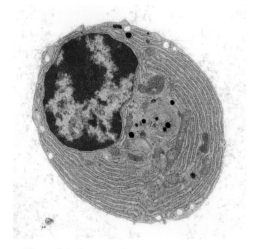

▲ Figure 7 A plasma cell

The activated helper T cells then bind to lymphocytes called B cells. Again, only B cells that have a receptor protein to which the antigen binds are selected and undergo the binding process. The helper T cell activates the selected B cells, both by means of the binding and by release of a signalling protein.

The role of plasma cells

Plasma cells secrete antibodies.

Plasma cells are mature B lymphocytes (white blood cells) that produce and secrete large number of antibodies during an immune response. Figure 7 shows a plasma cell. The cell's cytoplasm (orange) contains an unusually extensive network of rough endoplasmic reticulum (rER). rER manufactures, modifies and transports proteins, in this case, the antibodies. The cell produces a lot of the same type of protein meaning that the range of genes expressed is lower than a typical cell. This explains the staining pattern of the nucleus where dark staining indicates unexpressed genes.

Clonal selection and memory cell formation

Activated B cells multiply to form a clone of plasma cells and memory cells.

The activated B cells divide many times by mitosis, generating a clone of plasma cells that all produce the same antibody type. The generation of large numbers of plasma cells that produce one specific antibody type is known as clonal selection.

The antibodies are secreted and help to destroy the pathogen in ways described below. These antibodies only persist in the body for a few weeks or months and the plasma cells that produce them are also gradually lost after the infection has been overcome and the antigens associated with it are no longer present.

Although most of the clone of B cells become active plasma cells, a smaller number become memory cells, which remain long after the infection. These memory cells remain inactive unless the same pathogen infects the body again, in which case they become active and respond very rapidly. Immunity to an infectious disease involves either having antibodies against the pathogen, or memory cells that allow rapid production of the antibody.

The role of antibodies

Antibodies aid the destruction of pathogens.

Antibodies aid in the destruction of pathogens in a number of ways.

- **Opsonization:** They make a pathogen more recognizable to phagocytes so they are more readily engulfed. Once bound, they can link the pathogen to phagocytes.

- **Neutralization of viruses and bacteria:** Antibodies can prevent viruses from docking to host cells so that they cannot enter the cells.

- **Neutralization of toxins:** Some antibodies can bind to the toxins produced by pathogens, preventing them from affecting susceptible cells.

- **Activation of complement:** The complement system is a collection of proteins which ultimately lead to the perforation of the membranes of pathogens. Antibodies bound to the surface of a pathogen activate a complement cascade which leads to the formation of a "membrane attack complex" that forms a pore in the membrane of the pathogen allowing water and ions to enter into the cell ultimately causing the cell to lyse.

- **Agglutination:** Antibodies can cause sticking together or "agglutination" of pathogens so they are prevented from entering cells and are easier for phagocytes to ingest. The large agglutinated mass can be filtered by the lymphatic system and then phagocytized. The agglutination process can be dangerous if it occurs as a result of an incorrect blood transfusion.

Figure 8 summarizes some of the modes of action of antibodies.

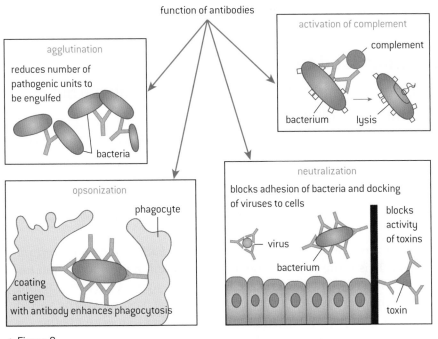

▲ Figure 8

Immunity

Immunity depends upon the persistence of memory cells.

Immunity to a disease is due either to the presence of antibodies that recognize antigens associated with the disease, or to memory cells that allow production of these antibodies. Immunity develops when the immune system is challenged by a specific antigen and produces antibodies and memory cells in response. Figure 9 distinguishes a primary immune response (launched the first time the pathogen infects

TOK

What can game theory tells us about the persistence of smallpox stockpiles?

Once wild smallpox had been eradicated there remained the challenge of what to do with samples of smallpox still in the hands of researchers and the military. Despite calls for the remaining stockpiles to be eradicated by the WHO, both the US and Russia have delayed complying with this directive.

Game theory is a branch of mathematics that makes predictions about human behaviour when negotiations are being undertaken. In terms of payoff, if one side reneges and the other proceeds on the basis of trust, the gain to the deal breaker is maximized. In this case, they are no longer threatened by the adversary but retain the ability to threaten. If both parties renege, the risk remains that the virus will be used as a weapon in both the first attack and in retaliation. Maximum net gain for all would involve both parties complying with the directive but this involves trust and risk taking.

the body) and the secondary immune response which is launched the second time the pathogen infects the body. Memory cells ensure that the second time an antigen is encountered, the body is ready to respond rapidly by producing more antibodies at a faster rate.

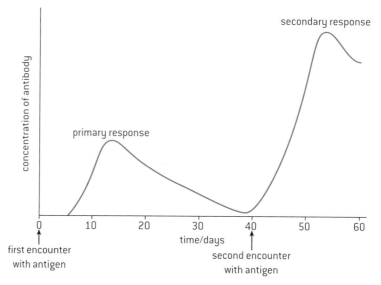

▲ Figure 9 The secondary immune response

Vaccines lead to immunity

Vaccines contain antigens that trigger immunity but do not cause the disease.

A vaccine is introduced into the body, usually by injection. The vaccine may contain a live attenuated (weakened) version of the pathogen, or some derivative of it that contains antigens from the pathogen. This stimulates a primary immune response. If the actual microorganism enters the body as a result of infection, it will be destroyed by the antibodies in a secondary immune response.

Figure 10 shows a phagocyte engulfing a *Mycobacterium bovis* bacterium (orange). This is the strain of the bacterium used in the vaccination for tuberculosis (TB). The bacteria are live but attenuated (weakened) and not as pathogenic as their relative *Mycobacterium tuberculosis*. The vaccine primes the immune system to produce antibodies that act on both species of bacteria, without causing the disease, so that it responds more rapidly if infected with *Mycobacterium tuberculosis* (TB) bacteria.

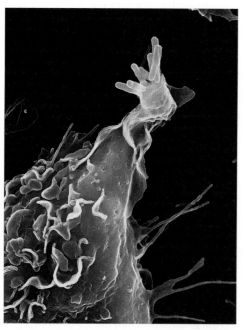

▲ Figure 10

 Ethical considerations of Jenner's vaccine experiments

Consider ethical implications of research: Jenner tested his vaccine for smallpox on a child.

Edward Jenner was an 18th century scientist who noted that a milkmaid claimed that because she had caught the disease cowpox

she would never develop smallpox. He infected an eight-year-old boy with cowpox. After a brief illness, the boy recovered. Jenner then purposely infected the boy with smallpox to confirm that he had the ability to resist the disease.

He was the first person to use human beings as research subjects in testing a vaccine. He did not do any preliminary laboratory research nor any preliminary animal studies before experimenting with human beings, his subject was a small child well below the age of consent, and he deliberately infected him with an extremely virulent, often fatal, disease-causing agent.

Jenner's experiments were performed well before the formulation of any statements of ethical principles for the protection of human research subjects. The Nuremberg Trials condemned medical experiments on children. These trials that followed the Second World War resulted in the Nuremberg Code for the protection of research subjects, and later the World Health Organization's International Ethical Guidelines for Biomedical Research Involving Human Subjects (1993). Jenner's experiments would not be approved by a modern ethical review committee.

 ## The eradication of smallpox

Smallpox was the first infectious disease of humans to have been eradicated by vaccination.

The efforts to eradicate smallpox are an example of the contributions that intergovernmental organizations can make to address issues of global concern. The first such effort was launched in 1950 by the Pan American Health Organization. The World Health Assembly passed a resolution in 1959 to undertake a global initiative to eradicate smallpox. It met with mixed success until a well-funded Smallpox Eradication Unit was established in 1967.

The last known case of wild smallpox was in 1977 in Somalia, though there were two accidental infections after this. The campaign was successful for several reasons:

- Only humans can catch and transmit smallpox. There is no animal reservoir where the disease could be maintained and re-emerge. This is the reason a yellow fever eradication effort failed in the early 1900s.

- Symptoms of infection emerge quite quickly and are readily visible allowing teams to "ring vaccinate" all of the people who might have come in contact with the afflicted person. In contrast, efforts to eradicate polio have been hampered because infected persons do not always present readily recognized symptoms.

- Immunity to smallpox is long-lasting unlike such conditions as malaria where reinfection is more common.

 ## Vaccines and epidemiology

Analysis of epidemiological data related to vaccination programmes.

Epidemiology is the study of the distribution, patterns and causes of disease in a population. The spread of disease is monitored in order to predict and minimize the harm caused by outbreaks as well as to determine the factors contributing to the outbreak. Epidemiologists would be involved in planning and evaluating vaccination programmes.

An effort to achieve the global eradication of polio was begun in 1988, as a combined effort between the World Health Organization (WHO), UNICEF and the Rotary Foundation. Similarly, UNICEF is leading a worldwide initiative to prevent tetanus through vaccination.

A small number of polio cases are the result of a failure in vaccination programmes. Figure 11 shows the incidence of "wild" rather than vaccine-induced polio cases in India over a seven-year period. Epidemiologists would investigate to determine the causes of the two peaks in numbers. Figure 12 shows the geographic distribution of polio cases over a 13-year period in India. Epidemiologists would use information about geographic distribution to determine origins of outbreaks so they could focus resources on those areas. They could track incidence to determine the effectiveness of reduction campaigns. It is heartening to know that by 2012, India had been declared polio-free.

The concern is that polio-free countries can still see some polio cases if infected individuals cross borders.

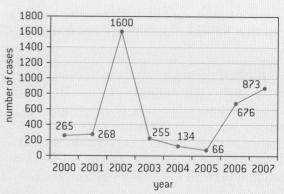

▲ Figure 11

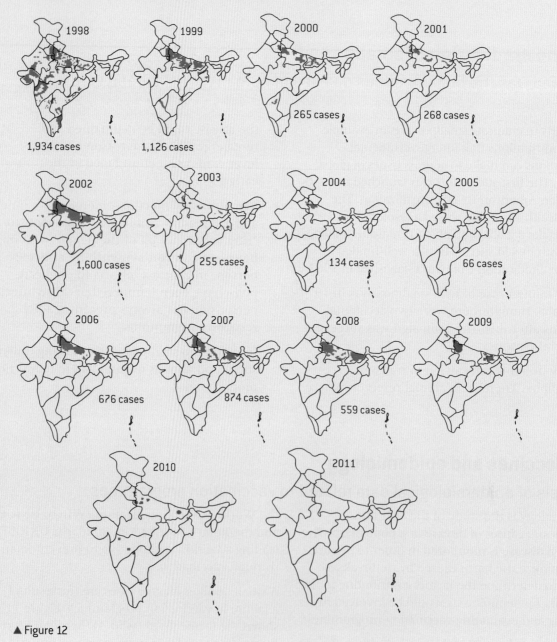

▲ Figure 12

Data-based questions: Polio incidence in 2012

Figure 13 provides data about polio incidence in the three countries where wild polio was still endemic as of mid-2012.

1 Define the term "endemic" (1)

2 Identify the three countries where polio was still endemic as of mid-2012. (1)

3 Identify the strain of polio virus which is the most prevalent. (1)

4 Identify one country where the situation appears to have improved between 2011 and 2012. (2)

5 Given that in 1988 there were an estimated 350,000 cases of polio globally, discuss the success of the polio eradication programme. (5)

6 Suggest some of the challenges an epidemiologist might face in gathering reliable data. (5)

7 Research to find the status of polio eradication in these countries.

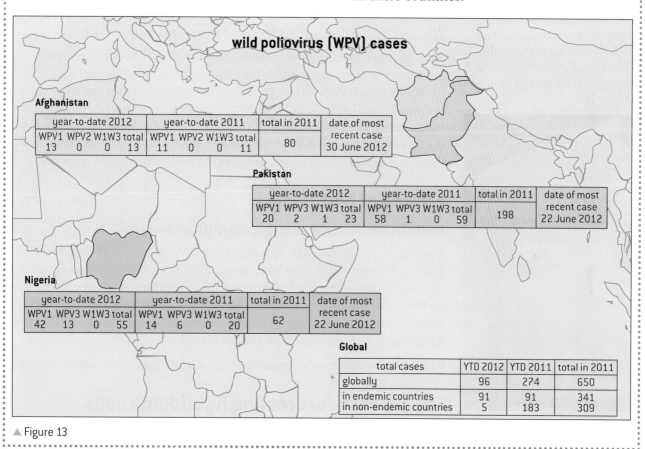

wild poliovirus (WPV) cases

Afghanistan

year-to-date 2012				year-to-date 2011				total in 2011	date of most recent case
WPV1	WPV2	W1W3	total	WPV1	WPV2	W1W3	total	80	30 June 2012
13	0	0	13	11	0	0	11		

Pakistan

year-to-date 2012				year-to-date 2011				total in 2011	date of most recent case
WPV1	WPV3	W1W3	total	WPV1	WPV3	W1W3	total	198	22 June 2012
20	2	1	23	58	1	0	59		

Nigeria

year-to-date 2012				year-to-date 2011				total in 2011	date of most recent case
WPV1	WPV3	W1W3	total	WPV1	WPV3	W1W3	total	62	22 June 2012
42	13	0	55	14	6	0	20		

Global

total cases	YTD 2012	YTD 2011	total in 2011
globally	96	274	650
in endemic countries	91	91	341
in non-endemic countries	5	183	309

▲ Figure 13

Zoonosis are a growing global health concern

Pathogens can be species-specific although others can cross species barriers.

Pathogens are often highly specialized with a narrow range of hosts. There are viruses that are specific to birds, pigs and bacteria for example. There are bacterial pathogens that only cause disease in humans. Humans are the only known organism susceptible to such pathogens as syphilis, polio and measles, but we are resistant to canine distemper virus, for example. The bacterium *Mycobacterium tuberculosis* does not cause disease in frogs because frogs rarely reach the 37 °C temperature necessary to support the proliferation of the bacterium. Rats injected with the diphtheria toxin do not become ill because their cells lack the receptor that would bring the toxin into the cell.

▲ Figure 14 A thermal scanning camera is being used to monitor the skin temperature of passengers arriving at Nizhny Novgorod airport, in Russia. Raised skin temperature can be an indicator of fever from illnesses. Such cameras have been used widely to screen for possible carriers of various possible zoonotic epidemic influenzas such as bird flu and swine flu

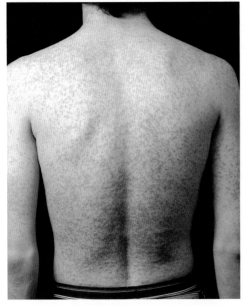

▲ Figure 15 The rash across the body of this male patient is due to the release of excessive histamines in response to taking Amoxicillin (penicillin) antibiotic

A zoonosis is a pathogen which can cross a species barrier. This is an emerging global health concern. Bubonic plague, Rocky Mountain spotted fever, Lyme disease, bird flu and West Nile virus are all zoonotic diseases. The major factor contributing to the increased appearance of zoonotic diseases is the growth of contact between animals and humans by such means as humans living in close contact with livestock or disruption of habitats.

For example, in the late 1990s in Malaysia, intensive pig farming in the habitat of bats infected with the Nipah virus eventually saw the virus move from the bats to the pigs to the humans and resulted in over 100 human deaths.

The immune system produces histamines

White cells release histamine in response to allergens.

Mast cells are immune cells found in connective tissue that secrete histamine in response to infection. Histamine is also released by basophils which circulate in the blood. Histamine causes the dilation of the small blood vessels in the infected area causing the vessels to become leaky. This increases the flow of fluid containing immune components to the infected area and it allows some of the immune components to leave the blood vessel resulting in both specific and non-specific responses.

Effects of histamines

Histamines cause allergic symptoms.

Histamine is a contributor to a number of symptoms of allergic reactions. Cells in a variety of tissues have membrane-bound histamine receptors. Histamine plays a role in bringing on the symptoms of allergy in the nose (itching, fluid build-up, sneezing, mucus secretion and inflammation). Histamine also plays a role in the formation of allergic rashes and in the dangerous swelling known as anaphylaxis.

To lessen the effects of allergic responses, anti-histamines can be taken.

The process for creating hybridoma cells

Fusion of a tumour cell with an antibody-producing plasma cell creates a hybridoma cell.

Monoclonal antibodies are highly specific, purified antibodies that are produced by a clone of cells, derived from a single cell. They recognize only one antigen.

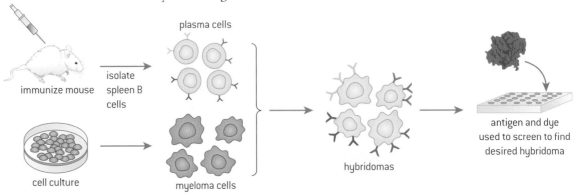

▲ Figure 16

To produce the clone of cells that will manufacture a monoclonal antibody, the antigen recognized by the antibody is injected into a mouse, or other mammal. In response to this challenge, the mouse's immune system makes plasma B cells that are capable of producing the desired antibody. Plasma cells are removed from the spleen of the mouse. They will be of many different types with only some producing the desired antibody.

The B cells are fused with cancer cells called myeloma cells. The cells formed by fusion of plasma B cells and myeloma cells are called hybridoma cells.

Production of monoclonal antibodies
Monoclonal antibodies are produced by hybridoma cells.

Because the full diversity of B cells are fused with the myeloma cells, many different hybridomas are produced and they are individually tested to find one that produces the required antibody.

Once identified, the desired hybridoma cell is allowed to divide and form a clone. These cells can be cultured in a fermenter where they will secrete huge amounts of monoclonal antibody. Figure 17 shows a 2000-litre fermenter used in the commercial production of monoclonal antibodies. The hybridoma cell is multiplied in the fermenter to produce large numbers of genetically identical copies, each secreting the antibody produced by the original lymphocyte.

Monoclonal antibodies are used both for treatment and diagnosis of diseases. Examples include the test for malaria that can be used to identify whether either humans or mosquitoes are infected with the malarial parasite, the test for the HIV pathogen or the creation of antibodies for injection into rabies victims.

 Pregnancy tests employ monoclonal antibodies

Monoclonal antibodies to hCG are used in pregnancy test kits.

Monoclonal antibodies are used in a broad range of diagnostic tests, including tests for HIV antibodies and for an enzyme released during heart attacks. Pregnancy test kits are available that use monoclonal antibodies to detect hCG (human chorionic gonadotrophin). hCG is uniquely produced during pregnancy by the developing embryo and later the placenta. The urine of a pregnant woman contains detectable levels of hCG.

Figure 18 shows how the pregnancy test strip works. At point C, there are antibodies to hCG immobilized in the strip. At point B there are free antibodies to hCG attached to a dye. At point D there are immobilized antibodies that bind to the dye-bearing antibodies. Urine applied to the end of a test strip washes antibodies down the strip.

▲ Figure 17

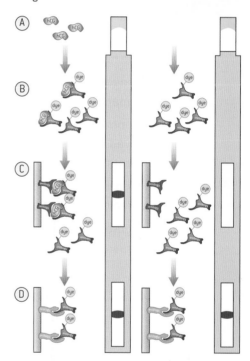

▲ Figure 18

Activity

1 Explain how a blue band appears at point C if the woman is pregnant. [3]

2 Explain why a blue band does not appear at point C if the woman is not pregnant. [3]

3 Explain the reasons for the use of immobilized monoclonal antibodies at point D, even though they do not indicate whether a woman is pregnant or not. [3]

11.2 Movement

Understanding

→ Bones and exoskeletons provide anchorage for muscles and act as levers.

→ Movement of the body requires muscles to work in antagonistic pairs.

→ Synovial joints allow certain movements but not others.

→ Skeletal muscle fibres are multinucleate and contain specialized endoplasmic reticulum.

→ Muscle fibres contain many myofibrils.

→ Each myofibril is made up of contractile sarcomeres.

→ The contraction of the skeletal muscle is achieved by the sliding of actin and myosin filaments.

→ Calcium ions and the proteins tropomyosin and troponin control muscle contractions.

→ ATP hydrolysis and cross-bridge formation are necessary for the filaments to slide.

 Applications

→ Antagonistic pairs of muscles in an insect leg.

 Skills

→ Annotation of a diagram of the human elbow.

→ Drawing labelled diagrams of the structure of a sarcomere.

→ Analysis of electron micrographs to find the state of contraction of muscle fibres.

 Nature of science

→ Fluorescence was used to study the cyclic interactions in muscle contraction.

Bones and exoskeletons anchor muscles

Bones and exoskeletons provide anchorage for muscles and act as levers.

Exoskeletons are external skeletons that surround and protect most of the body surface of animals such as crustaceans and insects. Figure 1 shows a scanning electron micrograph of a spider next to exoskeletons that have been moulted.

Bones and exoskeletons facilitate movement by providing an anchorage for muscles and by acting as levers. Levers change the size and direction of forces. In a lever, there is an effort force, a pivot point called a fulcrum and a resultant force. The relative positions of these three determine the class of lever.

In figure 2, the diagram shows that when a person nods their head backward, the spine acts as a first-class lever, with the fulcrum (F) being found between the effort force (E) provided by the *splenius capitis* muscle and the resultant force (R) causing the chin to be extended.

The grasshopper leg acts as a third-class lever as the fulcrum is at the body end and the effort force is between the fulcrum and the resultant force.

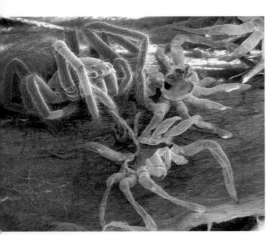
▲ Figure 1

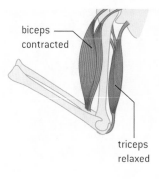

Muscles are attached to the insides of exoskeletons but to the outside of bones.

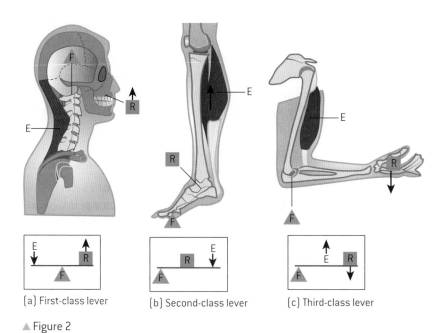

(a) First-class lever

(b) Second-class lever

(c) Third-class lever

▲ Figure 2

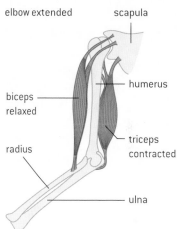

▲ Figure 3 The biceps and triceps are antagonistic muscles

Skeletal muscles are antagonistic

Movement of the body requires muscles to work in antagonistic pairs.

Skeletal muscles occur in pairs that are antagonistic. This means that when one contracts, the other relaxes. Antagonistic muscles produce opposite movements at a joint. For example, in the elbow, the triceps extends the forearm while the biceps flex the forearm.

Data-based questions: Flight muscles

In one research project, pigeons (*Columba livia*) were trained to take off, fly 35 metres and land on a perch. During the flight the activity of two muscles, the sternobrachialis (SB) and the thoracobrachialis (TB), was monitored using electromyography. The traces are shown in figure 4. The spikes show electrical activity in contracting muscles. Contraction of the sternobrachialis causes a downward movement of the wing.

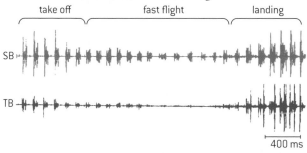

▲ Figure 4 Electrical activity in the sternobrachialis (SB) and the thoracobrachialis (TB) muscles during flight of a pigeon

1 Deduce the number of downstrokes of the wing during the whole flight. [1]

2 Compare the activity of the sternobrachialis muscle during the three phases of the flight. [3]

3 Deduce from the data in the electromyograph how the thoracobrachialis is used. [1]

4 Another muscle, the supracoracoideus, is antagonistic to the sternobrachialis. State the movement produced by a contraction of the supracoracoideus. [1]

5 Predict the pattern of the electromyograph trace for the supracoracoideus muscle during the 35-metre flight. [2]

An insect leg has antagonistic muscles

Antagonistic pairs of muscles in an insect leg.

The grasshopper, like all insects, has three pairs of appendages. The hindlimb of a grasshopper is specialized for jumping. It is a jointed appendage with three main parts. Below the joint is referred to as the tibia and at the base of the tibia is another joint below which is found the tarsus. Above the joint is referred to as the femur. Relatively massive muscles are found on the femur.

When the grasshopper prepares to jump, the flexor muscles will contract bringing the tibia and tarsus into a position where they resemble the letter "Z" and the femur and tibia are brought closer together. This is referred to as flexing. The extensor muscles relax during this phase. The extensor muscles will then contract extending the tibia and producing a powerful propelling force.

▲ Figure 6 Composite high-speed photograph of a grasshopper (Order Orthoptera) jumping from the head of a nail

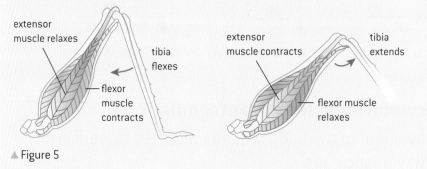

▲ Figure 5

The human elbow is an example of a synovial joint

Annotation of a diagram of the human elbow.

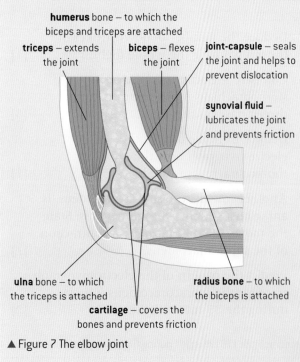

humerus bone – to which the biceps and triceps are attached

triceps – extends the joint

biceps – flexes the joint

joint-capsule – seals the joint and helps to prevent dislocation

synovial fluid – lubricates the joint and prevents friction

ulna bone – to which the triceps is attached

radius bone – to which the biceps is attached

cartilage – covers the bones and prevents friction

▲ Figure 7 The elbow joint

The point where bones meet is called a joint. Most joints allow the bones to move in relation to each other – this is called articulation. Most articulated joints have a similar structure, including cartilage, synovial fluid and joint capsule.

- Cartilage is tough, smooth tissue that covers the regions of bone in the joint. It prevents contact between regions of bone that might otherwise rub together and so helps to prevent friction. It also absorbs shocks that might cause bones to fracture.

- Synovial fluid fills a cavity in the joint between the cartilages on the ends of the bones. It lubricates the joint and so helps to prevent the friction that would occur if the cartilages were dry and touching.

- The joint capsule is a tough ligamentous covering to the joint. It seals the joint and holds in the synovial fluid and it helps to prevent dislocation.

Different joints allow different ranges of movement

Synovial joints allow certain movements but not others.

The structure of a joint, including the joint capsule and the ligaments, determines the movements that are possible. The knee joint can act as a hinge joint, which allows only two movements: flexion (bending) and extension (straightening). It can also act as a pivot joint when flexed. The knee has a greater range of movement when it is flexed than when it is extended. The hip joint, between the pelvis and the femur, is a ball and socket joint. It has a greater range of movement than the knee joint in that it can flex and extend, rotate, and move sideways and back. This latter type of movement is called abduction and adduction.

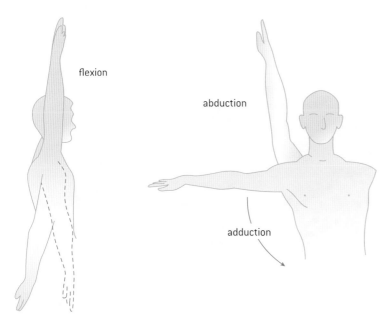

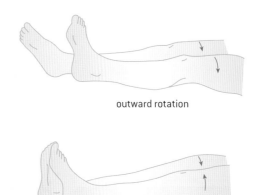

▲ Figure 8 Range of motion at the shoulder

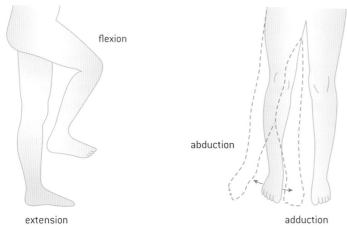

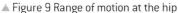

▲ Figure 9 Range of motion at the hip

Structure of muscle fibres

Skeletal muscle fibres are multinucleate and contain specialized endoplasmic reticulum.

The muscles that are used to move the body are attached to bones, so they are called skeletal muscles. When their structure is viewed using a microscope, stripes are visible. They are therefore also called striated muscle. The two other types of muscle are smooth and cardiac.

Striated muscle is composed of bundles of muscle cells known as muscle fibres. Although a single plasma membrane called the sarcolemma surrounds each muscle fibre, there are many nuclei present and muscle fibres are much longer than typical cells. These features are due to the fact that embryonic muscle cells fuse together to form muscle fibres. Figure 10 shows a muscle fibre.

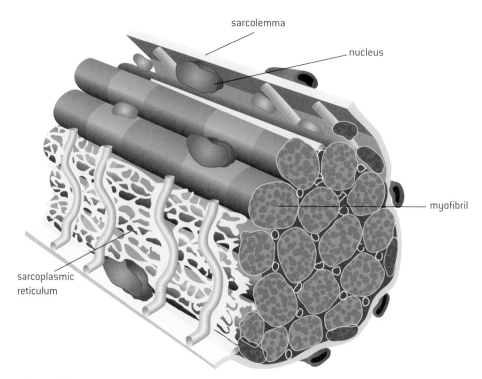

▲ Figure 10

A modified version of the endoplasmic reticulum, called the sarcoplasmic reticulum, extends throughout the muscle fibre. It wraps around every myofibril, conveying the signal to contract to all parts of the muscle fibre at once. The sarcoplasmic reticulum stores calcium. Between the myofibrils are large numbers of mitochondria, which provide ATP needed for contractions.

Myofibrils

Muscle fibres contain many myofibrils.

Within each muscle fibre there are many parallel, elongated structures called myofibrils. These have alternating light and dark bands, which give striated muscle its stripes. In the centre of each light band is a disc-shaped structure, referred to as the Z-line.

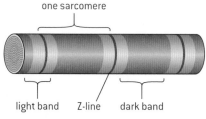

▲ Figure 11 The ultrastructure of the muscle fibre

Structure of myofibrils

Each myofibril is made up of contractile sarcomeres.

The micrograph in figure 13 shows a longitudinal section through a myofibril. A number of repeating units that alternate between light and dark bands are visible. Through the centre of each light area is a line called the Z-line. The part of a myofibril between one Z-line and the next is called a sarcomere. It is the functional unit of the myofibril.

The pattern of light and dark bands in sarcomeres is due to a precise and regular arrangement of two types of protein filament – thin actin filaments and thick myosin filaments. Actin filaments are attached to a Z-line at one end. Myosin filaments are interdigitated with actin filaments at both ends and occupy the centre of the sarcomere. Each myosin filament is surrounded by six actin filaments and forms cross-bridges with them during muscle contraction.

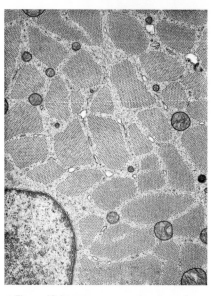

▲ Figure 12 A transverse section through a skeletal muscle fibre showing numerous myofibrils. A nucleus is shown in the bottom left

 The sarcomere

Drawing labelled diagrams of the structure of a sarcomere.

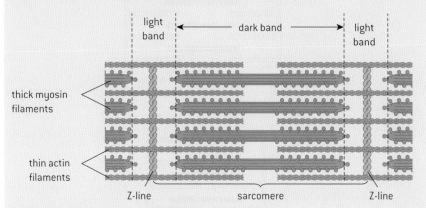

thick myosin filaments

thin actin filaments

light band — dark band — light band

Z-line   sarcomere   Z-line

▲ Figure 14 The structure of a sarcomere

When constructing diagrams of a sarcomere, ensure to demonstrate understanding that it is between two Z-lines. Myosin filaments should be shown with heads. Actin filaments should be shown connected to Z-lines. Light bands should be labelled around the Z-line. The extent of the dark band should also be indicated.

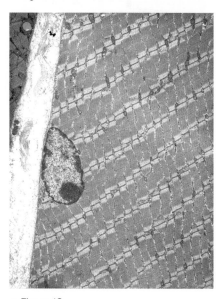

▲ Figure 13

Data-based questions: Transverse sections of striated muscle

The drawings in figure 15 show myofibrils in transverse section.

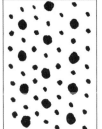 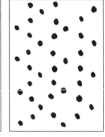

▲ Figure 15

1 Explain the difference between a transverse and a longitudinal section of muscle. [2]

2 Deduce what part of the myofibril is represented by the drawings as small dots. [2]

3 Compare the pattern of dots in the three diagrams. [3]

4 Explain the differences between the diagrams in the pattern of dots. [3]

Mechanism of skeletal muscle contraction

The contraction of the skeletal muscle is achieved by the sliding of actin and myosin filaments.

During muscle contraction, the myosin filaments pull the actin filaments inwards towards the centre of the sarcomere. This shortens each sarcomere and therefore the overall length of the muscle fibre (see figure 16).

The contraction of skeletal muscle occurs by the sliding of actin and myosin filaments. Myosin filaments cause this sliding. They have heads that can bind to special sites on actin filaments, creating cross-bridges, through which they can exert a force, using energy from ATP. The heads are regularly spaced along myosin filaments and the binding sites are regularly spaced along the actin filaments, so many cross-bridges can form at once (see figure 17).

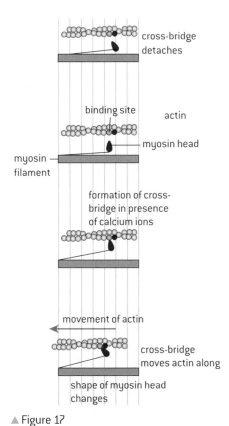

▲ Figure 17

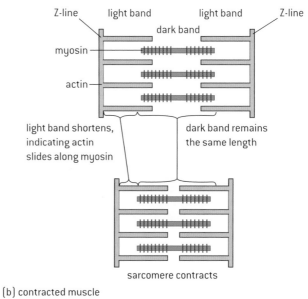

▲ Figure 16 Diagram of relaxed and contracted sarcomeres

Determining the state of skeletal muscle contraction

Analysis of electron micrographs to find the state of contraction of muscle fibres.

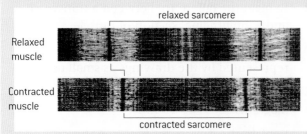

▲ Figure 18 Electron micrograph of relaxed and contracted sarcomeres

In a relaxed sarcomere, the Z-lines are farther apart, the light bands are wider and overall the sarcomere is longer. In the centre of the sarcomere, there is another line called the M-line. In a relaxed sarcomere, there is a more visible light band on either side of the M-line.

The control of skeletal muscle contraction

Calcium ions and the proteins tropomyosin and troponin control muscle contractions.

In relaxed muscle, a regulatory protein called tropomyosin blocks the binding sites on actin. When a motor neuron sends a signal to a muscle fibre to make it contract, the sarcoplasmic reticulum releases calcium ions. These calcium ions bind to a protein called troponin which causes tropomyosin to move, exposing actin's binding sites. Myosin heads then bind and swivel towards the centre of the sarcomere, moving the actin filament a small distance.

The role of ATP in the sliding of filaments

ATP hydrolysis and cross-bridge formation are necessary for the filaments to slide.

For significant contraction of the muscle, the myosin heads must carry out this action repeatedly. This occurs by a sequence of stages:

- ATP causes the breaking of the cross-bridges by attaching to the myosin heads, causing them to detach from the binding sites on actin.

- Hydrolysis of the ATP, to ADP and phosphate, provides energy for the myosin heads to swivel outwards away from the centre of the sarcomere – this is sometimes called the cocking of the myosin head.

- New cross-bridges are formed by the binding of myosin heads to actin at binding sites adjacent to the ones previously occupied (each head binds to a site one position further from the centre of the sarcomere).

- Energy stored in the myosin head when it was cocked causes it to swivel inwards towards the centre of the sarcomere, moving the actin filament a small distance. This sequence of stages continues until the motor neuron stops sending signals to the muscle fibre. Calcium ions are then pumped back into the sarcoplasmic reticulum, so the regulatory protein moves and covers the binding sites on actin. The muscle fibre therefore relaxes.

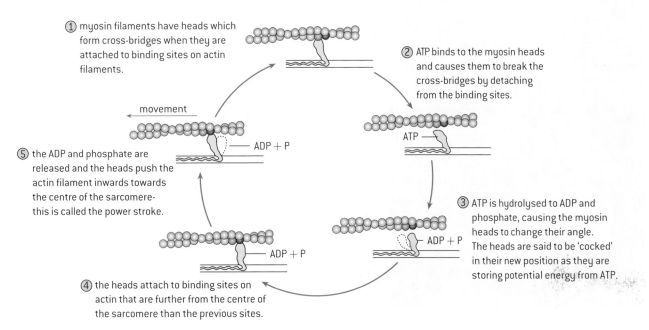

① myosin filaments have heads which form cross-bridges when they are attached to binding sites on actin filaments.

② ATP binds to the myosin heads and causes them to break the cross-bridges by detaching from the binding sites.

movement

⑤ the ADP and phosphate are released and the heads push the actin filament inwards towards the centre of the sarcomere- this is called the power stroke.

ADP + P

ATP

③ ATP is hydrolysed to ADP and phosphate, causing the myosin heads to change their angle. The heads are said to be 'cocked' in their new position as they are storing potential energy from ATP.

ADP + P

ADP + P

④ the heads attach to binding sites on actin that are further from the centre of the sarcomere than the previous sites.

▲ Figure 19

The use of fluorescence to study contraction

Fluorescence has been used to study the cyclic interactions in muscle contraction.

Fluoresence is the emission of electromagnetic radiation, often visible light, by a substance after it has been illuminated by electromagnetic radiation of a different wavelength. The fluorescence can often be detected in a light microscope and captured on film for later analysis.

Some of the classic experiments in the history of muscle research have depended on fluorescence. The coelenterate *Aequorea victoria* (figure 20) produces a calcium-sensitive bioluminescent protein, aequorin. Scientists studied the contraction of giant single muscle fibres of the acorn barnacle *Balanus nubilus* by injecting samples of the muscle with aequorin. When muscles were stimulated to contract in the study, initially there was strong bioluminescence coinciding with the release of Ca^{2+} from the sarcoplasmic reticulum. The light intensity began to decrease immediately after the cessation of the stimulus.

In another experiment, researchers cut apart *Nitella axillaris* cells. These cells are unique in that they have a network of actin filaments underlying their membranes. Researchers attached fluorescent dye to myosin molecules in an effort to show that myosin can "walk along" actin filaments.

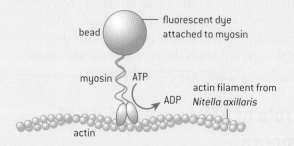

With this technique, the researchers were able to demonstrate the ATP-dependence of myosin-actin interaction.

The graph in figure 21 shows the velocity of myosin molecules as a function of ATP concentration.

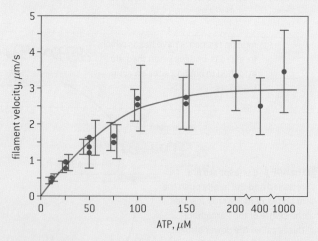

▲ Figure 21

▲ Figure 20 *Aequorea victoria*

11.3 The kidney and osmoregulation

Understanding

→ Animals are either osmoregulators or osmoconformers.

→ The Malpighian tubule system in insects and the kidney carry out osmoregulation and removal of nitrogenous wastes.

→ The composition of blood in the renal artery is different from that in the renal vein.

→ The ultrastructure of the glomerulus and Bowman's capsule facilitate ultrafiltration.

→ The proximal convoluted tubule selectively reabsorbs useful substances by active transport.

→ The loop of Henlé maintains hypertonic conditions in the medulla.

→ The length of the loop of Henlé is positively correlated with the need for water conservation in animals.

→ ADH controls reabsorption of water in the collecting duct.

→ The type of nitrogenous waste in animals is correlated with evolutionary history and habitat.

Applications

→ Consequences of dehydration and overhydration.

→ Treatment of kidney failure by hemodialysis or kidney transplant.

→ Blood cells, glucose, proteins and drugs are detected in urinary tests.

Skills

→ Drawing and labelling a diagram of the human kidney.

→ Annotation of diagrams of the nephron.

Nature of science

→ Curiosity about particular phenomena: investigations were carried out to determine how desert animals prevent water loss in their wastes.

Different responses to changes in osmolarity in the environment

Animals are either osmoregulators or osmoconformers.

Osmolarity refers to the solute concentration of a solution. Many animals are known as osmoregulators because they maintain a constant internal solute concentration, even when living in marine environments with very different osmolarities. All terrestrial animals, freshwater animals and some marine organisms like bony fish are osmoregulators. Typically these organisms maintain their solute concentration at about one third of the concentration of seawater and about 10 times that of fresh water.

Osmoconformers are animals whose internal solute concentration tends to be the same as the concentration of solutes in the environment.

Data-based questions

The striped shore crab *Pachygrapsus crassipes* (figure 1) is found on rocky shores over the west coast of North and Central America as well as in Korea and Japan. *P. crassipes* is often exposed to dilute salinities in tide pools and freshwater rivulets, but it only rarely encounters salt concentrations much higher than that of the ocean. Samples of crabs were placed in water concentrations of varying osmolarity and samples of blood were taken to determine osmolarity of the blood. In this experiment, the unit of osmolarity is measured in units based on freezing point depression. When solutes are added to water they disrupt hydrogen bonding. Freezing requires additional hydrogen bonding so adding solute lowers the freezing point. 2 *delta* is equivalent to about 100% ocean seawater, 0.2 *delta* is equivalent to about 10% ocean seawater, and 3.4 *delta* is equivalent to about 170% seawater.

1   Determine the solute concentration of crab blood at which the concentration of surrounding water is 1 *delta*.   (1)

2   Determine the range over which *P. crassipes* is able to keep its blood solute concentration fairly stable.   (1)

3   Predict what the graph would look like if *P. crassipes* was not able to osmoregulate.   (1)

4   Discuss whether *P. crassipes* is an osmoconformer or an osmoregulator.   (3)

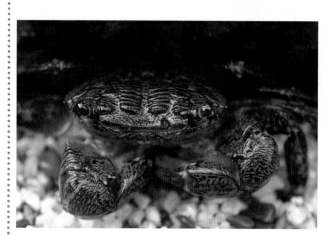

▲ Figure 1 The striped shore crab is exposed to varying salt concentrations in its habitat

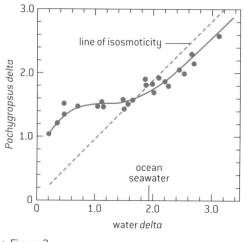

▲ Figure 2

The Malpighian tubule system

The Malpighian tubule system in insects and the kidney carry out osmoregulation and removal of nitrogenous wastes.

Arthropods have a circulating fluid, known as hemolymph, that combines the characteristics of tissue fluid and blood. Osmoregulation is a form of homeostasis whereby the concentration of hemolymph, or blood in the case of animals with closed circulatory systems, is kept within a certain range.

When animals break down amino acids, the nitrogenous waste product is toxic and needs to be excreted. In insects, the waste product is usually in the form of uric acid and in mammals it is in the form of urea.

Insects have tubes that branch off from their intestinal tract. These are known as Malpighian tubules. Cells lining the tubules actively transport ions and uric acid from the hemolymph into the lumen of the tubules. This draws water by osmosis from the hemolymph through the walls of the tubules into the lumen. The tubules empty their contents into the gut. In the hindgut most of the water and salts are reabsorbed while the nitrogenous waste is excreted with the feces.

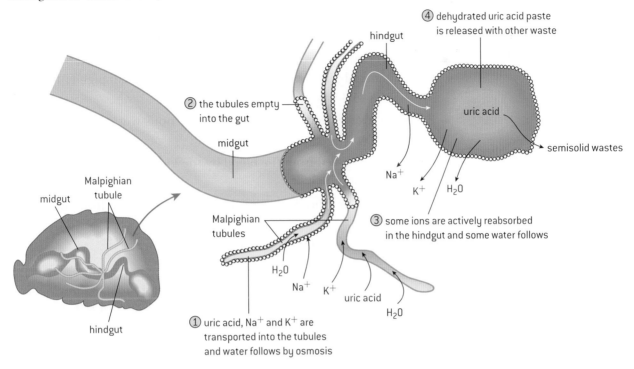

▲ Figure 3

⚠ Drawing the human kidney

Drawing and labelling a diagram of the human kidney.

When drawing a diagram of the kidney, the shape should be roughly oval with a concave side to which the renal artery and vein are attached. Drawings should clearly indicate the cortex shown at the edge of the kidney. It should be shown with a thickness of about $\frac{1}{5}$ the entire width. The medulla should be shown inside the cortex, with pyramids. The renal pelvis should be shown on the concave side of the kidney. The pelvis should drain into the ureter. The renal artery should have a smaller diameter than the renal vein.

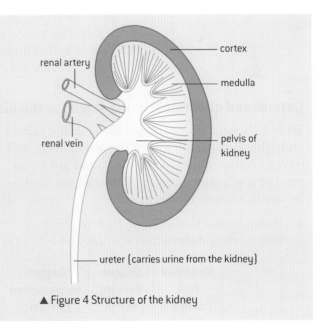

▲ Figure 4 Structure of the kidney

Comparing the composition of blood in the renal artery and the renal vein

The composition of blood in the renal artery is different from that in the renal vein.

Kidneys function in both osmoregulation and excretion. The kidneys are responsible for removing substances from the blood that are not needed or are harmful. As a result, the composition of blood in the renal artery, through which blood enters the kidney, is different from that in the renal vein, through which blood leaves.

Substances that are present in higher amounts in the renal artery than the renal vein include:

- Toxins and other substances that are ingested and absorbed but are not fully metabolized by the body, for example betain pigments in beets and also drugs.

- Excretory waste products including nitrogenous waste products, mainly urea.

Other things removed from the blood by the kidney that are not excretory products include:

- Excess water, produced by cell respiration or absorbed from food in the gut.

- Excess salt, absorbed from food in the gut.

These are not excretory products because they are not produced by body cells. Removal of excess water and salt is part of osmoregulation. While blood in the renal artery might contain a variable water or salt content, blood in the renal vein will have a more constant concentration because osmoregulation has occurred.

The kidneys filter off about one fifth of the volume of plasma from the blood flowing through them. This filtrate contains all of the substances in plasma apart from large protein molecules. The kidneys then actively reabsorb the specific substances in the filtrate that the body needs. The result of this process is that unwanted substances pass out of the body in the urine. These substances are present in the renal artery but not the renal vein.

Data-based questions: Blood supply to the kidney

Table 1 shows the flow rate of blood to the kidney and other organs, the rate of oxygen delivery and oxygen consumption. All of the values are given per 100 g of tissue or organ. The rates are for a person in a warm environment.

1 Compare the rate of blood flow to the kidney with flow to the other organs. [2]

	Blood flow rate (ml min⁻¹ 100 g⁻¹)	Oxygen delivery (ml min⁻¹ 100 g⁻¹)	Oxygen consumption (ml min⁻¹ 100 g⁻¹)
Brain	54.0	10.8	3.70
Skin	13.0	2.6	0.38
Skeletal muscle (resting)	2.7	0.5	0.18
Heart muscle	87.0	17.4	11.0
Kidney	420.0	84.0	6.80

▲ Table 1

2 Calculate the volume of oxygen delivered to the organs per litre of blood. [2]

3 In the brain, 34 per cent of the oxygen that is delivered is consumed. Calculate the same percentage for the other organs. [4]

4 Discuss the reasons for the difference between the kidney and the other organs in the volume of blood flowing to the organ, and the percentage of oxygen in the blood that is consumed. [4]

5 Some parts of the kidney have a high percentage rate of oxygen consumption, for example the outer part of the medulla. This is because active processes requiring energy are being carried out. Suggest one process in the kidney that requires energy. [1]

6 Predict, with a reason, one change in blood flow that would occur if the person were moved to a cold environment. [2]

A final set of differences between the composition of blood in the renal artery and the renal vein is due to the metabolic activity of the kidney itself. Blood leaving the kidney through the renal vein is deoxygenated relative to the renal artery because kidney metabolism requires oxygen. It also has a higher partial pressure of carbon dioxide because this is a waste product of metabolism. Even though glucose is normally filtered and then entirely reabsorbed, some glucose is used by the metabolism of the kidney and therefore the concentration is slightly lower in the renal vein compared to the renal artery.

Plasma proteins are not filtered by the kidney so should be present in the same concentration in both blood vessels. Presence in the urine indicates abnormal function. This is looked for during clinical examination of a urine sample.

The ultrastructure of the glomerulus

The ultrastructure of the glomerulus and Bowman's capsule facilitate ultrafiltration.

Blood in capillaries is at high pressure in many of the tissues of the body, and the pressure forces some of the plasma out through the capillary wall, to form tissue fluid.

In the glomerulus of the kidney, the pressure in the capillaries is particularly high and the capillary wall is particularly permeable, so the volume of fluid forced out is about 100 times greater than in other tissues. The fluid forced out is called glomerular filtrate. The composition of blood plasma and filtrate is shown in table 2. The data in the table shows that most solutes are filtered out freely from the blood plasma, but almost all proteins are retained in the capillaries of the glomerulus. This is separation of particles differing in size by a few nanometres and so is called **ultrafiltration.** All particles with a relative molecular mass below 65,000 atomic mass units can pass through. The permeability to larger molecules depends on their shape and charge. Almost all proteins are retained in the blood, along with all the blood cells.

	Content (per dm^{-3} of blood plasma)	
Solutes	plasma	filtrate
Na$^+$ ions (mol)	151	144
Cl$^-$ ions (mol)	110	114
glucose (mol)	5	5
urea (mol)	5	5
proteins (mg)	740	3.5

▲ Table 2

The structure of a section of the filter unit is shown in figure 6 and figure 7. Figure 6 is a coloured transmission electron micrograph (TEM) of a section through a kidney glomerulus showing its basement membrane (brown line running from top right to bottom left). The basement membrane separates the capillaries (the white space at the left is the lumen of a capillary). Note the gaps in the wall of the capillary which are referred to as fenestrations.

The smaller projections from the membrane are podocyte foot processes, which attach the podocytes (specialized epithelial cells) to the

Are there criteria that can be developed to justify the use of animals in research?

Figure 5 shows some of the techniques that have been used to investigate kidney function. The animals used include rats, mice, cats, dogs and pigs.

1 What are the reasons for carrying out kidney research?

2 What criteria should be used to decide if a research technique is ethically acceptable or not?

3 Apply your criteria to the three techniques outlined in figure 5 to determine whether they are ethically acceptable.

4 Who should make the decisions about the ethics of scientific research?

Living animal is anaesthetized and its kidney is exposed by surgery. Fluid is sampled from nephrons using micropipettes. Animal is then sacrificed so that the position of the sample point in the kidney can be located.

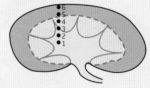

Animal is killed and kidneys are removed and frozen. Samples of tissue are cut from regions of kidney that can be identified. Temperature at which thawing occurs is found, to give a measure of solute concentration.

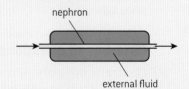

Animal is killed and kidneys are dissected to obtain samples of nephron. Fluids are perfused through nephron tissue, using experimental external fluids to investigate the action of the wall of the nephron.

▲ Figure 5

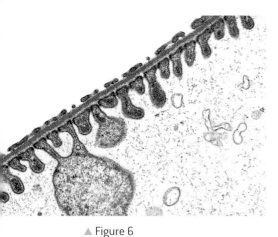

▲ Figure 6

membrane. The podocytes function as a barrier through which waste products are filtered from the blood.

There are three parts to the ultrafiltration system.

1 **Fenestrations** between the cells in the wall of the capillaries. These are about 100 nm in diameter. They allow fluid to escape, but not blood cells.

2 **The basement membrane** that covers and supports the wall of the capillaries. It is made of negatively-charged glycoproteins, which form a mesh. It prevents plasma proteins from being filtered out, due to their size and negative charges.

3 **Podocytes** forming the inner wall of the Bowman's capsule. These cells have extensions that wrap around the capillaries of the glomerulus and many short side branches called foot processes. Very narrow gaps between the foot processes help prevent small molecules from being filtered out of blood in the glomerulus.

If particles pass through all three parts they become part of the glomerular filtrate.

Figure 8 shows the relationship between the glomerulus and the Bowman's capsule.

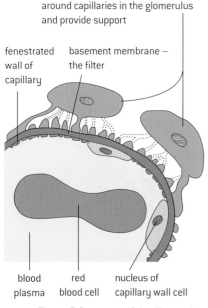

▲ Figure 7 Structure of the filter unit of the kidney

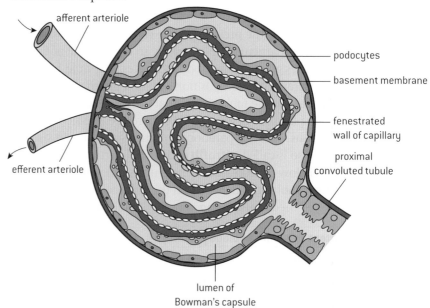

▲ Figure 8

─────────────────────────────────────

Data-based questions: Ultrafiltration of charged and uncharged dextrans

Dextrans are polymers of sucrose. Different sizes of dextran polymer can be synthesized, allowing their use to investigate the effect of particle size on ultrafiltration. Neutral dextran is uncharged, dextran sulphate has many negative charges, and DEAE is dextran with many positive charges.

Figure 9 shows the relationship between particle size and the permeability of the filter

unit of rat glomeruli. Animal experiments like this can help us to understand how the kidney works and can be done without causing suffering to the animals.

1 State the relationship between the size of particles and the permeability to them of the filter unit of the glomerulus. [1]

2 a) Compare the permeability of the filter unit to the three types of dextran. [3]

b) Explain these differences in permeability. [3]

3 One of the main plasma proteins is albumin, which is negatively charged and has a particle size of approximately 4.4 nm. Using the data in the graph, explain the diagnosis that is made if albumin is detected in a rat's urine. [3]

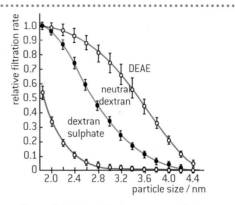

▲ Figure 9 Relationship between particle size of dextrans and filtration rate

The role of the proximal convoluted tubule

The proximal convoluted tubule selectively reabsorbs useful substances by active transport.

The glomerular filtrate flows into the proximal convoluted tubule. The volume of glomerular filtrate produced per day is huge – about 180 dm^{-3}. This is several times the total volume of fluid in the body and it contains nearly 1.5 kg of salt and 5.5 kg of glucose. As the volume of urine produced per day is only about 1.5 dm^3 and it contains no glucose and far less than 1.5 kg of salt, almost all of the filtrate must be reabsorbed into the blood. Most of this reabsorption happens in the first part of the nephron – the proximal convoluted tubule. Figure 10 shows this structure in transverse section. The methods used to reabsorb substances in the proximal convoluted tubule are described in table 3. By the end of the proximal tubule all glucose and amino acids and 80 per cent of the water, sodium and other mineral ions have been absorbed.

▲ Figure 10 Transverse section of the proximal convoluted tubule

Sodium ions: are moved by active transport from filtrate to space outside the tubule. They then pass to the peritubular capillaries. Pump proteins are located in outer membrane of tubule cells.
Chloride ions: are attracted from filtrate to space outside the tubule because of charge gradient set up by active transport of sodium ions.
Glucose: is co-transported out of filtrate and into fluid outside the tubule, by co-transporter proteins in outer membrane of tubule cells. Sodium ions move down concentration gradient from outside tubule into tubule cells. This provides energy for glucose to move at the same time to fluid outside the tubule. The same process is used to reabsorb amino acids.
Water: pumping solutes out of filtrate and into the fluid outside the tubule creates a solute concentration gradient, causing water to be reabsorbed from filtrate by osmosis.

▲ Table 3

Activity

The drawing below shows the structure of a cell from the wall of the proximal convoluted tubule. Explain how the structure of the proximal convoluted tubule cell, as shown in the diagram, is adapted to carry out selective reabsorption.

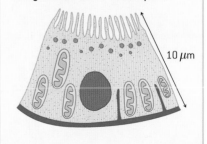

491

The nephron

Annotation of diagrams of the nephron.

The basic functional unit of the kidney is the nephron. This is a tube with a wall consisting of one layer of cells. This wall is the last layer of cells that substances cross to leave the body – it is an epithelium. There are several different parts of the nephron, which have different functions and structures (see figure 11):

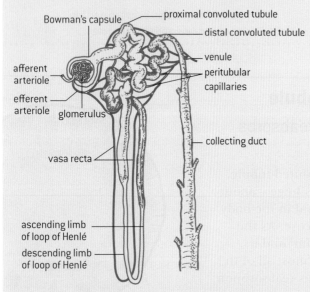

▲ Figure 11 The nephron and associated blood vessels. The human kidney contains about a million nephrons

- **Bowman's capsule** – a cup-shaped structure with a highly porous inner wall, which collects the fluid filtered from the blood.

- **Proximal convoluted tubule** – a highly twisted section of the nephron, with cells in the wall having many mitochondria and microvilli projecting into the lumen of the tube.

- **Loop of Henle** – a tube shaped like a hairpin, consisting of a descending limb that carries the filtrate deep into the medulla of the kidney, and an ascending limb that brings it back out to the cortex.

- **Distal convoluted tubule** – another highly twisted section, but with fewer, shorter microvilli and fewer mitochondria.

- **Collecting duct** – a wider tube that carries the filtrate back through the cortex and medulla to the renal pelvis.

- **Blood vessels** – associated with the nephron are blood vessels. Blood flows though them in the following sequence:

  - **Afferent arteriole** – brings blood from the renal artery.

  - **Glomerulus** – a tight, knot-like, high-pressure capillary bed that is the site of blood filtration.

  - **Efferent arteriole** – a narrow vessel that restricts blood flow, helping to generate high pressure in the glomerulus.

  - **Peritubular capillaries** – a low-pressure capillary bed that runs around the convoluted tubules, absorbing fluid from them.

  - **Vasa recta** – unbranched capillaries that are similar in shape to the loops of Henle, with a descending limb that carries blood deep into the medulla and an ascending limb bringing it back to the cortex.

  - **Venules** – carry blood to the renal vein.

The role of the loop of Henle

The loop of Henle maintains hypertonic conditions in the medulla.

The overall effect of the loop of Henle is to create a gradient of solute concentration in the medulla. The energy to create the gradient is expended by wall cells in the ascending limb. Here sodium ions are pumped out of the filtrate to the fluid between the cells in the medulla – called the interstitial fluid. The wall of the ascending limb is unusual in that it is impermeable to water, so water is retained in the filtrate, even though the interstitial fluid is now hypertonic relative to the filtrate; i.e., it has a higher solute concentration.

Normal body fluids have a concentration of 300 mOsm. The pump proteins that transfer sodium ions out of the filtrate can create a gradient of up to 200 mOsm, so an interstitial concentration of 500 mOsm is clearly achievable. The cells in the wall of the descending limb are permeable to water, but are impermeable to sodium ions. As filtrate flows down the descending limb, the increased solute concentration of interstitial fluid in the medulla causes water to be drawn out of the filtrate until it reaches the same solute concentration as the interstitial fluid. If this was 500 mOsm, then filtrate entering the ascending limb would be at this concentration and the sodium pumps could raise the interstitial fluid to 700 mOsm. Fluid passing down the descending limb would therefore reach 700 mOsm, and the sodium pumps in the ascending limb could cause a further 200 mOsm rise. The interstitial fluid concentration can therefore rise further and further, until a maximum is reached, which in humans is 1,200 mOsm.

This system for raising solute concentration is an example of a countercurrent multiplier system. It is a *countercurrent* system because of the flows of fluid in opposite directions. It is a countercurrent *multiplier* because it causes a steeper gradient of solute concentration to develop in the medulla than would be possible with a concurrent system. There is also a countercurrent system in the vasa recta. This prevents the blood flowing through this vessel from diluting the solute concentration of the medulla, while still allowing the vasa recta to carry away the water removed from filtrate in the descending limb, together with some sodium ions.

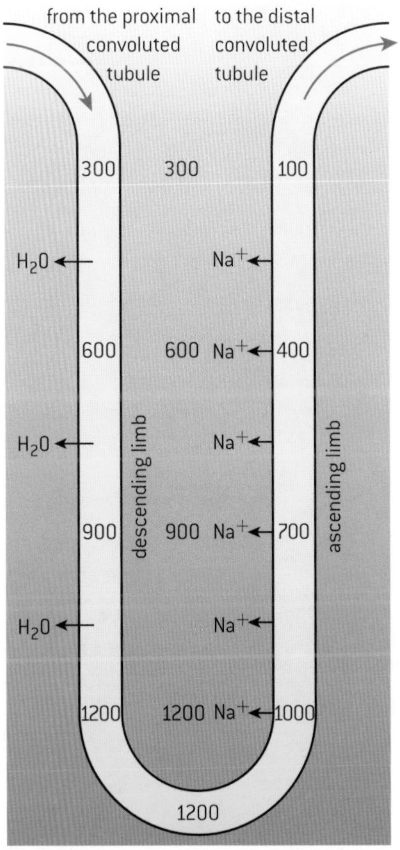

▲ Figure 12 Solute concentrations in the loop of Henle (in mOsm)

Some animals have relatively long loops of Henle

The length of the loop of Henle is positively correlated with the need for water conservation in animals.

The longer the loop of Henle, the more water volume will be reclaimed. Animals adapted to dry habitats will often have long loops of Henle. Loops of Henle are found within the medulla. In order to accommodate long loops of Henle, the medulla must become relatively thicker.

> **Data-based questions:** Medulla thickness and urine concentration
>
> Table 4 shows the relative medullary thickness (RMT) and maximum solute concentration (MSC) of the urine in mOsm for 14 species of mammal. RMT is a measure of the thickness of the medulla in relation to the overall size of the kidney. All the species in the table that are shown with binomials are desert rodents.
>
> 1 Discuss the relationship between maximum solute concentration of urine and the habitat of the mammal. [3]
>
> 2 Plot a scattergraph of the data in the table, either by hand or using computer software. [7]

493

3 **a)** Using the scattergraph that you have plotted, state the relationship between RMT and the maximum solute concentration of the urine. [1]

b) Suggest how the thickness of the medulla could affect the maximum solute concentration of the urine. [4]

Species	RMT	MSC (mOsm)
beaver	1.3	517
pig	1.6	1076
human	3.0	1399
dog	4.3	2465
cat	4.8	3122
rat	5.8	2465
Octomys mimax	6.1	2071
Dipodomys deserti	8.5	5597
Jaculus jaculus	9.3	6459
Tympanoctomys barrerae	9.4	7080
Psammomys obesus	10.7	4952
Eligmodontia typus	11.4	8612
Calomys mus	12.3	8773
Salinomys delicatus	14.0	7440

▲ Table 4

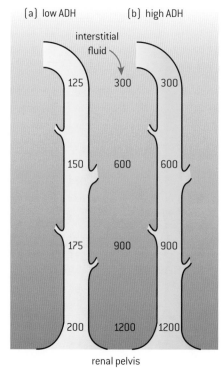

(a) low ADH   (b) high ADH

interstitial fluid

▲ Figure 13 Solute concentrations in the collecting duct

Function of ADH

ADH controls reabsorption of water in the collecting duct.

When filtrate enters the distal convoluted tubule from the loop of Henle, its solute concentration is lower than that of normal body fluids – it is hypotonic. This is because proportionately more solutes than water have passed out of the filtrate as it flows through the loop of Henle in the medulla.

If the solute concentration of the blood is too low, relatively little water is reabsorbed as the filtrate passes on through the distal convoluted tubule and the collecting duct. The wall of these parts of the nephron can have an unusually low permeability to water. A large volume of urine is therefore produced, with a low solute concentration, and as a result the solute concentration of the blood is increased (see figure 13a).

If the solute concentration of the blood is too high, the hypothalamus of the brain detects this and causes the pituitary gland to secrete a hormone – antidiuretic hormone or ADH. This hormone causes the walls of the distal convoluted tubule and collecting duct to become much more permeable to water, and most of the water in the filtrate is reabsorbed. This is helped by the solute concentration gradient of the medulla. As the filtrate passes down the collecting duct, it flows deep into the medulla, where the solute concentration of the interstitial fluid is high. Water continues to be reabsorbed along the whole length of the collecting duct and the kidney produces a small volume of concentrated urine (figure 13b). As result the solute concentration of the blood is reduced. The action of the kidney therefore helps to keep the relative amounts of water and solutes in balance at an appropriate level. This is called osmoregulation.

Data-based questions: ADH release and feelings of thirst

The plasma solute concentration, plasma antidiuretic hormone (ADH) concentration and feelings of thirst were tested in a group of volunteers. Figures 14 and 15 show the relationship between intensity of thirst, plasma ADH concentration and plasma solute concentration.

a) Identify the plasma ADH concentration at a plasma solute concentration of 300 mOsmol kg^{-1} using the line of best fit. [1]

b) Compare intensity of thirst and plasma ADH concentration. [1]

c) Outline what would happen to plasma solute concentration and ADH concentration if a person were to drink water to satisfy his/her thirst. [2]

d) State two reasons why a person's plasma solute concentration may increase. [2]

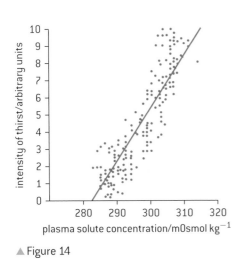

▲ Figure 14

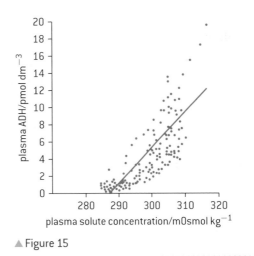

▲ Figure 15

Animals vary in terms of the type of nitrogenous waste they produce

The type of nitrogenous waste in animals is correlated with evolutionary history and habitat.

When animals break down amino acids and nucleic acids, nitrogenous waste in the form of ammonia is produced. Ammonia is highly basic and can alter the pH balance. It is also toxic as it is a highly reactive chemical. If the organism lives in a marine or freshwater habitat, such as fish, echinoderms or coelenterates, they can release the waste directly as ammonia as it can be easily diluted within that environment. Terrestrial

organisms will expend energy to convert ammonia to the less toxic forms of urea or uric acid depending on their habitats and evolutionary history. Marine mammals, despite their habitat, release urea because of their evolutionary history.

Some organisms like amphibians release the waste as ammonia when they are larva and after metamorphosis, release the waste as urea. Converting ammonia to urea requires energy and converting it to uric acid requires even more energy. The advantage of uric acid is that it is not water-soluble and therefore does not require water to be released. Birds and insects release their nitrogenous waste as uric acid. For birds, not having to carry water for excretion means less energy needs to be expended on flight.

Uric acid is linked to adaptations for reproduction. Nitrogenous wastes are released by the developing organism within eggs. Uric acid is released as it is not soluble and crystallizes rather than building up to toxic concentrations within the egg.

▲ Figure 16 The white paste in bird droppings is uric acid

Dehydration and overhydration

Consequences of dehydration and overhydration.

Dehydration is a condition that arises when more water leaves the body than comes in. It can arise from a number of factors including exercise, insufficient water intake or diarrhoea. It can lead to the disruption of metabolic processes.

One sign of dehydration is darkened urine due to increased solute concentration. Water is necessary to remove metabolic wastes so dehydration can lead to tiredness and lethargy due to decreased efficiency of muscle function and increased tissue exposure to metabolic wastes. Blood pressure can fall due to low blood volume. This can lead to increases in heart rate. Body temperature regulation may be affected because of an inability to sweat.

Overhydration is less common and occurs when there is an over-consumption of water. The result is a dilution of blood solutes. It might occur when large amounts of water are consumed after intense exercise without replacing the electrolytes lost at the same time. This makes body fluids hypotonic and could result in the swelling of cells due to osmosis. If this occurs, the most notable symptoms are headache and nerve function disruption.

Treatment options for kidney failure

Treatment of kidney failure by hemodialysis or kidney transplant.

Kidney failure can occur for a number of reasons but most commonly occurs as a complication from diabetes or chronic high blood pressure (hypertension) as a result of diabetes.

Figure 17 shows a patient undergoing renal dialysis (hemodialysis). The dialysis machine (artificial kidney) is on the left. Hemodialysis is required when the kidneys are no longer able to filter waste products from the blood properly. During the procedure, a steady flow of blood passes over an artificial semi-permeable membrane in the dialysis machine. The small waste products in the blood pass through the membrane, but the larger blood cells and proteins cannot. The purified blood is then returned to the patient via a vein. This procedure takes several hours.

An alternative to dialysis is a kidney transplant. In this treatment option, a kidney from one person is placed in the body of a person whose kidneys aren't functioning. The donor can either be living or deceased. A living donor is possible because a person can survive with one functional kidney. This approach can result in greater independence of movement and freedom to travel as compared

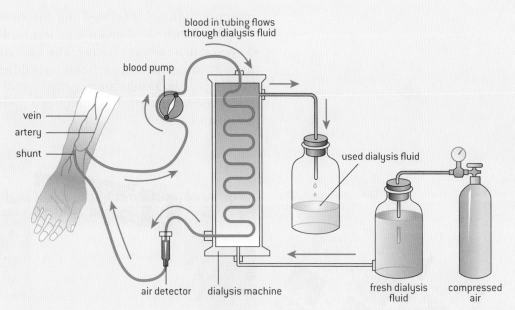

▲ Figure 17

to dialysis. Dialysis also carries with it the risk of infection and other complications.

A drawback to a transplant is that the recipient's body can reject the organ. Figure 19 is of a light micrograph through a transplanted kidney that has been rejected by the recipient's immune system. Numerous lymphocytes (with small dots) have infiltrated the kidney tissue.

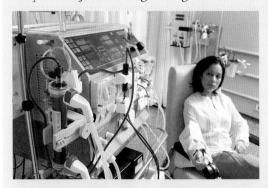

▲ Figure 18

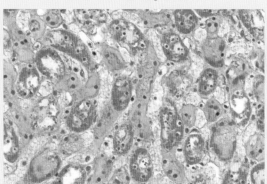

▲ Figure 19

Urinalysis

Blood cells, glucose, proteins and drugs are detected in urinary tests.

Urine is a product of osmoregulation, excretion and metabolism. These processes can be disrupted by illness or drug abuse. Urinalysis is a clinical procedure that examines urine for any deviation from normal composition.

Figure 20 shows a urine test strip being compared to the results chart on the testing kit bottle. This strip contains three test areas designed to change colour to indicate a positive or negative result after being dipped in urine.

The colours displayed can then be compared to a results chart on the testing kit. This test indicates the pH, protein level and glucose level in the urine. High levels of glucose and protein in the urine can be an indication of diabetes. High protein levels can indicate damage to the kidneys as these do not get through ultrafiltration in a healthy kidney. The strip in the picture is a normal negative result for protein and glucose.

presence of traces of banned and controlled drugs in urine. Figure 21 shows a drug test card being dipped into a sample of urine. The card contains five vertical strips that each test for a different drug. Here, the results are negative for all but the one second from left. This indicates a positive test for opiates.

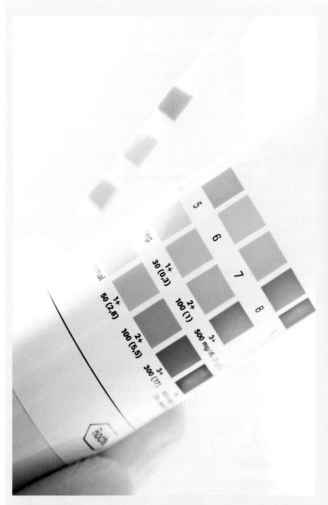

▲ Figure 20

The panel drug test also uses test strips based on monoclonal antibody technology to look for the

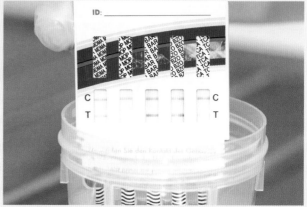

▲ Figure 21

Microscopic examination of urine is carried out to determine if cells are present, as under normal circumstances, these cells should not be present. Figure 22 shows white blood cells. The presence of 6–10 neutrophils (white blood cells with a nucleus visible) can be a sign of urinary tract infection. Figure 23 indicates the presence of red blood cells (erythrocytes) in the urine of this patient. This can be a sign that there is a kidney stone or a tumour in the urinary tract.

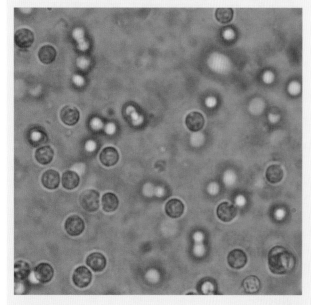

▲ Figure 22

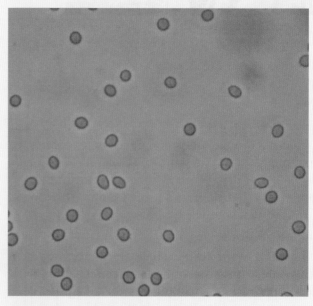

▲ Figure 23

11.4 Sexual reproduction

Understanding

→ Spermatogenesis and oogenesis both involve mitosis, cell growth, two divisions of meiosis and differentiation.

→ Processes in spermatogenesis and oogenesis result in different numbers of gametes with different amounts of cytoplasm.

→ Fertilization involves mechanisms that prevent polyspermy.

→ Fertilization in animals can be internal or external.

→ Implantation of the blastocyst in the endometrium is essential for the continuation of pregnancy.

→ hCG stimulates the ovary to secrete progesterone during early pregnancy.

→ The placenta facilitates the exchange of materials between the mother and embryo.

→ Estrogen and progesterone are secreted by the placenta once it has formed.

→ Birth is mediated by positive feedback involving estrogen and oxytocin.

Applications

→ The average 38-week pregnancy in humans can be positioned on a graph showing the correlation between animal size and the development of the young at birth for other mammals.

Skills

→ Annotation of diagrams of seminiferous tubule and ovary to show the stages of gametogenesis.

→ Annotation of diagrams of mature sperm and egg to indicate functions.

Nature of science

→ Assessing risks and benefits associated with scientific research: the risks to human male fertility were not adequately assessed before steroids related to progesterone and estrogen were released into the environment as a result of the use of the female contraceptive pill.

Similarities between oogenesis and spermatogenesis

Spermatogenesis and oogenesis both involve mitosis, cell growth, two divisions of meiosis and differentiation.

Oogenesis is the production of egg cells in the ovaries. Oogenesis starts in the ovaries of a female fetus. Germ cells in the fetal ovary divide by mitosis and the cells formed move to distribute themselves through the cortex of the ovary. When the fetus is four or five months old, these cells grow and start to divide by meiosis. By the seventh month, they are still in the first division of meiosis and a single layer of cells, called follicle cells, has formed around them. No further development takes place until after puberty. The cell that has started to divide by meiosis, together with the surrounding follicle cells, is called a **primary follicle.** There are about 400,000 in the ovaries at birth. No more primary follicles are produced, but at the start of each menstrual cycle a small batch are stimulated to develop by FSH. Usually only one goes on to become a **mature follicle,** containing a **secondary oocyte**.

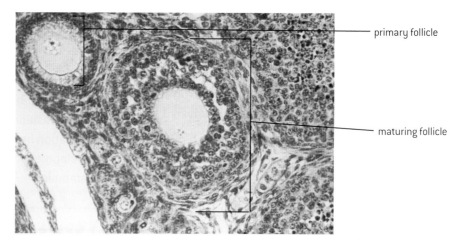

▲ Figure 1 Light micrograph of a section through tissue from an ovary, showing a primary follicle (left) and a maturing follicle (centre). Primary follicles contain a central oocyte (female germ cell, egg) surrounded by a single layer of follicle cells. A mature ovarian follicle has many more follicle cells, outer and inner follicle cells and cavities, and the oocyte is now more fully developed compared to the primordial and primary stages

Spermatogenesis is the production of sperm. It happens in the testes, which are composed of a mass of narrow tubes, called **seminiferous tubules,** with small groups of cells filling the gaps between the tubules. These gaps are called interstices, so the cells in them are **interstitial cells**. They are sometimes called Leydig cells. The seminiferous tubules are also made of cells. The outer layer of cells is called the **germinal epithelium**. This is where the process of sperm production begins. Cells in various stages of sperm production are found inside the germinal epithelium, with the most mature stages closest to the fluid-filled centre of the seminiferous tubule. Cells that have developed tails are called **spermatozoa,** though this is almost always abbreviated to sperm. Also in the wall of the tubule are large nurse cells, called **Sertoli cells**. Figure 3 shows a small area of testis tissue, in which the structures described above can be seen.

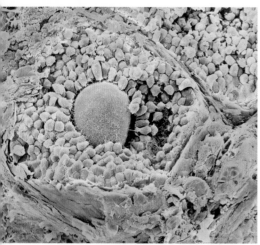

▲ Figure 2 Coloured scanning electron micrograph (SEM) of ovary tissue, showing two secondary follicles. A secondary oocyte (pink) is seen at the centre of one follicle. Follicles are surrounded by two types of follicle cells (coloured blue and green). Between the follicle cells a space develops (at centre right, coloured brown), into which follicular fluid is secreted. The amount of fluid will increase significantly as the follicle matures

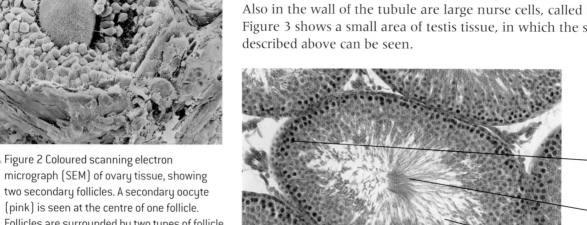

▲ Figure 3 Transverse section through a seminiferous tubule

Diagrams of a seminiferous tubule and the ovary

Annotation of diagrams of seminiferous tubule and ovary to show the stages of gametogenesis.

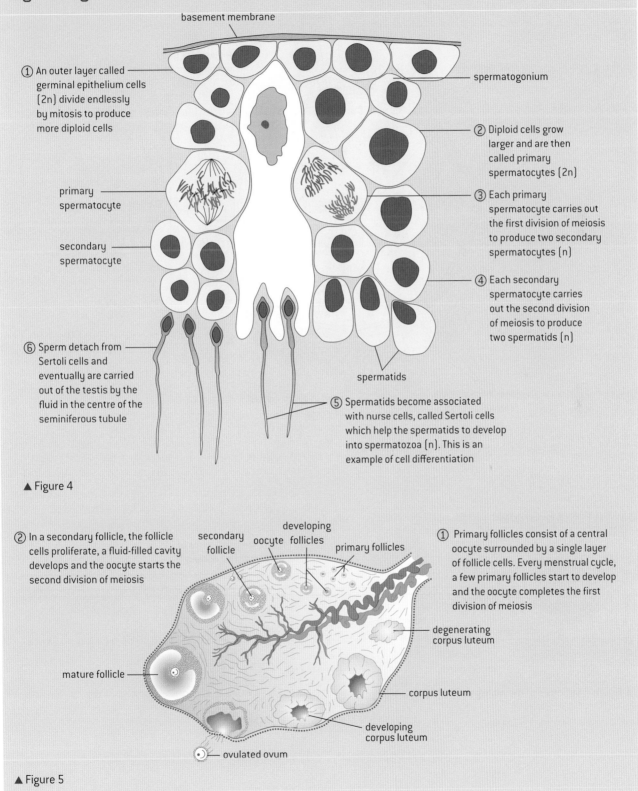

basement membrane

① An outer layer called germinal epithelium cells (2n) divide endlessly by mitosis to produce more diploid cells

spermatogonium

② Diploid cells grow larger and are then called primary spermatocytes (2n)

primary spermatocyte

③ Each primary spermatocyte carries out the first division of meiosis to produce two secondary spermatocytes (n)

secondary spermatocyte

④ Each secondary spermatocyte carries out the second division of meiosis to produce two spermatids (n)

⑥ Sperm detach from Sertoli cells and eventually are carried out of the testis by the fluid in the centre of the seminiferous tubule

spermatids

⑤ Spermatids become associated with nurse cells, called Sertoli cells which help the spermatids to develop into spermatozoa (n). This is an example of cell differentiation

▲ Figure 4

② In a secondary follicle, the follicle cells proliferate, a fluid-filled cavity develops and the oocyte starts the second division of meiosis

secondary follicle

developing oocyte follicles

primary follicles

① Primary follicles consist of a central oocyte surrounded by a single layer of follicle cells. Every menstrual cycle, a few primary follicles start to develop and the oocyte completes the first division of meiosis

degenerating corpus luteum

corpus luteum

developing corpus luteum

mature follicle

ovulated ovum

▲ Figure 5

(A) Diagrams of sperm and egg

Annotation of diagrams of mature sperm and egg to indicate functions.

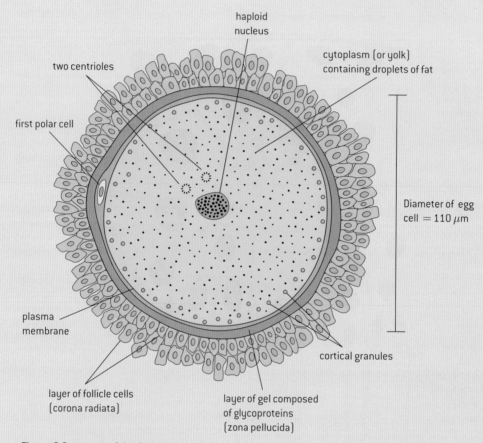

haploid nucleus

two centrioles

cytoplasm (or yolk) containing droplets of fat

first polar cell

Diameter of egg cell = 110 μm

plasma membrane

cortical granules

layer of follicle cells (corona radiata)

layer of gel composed of glycoproteins (zona pellucida)

▲ Figure 6 Structure of the female gamete

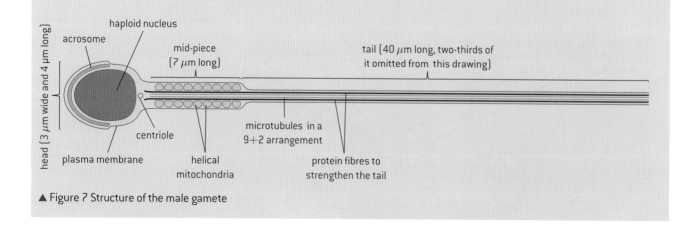

haploid nucleus

acrosome

mid-piece (7 μm long)

tail (40 μm long, two-thirds of it omitted from this drawing)

head (3 μm wide and 4 μm long)

centriole

microtubules in a 9+2 arrangement

plasma membrane

helical mitochondria

protein fibres to strengthen the tail

▲ Figure 7 Structure of the male gamete

Data-based questions: Sizes of sperm

Sperm tails have a 9 + 2 arrangement of microtubules in the centre, with thicker protein fibres around. Table 1 shows the structure of sperm tails of eight animals in transverse section, with the tail lengths and the cross-sectional area of the protein fibres.

1 Draw a graph of tail length and cross-sectional area of protein fibres in the eight species of animal. [4]

2 Outline the relationship between tail length and cross-sectional area of protein fibres. [2]

3 Explain reasons for the relationship. [2]

4 Discuss whether there is a relationship between the size of an animal and the size of its sperm. [2]

	chinese hamster	rat	guinea pig	hamster	bull	mouse	human	sea urchin
cross-sectional area of fibrous sheaths / μm^2	0.22	0.16	0.13	0.11	0.08	0.04	0.02	0
length of sperm / μm	258	187	107	187	54	123	58	45

▲ Table 1

Differences in the outcome of spermatogenesis and oogenesis

Processes in spermatogenesis and oogenesis result in different numbers of gametes with different amounts of cytoplasm.

While there are similarities in spermatogenesis and oogenesis, there are differences that are necessary to prepare the gametes for their different roles. Each mature sperm consists of a haploid nucleus, a system for movement and a system of enzymes and other proteins that enable the sperm to enter the egg. Each complete meiotic division results in four spermatids. The process of sperm differentiation eliminates most of the cytoplasm, whereas the egg must increase its cytoplasm.

All of the requirements for beginning the growth and development of the early embryo must be present in the egg. In females, the first division of meiosis produces one large cell and one very small cell (figure 8). The small cell is the first polar body which eventually degenerates. The large cell goes on to the second division of meiosis, completing it after fertilization. Again one large cell and one very small cell are produced. The small cell is the second polar body and it also degenerates and dies. Only the large cell, which is the female gamete, survives. The result is that the egg is much larger than the sperm cell. Figures 6 and 7 show the differences in structure. Note that the scale bars indicate that the sperm and egg are drawn to different scale and that the egg is much larger than the sperm.

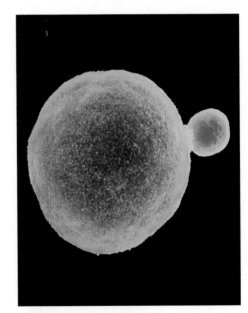

▲ Figure 8 The micrograph shows a primary oocyte split into two cells, known as the secondary oocyte (green) and the first polar body (yellow)

The process of egg formation happens once per menstrual cycle in humans and usually only one egg cell per cycle is produced. During the years from puberty to the menopause only a few hundred female gametes are likely to be produced.

From puberty onwards, the testes produce sperm continuously. At any time, there are millions of sperm at all stages of development.

Preventing polyspermy

Fertilization involves mechanisms that prevent polyspermy.

Fertilization is the union of a sperm and an egg to form a zygote.

The membranes of sperm have receptors that can detect chemicals released by the egg, allowing directional swimming towards the egg. Figure 9 illustrates that multiple sperm arrive at the egg. Once the egg is reached, a number of events take place (see figure 10). These events are designed to result in the union of a single sperm with the egg. The events are also designed to prevent multiple sperm entering, known as polyspermy.

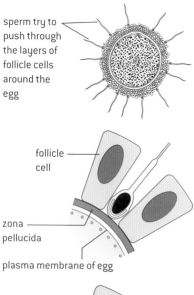

sperm try to push through the layers of follicle cells around the egg

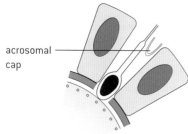

follicle cell

zona pellucida

plasma membrane of egg

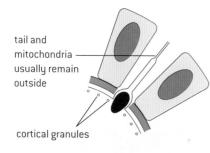

acrosomal cap

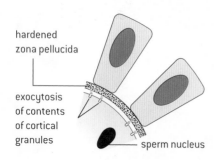

tail and mitochondria usually remain outside

cortical granules

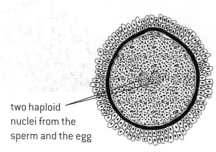

hardened zona pellucida

exocytosis of contents of cortical granules

sperm nucleus

two haploid nuclei from the sperm and the egg

▲ Figure 10 Stages in fertilization

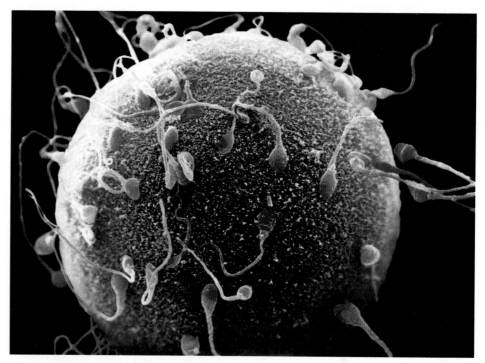

▲ Figure 9 Micrograph of egg surrounded by sperm

1 The acrosome reaction

The **zona pellucida** is a coat of glycoproteins that surrounds the egg. The **acrosome** is a large membrane-bound sac of enzymes in the head of the sperm. In mammals, the sperm binds to the zona pellucida and the contents of the acrosome are released. The enzymes from it digest the zona pellucida.

2 Penetration of the egg membrane

The acrosome reaction exposes an area of membrane on the tip of the sperm that has proteins that can bind to the egg membrane. The first sperm that gets through the zona pellucida therefore binds and the membranes of sperm and egg fuse together. The sperm nucleus enters the egg cell. This is the moment of fertilization.

3 The cortical reaction

Not only does the sperm bring male genes, it also causes the activation of the egg. The first effect of this is on the **cortical granules** – vesicles located near the egg membrane. There are thousands of these vesicles and when activation of the egg has taken place their contents are released from the egg by exocytosis. In mammals, the cortical vesicle enzymes result in the digestion of binding proteins so that no further sperm can bind. The enzymes also result in a general hardening of the zona pellucida.

▲ Figure 11 Breeding pair of *Anomalochromis thomasi* cichlids. The female (bottom) is laying eggs on a rock with the male in close proximity

Internal and external fertilization

Fertilization in animals can be internal or external.

Aquatic animals often release their gametes directly into water in a process that will lead to fertilization outside of the female's body. Such animals often have behaviours that bring eggs into proximity with sperm (see figure 11). External fertilization has several risks including predation and the susceptibility to environmental variation such as temperature and pH fluctuations and more recently, pollution.

Terrestrial animals are dependent on internal fertilization. Otherwise, gametes would be at risk of drying out. Internal fertilization also ensures sperm and ova are placed in prolonged close proximity to each other. Marine mammals which have reinvaded aquatic habitats still use internal fertilization. Once the eggs are fertilized, the developing embryo can be protected inside the female.

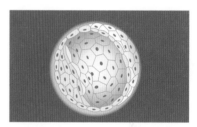

▲ Figure 12 Blastocyst

Implantation of the blastocyst

Implantation of the blastocyst in the endometrium is essential for the continuation of pregnancy.

After fertilization in humans, the fertilized ovum divides by mitosis to form two diploid nuclei and the cytoplasm of the fertilized egg cell divides equally to form a two-cell embryo. These two cells replicate their DNA, carry out mitosis and divide again to form a four-cell embryo. The embryo is about 48 hours old at this point. Further cell divisions occur, but some of the divisions are unequal and there is also migration of cells, giving the embryo the shape of a hollow ball. It is called a **blastocyst** (figure 12). At 7 days old the blastocyst consists of about 125 cells and it has reached the uterus, having been moved down the oviduct by the cilia of cells in the oviduct wall. At this age the zona pellucida, which has surrounded and protected the embryo, breaks down. The blastocyst has used up the reserves of the egg cell and needs an external supply of food. It obtains this by sinking into the endometrium or uterus

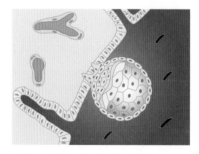

▲ Figure 13 Implantation of the blastocyst

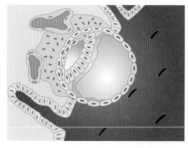

▲ Figure 14 Growth and differentiation of the early embryo

lining in a process called **implantation** (figure 13). The outer layer of the blastocyst develops finger-like projections allowing the blastocyst to penetrate the uterus lining. They also exchange materials with the mother's blood, including absorbing foods and oxygen. The embryo grows and develops rapidly and by eight weeks has started to form bone tissue. It is then considered to be a fetus rather than an embryo. It is recognizably human and soon visibly either male or female.

Role of hCG in early pregnancy

hCG stimulates the ovary to secrete progesterone during early pregnancy.

Pregnancy depends on the maintenance of the endometrium, which depends on the continued production of progesterone and estrogen. In part these hormones prevent the degeneration of the uterus lining which is required to support the developing fetus. Early in pregnancy the embryo produces human chorionic gonadotropin – hCG. This hormone stimulates the corpus luteum in the ovary to continue to secrete progesterone and estrogen. These hormones stimulate the continued development of the uterus wall, which supplies the embryo with everything that it needs.

Materials exchange by the placenta

The placenta facilitates the exchange of materials between the mother and embryo.

Humans are placental mammals. There are two other groups of mammals: the monotremes lay eggs and the marsupials give birth to relatively undeveloped offspring that develop inside a pouch. By the stage when a marsupial would be born, a human fetus has developed a relatively complex placenta and so can remain in the uterus for months longer. The placenta is needed because the body surface area to volume ratio becomes smaller as the fetus grows larger.

The placenta is made of fetal tissues, in intimate contact with maternal tissues in the uterus wall. The fetus also develops membranes that form the amniotic sac. This contains amniotic fluid, which supports and protects the developing fetus.

The basic functional unit of the placenta is a finger-like piece of fetal tissue called a placental villus. These villi increase in number during pregnancy to cope with the increasing demands of the fetus for the exchange of materials with the mother. Maternal blood flows in the inter-villous spaces around the villi (figure 15). This is a very unusual type of circulation as elsewhere blood is almost always confined in blood vessels. Fetal blood circulates in blood capillaries, close to the surface of each villus. The distance between fetal and maternal blood is therefore very small – as little as 5 μm. The cells that separate maternal and fetal blood form the **placental barrier**. This must be selectively permeable, allowing some substances to pass, but not others (figure 16).

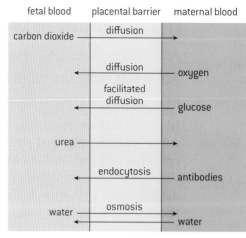

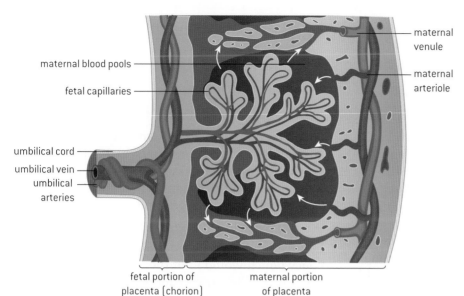

fetal portion of placenta (chorion)   maternal portion of placenta

▲ Figure 15

▲ Figure 16 Exchange processes in the placenta

Release of hormones by the placenta

Estrogen and progesterone are secreted by the placenta once it has formed.

By about the ninth week of pregnancy, the placenta has started to secrete estrogen and progesterone in large enough quantities to sustain the pregnancy, and the corpus luteum is no longer needed for this role. There is a danger of miscarriage at this stage of pregnancy if this switch-over fails.

Data-based questions: Electron micrograph of placenta

Figure 17 shows a small region at the edge of a placental villus. The magnification is × 17,000.

1   a)   Identify the structures that are visible in the upper part of the micrograph.   [1]

    b)   Explain the functions of these structures.   [3]

2   In much of the area of the electron micrograph there are rounded structures, surrounded by a single membrane. These are parts of a system of tubules called the smooth endoplasmic reticulum (sER). Its function is the synthesis of lipids, including steroids. Suggest a function for the sER in the placenta.   [3]

3   Identify, with reasons, the structure in the lower left part of the micrograph.   [3]

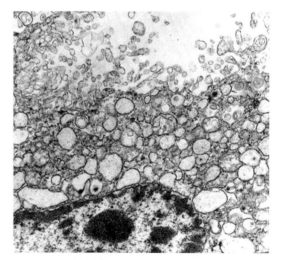

▲ Figure 17 Small region at the edge of a placental villus

Assessing risks of estrogen pollution

Assessing risks and benefits of scientific research: the risks to human male fertility were not adequately assessed before steroids related to progesterone and estrogen were released into the environment as a result of the use of the female contraceptive pill.

High levels of estrogen are present in pregnant women and inhibit FSH release. If women consume pills containing estrogen, then this would mimic pregnancy and inhibit the development of mature follicles thus preventing pregnancy. Ethinyl estradiol is a synthetic form of estrogen that was first introduced as a contraceptive in 1943. At the time, little thought was given to the idea that if a large number of women used this form of contraception, then levels of estrogen in bodies of water might be raised through sewage. It wasn't until the mid-1980s that the first reports of elevated contraceptive pill hormones present in water were reported. Since then, a number of problems have been attributed to estrogen pollution.

In 1992, a review article summarizing 61 different studies concluded that human male sperm counts have declined by 50% over the past 50 years.

In one of the largest studies of the problem, the UK government's Environment Agency found in 2004 that 86% of male fish sampled at 51 sites around the country were intersex, that is male fish showed signs of "feminization". However, there is limited scientific consensus that pollution with steroids related to estrogen and progesterone is the causative agent behind reduced male fertility.

In 2012 the European Commission proposed a policy which would limit the concentrations in water of a widely used contraceptive drug. This has sparked intense lobbying by the water and pharmaceutical industries, which say that the science is uncertain and the costs too high.

Upgrading the technology for wastewater treatment could eliminate most of the pollution. Researchers and policy experts suggest sharing the costs among all responsible parties, including the water and drug industries, and that some expense would be passed on to the public. The drugs are widely used in livestock, so preventing animals from urinating close to rivers could further reduce the amount of drugs leaking into surface waters.

Data-based questions: Estrogen pollution

Rivers vary in terms of the quantities of synthetic estrogen (E$_2$) found. A study was conducted to investigate the relationship between concentrations of synthetic estrogen in water and impacts on male fish from the genus *Rutilus* (roach) (see figure 18).

a) State the relationship between synthetic estrogen (E$_2$) and the appearance of oocytes in testes. [1]

b) Determine the mean percentage of male fish with oocytes in their testes at concentrations of estrogen greater than 10 ng/L. [2]

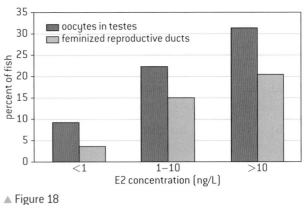

▲ Figure 18

Source: Jobling et al, *Environ Health Perspect.* 2006 April; 114(S-1): 32–39.

The role of hormones in parturition

Birth is mediated by positive feedback involving estrogen and oxytocin.

During pregnancy, progesterone inhibits secretion of oxytocin by the pituitary gland and also inhibits contractions of the muscular outer wall of the uterus – the myometrium. At the end of pregnancy, hormones produced by the fetus signal to the placenta to stop secreting progesterone, and oxytocin is therefore secreted.

Oxytocin stimulates contractions of the muscle fibres in the myometrium. These contractions are detected by stretch receptors, which signal to the pituitary gland to increase oxytocin secretion. Increased oxytocin makes the contractions more frequent and more vigorous, causing more oxytocin secretion. This is an example of a positive feedback system – a very unusual control system in human physiology. In this case it has the advantage of causing a gradual increase in the myometrial contractions, allowing the baby to be born with the minimum intensity of contraction.

Relaxation of muscle fibres in the cervix causes it to dilate. Uterine contraction then bursts the amniotic sac and the amniotic fluid passes out. Further uterine contractions, usually over hours rather than minutes, finally push the baby out through the cervix and vagina. The umbilical cord is broken and the baby takes its first breath and achieves physiological independence from its mother.

Data-based questions: Hormone levels during pregnancy

In the graph (figure 20), the thickness of the arrows indicates relative quantities.

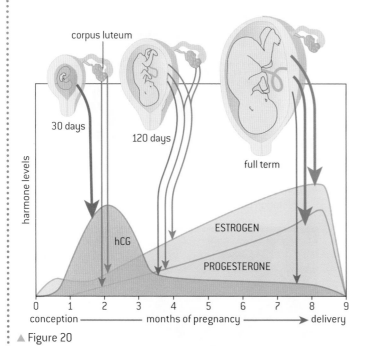

▲ Figure 20

1 Describe the changes over the course of a pregnancy in relative amounts and source of:

a) hCG [2]

b) estrogen [2]

c) progesterone [2]

2 Suggest reasons for the drop in hCG concentration after the second month of the pregnancy. [2]

3 Predict the consequences of the placenta failing to secrete estrogen and progesterone during a pregnancy. [2]

① Baby positions itself before birth so that its head rests close to the cervix

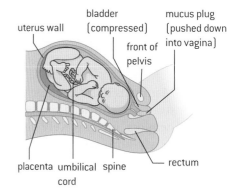

② Baby passes into vagina and amniotic fluid is released

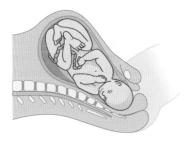

③ Baby is pushed out of mother's body

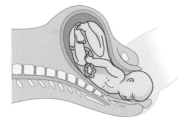

④ Placenta and umbilical cord are expelled from body

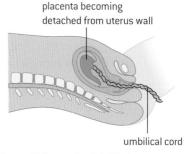

▲ Figure 19 Stages in childbirth

509

🌐 Gestation times, mass and growth, and development strategies

The average 38-week pregnancy in humans can be positioned on a graph showing the correlation between animal size and the development of the young at birth for other mammals.

Mammals differ in their growth and development strategies. Altricial species give birth to relatively helpless, incompletely developed offspring. Their newly-born young are relatively immobile, lack hair and are unable to obtain food on their own. At the opposite end of the spectrum are precocial mammals in which the offspring have open eyes, hair, are immediately mobile and are somewhat able to defend themselves against predators.

Mammals with a large body size are more likely to be precocial. This is correlated with a long gestation period.

Data-based questions: Gestation length and body mass

Figure 21 shows the relationship between gestation period and body mass for 429 placental mammal species subdivided into whether the species is described as altricial or precocial.

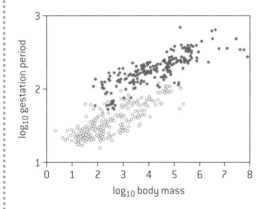

▲ Figure 21

▲ Figure 22 Laboratory mice are altricial. They have a gestation period of about 19 days

1 The solid dots and open dots are representative of two different growth and development strategies. Deduce which circles are used to represent precocial mammals. [2]

2 Outline the relationship between adult body mass and gestation period. [1]

3 Explain the relationship between body mass and the length of gestation. [3]

4 The mean length of human gestation is 283 days ($\log_{10} 283 = 2.45$) The mean body mass of an adult human is 65 kg ($\log_{10} 65 = 1.8$).

  (i) Determine the approximate location of humans on the graph. [1]

  (ii) Suggest reasons for humans being an outlier on this graph. [3]

▲ Figure 23 Elephant calves are born after a 22-month gestation period and they nurse for around three years. They are categorized as precocial. The African elephant is the largest and heaviest land animal alive today

Questions

1 Figure 24 shows how the surface pH of human skin varies between different areas of the body. It also shows differences between adults and newborn infants (neonates). Skin pH protects the skin from colonization by certain microorganisms.

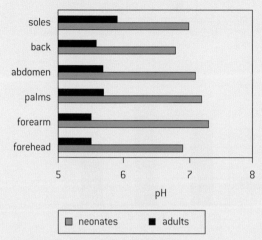

▲ Figure 24 How the surface pH of human skin varies between different areas of the body

a) Compare the skin pH of neonates and adults. [2]

b) Suggest how the adult skin pH might be established. [1]

c) Suggest why the use of soaps (which are basic) might have a more irritating effect on the skin of a neonate. [2]

d) Deduce how basic soaps might undermine the skin's defensive function. [2]

2 Figure 25 shows the ability of a calf (*Bos taurus*) to absorb antibodies after birth.

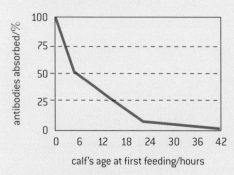

▲ Figure 25 The ability of a calf (*Bos taurus*) to absorb antibodies

a) Describe how the ability of a calf to absorb antibodies changes over the initial hours after birth. [2]

b) Suggest reasons for calves that have endured a long and difficult birth being more likely to suffer from infection. [2]

c) Predict how the concentration of antibodies might vary in the cow's colostrum over the first 24 hours after birth. [2]

d) Deduce the reasons for vaccinating sheep against pulpy kidney and other life-threatening diseases three weeks before lambs are due to be born. [2]

e) Explain which method of transport across membranes is likely to be used for absorption of antibodies in the stomach of newborn mammals. [2]

3 The blood glucose concentration of a person with untreated diabetes often rises to 300–500 mg per 100 ml of blood. It can even rise to concentrations above 1,000 mg per 100 ml. When the blood glucose level rises above 225 mg per 100 ml, glucose starts to appear in the urine. The volumes of urine produced become larger than normal, making the person dehydrated and thirsty.

a) Explain how glucose is completely reabsorbed from the glomerular filtrate of people who do not have diabetes. [3]

b) Explain why glucose is not all reabsorbed from the glomerular filtrate of diabetic patients. [4]

c) Suggest why untreated diabetics tend to pass large volumes of urine and often feel thirsty. [3]

4 Muscles often increase in mass if the amount that they are used increases. An experiment was performed to examine the effect of flight on muscle mass in European starlings (*Sturnus vulgaris*). Study birds were randomly assigned to three groups. Over 6 weeks, each group was subjected to 34 1-hour study periods. The exercise group was trained to fly for 1 hour by receiving food rewards. Control group 1 was allowed to feed freely but placed into cages that prevented flying. Control group 2 was fed the same food rewards at the same time as the exercise group, but was also placed into cages that prevented flying. Body mass was monitored before and during the experiment (see figure 26). At the end of the experiment,

the mean mass of the birds' pectoralis muscles was compared (figure 26).

a) Compare the changes in body mass in control group 2 and the exercise group.   [2]

b) Evaluate the claim that preventing exercise increases pectoralis muscle mass.   [3]

c) Suggest how the mass of the birds' pectoralis muscle could be determined.   [2]

d) One hypothesis that might be generated from this experiment would be that reducing motion in birds might lead to greater muscle mass per bird. Such knowledge might be used in the farming of poultry. Greater meat production per bird would result from the motion of the birds being restricted. Discuss the ethics of designing and carrying out experiments to test this hypothesis.   [3]

(a)
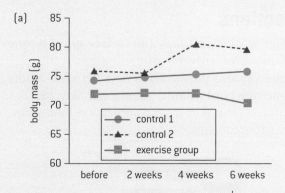

(b)
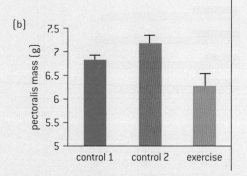

▲ Figure 26 The effect of exercise on body mass and muscle mass in starlings

Introduction

Neurobiology is the scientific study of the nervous system. Living organisms use their nervous system to detect and respond to changes in the environment. Communication between neurons can be altered through the manipulation of the release and reception of chemical messengers. Modification of neurons starts in the earliest stages of embryogenesis and continues to the final years of life. The parts of the brain specialize in different functions. Behaviour patterns can be inherited or learned. Natural selection favours types of behaviour that increase the chance of survival and reproduction.

A.1 Neural development

Understanding

→ The neural tube of embryonic chordates is formed by infolding of ectoderm followed by elongation of the tube.

→ Neurons are initially produced by differentiation in the neural tube.

→ Immature neurons migrate to a final location.

→ An axon grows from each immature neuron in response to chemical stimuli.

→ Some axons extend beyond the neural tube to reach other parts of the body.

→ A developing neuron forms multiple synapses.

→ Synapses that are not used do not persist.

→ Neural pruning involves the loss of unused neurons.

→ The plasticity of the nervous system allows it to change with experience.

Applications

→ Incomplete closure of the embryonic neural tube can cause spina bifida.

→ Events such as strokes may promote reorganization of brain function.

Skills

→ Annotation of a diagram of embryonic tissues in *Xenopus,* used as an animal model, during neurulation.

Nature of science

→ Use models as representations of the real world: developmental neuroscience uses a variety of animal models.

Animal models in neuroscience

Use models as representations of the real world: developmental neuroscience uses a variety of animal models.

Neuroscience is the branch of biology concerned with neurons and nervous systems. The aim of research in developmental neuroscience is to discover how nervous systems are formed as animals grow from embryo into adult. The aim of many neuroscientists is to understand and develop treatments for diseases of the nervous system, but many experiments are impossible to perform in humans for ethical reasons. Also, research into other animal species is usually easier because the nervous system develops more rapidly, is less complex and is easier to observe because the embryo develops externally rather than in a uterus.

For these reasons, even when researchers are trying to make discoveries about humans, they work with other species. A relatively small number of species is used for most of this research and these species are known as animal models:

- *Caenorhabditis elegans* (flatworm) because they have a low fixed number of cells as adults and mature very quickly.

- *Drosophila melanogaster* (fruit fly) because they breed readily, have only 4 pairs of chromosomes and mature very quickly.

- *Danio rerio* (zebrafish) because the tissues are almost transparent.

- *Xenopus laevis* (African clawed frog) because the eggs are large and easily manipulated.

- *Mus musculus* (house mouse) because after millennia living near people and their food, it shares many human diseases.

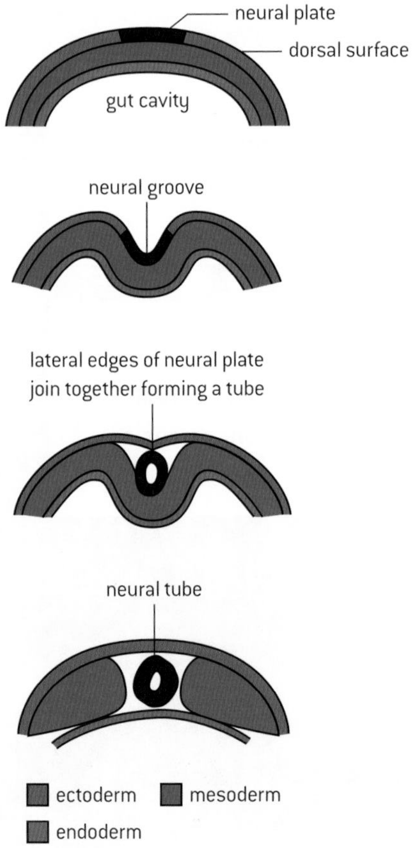

Development of the neural tube

The neural tube of embryonic chordates is formed by infolding of ectoderm followed by elongation of the tube.

All chordates develop a dorsal nerve cord at an early stage in their development. The process is called neurulation and in humans it occurs during the first month of gestation. An area of ectoderm cells on the dorsal surface of the embryo develops into the neural plate. The cells in the neural plate change shape, causing the plate to fold inwards forming a groove along the back of the embryo and then separate from the rest of the ectoderm. This forms the neural tube, which elongates as the embryo grows. The channel inside the neural tube persists as a narrow canal in the centre of the spinal cord.

Development of neurons

Neurons are initially produced by differentiation in the neural tube.

There are billions of neurons in the central nervous system (CNS), most of them in the brain. The origins of these neurons can be traced back to the early stages of embryonic development, when part of the ectoderm develops into neuro-ectodermal cells in the neural plate. Although not yet neurons, the developmental fate of these cells is now determined and it is from them that the nervous system is formed.

neural plate
dorsal surface
gut cavity

neural groove

lateral edges of neural plate join together forming a tube

neural tube

■ ectoderm ■ mesoderm
■ endoderm

▲ Figure 1 Stages in neurulation

The neural plate develops into the neural tube, with continued proliferation of cells by mitosis and differentiation along the pathways leading to the cells becoming functioning neurons. The mature CNS has far more neurons than are initially present in the embryonic neural tube, so cell proliferation continues in both the developing spinal cord and brain. Although cell division ceases before birth in most parts of the nervous system, there are many parts of the brain where new neurons are produced during adulthood.

Neurulation in *Xenopus*

Annotation of a diagram of embryonic tissues in *Xenopus*, used as an animal model, during neurulation.

The diagrams in figure 2 show five stages in the development of a *Xenopus* embryo, including the development of the neural tube. They show the notochord, a supportive structure that is present in all chordates during the early stages of embryonic development but which develops into the vertebral column in vertebrates. The notochord is part of the mesoderm of the embryo.

Make copies of the diagrams and annotate them to show these structures or stages:

- ectoderm, mesoderm and endoderm
- development of the neural tube
- wall of developing gut and gut cavity
- notochord
- developing dorsal fin.

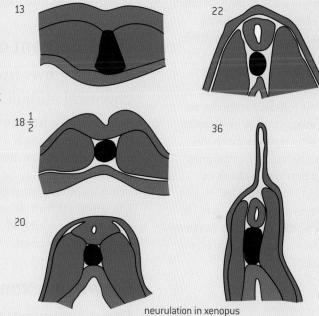

neurulation in xenopus

▲ Figure 2 Five stages of embryonic development in Xenopus from day 13 to day 36

Spina bifida

Incomplete closure of the embryonic neural tube can cause spina bifida.

In vertebrates, including all mammals, the spine comprises a series of bones called vertebrae. Each has a strong centrum that provides support and a thinner vertebral arch, which encloses and protects the spinal cord. The centrum develops on the ventral side of the neural tube at an early stage in embryonic development. Tissue migrates from both sides of the centrum around the neural tube and normally meets up to form the vertebral arch.

In some cases the two parts of the arch never become properly fused together, leaving a gap. This condition is called spina bifida. It is probably caused by the embryonic neural tube not closing up completely when it is formed from the neural groove. Spina bifida is commonest in the lower back. It varies in severity from very mild with no symptoms, to severe and debilitating.

515

Migration of neurons

Immature neurons migrate to a final location.

Neuronal migration is a distinctive feature of the development of the nervous system. The movement of the unicellular organism *Amoeba* is easy to observe under a microscope. Neural migration can occur by a similar mechanism. The cytoplasm and organelles in it are moved from the trailing end of the neuron to the leading edge by contractile actin filaments.

Migration of neurons is particularly important in brain development. Some neurons that are produced in one part of the developing brain migrate to another part where they find their final position. Mature, functional neurons do not normally move, though their axons and dendrites can often regrow if damaged.

Development of axons

An axon grows from each immature neuron in response to chemical stimuli.

An immature neuron consists of a cell body with cytoplasm and a nucleus. An axon is a long narrow outgrowth from the cell body that carries signals to other neurons. Only one axon develops on each neuron, but it may be highly branched. Many smaller dendrites that bring impulses from other neurons to the cell body may also develop. Chemical stimuli determine neuron differentiation when the axon grows out from the cell body and also the direction in which it grows in the developing embryo.

Growth of axons

Some axons extend beyond the neural tube to reach other parts of the body.

Axons grow at their tips. In some cases they are relatively short and make connections between neurons within the central nervous system, but other neurons develop very long axons which can reach any part of the body. Despite only being outgrowths of a single cell, axons can be more than a metre long in humans and many metres long in larger mammals such as blue whales. Axons carry impulses to other neurons or to cells that act as effectors – either muscle or gland cells.

As long as the cell body of its neuron remains intact, its axon may be able to regrow if severed or damaged in other ways outside the central nervous system. Regrowth rates can be as rapid as five millimetres per day so sensation or control of muscles can sometimes return over time after damage. Of course this recovery depends on the correct connections being re-established between an axon and the cells with which it should be communicating.

Development of synapses

A developing neuron forms multiple synapses.

The growth of an axon or dendrite is directed so that it reaches a cell with which it interacts. A synapse is then developed between the neuron and the other cell. The axons of motor neurons develop synapses with striated muscle fibres or gland cells for example. Synapse development involves special structures being assembled in the membranes on either side of the synapse and in the synaptic cleft between them.

The smallest number of synapses that a neuron could theoretically have is two – one to bring impulses from another cell and another to pass them on. In practice most neurons develop multiple synapses and some neurons in the brain develop hundreds, allowing complex patterns of communication.

Elimination of synapses

Synapses that are not used do not persist.

Many synapses are formed at an early stage of development, but new synapses can be formed at any stage of life. Synapses often disappear if they are not used. When transmission occurs at a synapse, chemical markers are left that cause the synapse to be strengthened. Synapses that are inactive do not have these markers so become weaker and are eventually eliminated. The maxim "use it or lose it" therefore describes synapses very well.

Neural pruning

Neural pruning involves the loss of unused neurons.

Measurements of the number of neurons have shown that there are more neurons in at least some parts of newborn babies' brains than in adults, which indicates that some neurons are lost during childhood. There is also evidence for the removal of dendrites and axon branches from some neurons. Neurons that are not used destroy themselves by the process of apoptosis. The elimination of part of a neuron or the whole cell is known as neural pruning.

Plasticity of the nervous system

The plasticity of the nervous system allows it to change with experience.

Connections between neurons can be changed by growth of axons and dendrites, by the establishment of new synapses and also by the elimination of synapses and pruning of dendrites, branches of axons or even whole neurons. This ability of the nervous system to rewire its connections is known as plasticity. It continues throughout life, but there is a much higher degree of plasticity up to the age of six than later.

The stimulus for a change in the connections between neurons comes from the experiences of a person and thus how their nervous system is used. Plasticity is the basis for forming new memories and also for certain forms of reasoning. It is also very important in repairing damage to the brain and spinal cord.

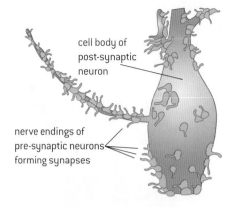

cell body of post-synaptic neuron

nerve endings of pre-synaptic neurons forming synapses

▲ Figure 3 Drawing based on an electron micrograph showing multiple synapses between pre-synaptic neurons and one post-synaptic neuron. Only the nerve endings of the pre-synaptic neurons are shown

Activity

Neural pruning in the mediodorsal thalamus

Newborn babies were found to have an estimated 11.2 million neurons in the mediodorsal nucleus of the thalamus, but in adult brains the estimated number was only 6.43 million. Assuming that no extra neurons were produced during childhood, what percentage of neurons disappears by neural pruning?

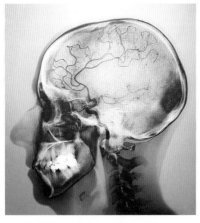

▲ Figure 4 Angiogram of the brain of a 48-year-old patient who had suffered a massive stroke. A middle cerebral artery has become blocked by a blood clot

🌐 Strokes

Events such as strokes may promote reorganization of brain function.

An ischemic stroke is a disruption of the supply of blood to a part of the brain. Most strokes are caused by a blood clot blocking one of the small vessels in the brain, but bleeding from a blood vessel is another cause. During a stroke part of the brain is deprived of sufficient oxygen and glucose. If cell respiration ceases in neurons, they become irreparably damaged and die.

Strokes vary greatly in severity. Many are so minor that the patient hardly notices. About one third of sufferers from major strokes make a full recovery and another third survive but are left with disability. In many cases recovery from strokes involves parts of the brain taking on new functions to supplement the damaged areas. Most recovery happens over the first six months after a major stroke and may involve relearning aspects of speech and writing, regaining spatial awareness and the ability to carry out skilled physical activities such as dressing or preparing food.

A.2 The human brain

Understanding

→ The anterior part of the neural tube expands to form the brain.

→ Different parts of the brain have specific roles.

→ The autonomic nervous system controls involuntary processes in the body using centres located in the medulla oblongata.

→ The cerebral cortex forms a larger proportion of the brain and is more highly developed in humans than other animals.

→ The human cerebral cortex has become enlarged principally by an increase in total area with extensive folding to accommodate it within the cranium.

→ The cerebral hemispheres are responsible for higher order functions.

→ The left cerebral hemisphere receives sensory input from sensory receptors in the right side of the body and the right side of the visual field in both eyes and vice versa for the right hemisphere.

→ The left cerebral hemisphere controls muscle activity in the right side of the body and vice versa for the right hemisphere.

→ Brain metabolism requires large energy inputs.

🌐 Applications

→ Visual cortex, Broca's area, nucleus accumbens as areas of the brain with specific functions.

→ Swallowing, breathing and heart rate as examples of activities coordinated by the medulla.

→ Use of the pupil reflex to evaluate brain damage.

→ Use of animal experiments, autopsy, lesions and fMRI to identify the role of different brain parts.

⚗️ Skills

→ Identification of parts of the brain in a photograph, diagram or scan of the brain.

→ Analysis of correlations between body size and brain size in different animals.

🧬 Nature of science

→ Use models as representations of the real world: the sensory homunculus and motor homunculus are models of the relative space human body parts occupy on the somatosensory cortex and the motor cortex.

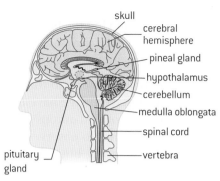

Development of the brain

The anterior part of the neural tube expands to form the brain.

During the development of vertebrate embryos a neural tube forms along the whole of the dorsal side, above the gut, near the surface. Most of the neural tube becomes the spinal cord, but the anterior end expands and develops into the brain as part of a process called cephalization, the development of a head. The human brain contains approximately 86 billion neurons (8.6×10^{10}).

The brain acts as the central control centre for the whole body, both directly from cranial nerves and indirectly via the spinal cord and numerous signal molecules carried by the blood. The advantage of having a brain is that communication between the billions of neurons involved can be more rapid than if control centres were more dispersed. The major sensory organs are located at the anterior end of vertebrates: the eyes, ears, nose and tongue.

▲ Figure 1 Diagram of the brain

Roles of the parts of the brain

Different parts of the brain have specific roles.

The brain has regions that are distinguishable by their shape, colour or by microscopic structure. These regions have different roles, identified by physiological research in humans and other mammals.

The **medulla oblongata** is used in autonomic control of gut muscles, breathing, blood vessels and heart muscle.

The **cerebellum** coordinates unconscious functions, such as posture, non-voluntary movement and balance.

The **hypothalamus** is the interface between the brain and the pituitary gland, synthesizing the hormones secreted by the posterior pituitary, and releasing factors that regulate the secretion of hormones by the anterior pituitary.

The **pituitary gland**: the posterior lobe stores and releases hormones produced by the hypothalamus and the anterior lobe produces and secretes hormones that regulate many body functions.

The **cerebral hemispheres** act as the integrating centre for high complex functions such as learning, memory and emotions.

 Methods of brain research

Use of animal experiments, autopsy, lesions and fMRI to identify the role of different brain parts.

Lesion studies gave the first useful information about brain functions. For example, in the 19th century, after the death and autopsy of a patient who could only say the word "Tan", the French neurologist Charcot found a single large tumour damaging the lower left side of the patient's brain. He deduced that this part of the brain is involved with speech. Another famous case was the railway construction worker Phineas Gage, who suffered severe damage to the frontal lobes of his brain in 1848 when an accident with explosives caused a large metal rod to pass through his forehead. He recovered from the wound but the brain damage radically and permanently altered his personality and particularly his capacity for social interaction.

Many lesions due to tumours, strokes or accidental damage have been investigated by carrying out an autopsy and relating the position of the lesion to observed changes in behaviour and capacities, but rather than wait for these fortuitous opportunities, some neuroscientists have studied experimental animals. Removal of parts of the skull gives access to the brain and allows experimental procedures to be performed. The brain itself does not feel pain – even today some forms of neurosurgery are performed on fully conscious patients. The effects of local stimulation in an animal's brain can be observed, as can long-term changes in the animal's temperament and capacities. There are widespread objections to such research, because of the suffering they may cause to the animal and because at the end the animal is often sacrificed, but the information obtained is useful to understanding, and therefore treating, conditions such as epilepsy, Parkinson's disease and multiple sclerosis. Increasingly genetic mutants and selective inactivation of genes, which are technically possible only in mice, are used to achieve similar experimental modification of brain structure and behaviour.

Magnetic resonance imaging (MRI) is a more modern and less controversial technique. Basic MRI is used to investigate the internal structure of the body, including looking for tumours or other abnormalities in patients. Figure 2 shows the results of an MRI scan of the upper part of a patient's body, including the head and brain.

A specialized version of MRI, called functional magnetic resonance imaging (fMRI) has been developed, which allows the parts of the brain

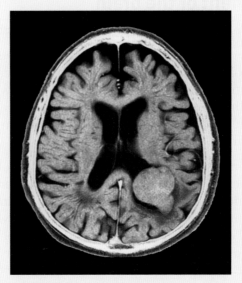

▲ Figure 2 Image of brain lesion

that are activated by specific thought processes to be identified. Active parts of the brain receive increased blood flow, often made visible by injecting a harmless dye, which fMRI records. The subject is placed in the scanner and a high-resolution scan of the brain is taken. A series of low-resolution scans is then taken while the subject is being given a stimulus. These scans show which parts of the brain are activated during the response to the stimulus.

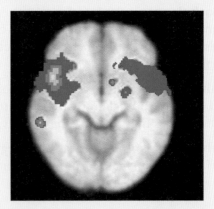

▲ Figure 3 fMRI scan of endometriosis pain

Examples of brain functions

Visual cortex, Broca's area, nucleus accumbens as areas of the brain with specific functions.

Each of the two cerebral hemispheres has a **visual cortex** in which neural signals originating from light sensitive rod and cone cells in the retina of the eyes are processed. Although

there is an initial stage in which a map of visual information is projected in a region called V1, the information is then analysed by multiple pathways in regions V2 to V5 of the visual

cortex. This analysis includes pattern recognition and judging the speed and direction of moving objects.

Broca's area is a part of the left cerebral hemisphere that controls the production of speech. If there is damage to this area an individual knows what they want to say and can produce sounds, but they cannot articulate meaningful words and sentences. For example, if we see a horse-like animal with black and white stripes, Broca's area allows us to say "zebra", but a person with a damaged Broca's area knows that it is a zebra but cannot say the word.

There is a **nucleus accumbens** in each of the cerebral hemispheres. It is the pleasure or reward centre of the brain. A variety of stimuli including food and sex cause the release of the neurotransmitter dopamine in the nucleus accumbens, which causes feelings of well-being, pleasure and satisfaction. Cocaine, heroin and nicotine are addictive because they artificially cause release of dopamine in the nucleus accumbens.

The autonomic nervous system

The autonomic nervous system controls involuntary processes in the body using centres located in the medulla oblongata.

The peripheral nervous system comprises all of the nerves outside the central nervous system. It is divided into two parts: the voluntary and the autonomic nervous systems. Involuntary processes are controlled by the autonomic nervous system, using centres in the medulla oblongata. The autonomic nervous system has two parts: sympathetic and parasympathetic. These often have contrary effects on an involuntary process. For example, parasympathetic nerves cause an increase in blood flow to the gut wall during digestion and absorption of food. Sympathetic nerves cause a decrease in blood flow during fasting or when blood is needed elsewhere.

 Activities coordinated by the medulla

Swallowing, breathing and heart rate as examples of activities coordinated by the medulla.

The first phase of swallowing, in which food is passed from the mouth cavity to the pharynx, is voluntary and so is controlled by the cerebral cortex. The remaining phases in which the food passes from the pharynx to the stomach via the esophagus, are involuntary and are coordinated by the swallowing centre of the medulla oblongata.

Two centres in the medulla control breathing: one controls the timing of inspiration; the other controls the force of inspiration and also active, voluntary expiration. There are chemoreceptors in the medulla that monitor blood pH. The carbon dioxide concentration in the blood is very important in controlling breathing rate, even more than oxygen concentration. If blood pH falls, indicating an increase in carbon dioxide concentration, breathing becomes deeper and/or more frequent.

The cardiovascular centre of the medulla regulates the rate at which the heart beats. Blood pH and pressure are monitored by receptor cells in blood vessels and in the medulla. In response to this information, the cardiovascular centre can increase or decrease the heart rate by sending signals to the heart's pacemaker. Signals carried from the sympathetic system speed up the heart rate; signals carried by the parasympathetic system in the vagus nerve slow the rate down.

Muscles in the iris control the size of the pupil of the eye. Impulses carried to radial muscle fibres by neurons of the sympathetic system cause them to contract and dilate the pupil; impulses carried to circular muscle fibres by neurons of the parasympathetic system cause the pupil to constrict.

The pupil reflex occurs when bright light suddenly shines into the eye. Photoreceptive ganglion cells in the retina perceive the bright light, sending signals through the optic nerve to the mid-brain, immediately activating the parasympathetic system that stimulates circular muscle in the iris, constricting the pupil and reducing the amount of light entering the eye, protecting the delicate retina from damage.

Doctors sometimes use the pupil reflex to test a patient's brain function. A light is shone into each eye. If the pupils do not constrict at once, the medulla oblongata is probably damaged. If this and other tests of brain stem function repeatedly fail, the patient is said to have suffered brain death. It may be possible to sustain other parts of the patient's body on a life support machine, but full recovery is extremely unlikely.

The cerebral cortex

The cerebral cortex forms a larger proportion of the brain and is more highly developed in humans than other animals.

The cerebral cortex is the outer layer of the cerebral hemispheres. Although it is only two to four millimetres thick, up to six distinctively different layers of neurons can be identified in sections studied under a microscope. It is has a highly complex architecture of neurons and processes the most complex tasks in the brain.

Only mammals have a cerebral cortex. Birds and reptiles have regions of the brain that perform a similar range of functions but they are structurally different, with cells arranged in clusters rather than layers. Among the mammals the cerebral cortex varies in size considerably. In humans it forms a larger proportion of the brain than in any other mammal.

The evolution of the cerebral cortex

The human cerebral cortex has become enlarged principally by an increase in total area with extensive folding to accommodate it within the cranium.

The cerebral cortex has become greatly enlarged during human evolution, and now contains more neurons than that of any other animal. There has been a modest increase in thickness, but the cortex is still only a few millimetres thick. The increase is due principally to an increase in total area and that necessitates the cortex becoming extensively folded during development. It is hard to measure, but the area is estimated to be about 180,000 mm² or 0.18 m². This is so large that the brain can only be accommodated inside a greatly enlarged cranium, forming the distinctive shape of the human skull.

Most of the surface area of the cerebral cortex is in the folds rather than on the outer surface. In contrast, mice and rats have an unfolded smooth cortex, but in cats there are some folds and elephants and dolphins have

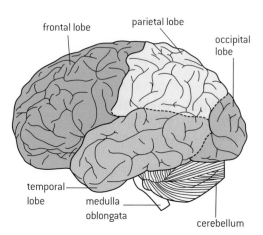
▲ Figure 4 The folded structure of the cerebral cortex, viewed from the left side. The four lobes are indicated

more. Among the primates, monkeys and apes show a range of cortex size and degree of folding, with larger sizes in primates that are more closely related to humans.

 Comparing brain size

Analysis of correlations between body size and brain size in different animals.

Scattergraphs show a positive correlation between body size and brain size in animals, but that the relationship is not directly proportional. The data-based questions below can be used to develop your skill in analysing this type of data.

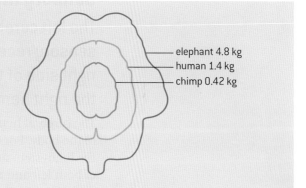

elephant 4.8 kg
human 1.4 kg
chimp 0.42 kg

Data-based questions: Brain and body size in mammals

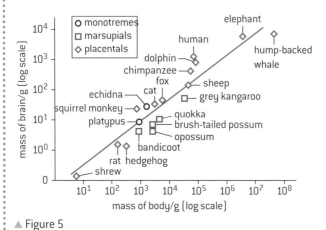

▲ Figure 5

The scattergraph in figure 5 shows the relationship between brain and body mass in species of placental, marsupial and monotreme mammal.

1 State the relationship between brain and body mass. [1]

2 Explain how the points on the scattergraph would have been arranged if brain mass was directly proportional to body mass. [2]

3 State which mammals have (a) the largest and (b) the smallest brain mass. [2]

4 Discuss the evidence provided by the scattergraph for the hypothesis that humans have the largest relative brain mass. [2]

5 Evaluate the hypothesis that marsupials have relatively small brains compared with other mammals. [2]

6 Suggest a reason for the researchers not including more data for monotremes in the scattergraph. [1]

Functions of the cerebral hemispheres

The cerebral hemispheres are responsible for higher order functions.

The cerebral hemispheres carry out the most complex of the brain's tasks. These are known as higher order functions and include learning, memory, speech and emotions. These higher order functions involve association of stimuli from different sources including the eye and ear and also from memories. They rely on very complex networks of neurons that are still only partially understood by neurobiologists. The most sophisticated thought processes such as reasoning, decision-making and planning occur in the frontal and prefrontal lobes of the cerebral cortex. Using these parts of the brain we can organize our actions in a logical sequence, predict their outcomes, develop a sense of right and wrong and be aware of our own existence.

Sensory inputs to the cerebral hemispheres

The left cerebral hemisphere receives sensory input from sensory receptors in the right side of the body and the right side of the visual field in both eyes and vice versa for the right hemisphere.

The cerebral hemispheres receive sensory inputs from all the sense organs of the body. For example, signals from the left ear pass to the left side of the brain stem and then on to the auditory cortex in the temporal lobe of both left and right hemispheres. Similarly, signals from the right ear pass to both hemispheres via the right side of the brain stem. Inputs from the skin, muscles and other internal organs pass via the spinal cord to the somatosensory area of the parietal lobe. Perhaps surprisingly, the impulses from each side cross in the base of the brain so that the left hemisphere receives impulses from the right side of the body and vice versa.

Inputs from the eye pass to the visual area in the occipital lobe, known as the visual cortex. Impulses from the right side of the field of vision in each eye are passed to the visual cortex in the left hemisphere, while impulses from the left side of the field of vision in each eye pass to the right hemisphere. This integration of inputs enables the brain to judge distance and perspective.

Motor control by the cerebral hemispheres

The left cerebral hemisphere controls muscle activity in the right side of the body and vice versa for the right hemisphere.

Regions in each of the cerebral hemispheres control striated ("voluntary") muscles. The main region is in the posterior part of the frontal lobe and is called the primary motor cortex. In this region there is a series of overlapping areas that control muscles throughout the body, from the mouth at one end of the primary motor cortex to the toes at the other end.

The primary motor cortex in the left hemisphere controls muscles in the right side of the body and that in the right side controls muscles in the left side of the body. So a stroke (or other brain damage) in the left side of the brain can cause paralysis in the right side of the body and vice versa.

 Homunculi

Use models as representations of the real world: the sensory homunculus and motor homunculus are models of the relative space human body parts occupy on the somatosensory cortex and the motor cortex.

Neurobiologists have constructed models of the body in which the size of each part corresponds to the proportion of the somatosensory cortex devoted to sensory inputs from that part. This type of model is called a sensory homunculus. Similar models have been constructed to show the proportion of the motor cortex that is devoted to control of muscles in each part of the body. These models are useful as they give a good impression of the relative importance given to sensory inputs from different parts of the body and to control of muscles in different parts.

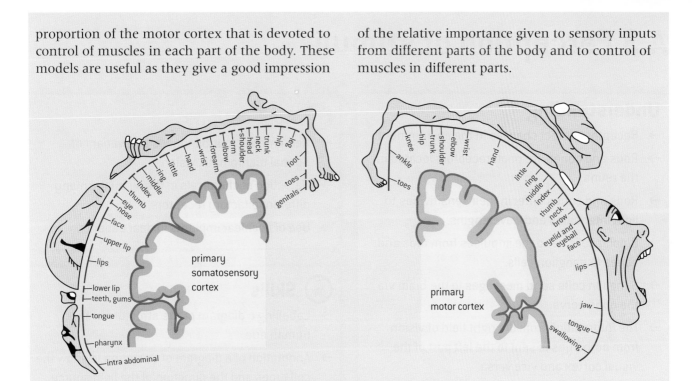

▲ Figure 6 Sensory homunculus (left) and motor homunculus (right)

Energy and the brain

Brain metabolism requires large energy inputs.

Energy released by cell respiration is needed to maintain the resting potential in neurons and to re-establish it after an action potential, as well as for synthesis of neurotransmitters and other signal molecules. The brain contains a huge number of neurons so it needs much oxygen and glucose to generate this energy by aerobic cell respiration. In most vertebrates the brain uses less than 10% of the energy consumed by basal metabolism but in the adult human brain it is over 20% and an even higher proportion in infants and small children.

A.3 Perception of stimuli

Understanding

→ Receptors detect changes in the environment.

→ Rods and cones are photoreceptors located in the retina.

→ Rods and cones differ in their sensitivities to light intensities and wavelengths.

→ Bipolar cells send the impulses from rods and cones to ganglion cells.

→ Ganglion cells send messages to the brain via the optic nerve.

→ The information from the right field of vision from both eyes is sent to the left part of the visual cortex and vice versa.

→ Structures in the middle ear transmit and amplify sound.

→ Sensory hairs of the cochlea detect sounds of specific wavelengths.

→ Impulses caused by sound perception are transmitted to the brain via the auditory nerve.

→ Hair cells in the semicircular canals detect movement of the head.

 Applications

→ Red-green colour-blindness as a variant of normal trichromatic vision.

→ Detection of chemicals in the air by the many different olfactory receptors.

→ Use of cochlear implants in deaf patients.

 Skills

→ Labelling a diagram of the structure of the human eye.

→ Annotation of a diagram of the retina to show the cell types and the direction of the light source.

→ Labelling a diagram of the structure of the human ear.

Nature of science

→ Understanding of the underlying science is the basis for technological developments: the discovery that electrical stimulation in the auditory system can create a perception of sound resulted in the development of electrical hearing aids and ultimately cochlear implants.

Sensory receptors

Receptors detect changes in the environment.

The environment, particularly its changes, stimulate the nervous system via sensory receptors. The nerve endings of sensory neurons act as receptors, for example touch receptors. In other cases there are specialized receptor cells that pass impulses to sensory neurons, as with the light-sensitive rod and cone cells of the eye. Humans have the following types of specialized receptor.

- Mechanoreceptors respond to mechanical forces and movements.

- Chemoreceptors respond to chemical substances.

- Thermoreceptors respond to heat.

- Photoreceptors respond to light.

Olfactory receptors

Detection of chemicals in the air by the many different olfactory receptors.

Olfaction is the sense of smell. Olfactory receptor cells are located in the epithelium inside the upper part of the nose. These cells have cilia which project into the air in the nose. Their membrane contains odorant receptor molecules, proteins which detect chemicals in the air. Only volatile chemicals can be smelled in air within the nose. Odorants from food in the mouth can pass through mouth and nasal cavities to reach the nasal epithelium.

There are many different odorant receptor proteins, each encoded by a different gene. In some mammals such as mice there are over a thousand different odorant receptors, each of which detects a different chemical or group of chemicals (though the exact mechanisms are still unclear in spite of intensive study). Each olfactory receptor cell has just one type of odorant receptor in its membrane, but there are many receptor cells with each type of odorant receptor, distributed though the nasal epithelium. Using these receptor cells most animals, including mammals, can distinguish a large number of chemicals in the air, or in water in the case of aquatic animals. In many cases the chemical can be detected in extremely low concentrations but the human sense of smell is very insensitive and imprecise compared to that of other animals.

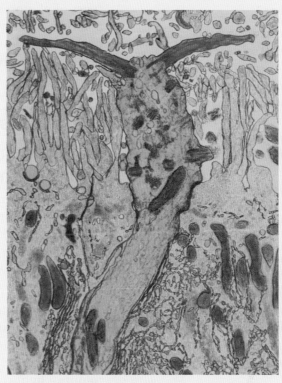

▲ Figure 1 Olfactory receptor cell (centre) with two of its cilia visible and also cilia in adjacent cells in the nasal epithelium

Structure of the eye

Labelling a diagram of the structure of the human eye.

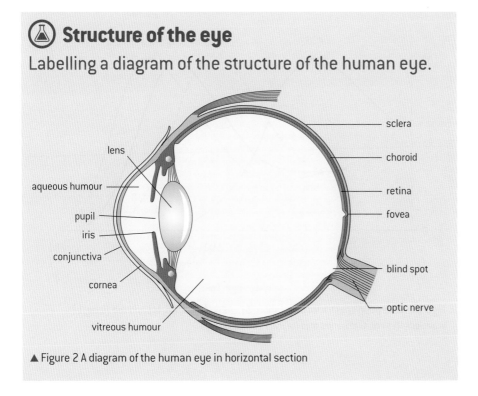

▲ Figure 2 A diagram of the human eye in horizontal section

Activity

Cataract surgery

Accumulation of metabolic wastes in the cells of the eye's lens gradually turns them yellow so blues fade. The difference in colour perception after a cataract operation is startling. Talk to a person, probably elderly, who has had cataract surgery to find out how it changed their colour perception.

Photoreceptors

Rods and cones are photoreceptors located in the retina.

Light entering the eye is focused by the cornea and the lens onto the retina, the thin layer of light-sensitive tissue at the back of the eye. Figure 5 shows the cell types in the retina. Two main types of photoreceptor are present in the human retina, rods and cones. Many nocturnal mammals have only rods and cannot distinguish colours. Rods and cones are stimulated by light and so together detect the image focused on the retina and convert it into neural signals.

Differences between rods and cones

Rods and cones differ in their sensitivities to light intensities and wavelengths.

Rods are very sensitive to light, so work well in dim light. In very bright light the pigment in them is temporarily bleached so for a few seconds they do not work. Rod cells absorb a wide range of visible wavelengths of light (see figure 3) but cannot respond selectively to different colours, so they give us black and white vision.

There are three types of cone, which absorb different ranges of wavelengths of light. They are named according to the colour that they absorb most strongly: red, blue or green. When light reaches the retina, the red, blue and green cones are selectively stimulated. By analysing the relative stimulation of each of the three cone types, the colour of light can be very precisely determined, though experiments show that people's perception of colour differs quite a lot. Cones are only stimulated by bright light and therefore colour vision fades in dim light.

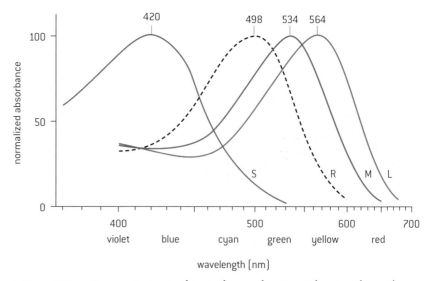

▲ Figure 3 Absorption spectra for blue (short, S), green (medium, M) and red (long, L) wavelength-sensitive cones and for rods (dotted line)

Red-green colour-blindness

Red-green colour-blindness as a variant of normal trichromatic vision.

Red-green colour-blindness is a common inherited condition in humans and some other mammals. It is due to the absence of, or a defect in, the gene for photoreceptor pigments essential to either red or green cone cells. Both genes are located on the human X chromosome so it is a sex-linked condition. The normal alleles of both genes are dominant and the alleles that cause red-green colour-blindness are recessive. Red-green colour-blindness is therefore much commoner among males, who have only one X chromosome, than females, and males inherit the allele that causes the condition from their mother.

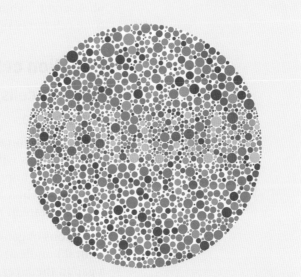

▲ Figure 4 Red and green colours cannot easily be distinguished by some males and fewer females

Structure of the retina

Annotation of a diagram of the retina to show the cell types and the direction of the light source.

The arrangement of the layers of cells in the retina may seem surprising. The light passes first through a layer of transparent nerve axons that carry impulses from the retina to the brain through the optic nerve, then through a layer of specialized "bipolar" neurons that process signals before they reach the optic nerve, and only then does the light reach the rod and cone cells. This is shown in figure 5.

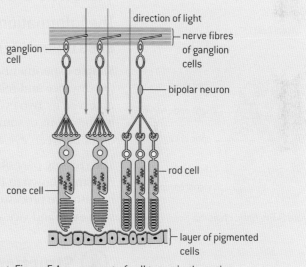

▲ Figure 5 Arrangement of cell types in the retina

Bipolar cells

Bipolar cells send the impulses from rods and cones to ganglion cells.

Rod and cone cells synapse with neurons called bipolar cells in the retina. If rod or cone cells are not stimulated by light they depolarize and release an inhibitory neurotransmitter onto a bipolar cell, causing it to become hyperpolarized and not transmit impulses to its associated retinal ganglion cell. When light is absorbed by a rod or cone cell it becomes hyperpolarized and stops sending inhibitory neurotransmitter to the bipolar cell. The bipolar cell can therefore depolarize, activating the adjacent ganglion cell.

Groups of rod cells send signals to the brain via a single bipolar cell, so the brain cannot distinguish which rod absorbed the light. The images transmitted to the brain by rods alone are lower resolution, like a grainy photograph, whereas those based on the cones are sharper because each cone cell sends signals to the brain via its own bipolar cell.

Ganglion cells

Ganglion cells send messages to the brain via the optic nerve.

Retinal ganglion cells have cell bodies in the retina with dendrites that form synapses with bipolar cells. Ganglion cells also have long axons along which impulses pass to the brain. Impulses are passed at a low frequency when the ganglion cell is not being stimulated and at an increased rate in response to stimuli from bipolar cells.

The axons of ganglion cells pass across the front of the retina to form a central bundle at the "blind spot", so called because their presence makes a gap in the layer of rods and cones. The axons of the ganglion cells pass via the optic nerve to the optic chiasma in the brain.

Vision in the right and left fields

The information from the right field of vision from both eyes is sent to the left part of the visual cortex and vice versa.

Simple experiments comparing vision with one eye or with both eyes show the distance and relative size of objects can be judged most precisely when observed by two eyes simultaneously. Stimuli from both eyes are integrated by the axons of some retinal ganglion cells crossing from one side to the other between eye and brain while other axons stay on the same side.

The crossing over of axons between left and right sides happens in the optic chiasma, shown in figure 6. As a result, the visual cortex in the right cerebral hemisphere processes visual stimuli from the left side of the visual field of both eyes, and vice versa for stimuli from the right side of the field of vision.

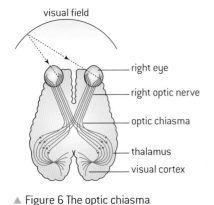

▲ Figure 6 The optic chiasma

🧪 Structure of the ear

Labelling a diagram of the structure of the human ear.

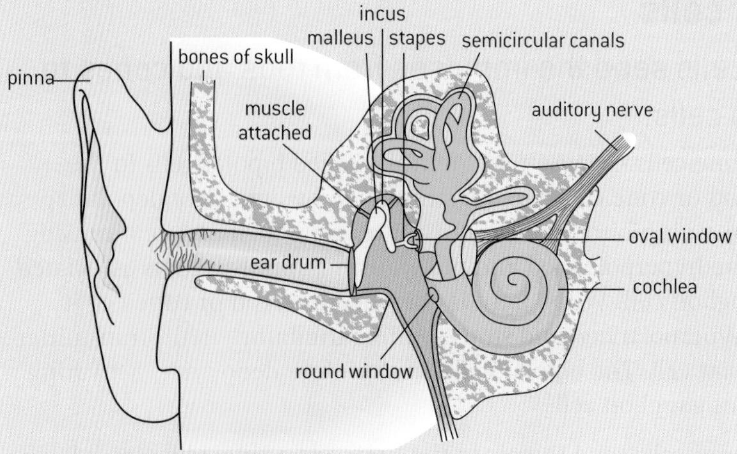

▲ Figure 7 The structure of the ear

The middle ear

Structures in the middle ear transmit and amplify sound.

The middle ear is an air-filled chamber between the outer ear and the inner ear. A thin, taut sheet of flexible tissue called the eardrum separates the middle ear from the outer ear. Two other thin sheets of tissue called the oval and round windows separate the middle ear from the inner ear.

Three tiny bones are in the middle ear, the malleus (hammer), incus (anvil) and stapes (stirrup), which articulate with each other to form a connection between the eardrum and the oval window. These bones, also called ossicles, transmit vibrations from the eardrum to the oval window, amplifying sound twentyfold because the oval window has a smaller area than the eardrum. During very loud sounds, the delicate sound-reception components of the ear are protected by contraction of the muscles attached to the bones in the middle ear, which weakens the connections between the ossicles and so damps the vibrations.

The cochlea

Sensory hairs of the cochlea detect sounds of specific wavelengths.

The cochlea is the part of the inner ear where vibrations are transduced into neural signals. It is a tubular, coiled, fluid-filled structure. Within the cochlea are layers of tissue (membranes) to which sensory cells are attached. Each of these cells has a bundle of hairs, stretching from one membrane to another. When vibrations are transmitted from the oval window into the cochlea, they resonate with the hair bundles of particular hair cells, stimulating these cells. Selective activation of different hair cells enables us to distinguish between sounds of different pitch.

The round window is another thin sheet of flexible tissue, located between the middle and inner ear. If it was stiff and indeformable, the oval window would not be able to vibrate, because the incompressible fluid in the cochlea would prevent it from moving. When vibrations of the oval window push the fluid in the cochlea inwards, the round window moves outwards, and when the oval window moves outwards, the round window moves inwards, enabling the oval window to transmit vibrations through the fluid in the cochlea.

The auditory nerve

Impulses caused by sound perception are transmitted to the brain via the auditory nerve.

When a hair cell in the cochlea is depolarized by the vibrations that constitute sounds, it releases neurotransmitter across a synapse, stimulating an adjacent sensory neuron. This triggers an action potential in the sensory neuron which propagates to the brain along the auditory nerve. The auditory nerve is one of the cranial nerves that serve the brain.

If humans are sensitive only to certain ranges of stimuli, what consequences or limitations might this have for the acquisition of knowledge?

Figure 8 shows the frequency sensitivity of six land mammals. The solid area shows where frequency sensitivity is best, while the lines indicate how much louder other frequencies need to be in order to be heard.

1 Does the world sound the same to any of the animals?

2 Which is the real world – the one we perceive or the world perceived by the bat?

3 Animals also differ considerably in their visual perception. Is what each animal sees what is really there, is it a construction of reality, or is reality a false concept?

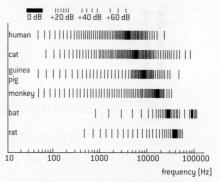

▲ Figure 8 Sensitivity of mammals to frequencies of sound

Cochlear implants

Use of cochlear implants in deaf patients.

Deafness has a variety of causes and in many cases a hearing aid that amplifies sounds can overcome the problem. However, if the hair cells in the cochlea are defective, such hearing aids do not help. In this case the best option, as long as the auditory nerve is functioning properly, is a cochlear implant. More than a quarter of a million people have had these devices implanted and although they do not fully restore normal hearing, they improve it and usually allow recognition of speech.

Cochlear implants consist of external and internal parts.

- The external parts are a microphone to detect sounds, a speech processor that selects the frequencies used in speech and filters out other frequencies, and a transmitter that sends the processed sounds to the internal parts.

- The internal parts are implanted in the mastoid bone behind the ear. They consist of a receiver that picks up sound signals from the transmitter, a stimulator that converts

these signals into electrical impulses and an array of electrodes that carry these impulses to the cochlea. The electrodes stimulate the auditory nerve directly and so bypass the non-functional hair cells.

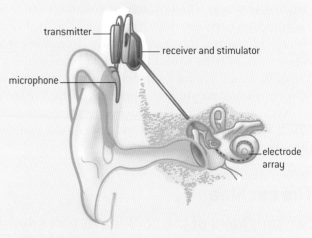

▲ Figure 8 Cochlear implant with microphone behind the ear connected to the transmitter and adjacent to this the internal receiver and stimulator, with electrodes leading to the auditory nerve that arises in the cochlea

The science behind cochlear implants

Understanding of the underlying science is the basis for technological developments: the discovery that electrical stimulation in the auditory system can create a perception of sound resulted in the development of electrical hearing aids and ultimately cochlear implants.

Research into artificial electrical stimulation of the cochlea began as early as the 1950s. Early attempts showed that it was possible to give some perception of sound to people who were severely or profoundly deaf due to non-functioning hair cells. Experiments with humans showed that electrical stimulation could be used to give perception of different frequencies of sound, as in music. Research continued and involved electronic engineers, neurophysiologists and clinical audiologists. An understanding of which frequencies are used to understand speech was used to develop speech processors for example.

During the 1970s early versions of cochlear implants were fitted to over a thousand patients. Since then research has led to huge technological developments in these devices with greatly improved outcomes for the increasing number of people that have had them fitted. Further improvements can be expected and although cochlear implants can never give severely or profoundly deaf people normal hearing, they can allow far better hearing than without this technology.

Detecting head movements

Hair cells in the semicircular canals detect movement of the head.

There are three fluid-filled semicircular canals in the inner ear. Each has a swelling at one end in which there is a group of sensory hair cells, with their hairs embedded in gel to form a structure called the cupula. When the head moves in the plane of one of the semicircular canals, the stiff wall of the canal moves with the head, but due to inertia the fluid inside the canal lags behind. There is therefore a flow of fluid past the cupula. This is detected by the hair cells, which send impulses to the brain.

The three semicircular canals are at right angles to each other, so each is in a different plane. They can therefore detect movements of the head in any direction. The brain can deduce the direction of movement by the relative amount of stimulation of the hair cells in each of the semicircular canals.

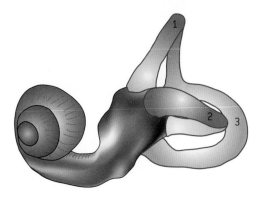

▲ Figure 9 Inner ear with cochlea (left) and semicircular canals (right): superior (1), lateral (2) and posterior (3)

A.4 Innate and learned behaviour (AHL)

Understanding

→ Innate behaviour is inherited from parents and so develops independently of the environment.

→ Autonomic and involuntary responses are referred to as reflexes.

→ Reflex arcs comprise the neurons that mediate reflexes.

→ Learned behaviour develops as result of experience.

→ Reflex conditioning involves forming new associations.

→ Imprinting is learning occurring at a particular life stage and is independent of the consequences of behaviour.

→ Operant conditioning is a form of learning which consists of trial and error experiences.

→ Learning is the acquisition of skill or knowledge.

→ Memory is the process of encoding, storing and accessing information.

Applications

→ Withdrawal reflex of the hand from a painful stimulus.

→ Pavlov's experiments into reflex conditioning in dogs.

→ The role of inheritance and learning in the development of birdsong.

Skills

→ Analysis of data from invertebrate behaviour experiments in terms of the effect on chances of survival and reproduction.

→ Drawing and labelling a diagram of a reflex arc for a pain withdrawal reflex.

Nature of science

→ Looking for patterns, trends and discrepancies: laboratory experiments and field investigations helped in the understanding of different types of behaviour and learning.

Innate behaviour

Innate behaviour is inherited from parents and so develops independently of the environment.

Animal behaviour is divided into two broad categories, innate and learned. The form of innate behaviour is unaffected by external influences that an animal experiences. It develops independently of the environment. For example, if an object touches the skin in the palm of a baby's hand, the baby grips the object by closing its fingers around it. This innate behaviour pattern, called the palmar grasp reflex, is seen in babies from birth until they are about six months old, whatever experiences the baby has.

Innate behaviour is genetically programmed, so it is inherited. It can change through evolution if there is genetically determined variation in behaviour and natural selection favours one behaviour pattern over others, but the rate of change is much slower than with learned behaviour.

 ## Research methods in animal behaviour

Looking for patterns, trends and discrepancies: laboratory experiments and field investigations helped in the understanding of different types of behaviour and learning.

The scientific study of animal behaviour became established as a significant branch of biology in the 1930s. Before then naturalists observed the behaviour of animals in natural habitats but had rarely analysed it scientifically. Two general types of methodology have since been used: laboratory experiments and field investigations.

The advantage of laboratory experiments is that variables can be controlled more effectively and innate behaviour in particular can be investigated rigorously. The disadvantage is that animal behaviour is an adaptation to the natural environment of the species and animals often do not behave normally when removed from that environment, especially with learned behaviour.

 ## Invertebrate behaviour experiments

Analysis of data from invertebrate behaviour experiments in terms of the effect on chances of survival and reproduction.

Many invertebrates have relatively simple behaviour patterns, so they can be studied more easily than mammals, birds or other vertebrates. A stimulus can be given and the response to it observed. Repeating the stimulus with a number of individuals allows quantitative data to be obtained and tests of statistical significance to be done. Once the response to a stimulus has been discovered, it may be possible to deduce how the response improves animals' chances of survival

and reproduction and thus how it evolved by natural selection as an innate behaviour pattern.

Many different invertebrates can be used in experiments. Planarian flatworms, woodlice, blowfly larvae, snails and beetles are often used. Some species can be purchased from suppliers but it is also possible to use invertebrates from local habitats. These should be kept for a short time only, protected from suffering during the

experiments and then returned to their habitat. Endangered species should not be used.

Two types of behaviour involving movement could be investigated:

- Taxis is movement towards or away from a directional stimulus. An example is movement of a woodlouse or slater away from light.

- Kinesis also involves movement as a response, but the direction of movement is not influenced by the stimulus. Instead, the speed of movement or the number of times the animal turns is varied. An example is slower movement, with more frequent turning, when woodlice are transferred from drier to more damp conditions.

Stages in designing an investigation:

1 Place the animals in conditions that are similar to the natural habitat.

2 Observe the behaviour and see what stimuli affect movement.

3 Choose one stimulus that appears to cause a taxis or kinesis.

4 Devise an experiment to test responses to the stimulus.

5 Ensure that other factors do not have an effect on the movement.

6 Decide how to measure the movement of the invertebrates.

Reflexes

Autonomic and involuntary responses are referred to as reflexes.

A stimulus is a change in the environment, either internal or external, that is detected by a receptor and elicits a response. A response is a change in an organism, often carried out by a muscle or a gland. Some responses happen without conscious thought and are therefore called involuntary responses. Many of these are controlled by the autonomic nervous system. These autonomic and involuntary responses are known as reflexes.

A reflex is a rapid unconscious response to a stimulus. The pupil reflex is an example: in response to the stimulus of bright light, the radial muscles in the iris of the eye contract, constricting the pupil. This involuntary response is carried out by the autonomic nervous system.

Reflex arcs

Reflex arcs comprise the neurons that mediate reflexes.

All reflexes start with a receptor that perceives the stimulus and ends with an effector, usually a muscle or gland, which carries out the response. Linking the receptor to the effector is a sequence of neurons, with synapses between them. The sequence of neurons is known as a reflex arc. In the simplest reflex arcs there are two neurons: a sensory neuron to carry impulses from the receptor to a synapse with a motor neuron in the spinal cord and a motor neuron to carry impulses on to the effector. Most reflex arcs contain more than two neurons, as there are one or more relay neurons connecting the sensory neuron to the motor neuron.

Activity

Reflex speed

The withdrawal reflex takes less than a tenth of a second. Reaction times that involve more complex processing take longer. Use online tests if you want to assess your reaction time, using the search term reflex test to find them.

 ## The withdrawal reflex

Withdrawal reflex of the hand from a painful stimulus.

The pain withdrawal reflex is an innate response to a pain stimulus. For example if we touch a hot object with the hand, pain receptors in the skin of the finger detect the heat and activate sensory neurons, which carry impulses from the finger to the spinal cord via the dorsal root of a spinal nerve. The impulses travel to the ends of the sensory neurons in the grey matter of the spinal cord where there are synapses with relay neurons. The relay neurons have synapses with motor neurons, which carry impulses out of the spinal cord via the ventral root and to muscles in the arm. Messages are passed across synapses from motor neurons to muscle fibres, which contract and pull the arm away from the hot object.

 ## Neural pathways in a reflex arc

Drawing and labelling a diagram of a reflex arc for a pain withdrawal reflex.

Figure 1 shows the reflex arc for the pain withdrawal reflex.

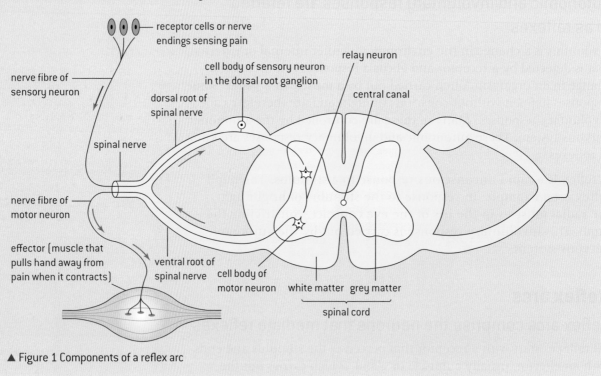

▲ Figure 1 Components of a reflex arc

Learned behaviour

Learned behaviour develops as result of experience.

Offspring inherit the capacity or propensity to acquire new patterns of behaviour during their life, as a result of experience. This is known as learned behaviour. Offspring learn behaviour patterns from their parents, from other individuals and from their experience of the

environment. For example, human offspring inherit the capacity to learn a language. The language that they learn is usually that of their biological parents, but not if they are adopted by adults who speak a different language. The ability to make sense of vocal patterns and then make them oneself is innate but the specific language spoken is learned.

 Development of birdsong

The role of inheritance and learning in the development of birdsong.

Birdsong has been investigated intensively in some species and evidence has been found for it being partly innate and partly learned. All members of a bird species share innate aspects of song, allowing each individual to recognize other members of the species. In many species, including all passerines, males learn mating calls from their father. The learned aspects introduce differences, allowing males to be recognized by their song and in some species mates to be chosen by the quality of their singing.

Data-based questions: Birdsong – innate or learned?

The sonograms in figure 2 are a visual representation of birdsong, with time on the x-axis and frequency or pitch on the y-axis.

1 Compare sonograms I and II, which are from two populations of white-crowned sparrows (*Zonotrichia leucophrys*). [2]

2 Sonogram III is from a white-crowned sparrow that was reared in a place where it could not hear any other birdsong.

   a) Compare sonogram III with sonograms I and II. [2]

   b) Discuss whether the song of white-crowned sparrows is innate, learned or due to both innate factors and learning. [3]

3 In 1981 Martin Morton and Luis Baptista published a very unusual discovery – a white-crowned sparrow had learned to imitate the song of another species. Sonogram IV is from a strawberry finch (*Amandava amandava*). Sonogram V is from a white-crowned sparrow that had been hand-reared by itself until it was 46 days old and then placed in an aviary with other white-crowned sparrows and a strawberry finch.

   a) Compare sonogram V with sonogram IV. [2]

   b) Compare sonogram V with sonograms I and II. [2]

c) Suggest two reasons why birds rarely imitate other species. [2]

d) Discuss whether Morton and Baptista's observation is evidence for innate or learned development of birdsong. [2]

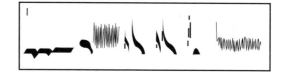

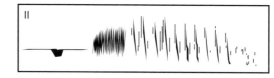

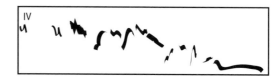

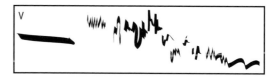

▲ Figure 2 Sonograms of birdsong

Innate and learned behaviour thus both depend on genes, but whereas the development of learned behaviour develops as a result of experience, innate behaviour is independent of it.

Reflex conditioning

Reflex conditioning involves forming new associations.

Several different types of learning have been defined. One of these, called reflex conditioning, was investigated by the Russian physiologist Ivan Pavlov, using dogs. Reflex conditioning involves forming new associations by establishing new neural pathways in the brain. Conditioned reflexes are used extensively in animal behaviour and can greatly increase survival chances.

For example, birds have an innate reflex to avoid foods with a bitter taste–this is an unconditioned reflex, but they have to learn which insects are likely to have that taste. If a bird tries to eat an insect with warning coloration of black and yellow stripes, for example, and finds that the insect tastes unpleasant, it develops an association between black and yellow stripes and bitter taste and therefore avoids all insects with such a colour pattern. In some cases the smell of the distasteful insect has to be combined with its coloration to cause avoidance.

▲ Figure 3 Monarch butterfly caterpillars ingest toxins (cardenolide aglycones) from the milkweed plants that they eat, making them distasteful to birds

🌐 Pavlov's experiments

Pavlov's experiments into reflex conditioning in dogs.

The 19th century Russian physiologist Pavlov developed apparatus to collect saliva from the mouth of his experimental dogs. He found that saliva was secreted in response to the sight or smell of food. These types of stimulus, to which all dogs respond without learning, are called unconditioned stimuli and the secretion of saliva that results is the unconditioned response.

Pavlov observed that after a while the dogs were starting to secrete saliva before they received the unconditioned stimulus. Something else had become a stimulus that allowed the dogs to anticipate the arrival of food. He found that the dogs could learn to use a variety of signals in this way, including the ringing of a bell, the flashing of a light, a metronome ticking or a musical box playing. These are examples of conditioned stimuli and the secretion of saliva that these stimuli elicit is the conditioned response. Pet dogs and children also quickly learn indicators that they will soon be fed.

▲ Figure 4 Pavlov's dogs

Imprinting

Imprinting is learning occurring at a particular life stage and is independent of the consequences of behaviour.

The word imprinting was first used in the 1930s by Konrad Lorenz to describe a type of learning. Imprinting can only occur at a particular stage of life and is the indelible establishment of a preference or stimulus that elicits behaviour patterns, often but not always, of trust and recognition. The example that was made famous by Lorenz was in greylag geese. Eggs are normally incubated by their mother so that she is the first large moving object that the hatchlings see. The young birds then follow their mother around during the first few weeks of life. She leads them to food and protects them.

Lorenz showed that young geese that are hatched in an incubator and who do not encounter their mother attach themselves to another large moving object and follow it around. This can be a bird of another species, Lorenz's boots or even an inanimate moving object. This attachment is what Lorenz called imprinting. The critical period in greylag geese when imprinting occurs is 13–16 hours after hatching. A distinctive feature of imprinting is that it is independent of the consequences of the behaviour – in experiments animals remain imprinted on something even if it does not increase their chance of survival.

▲ Figure 5 Young geese imprinted on their mother

Operant conditioning

Operant conditioning is a form of learning which consists of trial and error experiences.

Operant conditioning is sometimes explained in simple terms as learning by trial and error. It is a different form of learning from reflex conditioning. Whereas reflex conditioning is initiated by the environment imposing a stimulus on an animal, operant conditioning is initiated by an animal spontaneously testing out a behaviour pattern and finding out what its consequences are. Depending on whether the consequences are positive or negative for the animal or its environment, the behaviour pattern is either reinforced or inhibited.

Lambs learn not to touch electric fencing by operant conditioning. They explore their environment and if electric fencing is used to enclose their flock, lambs sooner or later touch it, probably with their nose. They receive a painful electric shock and through operant conditioning they avoid touching the fence in the future.

Learning

Learning is the acquisition of skill or knowledge.

The behaviour of animals changes during their lifetime. In a few cases behaviour patterns are lost, for example the palmar grasp reflex and other primitive reflexes in human babies. Far more commonly animals acquire types of behaviour pattern during their

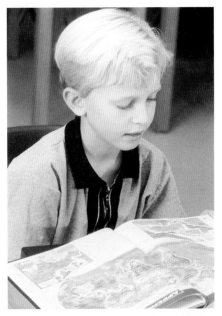

▲ Figure 6 Learning starts in children but is a lifelong process due to neural plasticity

lives. In some cases these behaviour changes are a natural part of growth and maturation, such as the behaviour changes that occur during puberty in humans. In other cases the modification of behaviour is acquired by learning – the behaviour does not develop unless it is learned.

Motor skills such as walking, talking or playing the violin are learned. Knowledge also has to be learned. For example the rainforest tribes learn the types of tree that can provide food or other useful materials and they also learn the location in the forest of individual trees of the useful types. Learning is a higher order function of the brain and humans have a greater capacity to learn than any other species. The degree of learning during an animal's lifetime is dependent on their longevity as well as their neural capacity. Social animals are more likely to learn from each other.

Memory

Memory is the process of encoding, storing and accessing information.

Memory is one of the higher order functions of the brain. Encoding is the process of converting information into a form in which it can be stored by the brain. Short-term memory lasts up to about a minute and may or may not lead to long-term memory, which can be retained for indefinite periods of time. Accessing is the recall of information so that it can be used actively in the thought processes of the brain.

Different parts of the brain have a role in the encoding, storage and accessing of memory. The importance of the hippocampus was strikingly demonstrated in 1953 when a patient called Henry Molaison had the amygdala and a section of hippocampus from both of his cerebral hemispheres removed in an experimental attempt to cure epilepsy. He immediately became incapable of making new memories unless they were procedural and his recall of memories formed during the eleven years before the surgery was also impaired. Recent research into the role of the hippocampus has shown that experiences cause large numbers of new synapses to be formed, which are then gradually pruned to refine the memory of the experience and allow it to be recalled when it is relevant and not at other times.

A.5 Neuropharmacology (AHL)

 Understanding

→ Some neurotransmitters excite nerve impulses in post-synaptic neurons and others inhibit them.

→ Nerve impulses are initiated or inhibited in post-synaptic neurons as a result of summation of all excitatory and inhibitory neurotransmitters received from pre-synaptic neurons.

→ Many different slow-acting neurotransmitters modulate fast synaptic transmission in the brain.

→ Memory and learning involve changes in neurons caused by slow-acting neurotransmitters.

→ Psychoactive drugs affect the brain by either increasing or decreasing post-synaptic transmission.

→ Anaesthetics act by interfering with neural transmission between areas of sensory perception and the CNS.

→ Stimulant drugs mimic the stimulation provided by the sympathetic nervous system.

→ Addiction can be affected by genetic predisposition, social environment and dopamine secretion.

 Applications

→ Effects on the nervous system of two stimulants and two sedatives.

→ The effect of anaesthetics on awareness.

→ Endorphins can act as painkillers.

 Skills

→ Evaluation of data showing the impact of MDMA (ecstasy) on serotonin and dopamine metabolism in the brain.

Nature of science

→ Assessing risk associated with scientific research: patient advocates will often press for the speeding up of drug approval processes, encouraging more tolerance of risk.

Excitatory and inhibitory neurotransmitters

Some neurotransmitters excite nerve impulses in post-synaptic neurons and others inhibit them.

The basic principles of synaptic transmission were described in sub-topic 6.5: neurotransmitter is released into the pre-synaptic neuron when a depolarization of the pre-synaptic neuron reaches the synapse. The neurotransmitter depolarizes the post-synaptic neuron by binding to receptors in its membrane. Excitatory neurotransmitters excite the post-synaptic neuron for periods ranging from a few milliseconds to many seconds, producing depolarization that may be sufficient to trigger action potentials.

Some neurotransmitters have a different effect – they inhibit the formation of action potentials in the post-synaptic neuron because the membrane potential becomes more negative when the neurotransmitter binds to the post-synaptic membrane. This hyperpolarization makes it more difficult for

541

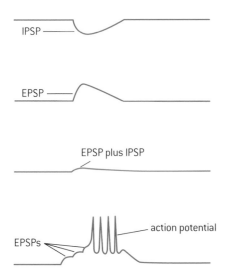

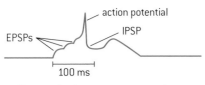

▲ Figure 1 Excitatory post-synaptic potentials (EPSP), inhibitory post-synaptic potentials (IPSP)

the post-synaptic neuron to reach the threshold potential so nerve impulses are inhibited. Inhibitory neurotransmitters are small molecules that are inactivated by specific enzymes in the membrane of the post-synaptic neuron.

Summation

Nerve impulses are initiated or inhibited in post-synaptic neurons as a result of summation of all excitatory and inhibitory neurotransmitters received from pre-synaptic neurons.

More than one pre-synaptic neuron can form a synapse with the same post-synaptic neuron. Especially in the brain, as there are hundreds or even thousands of pre-synaptic neurons! Usually a single release of excitatory neurotransmitter from one pre-synaptic neuron is insufficient to trigger an action potential. Either one pre-synaptic neuron must repeatedly release neurotransmitter, or several adjacent pre-synaptic neurons must release neurotransmitter more or less simultaneously. The additive effect from multiple releases of excitatory neurotransmitter is called summation.

Some pre-synaptic neurons release an inhibitory rather than an excitatory neurotransmitter. Summation involves combining the effects of excitatory and inhibitory neurotransmitters. Whether or not action potentials form in a post-synaptic neuron depends on the balance between the effects of the synapses that release excitatory and inhibitory neurotransmitters and therefore whether the threshold potential is reached. This integration of signals from many different sources is the basis of decision-making processes in the central nervous system.

Slow and fast neurotransmitters

Many different slow-acting neurotransmitters modulate fast synaptic transmission in the brain.

The neurotransmitters so far described have all been fast-acting, with the neurotransmitter crossing the synapse binding to receptors less than a millisecond after an action potential has arrived at the pre-synaptic membrane. The receptors are gated ion-channels, which open or close in response to the binding of the neurotransmitter, causing an almost immediate but very brief change in post-synaptic membrane potential.

Another class of neurotransmitter is slow-acting neurotransmitters or neuromodulators which take hundreds of milliseconds to have effects on post-synaptic neurons. Rather than having an effect on a single post-synaptic neuron they may diffuse through the surrounding fluid and affect groups of neurons. Noradrenalin/norepinephrine, dopamine and serotonin are slow-acting neurotransmitters.

Slow acting neurotransmitters do not affect ion movement across post-synaptic membranes directly, but instead cause the release of secondary messengers inside post-synaptic neurons, which set off sequences of intracellular processes that regulate fast synaptic transmission. Slow

acting neurotransmitters can modulate fast synaptic transmission for relatively long periods of time.

Memory and learning

Memory and learning involve changes in neurons caused by slow-acting neurotransmitters.

Psychologists have studied learning and memory for decades but it is only relatively recently that neurobiologists have been able to study these processes at the level of the synapse. Slow-acting neurotransmitters (neuromodulators) have a role in memory and learning. They cause the release of secondary messengers inside post-synaptic neurons that can promote synaptic transmission by mechanisms such as an increase in the number of receptors in the post-synaptic membrane or chemical modification of these receptors to increase the rate of ion movements when neurotransmitter binds.

The secondary messengers can persist for days and cause what is known as long-term potentiation (LTP). This may be central to the synaptic plasticity that is necessary for memory and learning. Even longer-term memories may be due to a remodelling of the synaptic connections between neurons. The learning of new skills has been shown to be linked to the formation of new synapses in the hippocampus and elsewhere in the brain.

Psychoactive drugs

Psychoactive drugs affect the brain by either increasing or decreasing post-synaptic transmission.

The brain has many synapses, perhaps as many as 10^{16} in children. These synapses vary in their organization and use a wide variety of neurotransmitters. Over a hundred different brain neurotransmitters are known. Psychoactive drugs affect the brain and personality by altering the functioning of some of these synapses. Some drugs are excitatory, because they increase post-synaptic transmission. Others are inhibitory because they decrease it.

Examples of excitatory drugs:

- Nicotine contained in cigarettes and other forms of tobacco, derived from the plant *Nicotiana tabacum*.

- Cocaine extracted from the leaves of a Peruvian plant, *Erythroxylon coca*.

- Amphetamines, a group of artificially synthesized compounds.

Examples of inhibitory drugs:

- Benzodiazepines, a group of compounds including Valium that are synthesized artificially.

- Alcohol in the form of ethanol, obtained by fermentation using yeast.

- Tetrahydrocannabinol (THC) obtained from the leaves of the *Cannabis sativa* plant.

 Endorphins

Endorphins can act as painkillers.

Pain receptors in the skin and other parts of the body detect stimuli such as the chemical substances in a bee's sting, excessive heat or the puncturing of skin by a hypodermic needle. These receptors are the endings of sensory neurons that convey impulses to the central nervous system. When impulses reach sensory areas of the cerebral cortex we experience the sensation of pain. Endorphins are oligopeptides that are secreted by the pituitary gland and act as natural painkillers, blocking feelings of pain. They bind to receptors in synapses in the pathways used in the perception of pain, inhibiting synaptic transmission and preventing the pain being felt.

Ecstasy

Evaluation of data showing the impact of MDMA (ecstasy) on serotonin and dopamine metabolism in the brain.

Data-based questions: Effects of ecstasy on the striatum

The graphs in figure 2 show the results of an experiment in which mice were treated with MDMA (ecstasy) and levels of dopamine and serotonin were measured in the striatum of their brains. Two doses of MDMA were used and also saline (no MDMA). Wild-type mice were used and also three strains of knockout mice that lacked genes for making the dopamine transporter protein (DAT-KO), the serotonin transporter (SERT-KO) or both transporters (DAT/SERT-KO). The graphs show the levels of dopamine and serotonin in the three-hour period after MDMA had been administered.

Questions

1   Describe the trends in dopamine level wild-type mice in the three hours after administration of 10 mg of MDMA.   [3]

2   Discuss the evidence from the data for the hypothesis that MDMA has a greater effect on serotonin level than dopamine level in the wild-type mice.   [3]

3   a) Distinguish between the results for the wild-type mice and the DAT-KO mice. [2]

    b) Discuss whether these differences are statistically significant or not.   [2]

4   Distinguish between the results for the DAT-KO mice and the SERT-KO mice.   [2]

5   Explain the results for the DAT/SERT-KO mice.   [2]

6   Suggest one benefit of using knockout mice in this experiment.   [1]

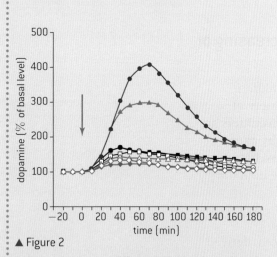

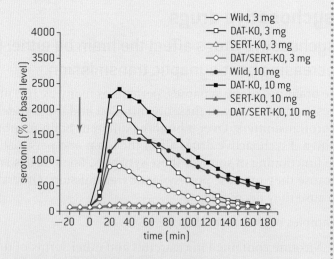

▲ Figure 2

Reference: Hagino et al, Effects of MDMA on Extracellular Dopamine and Serotonin Levels in Mice Lacking Dopamine and/or Serotonin Transporters, *Curr. Neuropharmacol.* 2011 March; 9(1): 91–95.

Anaesthetics

Anaesthetics act by interfering with neural transmission between areas of sensory perception and the CNS.

Anaesthetics cause a reversible loss of sensation in part or all of the body. Local anaesthetics cause an area of the body to become numb, for example the gums and teeth during a dental procedure. General anaesthetics cause unconsciousness and therefore a total lack of sensation.

Anaesthetics are chemically varied and work in a variety of ways. Many of them affect more than just the sense organs and can also inhibit signals to motor neurons and other parts of the nervous system so they should only ever be administered by highly trained medical practitioners.

 Anaesthetics and awareness

The effect of anaesthetics on awareness.

A patient who has been given a general anaesthetic normally has no awareness of the surgical or other procedures that they are undergoing because they are totally unconscious. There are some procedures where it is either not necessary or is undesirable for the patient to be unconscious. For example patients are kept partially conscious during some operations to remove brain tumours, so that the effects on the brain can be monitored.

There have been some cases of patients retaining some awareness during operations, when they have not been given a high enough dose of anaesthetic. The patient may or may not feel pain. The risk of awareness is highest in operations such as emergency caesarean sections in which it is best for the mother and child for the dose of anaesthetic to be minimized, although in these procedures a spinal block is almost always now used rather than a general anaesthetic, so the patient is awake and breathing is normal, but pain sensation cannot get beyond the spinal cord.

 Drug testing

Assessing risk associated with scientific research: patient advocates will often press for the speeding up of drug approval processes, encouraging more tolerance of risk.

There are strict protocols for testing new drugs with several phases that establish two things – an appropriate dose and route of administration that make the drug effective, and that its side-effects are minor and infrequent enough for the drug to be regarded as safe. These tests take many years to complete but approval for the introduction of a new drug is only given once the tests have been rigorously carried out. There have been some trials where the difference between the control group of patients given a placebo and the group given the new drug is so great that it seems unethical to deny the control group the treatment. It therefore seems reasonable to abandon the trials and introduce the drug immediately. The danger of this policy is that harmful side-effects may only then be discovered when large numbers of patients have been given the new drug.

There have been some cases where groups of patients have campaigned for a new drug to be introduced before it has been fully tested. This may be acceptable with terminal diseases such as AIDS or certain forms of heart disease where the patient may regard any level of risk acceptable given the certainty of death without treatment. It is unlikely to be acceptable with non-critical illnesses where the risks from using a drug that has not been fully tested are too great compared with the risks associated with the disease remaining untreated.

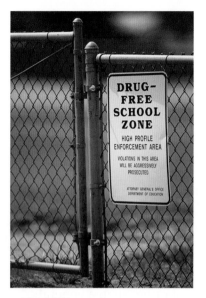

▲ Figure 3 Drug enforcement measures near a school

Stimulant drugs

Stimulant drugs mimic the stimulation provided by the sympathetic nervous system.

Stimulants are drugs that promote the activity of the nervous system. They make a person more alert, energetic and self-confident. They also increase heart rate, blood pressure and body temperature. The effects of stimulant drugs match those of the sympathetic nervous system. This is because stimulant drugs act by a variety of mechanisms to make the body respond as though it had been naturally stimulated by the sympathetic nervous system.

Some mild stimulants are present in foods and drinks, for example caffeine in tea and coffee and theobromine in chocolate. Doctors sometimes prescribe stronger stimulants to treat conditions such as clinical depression and narcolepsy. Stimulant drugs are also sometimes used against medical advice. Examples include cocaine, amphetamines and nicotine in cigarettes.

Examples of stimulants and sedatives

Effects on the nervous system of two stimulants and two sedatives.

Pramipexole mimics dopamine and binds to dopamine receptors in post-synaptic membranes at dopaminergic synapses. Whereas some drugs that mimic neurotransmitters are antagonists because they block synaptic transmission, pramipexole is an agonist because it has the same effects as dopamine when it binds. Pramipexole is used during the early stages of Parkinson's disease to help to reduce the effects of insufficient dopamine secretion that characterize this disease. It has also sometimes been used as an anti-depressant.

Cocaine also acts at synapses that use dopamine as a neurotransmitter. It binds to dopamine reuptake transporters, which are membrane proteins that pump dopamine back into the pre-synaptic neuron. Because cocaine blocks these transporters, dopamine builds up in the synaptic cleft and the post-synaptic neuron is continuously excited. Cocaine is therefore an excitatory psychoactive drug that gives feelings of euphoria that are not related to any particular activity.

Diazepam (Valium) binds to an allosteric site on GABA receptors in post-synaptic membranes.

GABA (γ-amino butyric acid) is an inhibitory neurotransmitter and when it binds to its receptor a chloride channel opens, causing hyperpolarization of the post-synaptic neuron by entry of chloride ions. When diazepam is bound to the receptor the chloride ions enter at a greater rate, inhibiting nerve impulses in the post-synaptic neuron. Diazapam is therefore a sedative. It can reduce anxiety, panic attacks and insomnia and it is also sometimes used as a muscle relaxant.

THC (Tetrahydrocannabinol) is present in cannabis. It binds to cannabinoid receptors in pre-synaptic membranes. Binding inhibits the release of neurotransmitters that cause excitation of post-synaptic neurons. THC is therefore an inhibitory psychoactive drug and a sedative. Cannabinoid receptors are found in synapses in various parts of the brain, including the cerebellum, hippocampus and cerebral hemispheres. The main effects of THC are disruption of psychomotor behaviour, short-term memory impairment, intoxication and stimulation of appetite.

Drug addiction

Addiction can be affected by genetic predisposition, social environment and dopamine secretion.

The American Psychiatric Association has defined addiction as: *"a chronically relapsing disorder that is characterized by three main elements: (a) compulsion to seek and take the drug, (b) loss of control in limiting intake and (c) emergence of a negative emotional state when access to the drug is prevented."* Only certain drugs cause addiction and usually repeated use over a prolonged period of time is needed. With a few drugs, addiction can develop more rapidly. The causes of addiction are clearly not simple and three areas need to be considered.

1   Some people seem much more vulnerable to addiction than others because of their genes. This is known as genetic predisposition. One example is the gene, DRD2, which codes for the dopamine receptor protein. There are multiple alleles of this gene and a recent study showed that people with one or more copies of the A1 allele consumed less alcohol than those homozygous for the A2 allele.

2   Addiction is more prevalent in some parts of society than others because the social environment greatly affects the likelihood of taking drugs and becoming addicted. Peer pressure, poverty and social deprivation, traumatic life experiences and mental health problems all contribute. Cultural traditions are very important and help to explain why different drugs cause problems in different parts of the world.

3   Many addictive drugs, including opiates, cocaine, nicotine and alcohol affect dopamine secreting synapses. Dopamine secretion is associated with feelings of well-being and pleasure. Addictive drugs cause prolonged periods with high dopamine levels in the brain. This is so attractive to the drug user that they find it very difficult to abstain.

▲ Figure 4 Alcohol is an addictive drug but is legal in many counties

A.6 Ethology (AHL)

Understanding

→ Ethology is the study of animal behaviour in natural conditions.

→ Natural selection can change the frequency of observed animal behaviour.

→ Behaviour that increases the chances of survival and reproduction will become more prevalent in a population.

→ Learned behaviour can spread through a population or be lost from it more rapidly than innate behaviour.

Nature of science

→ Testing a hypothesis: experiments to test hypotheses on the migratory behaviour of blackcaps have been carried out.

Applications

→ Migratory behaviour in blackcaps as an example of the genetic basis of behaviour and its change by natural selection.

→ Blood sharing in vampire bats as an example of the evolution of altruistic behaviour by natural selection.

→ Foraging behaviour in shore crabs as an example of increasing chances of survival by optimal prey choice.

→ Breeding strategies of hooknoses and jacks in coho salmon populations as an example of behaviour affecting chances of survival and reproduction.

→ Courtship in birds of paradise as an example of mate selection.

→ Synchronized oestrus in female lions in a pride as an example of innate behaviour that increases the chances of survival and reproduction of offspring.

→ Feeding on cream from milk bottles in blue tits as an example of the development and loss of learned behaviour.

Ethology

Ethology is the study of animal behaviour in natural conditions.

Animals are adapted to their natural habitat in their behaviour. If we remove them from this habitat and place them in a zoo or laboratory, animals may not behave normally because they may not receive the same stimuli as in their natural habitat. For this reason it is best whenever possible to carry out research into animal behaviour in their natural habitat rather than in an artificial environment. The study of the actions and habits of animals in their natural environment is called ethology.

Natural selection and animal behaviour

Natural selection can change the frequency of observed animal behaviour.

Natural selection is the theme that runs through the whole of modern biology, including ethology. It adapts species to all aspects of their environment. Adaptation extends over the whole range of animal characteristics, from the structure of a single molecule such as hemoglobin to the patterns of behaviour in a species.

Animal behaviour has been observed to change rapidly in some cases. House finches *Carpodacus mexicanus* are an example. In California the native population is sedentary – the birds remain in the same area throughout the year. A small number were illegally released in the 1940s in New York City and spread though the eastern United States, and within twenty years migratory behaviour was observed. The frequency of this behaviour rose to more than 50% of the population, presumably as a result of natural selection.

The mechanism of natural selection

Behaviour that increases the chances of survival and reproduction will become more prevalent in a population.

Natural selection works in same way for animal behaviour as for other biological characteristics. Individuals with the best-adapted actions and responses to the environment are most likely to survive and produce offspring. If behaviour is genetically determined, rather than learned, it can be inherited by offspring.

The breeding season of the great tit *Parus major* illustrates how behaviour evolves by natural selection, often as a response to environmental changes. This bird lives in woodland and feeds its young on caterpillars and other insects. The availability of this food rises to a peak in spring soon after the new leaves on trees have grown. Due to global warming, the time of peak availability has become earlier. The timing of nesting and egg laying varies within narrow limits within the population. Researchers have shown that birds that lay their eggs a few days earlier than the mean date have more success in rearing young. According to natural selection, the mean date of egg laying should evolve to be earlier and researchers found this prediction fulfilled.

 Breeding strategies in salmon

Breeding strategies of hooknoses and jacks in coho salmon populations as an example of behaviour affecting chances of survival and reproduction.

Coho salmon *Oncorhynchus kisutch* breed in rivers that discharge into the North Pacific Ocean, including those on the west coast of North America.

The adults die after breeding and the young live for about a year in the river and then migrate to the ocean where they remain for several years before returning to spawn. There are two breeding strategies among males. Hooknoses fight each other for access to females laying eggs, with the winner shedding sperm over the eggs to fertilize them. Jacks usually avoid fights and instead sneak up on females and attempt to shed sperm over their eggs before being noticed.

Obervations on individually identified fish, usually identified by a tag, show that whether a male becomes a jack or a hooknose depends on his growth rate. Males that grow rapidly are able to return to breed two years after they were spawned and are jacks. Males that grow less rapidly remain in the ocean for one year longer, but are then significantly larger and are hooknoses. The smaller jacks are more likely to reproduce by the sneaking strategy than by fighting the larger hooknoses. The larger

hooknoses are unlikely to sneak up on a female without being noticed so they must fight other hooknoses and fend off jacks if they are to be successful in breeding.

▲ Figure 1 Brown bear catching salmon as they swim upstream to breed

🌐 Synchronized oestrus

Synchronized oestrus in female lions in a pride as an example of innate behaviour that increases the chances of survival and reproduction of offspring.

Female lions remain in the group (pride) into which they were born, but male lions are expelled from the pride when they are about three years old. Males can only breed if, when fully grown adults, they overcome the dominant male in another pride by fighting. Within two or three years of taking over a pride of females, the breeding male is likely to be replaced by a younger rival. When a new dominant male takes over a pride he may kill all the suckling cubs, thus making the females come into oestrus more quickly so the male can then mate with them to father his own cubs. Females protect their cubs from marauding males, sometimes leading to fierce fights, but accept his sexual advances after he has taken over the pride. Sometimes two or more closely related young males together fight for dominance of another group. This increases their chance of success, especially if they are fighting a single dominant male.

Females can only breed when they come into oestrus. All females in a pride tend to come into

oestrus at the same time. This behaviour has several advantages: the females have their cubs at the same time so are all lactating while the cubs are suckling, so they can suckle each other's cubs when they are hunting, increasing the cubs' chance of survival. Also a group of male cubs of the same age are ready to leave the pride at the same time so can compete for dominance of another pride more effectively.

▲ Figure 2 Lions in a group known as a pride

Blackcap migration

Migratory behaviour in blackcaps as an example of the genetic basis of behaviour and its change by natural selection.

The blackcap *Sylvia atricapilla* breeds during the northern summer. Until relatively recently, populations of blackcaps that breed in Central Europe including Germany almost all migrated to Spain and Portugal for the winter, where the weather is warmer and the availability of food is greater. During the second half of the 20th century a few blackcaps from the population in Germany were found to be migrating to Britain and Ireland instead. The numbers of blackcaps overwintering in Britain rose rapidly to more than 10%.

There are several possible reasons for this change in migration behaviour. Global warming has led to winters being warmer in Britain so the long migration to Spain is not necessary. Many people in Britain feed wild birds in winter which may facilitate survival of overwintering blackcaps more than in Spain. In winter the minimum day length in Britain is shorter than in Spain, which may prompt earlier migration to breeding grounds. Blackcaps that arrive earlier take the best territories – another advantage of overwintering in Britain.

 Experiments with migrating blackcaps

Testing a hypothesis: experiments to test hypotheses on the migratory behaviour of blackcaps have been carried out.

In ethology as in other branches of science it is essential to test hypotheses and either obtain evidence for them or prove them to be false. The adaptive value of behaviour patterns have sometimes been assumed without evidence. These accounts of evolution are known as "Just So Stories" after Rudyard Kipling's children's story book. However intuitively obvious a hypothesis about the evolution of a behaviour pattern, it is only a just so story until tested.

Hypotheses about evolutionary changes in blackcap migration have been rigorously tested. For example, the hypothesis that the direction of migration is genetically determined has been tested. Eggs were collected in Germany from parent birds that had migrated to Britain in the previous winter and from parents who had migrated to Spain. The young were reared without their parents so that they could not learn from them and when they migrated the direction was recorded. Birds whose parents had migrated to Britain tended to fly west, wherever they were reared, and birds whose parents had migrated to Spain tended to fly south-west. They therefore responded to

migratory stimuli in the same way as their parents, indicating that the direction of migration is genetically determined, and can thus be subject to long-term evolutionary change under natural selection.

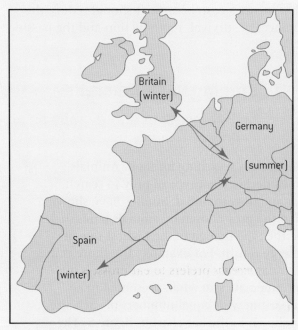
▲ Figure 3 Migration of blackcaps

Vampire bats

Blood sharing in vampire bats as an example of the evolution of altruistic behaviour by natural selection.

Female vampire bats *Desmodus rotundus* live in colonies of 8–12 individuals, with the same individuals roosting together for several years. Their diet is about 25ml of vertebrate blood, usually mammalian, each night. If a bat fails to feed for two or three consecutive nights they risk death from starvation. However, this rarely happens because when the bats return to the roost at the end of the night, those that have fed regurgitate blood for those that have not.

This behaviour pattern is a rare example of altruism. It fulfils two necessary criteria:

- there may be siblings or mothers with daughters in a group but tests have shown that there are also unrelated females who also share blood, so blood sharing is not kin-selection;

- giving blood to an individual who has not fed incurs a cost to the giver because some of their daily diet is lost, so blood sharing is not merely cooperation – it is genuine altruism.

The evolution of altruism is an interesting conundrum: we might not expect natural selection to promote the evolution of behaviour that incurs a cost, because it should reduce the chances of survival, reproduction and the passing

on of genes of the altruistic animal. Blood sharing is an example of reciprocal altruism. Individual A gains a benefit from giving blood to Individual B because Individual B survives and can share blood on a later night if Individual A fails to feed. It only occurs in stable groups of females that roost together regularly and as it aids the chances of survival and reproduction of all of the members of such groups, natural selection favours it.

▲ Figure 4 Vampire bats show reciprocal altruism by blood sharing

Foraging in shore crabs

Foraging behaviour in shore crabs as an example of increasing chances of survival by optimal prey choice.

Foraging is searching for food. Animals must decide what type of prey to search for and how to find it. Studies have shown that the prey chosen by animals tends to be the type that gives the highest rate of energy return. For example, the shore crab *Carcinus moenas* prefers to eat mussels of intermediate size when presented in an aquarium with equal numbers of each size, as shown in the bar chart in figure 5. The graph in figure 5 shows that mussels of intermediate size are the most profitable in terms of the energy yield per second of time spent breaking open the shells.

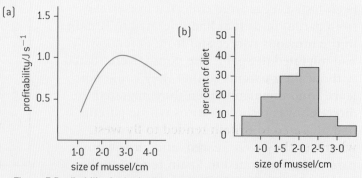

▲ Figure 5 Profitability in Joules per second and percentage in diet of mussels of different size

 Courtship in birds of paradise

Courtship in birds of paradise as an example of mate selection.

Some animals have anatomical features that seem to the human eye to be excessive, for example the tail feathers of the peacock. Other animals have behaviour patterns that seem bizarre. The plumage and courtship displays of male birds of paradise are examples of both of these types of exaggerated trait. There are about forty species of bird of paradise living on New Guinea and other nearby islands. The males have very showy plumage with bright coloration and elongated or elaborate tail feathers that are of no use in flying. The females, which build the nest, incubate the eggs and rear the young, are relatively drab.

Males in many species of bird of paradise have a complicated and eye-catching courtship dance that they use to try to attract females. In some species the males all gather at a site called a lek and females select a mate from among the males displaying. The coloured plumage and courtship dances of birds help to avoid interspecific hybridization by allowing females to determine if a male belongs to their species, but this could be achieved in much more subtle ways than those

used by birds of paradise and biologists have long speculated on the reasons for exaggerated traits.

Darwin explained them in terms of mate selection – females prefer to mate with males that have exaggerated traits. The reason may be that these traits indicate overall fitness. If a bird of paradise has enough energy to grow and maintain the elaborate plumage and repeatedly to carry out very vigorous courtship displays it indicates that the male must have fed efficiently. If it can survive in the rainforest with the encumbrance of its tail feathers and with bright plumage that makes it visible to predators, it is probably well adapted in other ways and is therefore a good mate to choose. Over the generations females that selected males with showier plumage and more spectacular courtship dances have produced offspring fathered by males with greater overall fitness. Natural selection has therefore caused these traits to become exaggerated.

An example of a male bird of paradise can be seen in sub-topic 4.1.

Changing learned and innate behaviour

Learned behaviour can spread through a population or be lost from it more rapidly than innate behaviour.

Some patterns of behaviour are entirely innate, for example the withdrawal reflex, so are programmed into an animal's genes. They can happen immediately in an individual without any period of learning. However, they can only be modified by natural selection relatively slowly because there must be variation in the alleles that affect the behaviour and a change in allele frequencies in the population due to one behaviour pattern increasing chances of survival and reproduction over the other patterns of behaviour.

Other patterns of behaviour are either partially or entirely learned – although these take longer to develop in an individual, they do not involve changes in allele frequency and can spread in a population relatively rapidly as one individual learns from another. Chimpanzees show many examples of tool use that are learned, with considerable variation between groups of chimpanzees in the types of tool used. If one individual discovers a new use of an object as a tool, others can learn it quickly. However, learned behaviour can also disappear from a population rapidly. An example is blue tits feeding on cream from milk bottles.

TOK

Why are scientists sometimes suspicious of evidence based on amateur observations rather than on numerical data from controlled experiments?

With respect to the observations of the changes in the behaviour of blue tits and milk bottles, an article appeared in 1952 in the journal Nature:

"Although no experimental analysis of the behaviour involved in the opening of milk bottles has yet been made, further observations in the field enable the discussion to be carried further."

Blue tits and cream

Feeding on cream from milk bottles in blue tits as an example of the development and loss of learned behaviour.

Blue tits *Cyanistes caeruleus* were first observed pecking through the aluminium foil caps of milk bottles left outside houses, to drink the cream, in the 1920s in Southampton, England. This behaviour was observed soon afterwards 150 kilometres away – far further than blue tits normally fly. Amateur birdwatchers followed the rapid spread of the behaviour, in both blue tits and great tits, across Europe to the Netherlands, Sweden and Denmark.

German occupation of the Netherlands during the Second World War stopped deliveries of milk for eight years – five years longer than the maximum life of a blue tit. However, within months of the resumption of deliveries, blue tits throughout the Netherlands were pecking through the bottle tops. The rapid spread of this behaviour pattern shows that it must be due to learned rather than innate behaviour.

Newspaper articles recently reported that blue tits had stopped feeding on cream from milk bottles.

Much less milk is now delivered to doorsteps because milk in supermarkets is cheaper. Also skimmed milk, without cream at the top, has become popular with humans. This may explain why blue tits have not recently been observed pecking through bottle tops.

▲ Figure 6 Blue tit pecking through milk bottle cap

Questions

1 When birds are in danger of attack by predators, they sometimes sleep with one eye open and one eye closed. Neurobiologists investigated this behaviour pattern using mallard ducks (*Anas platyrhynchos*). Video recordings were made of groups of four sleeping birds, arranged in a row. The birds at the ends of the row were more vulnerable to predator attacks and kept one eye open 150% more of the time than the two birds in the centre of the row.

Electroencephalograph (EEG) recordings were made to monitor the brain state of the birds at the ends of the rows. A region of the brain which indicates whether the bird is asleep or awake was monitored in each of the left and right cerebral hemispheres. EEG recordings were made when the birds were sleeping with both eyes closed, when the birds had both eyes open and also when they had one eye open. These results are shown in the bar chart below, as a percentage of the activity of the brain region when the birds were sleeping with both eyes closed.

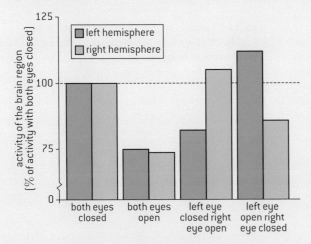

Source: Rattenborg, et al., *Nature*, 1999, 397, pages 397–398

a) State the effect of opening both eyes on activity in the region of the brain that was being monitored. [1]

b) (i) Using the data in the bar chart, deduce the effect on the two cerebral hemispheres of opening only the right eye. [2]

   (ii) Determine which hemisphere is more awake when the right eye is open. [1]

   (iii) Using the data in the bar chart, deduce how the left and right eyes and left and right hemispheres are connected. [1]

c) Suggest **two** advantages to birds of keeping one eye open during sleep. [2]

2 Alzheimer's disease (AD) is characterized by increasing dementia (mental and emotional deterioration) in affected persons.

Evidence from the *post-mortem* (after death) analysis of the brains of affected patients has revealed two abnormalities. Affected persons show a change in the concentration of nerve growth factor (NGF) in a region of the brain known as the cortex. The brains of affected patients also have plaques. These are accumulations of insoluble material in and around cells.

A study was carried out to measure the *post-mortem* NGF concentrations in two regions of the cortex, the temporal cortex and the frontal cortex. Three groups of people were compared:

- AD patients
- pre-AD patients with plaques but no dementia
- a control group with no plaques and no dementia.

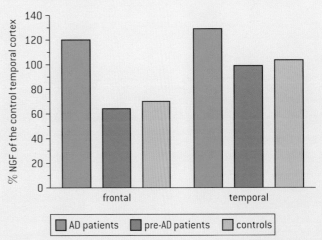

Source: R Hellweg et al., (1999), *International Journal of Development Neuroscience*, 16, (7/8), pages 787–794

a) Compare the data for the two regions of the cortex. [3]

b) Calculate the increase in percentage NGF in the frontal cortex of AD patients compared to the control group. [1]

c) Suggest what happens to the quantity of NGF in the cortex as the disease progresses. [2]

3 Many animal species use long-range calls to establish their use of space and their relationships with members of their own and other species. Most of the calls of the African Savanna elephant (*Loxodonta africana*) are below the range of human hearing. The area in which the elephants can detect the calls is known as the calling area. On any given day, the calling area undergoes expansions and contractions. The diagrams on the right show the calling area (solid line) of elephants in the Etosha National Park at different times of the day. The position of the calling elephants is the centre of the diagram. Circular rings depict distance (in km). The wind speed (in m s^{-1}) and direction are shown with an arrow. If there is no arrow on the diagram it shows there was no wind.

a) Identify the time of the day when the calling area was greatest. [1]

b) Identify the wind speed at 08:00h. [1]

c) Compare the calling area at 17:00h with 18:00h. [2]

d) Discuss the relationship between the wind and the calling area. [3]

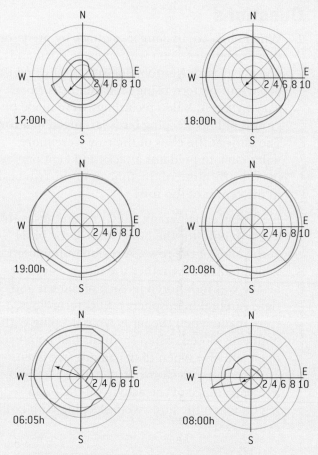

Source: D Larom, et al., *Journal of Experimental Biology* (1997), 200, page 421–431. Reprinted with the permission of the Company of Biologists

B BIOTECHNOLOGY AND BIOINFORMATICS

Introduction

Biotechnology is the use of organisms, especially microorganisms to perform industrial processes. The organisms used may be genetically modified to make them more suitable. Crops can be modified to increase yields and to obtain novel products. Biotechnology can be used in the prevention and mitigation of contamination from industrial, agricultural and municipal wastes. Biotechnology can also be used in the diagnosis and treatment of disease. Bioinformatics is the use of computers to analyse sequence data in biological research.

B.1 Microbiology: organisms in industry

Understanding

→ Microorganisms are metabolically diverse.

→ Microorganisms are used in industry because they are small and have a fast growth rate.

→ Pathway engineering optimizes genetic and regulatory processes within microorganisms.

→ Pathway engineering is used industrially to produce metabolites of interest.

→ Fermenters allow large-scale production of metabolites by microorganisms.

→ Fermentation is carried out by batch or continuous culture.

→ Microorganisms in fermenters become limited by their own waste products.

→ Probes are used to monitor conditions within fermenters.

→ Conditions are maintained at optimal levels for the growth of the microorganisms being cultured.

Applications

→ Deep-tank batch fermentation in the mass production of penicillin.

→ Production of citric acid in a continuous fermenter by *Aspergillus niger* and its use as a preservative and flavouring.

→ Biogas is produced by bacteria and archaeans from organic matter in fermenters.

Skills

→ Gram staining of Gram-positive and Gram-negative bacteria.

→ Experiments showing zone of inhibition of bacterial growth by bactericides in sterile bacterial cultures.

→ Production of biogas in a small-scale fermenter.

Nature of science

→ Serendipity has led to scientific discoveries: the discovery of penicillin by Alexander Fleming could be viewed as a chance occurrence.

▲ Figure 1 *Penicillium* mold growing on an orange. The antibiotic penicillin is derived from this microorganism

▲ Figure 2 Microalgae production for biofuels. Ponds being used to culture *Chlorella vulgaris* microalgae as a source of biofuel. The carbon dioxide is pumped into ponds (seen here) to promote photosynthesis and therefore growth of the algae

Metabolic diversity

Microorganisms are metabolically diverse.

Microorganisms occupy a number of niches in ecosystems. In order to serve their ecological role, they require certain metabolic pathways that correspond to their role.

Saprotrophs release nutrients trapped in detritus and make it available to ecosystems. As saprotrophs, bacteria and fungi compete with one another for food sources. Many fungi release anti-bacterial antibiotics into the environment in an effort to limit interspecific competition.

Other microorganisms act as producers. Cyanobacteria (blue-green algae) and protoctists such as algae and *Euglena* are photosynthetic. They produce carbohydrates by fixing carbon dioxide in the Calvin cycle.

Other microorgansims act as heterotrophs. Yeast such as *Saccharomyces cerevisiae* carry out anaerobic respiration producing alcohol and carbon dioxide by a pathway known as alcoholic fermentation.

The bacteria *Rhizobium* and *Azotobacter* can fix nitrogen and convert it to a form that living things can use. Bacteria such as *Nitrobacter* and *Nitrosomonas* can use inorganic chemicals as energy sources. They are known as chemoautotrophs.

Humans have been able to take advantage of the metabolic pathways of microorganisms in biotechnology applications.

The advantages of using microorganisms in biotechnology

Microorganisms are used in industry because they are small and have a fast growth rate.

Humans have been exploiting the metabolism of microorganisms throughout history for example in the production of food such as yogurt, bread, wine and cheese.

More recently, industrial biotechnology has increased the number of metabolic pathways exploited for drug and fuel production as well as additional applications involving genetically modified microbes.

Industrial biotechnology takes advantage of the facts that microorganisms are small and reproduce at a fast rate. They can be grown on a range of nutrient substrates and can produce a range of products. Conditions can be easily monitored in an industrial setting and maintained at optimum levels.

Pathway engineering

Pathway engineering is used industrially to produce metabolites of interest.

Traditionally either through selective breeding or genetic modification, microorganisms used in biotechnology applications were selected because they were the variants that provided the maximum yield of a desired metabolite. What this didn't take into account was the possibility that there were points in the metabolic pathway that constrained yields to the point where actual yields were much lower than theoretical yields.

What distinguishes pathway engineering from traditional methods is the use of detailed knowledge and analysis of the cellular system of metabolic reactions. This allows scientists to direct changes at multiple points to improve yields of metabolites of interest. This can include extending the range of substrates, elimination of by-products that slow the process down and extension of the range of products.

Pathway engineering uses knowledge of metabolic pathways to increase yields

Pathway engineering optimizes genetic and regulatory processes within microorganisms.

Pathway engineering is a technique that analyses the metabolic pathway of a particular microorganism to determine the "bottleneck" points of the pathway that constrain the production of the desired compound. Researchers can then address the constraint using genetic modification.

For example, the yeast *Saccharomyces cerevisiae* occurs naturally on the skin of grapes. The fermentation of grapes is carried out by *S. cerevisiae* with the desired end product being ethanol. Maintaining the correct pH is important in wine production. Malate is a metabolite that appears during wine making. Its degradation is essential for the deacidification of grapes. However, malate permease, a membrane protein necessary for the transport of malate into cells is not present in *S. cerevisiae*. Further, while *S. cerevisiae* has an enzyme that can degrade malate, it was found to be relatively inefficient.

The gene for MAE2, a highly efficient malate degrading enzyme from *Lactococcus lactis* was inserted into *S. cerevisiae* along with the gene for malate permease from the yeast *Schizosaccharomyces pombe*. The ability of transgenic *S. cerevisiae* to undertake more efficient malate degradation was successfully achieved.

Fermenters in industry

Fermenters allow large-scale production of metabolites by microorganisms.

Technically, fermentation refers to the anaerobic generation of ATP from glucose that generates characteristic end products such as alcohol and lactic acid. With respect to biotechnology, microbiologists have a broad interpretation of the term fermentation; i.e., the word refers to the processes involved in the large-scale culture of microorganisms to produce metabolites of interest.

A fermenter is often a large stainless steel vessel filled with sterile nutrient medium. The medium is inoculated with the desired microorganism. An impeller is a rotating set of paddles that mixes the medium preventing sedimentation. Gas is bubbled through if the desired metabolic process is aerobic. A pressure gauge detects gas build-up and

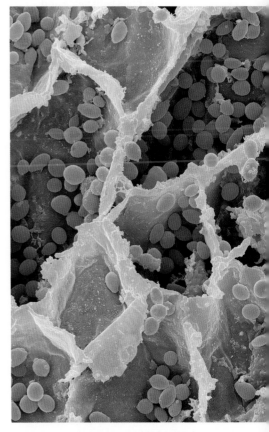

▲ Figure 3 Coloured scanning electron micrograph (SEM) of naturally occuring yeast cells (red) on the skin of a grape. In the processing of the grapes to make wine, the presence of the yeast is essential for the fermentation of the grapes that is part of the wine making process

allows waste gases to escape. Conditions within the vessel are monitored by probes. Because heat can build up as a waste product of metabolism, a cooling jacket surrounds the reaction vessel with cooling water flowing through it. Once the medium is used up, new medium can be added. Product removal may also occur leading out of the vessel.

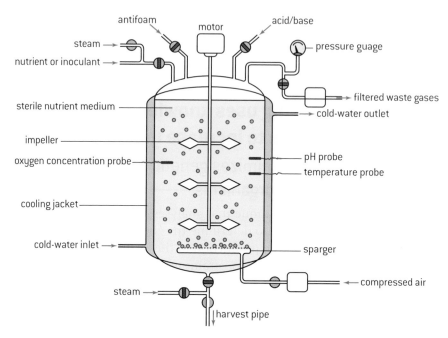

▲ Figure 4 A fermenter

There are two approaches to industrial fermentation

Fermentation is carried out by batch or continuous culture.

Mass culture of microorganisms is carried out in two ways in industry.

Batch culture is used for producing secondary metabolites; i.e., those which are not essential for the growth of the culture. In this case, inoculation of the medium is followed by the culture passing through all of the stages of the sigmoid growth curve. To begin the process, a fixed volume of medium is added to the closed fermentation vat. After inoculation, no further nutrients or microorganisms are added during the incubation period. The products are extracted only when they reach a high enough concentration.

In continuous culture, nutrients are added and products harvested at a constant rate. Conditions are monitored closely and efforts are made to keep conditions at a constant level so the process can continue over a long period.

Factors limiting industrial fermentation

Microorganisms in fermenters become limited by their own waste products.

A number of abiotic factors set limits to the activity of microorganisms in fermentation tanks. These can be due to the consumption of raw materials by the microorganism or by the production of waste products due to their activities.

- Carbon dioxide production can lower the pH affecting enzyme activity.

- Gas production can lead to pressure build-up possibly affecting reaction rates.

- Alcoholic fermentation can yield levels of alcohol which have an osmotic effect on cells.

- Oxygen levels can be depleted due to cellular respiration.

- Heat as a waste product of metabolism can raise the temperature of the reaction vessel.

Probes monitor conditions within fermenters

Probes are used to monitor conditions within fermenters.

In figure 5, oxygen concentration, pH, volume, foam levels and temperature probes are shown. These are the most commonly monitored variables in fermentation tanks. Computer-based probes gather data on these conditions. In batch fermentation, they can provide an indication of the stage of the production process. In continuous fermentation, they can signal to a technician actions to be taken to keep conditions within the favourable range.

▲ Figure 5 A system of probes is connected to the fermenter to monitor conditions within the vessel

Maintaining optimum conditions within fermenters

Conditions are maintained at optimal levels for the growth of the microorganisms being cultured.

Conditions are more likely to be monitored and kept at optimal levels in continuous culture. Such conditions include water content, temperature, pH, macro- and micro-nutrient levels, levels of waste products, cell density, dissolved oxygen content, dissolved carbon dioxide content, culture volume and culture mixing. The optimum level of each variable depends on the species.

The level of a variable is often influenced by a number of variables and is therefore important to monitor constantly. Consider the example of oxygen. Species differ in their tolerance of low oxygen. *Penicillium* is less tolerant of low oxygen than *Saccharomyces*. When concentrations of dissolved oxygen go below a critical value, then it becomes limiting. Dissolved oxygen is affected by the temperature and the nutrients being oxidized by the organism. Adding oxygen by aeration to a culture is not a simple matter as it generates foam which can limit production. Anti-foaming agents are often added to the reaction vessel.

 Deep-tank fermentation

Deep-tank batch fermentation in the mass production of penicillin.

In the early 20th century, efforts were concerted to find ways to mass produce penicillin. Initial experiments showed that *Penicillium notatum* grew best in shallow pans due to the need for aeration.

However, this did not produce significant enough yields to meet the demands for treatment of the casualties of World War II. Large-scale production was facilitated by deep-tank fermentation. This employed both a source of oxygen bubbled in to the tank and paddles to distribute the oxygen. The nutrient source for the *Penicillium* is corn steep liquor. This is the liquid produced by warming a vat of corn in water near 50 °C for two days.

Antibiotics are secondary metabolites in the sense that they are produced at a certain point in the life cycle of the microbe under certain conditions.

In the case of penicillin, optimum conditions are about 24 °C, slightly basic pH and a good oxygen supply. The product typically starts being formed about 30 hours after the start of the batch culture as nutrient concentrations begin to decline and continues for about six days after which the fermenter has to be drained and the liquid filtered. Using solvents, a crystalline precipitate is generated from the filtered liquid.

Industrial production of citric acid

Production of citric acid in a continuous fermenter by Aspergillus niger and its use as a preservative and flavouring.

Citric acid is an important food additive, both as a flavour enhancer and a preservative. Industrial production of citric acid relies on the fungus *Aspergillus niger*. While the greatest fraction of industrially produced citric acid is produced by batch fermentation, continuous fermentation has also been attempted. The optimal conditions for citric acid production are high dissolved oxygen concentration, high sugar concentration, an acidic pH and a temperature of about 30 °C. Citric acid is

produced in the Krebs cycle and so is referred to as a primary metabolite. If the culture medium is under-supplied with certain minerals such as iron, citric acid builds up in the reaction vessel.

After the contents of the fermentation vessel are filtered out, calcium hydroxide is added to the filtrate and solid calcium citrate precipitates out of solution. It can then be further treated chemically to yield citric acid.

Gram staining

Gram staining of Gram-positive and Gram-negative bacteria.

A traditional test used to classify bacteria is whether they are Gram-negative or Gram-positive, based on how they react to Gram-staining. The cell wall of Gram-positive bacteria consists of a thick layer of

peptidoglycan (a polymer consisting of amino acids and sugars). The greatest fraction of the Gram-positive cell wall is composed of peptidoglycan.

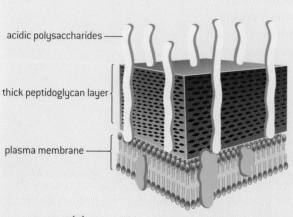

acidic polysaccharides

thick peptidoglycan layer

plasma membrane

(a) gram-positive: thick cell wall, no outer envelope

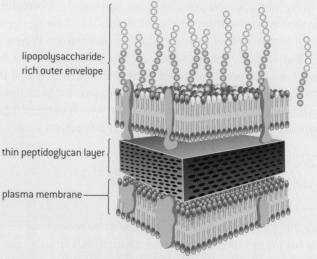

lipopolysaccharide-rich outer envelope

thin peptidoglycan layer

plasma membrane

(b) gram-negative: thinner cell wall, with outer envelope

▲ Figure 6

The cell wall of Gram-negative bacteria is much thinner – only about 20% peptidoglycan (see figure 6). Crystal violet binds to the outer membrane in Gram-negative bacteria and when alcohol is added, it washes away the outer membrane and the crystal violet stain with it. In contrast, the crystal violet binds to the multiple layers within the thick peptidoglycan layer which is not washed away by the alcohol and thus the colour persists.

Activity:

Gram-staining procedure

1 Prepare smears of *Bacillus cereus*, *Streptococcus fecalis*, *Escherichia coli* and *Micrococcus luteus*. Fix these preparations by heating over a bunsen burner.

2 Stain with crystal violet for about 30 seconds.

3 Rinse with water, then cover with Gram's iodine. Allow stain to act for about 30 s.

4 Rinse with water, then decolorize with 95% alcohol for 10–20 s.

5 Rinse with water, then counterstain with safranin for 20–30 s.

6 Rinse with water and blot dry. Gram-negative bacteria will be pink. Gram-positive bacteria will be blue or violet.

7 Depending on local restrictions, you might choose to examine prepared slides of Gram-negative and Gram-positive bacteria.

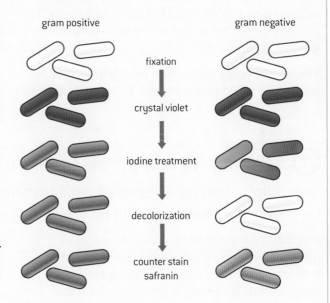

🌐 Biogas production

Biogas is produced by bacteria and archaeans from organic matter in fermenters.

Biogas refers to the combustible gas produced from the anaerobic breakdown of organic matter such as manure, waste plant matter from crops and household organic waste. Depending on the construction of the fermenter, biogas is mostly methane with some carbon dioxide, though other gases may be present.

Three different communities of anaerobic microbes are required. The first group convert the raw organic waste into a mixture of organic acids, alcohol, hydrogen and carbon dioxide. The second group use the organic acids and alcohol from the first stage to produce acetate, carbon dioxide and hydrogen. These first two communities are Eubacteria. The last group are Archaea called methanogens. The methanogens produce methane by one of the following two reactions:

$$CO_2 + 4H_2 \rightarrow CH_4 + 2H_2O$$

(reduction of carbon dioxide to methane)

$$CH_3COOH \rightarrow CH_4 + CO_2$$

(splitting ethanoic acid to form methane and carbon dioxide)

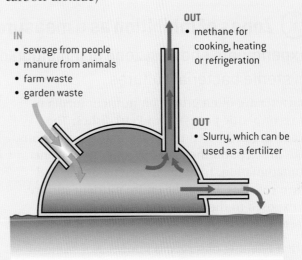

▲ Figure 7 Methane generator. Conditions inside must be anaerobic

563

Producing biogas in the classroom

Production of biogas in a small-scale fermenter.

Figure 8 shows an example of a set-up of a biogas generator. Mylar balloons are the ones that are commonly filled with helium as party balloons. The feedstock bottle should be plastic rather than glass due to the risk of explosion. The tube clamps can be used to prevent gas leakage when the balloon is to be disconnected from the set-up. The balloon should be sealed to the glass tube with insulating tape.

The rate of biogas generated by different feedstocks could be compared. Relative quantities of organic waste and water could be compared in terms of rate of biogas production.

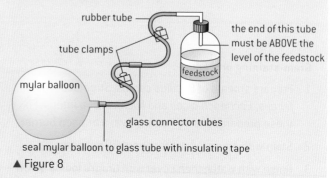
▲ Figure 8

Serendipity and the discovery of penicillin

Serendipity has led to scientific discoveries: the discovery of penicillin by Alexander Fleming could be viewed as a chance occurrence.

Serendipity is defined as a lucky accident or the situation where something good or useful is revealed when it was not being specifically searched for. However, it is only useful if the observer recognizes its value.

Alexander Fleming was a Scottish medical doctor and scientist who spent the early part of his career searching for anti-bacterial agents. In 1928, he was investigating the properties of the bacterium *Staphylococcus*. After returning from an extended holiday, he noticed that one of his bacterial plates was contaminated with fungus and the zone around the fungus on the plate appeared to have no bacteria while further away from the fungus, bacteria grew on the plate. Fleming was wise enough to connect the unexpected observation to his earlier studies of anti-bacterial agents.

He proceeded to grow the mold in pure culture and then test it on a number of pathogenic bacteria and discovered that it had an antibiotic effect on several species.

Zones of inhibition as a measure of bactericide effectiveness

Experiments showing zone of inhibition of bacterial growth by bactericides in sterile bacterial cultures.

Bacteria are often grown on nutrient media in glass or plastic plates called Petri dishes. The plates are incubated under laboratory conditions. Lids are kept on the plates in order to prevent contamination. Individual bacteria divide and form colonies, but if the entire nutrient surface is exposed to the bacterium, then a "lawn" of bacteria is grown.

What Fleming observed is known as a "zone of inhibition"; that is, a region on a bacterial lawn

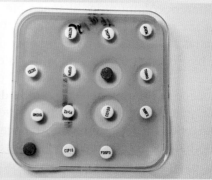
▲ Figure 9

where an anti-bacterial effect prevents the growth of bacteria. The consequence is an often circular-shaped disc region. The diameter of the zone of inhibition is a measure of the strength of the anti-bacterial agent. In figure 9 a plate inoculated with *Pseudomonas aeruginosa* bacteria had various types of antibiotic discs placed on the surface to determine which is the most effective. This is a species of bacteria that rarely infects healthy individuals but is a major cause of infections acquired by people in hospitals.

This technique can be modified by students to investigate the effectiveness of various anti-bacterial agents. Absorbent filter paper can be cut into disc shapes by a hole puncher. The discs can be soaked in disinfectants, for example, and placed on to a plate that has been inoculated with bacteria.

B.2 Biotechnology in agriculture

Understanding

→ Transgenic organisms produce proteins that were not previously part of their species' proteome.
→ Genetic modification can be used to overcome environmental resistance to increase crop yields.
→ Genetically modified crop plants can be used to produce novel products.
→ Bioinformatics plays a role in identifying target genes.
→ The target gene is linked to other sequences that control its expression.
→ An open reading frame is a significant length of DNA from a start codon to a stop codon.
→ Marker genes are used to indicate successful uptake.
→ Recombinant DNA must be inserted into the plant cell and taken up by its chromosome or chloroplast DNA.
→ Recombinant DNA can be introduced into whole plants, leaf discs or protoplasts.
→ Recombinant DNA can be introduced by direct physical and chemical methods or indirectly by vectors.

Applications

→ Use of tumour-inducing (Ti) plasmid of *Agrobacterium tumefaciens* to introduce glyphosate resistance into soybean crops.
→ Genetic modification of tobacco mosaic virus to allow bulk production of Hepatitis B vaccine in tobacco plants.
→ Production of Amflora potato (*Solanum tuberosum*) for paper and adhesive industries.

Skills

→ Evaluation of data on the environmental impact of glyphosate-tolerant soybeans.
→ Identification of an open reading frame (ORF).

Nature of science

→ Assessing risks and benefits associated with scientific research: scientists need to evaluate the potential of herbicide resistant genes escaping into the wild population.

Transgenic organisms

Transgenic organisms produce proteins that were not previously part of their species' proteome.

The complete set of proteins that a cell or organism can make is referred to as its proteome. Proteins are key components of a cell's structure and carry out most cellular functions. Sometimes genetic engineers seek to extend the proteome of an organism for the purposes of a biotechnological application. If the addition to the proteome is due to the addition of a gene from a different organism, then the modified organism is said to be transgenic.

Figure 1 shows glo-fish©, the first genetically modified organism to be sold as a pet. These transgenic fish have had the gene for the production of green fluorescent protein introduced into their genome. The original organism that was the source of the gene was *Aequorea victoria*, a jellyfish.

The protein SRY is a transcription factor which triggers the expression of genes that lead to the development of male characteristics. Figure 2 shows a transgenic female mouse (on the right) that has been genetically modified to express the protein SRY within its proteome. It has caused the female mouse to develop the same genetalia as the male on the left.

▲ Figure 1

▲ Figure 2

Genetically modified crop plants

Genetically modified crop plants can be used to produce novel products.

A novel product refers to the presence of a protein or phenotype that was not previously found in the species.

The production of "golden rice" involved the introduction into rice of three genes, two from daffodil plants and one from a bacterium, so that the orange pigment β-carotene is produced in the rice grains. β-carotene is a precursor to vitamin A. The development of golden rice was intended as a solution to the problem of vitamin A deficiency, which is a significant cause of blindness among children globally.

Corn has been genetically modified to produce the CRY toxin due to the insertion of a gene from *Bacillus thuringiensis*. As a consequence, the plant becomes unpalatable to the European corn borer, an insect pest that significantly reduces crop yields.

Overcoming environmental resistance in crops

Genetic modification can be used to overcome environmental resistance to increase crop yields.

Limiting factors affecting crop plant growth can be biological or non-biological.

Biotic factors include competition from weed species, predation by insects and infection by pathogens.

Resistance to the herbicide glyphosate has been introduced to crop plants such as soybeans as part of a strategy for reducing competition with weeds.

The introduction of genes for the production of Bt toxin into corn is part of a strategy for reducing predation by insects such as the western corn rootworm. Non-transgenic roots will suffer considerable damage from pests, but transgenic roots suffer little damage as they have resistance to the rootworm due to the expression of the Bt toxin.

In Hawaii, researchers genetically modified papaya plants to be resistant to papaya ringspot virus by leading the plant to express the gene for the virus coat triggering a protective response to the virus.

Abiotic factors that limit crop growth include such factors as drought, frost, low soil nitrogen and high soil salinity.

DroughtGard® maize contains the gene for "cold shock protein B" (*cspB*) from the bacterium *Bacillis subtilis* that enables it to retain water during drought conditions.

A gene from Thale cress (*Arabidopsis*), AtNHXI, codes for the production of a membrane protein that captures excess sodium into plant vacuoles. Peanut plants have been genetically modified to express this gene allowing them to grow in saline soils that would otherwise limit crop output.

Components of the gene construct

The target gene is linked to other sequences that control its expression.

To carry out genetic modification, more than the gene must be inserted. Additional sequences are necessary to control the expression of the gene.

Most commonly, sequences such as a eukaryotic promoter must be added upstream in the construct and a eukaryotic terminator sequence must be included downstream in the construct. The construct also often contains a second gene called a recognition sequence which allows engineers to confirm that the construct has been taken up by the host DNA and is being expressed.

In some cases, specific additional sequences have to be added. Consider the example of genetically modifying sheep to express human proteins such as alpha-1-antitrypsin in the sheep's milk. In this case, a specific promoter sequence that will ensure that the gene is expressed in milk is necessary in creating the gene construct. In addition, a signal sequence has to be added to ensure that the protein is produced by ribosomes on the endoplasmic reticulum rather than by ribosomes that are free in the cytoplasm. This is to ensure that the alpha-1-antitrypsin protein is secreted by the mammary cells rather than released intracellularly.

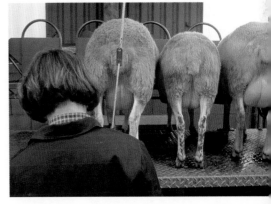

▲ Figure 3 Transgenic sheep, awaiting milking. The sheep are offspring of ewes which have a human gene responsible for the production of the protein alpha-1-antitrypsin (A1AT) incorporated into their DNA. A1AT is produced in mammary cells, and secreted in the sheep's milk. The A1AT can then be isolated and used to treat hereditary A1AT deficiency in humans, which leads to the lung disease emphysema

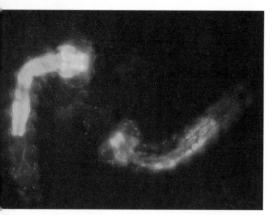

▲ Figure 4

Marker genes

Marker genes are used to indicate successful uptake.

In addition to the target gene, an additional gene is often added to provide some way to indicate that uptake of the target gene has occured. Some markers are based on artificial selection. In this case, the gene is called a selectable marker. The marker gene often confers antibiotic resistance. Those bacteria that have taken up the marker gene and the target gene construct will survive exposure to antibiotic. These can then be cultured separately.

The gene for the production of green fluorescent protein (GFP) is also used as a marker. Figure 4 shows mosquitos that have been genetically modified to resist being hosts for the malarial parasite. The donor gene has been linked to the gene for GFP so that the transgenic mosquitos can be detected under a microscope.

Recombinant DNA

Recombinant DNA must be inserted into the plant cell and taken up by its chromosome or chloroplast DNA.

Recombinant DNA is a molecule that has been manipulated so that it contains sequences from two or more sources.

In order to create a transgenic organism, the recombinant DNA must be taken up by the host cell.

In order for the gene to be expressed, it has to be taken up into a chromosome. In plant cells, it can also be taken up by a chloroplast. This process of uptake and expression of the donor DNA is called transformation. The new genes can be inserted into the DNA of the chloroplasts. The major advantage of inserting into chloroplasts is that the chloroplast DNA is not transmitted through pollen which prevents gene flow from the genetically modified plant to other plants. Transformation usually requires the use of a vector.

Different targets for genetic transformation

Recombinant DNA can be introduced into whole plants, leaf discs or protoplasts.

Once the transgene has been introduced into the host cell, the production of a whole plant from the transformed cell has to be performed.

Protoplasts are plant cells that have had their cell walls removed. Transformation by *Agrobacterium* was initially attempted on protoplasts. While this was somewhat successful, the difficulty of obtaining sufficient high quality protoplasts combined with the difficulty of growing whole plants from protoplasts led to the search for other methods.

The leaf disc methods involves incubating leaf cut-outs with *Agrobacterium* containing a plasmid with the target gene along with an antibiotic resistance gene. The leaf discs are then transferred to a plate containing two different antibiotics which ensures that only transformed cells will grow. The transformed cells are then cultured and treated in such a way so that roots and shoots develop from the discs.

Different methods of genetic transformation

Recombinant DNA can be introduced by direct physical and chemical methods or indirectly by vectors.

Genes can be introduced into plants in a number of different ways including microinjection, electroporation, virus infection, ballistic incorporation and incubation with *Agrobacterium tumefaciens*.

Incubating host cells at cold temperatures in calcium chloride solution and then heat shocking the solution is a chemical method that was one of the original methods for transforming cells.

Electroporation is a physical method that involves applying an external electric field that leads to the formation of temporary pores in the cell membrane allowing recombinant DNA to get into a cell.

Microinjection is another physical method of introducing genes. A pipette is used to aspirate and hold a cell in a fixed position while a needle is used to inject genes of interest.

In biolistics, metal particles coated with the gene of interest are fired at an entire plant.

A vector is a virus, a plasmid or some other biological agent that transfers genetic material from one cell to another. In the next section the use of the Ti plasmid vector is explained. On page 570 the use of a virus as a vector is explained.

 The use of Ti plasmid as a vector

Use of tumour-inducing (Ti) plasmid of *Agrobacterium tumefaciens* to introduce glyphosate resistance into soybean crops.

One way to introduce transgenes into plants is to use *Agrobacterium tumefaciens*. This is a species of bacteria that has a plasmid, called the Ti plasmid, that causes tumours in the plants it infects.

The glyphosate resistance gene is inserted into the Ti plasmid along with an antibiotic resistance gene. The construct is then re-inserted into an *A. tumefaciens* bacterium. Plant cells are then exposed to the transgenic bacterium and cultured on a plate containing antibiotic. The only plant cells that grow are those that have taken up the plasmid. The others are killed by antibiotic.

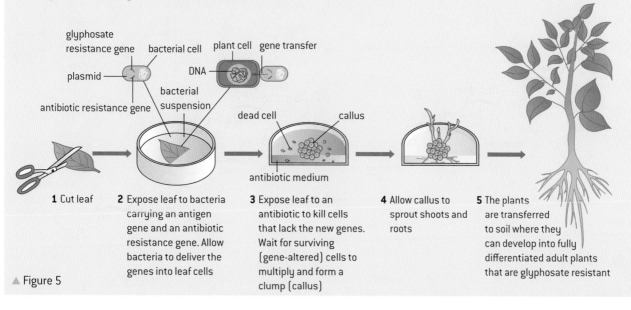

▲ Figure 5

1 Cut leaf

2 Expose leaf to bacteria carrying an antigen gene and an antibiotic resistance gene. Allow bacteria to deliver the genes into leaf cells

3 Expose leaf to an antibiotic to kill cells that lack the new genes. Wait for surviving (gene-altered) cells to multiply and form a clump (callus)

4 Allow callus to sprout shoots and roots

5 The plants are transferred to soil where they can develop into fully differentiated adult plants that are glyphosate resistant

Edible viruses

Genetic modification of tobacco mosaic virus to allow bulk production of Hepatitis B vaccine in tobacco plants.

Vaccination programmes are often impacted by lack of access to remote areas as well as the challenge of refrigerating vaccine preparations. One initiative has been to develop edible vaccines by incorporating antigens into plant matter. One attempt involved genetically modifying tobacco mosaic virus with antigens from the Hepatitis B virus and then infecting tobacco plants.

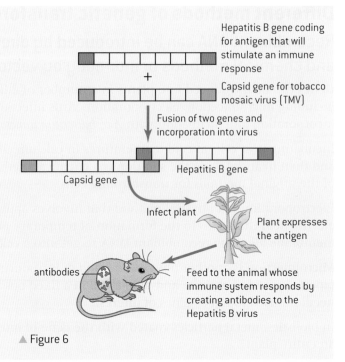

▲ Figure 6

Potatoes modified to produce starch containing only amylopectin

Production of Amflora potato (*Solanum tuberosum*) for paper and adhesive industries.

Potatoes are used in industry as a source of starch. Starch can be used for a number of purposes including use as an adhesive. Normally, potato starch consists of two different types of starch polymers (see figure 7). 80% of potato starch consists of the branched chain amylopectin and 20% is the straight chain form, amylose.

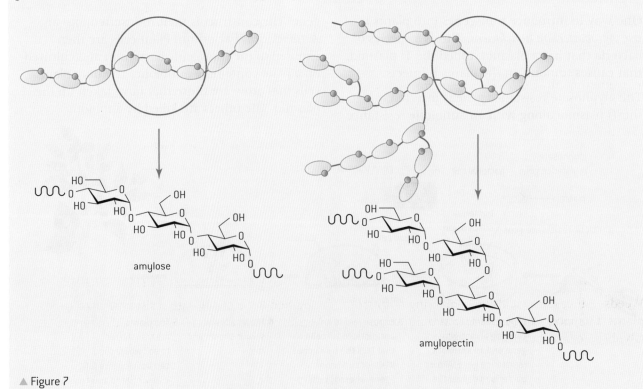

amylose

amylopectin

▲ Figure 7

When a starch mixture is heated and then cooled, it tends to form a gel which is undesirable for some applications such as paper manufacturing and adhesive production. To prevent this, conventional methods use chemical treatment to remove the amylose.

The company BASF produced a genetically modified potato where one of the genes involved in the production of amylose was deactivated. The gene product was "granule bound starch synthase". The method used was antisense technology. This involves inserting a version of the gene that is inverted such that it produces the antisense mRNA. The result would be that the normal sense strand would be produced as well as the antisense strand. The two would bind and the double stranded mRNA molecule gets degraded rather than being translated to form protein (figure 8).

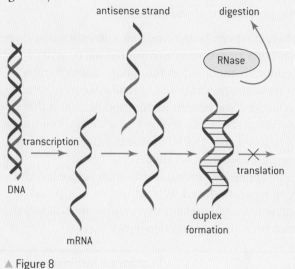

▲ Figure 8

 ## Assessing risks of transgenes entering wild populations

Assessing risks and benefits associated with scientific research: scientists need to evaluate the potential of herbicide resistant genes escaping into the wild population.

Gene flow is the movement of genes or genetic material from one population to another. In plant populations, it can occur through the transfer of pollen between related species.

Herbicide resistant genetically modified (GM) crops are the most common type of genetically modified crop grown.

The potential flow of transgenes from the GM crop to non-GM crops and from the GM crop to wild weed populations is an economic concern. If the transgene becomes expressed in the wild population, then controlling the effect of the weed population within a crop area would become difficult. If the transgene is for insect resistance, then ecological balance could be disrupted.

Assessing the risk requires estimating how frequently gene flow occurs, determining whether the transgene becomes expressed and determining the changes to the phenotype of the plant. One strategy for reducing risk is to incorporate "mitigator" genes with the transgene which is designed to reduce the success of any hybrid plants that might be accidentally created. Another strategy is to transform chloroplasts rather than nuclear DNA as the chloroplast DNA is not expressed in pollen.

 ## Evaluating the environmental impact of a GM crop

Evaluation of data on the environmental impact of glyphosate-tolerant soybeans.

Weeds reduce crop yields by competing with crop plants for space, sunlight, water and nutrients. Glyphosate is a chemical that kills a very broad spectrum of plants. Soybeans as well as a number of other crop species have been genetically modified to be glyphosate resistant allowing farmers to use a single broad-spectrum herbicide.

There are two potential environmental aspects to consider: the environmental risks of the genetic modification of a crop plant and the environmental risks of the widespread use of glyphosate as an herbicide that is encouraged by the prevalence of the GM crop.

There has been broad academic consensus that there has been at least some environmental benefit of genetically modified glyphosphate-tolerant crops in replacing previous widespread systems of weed reduction. The benefit to this modification is that crop weeds can be controlled by herbicide without the risk of reduced crop yields because the crop is resistant to the herbicide. Further the level of herbicide required to be applied is lower (table 1) than before the GM crop was introduced. While the data is disputed, many researchers claim that glyphosate is nearly the least toxic pesticide used in agriculture.

Geographic region	% change in herbicide use in comparison to non-GM crops in 1997
Heartland	−23%
Northern Crescent	−15%
Mississippi Portland	−11%
Southern Seaboard	−51%

▲ Table 1 Percentage reduction in the amount of herbicide applied in genetically modified crops over traditional crops in various regions of the US

Tillage is the practice of turning over the soil and has been commonly practised as a component of weed management strategies. The loss of top soil and erosion is one of the consequences of tillage. Glyphosate and glyphosate-resistant crops have enabled significantly less tillage and therefore preserved soil fertility. This has reduced the fossil fuel use required for tillage and reduced the need for inputs required to supplement soil fertility. Figure 9 shows the growth in the area cultivated with GM soybeans in Argentina with a corresponding growth in no-till agriculture.

Glyphosate resistance (GR) is under intense selection pressure given the widespread use of the crop and the reduced use of other herbicides. The environmental consequences of resistant weeds will include reduced crop yields for the same inputs and the need to increase the use of tillage and alternative herbicide formulations.

A review conducted in 2002 by the European Union reached the conclusion that there was little data to support claims of health impacts of glyphosate on humans. Some studies suggest that other components of the herbicide mixture used in combination with glyphosate did have environmental impacts. The Australian government has banned the use of some formulations of glyphosate near water.

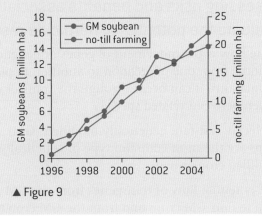

▲ Figure 9

Open reading frames

An open reading frame is a significant length of DNA from a start codon to a stop codon.

When the DNA of an organism has been sequenced, researchers will then look for the location of genes. The starting point for this search is to look for open reading frames.

The search for open reading frames (ORF) depends on the following concepts:

- There are 64 triplets of bases that are called codons.
- 61 codons are used to code for an amino acid.
- There are 3 three stop codons (TAA, TAG and TGA) that signal the end of an open reading frame.

- There is one start codon (ATG) that signals the start of an open reading frame and also codes for an amino acid.

Open reading frames in DNA are identified by searching for base sequences long enough to code for the amino acids in a polypeptide between a start codon and one of the three stop codons. In other words, they look for sequences where stop codons are absent. Researchers usually look for a base sequence long enough to code for one hundred amino acids. The average size of an ORF in *E. coli* is 317 amino acids long.

 Identifying open reading frames

Identification of an open reading frame (ORF).

A short base sequence is shown below.

AATTCATGTTCGTCAATCAGCACCTTTGTGGTTC
TCACCTCGTTGAAGCTTTGTACCTTGTTTGCGGT
GAACGTGGTTTCTTCTACACTCCTAAGACTTAA
TAGCCTGGTG

1   Find the first start codon and the first stop codon after it in the sequence.

  a) State how many bases there are before the start codon. [1]

  b) State how many codons there are in the open reading frame that you have found. [1]

  c) Calculate how many amino acids are encoded in this open reading frame. Show how you worked out your answer. [3]

2   Researchers need to distinguish between open reading frames that code for polypeptides and random base sequences in the genome that by chance have start codons followed by an extended sequence without a stop codon.

  a) Calculate the percentage chance of finding a start codon in a piece of DNA with a random sequence of ten base pairs. [2]

  b) If the start codon is found in a random base sequence, calculate the percentage chance that the next triplet of bases codes for an amino acid. [1]

  c) Calculate the percentage chance that the next 100 triplets all code for amino acids. [2]

Data-based questions: Determining an open reading frame

Once the sequence of bases in a piece of DNA has been determined, a researcher may want to locate a gene. To do this, computers search through the sequences looking for open reading frames. An open reading frame is one that is uninterrupted by stop sequences and could therefore code for the production of a protein. The stop codons are UGA, UAA and UAG.

1   State the number of codons in the genetic code. [1]

2   Determine the fraction of codons that are stop codons in the genetic code. [2]

3   In table 2, the codons could start with the first, second or third base. These correspond to three different reading frames (RF1, RF2 or RF3). Determine which of the reading frames, 1, 2 or 3, might be an open reading frame. [2]

DNA 3'	A T T A A C T A T A A A G A C T A C A G A G A G G G C T A G T A C
mRNA 5' RF1	U A A U U G A U A U U U C U G A U G U C U C U C C C G A U C A U G
RF2	A A U U G A U A U U U C U G A U G U C U C U C C C G A U C A U G
RF3	A U U G A U A U U U C U G A U G U C U C U C C C G A U C A U G

▲ Table 2

573

Activity

Alcanivorax borkumensis is a rod-shaped bacterium that utilizes oil as an energy source. It is relatively uncommon but quickly dominates the marine microbial ecosystem after an oil spill. Scientists sequenced the genome of this bacterium in an effort to identify the genetic aspects of its oil digesting ability. The entire genome can be accessed from the database GenBank.

Visit GenBank and search by genome to locate the genome of this organism. Click on FASTA to identify the organism's GI number. It is listed in the title. (GI number #110832861). View the genome.

Go to the open reading frame finder (http://www.ncbi.nlm.nih.gov/projects/gorf/). Enter the GI number and specify the range of bases that you are going to search.

Perhaps as a class, the genome can be divided up into 2000 bp pieces. Share information with one another about the open-reading frames identified.

TOK

What knowledge issues are created by the rapid growth in the amount of available data and information?

The technology of DNA sequencing and bioinformatics has evolved at a rapid pace. In 2009, the biggest problem for researchers was developing solutions to improve the sequencing of DNA. Time and cost limited the production of DNA sequence information. By 2013, researchers can sequence a whole human genome within a single day. The challenge has now shifted from sequencing DNA to managing and analysing the extraordinary volume of sequence data that is being produced. It has been estimated that five months of analysis are needed for every month's worth of data generated.

Identifying target genes

Bioinformatics plays a role in identifying target genes.

Bioinformatics is the use of computers to investigate biological phenomenon. Open reading frames are identified by subjecting genomic information held in a database to searches to find extended sequences without stop codons.

Once an open reading frame is identified, a BLAST search can be conducted. The acronym refers to **B**asic **L**ocal **A**lignment **S**earch **T**ool. A BLASTn search would search through databases to determine if an open reading frame with a similar nucleotide sequence existed in another species. A BLASTx search would search a protein database based on the translated sequence of the open reading frame.

Alternatively, if a researcher has found a protein and wants to determine the location of a gene, they can conduct a tBLASTn search using a computer search of multiple genomes using the translated sequence to search for potential genes that could have been transcribed to produce the protein. All three methods play a role in identifying target genes.

B.3 Environmental protection

Understanding

→ Responses to pollution incidents can involve bioremediation combined with physical and chemical procedures.

→ Microorganisms are used in bioremediation.

→ Some pollutants are metabolized by microorganisms.

→ Cooperative aggregates of microorganisms can form biofilms.

→ Biofilms possess emergent properties.

→ Microorganisms growing in a biofilm are highly resistant to antimicrobial agents.

→ Microorganisms in biofilms cooperate through quorum sensing.

→ Bacteriophages are used in the disinfection of water systems.

 Applications

→ Degradation of benzene by halophilic bacteria such as *Marinobacter*.

→ Degradation of oil by *Pseudomonas*.

→ Conversion by *Pseudomonas* of methyl mercury into elemental mercury.

→ Use of biofilms in trickle filter beds for sewage treatment.

 Skills

→ Evaluation of data or media reports on environmental problems caused by biofilms.

 Nature of science

→ Developments in scientific research follow improvements in apparatus: using tools such as the laser scanning microscope has led researchers to deeper understanding of the structure of biofilms.

Methods used to address pollution incidents

Responses to pollution incidents can involve bioremediation combined with physical and chemical procedures.

When chemicals are released to the environment by accident or through carelessness, the result can be significant in terms of ecological disruption. Bioremediation is the use of microbes to remove environmental contaminants from water or soil. In this section, we consider bioremediation strategies for addressing pollutants such as benzene, petroleum oil, heavy metals and sewage.

Not all pollution incidents can be addressed solely through bioremediation. Bioremediation may be undesirable in the case of heavy metals because these need to be removed from the food chain. In such cases phytoremediation, which relies on plants, might be employed. The heavy metals can bioaccumulate in the biomass of the crop. The crop can then be incinerated to concentrate the metal and then the metal can either be recycled or properly contained.

There are a number of physical and chemical procedures that can be combined with bioremediation to respond to pollution incidents.

- Physical methods for oil spills include the use of scrubbers, detergents and dispersants.

- Chemical-contaminated soil can be removed and incinerated to degrade volatile organic chemicals.

- Soil can be removed, crushed, sifted and then suspended in water that includes chemicals that will aid in dissolving the chemicals into the water. The chemical-contaminated water can then be purified separately.

- Oxidizing chemicals such as ozone and peroxide are sometimes injected into soils to accelerate the destruction of toxic organic compounds.

Microorganisms have properties that make them useful for bioremediation

Microorganisms are used in bioremediation.

Bacteria and archaeans are useful in bioremediation because they can multiply very quickly by binary fission and they are varied in their metabolism. They carry out a wider range of chemical reactions, especially inorganic reactions, than any other group of organisms. There is often a species of prokaryote that will perform the necessary reaction in a bioremediation process.

Figure 1 shows a biopile. This is a method for addressing pollution in soil. A bulking agent such as compost, hay or other nutrient source is dug into the piles and the piles are constantly watered. The microbial community which flourishes digests the contaminants.

Bioremediation relies on microorganism metabolism

Some pollutants are metabolized by microorganisms.

Microorganisms can use pollutants as energy sources, carbon sources and electron acceptors in cellular respiration.

The bacterium *Dehalococcoides ethenogenes* (shown in red in figure 2) has been used to break down chlorinated solvents in soil. It uses the chlorine compounds as electron acceptors in cellular respiration.

The bacterium *Geobacter sulfurreducens* uses uranium as an electron acceptor converting it from a soluble to an insoluble form, which allows the uranium to settle out and be collected.

Figure 3 shows the bacterium *Acidovorax sp.* (yellow) partially coated with iron (orange). This bacterium is able to precipitate iron and arsenic out of the soil and bind it. Due to this, it is being investigated as a means of reducing the amount of arsenic present in rice fields.

▲ Figure 1 Soil undergoing bioremediation at Fawley Refinery, an oil refinery and chemical plant located in Fawley, Hampshire, UK

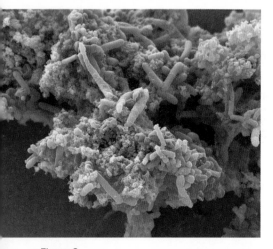

▲ Figure 2

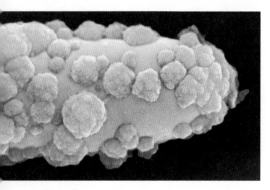

▲ Figure 3

Microorganisms can form biofilms

Cooperative aggregates of microorganisms can form biofilms.

A biofilm is a colony that coats a surface as a consequence of cooperation between individual cells. Members of a biofilm colony secrete signalling molecules that recruit independent, or planktonic, cells into the colony. They also secrete molecules that facilitate the aggregate adhering to the surface and facilitate individual cells sticking together. On their cell membranes, cells produce protein channels that facilitate the exchange of materials with other members of the colony. While biofilms normally form on solid surfaces, they can form on the surface of fluids. Sometimes, they can be composed of a community of organisms including bacteria, archaea, protozoa, algae and fungi. Dental plaque is a biofilm that can contain up to 500 taxa of microorganisms while the biofilm that forms in the lungs of patients afflicted with cystic fibrosis is often composed of a single species: *Pseudomonas aeruginosa*.

Figure 4 shows a magnified view of a bristle from a used toothbrush. The surface of the bristle is covered in a biofilm of cooperating bacteria. Figure 5 shows a biofilm inside a catheter. A catheter is a tube used in medical treatment to drain urine or maintain a connection to the bloodstream. The centre part is meant to be hollow but is covered in a white-coloured biofilm.

▲ Figure 4 Biofilm on the bristle of a used toothbrush

Emergent properties

Biofilms possess emergent properties.

Properties that emerge from the interaction of the members of a collective that are not present in the single cell form are referred to as emergent properties.

In biofilms, the ability of the cells to self-organize into a complex structure is an emergent property. Members of the colony secrete a chemical known as exopolysaccharide (EPS) that forms into a matrix that holds the colony together and protects it. This matrix is an emergent property.

Increased resistance to antibiotics; signalling between members of the colony; increased virulence; the formation of channels for water flow inside the colony; and the ability of cells to use the matrix to move leading to the colony itself moving are all considered emergent properties.

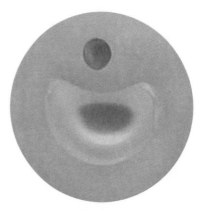

▲ Figure 5 Biofilm formed on the inside of a catheter

Biofilms resist antimicrobial agents

Microorganisms growing in a biofilm are highly resistant to antimicrobial agents.

Hospital acquired infections, or nosocomial infections, are commonly caused by biofilms. Increased resistance to antibiotics sometimes occurs in biofilms and is of concern to infection control officers within hospitals.

There are a number of proposed mechanisms for biofilm antibiotic resistance. In part, the resistance is due to the exopolysaccharide (EPS) matrix providing a physical barrier to the entry of the antibiotic into the colony.

Antibiotics often act on mechanisms that inhibit cell division. In some biofilms, limited supplies of nutrients leads to a suppression of the collective division rate which minimizes the effect antibiotics can have. This can especially be true of individuals deeper into the colony.

Quorum sensing

Microorganisms in biofilms cooperate through quorum sensing.

Quorum sensing is a system of behaviours that are triggered as a function of population density. It is observed in a diverse range of organisms.

In bacteria that form biofilms, gene expression can be affected by population density. Signalling molecules released by one cell bind to receptor molecules on another cell and lead to the expression of genes that are likely to facilitate the development of the biofilm. When population density is low, the concentration of the signalling molecule is low and is insufficient to trigger coordinated behaviour. When the population passes a threshold level; i.e., when the quorum is achieved, the concentration of the signalling molecule reaches a critical concentration and the behaviour becomes coordinated.

The pathogen *Pseudomonas aeruginosa* uses quorum sensing to coordinate movement, EPS production, cell aggregation and the formation of biofilms.

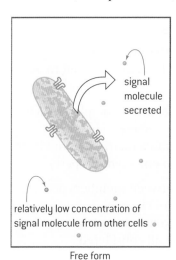
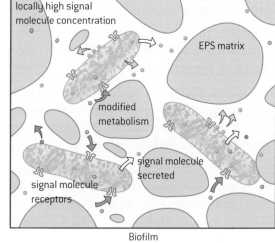

Free form

Biofilm

▲ Figure 6

Using viruses to kill bacteria in water systems

Bacteriophages are used in the disinfection of water systems.

When bacteria produce a biofilm, they can be difficult to eradicate. The control of biofilms within water systems is essential.

Some of the damage that can be done includes:

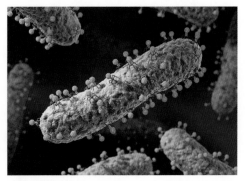

▲ Figure 7 Bacteriophages (pink) shown infecting a population of bacteria shown as green

- Anaerobic sulphate reducing bacteria produce sulphuric acid which can corrode pipes.

- Biofilms can affect heat exchange in systems where the release of waste heat to the environment is important.

- A proliferating biofilm can reduce the diameter of a pipe. This results in frictional drag, which lowers water pressure which leads to a need for increased pumping power.

Bacteria can be difficult to kill when they form a biofilm. The outer layer of bacteria in these biofilms can be killed by disinfectant, but the inner bacteria are sheltered.

Viruses solve this problem because they spread through the entire biofilm community. Viruses which attack bacteria are known as bacteriophages and they are specific to certain bacteria.

One study achieved the greatest success in killing biofilms by using a combination of bacteriophages and chlorine. An initial treatment with viruses followed by chlorine killed 97 percent of biofilms within five days of exposure while chlorine alone removed only 40 percent.

In addition, there may be specific pathogenic bacteria that are living in the bacterial community, such as coliform bacteria. Bacteriophages that are specific to the pathogen can be added to ensure reduction of the particular pathogen. The T4 bacteriophage is specific to *E. coli*.

🌐 Bioremediation in saline conditions

Degradation of benzene by halophilic bacteria such as *Marinobacter*.

The production of oil in marine environments generates volumes of saline (salty) wastewater that is contaminated with hydrocarbons such as benzene and toluene. Benzene (figure 8) is of particular concern as it can persist in the environment for a long time, is moderately soluble in water and is carcinogenic; i.e., it can lead to cancer. Bioremediation becomes difficult in this case as the salt content may be so high in the waste water that it kills most populations of bacteria.

Some archaea are adapted to living in extreme environments such as highly saline water (figure 9). They are referred to as halophiles. This adaptation has been useful in the bioremediation of saline wastewater. One species of halophilic archaea, *Marinobacter hydrocarbonoclasticus* has been shown to be able to fully degrade benzene.

▲ Figure 9 The colour in this salt pan pool is a indicator of the presence of a population of halophilic bacteria

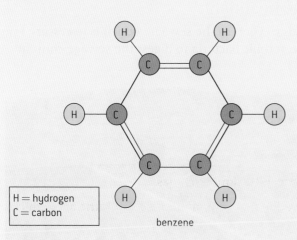

H = hydrogen
C = carbon

benzene

▲ Figure 8 Benzene molecule

Bioremediation of oil spills

Degradation of oil by *Pseudomonas*.

In natural environments, some petroleum seeps through cracks and vents in the ocean floor. Some members of the genus *Pseudomonas* thrive in these communities as they can use crude oil as an energy and carbon source. Clean-up at oil spills will often involve seeding the spill with *Pseudomonas*. These microbes also require substances such as potassium and urea as nutrients to metabolize the oil at a faster rate. These nutrients are often sprayed on to an oil spill to aid the bacteria in their work. Figure 10 shows a population of bacteria degrading a droplet of oil suspended in water.

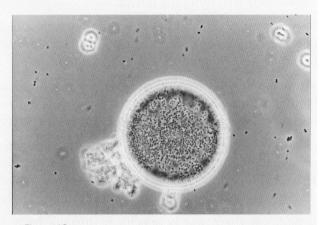

▲ Figure 10

Bioremediation of methyl mercury

Conversion by *Pseudomonas* of methyl mercury into elemental mercury.

Mercury ends up in garbage dumps as a component of some paints and some types of light bulbs. Elemental mercury is converted in this environment to the highly toxic organic methyl mercury by the bacterium *Desulfovibrio desulfuricans*. This form of mercury more easily enters food chains because it adheres to cell membranes and can dissolve in the cell membrane. It can bioaccumulate within the biomass of organisms and it can biomagnify up the food chain.

The bacterium *Pseudomonas putida* can convert the methyl mercury to methane and the mercury ion. Other bacteria then use the soluble mercury ion as an electron acceptor resulting in insoluble elemental mercury being reformed.

In a bioreactor, such elemental mercury can be separated from waste water as it is insoluble and will sink due to its density.

Biofilms used in trickle filter beds

Use of biofilms in trickle filter beds for sewage treatment.

The consequence of not treating sewage and allowing it to flow into watercourses is nutrient enrichment, or eutrophication of bodies of water. This favours algal blooms. When the mats of algae die, it leads to a loss of oxygen, because of bacterial activity on the dead organic matter. This is called biological oxygen demand.

Many sewage treatment plants make use of biofilms to address eutrophication. A trickling filter system has a rock bed that can be up to 2 metres deep. The rocks are colonized by a biofilm of aerobic bacteria. Sewage water is sprayed onto the rocks. The process of spraying adds oxygen to the sewage, which is necessary for the aerobic bacteria to digest the sewage content.

▲ Figure 11

Media reports on biofilms

Evaluation of data or media reports on environmental problems caused by biofilms.

Biofilms are commonly featured in the media as they have a number of novel and interesting properties. They are employed as innovative solutions to problems. At the same time they have been implicated in a number of environmental issues:

Virginia Tech scientists have provided new evidence that biofilms – bacteria that adhere to surfaces and build protective coatings – are at work in the survival of the human pathogen *Salmonella*.

One out of every six Americans becomes ill from eating contaminated food each year, with over a million illnesses caused by *Salmonella* bacteria, according to the Centers for Disease Control and Prevention. Finding out what makes *Salmonella* resistant to antibacterial measures could help curb outbreaks.

Researchers affiliated with the Fralin Life Science Institute discovered that in addition to protecting *Salmonella* from heat-processing and sanitizers such as bleach, biofilms preserve the bacteria in extremely dry conditions, and again when the bacteria are subjected to normal digestive processes.

Outbreaks of *Salmonella* associated with dried foods such as nuts, cereals, spices, powdered milk and pet foods have been associated with over 900 illnesses in the last five years. These foods were previously thought to be safe because the dry nature of the product stops microbial growth.

"Most people expect to find *Salmonella* on raw meats but don't consider that it can survive on fruits, vegetables or dry products, which are not always cooked," said Ponder.

In moist conditions, *Salmonella* thrive and reproduce abundantly. If thrust into a dry environment, they cease to reproduce, but turn on genes which produce a biofilm, protecting them from the detrimental environment.

Researchers tested the resilience of the *Salmonella* biofilm by drying it and storing it in dry milk powder for up to 30 days. At various points it was tested in a simulated gastrointestinal system. *Salmonella* survived this long-term storage in large numbers but the biofilm *Salmonella* were more resilient than the free-floating cells treated to the same conditions.

The bacteria's stress response to the dry conditions also made them more likely to cause disease. Biofilms allowed the *Salmonella* to survive the harsh, acidic environment of the stomach, increasing its chances of reaching the intestines, where infection results in the symptoms associated with food poisoning.

This research may help shape Food and Drug Administration's regulations by highlighting the need for better sanitation and new strategies to reduce biofilm formation on equipment, thus hopefully decreasing the likelihood of another outbreak.

Source: http://www.sciencedaily.com/releases/2013/04/130410154918.htm

---

Activity

Choose one or more of the following environmental issues related to biofilms. Create a brief research report outlining the scope of the problem. Ensure that you include the role of biofilms. Evaluate possible solutions to the problems caused by the biofilm.

- The role of biofilms in increasing biological oxygen demand in eutrophic bodies of water.

- The development of biofilms on equipment and piping systems in industry such as paper making facilities.

- The development of biofilms in clean water pipes at water treatment facilities.

- The binding of positively charged heavy metals to negatively charged biofilms.

- The sequestering of toxins within the biofilm.

Laser microscopes have enhanced our knowledge of biofilms

Developments in scientific research follow improvements in apparatus: using tools such as the laser scanning microscope has led researchers to deeper understanding of the structure of biofilms.

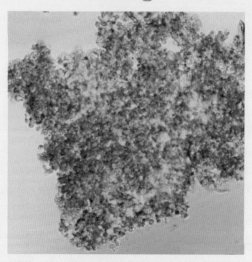
▲ Figure 12

Biofilms have a complex structure. The position of individual cells in relation to one another and the EPS matrix influences roles and functions. Three-dimensional visualization of living cells serving different functions can be carried out using a laser-scanning microscope in combination with dyes. This technique allows direct observation of the biofilm without disrupting its structure.

Figure 12 shows an image of a fragment of biofilm extracted from amniotic fluid. The image was generated using a laser scanning microscope. Red dots indicate EPS, green dots indicate bacteria and grey dots represent host cells.

B.4 Medicine (AHL)

Understanding

→ Infection by a pathogen can be detected by the presence of its genetic material or by its antigens.

→ Predisposition to a genetic disease can be detected through the presence of markers.

→ DNA microarrays can be used to test for genetic predisposition or to diagnose the disease.

→ Metabolites that indicate disease can be detected in blood and urine.

→ Tracking experiments are used to gain information about the localization and interaction of a desired protein.

→ Biopharming uses genetically modified animals and plants to produce proteins for therapeutic use.

→ Viral vectors can be used in gene therapy.

Applications

→ Use of PCR to detect different strains of influenza virus.

→ Tracking tumour cells using transferrin linked to luminescent probes.

→ Biopharming of antithrombin.

→ Use of viral vectors in the treatment of Severe Combined Immunodeficiency (SCID).

Skills

→ Analysis of a simple microarray.

→ Interpretation of the results of an of ELISA diagnostic test.

Nature of science

→ Developments in scientific research follow improvements in technology: innovation in technology has allowed scientists to diagnose and treat diseases.

Innovations in diagnostic techniques

Developments in scientific research follow improvements in technology: innovation in technology has allowed scientists to diagnose and treat diseases.

To be useful, new methods used to diagnose a disease must be accurate and preferably simple to use. They should provide a result that is timely and increases the time to carry out treatment in such a way that long-term complications do not result. In the case of infectious diseases, faster and more accurate diagnosis can lead to treatment which prevents the spread of the pathogen.

Infection by parasites has often been diagnosed by microscopic analysis to look for the presence of the organism or evidence of its activity.

Diagnosis by bacterial infection has traditionally been done by collecting samples of urine or stool, or swabs can be taken from an infected site. If

bacterial infection exists, the sample can be plated on culture media to look for the growth of the kind of bacterial colonies which characterize a certain disease. The limitation of this procedure is that sometimes different microorganisms present in the same way. Further, some pathogens are difficult or slow to culture.

Diagnosis of genetic diseases has traditionally been carried out by reviewing a combination of clinical observation and searching for the presence of high levels of unusual metabolites in the urine or blood.

Improvements in methods of diagnosis have increased the specificity, the speed and the reliability of diagnosis.

High levels of metabolites can indicate disease

Metabolites that indicate disease can be detected in blood and urine.

"Inborn errors of metabolism" is a term applied to a broad group of genetically inherited disorders that affect metabolism. The majority of these diseases are due to mutations in single genes that code for enzymes often resulting in a non-functional enzyme. This results in a build-up of substances which are toxic or a shortage of important molecules necessary for normal function leading to secondary symptoms. Table 1 shows three such diseases and the metabolites that are detected in blood and urine when an individual is affected.

Newborn infants are subjected to a heel prick test to detect phenylketonuria (PKU), in which a blood sample is taken from the heel of the foot. If the child is affected, there will be elevated levels of phenylpyruvate in the blood indicating the child lacks an enzyme for converting the amino acid phenylalanine to tyrosine. If diagnosed quickly enough, diet modification can prevent severe consequences for the child.

Disease	Metabolic pathway that is not functioning	Metabolite that is detected
Lesch–Nyhan syndrome	Production of purines	Uric acid crystals in the urine
Alkaptonuria	Breakdown of the amino acid tyrosine	High levels of homogentisic acid detected in both the urine and the blood by thin layer chromatography and paper chromatography
Zellweger syndrome	Assembly of peroxisomes (organelles essential for the degradation of long chain fatty acids)	Elevated very long chain fatty acids in the blood

▲ Table 1

Indicators of infection by a pathogen

Infection by a pathogen can be detected by the presence of its genetic material or by its antigens.

Modern molecular methods have the advantage of being much better at discriminating between pathogens. They can be automated to speed up the process and they don't present the challenge of having to culture the pathogen separately.

The **E**nzyme-**L**inked **I**mmuno**s**orbent **A**ssay (ELISA) detects the presence of antibodies to pathogens. The challenge with this diagnostic test is that it is usually only effective once the patient has developed an immune response to the pathogen resulting in the production of antibodies. Recent versions of the ELISA test for the antigen directly such as the p24 antigen from the HIV virus.

PCR can be used to detect the genetic material of a pathogen. If primers that have the same nucleotide sequence as the genetic material of the

 The ELISA test

Interpretation of the results of an ELISA diagnostic test.

An ELISA test can be used to detect the presence of infection by a pathogen. The test works by testing for the presence of antibodies to the antigens of the pathogen. Alternatively, it can test for the antigen directly.

Figure 1 shows the basis of a positive test for HIV.

A capture molecule is fixed to a surface. In the figure, these capture molecules are antibodies to the HIV p24 capsid protein.

The sample to be tested is exposed to the capture surface. Because the target molecules are present in a positive test, they bind to the capture molecules. Next a free version of the capture molecule is added. This version of the capture molecule is

linked to an enzyme. The solution is rinsed. In a negative test, this would wash away the free version of the capture molecule. In a positive test, they bind to the target molecule and they are not washed away. The last step is to add the substrate of the enzyme which changes colour when acted upon by the enzyme. A positive test is therefore indicated by a coloured solution (see figure 2).

Figure 2 shows a tray of wells containing human blood serum from different individuals being tested for antibodies to the hepatitis C virus. Wells which remain uncoloured are negative. Those that change colour to yellow/orange are positive and confirm that the patient has antibodies for hepatitis C virus.

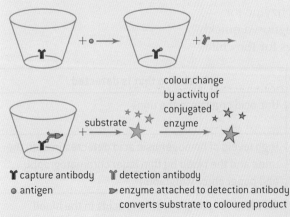

capture antibody   detection antibody
antigen   enzyme attached to detection antibody converts substrate to coloured product

▲ Figure 1 Steps in a positive ELISA test

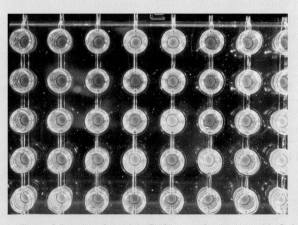

▲ Figure 2 Results of multiple ELISA tests for the Hepatitis C virus

pathogen are added to a sample from the patient, then amplification will only occur if the genetic material of the pathogen is present.

Another way to detect the presence of a pathogen is to use DNA probes in a microarray. These can be used to detect mRNA sequences complementary to the pathogen in samples from a patient.

Activity

Figure 3 shows a standard curve that relates quantity of antigen present in the test serum to optical density, a measure of the colour of solution. The darker the colour, the higher the optical density.

1   Explain how the standard curve could be used.   [2]

2   Determine the concentration of antigen present at an optical density of 1.0.   [1]

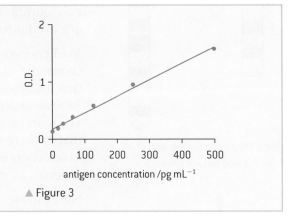

▲ Figure 3

PCR as a diagnostic tool

Use of PCR to detect different strains of influenza virus.

There are a number of clinical signs and tests that can indicate infection by an influenza virus. For some people, infection with more serious strains such as swine flu needs to be diagnosed quickly. This includes such patients as pregnant women, elderly patients or patients whose immune system is compromised, as the infection can result in death. Further, some strains can produce more serious side effects. In addition, rapid detection can prevent a serious epidemic. The PCR test is most likely to be able to identify the specific strain of the virus that infects a person.

Because the influenza virus is an RNA virus, a variation of PCR called reverse transcription polymerase chain reaction (RT-PCR) is used. Reverse transcriptase will produce a DNA molecule from an RNA template called cDNA.

The first step involves purifying mRNA from cells of an infected patient. The mRNA extract is converted into cDNA. Then primer sequences specific to the strain of influenza virus being tested for are added. If the influenza primers bind to sequences in the cDNA, this means that the

mRNA being sought was present in the original sample and the cDNA will be amplified. A recent modification is to include fluorescent dyes into the sample that bind specifically to double-stranded DNA. As the quantity of double-stranded DNA increases, fluorescence will be detected indicating a positive test.

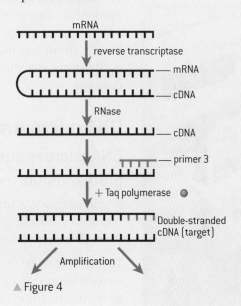

▲ Figure 4

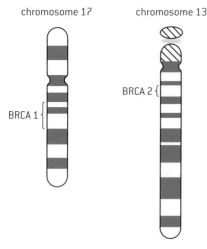

▲ Figure 5 Chromosomal location of the BRCA 1 and BRCA 2 genes

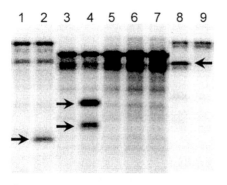

▲ Figure 6

▲ Figure 7 A DNA microarray cartridge being loaded into a machine that will be used to analyse the results from this test

Genetic markers

Predisposition to a genetic disease can be detected through the presence of markers.

Genetic markers are particular alleles which are associated with a predisposition to having a genetic disease. They can be single nucleotide polymorphisms or tandem repeats. Detection of the marker can be achieved through such methods as PCR, and DNA profiling.

Markers may be part of a coding or non-coding sequence; i.e., they may contribute to the disease or they may be genetically linked to the gene that influences the condition. To be useful, non-coding markers need to lie near to the defective gene to avoid being separated by crossing over. The marker should be an allele for which the population is polymorphic; that is, there should be a number of possible genotypes at the locus. Researchers look for alleles which are found more frequently than expected by chance in those people affected by the disease.

For example, mutations in the BRCA 1 and BRCA 2 genes indicate an increased risk of breast cancer and ovarian cancer in women and the gene itself contributes to the onset of the cancer. The genes are found on chromosome 17 and chromosome 13 respectively.

There are different alleles of BRCA mutations. Figure 6 shows the separation of proteins by electrophoresis. In this case, radioactively labelled amino acids were supplied during protein synthesis and the products were separated by electrophoresis and photographed using film that detects radioactivity. The arrows indicate the various types of marker proteins produced by different mutations of the BRCA 1 gene. The presence of such marker proteins in a blot from an individual would indicate a predisposition to cancer.

For diseases which are linked to a single gene, the marker has more predictive power. Where diseases are strongly influenced by the environment or are polygenic, particular markers have less predictive power, though considerable progress has been made recently in establishing statistical probabilities from more complex inheritance patterns.

DNA microarrays

DNA microarrays can be used to test for genetic predisposition or to diagnose the disease.

A microarray is a small surface that has a large range of DNA probe sequences adhering to its surface. Microarrays can be used to test for expression of a very large number of DNA sequences simultaneously.

The sample to be tested is the mRNA being expressed by a cell. cDNA is formed from the mRNA using reverse transcriptase. At the same time as synthesis, fluorescent dyes are linked to the cDNA. The microarray is exposed to the cDNA sample long enough for any complementary sequences to bind to the fixed probes and then the chip is rinsed. The chip is then exposed to laser light which will cause the fluorescent probes to give off light where there has been hybridization between the cDNA and the DNA probes within the chip. The brighter the light, the higher the level of gene expression in that region.

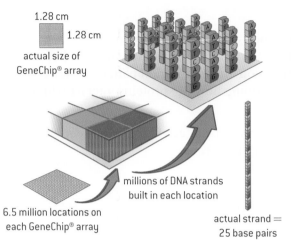
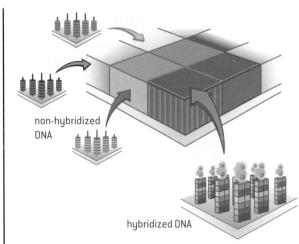

1.28 cm

1.28 cm

actual size of GeneChip® array

6.5 million locations on each GeneChip® array

millions of DNA strands built in each location

actual strand = 25 base pairs

non-hybridized DNA

hybridized DNA

▲ Figure 8

🧪 Interpreting a microarray

Analysis of a simple microarray.

As an example of the use of a microarray, an experimenter may want to assess the range and level of gene expression in a cancerous cell. They would extract mRNA from control cells and produce labelled cDNA from this sample. They would modify this cDNA with a green fluorescent dye. They would then extract mRNA from cancerous cells, produce cDNA and label it with red dye. They would then expose the microarray chip to both samples, allow time for hybridization and then wash the chip to remove unhybridized cDNA. The chip would then be exposed to fluorescent light. The part of the chip where green light is observed indicates sequences being expressed in the control only. The part of the chip where there is red light is where the sequences are being expressed by the cancerous cells only. Yellow light, which is a combination of green and red light, corresponds to regions where both types of cells are expressing the sequence.

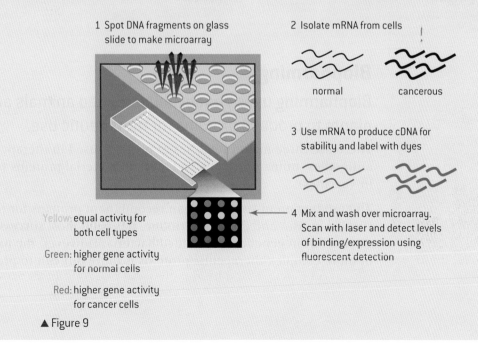

1 Spot DNA fragments on glass slide to make microarray

2 Isolate mRNA from cells

normal   cancerous

3 Use mRNA to produce cDNA for stability and label with dyes

4 Mix and wash over microarray. Scan with laser and detect levels of binding/expression using fluorescent detection

Yellow: equal activity for both cell types

Green: higher gene activity for normal cells

Red: higher gene activity for cancer cells

▲ Figure 9

587

Protein tracking experiments

Tracking experiments are used to gain information about the localization and interaction of a desired protein.

Proteins circulating in the blood can be traced if radioactive probes are attached to them. Such tracking experiments can allow researchers to follow distribution and localization patterns. They can also allow researchers to determine how the proteins interact with the target tissue.

Radioactive atoms or molecules can be attached to the proteins and their distribution can be tracked with PET scans.

Tracking experiments involving transferrin
Tracking tumour cells using transferrin linked to luminescent probes.

Transferrin is a molecule which binds iron. It is taken up by more by tumour cells than by surrounding cells.

Figure 10 shows a sequence of photos taken using luminescent dyes linked to transferrin molecules. The experiment is being used to study receptor-mediated endocytosis in lymphoma cells. At zero minutes, the transferrin is shown bound to receptors on the surface of cells. The dots represent the fluorescent dye. The bottom image shows some of the receptor–transferrin complexes having entered the cell. Once they have delivered their load of iron, the receptor–transferrin complex is recycled to the cell-surface membrane (top right).

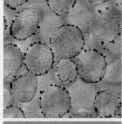

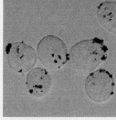

▲ Figure 10

Biopharming

Biopharming uses genetically modified animals and plants to produce proteins for therapeutic use.

There are three main categories of proteins used in therapy: antibodies, human proteins and viral or bacterial proteins (used in vaccines).

The production of simple human recombinant proteins for therapy, such as insulin and growth hormone, has been most successfully carried out in genetically modified bacteria. However, the production of more complex therapeutic proteins is more difficult to produce in these living systems. Prokaryotic systems do not carry out the required post-translational modification such as the addition of sugars. Sometimes, only mammal cells are capable of performing these modifications.

Producing these proteins in transgenic farm animals addresses the post-translational modification problem. Some domestic varieties of cows, sheep and goats have been selectively bred to produce high yields of milk. Lactating female animals have been engineered to secrete recombinant proteins into their milk. The combination of these two factors means a small herd of animals can yield a relatively large mass of therapeutic protein.

Plant-made therapeutic proteins have been made using whole plants and plant cell cultures. In May 2012, the first plant-made human therapeutic protein was approved for use in humans by the US Food and Drug Administration (FDA) as enzyme-replacement therapy to treat the symptoms of Gaucher's disease.

Biopharming to produce ATryn

Biopharming of antithrombin.

Antithrombin deficiency is a condition that puts patients at risk of blood clots during childbirth and surgery. ATryn is the commercial name of antithrombin that has been produced in the mammary glands of genetically modified goats.

To achieve this genetic modification, the gene of interest and specific additional sequences have to be added. A specific promoter sequence that will ensure that the gene is expressed in milk is necessary in creating the gene construct. In addition, a signal sequence has to be added to ensure that the protein is produced by ribosomes on the endoplasmic reticulum rather than by ribosomes that are free in the cytoplasm. This is to ensure that the antithrombin protein is secreted by the mammary cells rather than released intracellularly.

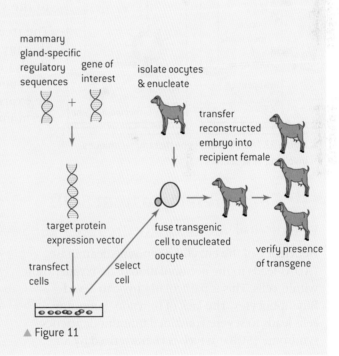

▲ Figure 11

Gene therapy

Use of viral vectors in gene therapy.

Some inherited diseases are caused by a defective gene, that results in the lack of a particular enzyme or protein. Cystic fibrosis is one such disease. It is caused by the lack of cystic fibrosis transmembrane protein (CFTP). This protein normally transports chloride ions out of cells and into mucus. The chloride ions draw

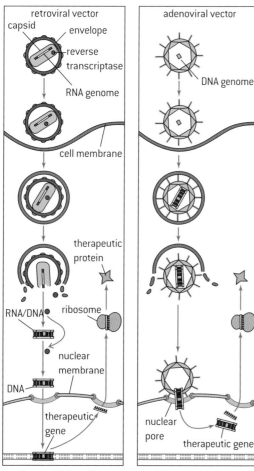

retroviral vector

capsid
envelope
reverse transcriptase
RNA genome

cell membrane

therapeutic protein

RNA/DNA   ribosome

DNA

nuclear membrane

therapeutic gene

adenoviral vector

DNA genome

nuclear pore

therapeutic gene

▲ Figure 12 Two different gene therapy techniques involving viral vectors

water out of the cells and make mucus watery. Cystic fibrosis patients suffer from thick mucus, which builds up in the airways.

Gene therapy may offer a cure for inherited diseases like cystic fibrosis. In gene therapy, working copies of the defective gene are inserted into a person's genome. To do this, a gene delivery system, or vector, is needed. Figure 12 shows two different ways of using viruses as vectors. The viral genome is altered so that the particles are not virulent. The therapeutic gene is then inserted into the virus.

Viruses that contain double-stranded (ds) DNA, such as adenovirus, cannot cause the problems found with retroviruses because the viral DNA is not inserted into the genome. However, the therapeutic gene is not passed on to the next generation of cells, so treatment has to be repeated more frequently. A challenge of using viruses as vectors is that the host may develop immunity to the virus.

The treatments described above are called somatic therapy, because the cells being altered are somatic (body) cells. An alternative method would be to inject therapeutic genes into egg cells. The missing gene would be expressed in all cells of the organism. This is called germ line therapy.

🌐 Gene therapy to treat SCID

Use of viral vectors in the treatment of Severe Combined Immunodeficiency (SCID).

Deficiency of the enzyme adenosine deaminase (ADA) leads to the accumulation of deoxyadenosine within cells. This is particularly toxic to T and B lymphocytes. The lack of functional immune cells leads to severe combined immunodeficiency syndrome (SCID) which is characterized by an inability to fight off the simplest of infections. ADA deficiency was the first condition successfully treated by gene therapy.

The steps involved in the successful therapy included:

- Removing ADA deficient lymphocytes from the patient with SCID.

- Culturing the cells in vitro.

- Infecting the cultured cells with genetically modified retrovirus containing the gene that can produce functional ADA.

- Delivering the modified lymphocytes by transfusion back into the patient.

The effect lasted for four years after the start of gene therapy in one patient.

B.5 Bioinformatics (AHL)

Understanding

→ Databases allow scientists easy access to information.

→ The body of data stored in databases is increasing exponentially.

→ BLAST searches can identify similar sequences in different organisms.

→ Gene function can be studied using model organisms with similar sequences.

→ Sequence alignment software allows comparison of sequences from different organisms.

→ BLASTn allows nucleotide sequence alignment while BLASTp allows protein alignment.

→ Databases can be searched to compare newly identified sequences with sequences of known function in other organisms.

→ Multiple sequence alignment is used in the study of phylogenetics.

→ EST is an expressed sequence tag which can be used to identify potential genes.

Applications

→ Use of knockout technology in mice to determine gene function.

→ Discovery of genes by EST data mining.

Skills

→ Explore the chromosome 21 in databases (for example in Ensembl).

→ Use of software to align two proteins.

→ Use of software to construct simple cladograms and phylograms of related organisms using DNA sequences.

Nature of science

→ Cooperation and collaboration between groups of scientists: databases on the internet allow scientists free access to information.

The role of databases in genetic research

Databases allow scientists easy access to information.

A database is a structured collection of information stored on a computer. It can include data in a range of formats including qualitative information, articles, images or quantitative information.

Types of databases used in bioinformatics include:

- Nucleotide sequence databases such as EMBL (The European Molecular Biology Laboratory).

- Protein sequence databases such as SwissProt.

- Three-dimensional structure databases such as PDB (Protein Data Bank).

- Microarray databases such as ArrayExpress which contain information about the level and types of mRNA expressed in different cells.

- Pathway databases which contain information about enzymes and reactions and can be used to model metabolic pathways. An example of such a database is KEGG (Kyoto Encyclopedia of Gene and Genomes).

591

TOK

To what extent does scientific research require regulation? If it does require regulation who should administer the regulations?

In 1999 a patient died as a result of participation in clinical trials for gene therapy. He suffered from ornithine transcarbamylase deficiency, or OTC, a liver disease marked by an inability to metabolize ammonia. Ammonia is a waste product of amino acid metabolism. He had been able to survive up to that point because of dietary modification and medication. The trial he participated in involved being injected with adenoviruses carrying the gene for transcarbamylase. He died within days due to a strong immune response to the viral vector. An investigation concluded that the scientists involved in the trial violated several rules of conduct.

- Four other patients who had received the treatment had reactions that were deemed so severe that the trial should have ended.

- The informed consent forms did not include information about primates that had died in similar trials.

- The patient had levels of ammonia that were so high he should have been excluded from the study.

- A principal investigator of the study had a major interest in the outcome of the trial as he held patents on the OTC treatment.

From *Welcome to the Genome* by Bob De Salle and Michael Yudell

1   Explain what is meant by informed consent.

2   **a)** Suggest what policy instruments might be put in to place to prevent such occurrences.

    b) Who should administer these policies – governments, other scientists or research institutions?

Hypothesis testing is increasingly possible by extracting data from a database rather than the researcher collecting the data directly for themselves.

A researcher can employ a database to do a number of tasks:

- add the results of their research for others to access

- extract subsets of data

- query the database by searching for a particular piece of data.

Growth in information housed in databases

The body of data stored in databases is increasing exponentially.

Advances in technology have meant that the rate of creation and publication of data is increasing. Advances in genome sequencing technology, microarrays, 3-D modelling programmes and computing power have resulted in a number of large-scale collaborative research projects which have generated an exponential growth in data housed in databases. One research report tracked the growth in information in bioinformatics databases and concluded that it has a doubling time of between 12 and 24 months.

Access to information issues in bioinformatics

Cooperation and collaboration between groups of scientists: databases on the internet allow scientists free access to information.

Most people presume that collaboration and cooperation between researchers characterizes the scientific endeavour. Most of the important bioinformatics databases are public and freely accessible to all researchers. Often, once data is added to one database, it is immediately synchronized with data in other databases. Such open access and synchronization facilitates collaboration and a spirit of cooperation.

One view is that the commercialization of bioinformatics databases is a threat to this spirit.

Some researchers working in private companies do not post their sequence information because of the need to make a profit. Some databases that have been public in the past have been taken over by for-profit companies who have started to charge for access to sequence information. Two examples are the *Saccharomyces cerevisiae* (yeast) and *Caenorhabditis elegans* (soil roundworm) databases, two of the most widely studied eukaryote model organisms. This was controversial as some of the information in the databases was derived from published studies and personal communications.

The academic journal *Science* twice created controversy due to the competing imperatives of public and private science. In 2001, the journal published the company Celera's version of the sequence of the human genome sequence while allowing the company to house the sequence on their own database. In 2002, *Science* published a version of the rice genome while allowing the company Syngenta to keep the data on their own private database. These two papers broke the industry standard of the previous 20 years that had seen data being published on the public database GenBank. It also did not comply with a second tradition of publication that was much more longstanding. Traditionally data supporting published reports has always been assumed to be published and therefore freely available to the scientific community so that, at a minimum, verification was possible.

Bioinformatics

BLAST searches can identify similar sequences in different organisms.

Once a researcher first identifies a sequence of interest by sequencing a protein, identifying an open reading frame or finding high levels of a certain type of mRNA within a cell, their next step would be to conduct a BLAST search.

The acronym refers to **B**asic **L**ocal **A**lignment **S**earch **T**ool. The tool finds regions of similarity between sequences. The computer program compares protein or nucleotide sequences housed in databases and carries out statistical calculations to determine matches with other sequences.

There are three main nucleotide databases: GenBank, EMBL and DDJB. Two of the most important protein sequence databases are PIR International and SwissProt.

BLASTn and BLASTp searches

BLASTn allows nucleotide sequence alignment while BLASTp allows protein alignment.

A researcher can identify open reading frames in nucleotide sequences. Once an open reading frame is identified, a BLASTn search can be conducted which involves searching through nucleotide databases to determine if a similar open reading frame exists in another species.

A BLASTp search uses a protein sequence to search a protein database.

A BLASTx searches a protein database based on the translated sequence of an open reading frame.

Alternatively, if a researcher has found a protein and wants to determine the location of a gene, they can conduct a tBLASTn search using a computer

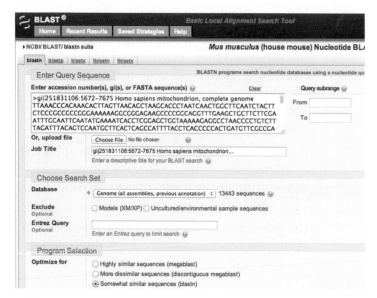

▲ Figure 1

search of multiple genomes using the translated sequence to search for potential genes that could have been transcribed to produce the protein.

Figure 1 shows a BLASTn search that is about to be conducted. A sequence from human mitochondrial DNA has been entered to search for similar sequences in mouse DNA.

Matching new sequences with those found in databases

Databases can be searched to compare newly identified sequences with sequences of known function in other organisms.

If a researcher has a sequence of unknown function, they can search a database to determine if a similar sequence has been identified in another organism.

If the sequence were a protein sequence, they could conduct a BLASTp search. The outcome would allow the researcher to determine if a protein of similar sequence exists in another organism and what its function might be.

If the researched sequence were a nucleotide sequence, they might conduct a BLASTn search to determine if a similar sequence of known function exists in another organism or a BLASTx search to see if a gene product of a similar sequence has been identified in another organism.

🌐 Knockout mice

Use of knockout technology in mice to determine gene function.

One method of determining gene function is to genetically modify mice by "knocking out" a gene. This involves replacing the functional sequence with a non-functional sequence within stem cells and then fusing the stem cells with an embryo. The resulting mouse is a chimera. The chimeras are mated with normal mice. Heterozygotes are interbred until a purebred knockout mouse is generated.

The loss of the activity of the gene will often lead to a detectable change in the phenotype of the mouse. This allows researchers to determine the likely function of the gene.

The gene for the production of the hormone leptin was knocked out by introducing a point mutation. Figure 2 shows a wild type mouse on the right and an obese (ob/ob) knockout mouse on the left. This is part of the evidence that leptin plays a role in regulating fat deposition and energy metabolism.

▲ Figure 2

Model organisms

Gene function can be studied using model organisms with similar sequences.

A model organism is a species that has been extensively studied based on the assumption that discoveries made in the model organism will have relevance to other organisms. Some of the most extensively studied model organisms are *Caenorhabditis elegans* (a soil roundworm), *Mus musculus* (the common house mouse), *Drosophila melanogaster* (the fruit fly), *Arabidopsis thaliana* (a plant with the common name Thale cress), *E. coli* and *Saccharomyces cerevisiae* (yeast).

The genomes of these organisms have been sequenced as has the genome of humans. Across the diversity of life, there are some conserved metabolic pathways and some conserved genetic sequences. Model organisms can be used as living, or *in vivo*, models of diseases related to these conserved pathways or diseases related to mutations in sequences.

Such studies might not be feasible or might be unethical in humans.

Computer-based sequence alignment

Sequence alignment software allows comparison of sequences from different organisms.

Sequences that are similar between organisms suggest evolutionary relationships. The greater the similarity, the closer the relationship. Visual comparison is possible when comparing two relatively short sequences, but comparing longer sequences or multiple sequence comparison relies on the use of computer algorithms.

There are a number of software programmes used to carry out sequence alignment including ClustalW and MUSCLE. Increasingly alignments can be carried out using web-based interfaces. For example the BLAST search web page of the National Centre of Biotechnology Information (NCBI) carries out the alignment of two sequences and the ClustalOmega web page will carry out multiple sequence alignment.

Figure 3 shows a DNA sequence alignment of nine different organisms generated using the programme ClustalX.

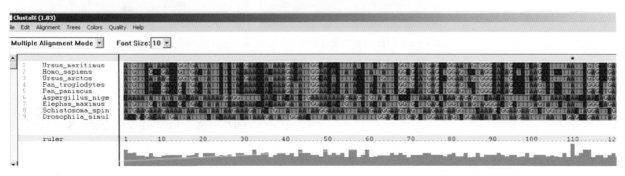

▲ Figure 3

Sequence alignment tools often start with the default of searching for global relationships over the entire length of the sequence. However, in terms of functions, two proteins might share a few common areas that are closely linked to a common function with other regions having little or no homologous areas. For this reason, alignment tools offer a choice between local or global alignment.

Using BLAST to align two proteins

Use of software to align two proteins.

There are a number of applications for aligning two protein sequences. The following instructions are for using the BLAST sequence alignment tool at the NCBI website (http://www.ncbi. nlm.nih.gov/protein/). In this example, we will conduct a sequence alignment using the cytochrome oxidase (cox1) protein for two species of primates called tarsiers. Horsfield's tarsier, variously classified as *Cephalopachus bancanus* and *Tarsius bancanus*, is a threatened species that lives in Borneo and Sumatra. The cox1 sequence for this tarsier will be compared to the sequence of the same gene for the Philippine tarsier (*Tarsius syrichta*). There is some uncertainty over the classification of Horsfield's tarsier and it is this type of sequence comparison which is often used to resolve this kind of controversy.

▲ Figure 4 Horsfield's tarsier

▲ Figure 5 Philippine tarsier

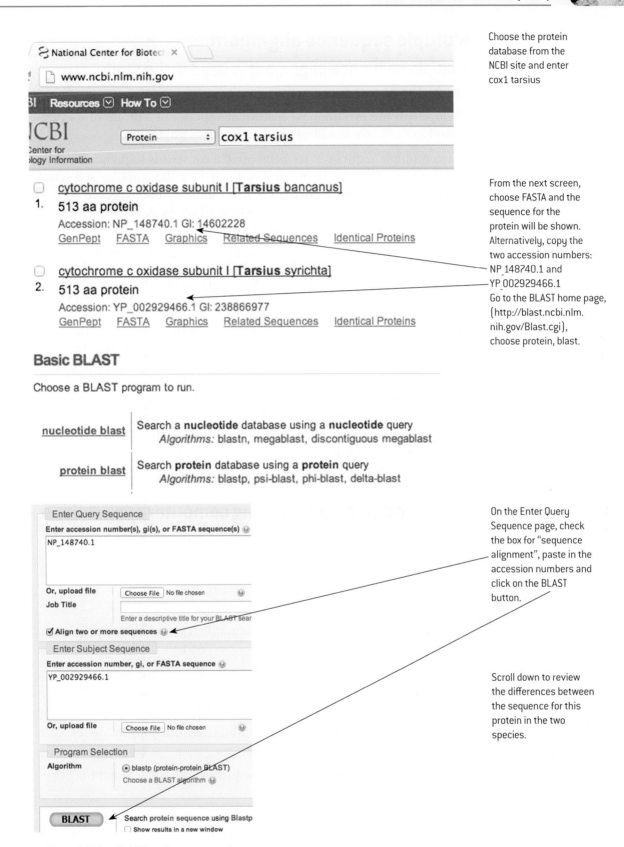

Choose the protein database from the NCBI site and enter cox1 tarsius

From the next screen, choose FASTA and the sequence for the protein will be shown. Alternatively, copy the two accession numbers: NP_148740.1 and YP_002929466.1 Go to the BLAST home page, (http://blast.ncbi.nlm. nih.gov/Blast.cgi), choose protein, blast.

On the Enter Query Sequence page, check the box for "sequence alignment", paste in the accession numbers and click on the BLAST button.

Scroll down to review the differences between the sequence for this protein in the two species.

▲ Figure 6 Using BLAST to align two proteins

Multiple sequence alignment

Multiple sequence alignment is used in the study of phylogenetics.

Phylogeny is the evolutionary history of a species or a group of species. A phylogenetic tree is a diagram that describes phylogeny.

When multiple sequences are compared, a consensus sequence is often identified based on the amino acid or nucleotide that appears at a certain position in the aligned sequences. As an example, if you aligned six sequences and the nucleotides at position 10 are G, A, G, G, C and G, then the consensus sequence will have a G at position 10.

Similarities in sequences can be caused by actual evolutionary relationships in which case the sequence similarities are said to homologous. Alternatively sequences which are the same by chance are referred to as analogous.

Most homologous sequences have many positions where mutations have occurred several times. However, not all mutations have the same effect. A mutation in a coding region which results in a change in amino acid sequence is less likely to persist in a population. The probability of a match by chance is higher for DNA sequences than it is for protein sequences. Nonetheless, computer based algorithms have been developed that can use sequence alignments to suggest evolutionary relationships.

Constructing phylograms and cladograms using computer applications

Use of software to construct simple cladograms and phylograms of related organisms using DNA sequences.

A phylogenetic tree that is created using the cladistics methods discussed in sub-topic 5.4 is a cladogram. This type of tree only shows a branching pattern and the length of its branch spans do not represent time or the relative amount of change that occurs along a branch. A phylogram is a phylogenetic tree that has branch lengths that are proportional to the amount of character change (see figure 8).

The following activity requires the use of two types of software: ClustalX and PhyloWin. This activity is based on an activity developed by the American Museum of Natural History. In this activity, we will conduct multiple sequence alignment for the gene for cytochrome oxidase for a number of primate species.

1   Visit the NCBI website and choose gene.

2   Search 'cox1 primate'.

3   Choose the species that you want to be included in the tree.

4   Under the Genomic regions, transcripts and products, choose 'FASTA'.

Highlight all of the DNA sequence including the title (for example '>gi|196123578:5667-7670 Homo sapiens neanderthalensis'.)

5   Open either Notepad from your PC or TextEdit on a Mac.

6   Paste your sequence into the text editing document.

7   Repeat with several other sequences from different organisms.

8   Edit the titles but remember to include the '>' symbol and to separate words in the title with an underscore. For example: >Homo_sapiens_neanderthalensis.

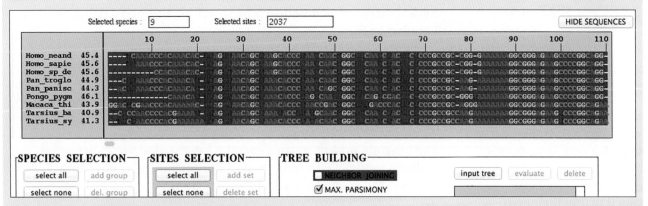

▲ Figure 7 A screen capture of an image generated using PhyloWin

9   When your document is complete, there is an extra step for Mac users. Under the 'Format' menu choose 'make plain text'.

10 Save your file as sequences.fasta

11 Open the ClustalX software.

12 Under the 'File' menu, chose load sequences.

13 Browse your files and open the 'sequences.fasta' file.

14 Once the sequences are loaded, choose 'do complete alignment' under the 'Alignment' menu. Make a note of where the output file 'sequences.fasta.aln' is saved.

15 Open PhyloWin.

16 Browse for the 'sequences.fasta.aln' sequence.

17 In the tree building window, choose max. parsimony. There are a number of possible ways the sequences could have ended up the way they have. For example, a change between a cytosine to a thymine back to a cytosine back to a thymine is a possible series of events, but maximum parsimony presumes that the change was simply cytosine to thymine. This means that you are choosing the tree that involves the least evolutionary change. Figure 8 is one possible phylogenetic tree showing the evolutionary relationship of the nine primates.

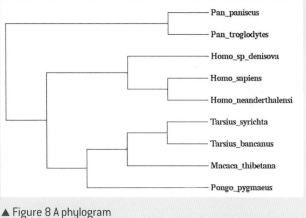

▲ Figure 8 A phylogram

Expressed sequence tags

EST is an expressed sequence tag which can be used to identify potential genes.

If a gene is being expressed, then mRNA transcribed from that gene will be found within the cell. The mRNA can then be used to search for the gene that produced it using expressed sequence tags (ESTs).

Scientists use the mRNA along with the enzyme reverse transcriptase to produce cDNA. This cDNA will not have any introns in it. Scientists use this DNA to synthesize ESTs. These are short sequences of DNA about 200 to 500 nucleotides long that are generated from both the 5′ end and the 3′ end. The 5′ end tends to have a sequence that is conserved across species and within gene families. The 3′ end is more likely to be unique to the gene.

Using ESTs to locate genes

Discovery of genes by EST data mining.

Because of their usefulness and the ease with which they are generated, a very large number of ESTs have been generated. The sequences have been deposited within their own database called dbEST. This database contains ESTs from over 300 organisms. Scientists can conduct a BLAST search once they have an EST to determine if it matches a DNA sequence from a known gene with an identified function.

The location of the gene within the genome then can be located through physical mapping techniques or by searching through database libraries of known ESTs.

 ## Exploring chromosome 21

Explore the chromosome 21 in databases (for example in Ensembl).

The Ensembl project collates genome information for 75 organisms. It allows for detailed exploration of the coding and non-coding sequences of each of the chromosomes from these species.

Chromosome 21 is the shortest human chromosome and perhaps best known for Downs syndrome, or trisomy 21.

Activity

To explore the information available about chromosome 21, visit the website of the web-based database Ensembl (www.ensembl.org).

The first column shows the position of the centromere as being 'acrocentric' which means to one side of the middle.

1 Click on a red bar to go to a detailed view of that coding region.

2 Click on three protein coding regions to determine the gene that they code for.

The search can be refined by looking at protein coding regions to determine what genes have a locus on the chromosome.

Visually, it can be seen that there the 'q' arm (the longer arm) has far more coding sequences per unit length that have been discovered.

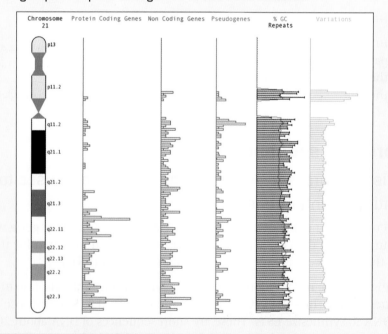

Questions

1 Release of sewage in marine waters is a common practice but it can cause water contamination with pathogens. A series of experiments were conducted to compare inactivation rates of two different groups of microbes with different sunlight exposures. One group were fecal coliform bacteria and the other were coliphage viruses. Experiments were conducted outdoors using 300-litre mixtures of sewage-seawater in open-top tanks.

A two-day experiment was carried out with untreated sewage added to seawater. Both days were sunny with no clouds. The figure below shows the inactivation of the microbes in seawater as a function of the cumulative amount of sunlight and time. The survival curves of the two microbes are plotted against sunlight exposure (lower x axis) during daylight periods and against time during the overnight period (upper x axis). The y axis gives counts of bacteria and viruses per 100 ml.

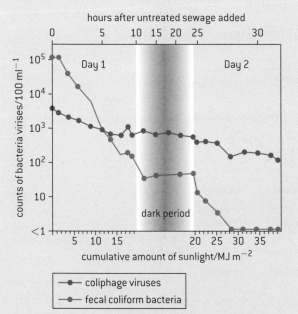

Source: adapted from L W Sinton, et al., (1999), Applied and Environmental Microbiology, 65 (8), pages 3605–3613

a) Identify the time at which fecal coliform bacteria counts fell below 1 unit per 100 ml. [1]

b) Deduce, using the data in the graph, the effect of sunlight on

(i)   fecal coliform bacteria. [2]

(ii) coliphage viruses. [2]

c) For an accidental sewage spill, suggest, giving a reason, which of the two microbes may be most useful as a fecal indicator two days after the spill. [1]

2 Wastewater from factories producing polyester fibres contains high concentrations of the chemical terephthalate. Removal of this compound can be achieved by certain bacteria. The graph below shows the relationship between breakdown of terephthalate and conversion into methane by these bacteria in an experimental reactor.

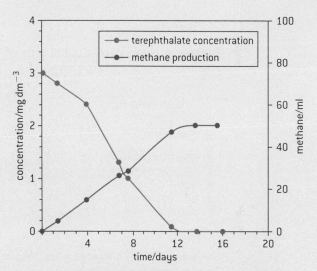

Source: Jer-Horng Wu, Wen-Tso Liu, I-Cheng Tseng, and Sheng-Shung Cheng, "Characterization of microbial consortia in a terephthalate-degrading anaerobic granular sludge system", Microbiology, Volume 147 (2001), pp. 373–382, © Society for General Microbiology. Reprinted with permission.

a) The reactor has a volume of 12 litres. Calculate the initial amount of terephthalate in the reactor. [1]

b) Describe the relationship between terephthalate concentration and methane production. [2]

c) Suggest which bacteria can be used for the degradation of terephthalate. [1]

d) Evaluate the efficiency of the terephthalate breakdown into methane. [2]

3 The evolution of hemoglobin molecules has been studied extensively by comparing the amino acid sequences in both myoglobin and hemoglobin. Myoglobin is used for oxygen storage while hemoglobin is used for oxygen transport. Ancient prehistoric animals had a single chain of simple globin for oxygen storage and transport. About 500 million years ago, a gene duplication event occurred and one copy became the present day myoglobin and the other evolved into an oxygen transport protein that gave rise to the present day hemoglobin.

The following figures are phylogenetic trees of hemoglobin in different organisms.

a) State how many years ago the hemoglobin split into α chains and β chains. [1]

b) Estimate the number of gene duplication events that have occurred from the simple globin. [1]

c) Using figure B, compare the phylogenetic relationship of myoglobin with vertebrate and invertebrate hemoglobin. [1]

d) Suggest a reason for the difference in function of hemoglobin between plants and animals. [1]

e) Explain why changes observed in the sequence of amino acids may lead to an underestimate of the actual number of mutations. [2]

Figure A Note: each shaded area of the chromosomes below represents a gene.

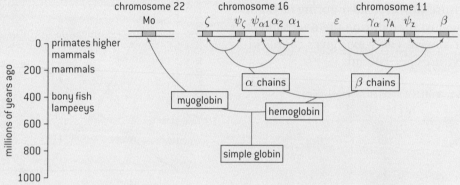

Source: Adapted from C K Mathews, K E van Holde and K G Ahern (2000), *Biochemistry*, 3rd edition, Benjamin Cummings, page 241

Figure B

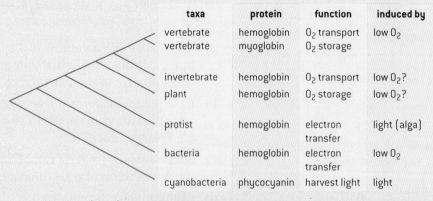

taxa	protein	function	induced by
vertebrate	hemoglobin	O₂ transport	low O₂
vertebrate	myoglobin	O₂ storage	
invertebrate	hemoglobin	O₂ transport	low O₂?
plant	hemoglobin	O₂ storage	low O₂?
protist	hemoglobin	electron transfer	light (alga)
bacteria	hemoglobin	electron transfer	low O₂
cyanobacteria	phycocyanin	harvest light	light

Source: R Hardison (1999), *American Scientist*, 87, pages 126–137

C ECOLOGY AND CONSERVATION

Introduction

Ecology is research into relationships between organisms and their natural environment. It underpins conservation measures that are aimed at ensuring the survival of as much of the Earth's biodiversity as possible. Community structure is an emergent property of an ecosystem. Changes in community structure affect and are affected by organisms. Human activities impact on ecosystem function. For this reason, entire communities need to be the target of conservation in order to preserve biodiversity.

C.1 Species and communities

Understanding

→ The distribution of species is affected by limiting factors.

→ Community structure can be strongly affected by keystone species.

→ Each species plays a unique role within a community because of the unique combination of its spatial habitat and interactions with other species.

→ Interactions between species in a community can be classified according to their effect.

→ Two species cannot survive indefinitely in the same habitat if their niches are identical.

Applications

→ Distribution of one animal and one plant species to illustrate limits of tolerance and zones of stress.

→ Local examples to illustrate the range of ways in which species can interact within a community.

→ The symbiotic relationship between zooxanthellae and reef-building coral reef species.

Nature of science

→ Use models as representations of the real world: zones of stress and limits of tolerance graphs are models of the real world that have predictive power and explain community structure.

Skills

→ Analysis of a data set that illustrates the distinction between fundamental and realized niche.

→ Use of a transect to correlate the distribution of plant or animal species with an abiotic variable.

Limiting factors

The distribution of species is affected by limiting factors

A limiting factor is the factor that is most scarce in relation to an organism's needs.

Plant distributions are affected by abiotic variables: temperature, water availability, light intensity, soil pH, soil salinity and availability of mineral nutrients. Every plant species has a range of tolerance for each of these factors, and is excluded from areas that are outside the range for one or more of the factors. For example, plant species from the tropics are not adapted to survive frosts so would not survive in northern regions. Plants from these northern regions have chemicals in their cells that act like anti-freeze and prevent frost damage from ice crystal formation. However, these northern plant species are not adapted to grow in the tropics. They would transpire excessively and their method of photosynthesis would be very inefficient at high temperatures.

Animal distributions are affected by temperature, water, breeding sites, food supply and territory. Extremes of temperature require special adaptations. For example, the large ears of elephants are adaptations for dissipating heat. This allows them to live in hot environments. Some animals have adaptations for living in arid conditions. For example, the kidneys of desert rats have longer loops of Henlé.

Many species of animals need a special type of breeding site and can only live in areas where these sites are available. The natterjack toad (*Epidalea calamita*) is native to sandy and heathland areas of northern Europe. The pools in which they lay their eggs need to have a very slight slope with sparse vegetation on the banks and in the water.

Some species of animals establish and defend territories either for breeding or for feeding. Some animal species have very specific food requirements, for example the leaves of only one plant species, which limits their distribution.

Food supply can affect animal distribution. Birds of temperate regions migrate because of the diminished food supply brought on by winter and also to escape cold weather. Tropical birds migrate because of the diminished food supply brought on by the dry season.

Using transects

Use of a transect to correlate the distribution of plant or animal species with an abiotic variable.

A sample is random if every member of a population has an equal chance of being selected in the sample. A transect is a method used to ensure that there is not bias in a student's selection of a sample and it can be used to correlate the distribution of a plant or animal species with an abiotic variable. For example, a transect from an area of grassland into woodland would reveal changes in distribution associated with light intensity and other variables.

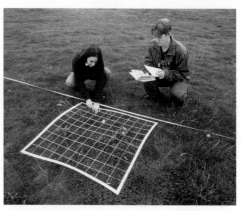

There are a number of types of transects including:

- Line transects where a tape is laid along the ground between two poles. In a line transect, sampling can be confined to describing all of the organisms that touch the line or distance of samples from the line can be recorded.

- Belt transects are where sampling is carried out between two lines separated by a fixed distance such as 0.5 metres or 1.0 metres.

- Point transects are used in studies of bird populations. Randomly selected points are selected and the researcher stands at that point and makes observations within a certain radius of the point.

▲ Figure 1 Students carrying out a survey of the plants in a grassy area by combining two different methods. The quadrat (square grid) is being used at intervals along a line transect (straight line). This is sometimes referred to as an interrupted belt transect

Data-based questions: Intertidal zonation

The kite diagram (figure 2) illustrates the distribution of common intertidal species 300 m south of Bembridge Lifeboat Station on the Isle of Wight, UK. The thickness of the shaded region indicates whether the organism is abundant, common, frequent, occasional or rare according to the key (abundance scale).

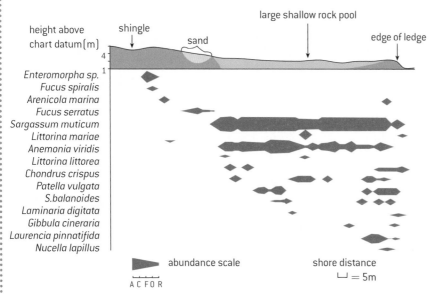

▲ Figure 2 Species abundance as a function of distance from the shore

1 Examine the kite diagram and explain the methods used to collect the data. [3]

2 State the species that is most abundant in the survey area. [1]

3 Using the scale bar, determine the length of the large shallow rock pool. [2]

4 Deduce one species adapted to: a) shingle b) sand c) rock pools. [3]

5   Several species are only found near the lower edge of the intertidal zone. Suggest reasons for them being absent from the upper parts of the intertidal zone.   [4]

6   Using the data in the kite diagram, predict two species that are adapted to the same abiotic environment.   [2]

 Ecological models

Use models as representations of the real world: zones of stress and limits of tolerance graphs are models of the real world that have predictive power and explain community structure.

Figure 4 is a model of how environmental gradients affect population levels of a species. The range of values of a biotic or abiotic factor that is tolerated is a characteristic of a species, but within a population there is variability. Some members of the population are more tolerant of extreme conditions than others and the limits of tolerance and where the zones of stress start is sometimes difficult to quantify. Another limitation of the model is that it is often presented as a symmetrical graph, when a shortage may have a more acute affect than an abundance or vice versa. For example, there is often an upper limit of a toxin that can be tolerated but often no lower limit. Consider the effect of water depth on the broadleaf cat-tail (*Typha latifolia*) from Michigan state in the US. It can grow out of water, but

appears to prefer a water depth of approximately 20 to 60 cm. Further increases of depth cause a precipitous decline in the dry biomass of the plant (see figure 3).

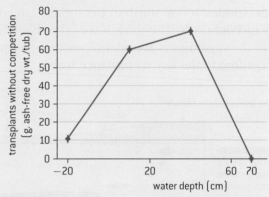

▲ Figure 3

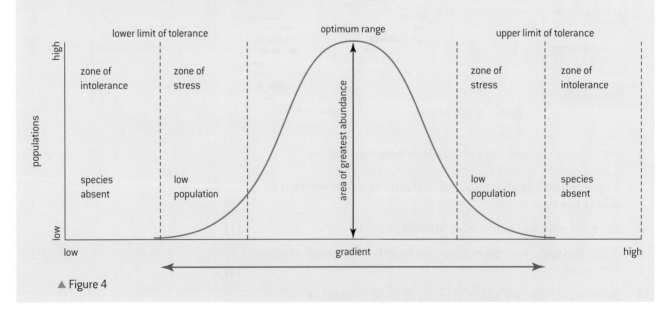

▲ Figure 4

Application of an ecological model

Distribution of one animal and one plant species to illustrate limits of tolerance and zones of stress.

Data-based questions

In the graph in figure 5, the relative shoot mass is shown for two plants in increasing concentrations of NaCl: sea blite (*Suaeda maritima*) is represented by the green line and the salt water cress (*Eutrema halophilum*) is represented by the red line.

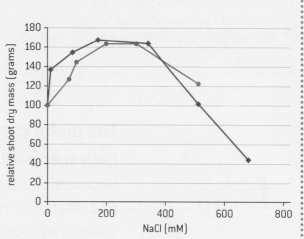

▲ Figure 5

Use the graph to suggest the value of the following:

1. The optimum range of NaCl concentration for both plants [1]

2. The starting value of the lower zone of stress. [1]

3. Explain why it is difficult to determine the limits of tolerance for the two plant species from the data given. [3]

Data-based questions: Maintaining conditions for aquarium fish

Ornamental fish for decorative aquariums are sometimes captured from wild populations in the Amazon and exported. One study found that between 30 and 70% of the fish captured die before being delivered to the final consumer. The cardinal tetra (*Paracheirodon axelrodi*) is the most popular ornamental fish in terms of export demand. According to one case study, four out of five fish imported from Brazil into the US were lost due to mortality.

Maintaining the water quality within the range tolerated by the fish is important for minimizing mortality.

Table 1 shows the lower and upper lethal limit of temperature at which 50% of fish do not survive (LT_{50}), the lower and upper lethal pH (LC_{50}), and the upper LC_{50} of ammonia and nitrite.

	Tolerance				
LT_{50}		LC_{50}			
Low temp.	High temp.	Acid pH	Alkaline pH	Ammonia	Nitrite
19.6 °C	33.7 °C	2.9	8.8	23.7 mg/L	1.1 mg/L

▲ Table 1 Lethal lower and higher temperatures (LT_{50}), and lethal concentrations (LC_{50}) of acid and alkaline pH, total ammonia and nitrite to cardinal tetra (*Paracheirodon axelrodi*)

1   Sketch a possible zone of tolerance graph for temperature and another one for pH for the cardinal tetra.

2   Sketch an upper zone of tolerance graph for ammonia and nitrite.

3   Use your models to suggest the optimum values of these water qualities to recommend to shipping agents who handle the cardinal tetra.

▲ Figure 6 Cardinal tetras (*Paracheirodon axelrodi*) in an aquarium

The niche concept

Each species plays a unique role within a community because of the unique combination of its spatial habitat and interactions with other species.

Within an ecosystem, each species fulfills a unique role, called its ecological niche. This includes the spatial habitat (where the species lives in the ecosystem), how the species obtains its food and the interactions with other species. For a species to be able to inhabit an area, there must be a suitable habitat where the abiotic variables which influence its survival are within the zone of tolerance. It must be able to obtain or synthesize food, and the other species with which it needs to interact must also be present.

Competitive exclusion principle

Two species cannot survive indefinitely in the same habitat if their niches are identical.

In the 1930s the Russian scientist Carl Friedrich Gauss investigated competition between two species of *Paramecium: P. caudatum* and *P. aurelia*. Gauss quantified biomass by estimating the volume of the paramecia. When cultured separately, under ideal laboratory conditions, both thrived. When cultured together, the numbers of both were reduced, but *P. caudatum* was reduced disproportionately.

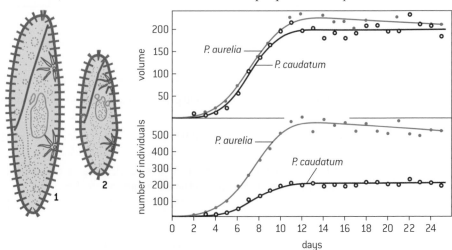

▲ Figure 7 *P. caudatum* has a higher volume than *P. aurelia*

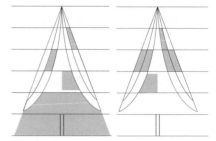

The bay-breasted warbler and the yellow-rumped warbler are migratory birds that appear to occupy the same niche, as they feed on similar prey items and can be found foraging together on the same tree. Figure 8 illustrates observed parts of the tree where each warbler most commonly forages. Note that the birds feed in such a way as to avoid competition with one another.

It appears from these observations that two species cannot coexist in the same habitat if their niches completely overlap. This is known as the competitive exclusion principle. Either one species will lead to the decline and extirpation of the other, or one or both of the competitors will narrow their niches to avoid competition.

Yellow-rumped warbler   Bay-breasted warbler

▲ Figure 8 The yellow-rumped warbler and bay-breasted warbler have similar prey yet they tend to forage in different parts of the same tree to avoid competition

Fundamental and realized niches

Analysis of a data set that illustrates the distinction between fundamental and realized niche.

The fundamental niche of a species is the potential mode of existence, given the adaptations of the species. It refers to the broadest range of habitats it can occupy and roles it can fulfill. The realized niche is the actual mode of existence, which results from the combination of its adaptations and competition with other species.

Data-based questions: Competitive exclusion in cat-tails

Figure 10 shows the distribution of two species of wetland plant over a range of depths of water. The plants are called cat-tails – *Typha latifolia* and *Typha angustifolia*. Negative depth means out of the water. The upper graph shows the situations where both species occur in a natural habitat together. The lower graph shows distributions in situations where the two species are grown separately.

1 Compare and contrast the distribution of *T. angustifolia* in the presence and absence of *T. latifolia*. [3]

2 With respect to this data, explain the concept of fundamental and realized niche. [4]

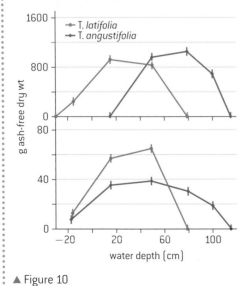

▲ Figure 10

▲ Figure 9 Yellow-rumped (top) and bay-breasted warblers

Data-based questions: Character displacement in ants

It has been suggested that competition between species may not only restrict a species to a narrower niche, but cause a change in some physical characteristics too. This is character displacement – the character changes when competition occurs. An example of character displacement is seen in seed-eating ants from the southwestern United Sates. The size of the mandibles (chewing mouth parts) of *Veromessor pergandei* determines the size of seeds they eat. The histograms in figure 11 show the number of *V. pergandei* in each frequency class of mandible size from different locations. The names of other seed-eating ants found in each habitat are included showing their mean mandible size.

1   Name the ant species with the smallest mean mandible size. (1)

2   Compare and contrast the frequency distribution of *Veromessor pergandei* mandible size between the four areas. (3)

3   Suggest what might be the fundamental niche of *Veromessor pergandei* in terms of size of seed eaten. (2)

4   Evaluate the hypothesis that the presence of multiple competitors decreases the variability of mandible size of *Veromessor pergandei*. (3)

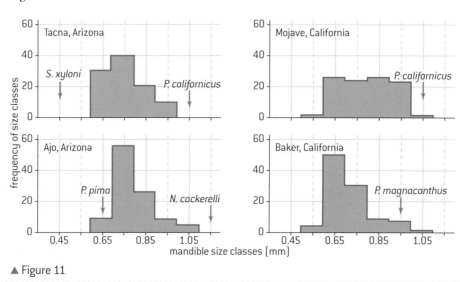

▲ Figure 11

Interspecific interactions

Interactions between species in a community can be classified according to their effect.

Within ecosystems, the interactions between species are complex. Five common types of interaction are described here. Competition occurs when two species require the same resource and the amount obtained by one species reduces the amount available to the other. Bracken and bluebells compete for light, but bluebells minimize the competition by starting into growth earlier than the taller-growing bracken. Red and grey squirrels compete for food in Britain where they occur together, with the grey squirrels usually obtaining so much more food that the red squirrels disappear.

Herbivory is primary consumers feeding on producers. Bison feed on grasses and limpets feed on algae growing on rocks in the intertidal zones of rocky shores.

Predation involves a consumer feeding on another consumer for example the bay-breasted warbler, which winters in Guatemala, feeds on insects including dragonflies and the dingo in New South Wales feeds on the red kangaroo.

Parasitism is when one organism feeds off another but does not normally kill it. The predatory organism in this case is termed a parasite and the prey a host. The host is harmed and the parasite benefits. North American bighorn sheep are frequently parasitized by the lungworm *Prolostrongylus stilesi*. Protists of the genus *Schistosoma* use people as hosts.

In mutualism, two species live in a close association where both organisms benefit from the association. Many mammals that consume grass have bacteria living in their guts, which digest cellulose in the grass. Many flowering plants and their insect or mammal pollinators have mutualistic relationships.

In commensalism, one organism benefits and the other is neither harmed nor helped. A broad category of plants called epiphytes live on other plants and rely on them for support but do not normally get their nutrition from their host. Examples include many different kinds of moss.

 ### The role of zooxanthellae in ecosystems

The symbiotic relationship between zooxanthellae and reef-building coral reef species.

Most corals that build reefs contain mutualistic photosynthetic algae called zooxanthellae. The coral provides the alga with a protected environment and a substrate that can hold it in place for photosynthesis to occur. The zooxanthellae provide the coral with molecules such as glucose and amino acids. The association ensures the recycling of nutrients which are in short supply in tropical waters.

Zooxanthellae are responsible for the unique coloration of many corals and make coral reefs one of the most biologically productive ecosystems.

 ### Local examples of interspecific interactions

Local examples to illustrate the range of ways in which species can interact within a community.

Below are examples of the different types of interaction between organisms found in and around the Bahamian island of New Providence.

Figure 12 shows dodder. This is a non-photosynthetic vine that invades plant tissues and obtains both nutrients and support from the plant. This is a form of parasitism.

Fire coral is a stinging coelerentate. The Hawkfish is immune to the effects of the coral and so gains protection from the coral without helping or harming it. This is an example of commensalism (see figure 13).

▲ Figure 12

▲ Figure 13

A major herbivore that consumes tissues of the buttonwood tree is the bagworm moth (*Biopsyche thoracia*) shown in figure 14.

▲ Figure 14

Visits to flowers by pollinating hummingbirds such as the Bahama woodstar is a form of mutualism. The bird gains nectar as a food source and the plant gets assistance with pollination.

▲ Figure 15 Bahama woodstar (*Calliphlox evelynae*)

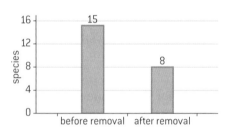

▲ Figure 16

Keystone species

Community structure can be strongly affected by keystone species.

A keystone species is one which has a disproportionate effect on the structure of an ecological community. Robert Paine was the first scientist to use the term related to his studies of the sea star *Pisaster*. He artificially removed the sea star from one part of the community and left the population intact in another area.

The following changes occurred as a consequence of the removal:

- The remaining members of the food web in the study area immediately began to compete with each other to occupy the newly available space and resources. Further, the sea star is an important predator of the species that eventually over-ran the study site.

- Within three months of the removal, the barnacle *Balanus glandula* had become dominant within the study area.

- Nine months later, *Balanus glandula* had been replaced by a population of another barnacle *Mitella* and the mussel *Mytilus*.

- Succession continued until *Mytilus* became the dominant species. The sea star is an important predator of *Mytilus*.

- Eventually the succession of species wiped out populations of benthic algae.

- Some species, such as the limpet, emigrated from the study area because of lack of food and/or space.

- Within a year of the starfish's removal, species diversity had decreased in the study area from fifteen to eight species (figure 16).

Other examples of keystone species include the sea otter, elephants, the mountain lion and the prairie dog.

▲ Figure 17 Ochre sea star *Pisaster ochraceus*

C.2 Communities and ecosystems

Understanding

→ Most species occupy different trophic levels in multiple food chains.

→ A food web shows all the possible food chains in a community.

→ The percentage of ingested energy converted to biomass is dependent on the respiration rate.

→ The type of stable ecosystem that will emerge in an area is predictable based on climate.

→ In closed ecosystems energy but not matter is exchanged with the surroundings.

→ Disturbance influences the structure and rate of change within ecosystems.

 ## Applications

→ Conversion ratio in sustainable food production practices.

→ Consideration of one example of how humans interfere with nutrient cycling.

 ## Skills

→ Comparison of pyramids of energy from different ecosystems.

→ Analysis of a climograph showing the relationship between temperature, rainfall and ecosystem type.

→ Construction of Gersmehl diagrams to show the inter-relationships between nutrient stores and flows between taiga, desert and tropical rainforest.

→ Analysis of data showing primary succession.

→ Investigation into the effect of an environmental disturbance on an ecosystem.

 ## Nature of science

→ Models are representations of the real world: pyramids of energy model the energy flow through ecosystems.

Trophic levels

Most species occupy different trophic levels in multiple food chains.

An organism's trophic level is its feeding position in a food chain. Because feeding relationships within an ecosystem are often web-like, an organism can occupy more than one trophic level. For example, the diet of an owl involves animals which occupy different trophic levels.

An owl pellet is a mass of undigested parts of the owl's diet which it regurgitates. The contents of the pellet can be used to gather information about the owl and its community without disturbing the bird. The contents might include such things as exoskeletons of insects, bones (including skulls), fur and claws.

If the species found within the pellet can be identified, their trophic level can be found. Alternatively, it may be possible to deduce the trophic

613

levels from the adaptations. The contents of a pellet often show that an owl has been feeding at more than one trophic level.

The three skulls in figure 1 are different species of rodent that might be found in an owl pellet. The dentition indicates whether the animal was a primary consumer, feeding on plant material, or a secondary or tertiary consumer, feeding on primary or secondary consumers.

When stating an organism's trophic level, reference needs to be made to a particular food chain.

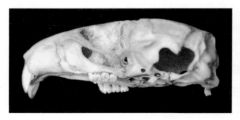  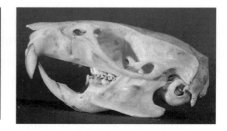

▲ Figure 1 Rodent skulls

Data-based questions: Fishing down marine food webs

Trophic levels can be represented by a number indicating the position of a species within an ecosystem. By definition, producers occupy the first trophic level (TL) and so have a TL of 1. For primary consumers TL = 2, and so on. The larger the number, the more energy-transfer steps between the organism and the initial fixing of the sun's energy. Trophic levels are not always stated as whole numbers. Fish and other animals that feed at more than one level often have estimated mean trophic levels.

One effect of commercial over-fishing is the reduction in the number of fish that feed at higher trophic levels, such as long-lived fish. The phrase "fishing down marine food webs" refers to the increased tendency for marine landings to consist of animals that feed at lower trophic levels (see figure 2).

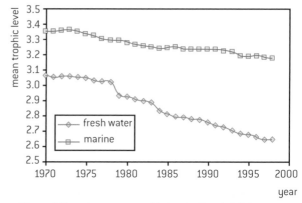

▲ Figure 2 How the mean trophic level of landed fish has changed over a 30-year period

1   Suggest a method that might be used to deduce the trophic level of a fish once it is captured. [2]

2   a)   Compare the changes in mean trophic level of landed fish from marine and freshwater fisheries since 1970. [3]

b)   Suggest why there is a difference in the two trends. [2]

3   Explain why the mean trophic level might increase with the age of an individual fish. [2]

4   Deduce the change in age of captured fish over the period shown. [2]

5   Explain two advantages of humans catching and consuming fish at a lower mean trophic level. [4]

Food webs

A food web shows all the possible food chains in a community.

Trophic relationships within ecological communities tend to be complex and web-like. This is because many consumers feed on more than one species and are fed upon by more than one species. A food web is a model that summarizes all of the possible food chains in a community.

Figure 3 shows a simplified food web for a pond.

When a food web is constructed, organisms at the same trophic level are often shown at the same level in the web. This isn't always possible, as some organisms feed at more than one trophic level.

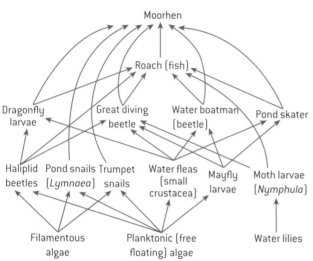

▲ Figure 3 A pond food web

🧬 Pyramids of energy as models

Models are representations of the real world: pyramids of energy model the energy flow through ecosystems.

A pyramid of energy is a type of bar chart that is used to show the relative amounts of energy flowing through each trophic level. The bars are horizontal and are arranged symmetrically. The lowest bar represents the production of the producers, either gross or net. The bar above represents the primary consumers, with the secondary consumers above that, and so on upwards. When constructing a pyramid of energy, every bar should be labelled and the units should be indicated – usually $kJ\ m^{-2}\ year^{-1}$. The same scale should used for each bar, if possible, though many pyramids in text books are not drawn to scale. The limitation of the model is that energy transfer rates can vary over seasons. Further, diet analysis is necessary for organisms occupying different trophic levels in different food chains. The percentage composition of their diets may vary according to season or opportunity.

Data-based questions: Energy pyramids

The diagram shows energy flow in a stream in Concord, Massachusetts. The debris consisted almost entirely of leaves and other parts of plants that dropped into the stream.

1   Explain how the heat shown in the diagram is produced. [2]

2   Calculate the net production of the herbivores (primary consumers). [1]

3   The amount of energy flowing to the herbivores is 2,300 kJ m^{-2} year^{-1}

a) State the amount of energy flowing to the primary consumers. [1]

b) Calculate the percentage of the energy flowing to the herbivores that flows on to the primary consumers. [2]

4   Construct an energy pyramid for four trophic levels in the stream. The gross production of producers is estimated to be 30,600 kJ m^{-2} year^{-1}. [4]

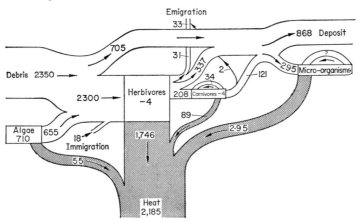

Food conversion ratios

Conversion ratio in sustainable food production practices.

The production of meat for consumption requires the animals to be fed. Feed conversion ratios refer to the quantity of dietary input in grams required to produce a certain quantity of body mass in livestock or fish.

For example, a feed conversion ratio of 1.2 means that 120 grams of feed are required to produce 100 grams of body mass.

Table 2 shows the feed conversion ratios of several animals farmed for human consumption as reported in one document. Such numbers vary significantly in the reported literature due to variation in the content of food used, feeding methods, the age of the animals and other variables.

The implication of these ratios for sustainability is that there are some dietary choices which are more sustainable. Lower feed inputs mean lower energy inputs for food production. Avoiding the consumption of meat means that there would be lower energy losses due to feed conversion.

The nature of the feed inputs are a consideration for sustainability. Consider the example of aquaculture operations to raise salmon. While farmed salmon is fed protein meal formed from other fish, livestock often consume plant matter. Fish farmers can use feed that is easier to digest so there is less fecal waste from fish. Feeding can be closely monitored to ensure that feed levels are adjusted to eliminate uneaten food. Uneaten food and fecal waste lower the carrying capacity of ponds used to raise fish and so more energy is required to produce the same quantity of fish.

Meat production	Feed conversion ratio estimate
Salmon	1.2
Beef	8.8
Pork	5.9
Chicken	1.9

▲ Table 2

The impact of climate on ecosystem type

The type of stable ecosystem that will emerge in an area is predictable based on climate.

Climate is a property that emerges from the interaction of a number of variables including temperature and precipitation.

Temperature influences the distribution of organisms. Temperature influences rates of cell respiration, photosynthesis, decomposition and transpiration and ultimately has an impact on productivity.

Precipitation also impacts productivity by influencing rates of photosynthesis and rates of decomposition. Information about the relative combinations of these two factors in an area can allow for predictions about what kind of stable ecosystem will exist in that area. High rainfall results in the formation of a forest, moderate or seasonal rains will result in grassland. Very little or no rainfall results in a desert.

Very high rainfall and high temperatures result in a tropical rainforest whereas very high rainfall and cooler temperatures result in a temperate rainforest.

 Interpreting a Whittaker climograph

Analysis of a climograph showing the relationship between temperature, rainfall and ecosystem type.

A climograph is a diagram which shows the relative combination of temperature and precipitation in an area. Figure 4 is a modified climograph first developed by the ecologist Robert Whittaker. It shows the most likely stable ecosystem that will emerge in an area under certain climactic conditions. The dashed line in the graph represents regions where the type of biome is strongly influenced by other factors such as fire, soil type, animal grazing and the seasonality of drought.

a) Determine the types of ecosystems which can exist where a mean annual precipitation level of 175 cm exists.

b) Determine the range of conditions under which a tropical rainforest will form.

c) List other variables that likely influence the type of stable ecosystem that forms.

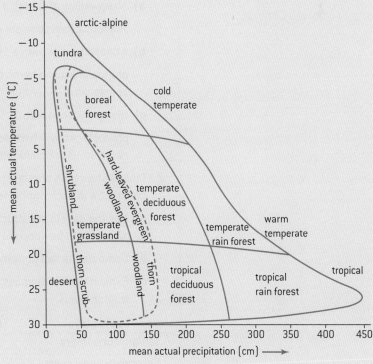

▲ Figure 4

Comparison of pyramids of energy from different ecosystems

The length of food chains is determined by the level of net primary productivity. The higher the productivity, the longer the food chains and the broader the trophic level at each stage of the pyramid. Figure 5 shows the differences in net productivity of different ecosystems.

Energy conversion efficiencies are affected by the organisms involved. For this reason, pyramids of energy differ between ecosystems.

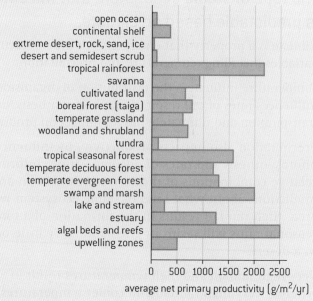

▲ Figure 5

Activity

Communities differ in their energy conversion efficiency.

1 For each of the following communities, construct energy pyramid models, drawn to scale, based on the energy conversion efficiency shown.

   a) an upwelling area with a 20% energy conversion efficiency.

   The food chain consists of phytoplankton→anchovy

   b) a coastal region with a 15% energy conversion efficiency

   phytoplankton→herbivorous zooplankton→carnivorous zooplankton→herring

   c) the open ocean with 10% energy conversion efficiency

   phytoplankton→herbivorous zooplankton→ carnivorous zooplankton→carnivorous fish→tuna

2 Table 3 shows the annual energy fixed in biomass in joules per square centimetre in each trophic level of two separate ecosystems.

   a) Use the data to construct two separate pyramids of energy. They should both be drawn to the same scale.

   b) Compare the two pyramids.

   c) Explain the low biomass and low numbers of organisms in higher trophic levels.

Trophic level	Cedar Bog Lake	Lake Mendota
Tertiary consumers	–	0.2
Secondary consumers	0.8	1.4
Primary consumers	3.6	35.1
Producers	27.1	104.4

▲ Table 3

Gersmehl nutrient cycle diagrams

Construction of Gersmehl diagrams to show the inter-relationships between nutrient stores and flows between taiga, desert and tropical rainforest.

A Gershmehl diagram is a model of nutrient storage and flow for terrestrial ecosystems. Figure 6 shows three Gersmehl diagrams for three different ecosystems. Figure 7 provides a detailed explanation of the diagram for a tropical rainforest. The model presumes three storage compartments: biomass, litter and the soil. Storage compartments, or pools, are represented by circles or ovals. Arrows represent nutrient flows, or fluxes. The thickness of the arrows represent rates of flow of nutrients. One arrow can represent more than one process.

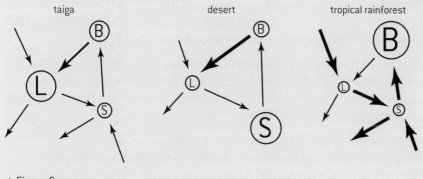

▲ Figure 6

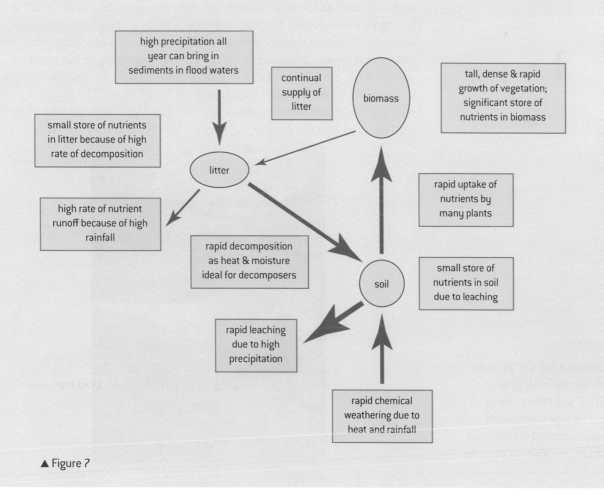

▲ Figure 7

Activity

Look at figure 6.

1 Identify the ecosystem type with the largest nutrient store being the soil.

2 Identify an ecosystem type where the rate of litter decomposition is low.

3 Identify arrows which could represent the following processes:

   a) run-off

   b) mineral absorption by plants

   c) regurgitation of an owl pellet

4 Compare the nutrient cycles of taiga, desert and tropical rainforests.

 Primary succession

Analysis of data showing primary succession.

Ecological successions are the changes that transform ecosystems over time. The changes involve both the species making up the community and their abiotic environment. They are the result of complex interactions between the community and the environment.

In an ecosystem, abiotic factors set limits to the distribution of living organisms, and the organisms have an effect on abiotic factors. Consider a forest next to an area of grassland. Relative to the grassland, the forest will have lower light intensity and be cooler and more humid, largely because of the presence of the trees. Leaf litter from the trees will increase the rate of water infiltration in the soil, increase the nutrient concentration of the soil and will directly and indirectly affect the aeration of the soil.

Communities of living organisms may change the abiotic factors to such an extent that the environment becomes limiting to some of them and other species join the community as they are better adapted. This happens during succession.

Two categories of succession are recognized: primary and secondary. Primary succession begins with an environment, such as a retreating glacier where living organisms have not previously existed. At the start of a primary succession, there are only organisms that can survive on rock surfaces, such as bacteria, lichens and mosses. Small amounts of soil are formed allowing small herbs to colonize and, as a deeper soil develops, successively larger plants colonize – tall herbs, shrubs and finally in most areas trees. Consumer populations will change with the plant populations, as will populations of decomposers and detritivores. Figure 8 shows a photo of the same area separated by 17 years. The top photo was taken in 1985, the bottom photo in 2002. The sign in the picture indicates that the area was covered in the ice of a retreating glacier in 1920.

Activity

Compare the site as shown in figure 8 in 1985 and 2002 and infer some of the changes that would have occurred to biotic and abiotic variables in the area.

▲ Figure 8

Data-based questions

Captain George Vancouver visited the area now known as Glacier Bay, Alaska in 1794. He made detailed notes regarding the position of the glaciers. This has allowed researchers to determine the time since the start of primary succession as the glacier retreated.

The first species to colonize the bare rock are bacteria, lichens and moss. Mountain avens (*Dryas drummondii*) is a flowering shrub that dominates after the moss stage. Deciduous alder trees (*Alnus sinuata*) invade next followed by the most stable ecosystem which is a spruce and hemlock forest.

Figure 9 shows the mean stem diameter and range of diameter of plants as a function of time since the tongue of the glacier covered the area at eight sites (E1-E8).

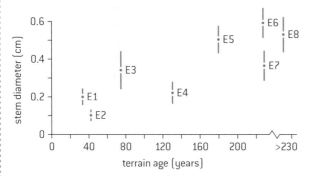

▲ Figure 9

1 a) Outline the changes in mean stem diameter with time. [2]

b) Explain the change in mean stem diameter. [2]

Figure 10 shows the number of species found in Glacier Bay, Alaska as a function of time since the glacier covered the area.

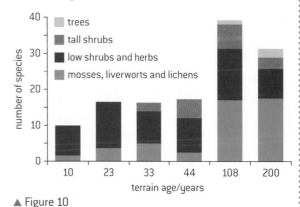

▲ Figure 10

2 a) Outline the changes in the number of species (species richness). [2]

b) Outline the changes in the relative numbers of species types (species evenness) [2]

Figure 11 shows changes in the properties of soil as the dominant plant species change.

3 a) Outline the changes in soil properties that are seen. [12]

b) Deduce the stage where the greatest changes in soil properties are observed. [2]

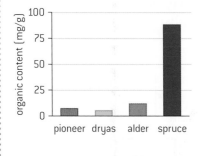
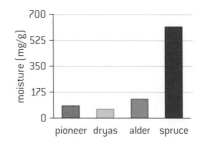
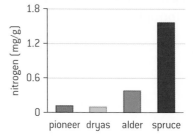
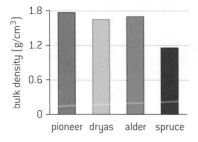
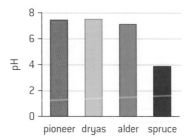

▲ Figure 11

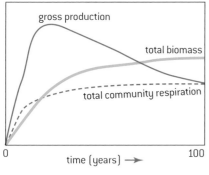

▲ Figure 12

Respiration rates and biomass accumulation

The percentage of ingested energy converted to biomass is dependent on the respiration rate.

Production in plants happens when organic matter is synthesized by photosynthesis. In animals it occurs when food is absorbed after digestion. Energy units are usually used for measuring production e.g. kilojoules. The amounts of energy are given per unit area, usually per m² and per year. Gross and net production values can be calculated using this equation:

$$\text{net production} = \text{gross production} - \text{respiration}$$

Gross production is the total amount of organic matter produced per unit area per unit time by a trophic level in an ecosystem.

Net production is the amount of gross production remaining after subtraction of the amount used for respiration by the trophic level.

In the early stages of primary production, the high availability of sunlight means that gross production is high and there is little total biomass in the community. As a result, the total amount of respiration to support the small biomass is low. As succession proceeds, the standing biomass increases and the total amount of respiration increases. Further, the amount of gross production begins to decline once all available spaces for stems become filled. An equilibrium is reached where the total community production to total community respiration (P/R) ratio equals 1. When this occurs, the ecosystem has reached a relatively stable stage.

Data-based questions: Calculating productivity values

The energy flow diagram in figure 13 is for a temperate ecosystem. It has been divided into two parts. One part shows autotrophic use of energy and the other shows heterotrophic use of energy. All values are kJ m⁻² yr⁻¹.

1 Calculate the net production of the autotrophs. [1]

2 Compare the percentage of heat lost through respiration by the autotrophs with that lost by the heterotrophs. [1]

3 Most of the heterotrophs are animals. Suggest one reason for the difference in heat losses between the autotrophs and animal heterotrophs. [1]

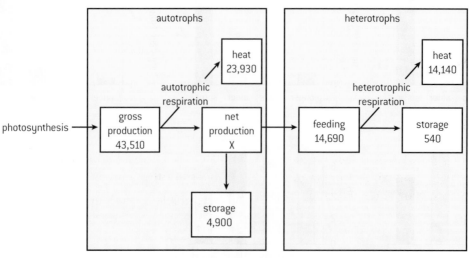

▲ Figure 13 An energy flow diagram for a temperate ecosystem

Secondary succession

Disturbance influences the structure and rate of change within ecosystems.

Secondary succession takes place in areas where there is already, or recently has been, an ecosystem. The succession is initiated by a change in conditions. Construction sites or roads might become disused and eventually plants grow up in the remains. Old-field succession occurs when an arable field or meadow (field of grassland) is abandoned. The lack of tillage or grazing initiates the succession. Figure 14 shows the sequence of communities following the abandonment of an arable field, with approximate timings. Examining the time scale in figure 14 indicates that the pace of change slows as succession proceeds. Close to the time of the disturbance, rates of system respiration and productivity increase rapidly and there is an accumulation of biomass. Species diversity increases close to the time of the disturbance. At the "climax" stage shown in the diagram, changes are still occurring, but they are slower and the ecosystem is viewed as being more stable and resistant to change at the climax stage.

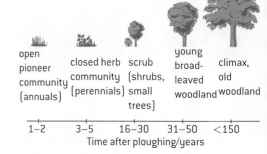

open pioneer community (annuals)	closed herb community (perennials)	scrub (shrubs, small trees)	young broad-leaved woodland	climax, old woodland
1–2	3–5	16–30	31–50	<150

Time after ploughing/years

▲ Figure 14 Secondary succession is the progression of communities where a pre-existing climax community has been disturbed but the soil is already developed

Data-based questions: Secondary succession

The three boxes in figure 15 show a field undergoing secondary succession. It is shown 5 years, 25 years and 30 years after the original disruption. Each numbered shape represents a distinct plant species.

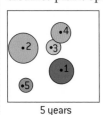 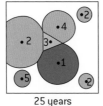 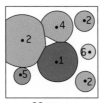

| 5 years | 25 years | 30 years |

▲ Figure 15 Community changes 5 years, 25 years and 30 years after an initial disturbance

1 Explain what happened to species number 3 during the succession. [2]

2 Deduce the changes that might have been occuring in

   a) gross plant production [1]

   b) species diversity [1]

   c) soil depth [1]

   d) amounts of minerals being recycled. [1]

3 Predict, with reasons, the species composition of this area after 50 years. [2]

🧪 Studying secondary succession

Investigation into the effect of an environmental disturbance on an ecosystem.

In your local area there may be opportunities to investigate secondary succession. Abandoned fields, wooded areas with disused roads and fields recovering from a fire are all examples of sites where succession can be studied. A number of possible variables that can be studied include:

- Species diversity
- Stem density
- Above ground biomass

- Leaf area index
- Volume of leaf litter
- Water cycle variables including infiltration rates and run-off rates
- Soil variables including soil structure, soil moisture, soil nutrient levels and compaction levels
- Light levels
- Bulk soil density.

623

▲ Figure 17 Biosphere 2 rainforest building near Tucson, Arizona. Biosphere 2 was created to explore the possible use of closed biospheres in space exploration

Closed ecosystems

In closed ecosystems energy but not matter is exchanged with the surroundings.

There are three categories of systems that can be modelled (figure 16). Open systems exchange matter and energy with their surroundings. Closed systems, such as the mesocosm that you constructed as part of your study of the core and Biosphere 2 (figure 17), exchange energy but not matter with the surroundings. Isolated systems, which are largely theoretical, exchange neither matter nor energy with their surroundings. Ecological systems exist along the continuum. Natural systems exchange both matter and energy with their surroundings and so are categorized as open systems. In undisturbed systems, the rate of exchange of matter with the surroundings occurs most notably due to the water cycle and nutrient cycles that have a gaseous phase. Human interference increases the exchange of matter either through harvesting of crops or addition or depletion of nutrients.

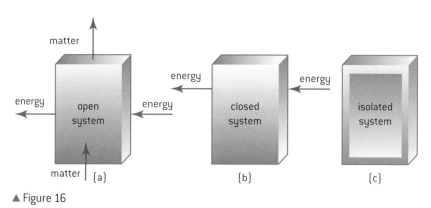

▲ Figure 16

Disruptions to nutrient cycling

Consideration of one example of how humans interfere with nutrient cycling.

While natural systems usually exchange matter with their surroundings, especially with respect to the water cycle and all nutrient cycles that have a gaseous phase, human activity can accelerate nutrient flows into and out of ecosystems.

Agriculture is an example of a human activity which interferes with nutrient cycling. The harvesting of crops and transport of the products out of the area where they were grown depletes the area of the nutrients that are locked in to the biomass of the crop. As a consequence,

regular inputs of nutrients must be added to the soil so that agriculture can continue in the area. Phosphate and nitrogen are key components of fertilizer.

Phosphate is mined, converted to fertilizer and shipped around the world for use in agriculture.

Nitrogen produced from gaseous N_2 in the Haber process has significantly increased inputs into the nitrogen cycle that would not occur naturally. Runoff from agricultural fields results in build-up in waterways and leads to eutrophication.

C.3 Impacts of humans on ecosystems

Understanding

→ Introduced alien species can escape into local ecosystems and become invasive.

→ Competitive exclusion and the absence of predators can lead to reduction in the numbers of endemic species when alien species become invasive.

→ Pollutants become concentrated in the tissues of organisms at higher trophic levels by biomagnification.

→ Macroplastic and microplastic debris has accumulated in marine environments.

Applications

→ Study of the introduction of cane toads in Australia and one other local example of the introduction of an alien species.

→ Discussion of the trade-off between control of the malarial parasite and DDT pollution.

→ Case study of the impact of marine plastic debris on Laysan albatrosses and one other named species.

Nature of science

→ Assessing risks and benefits associated with scientific research: the use of biological control has associated risk and requires verification by tightly controlled experiments before it is approved.

Skills

→ Analysis of data illustrating the causes and consequences of biomagnification.

→ Evaluation of eradication programmes and biological control as measures to reduce the impact of alien species.

Alien and invasive species

Introduced alien species can escape into local ecosystems and become invasive.

Human activity often results in an organism being introduced to an area where it did not previously occur. Species that are native to an area are referred to as endemic whereas species that are not native but are introduced by humans are referred to as alien species.

The impacts of an alien species are usually only significant if it increases in number and spreads rapidly. Species that do this are described as invasive. Many alien species are invasive, because the normal limiting factors in their original habitat are missing. The predators, diseases and vigorous competitors that controlled numbers in its native habitat are usually absent.

Many of these invasive species have significant effects on the ecosystems where they are released. Rats introduced to New Zealand contributed to the extinction of ground nesting bird species by predating their eggs. The signal crayfish, introduced from North America to Britain, competes with

625

the native white-clawed crayfish and has also brought a disease (crayfish plague) to which the native crayfish are not resistant, so large numbers have been killed. The floating fern, *Salvinia molesta,* has spread over the surfaces of many lakes in the tropics, eliminating the native aquatic plants by competition. Most of the impacts of alien species are harmful, particularly the excessive predation of native species, and interspecific competition due to niche overlap with native species.

Alien species compete with endemic species

Competitive exclusion and the absence of predators can lead to reduction in the numbers of endemic species when alien species become invasive.

An alien species can become so reproductively successful and aggressive that it dominates the new ecosystem and poses a serious threat to biodiversity. Organisms that are endemic to an area may occupy similar niches as the alien species. The competitive exclusion principle predicts that two species with overlapping niches cannot continue occupying overlapping niches indefinitely.

One consequence of the competition might be that either or both species may occupy smaller realized niches. If the alien species lacks predators it may be able to out-compete native species and become invasive. In the UK, the invasive Eastern grey squirrel (*Sciurus carolinensis*) occupies a similar niche to the native red squirrel (*Sciurus vulgaris*) and is a superior competitor. Whenever grey squirrels are introduced to an area, the native red squirrel population is reduced and often extirpated.

Alternatively, the ability of a new ecosystem to resist an alien species can prevent it from becoming invasive. The western thrips (*F. occidentalis*) is an invasive insect pest that has spread from the Western United States to many parts of the globe but has been less successful at colonizing the Eastern United States. One hypothesis is that it is being competitively excluded by a thrips species that is native to the Eastern US (*F. tritici*).

▲ Figure 1 Red squirrel (*Sciurus vulgaris*)

▲ Figure 2 Eastern grey squirrel (*Sciurus carolinensis*)

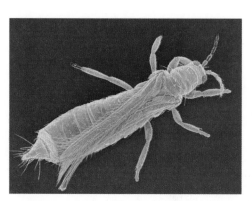

▲ Figure 3 *F. occidentalis*

🧬 The risk of biological control

Assessing risks and benefits associated with scientific research: the use of biological control has associated risk and requires verification by tightly controlled experiments before it is approved.

In some cases, biological control can be used to limit an invasive species. This involves introducing natural predators of the invasive species to limit its spread.

Figure 4 shows the aquatic plant taxifolia (*Caulerpa taxifolis*) being consumed by the violet sea slug (*Flabellina affinis*). Taxifolia is used as an ornamental plant in aquariums but it has become highly invasive in a number of areas. This is due in part to the toxin that the plant contains that allows it to resist predation. The violet sea slug is immune to these toxins and it has been suggested that it can be used as a form of biological pest control, but there are concerns that it could become invasive itself.

A risk-averse approach to introducing a natural enemy to control an invasive species involves holding the natural enemy in an approved facility that ensures it is quarantined. This prevents escape until research can establish that the natural enemy will have minimal negative impact in the new area where it is released.

▲ Figure 4

🌐 Case studies of introduced alien species

Study of the introduction of cane toads in Australia and one other local example of the introduction of an alien species.

Biological control can go wrong. One of the most troublesome invasive alien species, ironically, was originally introduced with the aim of it acting as a biological control. The cane toad (*Bufo marinus*) was introduced to Australia to control the cane beetle (*Dermoleida albohirtum*). The toad is a native of Central and South America. Unfortunately, the toad has become a generalist predator and a competitor for food resources. Research suggests that it has its biggest ecological impact on predators that consume the toad. The toad has a gland behind its head that can release a toxin when it is disturbed. The toad is lethal to consume for many predators because of the release of the toxin.

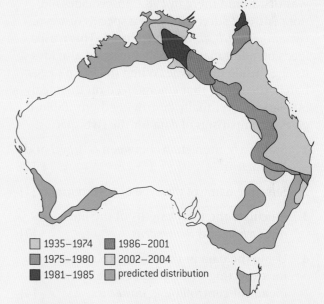

☐ 1935–1974	☐ 1986–2001
☐ 1975–1980	☐ 2002–2004
☐ 1981–1985	☐ predicted distribution

▲ Figure 5

Figure 5 shows a map of Australia that indicates changes in the distribution of the cane toad since its introduction in 1935.

The zebra mussel (*Dreissena polymorpha*) is an invasive species of the North American Great Lakes system that is native to the Black Sea and the Caspian Sea.

Empty cargo ships often fill their hulls with water as ballast to help them remain stable when they are not carrying cargo. When this ballast water is transported to other locations on the planet, it can lead to invasions of alien species if precautions are not taken when the ballast water is released. The zebra mussel is believed to have been transported to North America in ballast water.

▲ Figure 6 A section through a water pipe which has been opened to reveal an interior clogged with zebra mussels, *Dreissena polymorpha*

Since their introduction to the Great Lakes, zebra mussels have spread to many North American river systems. They have covered the undersides of docks and boats. Populations can grow so dense that they block pipes, clog municipal water systems and interfere with hydroelectric power generation. One proposed biological control mechanism of the zebra mussel is the use of *Pseudomonas fluorescens*, a bacterium that produces a toxin that appears to selectively affect the zebra mussel.

Evaluation of methods to control alien species

Evaluation of eradication programmes and biological control as measures to reduce the impact of alien species.

Data-based questions: The mango mealy bug

The mango mealy bug (*Rastrococcus invadens*) became a serious pest of crops in Ghana after it was first found there in 1982. Insecticides did not control it successfully, so a species of wasp, *Gyranusoidea tebygi*, was introduced. This wasp feeds on adult mango mealy bugs and also lays its eggs inside the larvae. When the wasp's eggs hatch, the larvae that emerge feed on the mealy bug larvae, eventually destroying them. Table 1 gives a comparison of data for the two species.

1 State two interactions between *G. tebygi* and *R. invadens*. [2]

2 State the name of this type of pest control. [1]

3 Calculate the total lengths of the life cycles of *G. tebygi* and *R. invadens*. [2]

4 Using the data in the table, explain reasons for expecting *G. tebygi* to be an effective means of controlling *R. invadens*. [4]

5 Suggest one risk of releasing *G. tebygi* in Ghana. [1]

	R. invadens	*G. tebygi*
Time between egg hatching and adult emerging/days	61	25
Time between adult female emerging and starting to reproduce/days	16	2
Mean number offspring produced per female per day	2.4	4.4
Percentage of offspring that are female	15	75

▲ Table 1

Purple loosestrife is a European plant that has become invasive in North America. It is often discussed in the context of successful biological control programmes. Two members of the beetle genus *Galerucella* have been introduced to eat the plant and this has been used to limit its spread as it specializes in feeding on the loosestrife.

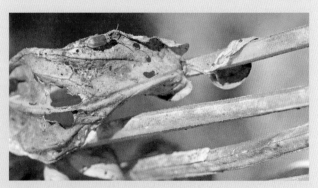

▲ Figure 7 Purple loosestrife plant showing feeding damage by *Galerucella*

Data-based questions: Control of purple loosestrife

Galerucella beetles were first introduced to several sites in Connecticut in 1996 and were released annually to the same areas in 1997 and 1998. The effects on purple loosestrife plants were assessed by looking for feeding damage (figure 8) and height of plants (figure 9).

The effects were determined as a function of distance from the release site.

a) Outline the relationship between feeding damage and time since first release. [1]

b) Outline the relationship between plant height and introduction of *Gallerucella*. [1]

c) Discuss the effect of biological control of loosestrife by *Galerucella*. [4]

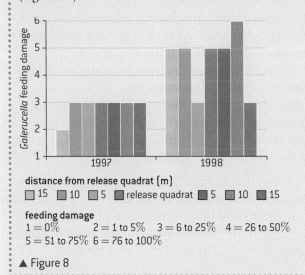

distance from release quadrat (m)

☐ 15 ☐ 10 ☐ 5 ■ release quadrat ■ 5 ■ 10 ■ 15

feeding damage
1 = 0% 2 = 1 to 5% 3 = 6 to 25% 4 = 26 to 50%
5 = 51 to 75% 6 = 76 to 100%

▲ Figure 8

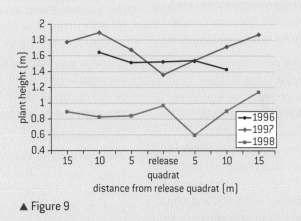

▲ Figure 9

Biological control is not the only mechanism by which invasive species can be controlled. Eradication programmes involve application of herbicides and selective harvesting of invasive plants as well as trapping and culling of invasive animals. Some of the most successful examples have been the removal of invasive mammal species from islands. Eradication techniques have improved and increasingly large islands can be cleansed of the invader. For example in New Zealand eradication of invasive Norway rats (*Rattus norvegicus*) has been achieved for all islands up to 10,000 ha (see figure 10).

The requirements for a successful eradication programme include removing the invader faster than it can reproduce, an ongoing commitment to see the eradication through

to completion, the support of local communities and preventing re-invasion. Knowledge of the ecology of the invasive species is important. Removal of the invader may cause an explosion in the prey species for example.

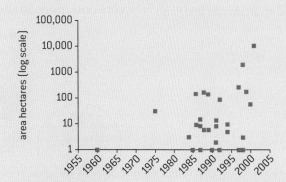

▲ Figure 10

To what extent does amphiboly obscure the communication of scientific information?

Amphiboly is an error in reasoning that applies to situations where words are treated as being synonymous when they are not. For example the term "mercury pollution" is often used to indicate both elemental mercury and compounds containing mercury such as methyl mercury. The latter form is far more harmful to ecological systems.

The consensus definition of biomagnification is that it is the transfer through food chains of foreign chemicals that can accumulate in tissues. The result is higher concentrations in organisms higher up the food chain. One study analysed the contents of 148 papers with the term "biomagnification" in the title and found that less than half were using the term according to the consensus definition. Many of the papers were equating biomagnification with bioconcentration which is the process of uptake of chemicals from the surrounding water. The consequence was a lack of recognition of distinct mechanisms that occur in aquatic invertebrates and fish.

Biomagnification

Pollutants become concentrated in the tissues of organisms at higher trophic levels by biomagnification.

Some toxins build up in the body of organisms, particularly if the toxin is fat-soluble and not easily excreted. This is known as bioaccumulation. For example, organic compounds containing mercury, such as methyl mercury, are more likely to be stored in fat tissue than metallic mercury.

Biomagnification is the process by which chemical substances become more concentrated at each trophic level. At each stage in a food chain, the predator will accumulate higher concentrations of the toxin than its prey. This is because the predator consumes large quantities of prey during its lifetime and bioaccumulates the toxins that they contain. Some organisms have greater concentrations of body lipids and so the accumulation is not even across trophic levels. Sometimes, the toxin can be taken up directly from the abiotic environment rather than entering through the food chain.

The concentration of toxins in the highest trophic levels may be lethal, even when the concentrations in organisms at the start of the food chain were very low. Bioaccumulation and biomagnification of toxins such as DDT caused catastrophic falls in the populations of some birds of prey such as peregrine falcons and ospreys in the 1950s and 1960s.

Figure 11 shows the concentrations of PCBs in an aquatic food chain in the Great Lakes. These chemicals were used as insulators in electrical devices and as flame-retardants. It was demonstrated in the 1950s that moderate doses killed experimental rats and manufacture of them stopped in the 1970s. However, PCBs have persisted in the environment and biomagnification can result in organisms in higher trophic levels having concentrations up to 10 million times greater than the concentrations in the water.

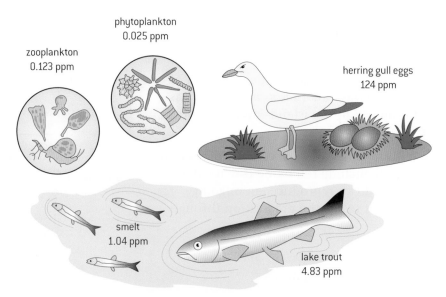

phytoplankton
0.025 ppm

zooplankton
0.123 ppm

herring gull eggs
124 ppm

smelt
1.04 ppm

lake trout
4.83 ppm

▲ Figure 11 The concentration of pollutants called PCBs in each level of the North America Great Lakes aquatic food chain (in parts per million, ppm)

The causes and consequences of biomagnification

Analysis of data illustrating the causes and consequences of biomagnification.

Data-based questions: Biomagnification of caesium

In addition to nutrients, other atmospheric elements may also enter the ecosystem. Radioactive caesium-137 was released into the atmosphere by atomic bomb tests in 1961. The caesium-137 was deposited in the soil and on to plants. Figure 12 shows the amount of radioactivity found in the tissues of lichens (an alga and a fungus growing together), caribou (a member of the deer family) and Inuit people in the Anaktuvuk Pass of Alaska.

1   The three organisms form a food chain. Deduce the trophic level of

   a) lichens   [1]

   b) Inuit.   [1]

2   Describe the level of caesium-137 in the Inuit from June 1962 through to December 1964.   [2]

3   **a)** Identify the time of year when caesium-137 concentrations were highest in

     (i) caribou   [1]

     (ii) Inuit.   [1]

   b) Explain the annual variation in caesium-137 concentrations in caribou and Inuit.   [2]

4   Predict, with reasons, whether the concentrations of caesium-137 would have dropped to zero in caribou and Inuit by the end of 1966.   [2]

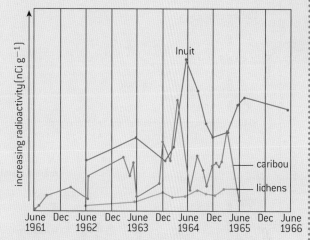

▲ Figure 12 Radioactivity levels in various trophic levels in an Alaskan community after an atmospheric atomic bomb test

Data-based questions

Biomagnification is the increase in concentration between trophic levels. The process of biomagnification for the same chemical may differ between aquatic and terrestrial food webs or between marine mammals and marine gill breathing animals (figure 13). Certain chemicals that are moderately lipid soluble but can still dissolve in water can be eliminated into water by gill breathing organisms but are not eliminated into the air by lung breathing organisms.

1   Determine the trophic level of sculpin. [1]

2   Explain how it is possible to have a non-whole number trophic level.   [2]

3   Outline the relationship between PCB concentration and trophic level in the terrestrial food web.   [2]

4   Deduce the food web where β-HCH does not biomagnify.   [2]

5   Compare the concentration of β-HCH at the third trophic level in the terrestrial food web and in the marine mammalian food web. [2]

6.   Explain the differences in biomagnification of β-HCH in the terrestrial food web and the marine mammalian food web. [3]

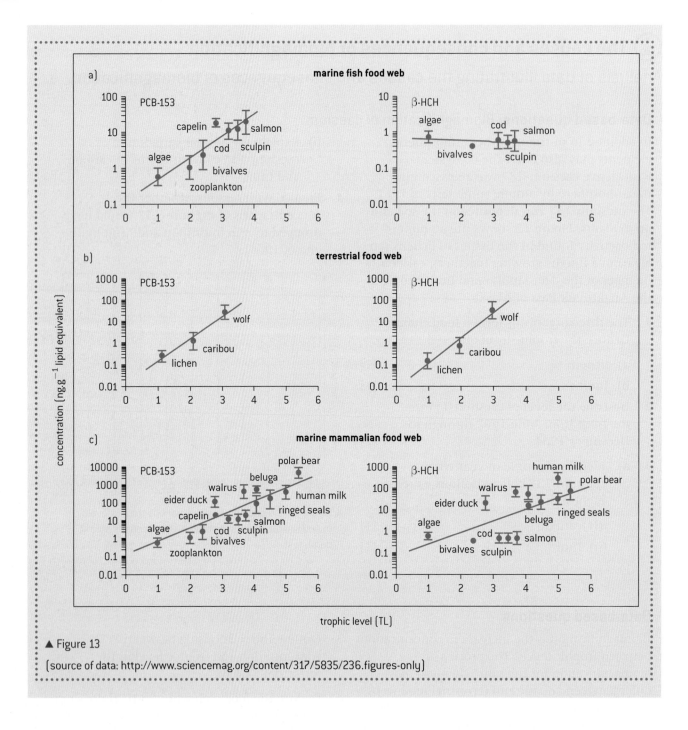

▲ Figure 13

(source of data: http://www.sciencemag.org/content/317/5835/236.figures-only)

The benefits and risks of DDT use

Discussion of the trade-off between control of the malarial parasite and DDT pollution.

DDT (**d**ichloro**d**iphenyl**t**richloroethane) is an insecticide that was used widely in the mid-20th century, first to control disease vectors such as ticks and mosquitos during and immediately after the Second World War and later on as an agricultural insecticide.

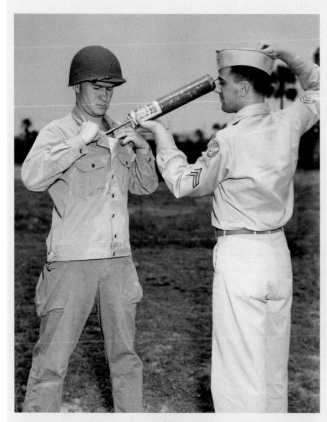

▲ Figure 14 A soldier demonstrating the DDT spraying equipment used to control the lice which carried the bacterium that causes typhus

DDT was made famous by the writing of Rachel Carson. In her book *Silent Spring* she expressed concern about how the indiscriminate use of DDT was leading to ecological effects. As DDT biomagnified up food chains, it led to birds of prey having thin-shelled eggs and consequently leading to their failure to reproduce successfully. As a consequence, DDT was banned for agricultural use under the terms of the Stockholm Convention on persistent organic chemicals, though residual indoor spraying to control mosquitos was permitted. This use remains controversial. The World Health Organization endorses the use of DDT for this purpose.

Where the use of DDT was discontinued for malaria vector control in an area, rates of malaria climbed. Substitute strategies to spraying DDT were attempted but were not as successful. Many countries that had banned the use of DDT completely reapproved its use for residual indoor spraying.

Concerned scientists argue that DDT may have a variety of human health effects, including reduced fertility, genital birth defects, cancer and damage to developing brains. Its metabolite, DDE, can block male hormones. The pesticide accumulates in body fat, and in breast milk, and evidence that it persists in the environment for decades is strong.

With its strong record in reducing malaria cases and the WHO endorsement, the use of DDT is increasing globally. Pressure is mounting for governments and intergovernmental organizations to rely on DDT only as a last resort and the call is growing for the development of an alternative form of malaria vector control.

Plastics in the ocean

Macroplastic and microplastic debris has accumulated in marine environments.

Plastic is a broad term that describes a number of different polymers that are used in a growing number of disposable consumer items. Some plastic enters the ocean through direct disposal from ships and platforms, but the majority comes from litter being blown into water systems.

Macroplastic is large visible debris including nets, buoys, buckets and trash that has not degraded (see figure 15).

Physical and chemical degradation of macroplastic results in microplastic fragments that are harder to see but are more omnipresent.

▲ Figure 15 Grey reef shark (*Carcharhinus amblyrhynchos*) with macroplastic sheeting in its mouth. It may have attempted to eat the plastic, mistaking it for prey

Ocean currents transport garbage to five concentration areas across the globe called gyres. These are very large "patches" within the ocean where circular currents concentrate the plastic waste (figure 16).

Some of the consequences of marine plastic pollution are:

- The degradation of the plastic at sea releases persistent organic chemicals into the ocean that can bioaccumulate and biomagnify.

- Plastics absorb other persistent organic chemicals and thus concentrate these toxins.

- Animals eat or become tangled in plastic pollution.

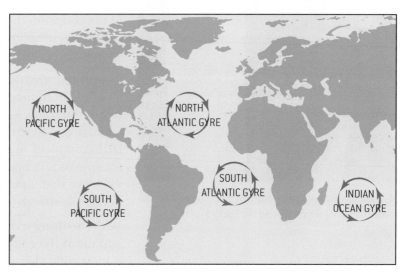

▲ Figure 16

🌐 Marine plastics and the Laysan albatross

Case study of the impact of marine plastic debris on Laysan albatrosses and one other named species.

Macroplastic is a concern for marine animals as it may be mistaken as a food source and ingested. The Laysan albatross is a large marine bird that nests on the island of Midway Atoll in the Pacific Ocean. The North Pacific Gyre contacts the shore of the island transporting large volumes of plastic onto its beaches. Parent Laysan albatrosses feed plastic pieces to chicks resulting in significantly higher mortality.

▲ Figure 17 Carcass of Laysan albatross chick that has been fed plastic debris

Data-based questions: Plastic ingestion by the leatherback turtle

The leatherback, *Dermochelys coriacea*, is a large sea turtle that preys on jellyfish. Figure 18 shows the results of a literature review of all reported cases of autopsies of dead leatherback turtles. The reported autopsies are grouped into five year "bins". The numbers labelling each point represent the number of autopsies reported in the literature for that five-year period. The percentage of turtles with plastic in their stomach is noted.

1   Identify the first year that an autopsy reported plastic being found in the stomach of a leatherback. [1]

2   Estimate the total percentage of leatherback turtles with plastic in their digestive tract over the past 50 years. [2]

3   Suggest a reason why such a large percentage of leatherback turtles ingest plastic. [2]

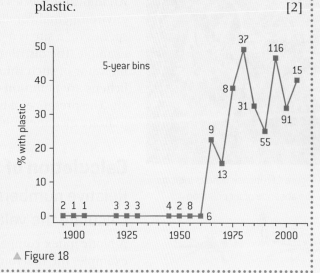

▲ Figure 18

C.4 Conservation of biodiversity

Understanding

→   An indicator species is an organism used to assess a specific environmental condition.

→   Relative numbers of indicator species can be used to calculate the value of a biotic index.

→   *In situ* conservation may require active management of nature reserves or national parks.

→   *Ex situ* conservation is the preservation of species outside their natural habitats.

→   Biogeographic factors affect species diversity.

→   Richness and evenness are components of biodiversity.

Applications

→   Case study of the captive breeding and reintroduction of an endangered animal species.

→   Analysis of the impact of biogeographic factors on diversity limited to island size and edge effects.

Nature of science

→   Scientists collaborate with other agencies: the preservation of species involves international cooperation through intergovernmental and non-governmental organizations.

Skills

→   Analysis of the biodiversity of two local communities using Simpson's reciprocal index of diversity.

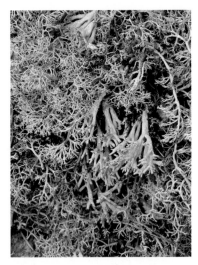

▲ Figure 1 Fruticose lichens indicate the absence of air pollution

1 Stonefly nymph (up to 30 mm)

2 Mayfly larva (up to 15 mm)

3 *Asellus* (freshwater louse) (up to about 12 mm)

4 Chironomid (bloodworm; a midge larva) (up to 20mm)

5 Rat-tailed maggot larva (up to 55 mm including tube)

6 *Tubifex* (sludge worm) (up to 40 mm)

▲ Figure 2 Benthic macroinvertebrates

Indicator species

An indicator species is an organism used to assess a specific environmental condition.

An indicator species is an organism that occurs only when specific environmental conditions are present. The presence or absence of these species on a site is a good indicator of environmental conditions. For example, the distribution of understorey plants in a forest is a good indicator of such things as soil fertility or water drainage. Fruticose lichens are pollution-intolerant, so their presence is an indicator of clean air. The presence of black greasewood (*Sarcobatus vermiculatus*) would indicate alkaline saline soils.

Calculation of a biotic index

Relative numbers of indicator species can be used to calculate the value of a biotic index.

A biotic index compares the relative frequency of indicator species. For example, the macroinvertebrate biotic index is a measure of stream health. The number of individuals of each indicator species in a sample is determined. Each number is multiplied by a pollution tolerance factor and a weighted average is determined. One possible biotic index multiplies the number of a certain kind or organism by its pollution tolerance rating. Each of these products is then added to the others and divided by the total number of organisms in the habitat.

Figure 2 shows six different benthic (bottom-dwelling) macroinvertebrates that are found in rivers. Benthic macroinvertebrates are useful indicators of stream health for a number of reasons: they live in the water for the duration of the aquatic stage of their life cycle and so they reflect conditions in the water over a period of time; they are easy to capture; and most importantly, they vary in their response to chemical and physical changes in their habitat.

In situ conservation

In situ conservation may require active management of nature reserves or national parks.

In situ conservation measures involve endangered species remaining in the habitat to which they are adapted. This allows the species to interact with other wild species, conserving more aspects of the organism's niche. Nature reserves are areas that are specially designated for the conservation of wildlife. Terrestrial, aquatic and marine nature reserves have all been established.

Establishment of a nature reserve is often not enough – active management is required.

This may involve:

- controlled grazing
- removal of shrubs and trees

- removal of alien plant species and culling invasive animals

- reintroduction of species that have become locally extinct

- re-wetting of wetlands

- limiting predators

- controlling poaching

- feeding the animals

- controlling access.

Ex situ conservation

Ex situ conservation is the preservation of species outside their natural habitats.

Ex situ conservation measures involve removal of organisms from their natural habitat.

Plant species can be grown in botanic gardens. The seeds of plants can be stored in seed banks at low temperatures, which maintain their viability for long periods.

Captive breeding of animals is sometimes used, followed by release of the captive-bred individuals into their natural habitats.

Ex situ conservation is used to back up *in situ* conservation measures, or where endangered species cannot safely remain in their natural habitats. Populations of endangered bird species in New Zealand have been moved to offshore islands to protect them from attacks by alien predators for example.

▲ Figure 3 White rhinoceros (*Ceratotherium simum*) horn removal, Umhlametsi Private Nature Reserve, South Africa. Horns are removed as an anti-poaching measure. The horn, which is made entirely from keratin protein, is removed with a chainsaw at ten centimetres above the base. Rhinoceros horns are used as ornaments and for some traditional medicines

Captive breeding to restore populations of endangered species

Case study of the captive breeding and reintroduction of an endangered animal species.

The peregrine falcon (*Falco peregrinus*) became endangered in parts of the United States, Canada and Europe because of the widespread use of DDT in the 1960s and 1970s. Biomagnification caused high levels of toxic metabolites in the fat tissue of the birds. This reduced the calcium content of eggshells. The thinner eggshells meant fewer successful hatching events occurred (see figure 4).

The captive breeding programme for peregrine falcons involved conservation workers collecting thin-shelled eggs from nests and replacing them with porcelain replicas. They incubated the thin-shelled eggs in a hatching facility under carefully controlled conditions ensuring greater frequency of successful hatching.

▲ Figure 4 Two peregrine falcon eggs. The egg on the left is normal. The one on the right has been weakened due to the mother's exposure to DDT

Fledglings were then replaced in the nest to be reared by wild peregrine falcons as well as by foster prairie falcons. Captive breeding pairs also generated eggs that were incubated artificially. Some fledglings were raised to the age of independence by providing them with food in "hack" boxes that were open to the wild for them to increasingly find food on their own.

In 1973, there were no breeding pairs of peregrine falcons detected in Southern Alberta, Canada and few detected in Northern Alberta. A captive breeding programme was initiated and the success of the programme is evident in figure 5.

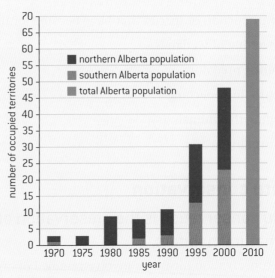

▲ Figure 5

Different groups work together to conserve biodiversity

Scientists collaborate with other agencies: the preservation of species involves international cooperation through intergovernmental and non-governmental organizations.

Scientific research can inform best practice but needs the support of agencies that can effect proposals for conservation. National governments can take enlightened steps when they set aside land as nature reserves.

Because threats to biodiversity extend beyond borders, conservation involves international cooperation. If one national government places restrictions on harvesting products based on endangered species, nationals from other countries might free ride on their restraint. Intergovernmental organizations such as the International Union for the Conservation of Nature (IUCN) can facilitate agreement between nations. The IUCN publishes the Red List of threatened species which publishes the status of the conservation of species. This organization facilitated the creation of the Convention on International Trade in Endangered Species (CITES) which is a treaty which regulates the international trade in specimens and products of wild plants and animals. This is done in an effort to ensure that trade does not threaten survival of the organism in the wild.

Non-governmental organizations (NGOs) such as the World Wide Fund for Nature (WWF) operate independently from any form of government. NGOs are typically non-profit groups that pursue wider social aims and lack political affiliation. The WWF is an example of an NGO. It raises funds for educational programmes and lobbying efforts.

▲ Figure 6 Teacher and school children holding tree saplings whilst listening to a World Wide Fund for Nature (WWF) conservation worker (left). These saplings will be planted as part of a tree planting in Sagarmatha National Park, Himalayas, Nepal. The scheme is aimed at replacing forest lost due to deforestation caused by tourism

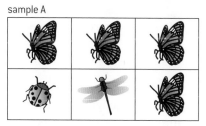

Components of biodiversity

Richness and evenness are components of biodiversity.

Biological diversity, or biodiversity, has two components. Richness refers to the number of different species present. In figure 7, sample A has greater richness as three species are present. Evenness refers to how close in numbers each species is. If two individuals were to be chosen from the sample, evenness would be an indicator of the likelihood that the two individuals would be from different species. By this standard, sample B is more "even".

sample A

sample B

▲ Figure 7

 Simpson's diversity index

Analysis of the biodiversity of two local communities using Simpson's reciprocal index of diversity.

The Simpson's reciprocal index quantifies biodiversity by taking into account richness and evenness. The greater the biodiversity in an area, the higher the value of D. The lowest possible defined value of D is 1 and would occur if the community contained only one species. The maximum value would occur if there was perfect evenness and would be equal to the number of species.

The formula for Simpson's reciprocal index of diversity is:

$$D = \frac{N(N-1)}{\Sigma n(n-1)}$$

D = diversity index, N = total number of organisms of all species found and n = number of individuals of a particular species.

Biogeography can influence diversity

Biogeographic factors affect species diversity.

The effectiveness of nature reserves at conserving biodiversity depends on their biogeographical features.

Large nature reserves are more effective than small ones at maintaining biodiversity. This is consistent with the island biogeography model if nature reserves are like islands. The larger an island is, the greater the biodiversity found there. The larger the area, the higher the population of a certain species that can be supported and the less likely that small populations will be extirpated by random events.

Connected nature reserves are more effective than isolated ones. If there are several small reserves near to one another, then corridors between them can increase effectiveness at preserving biodiversity. Even narrow wildlife corridors allow organisms to move between fragmented habitats, for example through tunnels under busy roads.

Because the ecology of the edges of ecosystems is different from the central areas, the shape of nature reserves is important. If the central area can be maximized and the total length of the perimeter can be minimized, then the reserve can preserve biodiversity better. A circular reserve would therefore be superior to an extended strip of land with the same total area.

Activity

Groups of students studied the species diversity of the beetle fauna found on two upland sites in Europe. The same number of students searched for a similar length of time in each of the two sites. The two sites were of equal area.

The number of individuals of the four species found at each site is given in the table below.

Species	Site A	Site B
Trichius fasciatus	10	20
Aphodius lapponum	5	10
Cicindela campestris	15	8
Stenus geniculatus	10	2

a) Calculate the reciprocal Simpson diversity index (D) for the beetle fauna of the two sites. [3]

b) Suggest a possible conclusion that can be formed. [2]

Source: International Baccalaureate November 07 exam.

639

Impact of island size and edge effects on diversity

Analysis of the impact of biogeographic factors on diversity limited to island size and edge effects.

Island biogeography studies have found that two important determinants of biodiversity on an island are:

- proximity to the mainland, and

- the area of the island.

These variables lead to a balance between colonization and extinction.

Proximity to the mainland will lead to new colonization.

If the island is small, rates of extinction will be higher. The total size of a population on an island with low area is more likely to be small and low in genetic diversity. Random accidents can have a large negative impact on a small island population.

Data-based questions: Island size and diversity

Figure 8 shows the relationship between species richness and island area for reptiles and amphibians in the West Indies. Figure 9 shows the relationship between species richness and island area for birds in the Sunda Islands. Both sets of data are from: R. H. MacArthur and E. O. Wilson, *The Theory of Island Biogeography*, Princeton University Press Princeton, N.J. (1967).

a) Estimate the number of bird species on an island of 10,000 square miles. [1]

b) Outline the relationship between the number of amphibian and reptile species and island area. [2]

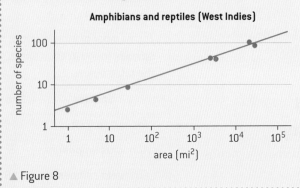

▲ Figure 8

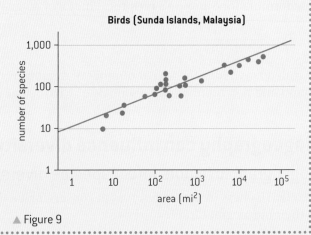

▲ Figure 9

Data-based questions: Forest size and songbird density

Figure 10 shows the probability of sighting three bird species whose preferred habitat is the forest interior away from forest edges. The dotted lines represent the range of probability.

a) (i) In a 10 ha forest, determine the range of probability of observing a wood thrush. [1]

   (ii) In a total of 20 point transects undertaken in a 3.2 ha forest, how many times is a red-eyed vireo likely to be observed. [2]

b) In a 32 ha forest, which is the

   (i) most likely species to be observed [1]

   (ii) least likely species to be observed. [1]

c) Based on this data, suggest with a reason, the minimum size a conservation area should be to preserve populations of woodland interior species. [2]

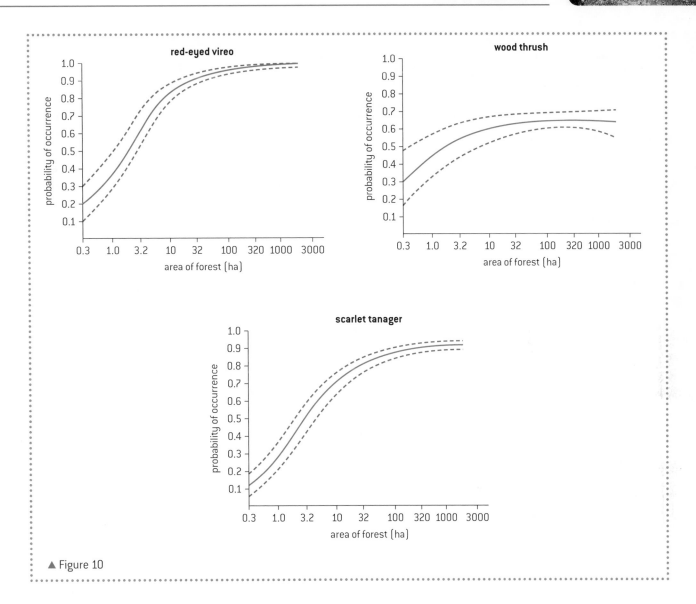

▲ Figure 10

C.5 Population ecology (AHL)

Understanding

→ Sampling techniques are used to estimate population size.

→ The exponential growth pattern occurs in an ideal, unlimited environment.

→ Population growth slows as a population reaches the carrying capacity of the environment.

→ The phases shown in the sigmoid curve can be explained by relative rates of natality, mortality, immigration and emigration.

→ Limiting factors can be top-down or bottom-up.

Applications

→ Evaluating the methods used to estimate the size of commercial stock of marine resources.

→ Use of the capture-mark-release-recapture method to estimate the population size of an animal species.

→ Discussion of the effect of natality, mortality, immigration and emigration on population size.

→ Analysis of the effect of population size, age and reproductive status on sustainable fishing practices.

→ Bottom-up control of algal blooms by shortage of nutrients and top-down control by herbivory.

Nature of science

→ Avoiding bias: a random number generator helps to ensure population sampling is free from bias.

Skills

→ Modelling the growth curve using a simple organism such as yeast or species of *Lemna*.

Estimating population size

Sampling techniques are used to estimate population size.

The simplest method of estimating population size or population density is to count the number of individuals in a given area. This is only feasible if the individuals are large and the area is small. For most other cases, ecologists use population sampling techniques. This requires the researcher to determine the population size in a small area and use this to estimate the entire population. This is referred to as population sampling. The sample is assumed to be representative of the entire population. Normally, several samples are taken to limit the effect of choosing a sample which is not representative.

Using a random number generator

Avoiding bias: a random number generator helps to ensure population sampling is free from bias.

In order for the sample to be representative of the entire population, the selection of the sample should be random. A random sample is one where every member of the population has an equal likelihood of being selected.

There are several methods of generating a random sample. A computer or graphic calculator can be used to generate a random sampie. The activity box describes a method that uses a graphic calculator to generate a random sample. Alternatively, a random number table can be used.

Activity

Using the Ti-84 to generate random numbers

A student has divided a 50 m² area into 25 quadrats and wants to select three quadrats randomly without bias. She decides to use her calculator to randomly select the quadrats for her.

For steps 1 to 3 below see screen 1

1 On the Ti-84, press the "MATH" button

2 Scroll over until the PRB button is highlighted

3 Push 5 so that "randInt" is selected.

For steps 4 and 5 below see screen 2

4 Type in 1 for the lowest number, 25 for the highest number and 3 for the number of quadrats. The comma button is found above the number 7.

5 Push enter. The calculator selects quadrat 8, 13 and 11. Note that three quadrats may not be sufficient to be representative of the whole area.

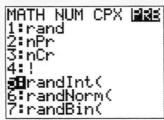

Screen 1

Screen 2

▲ Figure 1

The Lincoln index

Use of the capture-mark-release-recapture method to estimate the population size of an animal species.

One sampling technique used to determine population density is the Lincoln index or the capture-mark-release-recapture method.

1 Capture as many individuals as possible in the area occupied by the animal population, using netting, trapping or careful searching

e.g. careful searching for banded snails (*Cepaea nemoralis*)

2 Mark each individual, without making them more visible to predators.

e.g. marking the inside of the snail shell with a dot of non-toxic paint.

3 Release all the marked individuals and allow them to settle back into their habitat.

4 Recapture as many individuals as possible and count how many are marked and how many unmarked.

 24 marked

16 unmarked

5 Calculate the estimated population size by using the Lincoln index:

$$\text{population size} = \frac{n_1 \times n_2}{n_3}$$

$n_1 =$ number caught and marked initially
$n_2 =$ total number caught on the second occasion
$n_3 =$ number of marked individuals recaptured

 Figure 2

🌐 Estimating commercial fish populations

Analysis of the effect of population size, age and reproductive status on sustainable fishing practices.

Fish are an important food resource. Because they are an open access resource on the high seas, the incentives are limited for conservation.

An important component of managing fish is clear data about fish populations. The concept of maximum sustainable yield is related to the sigmoid growth curve. When the population size is low, the rate of population growth will increase until environmental resistance begins to limit population size. At point 2 in the graph in figure 3 the population is growing at its maximum rate. This is the point at which the maximum sustainable yield occurs. If fish were harvested at this rate, then fishing would be able to continue indefinitely.

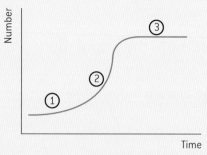

▲ Figure 3 Population growth curve

Figure 4 shows a graph of sustainable yield versus intensity of fishing. If there is no fishing, then the yield of fish would be zero. If there is very high-intensity fishing, it may be that the population of fish becomes extinct and there is no yield. The maximum of the curve in figure 4 would correspond to point 2 on the S-curve in figure 3.

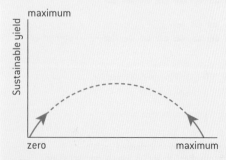

▲ Figure 4 Intensity of fishing

If a population is growing, then the relative number of younger fish will be higher. If a

population is in decline, then the proportion of older fish will be higher. In addition to population size information, the age structure of the population is important information when establishing sustainable levels of harvesting. Figure 5 shows a technician holding an otolith (ear bone) of a fish. The otolith contains rings, similar to the rings of a tree, that can be used to determine the age of the fish.

▲ Figure 5

There are a number of practices that are associated with sustainable fishing and each practice is informed by population size data. Most practices depend upon international cooperation.

- Restrictions exist on the catching of younger fish. Regulations or international agreements often restrict the net mesh size allowing younger fish to escape.

- Quotas are agreed upon for species with low stocks and moratoria declared on the fishing of all endangered species.

- Closed seasons are often declared to allow undisturbed breeding and exclusion zones are agreed upon in which all fishing is banned.

- Methods of fishing that are particularly damaging are often banned, for example drift nets, which catch many species of fish other than target species.

Evaluating methods of determining population size

Evaluating the methods used to estimate the size of commercial stock of marine resources.

The first stage in the conservation of fish is obtaining reliable estimates of fish stocks. This is difficult for marine species because most fish species are highly mobile and unevenly distributed, so random sampling methods are ineffective. Capture-mark-release-recapture methods are useful in lakes and rivers, but the numbers of recaptured fish are usually too small in the open ocean for reliable estimations. Fish can be temporarily stunned and then counted in lakes and rivers with an electric shock, but not in the ocean. Echo sounders can be used to estimate the size of shoals of fish, but many species do not form shoals. A common method for estimating stocks is based on data obtained from fish catches. The age structure of landed fish can be used to estimate population size. The numbers of the target species of fish of each age are counted. Spawning rates can be deduced, from which estimates of the total can be made. Violators of regulations designed to control the age of fish landed often do not report what they land or they dump the restricted fish as bycatch before landing so biased estimates might be formed by using age structure as a method of estimating fish stocks.

The "J" shaped population growth curve

The exponential growth pattern occurs in an ideal, unlimited environment.

If a population experiences ideal conditions, than the population will grow exponentially. A graph of population size over time will resemble a "J" shape (figure 6).

Figure 7 shows the population growth of a culture of the unicellular organism *Paramecium aurelia*, kept in controlled conditions, including a constant supply of food. The graph illustrates a pattern called the sigmoid, or S-shaped, growth curve.

The S-curve is representative of what happens when a population colonizes a new habitat. With a low level of environmental resistance, a population will grow exponentially initially. As the environment begins to offer resistance, the population reaches a transition point where the growth rate begins to slow down until it reaches the carrying capacity.

Factors that influence population size

The phases shown in the sigmoid curve can be explained by relative rates of natality, mortality, immigration and emigration.

With limited environmental resistance, a population will grow exponentially. At this stage birth rate (natality) is higher than death rate (mortality).

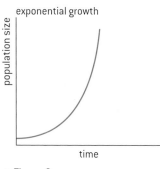

▲ Figure 6

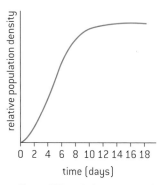

▲ Figure 7 Population growth of a culture of *Paramecium aurelia*

As population density increases, various density-dependent factors begin to limit population growth. Examples of such limiting factors include competition for resources, a build-up of the toxic by-products of metabolism, an increase in predation or an increase in the incidence of disease. The initial result is that natality slows in relation to mortality. This is the transition phase on the curve, when the slope begins to decrease. It is important to recognize that the population is still growing at this point. The plateau phase begins when mortality and natality rates equal out.

An important variable affecting population size is migration. Immigration increases the size of a population. As an example, an island that is close to the mainland will most likely have regular replenishment of a population through immigration.

Emigration decreases the size of a population. Emigration occurs when population members leave an area. The Norway lemming (*Lemmus lemmus*) is renowned for its emigration patterns from areas of high population density or poor habitat. This occurs in seasons with high population levels.

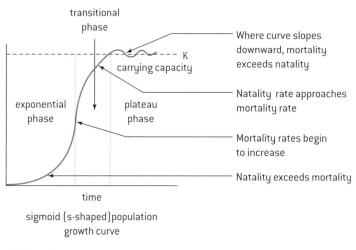

▲ Figure 8

Carrying capacity

Population growth slows as a population reaches the carrying capacity of the environment.

The maximum size of a population that an environment can support is its carrying capacity. It is represented by the variable "K". In the sigmoid growth pattern, when a population reaches its carrying capacity, the population will stop growing and natality and mortality will be equal. This is referred to as the plateau phase of the S-curve. At this point, an equilibrium population is often maintained.

However, some years can see a "boom" and "bust" pattern where populations far exceed the carrying capacity. Higher rates of mortality will return the population to the carrying capacity of the environment or the population may crash well below the carrying capacity.

Data-based questions

Two male and eight female ring-necked pheasants were released on Protection Island. Figure 9 shows how the population grew.

a) Explain the changes in population size:

    i)   in the first four years;   [3]

    ii)   from year 4 to year 6.   [3]

b) Predict with reasons what would have happened to the population after year 6.   [4]

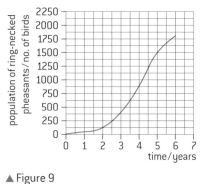

▲ Figure 9

Discussing the factors that influence population growth

Discussion of the effect of natality, mortality, immigration and emigration on population size.

The logistic growth curve is a model of population growth and is somewhat idealized.

There are a number of causes of mortality:

- senescence, or death from age related illness
- predation
- disease
- injury
- shortage of food or water
- density independent factors such as earthquakes, volcanic eruption, fire or storms.

The impact of mortality on population growth will depend on the age of the individual; i.e., whether the death occurred before or after reproductive age. While density dependent factors tend to affect the very young, the old and the weak individuals, density independent limiting factors affect all groups equally including the individuals who are at the peak of their reproductive potential.

Natality has the greatest impact on population size. Knowing the age structure is important when recording natality. Age and health status can both affect natality rates in different ways. Predicting whether natality rates will rise or fall depends on knowledge of population age structure.

If a population lacks genetic diversity, then environmental resistance can have a disproportionately negative effect on population; for example, a relatively disproportionate number might die in an epidemic. Immigration can diversify the gene pool and allow some members of the population to survive the selection pressure. In the case of the migrations found in Norway Lemmings, emigrating individuals are more likely to be individuals who are weaker and unable to defend territory; i.e., they have the lowest reproductive fitness.

Modelling population growth

Modelling the growth curve using a simple organism such as yeast or species of *Lemna*.

Population growth can be studied using model species such as yeast (*Saccharomyces cerevisiae*) or duckweed (*Lemna sp*).

Duckweed (*Lemna sp.*) are stemless water plants (see figure 10). Each plant grows one to four leaf-like structures called thalli. Duckweeds reproduce asexually by growing new thalli from older thalli and when they reach a certain size, they break free from the parent plant.

A number of experiments are possible:

- What is the carrying capacity of a given container?
- What conditions of light, nutrients or container surface area are ideal for population growth?

▲ Figure 10 Duckweed (stemless water plant) on pond together with pond skaters which also inhabit the surface film

647

Top-down and bottom-up limiting factors

Limiting factors can be top-down or bottom-up.

A limiting factor is an environmental selection pressure that limits population growth. There are two categories of limiting factor: top-down and bottom-up.

The population of organisms in an ecosystem can be influenced by the availability of resources such as nutrients, food and space. All such factors are referred to as bottom-up limiting factors.

Predation is referred to as a top-down limiting factor.

A keystone species exerts top-down influence on its community by preventing species at lower trophic levels from monopolizing critical resources, such as competition for space or food sources.

Case study of bottom-up and top-down limiting factors

Bottom-up control of algal blooms by shortage of nutrients and top-down control by herbivory.

The distinction between bottom-up control and top-down control can be illustrated by the example of free-living marine algae growing in coral reef communities. Free-living algae blooms can disrupt coral reef communities by blocking sunlight and preventing photosynthesis in the symbiotic zooxanthellae. Coral-reef ecosystems are generally nutrient-poor. This explains the selection pressure for the symbiotic relationship between zooxanthellae and coral. The absence of nutrients represents a bottom-up limiting factor to free-living algae population growth.

Coral are also populated by grazing fish which sets limits to the growth of free-living algae blooms. Parrotfish graze on free-living algae and thus have a top-down limiting effect (figure 11).

Nutrient enrichment through human activity, known as eutrophication, can have a bottom-up impact on the build-up of algae populations. Fishing practices which remove herbivorous fish from coral reefs can have a top-down impact on algae populations. Figure 12 summarizes the impacts on coral reef communities of either factor as well as both factors in combination.

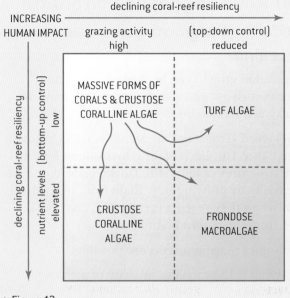

▲ Figure 12

▲ Figure 11 Parrotfish grazing on algae is an example of a top-down limiting factor

C.6 The nitrogen and phosphorus cycles (AHL)

Understanding

→ Nitrogen-fixing bacteria convert atmospheric nitrogen to ammonia.

→ *Rhizobium* associates with roots in a mutualistic relationship.

→ In the absence of oxygen denitrifying bacteria reduce nitrate in the soil.

→ Phosphorus can be added to the phosphorus cycle by application of fertilizer or removed by the harvesting of agricultural crops.

→ The rate of turnover in the phosphorus cycle is much lower than the nitrogen cycle.

→ Availability of phosphate may become limiting to agriculture in the future.

→ Leaching of mineral nutrients from agricultural land into rivers causes eutrophication and leads to increased biochemical oxygen demand.

Applications

→ The impact of waterlogging on the nitrogen cycle.

→ Insectivorous plants as an adaptation for low nitrogen availability in waterlogged soils.

Skills

→ Drawing and labelling a diagram of the nitrogen cycle.

→ Assess the nutrient content of a soil sample.

Nature of science

→ Assessing risks and benefits of scientific research: agricultural practices can disrupt the phosphorus cycle.

Nitrogen fixation

Nitrogen-fixing bacteria convert atmospheric nitrogen to ammonia.

The atmosphere is 78% nitrogen gas in the form of the diatomic molecule N_2. Nitrogen in this form cannot be taken up by plants.

Nitrogen would quickly become a limiting factor for ecosystems if it were not for the bacteria involved in the nitrogen cycle. The bacteria *Rhizobium* and *Azotobacter* can "fix" nitrogen gas and convert it to ammonia (NH_3), a form that living things can use. Once in this form, it can be absorbed by plants and then it can enter food chains. Other bacteria convert ammonia to nitrates, another bioavailable form of nitrogen.

Nitrogen fixation by *Rhizobium*

Rhizobium associates with roots in a mutualistic relationship.

The bacteria of the genus *Rhizobium* convert atmospheric nitrogen into a usable organic form. These bacteria are often not free-living but live in a close symbiotic association in the roots of plants such as the legume family. Because both organisms benefit, the symbiosis is an example of mutualism.

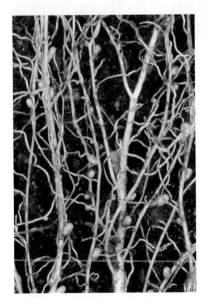

▲ Figure 1

Figure 1 is a photo of root nodules on the roots of white clover, *Trifolium repens*, caused by the nitrogen-fixing bacteria *Rhizobium trifolii*. The bacteria convert (fix) atmospheric nitrogen in the soil to ammonia (NH_3). The host plant cannot carry out this process itself, but it is vital for the production of amino acids, the building blocks of proteins. In return the plant passes carbohydrates produced during photosynthesis to the bacteria for use as an energy source.

Rhizobium infect the plant through root hairs, forming an infection thread, which conveys them from the entry point to the nodule site. Once they are within the nodule, they divide repeatedly and swell. The scanning electron micrograph in figure 2 shows that the outside of the nodule is composed of plant tissue.

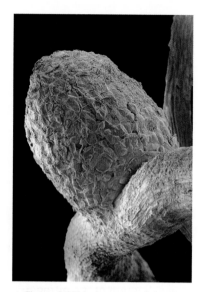

▲ Figure 2 SEM of a root nodule

Denitrification

In the absence of oxygen, denitrifying bacteria reduce nitrate in the soil.

The ammonia produced by nitrogen fixation is converted to nitrite (NO_2^-) by the genus of bacteria known as *Nitrosomonas*. These bacteria have a double membrane and use electrons gained from the oxidation of ammonia to produce energy. Nitrites are produced from this oxidation. The energy is used to fix carbon dioxide into organic carbon molecules. This means that *Nitrosomonas* are a genus of chemoautotrophs as they use the energy found in ammonia, an inorganic molecule.

Nitrites are converted to nitrates by the genus of bacteria known as *Nitrobacter* (figure 3). *Nitrobacter sp* are an example of chemoautotrophs as they derive energy from nitrites which are inorganic compounds. They oxidize nitrites as an energy source for carbon fixation and convert it to nitrates. Nitrate is a form of nitrogen that is bioavailable to plants.

Denitrification is the reduction of nitrate (NO_3^-) to nitrogen (N_2). Denitrifying bacteria, such as *Pseudomonas denitrificans* can use oxygen as an electron acceptor. However, when oxygen is in short supply, instead of using O_2 as an electron acceptor in electron transport, denitrifying bacteria use nitrate as an electron acceptor, releasing gaseous nitrogen as the product.

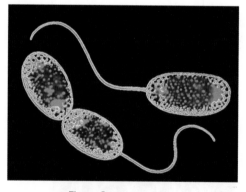

▲ Figure 3

When this occurs, the bioavailability of nitrogen within the ecosystem becomes reduced.

 Summary of the nitrogen cycle

Drawing and labelling a diagram of the nitrogen cycle.

The nitrogen cycle is an example of a nutrient cycle. When constructing nutrient cycle diagrams, also known as systems flow diagrams, three things are represented: pools, fluxes and processes. Compartments such as litter, biomass and the atmosphere are referred to as pools as they represent concentrations or stores of the nutrient. They are usually represented by shapes. In some systems flow diagrams, the size of the stock is indicated by varying the size of the shape.

Arrows are used to represent the direction of fluxes, or flows, of the nutrient. In some nutrient cycle diagrams, the thickness of arrows is used to indicate rates of flow.

Processes are usually written over flow arrows.

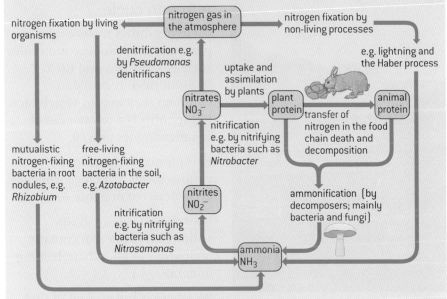

▲ Figure 4 The nitrogen cycle, showing the roles of *Rhizobium*, *Azotobacter*, *Nitrosomonas*, *Nitrobacter* and *Pseudomonas denitrificans*

 ## Waterlogging leads to denitrification

The impact of waterlogging on the nitrogen cycle.

Soil can become inundated by water through flooding or irrigation with poor drainage. The consequence is waterlogging. Oxygen is in short supply in waterlogged soils. This decreases available aeration and favours the process of denitrification by *Pseudomonas*.

Excessive irrigation can then lead to two problems related to the nitrogen cycle. If the excess water flows off the field and enters water courses as runoff, the nutrient enrichment of the body of water can lead to eutrophication, a problem discussed later in this sub-topic. Secondly, waterlogging can lead to the loss of bioavailable nitrogen through denitrification.

Carnivorous plants are adapted to low nitrogen soils

Insectivorous plants as an adaptation for low nitrogen availability in waterlogged soils.

Wetlands such as swamps and bogs have permanently waterlogged soils and would therefore have nitrogen-deficient soils. One adaptation of bog plants is to become "carnivorous" and obtain nitrogen through the extracellular digestion of animals.

In figure 5, the fly has been attracted by the droplets at the tips of the tentacles, which extend from the leaf surface. Insects stick to the tips of the longer tentacles, which bend over bringing them inward towards the shorter tentacles. Enzymes secreted by the tentacles digest the animal and the products are absorbed by the modified leaf. Note that the plant is not truly carnivorous as it obtains its energy and carbon from photosynthesis rather than the tissue of the animal.

▲ Figure 5 A hoverfly has been captured by the leaf of the carnivorous plant *Drosera rotundifolia*, the sundew

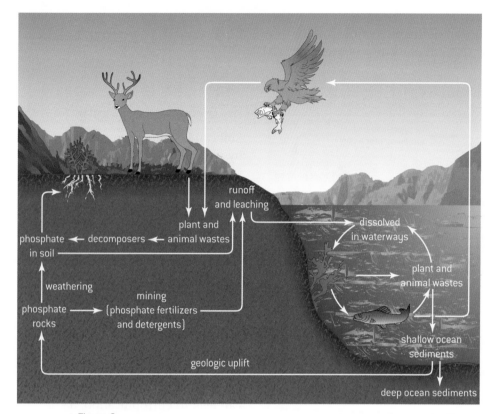

▲ Figure 6

The phosphorus cycle

The rate of turnover in the phosphorus cycle is much lower than the nitrogen cycle.

All living things require phosphorus to produce molecules such as ATP, DNA and RNA. Phosphorus is required to maintain skeletons in vertebrates. Phosphorus is a component of cell membranes as well.

Phosphorus cycles through various forms in a biogeochemical cycle depicted in figure 6.

Phosphorite (figure 7) is a sedimentary rock that contains high levels of phosphate-bearing minerals.

Weathering and erosion of such rocks releases phosphates into the soil. Phosphorus in the form of phosphates is readily taken up by plants, where it enters food chains.

The largest stocks of phosphate occur in marine sediments and in mineral deposits.

Turnover rate refers to the amount of phosphorus released from one stock to another per unit time. Phosphate is only slowly released to ecosystems by weathering and so has a relatively low turnover rate in comparison to nitrogen.

Effect of agriculture on soil phosphorus

Phosphorus can be added to the phosphorus cycle by application of fertilizer or removed by the harvesting of agricultural crops.

Human activity impacts the phosphorus cycle. Phosphate is mined and converted to phosphate-based fertilizer. The fertilizer is then transported great distances and applied to crops. Phosphorus in the biomass of crops is transferred from fields in one area to markets in other areas.

Waterlogging of soils in poorly drained irrigated crops can dissolve phosphate and bring it into solution. Runoff containing phosphate from fertilizer can contribute to fresh water eutrophication.

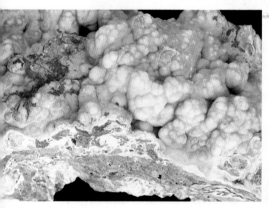

▲ Figure 7

Peak phosphorus

Availability of phosphate may become limiting to agriculture in the future.

The depletion of phosphate resources that can be mined is a concern because of the role it plays in fertilizer for modern intensive farming practices. Peak phosphorus is the point in time at which the maximum global phosphate production rate is reached and then begins to fall because of the depletion of reserves. Figure 8 shows a graph of world phosphate rock production against time from 1900 to 2009 obtained from the US Geological Survey. This graph suggests that the point of peak phosphorus production is approaching.

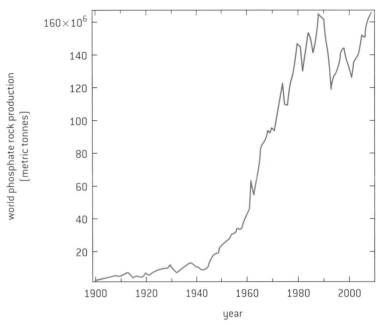

▲ Figure 8

There is a lack of agreement on the amount of available phosphate rock reserves though many agree that the problem will become acute within 50 to 100 years.

Without fertilizer, famine would most certainly result because yields per unit of farmland would plummet without the addition of fertilizer. There are no alternative sources of phosphate and no synthetic way of creating it, unlike ammonia which can be created by the industrial conversion of plentiful supplies of atmospheric nitrogen. Exploring alternative ways of conducting agriculture and conserving nutrients is one possible solution.

Eutrophication and biochemical oxygen demand

Leaching of mineral nutrients from agricultural land into rivers causes eutrophication and leads to increased biochemical oxygen demand.

When rain falls on agricultural land, water-soluble nutrients that have been added to crops such as phosphates and nitrates can dissolve in the water and the resulting runoff can enter water courses and streams. In addition to crops, nutrients from manure and urine of livestock can also contribute to nutrient enrichment of bodies of water.

▲ Figure 9 Algal bloom in marshland by the Thames Estuary, London, England

The nutrient enrichment of water is known as eutrophication. The nutrients favour the growth of algae leading to algal blooms (see figure 9). The algal blooms block light to the plants below.

When the mats of algae die and the plants below them die, it leads to a loss of oxygen, because of bacterial activity on the dead organic matter. This is called biological oxygen demand (BOD). The higher the BOD the more "anoxic" a body of water becomes and the more limiting the habitat becomes for certain fish species.

Eutrophication can also occur due to the release of untreated sewage.

Data-based questions: Sewage release into a river

Figure 10 shows changes in the biotic and abiotic factors at increasing distances from an untreated sewage outfall into a river.

1   Outline the relationship between distance from untreated sewage outfall and:

    a)   numbers of bacteria   [2]

    b)   oxygen concentrations   [2]

    c)   numbers of algae.   [2]

2   Explain the relationship between:

    a)   numbers of bacteria and concentrations of oxygen   [2]

    b)   numbers of algae and concentrations of nitrate   [2]

    c)   numbers of algae and concentrations of oxygen.   [2]

3   Predict, with reasons, the changes in BOD of water as it flows downstream from where untreated sewage enters the river.   [3]

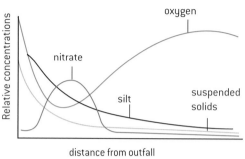

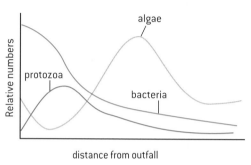

▲ Figure 10

Solutions to disruptions to the phosphorus cycle

Assessing risks and benefits of scientific research: agricultural practices can disrupt the phosphorus cycle.

Modern agriculture involves the harvesting of crops and shipping of the crops to markets outside of the ecosystem. As a consequence nutrients including phosphate are removed from the field in the crop biomass and have to be replaced by adding fertilizer. Demand for food due to population growth and increasing affluence leads to increasing pressure on the land. Higher levels of inputs are used to get more produce out of the same area of land. This is known as intensive agriculture and it requires even greater inputs of fertilizer. Concerns about the future limited availability of phosphate deposits that can be mined (see section on "peak phosphorus") as well as the fertilizer pollution (see section on "eutrophication") are leading scientists to search for solutions.

One possible solution to the supply problem is the recovery of phosphorus from sewage. People can excrete between 200 and 1,000 grams of phosphorus annually in urine. Many washing powders also contain phosphates that contributed to the phosphate load in sewage treatment plants, but alternative formulations of detergents have reduced this component of the problem.

One biological solution is the seeding of the sludge that settles in sewage treatment plants with particular groups of bacteria that selectively accumulate phosphorus. The bacterial/sludge mass can then be removed and used as fertilizer.

Phosphorus can also be removed by chemical precipitation with iron chloride or alum. This can lead to additional quantities of sludge and the chemicals are expensive, but this is a more straightforward solution than biological removal and gives higher yields of phosphate.

Livestock production can lead to an additional problem associated with the phosphorus cycle. Runoff that comes in contact with livestock farm manure might result in eventual phosphate pollution and eutrophication.

The development of genetically modified organisms is very expensive, so it is only likely to be done if there are clear benefits. The Enviropig™ has been proposed as a solution to the phosphate pollution problem. This pig has been genetically engineered with DNA from *E. coli* to produce phytase in its saliva. This enzyme digests normally insoluble phytate in pig feed, leading more phosphate to be absorbed by the pig and less to be released into the manure of the pig.

Benefits	Possible harmful effects
1. Less release of phosphorus to the environment in pig manure.	1. Traces of phytase in pork might cause allergies in human consumers.
2. Less risk of phosphorus deficiency in growing pigs.	2. Phytase gene might transfer to wild species by cross-breeding.
3. Less need to deplete world mineral phosphate reserves by its use as a dietary supplement in pig feed.	3. The genetic modification might cause suffering to the pigs in some way that is unexpected or difficult to detect.

 ## Soil testing

Assess the nutrient content of a soil sample.

Garden supply companies commonly sell soil quality assessment kits. The kits involve adding a chemical to a sample of soil that reacts with the nutrient in question if present. A colour is produced that can be visually compared to a key or the concentration can be quantified in a colorimeter (figure 11).

Soil nutrient deficiencies often produce characteristic symptoms in leaves. Figure 12 summarizes some of these characteristic signs.

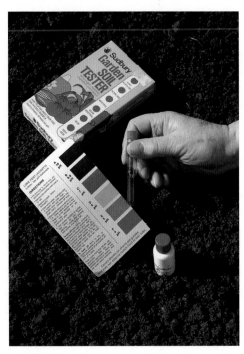

▲ Figure 11

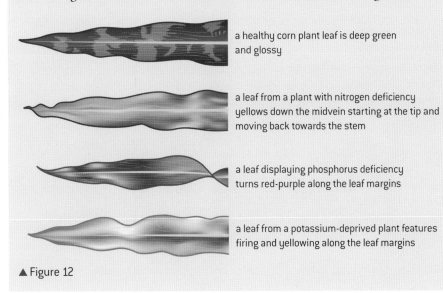

a healthy corn plant leaf is deep green and glossy

a leaf from a plant with nitrogen deficiency yellows down the midvein starting at the tip and moving back towards the stem

a leaf displaying phosphorus deficiency turns red-purple along the leaf margins

a leaf from a potassium-deprived plant features firing and yellowing along the leaf margins

▲ Figure 12

TOK

In what ways does technology enhance our ability to know the world?

The original method of colorimetry involved visually comparing the test sample to a known colour standard. However, the subjectivity of the perception of the experimenter, inconsistencies in the light sources, and the fading of colour standards limit the accuracy and reliability of results.

In 1931 the Commission Internationale de l'Eclairage developed a system to quantify the light that humans perceive by matching the three primary colours that make up all colours with three values, called the tristimulus values which approximately correspond to red, blue and green. They are also known as RBG values. Any visible colour can be quantified using these three values, and this has allowed for objective measuring and comparing of colours. A colorimeter or spectrophotometer can be used to measure the amount of light absorbed by a coloured sample in reference to a colourless sample or blank. Modern cell phone applications for measuring RBG values make quantification of colour even more accessible.

Questions

1 *Lecanora muralis* is a species of lichen that grows on walls and roofs in northwest Europe. In 1976 ecologists did a survey of the distribution of *L. muralis* in a sector of Leeds, an industrial city in the north of England. Wind direction in this area is variable and levels of air pollution decrease from the centre of the city outwards. *L. muralis* was found growing on three habitat types:

- sandstone blocks, used to build the tops of walls
- walls constructed using cement or concrete
- roofs made of asbestos cement.

Like many lichens, this species does not tolerate high levels of sulphur dioxide, an acidic gas that is a major component of acid rain. Acid rain can be neutralized by alkaline materials, including cement and concrete. The results of the survey are shown in the map below. *L. muralis* was found north of the lines shown on the map for each of the three types of habitat. The grid lines are 1 km apart.

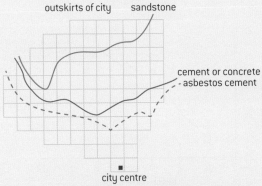

Source: Oliver Gilbert, *Lichens*, 2000, Harper Collins, page 56

a) **(i)** Deduce which habitat type allows *L. muralis* to tolerate the highest level of sulphur dioxide pollution. Give a reason for your answer. [2]

(ii) Suggest a reason for the differences in tolerance between the habitat types. [1]

b) Explain the value of a survey of this kind, especially if it is repeated at regular intervals. [3]

2 The Kluane boreal forest ecosystem project was a large-scale ten-year experimental manipulation of food and predators on an arctic ground squirrel (*Spermophilus parryii plesius*) population.

Three areas were set up:

- a food addition area
- a predator exclusion area
- a food addition area enclosed within a predator exclusion area.

The areas were monitored from 1986 to 1996. In spring 1996 all fences were dismantled and food addition was stopped.

As a further experiment, spring and summer mark-recapture population estimates of the squirrels were conducted from spring 1996 to spring 1998. The results for these two years are shown below. The areas are labelled according to the conditions imposed during the previous ten years.

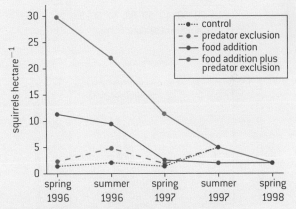

Source: Karels *et al.*, *Nature*, (2000), 408, pages 460–463

a) State the squirrel population in the food addition plus predator exclusion area in spring 1996. [1]

b) Describe the effect of ending food addition on the squirrel population. [2]

c) Scientists believed that the number of ground squirrels in the boreal forests was limited by an interaction between food and predators that acted primarily through changes in reproduction. Using the data, discuss this hypothesis. [3]

3 Destruction of subtidal and intertidal kelp (seaweed) and seagrass beds has been observed over a wide geographical range. Removal of sea urchins (*Strongylocentrotus sp.*) by experimental manipulation and accidental oil spill has resulted in the rapid development of marine vegetation. The presence and absence of kelp beds has a major effect on the structure of the marine community.

A survey was carried out of two of the western Aleutian Islands with and without sea otters (*Enhydra lutris*). Sea urchin size, density and biomass were measured. Densities and biomass were recorded per 0.25 m². Data was collected

657

from Amchitka Island (with sea otters) and Shemya Island (without sea otters).

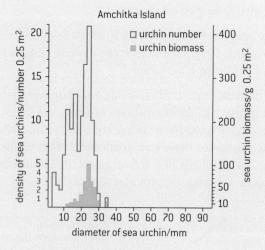

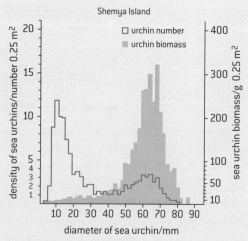

Source: reprinted with permission from J A Estes and J F Palmisano, *Science*, (1974), 185, pages 1058–1060, © copyright 1974 AAAS

a) (i) State the diameter of sea urchins with the most frequent biomass on Amchitka Island. [1]

   (ii) Suggest, giving a reason, which island would have the oldest sea urchins. [1]

b) Compare the sea urchin densities and biomass on Shemya Island and Amchitka Island. [2]

c) Deduce the trophic level of the sea urchins in this marine community. [1]

d) Explain the observed differences in sea urchin populations on the two islands. [2]

4 The horseshoe crab (*Limulus polyphemus*) lay their eggs in the sand on beaches in the intertidal zone. The nesting site is selected on the basis of distance above the mean high tide line, oxygen concentration and temperature of the sand. Egg development was assessed after 10 days and recorded as an arbitary unit; the

higher the value the more developed the eggs were. The "0 m" beach distance is based on the mean high tide line.

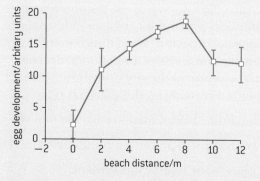

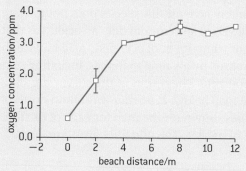

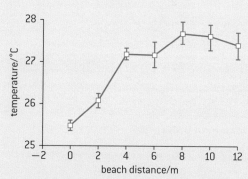

▲ Figure 3 from Penn and Brockmann. 1994. Biol. Bull. 187: 373–384. with permission from the Marine Biological Laboratory, Woods Hole, MA.

a) State the optimum distance above the high tide line for egg laying. [1]

b) Describe the effect of oxygen concentration and temperature on egg development. [2]

c) Scientists believe that egg development was influenced by oxygen concentration, temperature of the sand and distance from the mean high tidal line.

   (i) Evaluate this study with respect to these three factors. [2]

   (ii) State **one** other possible factor that might influence egg development. [1]

D HUMAN PHYSIOLOGY

Introduction

Health in humans depends on physiological mechanisms working efficiently. The study of disease both helps understand normal physiology and how treatments may be developed. Many physiological mechanisms are involved in homeostasis. To achieve a state of equilibrium, hormones need to be secreted at a variable rate.

A balanced diet is required with digestion regulated by both nervous and hormonal mechanisms. The chemical composition of the blood is regulated by the liver. Heart function is regulated by both internal and external factors. Red blood cells transport respiratory gases and respiratory gases influence blood pH.

D.1 Human nutrition

Understanding

→ Essential nutrients cannot be synthesized by the body, therefore they have to be included in the diet.

→ Dietary minerals are essential chemical elements.

→ Vitamins are chemically diverse carbon compounds that cannot be synthesized by the body.

→ Some fatty acids and some amino acids are essential.

→ Lack of essential amino acids affects the production of proteins.

→ Malnutrition may be caused by a deficiency, imbalance or excess of nutrients in the diet.

→ Appetite is controlled by a centre in the hypothalamus.

→ Overweight individuals are more likely to suffer hypertension and type II diabetes.

→ Starvation can lead to breakdown of body tissue.

Applications

→ Production of ascorbic acid by some mammals, but not others which need a dietary supply.

→ Cause and treatment of phenylketonuria (PKU).

→ Lack of Vitamin D or calcium can affect bone mineralization and cause rickets or osteomalacia.

→ Breakdown of heart muscle due to anorexia.

→ Cholesterol in blood as an indicator of the risk of coronary heart disease.

Skills

→ Determination of the energy content of food by combustion.

→ Use of databases of nutritional content of foods and software to calculate intakes of essential nutrients from a daily diet.

Nature of science

→ Falsification of theories with one theory being superseded by another: scurvy was thought to be specific to humans, because attempts to induce the symptoms in laboratory rats and mice were entirely unsuccessful.

Essential nutrients

Essential nutrients cannot be synthesized by the body, therefore they have to be included in the diet.

Nutrients are chemical substances, found in foods, that are used in the human body. Some nutrients are **essential** in the human diet, because foods are the only possible source of the nutrient. This includes some amino acids, some unsaturated fatty acids, some minerals, calcium, vitamins and water.

Other nutrients are **non-essential**, either because another nutrient can be used for the same purpose or because they can be made in the body from another nutrient. Glucose, starch and other carbohydrates are non-essential, because they are used in respiration to provide energy and lipids can be used instead.

Some essential nutrients are conditionally essential. In adults, vitamin K is produced by the metabolism of symbiotic bacteria in the intestine. Because infants do not have colonies of such bacteria at birth, they are often given a supplementary injection of vitamin K.

 Ascorbic acid is an essential nutrient in some animals

Production of ascorbic acid by some mammals, but not others which need a dietary supply.

Vitamin C is a compound called ascorbic acid. It is needed for the synthesis of the collagen fibres that form part of many tissues in the body, including skin and blood vessel walls. The vast majority of plants and animals, including most mammals, can synthesize vitamin C. The pathway by which it is synthesized in vertebrates is shown in figure 1 below.

Mutations that led to genes that no longer produce the protein necessary to make vitamin C have occurred several times in evolutionary history (see figure 2). In all cases studied so far, the inability to synthesis vitamin C is due to mutations in the GLO gene which codes for the production of the enzyme L-gulono-γ-lactone oxidase. In figure 1, this is the enzyme that catalyses the final reaction in the pathway.

A group of fish called teleost, or ray-finned, fish have lost the ability to produce vitamin C. Examples of fish from this diverse group include cod, salmon and herring. Most mammals can synthesize vitamin C: examples include carnivores such as dogs and cats. However, many primates including humans, chimpanzees and apes cannot synthesize vitamin C, though more primitive primates such as lorises and lemurs can. Only a few species of bats can synthesize vitamin C.

A variety of symptoms develop as a result of vitamin C deficiency, which are collectively known as scurvy. The symptoms of scurvy can be alleviated by intake of dietary sources of the compound.

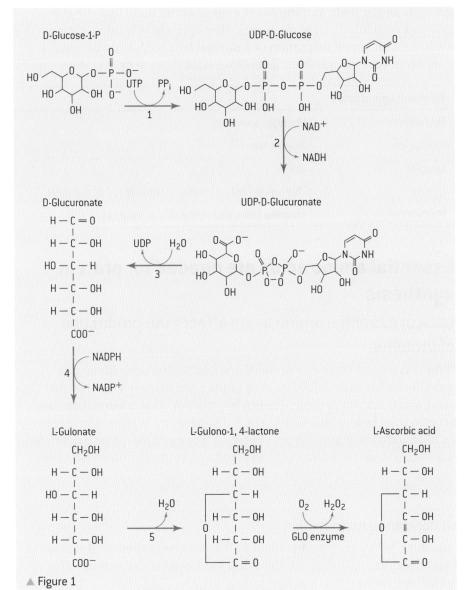

▲ Figure 1

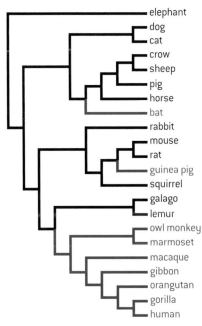

▲ Figure 2 Phylogenetic distribution of the ability to synthesize vitamin C in mammals. Lineages able to synthesize vitamin C are in black, those incapable are in purple

Essential fatty acids and amino acids

Some fatty acids and some amino acids are essential.

Of the 20 amino acids in proteins, about half are essential in humans, because they cannot be synthesized in sufficient quantities, but the other half can be made from other simpler nitrogen compounds. Threonine and arginine are conditionally essential. Threonine is an essential amino acid that can be synthesized by the body if phenylalanine is present. Sufficient arginine can normally be produced by a healthy individual. The synthesis pathway of arginine is not active in prematurely born infants and so they must obtain it through their diet.

There are some omega-3 and omega-6 fatty acids that are essential in the diet because they cannot be synthesized in the body. The "omega-3" and "omega-6" refers to the position of a double bond in relation to the end of the molecule. Alpha-linolenic acid and linoleic acid are used in the biosynthesis of a number of other compounds. They are needed

▲ Figure 3 Sunflower seeds (*Helianthus annuus*). Sunflower seeds can be eaten as a source of dietary linoleic acid

661

throughout the body, but the development of the brain and the eye involves particularly large quantities. However, there is little or no evidence that supplementation of a normal balanced diet with omega-3 fatty acids, for example from fish oils, enhances brain or eye development.

Essential amino acids	
Histidine	Phenylalanine
Isoleucine	Tryptophan
Leucine	Valine
Lysine	Threonine (only if phenylalanine is not in the diet)
Methionine	Arginine (required in the diet of infants)

Essential amino acids are needed for protein synthesis

Lack of essential amino acids affects the production of proteins.

If there is a shortage of one or more essential amino acids in the diet then the body cannot make enough of the proteins that it needs. This condition is known as protein deficiency malnutrition. Essential amino acids may be lacking due to an overall insufficiency of protein in the diet or to an imbalance in the types of protein. For example, protein deficiency malnutrition causes a lack of blood plasma proteins, with the result that

---

Data-based questions: Protein deficiency malnutrition

Figure 4 shows the incidence of stunting, wasting and developmental disability for eight regions of the world. The statistic used is YLD per 1,000, which is years lost to disease per thousand members of the population. The chart shows the results for males in the years 1990 and 2000. The figures for females showed the same trends.

1 **a)** Identify the region with the greatest evidence of protein deficiency malnutrition.

   b) Suggest reasons for this.

2 **a)** Determine the percentage difference in YLD for the year 2000 between the region with the highest rate and the region with the lowest rate.

   b) What could be done to narrow the difference?

3 **a)** Outline what the data reveals about the worldwide trend in protein deficiency malnutrition.

   b) Identify the regions of the world where the trend has been most pronounced.

   c) Suggest reasons for this.

4 Predict with a reason what the pattern might have been in 2010.

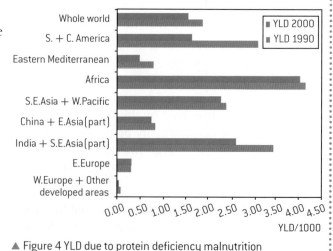

▲ Figure 4 YLD due to protein deficiency malnutrition

fluid is retained in tissues. This causes swelling (edema), which is often very obvious in the abdomen. Child development may be both mentally and physically retarded, with stunted growth and developmental disabilities.

Adults may undergo serious weight loss (wasting).

Essential minerals

Dietary minerals are essential chemical elements.

Minerals are needed in the diet in relatively small quantities – milligrams or micrograms per day rather than grams. They can be distinguished from vitamins by their chemical nature.

Minerals are chemical elements, usually in ionic form; for example, calcium is required in the diet in the form of Ca^{2+} ions. If any mineral is lacking from the diet, a deficiency disease results. The consequences of deficiency diseases can be serious, even though the quantities of the mineral needed in the diet are small. An example of this is the mineral iodine. It is needed by the thyroid gland for synthesis of the hormone thyroxin. This hormone stimulates the metabolic rate and ensures that enough energy is released in the body. A lack of iodine causes iodine deficiency disorder (IDD). If a pregnant woman has IDD, her baby may be born with permanent brain damage, and if children suffer from IDD after birth, their mental development and intelligence are impaired. Tens of millions of people worldwide have been affected in this way by IDD. Iodine supplementation can be done easily by adding the mineral to salt sold for human consumption. It costs about five cents per person to iodize salt and prevent IDD in a population. There are other types of nutrient supplementation that have considerable benefits, for little cost, in populations where deficiencies exist.

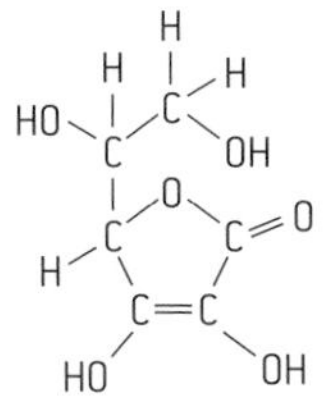

Vitamins

Vitamins are chemically diverse carbon compounds that cannot be synthesized by the body.

Vitamins are organic compounds that are needed in very small amounts because they cannot be synthesized by the body but must be obtained from the diet. They serve a variety of roles such as co-factors for enzymes, anti-oxidants and hormones. The word vitamin is derived from the words "vital amine" as the first vitamins to be discovered contained an amino group. Other vitamins discovered since do not necessarily contain an amino group such as vitamins A, C, D and E. Figure 5 shows just some of the range of structures of vitamins. Vitamin C is derived from a monosaccharide, vitamin A is hydrophobic and contains a hydrocarbon ring and chain. Vitamin B_2 contains nitrogen rings and is readily converted to the nucleotide FMN (flavin mononucleotide) by the addition of a phosphate to the carbohydrate within the molecule.

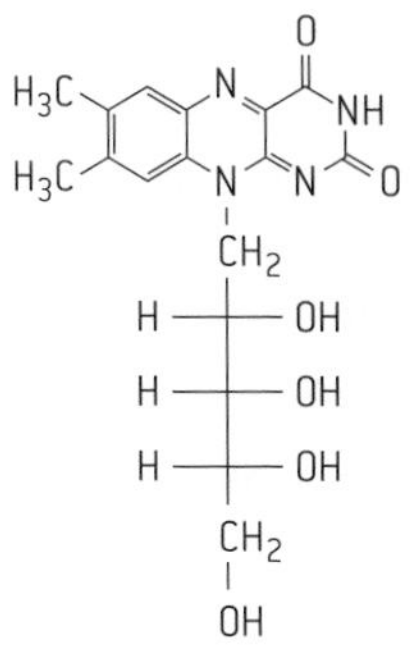

Vitamins are often broadly categorized as fat soluble and water soluble. The water soluble vitamins have to be constantly consumed and any excess is lost in urine. The fat soluble vitamins can be stored in the body.

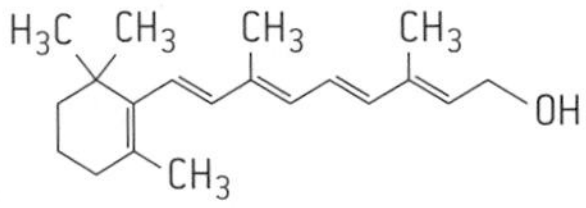

vitamin C

vitamin B_2

vitamin A

▲ Figure 5

To what extent should ethical constraints limit the pursuit of scientific knowledge?

During the Second World War, experiments were conducted both in England and in the US using conscientious objectors to military service as volunteers. The volunteers were willing to sacrifice their health to help extend medical knowledge. A vitamin C trial in England involved 20 volunteers. For six weeks they were all given a diet containing 70 mg of vitamin C. Then, for the next eight months, three volunteers were kept on the diet with 70 mg, seven had their dose reduced to 10 mg and ten were given no vitamin C. All of these ten volunteers developed scurvy. Three-centimetre cuts were made in their thighs, with the wounds closed up with five stitches. These wounds failed to heal. There was also bleeding from hair follicles and from the gums. Some of the volunteers developed more serious heart problems. The groups given 10 mg or 70 mg of vitamin C fared equally well and did not develop scurvy.

Experiments on requirements for vitamin C have also been done using real guinea pigs, which ironically are suitable because guinea pigs, like humans, cannot synthesize ascorbic acid. During trial periods with various intakes of vitamin C, concentrations in blood plasma and urine were monitored. The guinea pigs were then sacrificed and collagen in bone and skin was tested. The collagen in guinea pigs with restricted vitamin C had less cross-linking between the protein fibres and therefore lower strength.

1 Sometimes, people are paid to participate in medical experiments, such as new drug trials. What are some of the ethical issues associated with being paid to be a subject in an experiment?

2 For some drug trials, there is the potential for the subject to be harmed. What are the associated ethical issues related to risks to volunteers in drug trials?

3 Some experiments on humans were done against the subject's will or without the subject's knowledge. Once the data is generated from these experiments, it cannot become "unknown". What are the associated ethical issues with people other than the original experimenter using the information gathered under these conditions?

Water soluble vitamins	Fat soluble vitamins
C ascorbic acid	A
B_1 thiamin	
B_2 riboflavin	E
B_3 niacin	
B_5 pantothenic acid	K
B_6 pyridoxine	
B_7 biotin	
B_9 folic acid	D (conditionally essential)
B_{12} cobalamin	

Types of malnutrition

Malnutrition may be caused by a deficiency, imbalance or excess of nutrients in the diet.

Malnutrition is the outcome of a poor diet. Diets can be low in overall quantity with low protein and calorie content. They can be unbalanced and fail to provide essential nutrients or they can contain excess fats and refined carbohydrates. Malnutrition is often associated with poverty. Starvation is a consequence of a diet lacking in adequate protein and carbohydrates. Increasingly, obesity is observed in developing countries as well as in the lower socio-economic classes of developed nations as a consequence of unhealthy diets with excess fat and refined carbohydrates.

The appetite control centre

Appetite is controlled by a centre in the hypothalamus.

In the hypothalamus of the brain there is a centre that is responsible for making us feel satisfied when we have eaten enough food (satiated). It is called the appetite control centre. The small intestine secretes the hormone PYY3–36 when it contains food. The pancreas secretes insulin when the blood glucose concentration is high. Adipose tissue secretes the hormone leptin when amounts of stored fat increase. If the appetite control centre receives these hormones, it reduces the desire to eat. This helps to us to avoid health problems due to overeating, including excessive blood glucose levels and obesity.

Consequences of being overweight

Overweight individuals are more likely to suffer hypertension and type II diabetes.

Unhealthy diets with excess fat and refined carbohydrates have health consequences. Two examples of nutrition related diseases are diabetes and hypertension.

There are several diseases involving excessive excretion of urine, all of which are forms of diabetes. In the commonest form, sugar is present in the urine. This is **diabetes mellitus**, and it affects hundreds of millions of people worldwide. There are two ways in which this sort of diabetes can develop:

- Auto-immune destruction of insulin-secreting cells in the pancreas (type I diabetes).

- Decreased responsiveness of body cells to insulin due to "burn-out" (type II diabetes).

Prevalence rates of type II diabetes are rising rapidly in many countries. The study of the rates and distribution of a disease, to try to find its causes, is known as epidemiology. Epidemiological studies of type II diabetes have implicated increased blood concentrations of fatty acids, linked to the following risk factors:

- diets rich in fat and low in fibre

- obesity due to overeating and lack of exercise

- genetic factors which affect fat metabolism.

There is huge variation between ethnic groups in rates of type II diabetes, from less than 2 per cent in China to 50 per cent among the Pima Indians. The symptoms are not always recognized, so not all people with diabetes are diagnosed. The main symptoms are:

- elevated levels of blood glucose

- glucose in the urine – this can be detected by a simple test

- dehydration and thirst resulting from excretion of large volumes of urine.

Unless carefully managed, diabetes can cause other health problems to develop, several of which relate to the cardiovascular system:

- atherosclerosis (narrowing of arteries by fatty deposits)

- hypertension (raised blood pressure, discussed below)

- coronary heart disease (narrowing of the coronary arteries with the associated risk of heart attacks).

There also seems to be a link between these cardiovascular problems and blood lipid concentrations. Links have been suggested between high concentrations of cholesterol, high concentrations of LDL and low concentrations of HDL. There has been much controversy about the role of cholesterol, in particular in the development of coronary heart disease (CHD).

There is a clear correlation between excessive weight gain and hypertension, though the relationship is complex. Weight gain can increase the release of several hormones as well as cause changes in body physiology and anatomy all of which can lead to hypertension:

- weight gain leads to higher cardiac (heart) output which can raise blood pressure

- abdominal obesity can increase vascular resistance which can raise blood pressure

- weight gain is associated with arteries becoming stiffer and narrower which can raise blood pressure.

Hypertension can also be caused by high salt intake. Circulating salt has an osmotic effect.

Effects of starvation

Starvation can lead to breakdown of body tissue.

Starvation occurs due to the severe lack of intake of essential and non-essential nutrients. In the absence of dietary intake of energy sources, the body will first access glycogen stores. However, if no glucose is available, the body will break down its own muscle tissue to utilize the resulting amino acids as energy sources. The amino acids are sent to the liver where they are converted to glucose. This results in a loss of muscle mass. In figure 6, the child is suffering from marasmus. His thin limbs indicate that muscle tissue has been broken down as an energy source by his body.

▲ Figure 6

Anorexia

Breakdown of heart muscle due to anorexia.

The medical term anorexia means reduced appetite. Anorexia nervosa is a psychiatric illness, with causes that are complex. It involves voluntary starvation and loss of body mass. The amounts of carbohydrate and fat consumed are too small to satisfy the body's energy requirements, so protein and other chemicals in the body are broken down. There is wasting of muscles, resulting in loss of strength. Hair becomes thinner and can drop out. The skin becomes dry and bruises easily. A fine growth of body hair tends to develop. Blood pressure is reduced, with slow heart rate and poor circulation. In females, infertility is another common consequence, with no ovulation or menstrual cycles.

As body weight in a person with anorexia falls, not only is skeletal muscle digested, but heart muscle deteriorates. To some degree, the skeletal muscle mass reduces disproportionately faster than the cardiac mass. Lack of protein, electrolytes and micronutrients may result in the deterioration of muscle fibres. The lack of dietary intake also alters the electrolyte balance; i.e., concentrations of calcium, potassium and sodium. Both skeletal muscle and cardiac muscle do not contract normally under these circumstances. There is often reduced blood pressure, a slower heart rate and reduced heart output in patients.

Data-based questions: Changes in heart dimensions in patients with anorexia

The data in figure 7 shows the dimensions of various structures in subjects with normal diet and in patients with anorexia.

1   Calculate the percent change in the mean dimensions of

   a) the left ventricle

   b) the left ventricle wall

   c) the left atrium

   d) the base of the aorta.   [5]

2   Identify the part of the heart with the largest decrease in dimension due to anorexia.   [1]

3   Suggest what might be the symptoms of this change in the affected patient.   [3]

		Left ventricle	Left atrium	Base of aorta	Ventricle wall
Normal	mean	47 mm	29 mm	27 mm	9 mm
	range	(35–57)	(19–40)	(20–37)	(6–11)
Anorexia	mean	38 mm	26 mm	21 mm	8 mm
	range	(38–44)	(17–34)	(18–26)	(6–9)

▲ Figure 7

The guinea pig as a model organism for studying scurvy

Falsification of theories with one theory being superseded by another: scurvy was thought to be specific to humans, because attempts to induce the symptoms in laboratory rats and mice were entirely unsuccessful.

In 1907, two scientists, Holst and Frolisch, published a research paper on their success in developing an animal model for the study of scurvy. They caused scurvy by feeding guinea pigs (*Cavia porcellus*) with whole grains. They cured scurvy in the guinea pigs through dietary modification including feeding fresh cabbage and lemon juice. The ideas within their paper were somewhat unpopular with the scientific community as the concept of nutritional deficiencies was unheard of at the time. The use of the term "vitamin" did not begin until later.

Their animal model allowed for the systematic study of the factors that led to the scurvy, as well as the preventive value of different substances. Substituting guinea pigs for pigeons, an animal model that had been used in beriberi research, was a lucky coincidence, as the guinea pig was later shown to be among the very few mammals capable of showing scurvy-like symptoms, while pigeons, as seed-eating birds, were later shown to make their own vitamin C and could not develop scurvy.

▲ Figure 8 Dermatitis in a guinea pig fed exclusively on rabbit pellets. This is one of a number of symptoms of scurvy seen in guinea pigs with the disease

Phenylketonuria

Cause and treatment of phenylketonuria (PKU).

Phenylketonuria (PKU) is a genetic disease. It is caused by mutations of a gene coding for the enzyme that converts phenylalanine into tyrosine.

The mutations produce alleles of the gene that code for enzymes unable to catalyse the conversion reaction. Only one normal allele is needed for satisfactory conversion of phenylalanine to tyrosine, so this allele is dominant. The symptoms of PKU only occur in individuals with two recessive mutant alleles. Phenylalanine then accumulates in the body and there can also be a deficiency in tyrosine.

The consequences of PKU are potentially very serious. The high phenylalanine levels cause reduced growth of head and brain, with mental retardation of young children and severe learning difficulties, hyperactivity and seizures in older children. Other consequences are a lack of skin and hair pigmentation.

PKU babies are unaffected at birth because the mother's metabolism has kept phenylalanine and tyrosine at normal levels. This gives an opportunity for early diagnosis and treatment. A test should be carried out at about 24 hours after birth, by which time blood phenylalanine concentrations will have started to rise. Treatment involves eating a diet low in phenylalanine for the rest of the person's life. Meat, fish, nuts, cheese, peas and beans can only be eaten in small quantities. Tyrosine supplements may be needed. If a suitable diet is rigorously adhered to, the harmful consequences of PKU can be avoided.

phenylalanine

phenylalanine
hydroxylase

OH

tyrosine

▲ Figure 9 Synthesis of tyrosine from phenylalanine

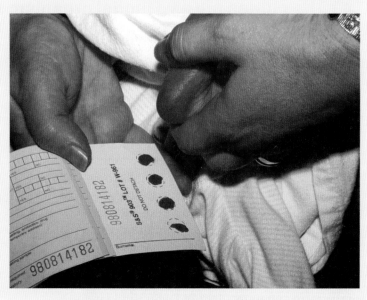

▲ Figure 10 A new born baby being tested for PKU using the Guthrie test

🌐 Vitamin D deficiency

Lack of Vitamin D or calcium can affect bone mineralization and cause rickets or osteomalacia.

Vitamin D is needed for calcium absorption from food in the intestines, so the symptoms of vitamin D deficiency are similar to those of calcium, with children developing the skeletal deformities known as rickets. Vitamin D does not fit the definition of a vitamin very well, as it can be synthesized in the skin. This only happens when sunlight, or another light source containing ultraviolet light with wavelengths in the range 290–310 nm, strikes the skin. If teenagers and adults spend enough time outside, with their skin uncovered, no vitamin D is required in the diet. Children, pregnant women and elderly people are recommended to eat 10 μg per day, to supplement the amount made in their skin. There are few dietary sources of vitamin D. Oily fishes including herring, mackerel, sardines and tuna are rich sources. Eggs and liver also contain some, and certain foods such as margarine and milk are artificially fortified with vitamin D.

Ultraviolet light has some harmful consequences, including mutations that can lead to skin cancer. Melanin in the skin intercepts and absorbs light, including the ultraviolet wavelengths. Dark skins therefore give good protection against cancer, but they also reduce vitamin D synthesis. In indigenous human populations, skin colour balances the twin risks of vitamin D deficiency and cancer or other damage due to ultraviolet light. After population migrations there can be problems. In the 1970s immigrants with dark skin from the Indian subcontinent living in the United Kingdom started to show symptoms of vitamin D deficiency. Immigrants from northern Europe with light skin living in Australia were found to have high rates of malignant melanoma. Australians with light skin were then advised to stay out of bright sunlight, cover their skin or apply sun-block creams.

Blood cholesterol and heart disease

Cholesterol in blood as an indicator of the risk of coronary heart disease.

Cholesterol is a normal component of plasma membranes in human cells, but nevertheless it has developed a reputation for being a harmful substance. This is because research has shown a correlation between high levels of cholesterol in blood plasma and an increased risk of coronary heart disease. Advice is often given to minimize dietary cholesterol intake, but it is not certain that this will actually lower the risk of coronary heart disease (CHD), for a variety of reasons.

- Much research has involved total blood cholesterol levels, but only cholesterol in LDL (low-density lipoprotein) is implicated in CHD.

- Reducing dietary cholesterol often has a very small effect on blood cholesterol levels and therefore presumably has little effect on CHD rates.

- The liver can synthesize cholesterol so dietary cholesterol is not the only source.

- Genetic factors are more important than dietary intake and members of some families have high cholesterol levels even with a low dietary intake.

- Drugs can be more effective at reducing blood cholesterol levels than reductions in dietary intake.

- There is a positive correlation between dietary intake of saturated fats and intake of cholesterol, so it is possible that saturated fats, not cholesterol, cause the increased risk of CHD in people with high cholesterol intakes.

Calorimetry

Determination of the energy content of food by combustion.

The determination of the energy content of a substance is called calorimetry. Figure 11 shows an experimental set-up for a simple calorimeter. It is based on the knowledge of the specific heat capacity of water. It takes 4.186 J of heat energy to raise the temperature of 1 gram of water by 1 degree Celsius.

Q = mass of water × specific heat capacity × change in temperature

The apparatus consists of a thermometer to detect the change in temperature and a vessel containing a known mass of water (1 ml of water has a mass of 1 g). The sample to be tested for its energy content is ignited and placed below the container containing the water and the temperature change is noted.

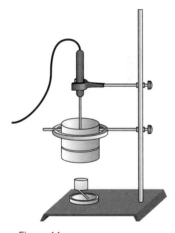

▲ Figure 11

Activity

Using the experimental results below, estimate the energy content of the nut per gram.

Sample data:

Mass of nut = 0.60 g

Volume of water = 25 ml

Initial water temperature = 20 °C

Final water temperature = 65 °C

Monitoring personal dietary intake

Use of databases of nutritional content of foods and software to calculate intakes of essential nutrients from a daily diet.

When nutritionists refer to a balanced diet, they are referring to a combination of foods that will provide essential and non-essential nutrients in the correct balance. Figure 12 shows a food wheel showing a healthy balanced diet. The wheel shows what proportion of the diet should be made up by each of the major food groups. Fresh fruit and vegetables should make up the largest part of the diet, followed by carbohydrates, proteins and then dairy products. Fats and sugars are on the chart, not because they are encouraged or required but because they should make up the smallest part of the diet.

Computer applications can be used to keep a record of the food consumed by an individual. Based on information entered in databases about the composition of the foods consumed, the nutrient intake of an individual can be tracked and compared to the recommended intake. Figure 13 shows an

image generated by entering a record of the contents of a typical breakfast into the free software "Super Tracker" available from the US Department of Agriculture (USDA).

▲ Figure 12

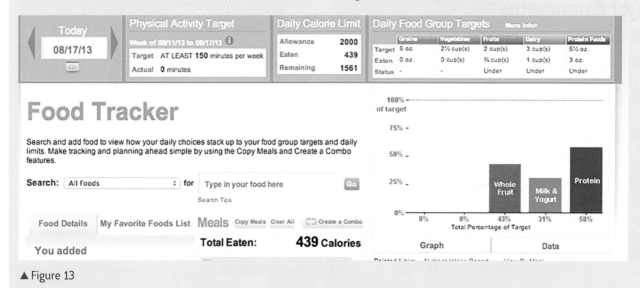

▲ Figure 13

D.2 Digestion

Understanding

→ Nervous and hormonal mechanisms control the secretion of digestive juices.

→ Exocrine glands secrete to the surface of the body or the lumen of the gut.

→ The volume and content of gastric secretions are controlled by nervous and hormonal mechanisms.

→ Acid conditions in the stomach favour some hydrolysis reactions and help to control pathogens in ingested food.

→ The structure of cells of the epithelium of the villi is adapted to the absorption of food.

→ The rate of transit of materials through the large intestine is positively correlated with their fibre content.

→ Materials not absorbed are egested.

 ## Applications

→ The reduction of stomach acid secretion by proton pump inhibitor drugs.

→ Dehydration due to cholera toxin.

→ *Helicobacter pylori* infection as a cause of stomach ulcers.

 ## Skills

→ Identification of exocrine gland cells that secrete digestive juices and villus epithelium cells that absorb digested foods from electron micrographs.

 ## Nature of science

→ Serendipity and scientific discoveries: the role of gastric acid in digestion was established by William Beaumont while observing the process of digestion in an open wound caused by gunshot.

Regulation of digestive secretions

Nervous and hormonal mechanisms control the secretion of digestive juices.

Under natural conditions, there are gaps between meals. Animals may go for long periods of time between meals. In order to conserve energy, animals do not have their digestive systems active constantly. In the case of the "fight or flight" response, conserving energy for muscle use creates the need to divert energy from the digestive process. In both cases, nerves and hormones ensure resources are devoted to digestion only when needed. Consider the example of gastric juice secretion in the stomach.

Regulation of gastric secretions

The volume and content of gastric secretions are controlled by nervous and hormonal mechanisms.

Both nerves and hormones are involved in controlling the secretion of digestive juices. Gastric juice secretion is an example of this. The sight or smell of food causes the brain to send nerve impulses via the vagus nerve from the medulla. Gland cells in the stomach wall are stimulated to secrete components of gastric juice. If chemoreceptors in the stomach wall detect peptides in the stomach contents or if stretch receptors detect distension of the stomach, impulses are sent to the brain. The brain responds by sending impulses via the vagus nerve to endocrine cells in the wall of the duodenum and the part of the stomach nearest to the duodenum, stimulating them to secrete gastrin. The hormone gastrin stimulates secretion of acid and pepsinogen by two types of exocrine gland cell in the stomach wall. Two other hormones, secretin and somatostatin, inhibit gastrin secretion if the pH in the stomach falls too low.

Exocrine glands

Exocrine glands secrete to the surface of the body or the lumen of the gut.

The passage through which food passes from mouth to anus is called the alimentary canal. Digestive juices are added to food in the alimentary canal at several points. Exocrine glands secrete the juices, including salivary glands, the pancreas, gland cells in the stomach wall and in the wall of the small intestine. The composition of the juices secreted by the glands is different, reflecting the processes that occur in each part of the alimentary canal (see table 1).

Digestive fluid	Source	Composition
saliva	salivary glands	water, electrolytes, salivary amylase, mucus, lysozyme
gastric juice	stomach	water, mucus, enzymes including pepsin, rennin and hydrochloric acid
pancreatic juice	pancreas	water, bicarbonate, enzymes including: amylase, lipase, carboxypeptidase, trypsinogen

▲ Table 1

Unlike endocrine glands, which secrete directly into the bloodstream, exocrine glands secrete into ducts. Figure 1 shows the arrangement of cells in part of an exocrine gland. Secretory cells are in groups around the duct branch. Each group of cells is called an acinus. The structure of the individual exocrine gland cells that secrete digestive enzymes is revealed in electron micrographs (figure 2). There is extensive endoplasmic reticulum for synthesis of enzymes. There are numerous mitochondria to provide ATP for protein synthesis and other cell activities. There are also large numbers of secretory vesicles containing enzymes. The process of exocytosis of these vesicles can sometimes be seen in progress where the plasma membrane of the cell is in contact with the duct.

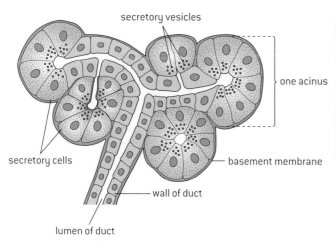

▲ Figure 1 An exocrine gland

secretory vesicles

one acinus

secretory cells

basement membrane

wall of duct

lumen of duct

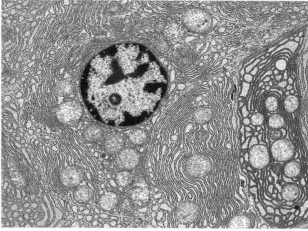

▲ Figure 2 An exocrine cell

Adaptations of the villus

The structure of cells of the epithelium of the villi is adapted to the absorption of food.

Figure 3 shows a longitudinal cross-section through the ileum, the site of a significant amount of the absorption that takes place in the small intestine.

The inner surface of the ileum has numerous folds. Each of the folds is covered in tiny projections called villi. Absorption takes place through the epithelial cells covering each villus.

- Each epithelial cell covering the villus adheres to its neighbours through tight junctions, which ensure that most materials pass into the blood vessels lining the villi through the epithelial cell.

- The cell surface membrane on the intestinal lumen side has a number of extensions called microvilli. The collection of microvilli on the intestinal side of the epithelial cells is termed the brush border. The function of the brush border is to increase the surface area for absorption.

- Relatively high amounts of ATP are required to drive active transport processes. Thus epithelial cells have large numbers of mitochondria.

- Pinocytic vesicles are often present in large numbers due to absorption of some foods by endocytosis.

- The surface facing the lumen of the intestine is referred to as the apical surface and the surface facing the blood vessels is referred to as the basal surface. These surfaces have different types of proteins involved in material transport.

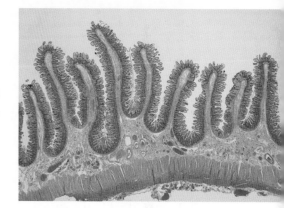

▲ Figure 3 Longitudinal section through the ileum wall

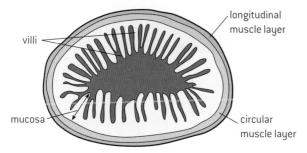

villi

longitudinal muscle layer

mucosa

circular muscle layer

▲ Figure 4 Transverse section of ileum

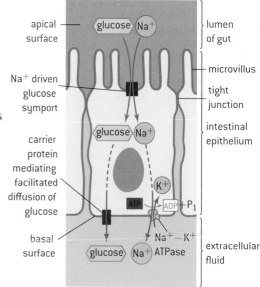

apical surface

glucose Na^+

lumen of gut

Na^+ driven glucose symport

microvillus

tight junction

carrier protein mediating facilitated diffusion of glucose

glucose Na^+

intestinal epithelium

K^+

ATP ADP P_1

basal surface

glucose Na^+

$Na^+ - K^+$ ATPase

extracellular fluid

▲ Figure 5

673

Identification of exocrine glands

Identification of exocrine gland cells that secrete digestive juices and villus epithelium cells that absorb digested foods from electron micrographs.

Data-based questions: Adaptations of villus epithelium cells

The electron micrograph shows part of two villus epithelium cells. False colour has been used to distinguish between some of the structures that are present.

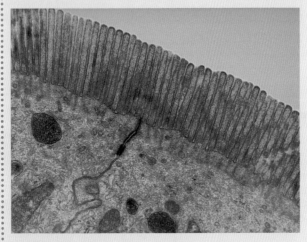

▲ Figure 6 Micrograph of villus epithelium cell

1 **a)** Identify the structures that have been coloured orange. [1]

b) Explain the function of these structures. [2]

c) Calculate the magnification of the electron micrograph, assuming that these structures are 0.85mm long. [3]

2 **a)** Identify which structures are mitochondria. [1]

b) Explain the need for large numbers of mitochondria in villus epithelium cells. [2]

3 Large numbers of vesicles are visible in the cytoplasm of the cells.

a) State the name of the process used to form these vesicles. [1]

b) Predict the contents of the vesicles. [2]

4 Part of the junction between the two cells has been coloured blue.

a) State the name of this structure. [1]

b) Explain its function. [2]

Figure 7 is an electron micrograph showing two elongated, acinar cells of the exocrine human pancreas. Arranged in rounded glands, these cells secrete an alkaline, enzyme-rich fluid into the duodenum via the small duct (in blue) at top of image. Acinar cells are often pyramidal-shaped cells. Vesicles and granules will often be found at the surface next to the duct. In this image, granules of pancreatic enzymes are being carried through the cytoplasm towards the duct at the top.

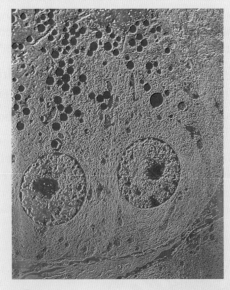

▲ Figure 7

 Discovering the chemical nature of digestion in the stomach

Serendipity and scientific discoveries: the role of gastric acid in digestion was established by William Beaumont while observing the process of digestion in an open wound caused by gunshot.

Alexis St. Martin was a Canadian fur trader who received a gunshot wound to his side. He survived the accident, but the wound healed in such a way that the inside of his stomach could be seen from the outside. William Beaumont, the surgeon who first treated the wound, used the opportunity to study the process of digestion. He continued to conduct investigations over an eleven-year period. He published his results in 1833. Beaumont is credited with overturning the notion that digestive processes within the stomach are solely physical providing evidence through his experiments of the chemical nature of digestion.

The role of acid conditions in the process of digestion

Acid conditions in the stomach favour some hydrolysis reactions and help to control pathogens in ingested food.

Acid is secreted by the parietal cells of the stomach. The acid disrupts the extracellular matrix that holds cells together in tissues. It also leads to the denaturing of proteins, exposing the polypeptide chains so that the enzyme pepsin can hydrolyse the bonds within the polypeptides.

Pepsin is released by chief cells as the inactive pepsinogen. The acid conditions within the stomach convert the inactive pepsinogen to pepsin. This ensures that the cells that produce pepsinogen are not digested at the same time as the protein in the diet.

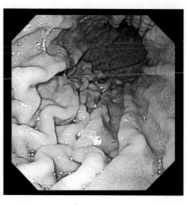

▲ Figure 8 Interior of stomach

Bacterial infection as a cause of ulcers

Helicobacter pylori infection as a cause of stomach ulcers.

Stomach ulcers are open sores, caused by partial digestion of the stomach lining by the enzyme pepsin and hydrochloric acid in gastric juice. Stomach cancer is the growth of tumours in the wall of the stomach. Until recently, emotional stress and excessive gastric juice secretion were believed to be a major contributory factor in the development of stomach ulcers, but a bacterium, *Helicobacter pylori*, has been shown to be a more significant cause. This bacterium also seems to be associated with stomach cancer.

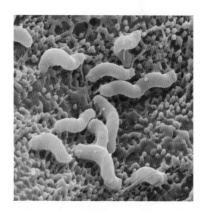

▲ Figure 9 *Helicobacter pylori* bacteria on the surface of the human gut. Colonies of *H. pylori* occur on the stomach mucous membrane in people who suffer from gastritis. This bacteria has been linked to stomach ulcer formation. *H. pylori* may also be a factor for gastric cancer as its presence increases the risk of stomach tumours

Proton pump inhibitors

The reduction of stomach acid secretion by proton pump inhibitor drugs.

There are several disease conditions of the stomach that are made worse by the release of acid. Stomach acid is corrosive so the body produces a natural mucus barrier which protects the lining of the stomach from being attacked by the acid.

In some people this barrier may have broken down allowing the acid to damage the stomach, causing bleeding. This is known as an ulcer. In others there may be a problem with the circular muscle at the top of the stomach that prevents fluid from escaping the stomach. If the muscle is not functioning, the acid escapes and irritates the esophagus. This is called "acid reflux" which can cause a symptom referred to as heartburn.

The production of the acidic environment within the stomach is achieved by a proton pump called the H^+, K^+-ATPase. This pump uses one ATP molecule to exchange two protons from the cytoplasm for two potassium ions in the lumen surrounding the parietal cell. One therapy that is increasingly prescribed for gastric diseases is proton pump inhibitors or PPIs.

PPIs bind irreversibly to a single pump. The effect on the overall acid production system is not permanent as the pumps are normally recycled and replaced with new pumps.

The PPIs are consumed in an inactive form. Acid conditions in the vicinity of the parietal cells convert them to the active form close to their target.

Egestion

Materials not absorbed are egested.

Dietary fibre is the edible parts of plants that are resistant to being digested and are not absorbed from the small intestine. Examples include cellulose and lignin. As a consequence, there is a fraction of ingested food which never leaves the digestive tube. In addition, secretion into the digestive tube occurs. Some of what is added is excretory products such as bilirubin from the breakdown of red blood cells. A large volume of water is added to the tube in the process of digestion by secretions in the mouth, stomach and small intestine, and has to be reclaimed in the large intestine. The excretory products, the unabsorbed water and undigested dietary fibre are egested as feces.

The role of dietary fibre

The rate of transit of materials through the large intestine is positively correlated with their fibre content.

Dietary fibre is material such as cellulose, lignin and pectin that cannot be readily digested. There are two categories of dietary fibre: soluble and insoluble. A healthy balanced diet contains fibre as it increases the bulk of material passing through the intestines and helps to prevent constipation as it draws water into the intestine. The higher the water content of the intestine, the faster the movement of fecal matter.

There are other possible benefits of fibre in the diet. The risk of various diseases of the large intestine may

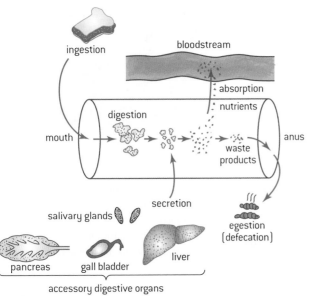

▲ Figure 10

be reduced, including bowel cancer, hemorrhoids and appendicitis. The presence of bulky material in the stomach and intestines may increase feelings of satiety, reducing the desire to eat and the risk of obesity. Absorption of sugars may be slowed down, helping to prevent the development of type II diabetes. Foods of plant origin contain dietary fibre, especially whole-grain bread and cereals, vegetables such as cabbage and salads such as celery. Foods made from cultured fungi (mycoprotein) also contain dietary fibre.

Data-based questions: Dietary fibre and mean residence time

Figure 11 shows the correlation between digestible matter content (meaning less dietary fibre) and mean residence time (the length of time in the intestine).

1   Using the curve, determine the digestible matter content of a feces which has a mean residence time of 40 hours.   [1]

2   Explain the relationship between digestibility and mean residence time.   [3]

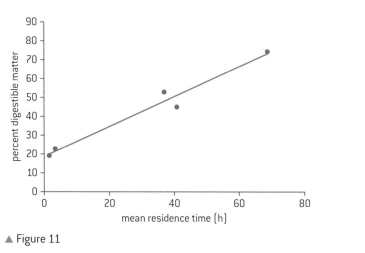

▲ Figure 11

🌐 Dehydration due to cholera

Dehydration due to cholera toxin.

Cholera is a disease caused by infection by the bacterium *Vibrio cholera*. The bacterium releases a toxin that binds to a receptor on intestinal cells. The toxin is then brought into the cell by endocytosis. Once inside the cell, the toxin triggers a cascade response that ultimately leads to the efflux of Cl^- and HCO_3^- ions from the cell into the intestine. Water follows by osmosis leading to watery diarrhoea. Water is drawn from the blood into the cells to replace the fluid loss from the intestinal cells. Quite quickly severe dehydration can result in death if the patient does not receive rehydration.

TOK

What role does conservatism play in science?

Thirty years ago, it was widely believed that emotional stress and lifestyle factors caused stomach ulcers. It is now recognized about 80 per cent of ulcers are caused by infection from *Helicobacter pylori*. The theory that ulcers were the consequence of an infection was put forward in the early 1980's by Barry Marshall and Robin Warren, two little-known Australian scientists. By the mid-1980s, they worked out an inexpensive treatment that cured about 75 per cent of patients. By 1988, they had shown definitively that antibiotics which killed *H. pylori* would cure ulcers for good. But the treatment did not become widely available until the early 1990s. Marshall attributes the slow take-up of their discovery to at least three different factors. The first problem is the inertia of existing beliefs. Doctors and drug companies had convinced themselves that they already knew the cause of ulcers: emotional stress. Marshall and Warren's infectious-agent theory had to displace the mindset. Also the blockbuster drugs of the time, Smith Kline Beecham's Tagamet and Glaxo's Zantac were both very good at putting ulcers into remission. The second problem lay in the way funding is allocated. Research grants are often awarded for three-year stints. When, in 1988, Marshall and Warren demonstrated that antibiotics could cure ulcers, many researchers who might have confirmed their result were already locked into research on acid-lowering drugs. Third, Marshall says that initially, they found it difficult for their publications to be noticed. Pharmaceutical companies fund an enormous amount of drug research in universities and hospitals. Pharmaceutical companies understandably tend to concentrate their efforts on conservative research that tends toward lucrative ongoing treatments rather than speculative ventures that might produce cheaper, permanent cures.

D.3 Functions of the liver

Understanding

→ The liver removes toxins from the blood and detoxifies them.

→ Components of red blood cells are recycled by the liver.

→ The breakdown of erythrocytes starts with phagocytosis of red blood cells by Kupffer cells.

→ Iron is carried to the bone marrow to produce hemoglobin in new red blood cells.

→ Surplus cholesterol is converted to bile salts.

→ Endoplasmic reticulum and Golgi apparatus in hepatocytes produce plasma proteins.

→ The liver intercepts blood from the gut to regulate nutrient levels.

→ Some nutrients in excess can be stored in the liver.

 Applications

→ Causes and consequences of jaundice.

→ Dual blood supply to the liver and differences between sinusoids and capillaries.

Nature of science

→ Educating the public on scientific claims: scientific studies have shown that high-density lipoprotein could be considered "good" cholesterol.

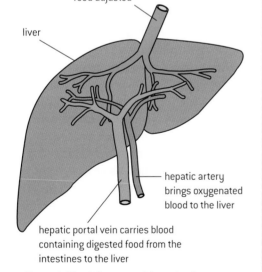

hepatic vein carries blood from the liver on to the heart with levels of food adjusted

liver

hepatic artery brings oxygenated blood to the liver

hepatic portal vein carries blood containing digested food from the intestines to the liver

▲ Figure 1 Blood flow to and from the liver

Blood supply to the liver

Dual blood supply to the liver and differences between sinusoids and capillaries.

Figure 1 illustrates the blood vessels that serve the liver. Blood arrives at the liver from two sources.

The hepatic artery branches off from the aorta bringing oxygen-rich blood from the heart.

Most of the blood circulating in the liver comes from the hepatic portal vein which brings blood from the stomach and the intestines to the liver. The blood can be rich in nutrients that have been absorbed from digested food depending on how recently the individual has eaten. Because the hepatic portal vein has travelled from the heart to the stomach or the intestine first, its oxygen content is relatively low.

Within the liver, the vein subdivides into divisions called sinusoids. Sinusoids are like capillaries but are wider and the walls are not continuously lined with cells (figure 2). This allows the blood flowing through to come in contact with the hepatocytes (liver cells), and also allows proteins such as albumin to enter and leave the blood. The hepatic artery subdivides into arterioles, which join with the sinusoids at various points, providing oxygenated blood. The sinusoids merge with venules that lead to the hepatic vein. This carries blood away from the liver to the vena cava.

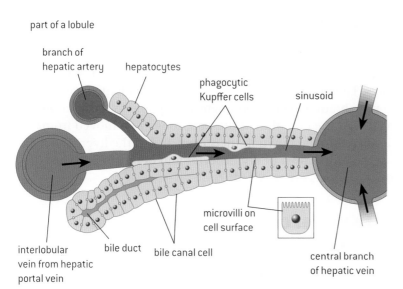

part of a lobule

branch of
hepatic artery   hepatocytes

phagocytic
Kupffer cells   sinusoid

microvilli on
cell surface

interlobular   bile duct   bile canal cell
vein from hepatic
portal vein

central branch
of hepatic vein

▲ Figure 2 Circulation within the liver

Processing of nutrients by the liver

The liver intercepts blood from the gut to regulate nutrient levels.

One of the main functions of the liver is to regulate the quantity of nutrients circulating in the blood. It plays a key role in the regulation of circulating glucose by either storing glucose as glycogen or breaking glycogen down to glucose. Because the body cannot store proteins or amino acids, excess quantities of these in the diet are broken down in the liver to be utilized as energy sources. The liver processes the resulting nitrogenous waste.

The liver is responsible for managing circulating lipids which arrive in a variety of forms. Some forms such as chylomicrons arriving from the intestine are broken down. The liver processes lipids in one form and distributes it in other forms. For example, very low density lipoproteins (VLDL) are synthesized in hepatocytes. Their purpose is to transport the triglycerides synthesized in the liver into blood plasma for storage or use in the body. Surplus cholesterol is converted into bile salts.

Storage of nutrients in the liver

Some nutrients in excess can be stored in the liver.

When levels of glucose are high, insulin is released. The insulin stimulates hepatocytes to take up the glucose and store it as glycogen. When the levels of blood glucose fall, hormones such as glucagon will be released. This release will result in the breakdown of glycogen, glycerol, amino acids and fatty acids in the liver releasing glucose to the blood stream.

Iron, retinol (vitamin A) and calciferol (vitamin D) are stored in the liver when in excess and released when there is a deficit in the blood.

Recycling of red blood cells

Components of red blood cells are recycled by the liver.

The typical lifespan of an erythrocyte (red blood cell) in an adult is about 120 days. The old and damaged erythrocytes undergo changes in their plasma membrane which make them susceptible to recognition by macrophages. At the end of their lifespan, they are removed from circulation and are broken down in the spleen and in the liver. The liver is involved in the breakdown of erythrocytes and hemoglobin. Most of the breakdown products are recycled.

The role of Kupffer cells in the breakdown of erythrocytes

The breakdown of erythrocytes starts with phagocytosis of red blood cells by Kupffer cells.

As red blood cells age, they swell and some are engulfed by Kupffer cells, which are macrophages which line the sinusoids in the liver. Inside the Kupffer cell, the hemoglobin molecule is split into globin chains and a heme group. Amino acids from the globin chains are recycled, while the heme group is further broken down into iron and bilirubin. The Kupffer cells release bilirubin to the blood. The iron is bound to transferrin and transported to the liver and spleen for storage or to the bone marrow to be used in the synthesis of new red blood cells.

Figure 3 is a scanning electron micrograph (SEM) of Kupffer cells in the liver. Here, within a sinusoid (coloured blue) of the liver, Kupffer cells (yellow, at right) are seen. Kupffer cells are capable of phagocytosis, whereby they trap and engulf foreign particles and substances. Note their long arm-like extensions of cytoplasm called filopodia.

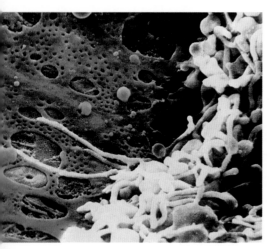

▲ Figure 3

The transport of iron to bone marrow

Iron is carried to the bone marrow to produce hemoglobin in new red blood cells.

Hemoglobin is synthesized in red blood cells. It is here that iron is added to the heme group.

Iron is essential for red blood cell function as it is a component of the hemoglobin molecule but it is toxic at high concentrations. When iron is absorbed from the intestine or when it is released during the breakdown of damaged red blood cells, it is transferred in the blood bound to a protein. The protein bound to iron is called transferrin. Cells have receptors for the transferrin molecule. Red blood cells are formed from stem cells in the bone marrow. Developing red blood cells have relatively high levels of transferrin receptors. Once bound, the receptor-iron complex enters the cell and the iron is either incorporated into the heme molecule or it is transferred to a storage molecule called ferritin.

Jaundice

Causes and consequences of jaundice.

When red blood cells are broken down in the liver and the spleen, hemoglobin is released. Macrophages digest the hemoglobin releasing heme and globin. The globin is digested into amino acids which are recycled. The heme group is further converted to iron and a yellow pigment called bilirubin. Bilirubin is also released from the breakdown of other proteins such as myoglobin and cytochrome. Any bilirubin produced outside of the liver is transported to the liver bound to the protein albumin.

Bilirubin is relatively insoluble, so in the liver, it is reacted with glucaronic acid to make it soluble. The water soluble form is secreted into passages called canaliculi along with water, electrolytes, bicarbonate, cholesterol, phospholipids and salts. This mixture is called bile. Figure 4 shows a micrograph of a human liver cell (red-brown), also known as hepatocyte, and a bile canaliculus (green). Hepatocytes secrete a green-brown fluid called bile. Bile is drained away from the liver through a dense network of bile canaliculi towards the gall-bladder. After a meal, bile is expelled from the gall-bladder and enters the duodenum where it plays an important role in the breakdown and digestion of fatty compounds.

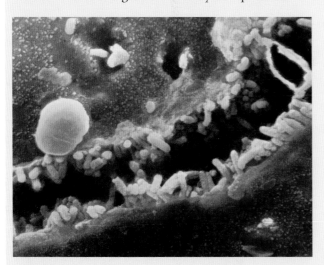

▲ Figure 4

When a disease interferes with the normal metabolism or excretion of bilirubin, it can build up in the blood. The result is a condition known as jaundice. The key symptom of jaundice is a yellowing of the eyes and of the skin. The normal concentration of bilirubin in blood plasma is 1.2 mg dl^{-1}. A concentration higher than 2.5 mg dl^{-1} results in jaundice.

Jaundice is a condition in which the skin and eyes are discoloured due to the deposition of excess bilirubin (pigment) in skin tissues. Figure 5 shows an 81-year-old man's hand (bottom) exhibiting jaundice as a result of taking the antibiotic Augmentin to treat a sinus infection. A normal person's hand is seen at top for comparison. Jaundice is not a disease in itself, but is a symptom of many disorders of the liver and biliary system. Treatment is aimed at correcting the underlying cause.

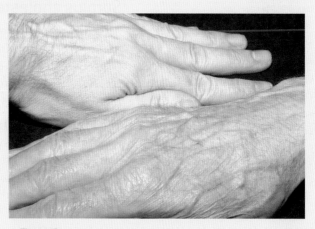

▲ Figure 5

Jaundice is seen in liver diseases such as hepatitis or liver cancer. It can occur due to obstruction of the bile duct by gallstones or pancreatic cancer. Jaundice in newborns is relatively common. There are different causes of newborn jaundice:

- Newborns have a relatively high turnover of red blood cells.

- A newborn's liver is often still developing and may not be able to process the bilirubin fast enough.

- Some newly born infants do not feed properly and the lack of intestinal contents means that excreted bilirubin can be reabsorbed.

Jaundice itself is not a disease. Treatment to remove bilirubin involves exposure to ultraviolet light either under a special "bili" lamp or by exposing the skin to the sun. Figure 6 shows a newborn baby undergoing bili therapy. The ultraviolet light converts the excess bilirubin into products that can be excreted. The eyes are covered to prevent any possible damage to the retina.

The underlying cause needs to be addressed to stop the jaundice. There can be significant consequences to extended periods of elevated serum bilirubin in infants including a form of neurological damage called kernicterus in which brain damage results in deafness and cerebral palsy. Adult patients with jaundice normally just experience itchiness.

▲ Figure 6

Conversion of cholesterol to bile salts

Surplus cholesterol is converted to bile salts.

Although cholesterol is absorbed from food in the intestine, a large quantity is synthesized each day by hepatocytes (liver cells). Cholesterol is a raw material needed for the synthesis of vitamin D as well as for the synthesis of steroid hormones. It is a structural component of membranes and it is used in the production of bile.

The liver regulates the amount of circulating lipids such as cholesterol and lipoproteins, either synthesizing them as required, or breaking them down and secreting cholesterol and phospholipids in the bile. The amount of cholesterol synthesized by the body varies to some degree with diet. Excess saturated fat in the diet increases the production of cholesterol.

Data-based questions: Lipase and bile

The graph in figure 7 shows the rate of bile flow into the gall bladder at different levels of bile salt secretion. The effect of a hormone, secretin, is also shown.

Look at figure 7.

1   **a)**   State the relationship between the rate of bile salt secretion and the rate of bile flow, without secretin.   [1]

    b)   Suggest the cause of this relationship.   [1]

2   Suggest when the rate of secretion of bile salt by liver cells needs to be highest.   [1]

3   Secretin causes HCO_3^- (hydrogen carbonate) ions to be secreted into the bile. Using only the data in the graph, outline the effect of secretin on bile flow.   [2]

4   The results in the graph show that in addition to HCO_3^- and bile salt, another solute is secreted into the bile. Explain how this conclusion can be drawn from the results in the graph.   [2]

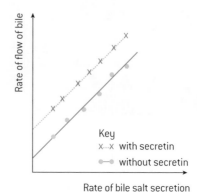

Key
x---x   with secretin
●—●   without secretin

Rate of bile salt secretion

▲ Figure 7 The effect of bile salts and secretin on the rate of flow of bile

 Claims about cholesterol

Educating the public on scientific claims: scientific studies have shown that high-density lipoprotein could be considered "good" cholesterol.

Lipids are transported in the blood in vesicles known as lipoproteins. There are five types of lipoproteins. Lipoproteins are composed of a hydrophilic exterior of phospholipids, proteins and cholesterol and a core of cholesterol and fats (triglycerides). Chylomicrons transport lipids from the intestine to the liver. Other lipoproteins are synthesized in the liver. Some of the lipoproteins change their density as molecules are selectively removed from them.

Data-based questions: Composition of various lipoproteins

Lipoprotein class	Density (g ml^{-1})	Diameter (nm)	% Protein	% Cholesterol	% Phospholipid	% Triglyeerides
HDL	1.063–1.210	5–15	33	30	29	8
LDL	1.019–1.063	18–28	25	50	21	4
IDL	1.006–1.019	25–30	18	29	22	31
VLDL	0.95–1.006	30–80	10	22	18	50
Chylomicrons	<0.95	100–1000	<2	8	7	84

▲ Table 1

Source of data: http://www.learn.ppdictionary.com/exercise_and_lipoproteins3.htm

1   State the relationship between density and

  i)   % triglycerides   [1]

  ii) % protein   [1]

  iii) % cholesterol.   [1]

2   Compare the cholesterol content of HDLs and IDLs.   [2]

3   Suggest, with a reason, why levels of HDLs are referred to as "good cholesterol" levels and levels of LDL are referred to as "bad cholesterol".   [3]

Production of plasma proteins by hepatocytes

Endoplasmic reticulum and Golgi apparatus in hepatocytes produce plasma proteins.

The rough endoplasmic reticulum of hepatocytes within the liver produce 90% of the proteins in blood plasma, including such proteins as fibrinogen and albumin. Albumin is a carrier protein that binds to such things as bilirubin. For this reason it is referred to as a transport protein, though it also plays a role in maintaining osmotic balance in the blood. Fibrinogen is a protein that is essential for clotting.

The fact that hepatocytes are actively involved in protein synthesis explains the characteristic appearance of hepatocytes. They show extensive networks of ER and Golgi body providing evidence of high levels of protein synthesis. It has been estimated that there are 13 million ribosomes attached to the ER of a typical liver cell.

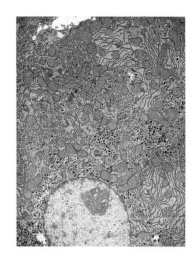

▲ Figure 8

683

Detoxification by the liver

The liver removes toxins from the blood and detoxifies them.

An important role of the liver is detoxification. Liver cells absorb toxic substances from the blood and convert them into non-toxic or less toxic substances, using a range of chemical conversions. For example, alcohol is converted into a less toxic substance by the enzyme ethanol dehydrogenase. The liver converts toxic ammonia into urea. The liver also works to detoxify biochemicals which are foreign to the organism's normal biochemistry such as poisons or drugs. One means by which the liver does this is to convert hydrophobic compounds into more easily excreted hydrophilic compounds.

D.4 The heart

Understanding

→ Structure of cardiac muscle cells allows propagation of stimuli through the heart wall.

→ Signals from the sinoatrial node that cause contraction cannot pass directly from atria to ventricles.

→ There is a delay between the arrival and passing on of a stimulus at the atrioventricular node.

→ This delay allows time for atrial systole before the atrioventricular valves close.

→ Conducting fibres ensure coordinated contraction of the entire ventricle wall.

→ Normal heart sounds are caused by the atrioventricular valves and semilunar valves closing causing changes in blood flow.

Applications

→ Use of artificial pacemakers to regulate the heart rate.

→ Use of defibrillation to treat life-threatening cardiac conditions.

→ Causes and consequences of hypertension and thrombosis.

Skills

→ Measurement and interpretation of the heart rate under different conditions.

→ Interpretation of systolic and diastolic blood pressure measurements.

→ Mapping of the cardiac cycle to a normal electrocardiogram (ECG) trace.

→ Analysis of epidemiological data relating to the incidence of coronary heart disease.

Nature of science

→ Developments in scientific research followed improvements in apparatus or instrumentation: the invention of the stethoscope led to improved knowledge of the workings of the heart.

Cardiac muscle cells

Structure of cardiac muscle cells allows propagation of stimuli through the heart wall.

Cardiac muscle tissue is unique to the heart. Like skeletal muscle, cardiac muscles are striated in appearance. The arrangement of the contractile proteins actin and myosin is similar to what is seen in skeletal muscle. However, cardiac muscle cells are shorter and wider than skeletal muscles and most commonly have just one nucleus per cell. Cardiac muscle contraction is not under voluntary control and many of the cardiac cells contract even in the absence of stimulation by nerves for the entire life of the organism. For these reasons, they have special structural features.

The cells are Y-shaped and are joined end to end in a complex network of interconnected cells. Where the end of one cell contacts the end of another cell, there is a specialized junction called an intercalated disc. This structure appears only in cardiac muscle. The intercalated disc consists of a double membrane containing gap junctions which provide channels of connected cytoplasm between the cells. This allows for the rapid movement of ions and a low electrical resistance. Being interconnected because of their Y-shapes and being electrically connected due to gap junctions allows a wave of depolarization to pass easily from one cell to a network of other cells leading to the synchronization of muscle contraction; that is, the network of cells contract as if it was one large cell.

Figure 1 shows a coloured transmission electron micrograph (TEM) of cardiac muscle fibrils (orange and blue). Mitochondria (red) supply the muscle cells with energy. The muscle fibrils, or myofibrils, are crossed by transverse tubules (narrow dark blue lines). These tubules mark the division of the myofibrils into contractile units (sarcomeres). In the centre is the intercalated disc (wavy dark blue line).

▲ Figure 1

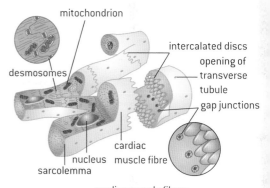

▲ Figure 2

The sinoatrial node

Signals from the sinoatrial node that cause contraction cannot pass directly from atria to ventricles.

The cardiac cycle is a repeating sequence of actions in the heart which result in the pumping of blood to the lungs and all other parts of the body. The cycle represents all of the events from the beginning of one heartbeat to the beginning of the next. Cardiologists refer to contraction of the heart's chambers as systole and relaxation as diastole. Figure 3 shows the sequence in which systole and diastole occur in the atria and ventricles. Figure 4 provides details of the events and pressure changes that occur in the stages in the cardiac cycle.

Within the wall of the right atrium, there is a collection of uniquely structured cardiac cells that spontaneously initiate action potentials without stimulation by other nerves. The initiation occurs rhythmically. This is the sinoatrial (or SA) node. The SA node is sometimes referred to as the pacemaker of the heart. Because gap junctions allow electric charges to flow freely between cells, the contraction which originated in the SA node spreads very rapidly across the entire atrium as if it were one cell. This causes the atria to undergo systole, i.e. they contract.

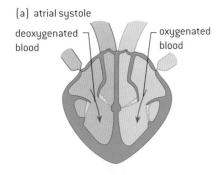

(a) atrial systole

deoxygenated blood

oxygenated blood

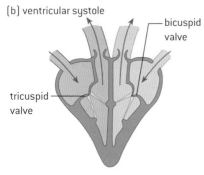

(b) ventricular systole

bicuspid valve

tricuspid valve

(c) diastole

semilunar valves

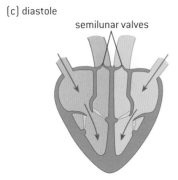

▲ Figure 3 The cardiac cycle

Signals from the sinoatrial node that cause contraction within the atria cannot pass directly from the atria to ventricles. Instead the signal from the SA node reaches the atrioventricular (AV) node. From there the signal spreads throughout the heart via specialized heart muscle tissue called Purkinje fibres. This signal causes the ventricles to undergo systole. This snaps the atrioventricular valves shut. After the ventricles are emptied the semilunar valves close.

The ventricles begin diastole, the atrioventricular valves open and the ventricles start filling with blood. Finally, all four chambers are in diastole and filling. When the atria are filled and the ventricles are 70 per cent filled, the cycle has ended.

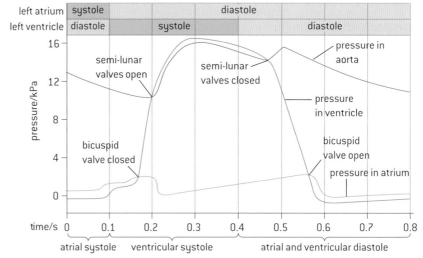

▲ Figure 4 The pressure changes inside the heart during the cardiac cycle

The atrioventricular node

There is a delay between the arrival and passing on of a stimulus at the atrioventricular node.

There are mechanisms in place to stagger the contraction of the atria and the ventricle. The fibres which connect the SA node to the AV node carry the action potential relatively slowly. There is a delay of approximately 0.12 s between arrival of the stimulus from the SA node and initiation of the impulse with the ventricles.

The cells of the AV node take longer to become excited than the cells of the SA node. There are a number of features of the AV node that lead to the delayed initiation of contraction of ventricles by the AV node.

- The AV node cells have a smaller diameter and do not conduct as quickly.

- There is a relatively reduced number of Na^+ channels in the membranes of AV node cells, a more negative resting potential and a prolonged refractory period within the cells of the AV node.

- There are fewer gap junctions between the cells of the AV node.

- There is relatively more non-conductive connective tissue in the node.

The delay in conduction

This delay allows time for atrial systole before the atrioventricular valves close.

The delay in the initiation of contraction caused by the AV node is important because it ensures that the atria contract and empty the blood they contain into the ventricles first before the ventricles contract. The contraction of ventricles causes the AV valves to snap shut, so that contraction of the ventricles too early would lead to too small a volume of blood entering the ventricles.

Coordination of contraction

Conducting fibres ensure coordinated contraction of the entire ventricle wall.

Once through the AV bundle, the signal must be conducted rapidly in order to ensure the coordinated contraction of the ventricle.

The atrioventricular bundle receives the impulse from the AV node and conducts the signal rapidly to a point where it splits into the right and left bundle branches. The bundle branches conduct the impulses through the wall between the two ventricles. At the base, or apex of the heart, the bundle branches connect to the Purkinje fibres which conduct the signal even more rapidly to the ventricles. These fibres have a number of modifications that facilitate them conducting signals at such a high speed:

- They have relatively fewer myofibrils.
- They have a bigger diameter.
- They have higher densities of voltage-gated sodium channels.
- They have high numbers of mitochondria and high glycogen stores.

The contraction of the ventricle begins at the apex.

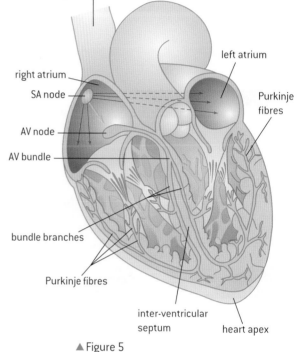

▲ Figure 5

 The invention of the stethoscope

Developments in scientific research followed improvements in apparatus or instrumentation: the invention of the stethoscope led to improved knowledge of the workings of the heart.

Stethoscopes are one of the most recognizable symbols of the medical profession. They were invented in the 19th century by Rene Laënnec though the original design has been significantly modified since then. Though not widely practised, practitioners would place their ears directly on the chest of patients to listen to the heart beat. In the 19th century, many patients were too obese for sounds to be heard by this method, washing was not the social norm and some patients were "infested with vermin" and if the patient was a female, modesty was an issue. While these variables were the main pressure behind the development of the tool, there were unintended benefits of the device. It became one of the first tools that allowed for the non-invasive investigation of internal anatomy. Different types of heart abnormalities result in different sounding heartbeats which can be detected through the stethoscope.

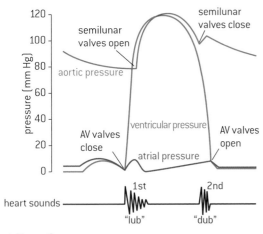

▲ Figure 6

Causes of the sound of the heartbeat

Normal heart sounds are caused by the atrioventricular valves and semilunar valves closing causing changes in blood flow.

A normal heartbeat has two sounds, both of which are caused by the closing of valves. When the atrioventricular valves snap shut, there is a "lub" sound. After the ventricles are emptied the semilunar valves close, causing the second sound, the "dub" sound.

▲ Figure 7

🧪 Variables affecting the heart rate

Measurement and interpretation of the heart rate under different conditions.

A number of variables that can influence heart rate can be assessed in the school laboratory setting. Some examples include types of exercise, intensity of exercise, recovery from exercise, relaxation, body position including lying down, breathing and breath holding, exposure to a cold stimulus and facial immersion in water.

Detecting heart rate can be done in a number of ways. Figure 7 shows how to detect the pulse of an artery in the wrist. The researcher uses fingers rather than the thumb. The side of the neck below the jaw has an artery where determining pulse is relatively easy. Data-logging equipment including hand-grip heart monitors, ear clips, EKG sensors and wrist watches can feed data into computers. The built-in cameras on some tablet computers can be used as a device to detect heart rate.

Data-based questions: Cold exposure and heart rate

The resting heart rate of a sample of students was determined through monitoring by a wrist band that measured heart rate. An ice pack was then placed on the forearm of these students for one minute. The heart rate was measured at the end of the one minute of cold exposure and then again at the end of each minute for two minutes of recovery.

1 Determine the mean resting heart rate. [1]

2 Calculate the percent decline in mean heart rate with cold exposure. [2]

3 Evaluate the conclusion that cold exposure suppresses heart rate. [2]

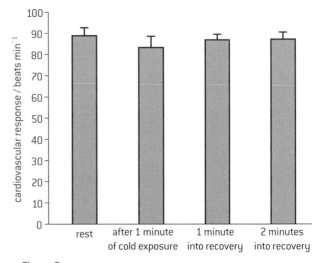

▲ Figure 8

Artificial pacemakers

Use of artificial pacemakers to regulate the heart rate.

Artificial pacemakers are medical devices that are surgically fitted in patients with a malfunctioning sinoatrial node, the part of the heart that initiates the heartbeat, or in patients with a block in the signal conduction pathway within the heart, which impairs the nerve impulses generated by the node. The purpose of the device is to maintain the rhythmic nature of the heart beat when the heart does not beat fast enough or when there is a fault in the heart's electrical conduction system.

Pacemakers can either provide a regular impulse or discharge only when a heartbeat is missed so that it beats normally. The most common, basic pacemaker monitors the heart's rhythm and when a heartbeat is not detected, the ventricle is stimulated with a low voltage pulse. More complex forms stimulate both the atria and the ventricles.

Figure 9 shows an X-ray of the chest of a male patient with a heart pacemaker (upper right). The heart is the blue mass at centre right, in between the lungs (white). The pacemaker has leads (running from upper right to lower centre) to supply regular electrical impulses to the heart.

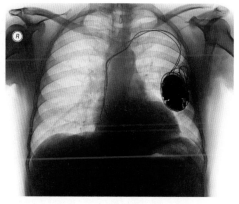

▲ Figure 9

Relating the cardiac cycle to the ECG trace

Mapping of the cardiac cycle to a normal electrocardiogram (ECG) trace.

Cardiac muscle contracts because it receives electrical signals. These signals can be detected and quantified using an electrocardiogram (ECG or EKG). Data-logging ECG sensors can be used to produce a pattern as shown in figure 10. The P-wave is caused by atrial systole, the QRS wave is caused by ventricular systole. The T-wave coincides with ventricular diastole. Interval analysis can be performed on the EKG signal, for example on the times between the beginning of P and Q (P–Q), QRS, and Q to the end of T (Q–T) intervals. The height of the R-wave can be compared when the body changes position from standing to lying down. The overall pattern can be compared before and after mild exercise.

Specialists can use changes to the size of peaks and lengths of intervals to detect heart pathology.

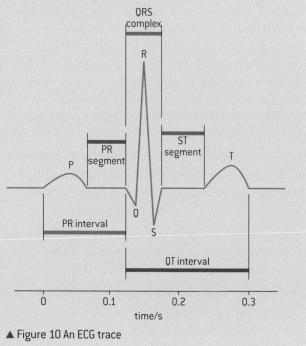

▲ Figure 10 An ECG trace

Explaining the use of a defibrillator

Use of defibrillation to treat life-threatening cardiac conditions.

Cardiac arrest occurs when the blood supply to the heart becomes reduced and heart tissues are deprived of oxygen. One of the first negative consequences of this is abnormalities in the cardiac cycle such as ventricular fibrillation. This is essentially the twitching of the ventricles due to rapid and chaotic contraction of individual muscle cells.

When "first responders" reach a scene where a victim is not breathing, they will apply the two paddles of a defibrillator to the chest of the patient, setting up a diagonal line between the two paddles with the heart in the middle. The device will first detect whether fibrillation is happening and if it is, an electric discharge is given off to restore a normal heart rhythm.

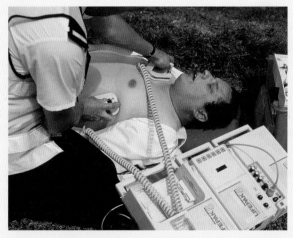

▲ Figure 11 First responders applying a defibrillator to the chest of a man who is undergoing cardiac arrest

Hypertension and thrombosis

Causes and consequences of hypertension and thrombosis.

Atherosclerosis is hardening of the arteries caused by the formation of plaques, or atheromas, on the inner lining of arteries (figure 12). Plaques are areas that are swollen and accumulate a diversity of debris. The plaques often develop because of high circulating levels of lipids and cholesterol. The plaques can reduce the speed at which blood moves through vessels. This can trigger a clot, or thrombosis, which can block the blood flow through the artery and deny the tissue access to oxygen. If this occurs on the surface of the heart, the consequence can be a myocardial infarction, or heart attack.

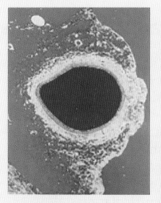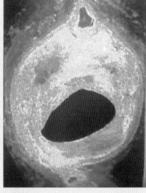

▲ Figure 12 A normal artery (top) can be compared to an artery where a plaque has formed (bottom)

Greater resistance to the flow of blood can slow the flow of blood. The result is greater pressure on the walls of arteries, also known as hypertension. Hypertension has a number of consequences.

- Damage to the cells that line arteries can cause a cascade of events that ultimately leads to the arteries becoming narrower and stiff.

- Constant high blood pressure can weaken an artery causing a section of the wall to enlarge and form a bulge called an aneurysm. An aneurysm can burst and cause internal bleeding. They can form in any artery in the body but are most common in the aorta.

- Chronic high blood pressure can lead to stroke by weakening blood vessels in the brain causing them to narrow, leak or rupture. It can also lead to blood clots in the arteries leading to the brain potentially causing a stroke.

- Chronic high blood pressure is one of the most common causes of kidney failure as it damages both the arteries leading to the kidney and the capillaries within the glomerulus.

There are a number of factors that are correlated with a greater incidence of thrombosis and hypertension.

- Having parents who have experienced heart attacks indicates a genetic precondition to either condition.

- Old age leads to less flexible blood vessels. In children, the normal ranges are lower than for adults.

- Risk in females increases post-menopause correlated with a fall in estrogen levels.

- Males are at greater risk compared with females correlated with lower levels of estrogen.

- Smoking raises blood pressure because nicotine causes vasoconstriction.

- A high-salt diet, excessive amounts of alcohol and stress are also correlated with hypertension.

- Eating too much saturated fat and cholesterol promotes plaque formation.

- Height affects blood pressure.

- Sedentary lifestyle, i.e. a lack of exercise is correlated with obesity and prevents the return of venous blood from the extremities leading to a greater risk of clot formation.

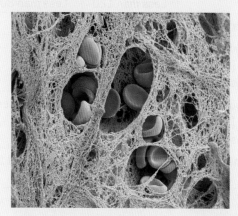

▲ Figure 13 A blood clot (thrombus) in the coronary artery, showing red blood cells (purple) in a fibrin mesh (threads). The coronary artery supplies blood to the heart

⚗ Interpreting blood pressure measurements

Interpretation of systolic and diastolic blood pressure measurements.

Blood pressure, or more accurately arterial pressure is the pressure that circulating blood puts on the walls of arteries. During each heartbeat, the pressure of blood within arteries varies from a peak during the ventricle systole to a minimum near the beginning of the cardiac cycle when the ventricles are filled with blood and are in systole.

Blood pressure measurements are often quoted in the pressure unit "mm Hg". An example blood pressure would be "120 over 80". The higher number refers to the pressure in the artery caused by ventricular systole and the lower number refers to the pressure in the artery due to ventricular diastole.

Figure 14 shows a pregnant woman having her blood pressure measured. Monitoring blood pressure during pregnancy is important. High blood pressure during pregnancy is called pre-eclampsia and it can be a life-threatening condition if it is not treated.

To measure blood pressure, a cuff is placed on the bicep and inflated so that it constricts the arm and

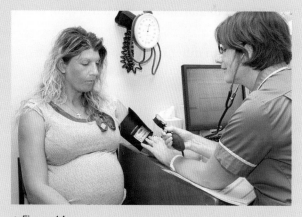

▲ Figure 14

prevents blood from entering the forearm. The cuff is slowly deflated and the nurse listens for the occurrence of a sound. This occurs when the cuff pressure is lowered below the systolic pressure. The sound is caused by the opening and closing of the artery. The cuff is further deflated until normal blood flow returns and there is no longer a sound. The absence of sound occurs when the cuff pressure is less than the diastolic pressure.

Blood pressure category	Systolic	Diastolic
Hypotension (low blood pressure)	90 or less	60 or less
Normal	Less than 120	Less than 80
Pre-hypertension	120–139	80–89
High blood pressure (Stage 1 hypertension)	140–159	90–99
High blood pressure (Stage 2 hypertension)	160 or higher	100 or higher
Hypertension crisis	Higher than 180	Higher than 110

▲ Table 1

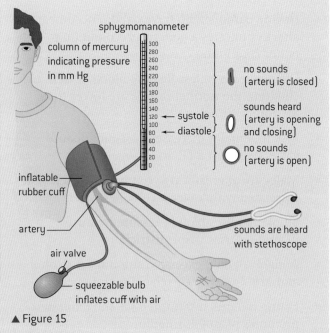

▲ Figure 15

🧪 Data relating to coronary heart disease

Analysis of epidemiological data relating to the incidence of coronary heart disease.

Coronary heart disease (CHD) refers to the damage to the heart as a consequence of reduced blood supply to the tissues of the heart itself. This is often caused by narrowing and hardening of the coronary artery.

Ethnic groups can differ in their predisposition to CHD because of differing diets and lifestyles.

Gender groups, age groups, groups that differ in their level of physical activity, groups with different genotypes, groups with differing medical histories – all can have different probabilities of experiencing CHD. Epidemiology is the study of the patterns, causes and effects of diseases in groups of individuals or populations.

Data-based questions: Hypertension

High blood pressure (hypertension) is a major risk factor for coronary heart diseases. In a major study, more than 316,000 males were followed for 12 years to investigate the effects of high blood pressure (BP). Figure 16 shows the relationship between systolic and diastolic blood pressure and the effect on the death rate per 10,000 persons year^{-1}.

1 Determine the death rate for a systolic blood pressure between 140 and 159 mmHg and a diastolic blood pressure between 75 and 79 mmHg. [1]

2 Describe the effect of systolic blood pressure and diastolic blood pressure on the death rate. [2]

3 Calculate the minimum difference between systolic and diastolic blood pressure where the death rate is highest. [1]

4 Evaluate the impact of differences between systolic and diastolic pressure on death rate. [3]

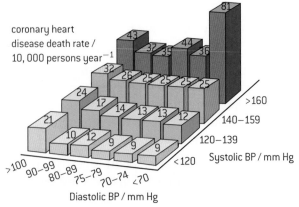

▲ Figure 16 The effect of blood pressure on coronary heart disease

Data-based questions: Cholesterol

Cholesterol and lipids are not soluble in the blood because blood is water-based. To solve this problem, lipids are transported in the blood in the form of lipoproteins called chylomicrons. The concentration of cholesterol in the blood as lipoproteins is a determining factor in the onset of coronary heart disease.

In 1998 the blood cholesterol level of 70,000 people in Mexico was measured. The people were divided into two age groups: 1 to 19 (young people) and 20 to 98 (adults). Mean blood cholesterol levels were calculated for the two age groups in each of the different states of Mexico.

Figure 17 shows the results. Each point on the graph shows the mean blood cholesterol level for the two age groups in one state.

1 State the relationship between cholesterol levels in young people and adults. [1]

2 Predict, using the data in the graph, how the blood cholesterol level usually changes over a lifetime. [2]

3 The maximum desirable blood cholesterol level is 200 mg 100cm^{-3} of blood. Suggest the implications of the survey of blood cholesterol levels for the population of Mexico. [3]

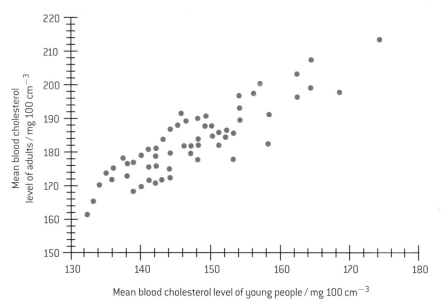

▲ Figure 17 Relationship between blood cholesterol in adults and blood circulation in adolescents in different Mexican states

693

D.5 Hormones and metabolism (AHL)

Understanding

→ Endocrine glands secrete hormones directly into the bloodstream.

→ Steroid hormones bind to receptor proteins in the cytoplasm of the target cell to form a receptor–hormone complex.

→ The receptor–hormone complex promotes the transcription of specific genes.

→ Peptide hormones bind to receptors in the plasma membrane of the target cell.

→ Binding of hormones to membrane receptors activates a cascade mediated by a second messenger inside the cell.

→ The hypothalamus controls hormone secretion by the anterior and posterior lobes of the pituitary gland.

→ Hormones secreted by the pituitary control growth, developmental changes, reproduction and homeostasis.

 Applications

→ Some athletes take growth hormones to build muscles.

→ Control of milk secretion by oxytocin and prolactin.

 Nature of science

→ Cooperation and collaboration between groups of scientists: the International Council for the Control of Iodine Deficiency Disorders includes a number of scientists who work to eliminate the harm done by iodine deficiency.

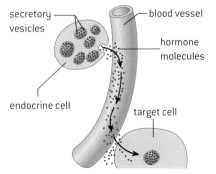

secretory vesicles

blood vessel

hormone molecules

endocrine cell

target cell

▲ Figure 1 Endocrine glands secrete chemical messages directly into the blood

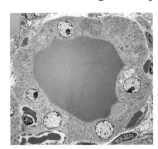

▲ Figure 2

Endocrine glands

Endocrine glands secrete hormones directly into the bloodstream.

Endocrine glands are structures that secrete chemical messages, called hormones, directly into the blood. These messages are transported to specific target cells (figure 1). Hormones can be steroids, proteins, glycoproteins, polypeptides, amines or tyrosine derivatives.

As an example, figure 2 shows a cross-section through a thyroid gland follicle. Thyroid hormones regulate the body's metabolism. The follicle consists of a layer of cells (pink) around a central storage chamber. The cells produce the thyroid hormones and secrete them into the central chamber where they are stored in a viscous fluid colloid (yellow). The follicle is surrounded by blood vessels (red), which transport the hormones around the body.

 Eradicating iodine deficiency

Cooperation and collaboration between groups of scientists: the International Council for the Control of Iodine Deficiency Disorders includes a number of scientists who work to eliminate the harm done by iodine deficiency.

Thyroid hormone refers to two similar hormones derived from tyrosine. Triiodothyronine (T_3) contains three iodine atoms and tetraiodothyronine (T_4) contains four iodine atoms. Correct functioning of the thyroid requires iodine in the diet. If there is dietary insufficiency, then there are a number of consequences including a condition known as goiter. The inability to produce the thyroid hormones because of the absence of iodine means that the hypothalamus and the anterior pituitary continuously stimulate the thyroid and enlargement of the thyroid results. Iodine deficiency during pregnancy can affect fetal nervous development and can lead to mental retardation in children.

The International Council for the Control of Iodine Deficiency Disorders (ICCIDD) is a non-governmental organization (NGO) that is partnered with intergovernmental organizations such as UNICEF and the WHO as well as national governments to eliminate iodine deficiency, mainly through a push for universal iodization of salt.

Since it was created the ICCIDD has partnered with academic institutions to produce reference works that guide national efforts to overcome iodine deficiency disorder.

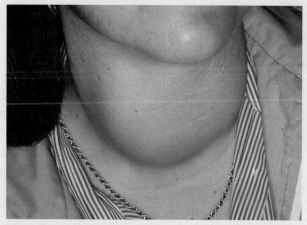

▲ Figure 3

The mechanism of action of steroid hormones

Steroid hormones bind to receptor proteins in the cytoplasm of the target cell to form a receptor–hormone complex.

Peptide hormones and lipid hormones differ in their solubility. This leads to different mechanisms of action. Both types of hormones act by binding to a receptor.

Steroid hormones can cross directly through the plasma membrane and the nuclear membrane and bind to receptors (see figure 4). An example is estrogen. The receptor–hormone complex then serves as a transcription factor, promoting or inhibiting the transcription of a certain gene.

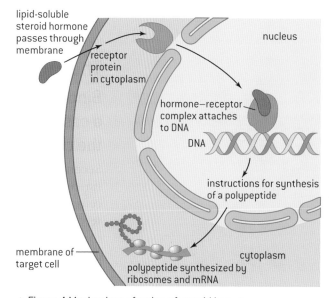

▲ Figure 4 Mechanism of action of steroid hormones

695

The receptor–hormone complex

The receptor–hormone complex promotes the transcription of specific genes.

The steroid hormone calciferol crosses the intestinal cell membrane and binds to a receptor within the nucleus of the cell. The receptor–hormone complex effects expression of the calcium transport protein calbindin in the small intestine which then allows for the absorption of calcium from the intestine.

Some steroids, such as cortisol, bind to receptors in the cytoplasm and the receptor–hormone complex then passes through the nuclear membrane into the nucleus to effect transcription.

The hormone may have different effects in different cells and it may even have an inhibitory effect. For example, when the steroid hormone cortisol binds to its receptor in the cytoplasm of a liver cell and enters the nucleus it activates many of the genes needed for gluconeogenesis; i.e., the conversion of fat and protein into glucose raising blood glucose. At the same time, it decreases the expression of the insulin receptor gene, preventing glucose from being stored in the cells and also raising blood glucose. In the pancreas, the cortisol–receptor complex inhibits the transcription of insulin genes.

Mechanism of action of peptide hormones

Peptide hormones bind to receptors in the plasma membrane of the target cell.

Protein hormones are hydrophilic so they cannot pass through the membrane directly. Instead they bind to surface receptors that can trigger a cascade reaction mediated by chemicals called second messengers.

The role of second messengers

Binding of hormones to membrane receptors activates a cascade mediated by a second messenger inside the cell.

Second messengers are small water soluble molecules that can quickly spread throughout the cytoplasm and relay signals throughout the cells. Calcium ions and cyclic AMP (cAMP) are the two most common second messengers. A large number of proteins are sensitive to the concentration of these molecules.

Epinephrine is a hormone that mediates the "fight or flight" response when released. When under threat, an organism needs a supply of blood glucose as an energy source. When epinephrine reaches the liver, it binds to a receptor called the G-protein couple receptor. Binding to the receptor activates the G-protein which uses guanosine triphosphate (GTP) as an energy source to activate the enzyme adenylyl cyclase. This converts ATP to cAMP. The cAMP then activates protein kinase enzymes which in turn activate the processes of glycogen breakdown and inhibit glycogen synthesis.

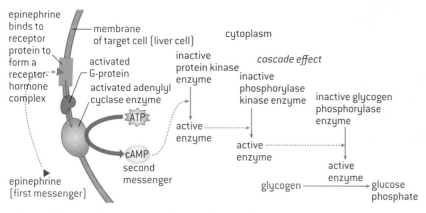

▲ Figure 5 Mechanism of action of adrenaline on a liver cell

Pituitary hormones

Hormones secreted by the pituitary control growth, developmental changes, reproduction and homeostasis.

The anterior pituitary synthesizes and secretes a number of hormones that control growth, reproduction and homeostasis. Examples include FSH and LH. The posterior pituitary gland secretes oxytocin and ADH, but these hormones are not produced there. Instead, they are synthesized in unusual cells called neurosecretory cells found in the hypothalamus. The hormones travel down the axons of the neurosecretory cells and are stored at the ends of the axons, until impulses pass down the axons from the hypothalamus, stimulating secretion.

The role of the hypothalamus

The hypothalamus controls hormone secretion by the anterior and posterior lobes of the pituitary gland.

Both the nervous system and the endocrine system play a role in homeostasis and in the control of other processes including reproduction. The hypothalamus links the nervous system to the endocrine system via the pituitary gland. There are two parts of the pituitary gland, which are effectively different glands, with a different mode of operation.

The role of the hypothalamus is to secrete releasing factors, which stimulate the secretion of the anterior pituitary gland's hormones. The releasing factors are carried from the hypothalamus to the anterior pituitary gland by a portal vein. This is an unusual type of blood vessel as it links two capillary networks – one in the hypothalamus which unites to form the portal vein and another in the anterior pituitary gland, from which blood flows on to the rest of the body (figure 6).

Negative feedback is involved in the control of secretion of many of the pituitary hormones. ADH can be used as an example. Blood solute concentration is monitored by osmoreceptors in the hypothalamus. If these receptors detect that the solute concentration is too high, impulses are sent

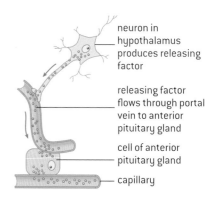

▲ Figure 6 A portal vein carries releasing factors to the anterior pituitary

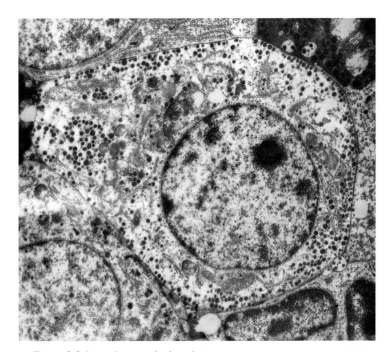

▲ Figure 8 Coloured transmission electron micrograph (TEM) of cells in the anterior pituitary gland, a hormone-secreting gland at the base of the brain. The cells' nuclei, which contain their genetic information, are purple. The cell at centre is a somatotroph cell, a secretory cell that has granules (red) containing hormones to be secreted in its cytoplasm (green). Somatotrophs secrete human growth hormone, which promotes growth and controls numerous metabolic processes

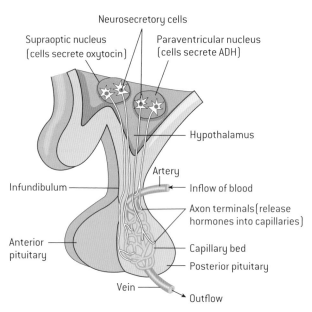

▲ Figure 7 The neurosecretory cells carry the hormones to the posterior pituitary

along the axons of neurosecretory cells, causing ADH secretion to increase. ADH acts on the kidney, as described in sub-topic 11.3. It causes blood solute concentration to decrease. If blood solute concentrations decrease too much, this is also detected by the osmoreceptors in the hypothalamus. Fewer or no impulses are sent via neurosecretory cells, so ADH secretion reduces or stops. This allows blood solute concentrations to rise.

Regulation of milk secretion
Control of milk secretion by oxytocin and prolactin.

A unique adaptation of mammals is the production of milk in mammary glands for feeding offspring. The production and secretion of milk is under hormonal control.

Prolactin is a hormone produced by the anterior pituitary found in a number of vertebrate species where it has a wide diversity of roles. It is not limited to mammals. However, in mammals it stimulates mammary glands to grow and it also stimulates the production of milk.

During pregnancy, high levels of estrogen increase prolactin production but inhibit the effects of prolactin on mammary glands. The abrupt decline in estrogen and progesterone following delivery removes this inhibition and the production of milk begins. However, the release of the milk after it is produced depends on the hormone oxytocin. Nursing by an infant stimulates the continued creation of prolactin. It also stimulates oxytocin release. Oxytocin stimulates the contraction of cells that surround the structures holding the milk leading to the ejection of the milk.

Oxytocin is produced by the neurosecretory cells of the hypothalamus and is stored in the posterior pituitary gland.

Injection of growth hormone by athletes

Some athletes take growth hormones to build muscles.

Growth hormone is another polypeptide hormone produced in the anterior pituitary. One of its main targets is receptors in liver cells. The binding of growth hormone stimulates the release of insulin-like growth factor which circulates in the blood and stimulates bone and cartilage growth. Growth hormone has a number of additional affects, one of which is increase in muscle mass. For this reason, it has been used as a performance enhancing drug. The availability of growth hormone has increased due to the development of genetically modified organisms that can produce it in large quantities.

Because there is a correlation between muscle size and strength, competitors in sports that require short bursts of explosive strength would benefit. While it is clear that it leads to greater muscle mass, the data is not clear that it leads to greater strength. Another claim is that it allows tired muscles to recover more quickly allowing an individual to train harder and more often. The scientific research on the topic suggests that the benefits provided in terms of enhanced performance are small or non-existent compared to the risks of injecting the hormone. For this reason, use of the drug is banned by most international sporting federations.

D.6 Transport of respiratory gases (AHL)

Understanding

→ Oxygen dissociation curves show the affinity of hemoglobin for oxygen.

→ Carbon dioxide is carried in solution and bound to hemoglobin in the blood.

→ Carbon dioxide is transformed in red blood cells into hydrogencarbonate ions.

→ The Bohr shift explains the increased release of oxygen by hemoglobin in respiring tissues.

→ Chemoreceptors are sensitive to changes in blood pH.

→ The rate of ventilation is controlled by the respiratory control centre in the medulla oblongata.

→ During exercise the rate of ventilation changes in response to the amount of CO_2 in the blood.

→ Fetal hemoglobin is different from adult hemoglobin allowing the transfer of oxygen in the placenta onto the fetal haemoglobin.

 ## Applications

→ Consequences of high altitude for gas exchange.

→ pH of blood is regulated to stay within the narrow range of 7.35 to 7.45.

→ Causes and treatments of emphysema.

 ## Skills

→ Analysis of dissociation curves for hemoglobin and myoglobin.

→ Identification of pneumocytes, capillary endothelium cells and blood cells in light micrographs and electron micrographs of lung tissue.

 ## Nature of science

→ Scientists have a role in informing the public: scientific research has led to a change in public perception of smoking.

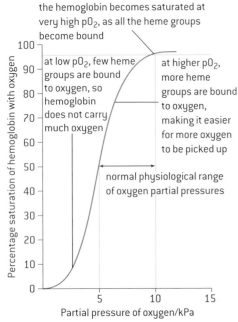

the hemoglobin becomes saturated at very high pO2, as all the heme groups become bound

at low pO2, few heme groups are bound to oxygen, so hemoglobin does not carry much oxygen

at higher pO2, more heme groups are bound to oxygen, making it easier for more oxygen to be picked up

normal physiological range of oxygen partial pressures

▲ Figure 1 Oxygen dissociation of hemoglobin

Oxygen dissociation curves

Oxygen dissociation curves show the affinity of hemoglobin for oxygen.

Hemoglobin is an oxygen transport protein in the blood. The degree to which oxygen binds to hemoglobin is determined by the partial pressure of oxygen (pO_2) in the blood. The oxygen dissociation curve shown in figure 1 describes the saturation of hemoglobin by oxygen at different partial pressures of oxygen.

Note the significant change in saturation over a narrow range of oxygen partial pressure. This narrow range typifies oxygen pressures surrounding cells under normal metabolism. At low pO_2, such as might occur in the muscles, O_2 will dissociate from hemoglobin. At high pO_2, such as might occur in the lungs, the hemoglobin will become saturated.

Carbon dioxide transport in the blood

Carbon dioxide is carried in solution and bound to hemoglobin in the blood.

Carbon dioxide is carried in three forms in blood plasma:

- dissolved as carbon dioxide;
- reversibly converted to bicarbonate (hydrogencarbonate) ions (HCO_3^-) that are dissolved in the plasma;
- bound to plasma proteins.

Table 1 shows the amounts of each form in arterial blood and in venous blood at rest and during exercise.

Form of transport	Arterial mmol⁻¹ blood	Venous mmol⁻¹ blood	
		Rest	Exercise
dissolved CO_2	0.68	0.78	1.32
bicarbonate ion	13.52	14.51	14.66
CO_2 bound to protein	0.3	0.3	0.24
total CO_2 in plasma	14.50	15.59	16.22
pH of blood	7.4	7.37	7.14

▲ Table 1 CO_2 transport in blood plasma at rest and during exercise

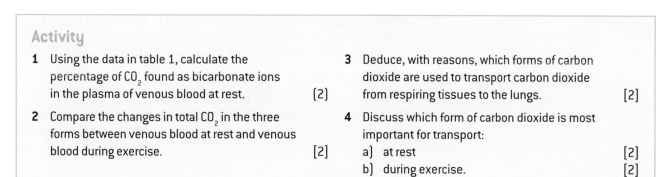

Activity

1 Using the data in table 1, calculate the percentage of CO_2 found as bicarbonate ions in the plasma of venous blood at rest. [2]

2 Compare the changes in total CO_2 in the three forms between venous blood at rest and venous blood during exercise. [2]

3 Deduce, with reasons, which forms of carbon dioxide are used to transport carbon dioxide from respiring tissues to the lungs. [2]

4 Discuss which form of carbon dioxide is most important for transport:
   a) at rest [2]
   b) during exercise. [2]

Conversion of carbon dioxide into hydrogen carbonate ions

Carbon dioxide is transformed in red blood cells into hydrogencarbonate ions.

The majority of carbon dioxide produced by the body during cellular respiration is converted to the more soluble and less toxic bicarbonate (hydrogencarbonate) ion. The reaction occurs inside red blood cells and is catalysed by the enzyme carbonic anhydrase.

$$CO_2 + H_2O \leftrightharpoons H_2CO_3 \leftrightharpoons H^+ + HCO_3^-$$

The two-sided arrows indicate that the reaction is reversible. In the tissues where carbon dioxide is generated, the reaction proceeds to the right; that is, more bicarbonate ion is generated as are H^+ ions. This lowers the pH of the blood. In the lungs, when carbon dioxide leaves the blood, the reaction is driven to the left and bicarbonate ion is converted to carbon dioxide.

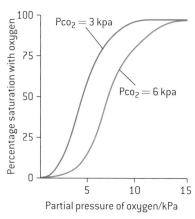

▲ Figure 2 The Bohr shift

The Bohr shift

The Bohr shift explains the increased release of oxygen by hemoglobin in respiring tissues.

Increased metabolism results in greater release of CO_2 into the blood, which lowers the pH of the blood. This increased acidity shifts the oxygen dissociation curve to the right, which results in a decreased affinity of the hemoglobin for oxygen; that is, a greater release of oxygen from hemoglobin at the same partial pressure of oxygen (see figure 2).

This is known as the Bohr shift. This ensures that respiring tissues have enough oxygen when their need for oxygen is greatest. Also, in the lungs, pCO_2 is lower, so saturation of hemoglobin can occur at lower partial pressures of oxygen.

Effect of CO_2 on ventilation rate

During exercise the rate of ventilation changes in response to the amount of CO_2 in the blood.

Exercise increases metabolism and leads to an increase in the production of CO_2 as a waste product of cellular respiration. Increased CO_2 causes blood pH to decrease because CO_2 dissolves in water to form carbonic acid

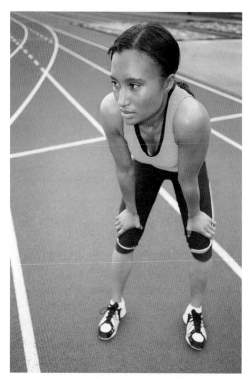

▲ Figure 3 Hyperventilation occurs following vigorous exercise as a mechanism to maintain blood pH by ridding the body of carbon dioxide

701

(H_2CO_3) which further dissociates into H^+ and HCO_3^-. Recall that high H^+ concentration means low pH. Chemoreceptors in the medulla, the aorta and the carotid artery are able to detect a change in blood carbon dioxide.

High levels of carbon dioxide in the blood trigger an increase in the ventilation rate in order to rid the body of the carbon dioxide build-up. Carbon dioxide diffuses into alveoli and ventilation expels the carbon dioxide from the body. This explains the hyperventilation that occurs in response to exercise.

Regulation of the ventilation rate

The rate of ventilation is controlled by the respiratory control centre in the medulla oblongata.

The rate of ventilation is regulated by the respiratory centre in the medulla oblongata of the brainstem. Two sets of nerves travel to the lungs from this centre: the intercostal nerves stimulate the intercostal muscles of the thorax and the phrenic nerves stimulate the diaphragm.

When the lungs expand due to stimulation from the nerves, stretch receptors in the walls of the chest and lungs send signals to the respiratory centre which triggers a cessation of the signals leading to inspiration until the animal exhales. Then a new signal is sent.

Chemoreceptors and blood pH

Chemoreceptors are sensitive to changes in blood pH.

If an increase in blood carbon dioxide or a drop in blood pH is detected, the chemoreceptors in the carotid artery and the aorta send a message to the breathing centre in the medulla oblongata. Nerve impulses are sent from the medulla to the diaphragm and the intercostal muscles causing them to increase the ventilation rate. This leads to an increased rate of gas exchange. There are also chemoreceptors in the medulla oblongata that can detect in an increase in blood carbon dioxide.

 Regulation of blood pH

pH of blood is regulated to stay within the narrow range of 7.35 to 7.45.

If the blood pH falls below 7.35, then chemoreceptors signal to the respiratory centre to increase the rate of ventilation. Hyperventilation withdraws carbon dioxide from the blood driving the carbonic acid reaction to the left. This withdraws hydrogen ions from the blood raising the pH.

$$CO_2 + H_2O \rightleftharpoons H_2CO_3 \rightleftharpoons H^+ + HCO_3^-$$

In the kidney, H^+ ions can be secreted into the urine bound to buffers to raise the pH. Greater amounts of bicarbonate will be reabsorbed from the tubules to neutralize the acid.

If the blood becomes too basic, then bicarbonate ions can be secreted into the distal convoluted tubule of the kidney.

Chemical buffers exist within the extracellular fluid and these can't remove the acids or bases, but they can minimize their effect.

Analysis of dissociation curves

Analysis of dissociation curves for hemoglobin and myoglobin.

Myoglobin is a specialized oxygen transport protein in muscles. It has a much higher affinity for oxygen and will only release its oxygen when the pO_2 is quite low, for example in the muscles during heavy exercise. The shapes of the two curves in figure 4 are different because hemoglobin has four chains with four heme groups, whereas myoglobin has one. The release of each O_2 from hemoglobin triggers a conformational change, which causes the hemoglobin to more rapidly release subsequent O_2 molecules.

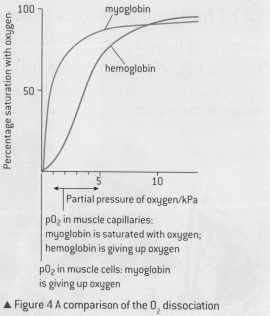

▲ Figure 4 A comparison of the O_2 dissociation curves of hemoglobin and myoglobin

Differences in oxygen affinity between fetal and adult hemoglobin

Fetal hemoglobin is different from adult hemoglobin allowing the transfer of oxygen in the placenta onto the fetal haemoglobin.

Figure 5 compares the oxygen dissociation curves of adult and fetal hemoglobin. Note that fetal hemoglobin has a higher affinity for O_2 at all partial pressures. This ensures that O_2 is transferred to the fetus from the maternal blood across the placenta.

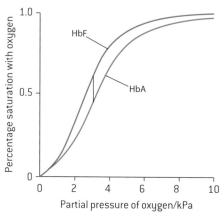

▲ Figure 5 A comparison of the O_2 dissociation curves of fetal and adult hemoglobin

 Gas exchange at high altitude

Consequences of high altitude for gas exchange.

At high altitude there is a low pO_2 in the air. Hemoglobin may not become fully saturated and as a consequence, the tissues may not be adequately supplied with oxygen. To some degree, human physiology can adapt to high altitude. Red blood cell production can increase, which increases the total amount of circulating hemoglobin. Ventilation rate increases to increase gas exchange. Muscles produce more myoglobin to ensure delivery of oxygen to the tissues. Populations living permanently at high altitude have greater mean lung surface area and larger vital capacities than people living at sea level. Their oxygen dissociation curve shifts to the right, encouraging release of oxygen into the tissue.

Changing attitudes to smoking

Scientists have a role in informing the public: scientific research has led to a change in public perception of smoking.

Figure 6 is a surprising picture which shows an athlete with a cigarette.

▲ Figure 6 British hurdler Shirley Strong

In the early part of the 20th century, there was a belief that tobacco smoking could improve ventilation. Doctors even prescribed smoking of medicine for such conditions as asthma.

In the 1930s and 1940s, smoking was common in both men and women. Even a majority of medical doctors smoked. At the same time, there was rising public concern about the health risks of smoking cigarettes. One response of tobacco companies was to devise advertising that featured images of physicians and scientists, to assure the consumer that their respective brands were safe.

As epidemiological evidence mounted, the US Surgeon General published a report in 1964 calling upon evidence from more than 7,000 scientific journal articles to link smoking to chronic bronchitis and several types of cancer.

The number of smokers is in steady decline in developed nations with nearly half of all living adults who have ever smoked having quit. This is a credit to public health departments pushing for policy measures informed by convincing scientific evidence.

 ## Emphysema

Causes and treatments of emphysema.

Emphysema is a lung condition in which the walls between individual alveoli break down leading to an increase in their size and therefore a reduction in the surface area for gas exchange which restricts oxygen uptake into the blood.

Figure 7 shows a computer tomography scan of the lungs with one of the characteristic indications of emphysema: large areas of trapped air that show up as transparent areas in images. This can cause the lungs to become trapped in "inspiration" position in the ventilation cycle. This is known informally as "barrel chest".

The main cause of emphysema is long-term exposure to airborne irritants, most commonly tobacco smoke, but possibly also air pollution, coal and silica dust.

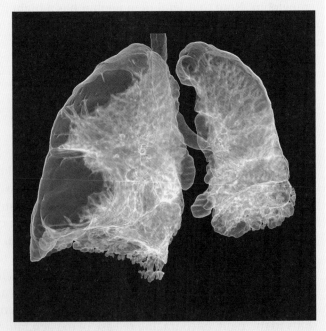

▲ Figure 7

The damage to lung tissue by smoke is due to three factors:

- Oxidation reactions produced by high concentrations of chemicals known as free radicals in tobacco smoke.

- Inflammation due to the body responding to the irritating particulates within smoke.

- Free radicals and other components of tobacco smoke impair the activity of the enzyme alpha-1-antitrypsin which would normally block the activity of proteases that degrade the proteins that maintain the elasticity of the lung.

A rare genetic cause of emphysema is a deficiency in the enzyme alpha-1-antitrypsin.

Emphysema can't be cured, but the symptoms can be alleviated and the spread of the disease can be checked by treatment. Figure 8 shows a man sitting in a chair at home, breathing oxygen through a tube to the nose. Beside him is oxygen administering equipment. Oxygen therapy supplies oxygen-enriched air to emphysema patients.

Patients are trained in breathing techniques that reduce breathlessness and improve the ability of the patient to exercise. Quitting smoking is essential so sometimes prescription medications can facilitate this process. Surgery is sometimes undertaken to reduce the volume of the lungs by removing damaged lung tissue. Lung transplants are also sometimes performed on patients suffering from emphysema.

▲ Figure 8

Interpreting micrographs of lung tissue

Identification of pneumocytes, capillary endothelium cells and blood cells in light micrographs and electron micrographs of lung tissue.

The wall of the alveolus is composed of two types of cells. 90% of the surface of the alveolus is composed of cells referred to as type 1 pneumocytes. They are extremely thin. Their primary purpose is gas exchange. The second type of cell forming the wall is the type 2 pneumocyte. These cells are covered in microvilli, are thicker and function to secrete surfactant, a substance that reduces surface tension, preventing the alveolus from collapsing.

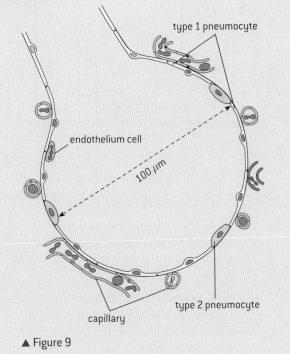

▲ Figure 9

Questions

1 A number of chemicals have been shown to cause tissue damage due to the production of free radicals. Free radicals are chemicals, such as superoxides and peroxides, which can react to damage DNA and lipids. Antioxidants produced by our body, such as reduced glutathione, combine with free radicals and decrease tissue damage. Reduced glutathione reacts with free radicals and in the process is converted to oxidized glutathione.

Recently dietary antioxidants such as lignins have also been shown to protect against tissue damage. Flaxseed is known to contain lignins but its antioxidant effects have yet to be evaluated. Research was done to see if flaxseed could help prevent damage to the liver by tetrachloromethane. Metabolism of tetrachloromethane by the liver leads to the formation of free radicals. Rats were pretreated by oral injection with flaxseed extract (+) or corn oil (−) (control) for three days and then injected with buffered saline solution (control) or tetrachloromethane. The glutathione levels were then measured.

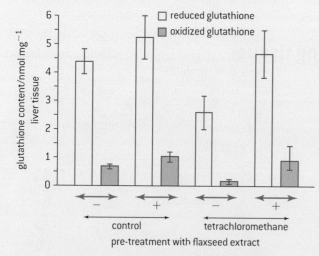

Source: Endoh, *et al J Vet Medical Science*, (2002), 64, page 761

a) (i) State the reduced glutathione content of liver tissue injected with tetrachloromethane with no flaxseed pretreatment. [1]

(ii) Calculate the total glutathione content (oxidized + reduced) in liver tissue treated with flaxseed extract but not injected with tetrachloromethane. [1]

b) Describe the effect of tetrachloromethane injection on total glutathione and reduced glutathione content in liver tissue without flaxseed pretreatment. [2]

c) Predict, using the data, the effect of using flaxseed extract in protecting liver tissue from damage due to tetrachloromethane. [3]

2 Blind mole rats (*Spalax ehrenbergi*) are adapted to live in underground burrows with very low oxygen conditions. Scientists compared blind mole rats and white rats in order to determine whether these adaptations are due to changes in their ventilation system.

Both types of rat were placed on a treadmill and the amount of oxygen consumed was measured at different speeds. This study was done under normal oxygen conditions and under low oxygen conditions. The results are shown in the scatter graph below.

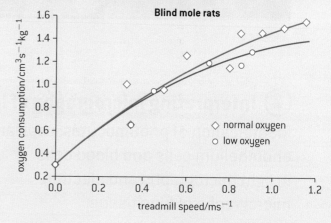

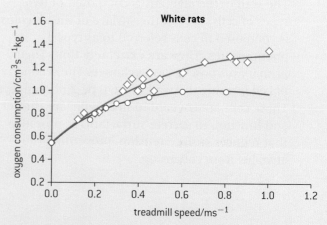

Source: Hans R. Widmer *et al.*, "Working underground: respiratory adaptations in the blind mole rat", *PNAS* (4 March 1997), vol. 94, issue 4, pp. 2062-2067, Fig. 1, © 2003 National Academy of Sciences, USA

a) Compare the oxygen consumption of blind mole rats and white rats when the treadmill is not moving. [1]

b) Compare the effect of increasing the treadmill speed on the oxygen consumption in both types of rats under normal oxygen conditions. [3]

c) Evaluate the effect of reducing the amount of oxygen available on both types of rat. [2]

The lungs of both types of rats were studied and the features important to oxygen uptake were compared. The results are shown in the bar chart below.

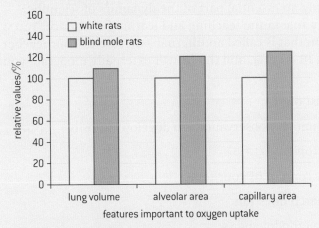

features important to oxygen uptake

Source: Hans R. Widmer *et al.*, "Working underground: respiratory adaptations in the blind mole rat", *PNAS* (4 March 1997), vol. 94, issue 4, pp. 2062-2067, Fig. 1, © 2003 National Academy of Sciences, USA

d) Using your knowledge of gaseous exchange in lungs, explain how these adaptations would help the blind mole rats to survive in underground burrows. [3]

e) Suggest how natural selection played an important part in the adaptations of blind mole rats. [3]

3 In the production of saliva, the acinar cells actively transport ions from the blood plasma into the ducts of the salivary gland resulting in water being drawn into the ducts. As this saliva moves down the duct, some ions are re-absorbed but the amount that can be re-absorbed depends on the rate of flow of saliva.

Graph A below shows how the composition of saliva varies depending on the rate of flow of saliva. Graph B shows the composition of blood plasma.

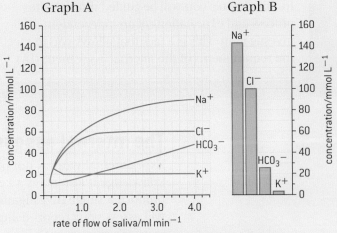

rate of flow of saliva/ml min^{-1}

Source: Jørn Hess Thaysen and Niels A. Thorn, Excretion of Urea, Sodium, Potassium and Chloride in Human Tears, *American Journal of Physiology*, 178: 160–164, 1954. American Physiological Society.

a) Using the data provided compare the concentration of ions in saliva produced at 4.0 ml min^{-1} with the concentration of those ions in the blood plasma. [2]

b) Outline the relationship between the concentration of Na$^+$ in saliva and the rate of flow of saliva. [2]

c) As the saliva moves down the ducts, Na$^+$ is re-absorbed into the blood plasma. Deduce, with a reason, the type of transport used to bring Na$^+$ back into the blood plasma. [1]

d) Suggest why the concentration of Na$^+$ varies with rate of flow. [2]

INTERNAL ASSESSMENT

Introduction

In this chapter you will be guided through the process and expectations of an independent investigation. This is a task that allows you to meet the requirements of the internal assessment (IA) component of the IB biology course. The aim is that you investigate something about the living world that genuinely interests you so that you can adopt a personal approach in your assessment.

The internal assessment

Experimental work is not only an essential part of the dynamics of scientific knowledge it also plays a key role in the teaching and learning of science. You will produce a single investigation that is called an internal assessment. Your teacher will assess your report using IB criteria, and the IB will externally moderate your teacher's assessment.

Your investigation will consist of:

- reading and making preliminary observations to develop curiosity

- selecting an appropriate topic

- researching the scientific content of your topic

- defining a workable research question

- adapting or designing a methodology

- obtaining, processing, and analysing data

- appreciating errors, uncertainties, and limits of data

- writing a scientific report 6–12 pages long receiving continued guidance from your teacher.

Planning and guidance

After your teacher introduces the idea of an internal assessment investigation, you will have an opportunity to discuss the topic of your investigation with your teacher. In dialogue with your teacher you can then select an appropriate topic, define a workable research question, and begin to do research into what is already known about your topic. You will not be penalized for seeking your teacher's advice.

It is your **teacher's responsibility** to provide you with a clear understand of the IA expectations, rules, and requirements. Your teacher will also provide you with continued guidance at all stages of your work. Your teacher will help you develop a topic, then a research question, and then an appropriate methodology. Your teacher will provide guidance as you work and they will read a draft of your report, making general suggestions for improvements or completeness. Your teacher will not, however, edit your report or give you a tentative grade for your report until it is finally completed. After this you are not allowed to make any changes.

It is **your responsibility** to appreciate the meaning of academic honesty, especially authenticity and intellectual property. You are also responsible for initiating your research question with the teacher, seeking help when in doubt, and demonstrating independence of thought and initiative in the design and implementation of your investigation. You are also responsible for meeting the deadlines set by your teacher.

Academic honesty

The IA is your responsibility, and it is your work. Plagiarism and copying others' work is not permissible. You must clearly distinguish between your own words and thoughts and those of others by the use of quotation marks (or another method like indentation) followed by an appropriate citation that denotes an entry in the bibliography.

Although the IB does not prescribe referencing style or in-text citation, certain styles may prove most commonly used; you are free to choose a style that is appropriate. It is expected that the minimum information included is: name of author, date of publication, title of source, and page numbers as applicable.

Types of investigations

After you have covered a number of biology syllabus topics and performed a number of hands-on experiments in class, you will be required to research, design, perform, and write up your own investigation. This project, known as an **internal assessment**, will count for 20% of your grade. You will have 10 hours of class time, you will consult with your teacher at all stages of your work, and you can research and write your report out of class. Your IA investigation cannot be used as part of a biology extended essay.

The variety and range of possible investigations is large, you could choose from:

Traditional hands-on experimental work. This could involve extending some of the protocols that you undertook as part of the syllabus or you might investigate in a practical way an experiment relevant to some of the concepts you have learned through the course.

Database investigations. A database is a mass of information that can be searched through the use of query. You may obtain data and process and analyse the information for your investigation. Examples might include GenBank, the Allele Frequencey Database (AlFreD) or the Audubon Christmas bird count.

Simulations and Models. It may not be feasible to perform some investigations in the classroom, but you may be able to find a computer simulation. The data from a simulation could then be processed and presented in such a way that something new is revealed. For example the Game of Life simulation allows the exploration of emergent properties.

Combinations of the above are also possible. The subject matter of your investigation is up to you. It may be something within the syllabus, something you have or will study, or it can be related to the syllabus or outside the syllabus. The depth of understanding should be, however, commensurate with the course you are taking. This means that your knowledge of IB biology (either SL or HL) will be sufficient to achieve maximum marks when assessed.

The assessment criteria

Your IA will be a single investigation and the report will be 6–12 pages long. The report should have an academic and scholarly presentation, and demonstrate scientific rigor commensurate with the course. There is the expectation of personal involvement and an understanding of biology, and there is also the expectation of setting your study within the known academic context. This means you need to research your topic and find out what is already known about it.

There are six assessment criteria, ranging in weight from 8–25% of the total possible marks. Each criterion reflects a different aspect of your overall investigation.

Criterion	Points	Weight
Personal engagement	0–2	8%
Exploration	0–6	25%
Analysis	0–6	25%
Evaluation	0–6	25%
Communication	0–4	17%
Total	**0–24**	**100%**

The IA grade will count for 20% of your total biology grade. The criteria are the same for standard and higher level students. We will now consider each criterion in detail.

PERSONAL ENGAGEMENT. *This criterion assesses the extent to which you engage with the investigation and make it your own. Personal engagement may be recognized in different attributes and skills. These include thinking independently and/or creatively, addressing personal interests, and presenting scientific ideas in your own way.*

For maximum marks under the personal engagement criterion, you must provide clear evidence that you have contributed significant thinking, initiative, or insight to your investigation. Your research question could be based upon something covered in class or an extension of your own interest.

For example, you may have a 'green thumb' and you enjoyed the practicals that you did with plants in class. You could turn your botany talents to growing a number of plants for your study. Personal significance, interest, and curiosity are expressed here.

You may also demonstrate personal engagement by showing personal input and initiative in the design, implementation, or presentation of the investigation. Perhaps you designed an improved method for measuring the rate of an enzyme controlled reaction or devised an interesting method for the analysis of data. You are not to simply perform a cookbook-like experiment or repeat an experiment that is commonly carried out in most practical work programmes without any modifications.

The **key** here is to be involved in your investigation, to contribute something that makes it your own.

Example

A personal approach to design

A student is interested in diving and wants to investigate the slowing of the heart when a diver holds their breath underwater (bradycardia). She takes her pulse on the surface and then after holding her breath for 30 s at the bottom of a swimming pool. There is a reduction in the pulse rate. She takes her pulse using the simple method of feeling her radial artery and counting for 30 s. The student wants to find out how rapidly the pulse falls, whether it falls suddenly or gradually and whether it stabilizes at a lower rate. She needs to monitor the heart rate continuously for one minute or more underwater. For this she needs an electronic probe, but the equipment that her teacher offers her has to be kept dry. If the student designs her own method for this, she will certainly have demonstrated a personal approach.

EXPLORATION. *This criterion assesses the extent to which you establish the scientific context for your work, state a clear and focused research question, and use concepts and techniques appropriate to the course you are studying. Where appropriate, this criterion also assesses awareness of safety, environmental, and ethical considerations.*

For maximum marks under the exploration criterion, your topic must be appropriately identified and a relevant and fully focused research question is described. Background information about your investigation must be appropriate and relevant, and the methodology must be appropriate to address your research question. Moreover, for maximum marks, your research must identify significant factors that may influence the relevance, reliability, and sufficiency of your data. Finally, if safety, environmental, and ethical considerations are relevant to your investigation, then your work must demonstrate a full awareness of these issues.

The **key** here is your ability to select, develop, and apply appropriate methodology and produce a scientific work.

ANALYSIS. *This criterion assesses the extent to which your report provides evidence that you have selected, processed, analysed, and interpreted the data in ways that are relevant to the research question and can support a conclusion.*

For maximum marks under the analysis criterion, your investigation must include sufficient raw data to support a detailed and valid conclusion to your research question. Your processing of data must be carried out with sufficient accuracy. Moreover, your report must show evidence of full and appropriate consideration of the impact of measurement uncertainty on your analysis. Finally, for maximum marks, you must correctly interpret your data, so that completely valid and detailed conclusions to the research questions can be deduced.

The **key** here is to make an appropriate and justified analysis of your data that is focused on your research question.

Is there any statistical hypothesis test you could do?

How much do the repeats vary? This indicates how reliable your evidence is.

Can the anomalous results be explained by mistake or are you unsure about the overall trend or pattern?

Are there any results or groups of results that do not fit the overall trend or pattern? These are often called anomalous results.

What trends or patterns are visible in the data? One example is a positive correlation. Can you show the trends more clearly by adding another type of chart?

What sort of chart or graph will display your data most clearly: a scatter graph, bar chart or pie chart, or other type of presentation? The aim is to make it easy for the reader to pick out the trends and patterns.

Example

Observations and questions

Two gerbils were being kept in a biology laboratory and the person who cleaned the laboratory started giving them a peanut each when she cleaned the lab each morning. After a few weeks, she noticed that when she entered the lab the gerbils came over to the front of the cage, stood on their back legs and waited for their peanut. When other people came into the lab at other times they did not do this.

This observation suggests some interesting questions.

1   Were the gerbils able to recognize the cleaner and if they were, what were the recognition features?

2   If clothes of different colour, but the same design were worn, was she still recognizable?

3   Was the time of day critical to recognizing the cleaner?

4   Could the gerbils predict the arrival of the cleaner before she actually came into the lab?

5   Could sight, sound and/ or smell be used for recognition?

A simple observation can lead to interesting and worthwhile questions.

EVALUATION. *This criterion assesses the extent to which your report provides evidence of evaluation of the investigation and results with regard to the research question and the wider world.*

For maximum marks under the evaluation criterion, you must state a detailed conclusion that is described and justified, that is entirely relevant to the research question, and fully supported by the data presented. You should make a comparison to the accepted scientific context if relevant. The strengths and weaknesses of your investigation, such as the limitation of data and sources of uncertainty, must be discussed and you must provide evidence of a clear understanding of the methodological issues involved in establishing your conclusion. This means not only identifying limitations, but also discussing the implications and consequences of these limitations. Finally, to earn maximum marks for evaluation, you must discuss realistic and relevant improvements and possible extensions to your investigation. The **key** here is different than the previous criterion for analysis. The focus of evaluation is to incorporate the methodology and to set the results within a genuine scientific context while making reference to your research topic.

Your evidence is strong if you answer "yes" to these questions	Your evidence is weak if you answer "yes" to these questions
Are your results consistent enough to give you reliable evidence to use to answer the research question?	Are your results variable or are there many anomalous results that can't easily be explained?
Was the design of your experiment successful so that only it gave precise and accurate results?	Were there faults in the experimental design which limited the precision or the accuracy?
Were all the variables controlled satisfactorily so that only the independent variable was varied?	Were there uncontrolled variables, which introduced uncertainties into your interpretation of the results?
Is there only one explanation that fits all the evidence and answers the research question?	Are there alternative explanations that would also fit the evidence and which you cannot refute?
Can you support each part of your answer to the research question with experimental evidence or by reference to other published data?	Are there parts of your answer to the research question which are unsubstantiated or uncertain and which need further investigation?

COMMUNICATION. *This criterion assesses whether the investigation is presented and reported in a way that supports effective communication of the focus, process and outcomes of the investigation.*

For maximum marks under the communication criterion, your report must be clear and easy to follow. Although your writing does not have to be perfect, any mistakes or errors should not hamper the reader's understanding of the focus, process, and outcomes of your investigation. Your report must be well structured and focused on the necessary information, the process and outcomes must be presented in a logical and coherent way. Your text must be relevant without wandering off onto tangential issues. Your use of specific biology terminology and conventions must be appropriate and correct. Graphs, tables, and images must all be appropriately presented. Your lab report should be 6–12 pages long. Excessive length (beyond 12 pages) will be penalized under the communication criterion.

The **key** here is to demonstrate a concise, logical, and articulate report, one that is easy to follow and is written in a scientific context.

INDEX

Page numbers in *italics* refer to data-based questions and questions at the end of each chapter.

719

LIBRARY; p416a: Taiz and Zeiger: Plant Physiology, Third Edition (2002); p416b: DR. RICHARD KESSEL & DR. GENE SHIH/VISUALS UNLIMITED, INC. /SCIENCE PHOTO LIBRARY; p416c: Trends in Plant Science, Volume 7, Issue 3, 1 March 2002, Pages 126-132/Science Direct; p417: D. Fischer; p419: Yuriko Nakao/ Reuters; p420a: POWER AND SYRED/SCIENCE PHOTO LIBRARY; p420b: DR KEITH WHEELER/SCIENCE PHOTO LIBRARY; p421a: STEVE GSCHMEISSNER/ SCIENCE PHOTO LIBRARY; p421b: DR KEITH WHEELER/SCIENCE PHOTO LIBRARY; p423a: Nature Reviews Genetics 4, 169-180 (March 2003)/Nature. com; p423b: DR JOHN RUNIONS/SCIENCE PHOTO LIBRARY; p423c: Biodisc; p425a: ANDREW LAMBERT PHOTOGRAPHY/SCIENCE PHOTO LIBRARY; p425b: POWER AND SYRED/SCIENCE PHOTO LIBRARY; p427a: www. wageningenur.nl; p427b: web.tiscali.it; p429a: InavanHateren/Shutterstock; p429b: Andrey Shtanko/Shutterstock; p432a: DR. JOHN BRACKENBURY/ SCIENCE PHOTO LIBRARY; p432b: NHPA/Photoshot; p433: MERLIN TUTTLE/ SCIENCE PHOTO LIBRARY; p434: American Institute of Biological Sciences; p436: Andrew Allot; p439: Arvind Balaraman/Shutterstock; p448a: DR JEREMY BURGESS/SCIENCE PHOTO LIBRARY; p448b: EYE OF SCIENCE/ SCIENCE PHOTO LIBRARY; p451: PNAS.org; p460: R A J Case, et al., (2005), Marine Ecology Progress Series, 201; p461a: Dr. Michael Mares/Sam Noble Museum; p461b: Brain Gratwicke; p462a: Annales Botanici Fennici/Finnish Zoological and Botanical Publishing Board; p462b: Annales Botanici Fennici/ Finnish Zoological and Botanical Publishing Board; p463a: Ed Reschke/ Photolibrary; p463b: Dr. Glyn Jenkins; p465: Four Oaks/Shutterstock; p466: AMI IMAGES/SCIENCE PHOTO LIBRARY; p467: REVY, ISM/SCIENCE PHOTO LIBRARY; p468: STEVE GSCHMEISSNER/SCIENCE PHOTO LIBRARY; p470: SCIENCE PHOTO LIBRARY; p474a: RIA NOVOSTI/SCIENCE PHOTO LIBRARY; p474b: SCIENCE PHOTO LIBRARY; p475: JAMES HOLMES/CELLTECH LTD/SCIENCE PHOTO LIBRARY; p476: THIERRY BERROD, MONA LISA PRODUCTION/ SCIENCE PHOTO LIBRARY; p478: DR. JOHN BRACKENBURY/ SCIENCE PHOTO LIBRARY; p481a: MICROSCAPE/SCIENCE PHOTO LIBRARY; p481b: MICROSCAPE/SCIENCE PHOTO LIBRARY; p482: www.mrothery. co.uk; p484: Dwight Smith/Shutterstock; p486: matt knoth from San Francisco, yesicannibus/Wikipedia; p490: SCIENCE PHOTO LIBRARY; p496: Carsten Medom Madsen/Shutterstock; p497a: AJ PHOTO/HOP AMERICAIN/ SCIENCE PHOTO LIBRARY; p497b: HERVE CONGE, ISM /SCIENCE PHOTO LIBRARY; p498a: SATURN STILLS/SCIENCE PHOTO LIBRARY; p498b: DR P. MARAZZI/SCIENCE PHOTO LIBRARY; p498c: Bobjgalindo/Wikipedia; p498d: Bobjgalindo/Wikipedia; p500a: PROF. R. WEGMANN/SCIENCE PHOTO LIBRARY; p500b: PROFESSOR P.M. MOTTA, G. MACCHIARELLI, S.A NOTTOLA/SCIENCE PHOTO LIBRARY; p500c: POWER AND SYRED/SCIENCE PHOTO LIBRARY; p503: PROF. P. MOTTA/DEPT. OF ANATOMY/UNIVERSITY "LA SAPIENZA", ROME/SCIENCE PHOTO LIBRARY; p504: D. PHILLIPS/ SCIENCE PHOTO LIBRARY; p505: ANGEL FITOR/SCIENCE PHOTO LIBRARY; p510a: Alis Leonte/Shutterstock; p510b: Four Oaks/Shutterstock; p513: OUP; p518: ZEPHYR/SCIENCE PHOTO LIBRARY; p520a: ZEPHYR/SCIENCE PHOTO LIBRARY; p520b: Nuffield Department of Obstetrics and Gynaecology; p527: Prof. P Motta/Dept. of Anatomy/University La sapimeza, Rome/SPL; p529: DAVID NICHOLLS/SCIENCE PHOTO LIBRARY; p538a: OUP; p538b: Kim Taylor/ Dorling Kindersley/Getty Images; p539: OUP; p540: OUP; p546: OUP; p547: OUP; p550a: OUP; p550b: OUP; p552: Tom Mchug/SPL; p554: Kim Taylor/ Dorling Kindersley/Getty Images; p557: GloFish®; p558a: Bios/Wikipedia; p558b: PASCAL GOETGHELUCK/SCIENCE PHOTO LIBRARY; p559: VIKTOR SYKORA/SCIENCE PHOTO LIBRARY; p561: www.nyp.edu.sg; p564: CC STUDIO/SCIENCE PHOTO LIBRARY; p566a: GloFish®; p566b: MEDICAL RESEARCH COUNCIL/SCIENCE PHOTO LIBRARY; p567: PHILIPPE PLAILLY/ SCIENCE PHOTO LIBRARY; p568: SINCLAIR STAMMERS/SCIENCE PHOTO LIBRARY; p576a: PAUL RAPSON/SCIENCE PHOTO LIBRARY; p576b: EYE OF SCIENCE/SCIENCE PHOTO LIBRARY; p576c: EYE OF SCIENCE/SCIENCE PHOTO LIBRARY; p577a: STEVE GSCHMEISSNER/SCIENCE PHOTO LIBRARY; p577b: Nature .com; p578: ANIMATED HEALTHCARE LTD/SCIENCE PHOTO LIBRARY; p579: VINCENT AMOUROUX, MONA LISA PRODUCTION/ SCIENCE PHOTO LIBRARY; p580a: JAMSTEC; p580b: John & Margaret Rostron; p582: American Journal of Obstetrics and Gynecology/Science Direct; p584: PHILIPPE PLAILLY/SCIENCE PHOTO LIBRARY; p586a: American Journal of Obstetrics and Gynecology, Volume 178, Issue 1, Part 1, January 1998, Pages 85-90 Ronald P Zweemer, Renée H.M. Verheijen, Johan J.P. Gille, Paul J. van Dies, Gerard Pals, Fred H. Menko/Science Direct; p586b: JAN VAN DE VEL/ REPORTERS/SCIENCE PHOTO LIBRARY; p588: Alan R Hibbs; p593: blast.ncbi. nlm.nih.gov; p594: OAK RIDGE NATIONAL LABORATORY/US DEPARTMENT OF ENERGY/SCIENCE PHOTO LIBRARY; p595: PhyloWin; p596a: FRANS LANTING, MINT IMAGES / SCIENCE PHOTO LIBRARY; p596b: Edwin Verin/ Shutterstock; p597: blast.ncbi.nlm.nih.gov; p599a: PhyloWin; p599b: PhyloWin; p600: doi: 10.1093/nar/gks1236 /Ensembl; p603: Shutterstock; p605: MARTYN F. CHILLMAID/SCIENCE PHOTO LIBRARY; p608: NATURE'S IMAGES/SCIENCE PHOTO LIBRARY; p609a: JIM ZIPP/SCIENCE PHOTO LIBRARY; p609b: JIM ZIPP/SCIENCE PHOTO LIBRARY; p611a: MarkMolander/ Wikipedia; p611b: DR GEORGE GORNACZ/SCIENCE PHOTO LIBRARY; p612a: Anurag Agrawal/ Cornell University; p612b: ANTHONY MERCIECA/SCIENCE PHOTO LIBRARY; p612c: © Pete Saloutos/Blend Images/Corbis; p614: Andrew Allot; p620: Jurg Alean/swisseduc.ch; p624: Dave G. Houser/Corbis p626a: DUNCAN SHAW/SCIENCE PHOTO LIBRARY; p626b: JOHN DEVRIES/SCIENCE PHOTO LIBRARY; p626c: DR DAVID FURNESS, KEELE UNIVERSITY/SCIENCE PHOTO LIBRARY; p627a: ALEXIS ROSENFELD/SCIENCE PHOTO LIBRARY; p627b: PETER YATES/SCIENCE PHOTO LIBRARY; p628: © Scientifica/ Visuals Unlimited/Corbis; p633a: Public Domain/Wikipedia; p633b: ALEXIS ROSENFELD/SCIENCE PHOTO LIBRARY; p634: © Tui De Roy/Minden Pictures/ Corbis; p636: BOB GIBBONS/SCIENCE PHOTO LIBRARY; p637a: LOUISE MURRAY/SCIENCE PHOTO LIBRARY; p637b: FRANS LANTING, MINT IMAGES / SCIENCE PHOTO LIBRARY; p638: DAVID WOODFALL IMAGES/SCIENCE PHOTO LIBRARY; p644: NOAA/Ocean Explorer; p647: Andrew Allot; p648: GEORGETTE DOUWMA/SCIENCE PHOTO LIBRARY; p649: DR JEREMY BURGESS/SCIENCE PHOTO LIBRARY; p650a: DR JEREMY BURGESS/SCIENCE PHOTO LIBRARY; p650b: PASIEKA/SCIENCE PHOTO LIBRARY; p651: CLAUDE NURIDSANY & MARIE PERENNOU/SCIENCE PHOTO LIBRARY; p652: DIRK WIERSMA/SCIENCE PHOTO LIBRARY; p653: ROBERT BROOK/SCIENCE PHOTO LIBRARY; p655: GEOFF KIDD/SCIENCE PHOTO LIBRARY; p659: MAXIMILIAN STOCK LTD/SCIENCE PHOTO LIBRARY; p661: GEOFF KIDD/ SCIENCE PHOTO LIBRARY; p666: CDC/SCIENCE PHOTO LIBRARY; p667: American Journal of Obstetrics and Gynecology, Volume 178, Issue 1, Part 1, January 1998, Pages 85-90 Ronald P Zweemer, Renée H.M. Verheijen, Johan J.P. Gille, Paul J. van Dies, Gerard Pals, Fred H. Menko Direct; p668: SIMON FRASER/SCIENCE PHOTO LIBRARY; p670a: MAXIMILIAN STOCK LTD/ SCIENCE PHOTO LIBRARY; 670b: www.supertracker.usda.gov/Foodtracker; p673a: Steve Gschmeissner/SPL; p673b: Innerspace Imaging/SPL; p674a: Dennis Kunkel/Photolibrary; p674b: CNRI/SCIENCE PHOTO LIBRARY PHOTO LIBRARY; p675a: David M Martin, Md/SPL; p675b: EYE OF SCIENCE/ SCIENCE PHOTO LIBRARY; p680: PROFESSORS P.M. MOTTA, T. FUJITA & M. MUTO /SCIENCE PHOTO LIBRARY; p681a: PROFESSORS P.M. MOTTA, T. FUJITA & M. MUTO /SCIENCE PHOTO LIBRARY; p681b: DR P. MARAZZI/ SCIENCE PHOTO LIBRARY; p682: RIA NOVOSTI/SCIENCE PHOTO LIBRARY; p683: MICROSCAPE/SCIENCE PHOTO LIBRARY; p685: THOMAS DEERINCK, NCMIR/SCIENCE PHOTO LIBRARY; p688: FAYE NORMAN/SCIENCE PHOTO LIBRARY; p689: GUSTOIMAGES/SCIENCE PHOTO LIBRARY; p690a: ADAM HART-DAVIS/SCIENCE PHOTO LIBRARY; p690b: Andrew Allot; p691a: STEVE GSCHMEISSNER/SCIENCE PHOTO LIBRARY; p691b: DR P. MARAZZI/SCIENCE PHOTO LIBRARY; p694: STEVE GSCHMEISSNER/SCIENCE PHOTO LIBRARY; p695: SCIENCE PHOTO LIBRARY; p698: STEVE GSCHMEISSNER/SCIENCE PHOTO LIBRARY; p701: GUSTOIMAGES/SCIENCE PHOTO LIBRARY; p704a: SSPL/Getty Images; p704b: DU CANE MEDICAL IMAGING LTD/SCIENCE PHOTO LIBRARY; p705: CONOR CAFFREY/SCIENCE PHOTO LIBRARY

Artwork by Six Red Marbles and OUP

The authors and publisher are grateful for permission to reprint extracts from the following copyright material:

p388: Dr.Carmen Mannella for her comment, reprinted by kind permission.

p581: Cecilia Elpi, 'Biofilm helps salmonella survive hostile conditions', 10 April 2013, reprinted by kind permission of Virginia Tech

Sources:

p309: Quotation from BBC News © BBC 2013

Although we have made every effort to trace and contact all copyright holders before publication this has not been possible in all cases. If notified, the publisher will rectify any errors or omissions at the earliest opportunity.

Links to third party websites are provided by Oxford in good faith and for information only. Oxford disclaims any responsibility for the materials contained in any third party website referenced in this work.